P9-CIV-631

North American Forest and Conservation History

North American Forest and Conservation History

A Bibliography

Ronald J. Fahl

PUBLISHED UNDER CONTRACT WITH THE
FOREST HISTORY SOCIETY, INC.

A.B.C. — Clio Press
Santa Barbara, California
Oxford, England

Library of Congress Cataloging in Publication Data

Fahl, Ronald J 1942-
North American forest and conservation history.
Includes index.
1. Forests and Forestry—North America—History—Bibliography. 2. Forest conservation—North America—History—Bibliography. I. Forest History Society. II. Title.
Z5991.F33 [SD140] 016.3337'5'0973 76-27306
ISBN 0-87436-235-0

American Bibliographical Center—Clio Press
2040 Alameda Padre Serra
Santa Barbara, California

European Bibliographical Center—Clio Press
Woodside House, Hinksey Hill
Oxford OX1 5BE, England

North American Forest and Conservation History was composed on a Compugraphic ACM9000 and a Variable Input Photocompositor, using a Baskerville text and display by Camera-ready Composition, Santa Barbara, California. Offsetting and bindery work were done by Edwards Brothers, Ann Arbor, Michigan. The text paper is a 45-pound Educator's Coated. The casesides are Kivar 9 which are applied over .080 binder's boards. Binding is Smyth sewn with 80-pound white endsheets, rounded and backed with super and liner. The text was designed by Shelly Lowenkopf, the cover art was designed, separated, and prepared by Jack Swartz.

Abbreviations Used in the Text

USDA — United States Department of Agriculture
USFS — United States Forest Service
USDI — United States Department of the Interior
GPO — Government Printing Office
OHI — Oral History Interview

Acknowledgment

The publisher acknowledges with thanks the contributions of the Author and the Forest History Society, for their supervision of the entire editorial process.

Contents

Introduction

Compiling a historical bibliography can become an obsessive and overwhelming task—completion often delayed or frustrated by the chimerical goal of producing a definitive listing. Even in a subject such as forest history, which the uninitiated might regard as a remote corner of the whole realm of knowledge, the quantity of literature is deceptively extensive. In this project, therefore, finite amounts of time and money necessitated restrictions upon the categories of literature included, the range of bibliographic searching, and the degree to which individual entries could be identified and annotated. This introduction is intended to clarify the reference work's limits and to explain the compromises involved in its compilation. It also serves the traditional purpose of crediting the chief sources of information and acknowledging those persons and institutions who assisted in its preparation.

For the purpose of this bibliography, forest history is broadly defined to encompass man's exploitation, utilization, and appreciation of the forest and its resources. The Forest History Society, sponsor of the research leading to this volume, has made the history of forestry, forest conservation, and the forest and wood industries central to its program for nearly thirty years. Geographically it has claimed North America as its province, although its quarterly *Journal of Forest History* occasionally publishes articles and book reviews on non-North American topics. The present work follows these general precedents but limits its scope to Canada, the United States, and dependencies. (A companion volume by Richard C. Davis is entitled *North American Forest History: A Guide to Archives and Manuscripts in the United States and Canada.)*

Although this volume offers a serviceable guide to light reading in the subject, the compiler's education and inclinations as a historian, the academic orientation of the sponsoring institution, and the nature of available library resources have combined to focus attention on the traditional historical writing produced by scholars. Despite the fact that it constitutes a minority of the bibliography's entries, this scholarly material has been examined more thoroughly and systematically than other categories of literature.

The chief criterion for including a work in this bibliography is that it be written as history—that it treat a subject which happened in the past—as distinguished from the great body of published writings about contemporaneous subjects which historians also find useful as primary sources. By way of example, this work does not encompass the technical literature of forestry and related sciences; it does not contain the contemporary reports of the forest industries; it does not include the descriptive or polemical writings of the conservation movement—except, in all of these cases, when such literature incorporates explicitly historical treatments of forest subjects.

These exceptions, it should be noted, constitute a sizeable minority of the entries. Much forestry literature, for example, commences with an obligatory historical review, or, in the case of a policy statement, may deliberately draw upon historical examples to bolster a thesis. The historian seeking to reconstruct a phase of the past may welcome such material, especially for topics which have not been previously illuminated by formal historical analysis. Similarly, a preservationist plea to save some endangered wilderness may also include a convenient historical introduction to the disputed area. On the same grounds even relatively superficial treatments of forest subjects in popular magazines might qualify for listing. Neither the nature nor the quality of scholarship determines inclusion in the bibliography, despite the stated emphasis on formal history. Subject matter is the overriding consideration: this work seeks to bring together from many academic disciplines and realms of experience references to man's deliberate recollections and reconstructions of North American forest history.

Several categories of writing on forest history merit further explanation. Forest industry trade journals are immensely valuable but underexploited sources. Especially in the early twentieth century, these weekly and semimonthly magazines reported broadly on industrial developments, trade association activities, and political issues of interest to a high-riding fraternity of lumbermen. They included, moreover, reliable historical articles and numerous reminiscences by old-timers of the forest industries. *American Lumberman, Canada Lumberman, Southern Lumberman,* and *Timberman* are notable examples of trade publications which consciously sought to register the history of lumbering and related enterprise in their respective spheres of influence. James E. Defebaugh, George H. Hotchkiss, Stanley F. Horn, and George M. Cornwall stand out as editors and publishers whose historical propensities have provided modern scholars with good clear lumber for constructing monographs in forest history. The entries to trade journal articles in this bibliography should be regarded only as samples of the genre. What is included reflects the incomplete holdings of the Forest History Society's library. *The Union List of Serials in Libraries of the United States and Canada* provides information on the availability of older trade journals.

Professional forestry journals, notably the *Journal of Forestry* and its Canadian counterpart, *Forestry Chronicle,* are also rich sources of forest history. Valuable for reconstructing both the scientific and political climates of forestry opinion throughout the twentieth century, they also contain historical articles by knowledgeable and prolific writers such as Henry Clepper and Samuel Trask Dana. Most of these articles are listed herein, but the serious historian of forestry is advised to repair to a good forestry library for a thorough examination of these journals, the various forestry school annuals, and other periodicals in the field. He will find that feature articles and obituaries provide a wealth of material on the profession's issues and leaders. For the early literature, Edward N. Munns' *A Selected Bibliography of North American Forestry,* 2 volumes, USDA Miscellaneous Publication No. 364 (Washington: GPO, 1940), is a useful aid.

Another fertile source for historians is the conservation magazine. *American Forests* and its predecessors have always featured historical treatments of subjects which continue to interest scholars today. *Audubon, Living Wilderness,* and *Sierra Club Bulletin* have likewise given an eye to forest history, albeit through glasses differently colored than those of the industrial and professional journals. Routine use of the *Reader's Guide to Periodical Literature* will turn up references to many articles which escaped inclusion here.

In the past century, a proliferation of local, state, regional, and national agencies have issued, in an almost geometric progression, mountains of documents and reports on natural resources. A small fraction of this printed matter is explicitly historical; much more is useful as primary source material. Only the most notable examples—or the most frequently cited by historians of governmental involvement in forestry and natural resources—were incorporated in this listing. The thorough and persistent researcher will want to seek out a large library, preferably a federal depository, and engage the talents of a documents librarian to unlock the mysteries of government cataloging. No matter how remote the subject, be assured that somewhere it is treated in a government publication. *Agriculture Index, Bibliography of Agriculture,* and *Forestry Abstracts* are useful guides to this and aforementioned categories of literature.

State and county histories, including the weighty albums of biographical sketches (or "mugbooks") so popular with former generations of successful pioneers and businessmen, should also be regarded as fruitful sources for the study of lumbering and its practitioners. While much local history is marked by a focus on "trees" rather than "forest," the researcher will profit nonetheless from a patient examination of such material. Only a relative handful of such items have been listed in this work, and the user is urged to search state and local historical bibliographies for additional references to this genre and to the untold other *festschriften* and miscellanies which have escaped notice here. Researchers also should be aware of the growing body of oral history interviews which relate to forest subjects. Published interviews and excerpts, primarily those produced by the Forest History Society, are listed herein.

Despite an original intent to list only published works, doctoral dissertations and master's theses are included in this bibliography. To exclude a scholarly composition of several hundred pages while including a brief reminiscence from a trade journal would violate the spirit of this reference work, especially in an age when xerographic, microfilming, and interlibrary loan services make available much of this quasi-published material at nominal cost. *Dissertation Abstracts International* and Warren F. Kuehl's two-volume index, *Dissertations in History,* provided information on most of those listed here. References to master's theses were gathered somewhat randomly.

The literature of historians has long escaped the comprehensive bibliographical control that distinguishes some academic disciplines and realms of knowledge and inquiry. The American Historical Association's annual *Writings on American History* once provided reasonably good coverage, but it fell more than a decade behind the outpouring of literature before being discontinued in its traditional format. The AHA recently has taken steps to bring its coverage of periodical literature up to date. A more useful guide, however, is the improved *America: History and Life,* an abstracting journal published by the American Bibliographical Center-Clio Press. Book reviews, book notices, and bibliographical listings in the standard national, regional, state, and specialty history journals are the most current sources of information on literature in American history. All of these resources were tapped in the course of this project.

Several specialized historical bibliographies, among the hundreds consulted, are deserving of specific mention: Eugene S. Ferguson, *Bibliography of the History of Technology* (Cambridge: Society for the History of Technology and M.I.T. Press, 1968); Henrietta M. Larson, *Guide to Business History* (Cambridge: Harvard University Press, 1948; Reprint, 1964); Robert W. Lovett, *American Economic and Business History Information Sources: An Annotated Bibliography of Recent Works* . . . . (Detroit: Gale Research, 1971); Robert G. Albion, *Maritime and Naval History: An Annotated Bibliography,* 4th ed. (Mystic, Connecticut: Marine Historical Association, 1972); Charles Haywood, *A Bibliography of North American Folklore and Folksong,* 2 volumes, 2nd ed. (New York: Dover Publications, 1961); Louis Kaplan, *A Bibliography of American Autobiographies* (Madison: University of Wisconsin Press, 1961); John T. Schlebecker, *Bibliography of Books and Pamphlets on the History of Agriculture in the United States, 1607-1967* (Santa Barbara: ABC-Clio Press, 1969); and Oscar O. Winther, *A Classified Bibliography of the Periodical Literature of the Trans-Mississippi West, 1811-1957* (Bloomington: Indiana University Press, 1961; Supplement, 1970). Two articles stand out as thoughtful evaluations of literature in conservation history: Gordon B. Dodds, "Conservation & Reclamation in the Trans-Mississippi West: A Critical Bibliography," *Arizona and the West* 13 (Summer 1971), 143-71, and Lawrence Rakestraw, "Conservation Historiography: An Assessment," *Pacific Historical Review* 41 (August 1972), 271-88. Needless to say, the bibliographical listings and essays in the academic histories of forestry, lumbering, and conservation are also rich with leads to all kinds of source material.

The most important repository of bibliographical information for this project was the collective research projects, library resources, and publications of the Forest History Society. Since its founding by historians, foresters, and businessmen in 1946, the Society has sought to advance knowledge of the field by identifying, collecting, and preserving archival and manuscript sources in cooperation with affiliated libraries and archival institutions. The first fruit of these efforts soon came in the form of graduate research, much of which was published in book and article form. Indeed, the *Journal of Forest History* (formerly *Forest History*), which has appeared continuously since 1957, published some of the new scholarship and sought to identify the rest by listing and/or reviewing in its pages the growing literature of forest and conservation history.

In the early 1960s the Society sponsored, with generous aid from the Weyerhaeuser Foundation and the Louis W. and Maud Hill Family Foundation, a bibliographic project with objectives similar to those which motivated this volume. Joseph A. Miller, formerly editor of *Forest History* and associate director of the Society, compiled a massive list of references, but his manuscript was neither indexed or published. The information assembled by Miller measurably assisted the compiler by reducing the load of bibliographic searching. The two projects differed fundamentally, however, in their philosophical conception, organization, and execution. All who use this volume profitably owe Miller a vote of thanks, but he bears no responsibility for its shortcomings.

Two bibliographies issued in limited numbers by the Forest History Society also were consulted throughout the course of the project. Judith A. Steen's *A Guide to Unpublished Sources for a History of the United States Forest Service* (1973) contains references to published as well as unpublished materials bearing on the field. Gerald R. Ogden's *The United States Forest Service: A Historical Bibliography, 1876-1972* (1973) lists over 7,000 published items.

The Society's specialized library has been the backbone of this project's bibliographic research. Despite gaps in its holdings, the collection has permitted first-hand examination of many of the book entries; its growing files of periodicals and article reprints from several disciplines have saved the compiler much time-consuming and expensive travel. Research in other institutions was limited to libraries at the University of California at Berkeley, the University of California at Santa Cruz, and the Oregon Historical Society in Portland.

An early decision was to organize the main body of the bibliography alphabetically by author. Given its interdisciplinary nature, the field of forest and conservation history does not lend itself neatly to subject classification or organization. A geographical arrangement of entries would offer the user many conveniences but would result in much overlap and added cost to an already expensive volume. Although alphabetical arrangement is handy for the author-minded researcher, the key to successful use of this bibliography is its subject index. The user should consult its introduction for general instructions and precautions.

The bibliography lists materials published as recently as mid-1975, including several items scheduled for appearance after the manuscript was completed. Given the variable time lag between publication and bibliographical listing in standard sources, however, references to some recent books and articles may not appear here. The compiler will continue his bibliographic effort with published works by listing and/or arranging for their review in the *Journal of Forest History.* A supplement to this bibliography may also appear at some future date. Users are encouraged, therefore, to notify the Forest History Society of omissions as well as newly published books and articles and recently completed dissertations and theses. In so doing, a community of cooperative forest historians will certainly provide mutual bibliographical benefits while advancing the identity and status of the field.

A grant from the National Endowment for the Humanities (RO-7813-73-329) in 1973 made possible this reference work. Under the grant provisions full project funding was contingent upon the receipt of donations to be matched by the NEH. The Forest History Society, the NEH, and the compiler are pleased to credit the following individuals and organizations whose generous gifts of money facilitated maximum project funding: Laird Norton Foundation, John M. Musser, Weyerhaeuser Foundation, Inc., Garrett Eddy, Gene C. Brewer, Susan L. Flader, Harold A. Miller, Nils B. Hult, James D. Bronson, J. Paul Neils, Joseph E. McCaffrey, John H. Hauberg, Edmund G. Hayes, George E. Lamb, The Langdale Company, George S. Kephart, Robert Noyes, and Hans Schneider. Others who supported the project were John H. Hinman, James W. Craig, E.R. Titcomb, F.P. Keen, Connwood, Inc., Philip H. Jones, G.E. Karlen, Selwyn J. Sharp, Hardy L. Shirley, Clare W. Hendee, Carl G. Krueger, Rolf B. Jorgensen, Gene S. Bergoffen, and H.E. Ruark. The Forest History Society thanks these donors, moreover, for their expressions of confidence in the worthiness of the project.

Work on the bibliography commenced in July 1973 and continued through August 1975. Several associates at the Forest History Society rendered valuable assistance at various stages of the project. Elwood R. Maunder, the Society's executive director since 1952, made the completion of this reference work both a private and public goal for many years. In addition to his part in securing the NEH grant and raising matching funds, he drew upon his broad familiarity with the literature of forest history to make substantive contributions and criticisms. George A. Garratt, former president of the Society, helped Maunder formulate the project proposal and solicit financial support. Maunder had the able support of Roberta M. Barker in administering financial aspects of the project.

Harold K. Steen, the Society's associate director for research and library services, assisted the compiler in shaping the project's objectives and methodology. His dual background in forestry and history enabled him to offer insightful suggestions and valuable counsel. Richard C. Davis, who assembled the companion guide to unpublished sources of North American forest history, collaborated with the compiler on a daily basis for the duration of the project. It profited from his knowledgeable criticism.

Joan Hodgson, head of the interlibrary loan service at the University of California, Santa Cruz, rendered much valuable assistance in locating and obtaining books and articles. Jill Feblowitz, Katrina Kuizenga, Sarah Silverman, and Pamela Mathis helped with the tedious but vital task of indexing. Pamela O'Neal, who typed the manuscript, brought to the project a discerning appreciation of the subject and a remarkable memory—attributes which permitted her to make useful editorial contributions. Kathryn A. Fahl was both a paid and volunteer assistant for over two years. An experienced bibliographic searcher, she was a versatile and dependable aide in every aspect of the work. More importantly, she nurtured and cheered her husband through the most difficult phases and acquiesced in the extensive evening and weekend efforts required to complete this volume.

The compiler alone, of course, accepts responsibility for any errors, omissions, and misinterpretations on the following pages.

Bibliography

Bibliography

A1 Abbot, Arthur P. *The Greatest Park in the World, Palisades Interstate Park: Its Purposes, History and Achievements.* New York: Historian Publishing Company, 1914. 64 pp. Illus., map. Palisades Interstate Park, New York–New Jersey.

A2 Abbot, Donald P. "Lancaster's Lodge." *Oregon Historical Quarterly* 75 (Sept. 1974), 277-81. Reminiscences of Samuel C. Lancaster, his recreational lodge in Oregon's Columbia Gorge, and fighting a forest fire in 1922.

A3 Abbot, M.F. "A Brief History of the Conservation Movement." *Twentieth Century Magazine* 1 (Mar. 1910), 537-43; 2 (May 1910), 143-48; 2 (June 1910), 228-34; 2 (Aug. 1910), 410-15; 2 (Sept. 1910), 511-15.

A4 Abbott, Carl. "The Plank Road Enthusiasm in the Antebellum Middle West." *Indiana Magazine of History* 67 (June 1971), 95-116.

A5 Abbott, Phyllis R. "The Development and Operation of an American Land System to 1800." Ph.D. diss., Univ. of Wisconsin, 1959. 456 pp.

A6 Aber, Ted, and King, Stella (Brooks). *The History of Hamilton County.* Lake Pleasant, New York: Great Wilderness Books, 1965. 1209 pp. Illus., maps. Includes some forest history of New York's Adirondack Mountains.

A7 Aberg, William J.P. *Conservation Reminiscences of Wm. J.P. Aberg.* OHI by Walter E. Scott and Bill Alderfer. Madison: Wisconsin Conservation Department and State Historical Society of Wisconsin, 1964. 22 pp. Processed. Reminiscences of his work with the Izaak Walton League, the National Wildlife Federation, the Wisconsin Conservation Commission, and other Wisconsin conservation organizations, since the 1920s.

A8 Abernethy, Thomas P. *Western Lands and the American Revolution.* New York: Appleton-Century, for the Institute for Research in the Social Sciences, University of Virginia, 1937. Reprint. Russell & Russell, 1964. xv + 413 pp. Illus., maps, bib. On the trans-Appalachian westward movement and its political consequences on the land policies of the colonies, British government, Continental Congress, and various states, from the mid-18th century to ca. 1800.

A9 Abramoske, Donald J. "The Public Lands in Early Missouri Politics." *Missouri Historical Review* 53 (July 1959), 295-305. Land policies and land reform proposals, 1819-1828.

A10 Abramoske, Donald J. "The Federal Land Leasing System in Missouri." *Missouri Historical Review* 54 (Oct. 1959), 27-38. From 1807 to 1829.

A11 Abrams, A. "Pulp and Paper, 1918-1938." *Chemical Industry* 42 (May 1938), 514-19.

A12 Abrey, Daniel. *Reminiscences of Daniel Abrey.* Corunna, Michigan: Sheardy, 1903. 195 pp. Autobiography of a Michigan lumberman, 1861-1901.

A13 Academie des Sciences, Paris. "Recently Acquired Botanical Documents." Ed. by Gilbert Chinard. *Proceedings of the American Philosophical Society* 101 (Dec. 1957), 508-22. Papers relating to the feasibility of obtaining timber in the United States for French naval use, and of naturalizing American trees in France, 1780s.

A14 Acker, W.B. *Memorandum History of the Department of the Interior.* Washington: GPO, 1913. 20 pp. Processed.

Ackerman, Edward A. See Whitaker, J. Russell, #W231

A15 Ackerman, Robert K. "South Carolina Colonial Land Policies." Ph.D. diss., Univ. of South Carolina, 1965. 207 pp.

A16 Adamovich, Laszlo, and Sziklai, Oszkar. *Foresters in Exile: The Sopron Forestry School in Canada.* Vancouver: University of British Columbia, Faculty of Forestry, 1970. ix + 67 pp. Illus., tables, apps. History of the Forest Engineering University of Sopron in Hungary and its exile after 1956 to the University of British Columbia.

A17 Adams, Alexander B. *John James Audubon: A Biography.* New York: G.P. Putnam's Sons, 1966. 510 pp. Illus., bib. Audubon (1785-1851), an artist and ornithologist, was a precursor of wildlife conservationists.

A18 Adams, Ansel. *There We Inherit: The Parklands of America.* San Francisco: Sierra Club, 1962. 103 pp. Illus., notes. Incidental history; largely a photographic tribute to the national parks.

A19 Adams, Bristow. "Cornell: An Appreciation." *Journal of Forestry* 35 (July 1937), 649-53. Forestry education at Cornell University since 1898.

A20 Adams, Kramer A. *Logging Railroads of the West.* Seattle: Superior Publishing Company, 1961. 144 pp. Illus., app., index. An illustrated history of logging railroads in eleven Western states since the 1880s. A useful roster of logging railroads is appended to the text.

A21 Adams, Kramer A. "Common Carriers in the Woods." *Forest History* 6 (Spring/Summer 1962), 10-11. On naming logging railroads in the West.

A22 Adams, Kramer A. *The Redwoods.* New York: Popular Library, 1969. 176 pp. Illus., bib., index. Includes some history of the lumber industry and the movement to preserve California redwoods.

A23 Adams, Kramer A. "Blue Water Rafting: The Evolution of Ocean Going Log Rafts." *Forest History* 15 (July 1971), 16-27. On log rafts and the Pacific lumber trade, 1880s-1940s.

A24 Adams, Sherman. *Firsthand Report: The Story of the Eisenhower Administration.* New York: Harper, 1961. 481 pp. Illus. Reprint. Westport, Connecticut: Greenwood Press, 1974. Adams, a forester, was an advisor to President Dwight D. Eisenhower in the 1950s.

A25 Adams, Sherman. "A Couple of Loggers." *Appalachia* 31 (Dec. 1965), 609-14. Reminiscences of Charlie Henderson, Leonard Thibodeau, and a logging camp in New Hampshire, 1926.

A26 Adams, W. Claude. "History of Papermaking in the Pacific Northwest." *Oregon Historical Quarterly* 52 (Mar. 1951), 21-37; 52 (June 1951), 83-100; 52 (Sept. 1951), 154-85. Individual companies and personnel in Oregon, Washington, and British Columbia, since 1866. The three installments were also issued under one cover in 1951 by Portland's Binfords & Mort.

A27 Adams, William R. "Florida Live Oak Farm of John Quincy Adams." *Florida Historical Quarterly* 51 (Oct. 1972), 129-42. On efforts to preserve live oak for naval purposes, 1830s.

A28 Addison, J.A., and Challenger, Jack. "History of the Power Saw." *Loggers Handbook* 9 (1949), 70-78. With capsule accounts of early chain saw manufacturers, since the 1930s.

A29 Adler, G. *Englands Versorgung mit Schiffbaumaterialien aus Englischen und Amerikanischen Quellen vornehmlich im 17. Jahrhundert.* Stuttgart, Germany: W. Kohlhammer, 1929. 118 pp. On 17th-century British shipbuilding and naval stores policy.

A30 Agee, James K. "Fire Management in the National Parks." *Western Wildlands* 1 (Summer 1974), 27-33. Includes some history of fire management philosophies and policies.

A31 Agerter, S.R., and Glock, W.S. *An Annotated Bibliography of Tree Growth and Growth Rings, 1950-1962.* Tucson: University of Arizona Press, 1965. 180 pp.

A32 Agnew, Dwight L. "The Government Land Surveyor as a Pioneer." *Mississippi Valley Historical Review* 28 (Dec. 1941), 369-82.

A33 Ahern, George P. *Deforested America: Statement of the Present Forest Situation in the United States.* Senate Document No. 216, 70th Congress, 2nd Session. Washington: GPO, 1929. 44 pp. Includes some historical references to forest depletion in each state and region. A privately printed version of 79 pages was issued in 1928.

A34 Ahern, George P. *Forest Bankruptcy in America: Each State's Own Story.* Washington: Green Lamp League, 1933. 319 pp. Bib., tables. A state by state survey of the forestry situation with some historical references.

Ahlgren, C.E. See Ahlgren, I.F., #A35

A35 Ahlgren, I.F., and Ahlgren, C.E. "Ecological Effects of Forest Fires." *Botanical Review* 26 (Oct.-Dec. 1960), 483-533. Includes some history of forest fires in the United States and abroad since 1884.

A36 Aiken, Samuel R. "The New-Found-Land Perceived: An Exploration of Environmental Attitudes in Colonial British America." Ph.D. diss., Pennsylvania State Univ., 1971. 323 pp.

A37 Aikens, Andrew J., and Proctor, Lewis A., eds. *Men of Progress.* Milwaukee: The Evening-Wisconsin Company, 1897. 640 pp. Illus. Biographical sketches of prominent Wisconsin lumbermen included.

Ainsworth, Charles P. See Wickstrom, George W., #W275

A38 Aitken, Hugh G.J. "The Changing Structure of the Canadian Economy, with Particular Reference to the Influence of the United States." In *The American Economic Impact on Canada.* Durham, North Carolina: Duke University Press, 1959. Pp. 3-35. On the displacement of Great Britain by the United States in the years since 1914 as the main buyer from, importer into, and foreign investor in Canada, and its effect upon the Canadian pulpwood and other extractive industries.

A39 Aitken, Hugh G.J. *American Capital and Canadian Resources.* Cambridge: Harvard University Press, 1961. xiv + 217 pp. Notes, tables, charts, app., index. Includes some account of American investment in Canadian raw materials and resource industries, among them lumber.

Alaska Travel Publications, eds. See Montague, Richard W., #M517

A40 *Albany Argus. The Albany Lumber Trade, Its History and Extent.* Albany, New York, 1872. 42 pp.

A41 Albason, William. "History of Willapa Harbor Lumbering: Three-Fourths of a Century Brings Many Changes in Ranks of Firms Which Pioneered This Section." *Timberman* 33 (June 1932), 20, 32. Pacific County, Washington.

A42 Albert, E.J. "Historical Background of the Delaware Valley Section." *TAPPI* 35 (May 1952), 148-150A. On the early meetings and individuals associated with this section of

TAPPI (Technical Association of the Pulp and Paper Industry).

A43 Albertson, Dean. "Roosevelt's Farmer: The Life of Claude R. Wickard." Ph.D. diss., Columbia Univ., 1955. 442 pp.

A44 Albertson, Dean. *Roosevelt's Farmer: Claude R. Wickard in the New Deal.* New York: Columbia University Press, 1961. Reprint. New York: DaCapo Press, 1975. xii + 424 pp. Illus., notes, index. Wickard served Presidents Franklin D. Roosevelt and Harry S. Truman as Secretary of Agriculture, 1940-1945.

A45 Albion, Robert G. "Forests and Sea Power: The Timber Problem of the English Navy from 1652 to 1862." Ph.D. diss., Harvard Univ., 1924.

A46 Albion, Robert G. *Forests and Sea Power: The Timber Problem of the Royal Navy, 1652-1862.* Harvard Economic Studies, Volume 29. Cambridge: Harvard University Press, 1926. Reprint. Hamden, Connecticut: Archon Books of Shoe String Press, 1965. xv + 485 pp. Illus., notes, apps., tables, bib., index. Includes several chapters on British forest and naval stores policy in North America, especially during the 18th century.

A47 Albion, Robert G., and Pope, Jennie Barnes. *The Rise of the New York Port (1815-1860).* New York: Scribner's, 1939. Reprint. New York: Newton Abbot, 1970. xiv + 485 pp. Illus., apps., bib., index. Includes some history of shipbuilding and of the trade in lumber and forest products.

A48 Albion, Robert G. "The Timber Problem of the Royal Navy, 1650-1860." *Mariner's Mirror* 38 (Feb. 1952). Includes English interest in American naval timber.

A49 Albright, Horace M., and Taylor, Frank J. *'Oh, Ranger!' A Book about the National Parks.* New York: Dodd, Mead & Company, 1941. xiv + 272 pp. Illus., index. Much description but also a chapter on the history of the national parks. Albright was director of the National Park Service from 1929 to 1933.

A50 Albright, Horace M. "John D. Rockefeller, Jr." *National Parks Magazine* 35 (Apr. 1961). On Rockefeller's philanthropy to national parks.

Albright, Horace M. See Miller, Leslie A., #M463

A51 Albright, Horace M., and Taylor, Frank A. "How We Saved the Big Trees." *Saturday Evening Post* 225 (Feb. 7, 1953), 31-32, 107-08. On the work of the U.S. National Park Service and the Save-the-Redwoods League in preserving coastal redwoods and groves of sequoia in California's Sierra Nevada.

Albright, Horace M. See Newhall, Nancy, #N88

A52 Albright, Horace M. "Harlan Page Kelsey." *National Parks Magazine* 33 (Feb. 1959), 12-13. On the career of Kelsey (1872-1958), including his promotion of national parks in North Carolina, Tennessee, Virginia, and New England.

A53 Albright, Horace M. *Great American Conservationists.* Horace M. Albright Conservation Lectures, No. 1. Berkeley: University of California, School of Forestry and Conservation, 1961. 30 pp.

A54 Albright, Horace M. "Harding, Coolidge, and the Lady Who Lost Her Dress." *American West* 6 (Sept. 1969), 24-32. Reminiscences by the former superintendent of Yellowstone National Park of famous visitors and problems of park administration during the 1920s.

A55 Albright, Horace M. *Origins of the National Park Service Administration of Historic Sites.* Philadelphia: Eastern National Park and Monument Association, 1971. 24 pp.

A56 Albright, Horace M. "The Great and Near-Great in Yellowstone." *Montana, Magazine of Western History* 22 (Summer 1972), 80-89. Albright was superintendent of Yellowstone National Park, 1919-1929, and director of the National Park Service, 1929-1933. Here he recollects noted park visitors, problems of civilian administration, and other events, 1913-1929.

A57 Alden, Vera. "A History of the Shelterbelt Project in Kansas." Master's thesis, Kansas State Univ., 1949.

Alderfer, Bill. See Aberg, William J.P., #A7

A58 Alexander, J.H.H. "Wisconsin Industry, Past and Present: The Golden Era of Lumbering." *Wisconsin Magazine* (Feb. 1930).

A59 Alexander, J.H.H. "Eighty-Four Years Ago Wisconsin Had but One Paper Mill." *Paper Industry* 14 (Dec. 1932), 67-73. On the growth of pulp and paper industry since the mid-19th century.

A60 Alexander, Marjorie. "Irasburg Mill." *Vermont Life* 6 (Spring 1952), 14-15. A sawmill on Vermont's Black River, since 1828.

A61 Alexander, R.H. "Growth of the Lumber Industry in British Columbia." *Canada Lumberman* 34 (Dec. 1, 1914), 34-35.

A62 Alexander, Robert R. "Thinning Lodgepole Pine in the Central Rocky Mountains." *Journal of Forestry* 58 (Feb. 1960), 99-104. On forest management in Colorado and Wyoming since 1924.

A63 Alexander, Thomas G. "The Federal Frontier: Interior Department Financial Policy in Idaho, Utah, and Arizona, 1863-1896." Ph.D. diss., Univ. of California, Berkeley, 1965. 437 pp. Includes some history of federal land surveys.

A64 Alexander, Thomas G. "The Powell Irrigation Survey and the People of the Mountain West." *Journal of the West* 7 (Jan. 1968), 48-54. Late 19th century.

A65 Alexander, Thomas G. "Senator Reed Smoot and Western Land Policy, 1905-1920." *Arizona and the West* 13 (Autumn 1971), 245-64. On his role in legislation affecting public lands and conservation.

Alexander, Thomas G. See Bloom, John Porter, #B317

A66 Allaby, Eric. *Shipbuilding in the Maritimes.* Toronto: Ginn and Company, 1970. 24 pp. Illus. Canadian Maritime provinces.

A67 Allan, A.M. "Some Lessons We Have Learned and Suggestions for the Future." *American Journal of Forestry* 1 (1883), 292-97.

A68 Allard, Dean C., Jr. "Spencer Fullerton Baird and the United States Fish Commission: A Study in the History of American Science." Ph.D. diss., George Washington Univ., 1967. 440 pp. Baird was commissioner, 1871-1887, and made many contributions to the conservation movement.

A69 Allard, Harry A., and Leonard, E.C. "The Canaan and the Stony River Valleys of West Virginia, Their Former Magnificent Spruce Forests, Their Vegetation and Floristics Today." *Castanea* 17 (Mar. 1952), 1-60. On the vegetation in part of Tucker County, West Virginia, denuded of its spruce forests by logging and fire between 1884 and 1924.

Allcorn, Bill. See Texas, General Land Office, #T45

A70 Allen, Alice Benson. *Simon Benson: Northwest Lumber King.* Portland: Binfords & Mort, 1971. 144 pp. Illus., map. Benson (1852-1942) was a Portland lumberman and philanthropist who introduced several technical innovations to Pacific Coast log rafting and to logging along the lower Columbia River.

A71 Allen, B.E. "Henry J. Malsberger." *Journal of Forestry* 48 (June 1950), 432-33. Sketch of Malsberger's career since 1925 with the Florida Forest Service and the Southern Pulpwood Conservation Association.

A72 Allen, Ben S. "Redwood Conservation Council's History Traced." *Western Conservation Journal* 14 (May-June 1957), 32, 67. Since 1949.

Allen, C.F.H. See Bullis, Raymond S., #B558

A73 Allen, Durward L. *Our Wildlife Legacy.* New York: Funk and Wagnalls, 1954. 422 pp. Illus., bib. Includes some history of wildlife conservation.

A74 Allen, Edward F. *A Guide to the National Parks of America.* New York: McBride, Nast, 1918. 338 pp. Illus., map. Incidental history.

A75 Allen, Edward Tyson. *Practical Forestry in the Pacific Northwest: Protecting Existing Forests and Growing New Ones, from the Standpoint of the Public and That of the Lumberman, with an Outline of Technical Methods.* Portland: Western Forestry and Conservation Association, 1911. 130 pp. Tables, apps. A text with some historical references.

A76 Allen, Edward Tyson. "America's Transition from Old Forests to New." *American Forestry* 29 (Feb. 1923), 67-71, 106; 29 (Mar. 1923), 163-68; 29 (Apr. 1923), 235-40; 29 (May 1923), 307-11. Some history of the lumber industry, forestry, and forest conservation prior to the 1920s.

A77 Allen, Edward Tyson. "Men, Trees and an Idea: The Genesis of a Great Fire Protective Plan." *American Forests and Forest Life* 32 (Sept. 1926), 529-32. On the origins of cooperative fire protection among Idaho lumbermen in 1906 and its subsequent expansion under the aegis of the Western Forestry and Conservation Association in the Pacific Northwest and California.

A78 Allen, G.F. *The Forests of Mount Rainier National Park.* Washington: USDI, 1916. 32 pp. Illus., index.

A79 Allen, G.S. "Forestry at the University of British Columbia." *Timber of Canada* 18 (No. 1, 1957), 27-32. Brief history of forestry education.

A80 Allen, J. Davis. "Twenty-five Years of Sawmilling in the South." *Southern Lumberman* (No. 603, 1906), 57-58.

A81 Allen, J. Davis. "The First Log Band Saw: An Historical Monograph." *Lumber World Review* 39 (July 10, 1920), 33-34. On band saw technology and evolution since 1870s.

A82 Allen, James B. *The Company Town in the American West.* Norman: University of Oklahoma Press, 1966. xvii + 205 pp. Illus., app., bib., notes, index. Includes a chapter on lumber company towns in the Pacific states and an extensive list of such towns in the appendix.

A83 Allen, James H. "History of Pulp and Paper in the South." *Paper Trade Journal* 105 (Oct. 28, 1937), 133-34.

A84 Allen, James H. "History of Making Pulp and Paper in the South." *Southern Pulp and Paper Journal* 1 (Dec. 1938), 9-13.

A85 Allen, John C. "A Half Century of Reforestation in the Tennessee Valley." *Journal of Forestry* 51 (Feb. 1953), 106-13. Includes some history of tree planting since 1890, with emphasis on reforestation efforts by the Tennessee Valley Authority since the 1930s.

A86 Allen, John M. *Indiana Pittman-Robertson Wildlife Restoration, 1939-1955.* Indianapolis: Indiana Division of Fish and Game, 1955. 240 pp.

A87 Allen, Milford F. "United States Government Exploring Expeditions and Natural History, 1800-1840." Ph.D. diss., Univ. of Texas, 1958. 252 pp.

A88 Allen, Richard S.; Gove, William; Maloney, Keith; and Palmer, Richard F. *Rails in the North Woods.* Lakemont, New York: North Country Books, 1972. On seven shortline railroads and their role in the lumber industry of Pennsylvania, New York, Vermont, and New Hampshire, since the mid-19th century.

A89 Allen, Ruth Alice. *East Texas Lumber Workers: An Economic and Social Picture, 1870-1950.* Austin: University of Texas Press, 1961. x + 239 pp. Illus., maps, tables, notes, bib., index. A historical study of the lumber industry and especially the social and economic condition of its working force in eastern Texas.

A90 Allen, Shirley W. "New Forests for Northern New York." *American Forests and Forest Life* 34 (Mar. 1928), 141-42, 181. On reforestation projects undertaken by the Northern New York Utilities Company, Watertown, New York, since 1914.

A91 Allen, Shirley W. "Conservation Aspects of the History of the Oregon and California Railroad Land Grant." Master's thesis, Iowa State College, 1929. Western Oregon.

A92 Allen, Shirley W. "E.T. Allen." *Journal of Forestry* 43 (Mar. 1945), 222-23. Edward Tyson Allen (1875-1942) was one of the first foresters in the Pacific Northwest, served as state forester of California, was an organizer of the Western Forestry and Conservation Association, and served as its forester and manager for over thirty years.

A93 Allen, Shirley W. *An Introduction to American Forestry.* American Forestry Series. 1938. Third edition. New York: McGraw-Hill, 1960. 466 pp. Illus., maps, tables, apps., index. A standard text containing many historical references to forests and forestry. G.W. Sharpe is coauthor of the third edition.

A94 Allen, Shirley W. *Conserving Natural Resources: Principles and Practice in a Democracy.* 1955. Third edition. New York: McGraw-Hill, 1966. 432 pp. Illus., maps, bib., index. A standard text which includes a chapter on forests and the national forest movement. The introduction presents a brief history of the conservation movement.

A95 Allerton, C.K. "Short History of Solid Fibreboard Cases." *Paper Making,* Special Annual Number (1946), 25-26.

A96 Alling & Cory Company. *One Hundred Years in the Paper Business, 1819-1919; Being a Brief History of the Founding of the Paper Business of the Alling and Cory Company, together with an Account of Its Growth during the Centenary of Its Existence.* Rochester, 1919. 21 pp. Illus., chart.

A97 Allison, Jack. "Development of Forest Fire Fighting Weapons." Master's thesis, Pennsylvania State College, 1933.

A98 Allison, John H., and Brown, R.M. *Management of the Cloquet Forest: Second Ten-Year Period.* Cloquet, Minnesota: University of Minnesota, Agricultural Experiment Station, 1945. 95 pp. Illus., tables. Includes some history of the Cloquet Experimental Forest in northern Minnesota, since 1910.

A99 Allison, John H. "Edward G. Cheyney." *Journal of Forestry* 46 (Sept. 1948), 692-93. On Cheyney's career in forestry education at the University of Minnesota, as member of the Minnesota Forestry Board, and as author of forestry texts and fiction, since 1896.

Allison, John H. See Dana, Samuel Trask, #D29

A100 Allison, John H. "St. Paul's Municipal Forest and Its 50 Years of Growth." *Ramsey County History* 1 (Spring 1964), 19-23. St. Paul, Minnesota.

A101 Allison, John H. "The Story of Samuel Green." *Conservation Volunteer* 30 (Sept.-Oct. 1967), 36-44; 30 (Nov.-Dec. 1967), 21-30. Green (1859-1910) was a pioneer forester, forestry educator, and conservationist in Minnesota.

A102 Allison, R.C. "Marketing of Lumber Produced by Sawmills in Pennsylvania." Master's thesis, Pennsylvania State Univ., 1960.

A103 Allyn, George W. *When Blue Earth County Was Young.* Madison Lake, Minnesota: N.p., 1919. 40 pp. Autobiography of Minnesota lumber dealer.

A104 Ambler, Charles Henry. *A History of Transportation in the Ohio Valley with Special Reference to Its Waterways, Trade, and Commerce from the Earliest Period to the Present Time.* Glendale, California: Arthur H. Clark Company, 1932. 465 pp. Illus., maps. Includes some history of shipbuilding and the lumber trade, chiefly on the Ohio River, before 1900.

A105 Ambridge, D.W. *Frank Harris Anson (1859-1923), Pioneer in the North.* New York: Newcomen Society in North America, 1952. 24 pp. Illus. Of the Abitibi Power & Paper Company, Ltd., Ontario.

A106 American Association for the Advancement of Science. *Memorial upon the Cultivation of Timber and the Preservation of Forests.* Washington, 1874. 6 pp. The memorial to Congress which resulted in its creation of a federal forestry agency in 1876.

A107 American Forestry Association. *The Progress of Forestry, 1945 to 1950.* Washington, 1952. vi + 90 pp. Map, tables, diags.

A108 American Forest Products Industries. *Progress in Private Forestry.* Washington, 1961. 49 pp. Illus., maps, tables, graphs. Includes a section on the history of forestry.

A109 American Forest Products Industries. *Government Land Acquisition: A Summary of Land Acquisition by Federal, State and Local Governments up to 1964.* Washington, 1965. iii + 127 pp. Tables. A largely statistical account.

A110 American Forestry Association. "The Progress of Forestry, 1945-1950." *American Forests* 57 (Oct. 1951), 45-52. Highlights of an AFA progress report on all phases of forestry and forest conservation.

A111 *American Forestry.* "The Walnut—Our National Tree?" *American Forestry* 26 (Aug. 1920), 461-71. Includes some history of its use since the 17th century.

A112 *American Forestry.* "Fifty Years of Arbor Days." *American Forestry* 28 (May 1922), 279-82. On the celebration of Arbor Day since its establishment in Nebraska by J. Sterling Morton in 1872.

A113 *American Forestry.* "Franklin B. Hough—A Tribute." *American Forestry* 28 (July 1922), 431-32. Hough (1822-1885) became the first federal forestry official in 1876 and served as head of the Division of Forestry, 1881-1883.

A114 *American Forests and Forest Life.* National Park Number. *American Forests and Forest Life* 35 (Aug. 1929), complete. Many articles on national parks and the National Park Service, with historical references.

A115 *American Forests.* National Forest Number. *American Forests* 36 (July 1930), complete. Many articles on the history of national forests and the USFS.

A116 *American Forests.* Anniversary Number. 41 (Sept. 1935), complete. Commemorates the 60th anniversary of the American Forestry Association. Includes many articles on the history of forest conservation. The issue also appeared as a book, *American Conservation in Picture and in Story,* edited by Ovid M. Butler in 1935.

A117 *American Forests.* Special Fire Prevention Number. *American Forests* 45 (Apr. 1939), complete. Many articles on forest fires and fire prevention, some with historical references.

A118 *American Forests.* "Pathfinders in Southern Forestry." *American Forests* 50 (May 1944), 340-42, 265. Brief evaluations of foresters W.W. Ashe, Charles H. Herty, Austin Cary, Henry E. Hardtner, and Wilbur R. Mattoon.

A119 *American Forests.* 75th Anniversary Issue. *American Forests* 56 (Oct. 1950), complete. Many articles (some listed below) on the history of the American Forestry Association, the forestry profession, and the conservation movement generally.

A120 *American Forests.* "Milestones Are Its History." *American Forests* 56 (Oct. 1950), 12-23. Chronicles major events in the history of forest conservation and the American Forestry Association since 1875.

A121 *American Forests.* "They Were Its Leaders." *American Forests* 56 (Oct. 1950), 18-19, 55-68. Biographical sketches and achievements of American Forestry Association presidents since 1875.

A122 *American Forests.* "The Printed Word." *American Forests* 56 (Oct. 1950), 24-25, 68. On the editorial history of the American Forestry Association's publications since 1875.

A123 *American Forests.* "Projects Widened Its Influence." *American Forests* 56 (Oct. 1950), 26-32. Chronicle of the American Forestry Association's action programs to "awaken the public to its forestry obligations," since 1917.

A124 *American Forests.* "Conservation Timetable." *American Forests* 56 (Oct. 1950), 33-54. Chronology of events in American forest conservation history since 1608.

A125 *American Forests.* Special Number ("AFA Salutes the Service"). *American Forests* 61 (Mar. 1955), complete. Commemorates the 50th anniversary of the USFS; includes many articles and features of a historical nature.

A126 *American Forests.* "Maryland's First State Forester." *American Forests* 62 (Oct. 1956), 38, 77-84. On Fred Wilson Besley (1872-1960) and his career since 1900—his assignments with the USFS in Kentucky, Texas, Nebraska, and Colorado; his service as state forester of Maryland, 1906-1942; his subsequent management of his own woodlands; and his book, *The Forests of Maryland* (1916).

A127 *American Forests.* "Ovid Butler, Pioneer in Forest Education." *American Forests* 66 (Mar. 1960), 27, 61. Butler (1881-1960) served with the USFS, 1907-1922, and as executive director of the American Forestry Association and editor of *American Forests,* 1922-1948. See editorial by James B. Craig, p. 9.

A128 *American Forests.* "A Capsule History of AFA." *American Forests* 66 (Aug. 1960), 2, 70-76. The American Forestry Association since 1875; separate texts in English, French, and Spanish.

A129 *American Forests.* "A Salute to Frank Hubachek at 80." *American Forests* 80 (Aug. 1974), 22-23, 48. Biographical and career sketch of a philanthropist, conservationist, and supporter of the Boundary Waters Canoe Area of the Superior National Forest, Minnesota.

A130 *American Lumberman.* "Yellow Pine Development." *American Lumberman* 53 (Jan. 7, 1899), 23. On the introduction of southern pine into northern markets in the 1880s.

A131 *American Lumberman.* "A Lumberman in the Senate." *American Lumberman* 53 (Feb. 11, 1899), 1. On Addison G. Foster, Tacoma lumberman (St. Paul and Tacoma Lumber Company) and U.S. senator from Washington, 1899-1905.

A132 *American Lumberman.* "Model Sawmill Plants, XVIII." *American Lumberman* 53 (Feb. 18, 1899), 59-62. On the F.B. Williams Cypress Company, Louisiana.

A133 *American Lumberman.* "James A. Tawney." *American Lumberman* 53 (Mar. 25, 1899), 1. U.S. representative from Minnesota, 1893-1911, and advocate of tariff protection on lumber.

A134 *American Lumberman.* "Frederick Weyerhaeuser." *American Lumberman* 54 (May 6, 1899), 1, 19. And the Mississippi River Logging Company, Rock Island, Illinois.

A135 *American Lumberman.* "James D. Lacey." *American Lumberman* 54 (May 13, 1899), 1, 19. Robinson & Lacey Company and James D. Lacey Company, brokers in southern and western timberlands since 1880.

A136 *American Lumberman.* "A Northwestern Lumberman." *American Lumberman* 54 (June 24, 1899), 1, 14. Sumner T. McKnight and the North Western Lumber Company, Eau Claire, Wisconsin, 1870s-1890s.

A137 *American Lumberman.* "Model Sawmill Plants, IX." *American Lumberman* 54 (June 24, 1899), 30-32. Ruddock Cypress Company, Ruddock, Louisiana.

A138 *American Lumberman.* "Manufacturer and Dealer." *American Lumberman* 55 (July 8, 1899), 1, 14. Samuel H. Fullerton, president of the Chicago Lumber and Coal Company, St. Louis, Missouri.

A139 *American Lumberman.* "Line Yard History and Methods." *American Lumberman* 55 (July 8, 1899), 13.

A140 *American Lumberman.* "The Hardwood Wholesaler." *American Lumberman* 55 (Aug. 5, 1899), 1, 26. Maurice Martin Wall of Buffalo Hardwood Lumber Company, Buffalo, New York.

A141 *American Lumberman* "A Climb to Success." *American Lumberman* 55 (Sept. 9, 1899), 1, 16. Charles A. Smith, owner of the C.A. Smith Lumber Company, Minneapolis, Minnesota.

A142 *American Lumberman.* "Model Sawmill Plants, XI." *American Lumberman* 55 (Sept. 16, 1899), 36-37. J.J. Newman Lumber Company, Hattiesburg, Mississippi.

A143 *American Lumberman.* "A Man." *American Lumberman* 55 (Sept. 23, 1899), 1, 30. Chauncey W. Griggs, Minnesota lumberman and partner with Addison G. Foster in the formation of the St. Paul & Tacoma Lumber Company, Tacoma, Washington, and twice a candidate for the U.S. Senate from Washington.

A144 *American Lumberman.* "Model Sawmill Plants, XII." *American Lumberman* 55 (Sept. 23, 1899), 28-29. Eastabuchie Lumber Company, Eastabuchie, Mississippi.

A145 *American Lumberman.* "A Worker and a Winner." *American Lumberman* 56 (Oct. 14, 1899), 1, 19. John Charles Turner of J.C. Turner Cypress Lumber Company, New York City.

A146 *American Lumberman.* "Model Sawmill Plants, XIII." *American Lumberman* 56 (Oct. 14, 1899), 28-29. Sierra Nevada Wood and Lumber Company, Overton, California.

A147 *American Lumberman.* "Model Sawmill Plants, XIV." *American Lumberman* 56 (Nov. 4, 1899), 29-30. Camp and Hinton Company, Lumberton, Mississippi.

A148 *American Lumberman.* "A Veteran Lumberman." *American Lumberman* 56 (Nov. 18, 1899), 1, 15. Alexander Stewart, owner of Alexander Stewart Lumber Company, Wausau and Merrill, Wisconsin. Stewart was a U.S. representative from Wisconsin, 1895-1901. See obituary, *ibid.* (June 1, 1912), 1.

A149 *American Lumberman.* "A Great and Model Lumber Plant." *American Lumberman* 56 (Dec. 2, 1899), 27-30. John Schroeder Lumber Company, Milwaukee, Wisconsin.

A150 *American Lumberman.* "The Maker of Opportunities." *American Lumberman* 56 (Dec. 9, 1899), 1, 20. Thomas Henry Shevlin, Minnesota lumberman with interests in Wisconsin and Ontario, 1890s.

A151 *American Lumberman.* "Model Sawmill Plants, XV." *American Lumberman* 56 (Dec. 9, 1899), 20-21. Stimson Mill Company, Ballard, Washington.

A152 *American Lumberman.* "Model Sawmill Plants, XVI." *American Lumberman* 57 (Jan. 13, 1900), 28-29. C.A. Smith Lumber Company, Minneapolis, Minnesota.

A153 *American Lumberman.* "Maryland's Lumberman Governor." *American Lumberman* 57 (Jan. 20, 1900), 27. John W. Smith of the Smith, Moore & Company, box manufacturers of Snow Hill, Maryland, elected governor of Maryland in 1900.

A154 *American Lumberman.* "Missouri and a Missourian." *American Lumberman* 57 (Jan. 20, 1900), 1, 19. Richard H. Keith, lumberman of Kansas City, Missouri.

A155 *American Lumberman.* "Model Sawmill Plants, XVII." *American Lumberman* 57 (Jan. 20, 1900), 28-29. Backus-Brooks Company, Minneapolis, Minnesota.

A156 *American Lumberman.* "A Business Evolution." *American Lumberman* 57 (Feb. 3, 1900), 1, 25. Robert Alexander Long, founder of the Long-Bell Lumber Company, Kansas City, Missouri.

A157 *American Lumberman.* "The Saw and Its Maker." *American Lumberman* 57 (Feb. 10, 1900), 1, 41. Elias C. Atkins, founder of a saw manufacturing firm, E.C. Atkins & Company, Indianapolis, Indiana.

A158 *American Lumberman.* "Nashville As a Hardwood Center." *American Lumberman* 57 (Feb. 17, 1900), 28-29. Brief sketches of five leading lumber companies in Nashville, Tennessee.

A159 *American Lumberman.* "Model Sawmill Plants, XVIII." *American Lumberman* 57 (Feb. 24, 1900), 28-29. Wheeler, Osgood and Company, Tacoma, Washington.

A160 *American Lumberman.* "Model Sawmill Plants, XIX." *American Lumberman* 57 (Mar. 10, 1900), 28-29. California Door Company, Oakland, California.

A161 *American Lumberman.* "Model Sawmill Plants, XX." *American Lumberman* 57 (Mar. 17, 1900), 36-37. Nebagamon Lumber Company, Lake Nebagamon, Wisconsin.

A162 *American Lumberman.* "From the Ranks to the Rank." *American Lumberman* 58 (Apr. 21, 1900), 1, 15. Charles M. Betts, wholesale lumberman of Philadelphia, Pennsylvania.

A163 *American Lumberman.* "A Great Business Organization." *American Lumberman* 58 (Apr. 21, 1900), 27-45. Includes some history of the Long-Bell Lumber Company, Kansas City, Missouri, and of affiliated firms.

A164 *American Lumberman.* "Model Sawmill Plants, XXI." *American Lumberman* 58 (Apr. 28, 1900), 27-31. Cummer Lumber Company, Jacksonville, Florida.

A165 *American Lumberman.* "A Pioneer Saw Maker." *American Lumberman* 58 (May 5, 1900), 1, 13. Joseph W. Branch and the Branch Saw Company, St. Louis, Missouri.

A166 *American Lumberman.* "The Growth of a Business." *American Lumberman* 58 (May 19, 1900), 28-29. Francis Beidler & Company, Chicago.

A167 *American Lumberman.* "The Man at the Lever." *American Lumberman* 58 (June 16, 1900), 1, 15. William Russell Pickering and the W.R. Pickering Lumber Company, Kansas City, Missouri.

A168 *American Lumberman.* "Model Sawmill Plants, XXII." *American Lumberman* 59 (July 21, 1900), 28-29. Pearl River Lumber Company, Brookhaven, Mississippi.

A169 *American Lumberman.* "Modern Selling Methods." *American Lumberman* 59 (Aug. 4, 1900), 1, 29. W.D. Johnston, president of the American Lumber and Manufacturing Company, Pittsburgh, Pennsylvania.

A170 *American Lumberman.* "The West Side Flume and Lumber Company, Carters, California." *American Lumberman* 59 (Sept. 29, 1900), 30-31.

A171 *American Lumberman.* "A Man Who Does Things." *American Lumberman* 60 (Oct. 13, 1900), 1, 26. Robert Laird McCormick, North Wisconsin Lumber Company, Hayward, Wisconsin, since 1881.

A172 *American Lumberman.* "A Worker and His Work." *American Lumberman* 60 (Oct. 27, 1900), 1, 19. Theodore

Stewart Fassett of Smith, Fassett & Company, a pioneer of the lumber industry in Buffalo and Tonawanda, New York.

A173 *American Lumberman*. "Model Sawmill Plants, XXIII." *American Lumberman* 60 (Nov. 10, 1900), 28-29. E.E. Jackson Lumber Company, Riderville, Alabama.

A174 *American Lumberman*. "The Story of William Cameron and of William Cameron and Co., Inc., Waco, Texas." *American Lumberman* 60 (Nov. 17, 1900), 43-59.

A175 *American Lumberman*. "Career of an Eastern Retailer." *American Lumberman* 60 (Dec. 8, 1900), 1, 21. George B. Swain of Swain & Jones Company, Newark, New Jersey, and an official of the New Jersey Lumbermen's Protective Association.

A176 *American Lumberman*. "Development of the Tensas Delta District of La." *American Lumberman* 60 (Dec. 29, 1900), 37-41. On the activities of lumbermen James D. Lacey, Wood Beal, and Victor Thrane in northeastern Louisiana.

A177 *American Lumberman*. "The Story of Calcasieu Yellow Pine." *American Lumberman* 61 (Jan. 5, 1901), 43-52. On the Bradley-Ramsay Lumber Company, Calcasieu Parish, Louisiana, since 1880.

A178 *American Lumberman*. "A Typical Southern Lumberman." *American Lumberman* 61 (Feb. 2, 1901), 1, 42. Biographical sketch of Peter G. Gates, who, with his brothers, held vast Southern pine lands and owned lumber companies in Arkansas, Louisiana, and Alabama.

A179 *American Lumberman*. "The Story of a Great Lumber Enterprise." *American Lumberman* 61 (Feb. 23, 1901), 27-30. Atlantic Coast Lumber Company, Georgetown, South Carolina.

A180 *American Lumberman*. "An Organization and a Man." *American Lumberman* 61 (Mar. 2, 1901), 1, 33. Richard S. White of the John C. Orr & Company, Brooklyn, New York, and president of the New York Lumber Trade Association.

A181 *American Lumberman*. "Inherited Ability: A Noble Family of Saw Makers." *American Lumberman* 61 (Feb. 23, 1901), 41; 62 (May 11, 1901), 36. Joshua Oldham & Sons, Inc., saw manufacturers of New York City since 1890.

A182 *American Lumberman*. "Decker Log Loader in the Market." *American Lumberman* 61 (Mar. 9, 1901), 32-33. Includes some history of the log loader and its manufacture by P.H. & F.M. Roots Company, Connersville, Indiana.

A183 *American Lumberman*. "Success in Youth." *American Lumberman* 62 (June 8, 1901), 1. Robert F. Whitmer, president of William Whitmer & Sons, Philadelphia, Pennsylvania, and Parsons Pulp and Paper Company, Parsons, West Virginia.

A184 *American Lumberman*. "Commercial Fraternity." *American Lumberman* 62 (June 29, 1901), 1, 34. Pendennis White of White, Gratwick & Company, North Tonawanda, New York.

A185 *American Lumberman*. "A Georgia Haunt of Red Cypress." *American Lumberman* 62 (June 29, 1901), 28-29. Red Cypress Lumber Company, Macon, Georgia.

A186 *American Lumberman*. "A Coast Type of the Modern Plant." *American Lumberman* 63 (July 13, 1901), 17-18. Far West Lumber Company, Tacoma, Washington.

A187 *American Lumberman*. "A Plain Lumberman Citizen." *American Lumberman* 63 (July 20, 1901), 1, 23. Alfred James Bond, lumberman of Bradford, Pennsylvania.

A188 *American Lumberman*. "R.H. Dowman and Louisiana Red Cypress." *American Lumberman* 63 (July 27, 1901), 27-34. President of several Louisiana companies.

A189 *American Lumberman*. "Maple Flooring Production." *American Lumberman* 64 (Dec. 14, 1901), 27-31. Cobbs and Mitchell, a flooring firm of Cadillac, Michigan.

A190 *American Lumberman*. "A Winner of the Forests." *American Lumberman* 64 (Dec. 21, 1901), 1, 24. On Jacob Cummer, a Michigan lumberman with interests in Florida and Virginia.

A191 *American Lumberman*. "The Acme of Maple Flooring Making." *American Lumberman* 66 (Apr. 5, 1902), 20-21. Mitchell Brothers, a flooring firm in Cadillac, Michigan.

A192 *American Lumberman*. "A Pioneer in Southern Pine." *American Lumberman* 66 (Apr. 19, 1902), 1, 49. William Grayson of Grayson-McLeod Company, St. Louis, Missouri.

A193 *American Lumberman*. "The Finer Manipulation of White Pine." *American Lumberman* 66 (Apr. 19, 1902), 20-21. Mershon, Schuette, Parker & Company, a wholesale lumber firm of Saginaw, Michigan.

A194 *American Lumberman*. "A Builder of Civilization." *American Lumberman* 66 (Apr. 26, 1902), 1, 44. On William Harris Laird of the Laird, Norton Company, Winona, Minnesota, since the 1850s.

A195 *American Lumberman*. "A Type of Southwestern Progress." *American Lumberman* 66 (May 17, 1902), 1, 23. Charles Warner Gates, president of the Crossett Lumber Company, Hamburg, Arkansas.

A196 *American Lumberman*. "The Commercial Forests of New Mexico." *American Lumberman* 66 (May 24, 1902), 35-42. Especially the white pine timberlands of the American Lumber Company of Chicago.

A197 *American Lumberman*. "Where Yellow Poplar Log and Lumber Stocks Are Found." *American Lumberman* 66 (May 24, 1902), 47-56. Yellow Poplar Lumber Company, Coal Grove, Ohio.

A198 *American Lumberman*. "One of the Largest Southern Pine Operations." *American Lumberman* 66 (May 31, 1902), 14-15. On the formation of the Jackson Lumber Company at Lockhart, Alabama, by Edward S. Crossett, J.W. Watzek, and Charles W. Gates.

A199 *American Lumberman*. "A Man Among Men." *American Lumberman* 66 (June 28, 1902), 1, 41. Charles Hebard, lumberman of Pennsylvania and Michigan, since 1853.

A200 *American Lumberman*. "Model Sawmill Plants, XXIV." *American Lumberman* 67 (July 5, 1902), 28-30. A.H. Stange Company, Merrill, Wisconsin.

A201 *American Lumberman.* "Midsummer Musings among the Memphis Hardwood Trade." *American Lumberman* 67 (Aug. 2, 1902), 31-41. The hardwood lumber industry in Memphis, Tennessee.

A202 *American Lumberman.* "A Half Century Personified." *American Lumberman* 67 (Aug. 16, 1902), 1, 39. Isaac Stephenson, prominent Wisconsin lumberman since the 1850s (N. Ludington & Company, Marinette; Peshtigo Lumber Company, Peshtigo; Menominee Boom Company), with redwood interests in California. Stephenson was U.S. representative from Wisconsin, 1883-1889, and U.S. senator, 1907-1915.

A203 *American Lumberman.* "Model Sawmill Plants, XV." *American Lumberman* 67 (Aug. 30, 1902), 28-29. East Union Lumber and Manufacturing Company, Brookhaven, Mississippi.

A204 *American Lumberman.* "Development of Broad Gaged Ideas." *American Lumberman* 67 (Sept. 6, 1902), 1, 24. Arthur Ross Rogers, Minnesota lumberman (C. A. Smith Lumber Company, Minneapolis), with interests in Oregon and California timberlands.

A205 *American Lumberman.* "A Plain Exponent of Commerce." *American Lumberman* 67 (Sept. 27, 1902), 1. Walter S. Eddy, lumberman, Saginaw, Michigan, with interests in Ontario and Tennessee, 1890s.

A206 *American Lumberman.* "America's Pioneer Hardwood Flooring Plant." *American Lumberman* 68 (Oct. 4, 1902), 27-32. T. Wilce Company, Chicago.

A207 *American Lumberman.* "A Forceful Quartet." *American Lumberman* 68 (Oct. 11, 1902), 1, 48. Brief history of the Blades Lumber Company, Elizabeth City, North Carolina.

A208 *American Lumberman.* "The Central Coal and Coke Company." *American Lumberman* 68 (Nov. 1, 1902), 47-70. Includes some history of the Kansas City company, with emphasis on its logging operations in Texas and Louisiana.

A209 *American Lumberman.* "A Great Eastern Michigan Flooring and Hardwood Lumber Concern." *American Lumberman* 68 (Nov. 8, 1902), 43-46. W.D. Young & Company, West Bay City, Michigan.

A210 *American Lumberman.* "Timber Resources of East Texas, Their Recognition and Development by John Henry Kirby through the Inception and Organization of the Kirby Lumber Company of Houston, Texas." *American Lumberman* 68 (Nov. 22, 1902), 43-78.

A211 *American Lumberman.* "The Rugged Sons of Maine." *American Lumberman* 68 (Dec. 6, 1902), 1, 45. On Joseph William Cochran, Maine-born lumberman with operations in Williamsport, Pennsylvania, 1860s-1880s, and later owner of the Keystone Lumber Company, Ashland, Wisconsin.

A212 *American Lumberman.* "A Story of the Old West." *American Lumberman* 69 (Jan. 24, 1903), 1, 69. On Edward Savage Crossett, Iowa lumberman who organized several companies in Arkansas, Florida, and Alabama.

A213 *American Lumberman.* "A Master of Opportunity." *American Lumberman* 69 (Feb. 14, 1903), 1, 50. On Nelson Augustus Gladding, affiliated with E.C. Atkins & Company, saw manufacturers of Indianapolis, Indiana, since the 1890s.

A214 *American Lumberman.* "Texas Enterprise as Exemplified at Orange." *American Lumberman* 69 (Feb. 21, 1903), 17-18. On the pulp and paper mill, logging operations, railroad facilities, and company town of the Orange Paper Company, Orange, Texas.

A215 *American Lumberman.* "The Industrial Army of America." *American Lumberman* 69 (Mar. 7, 1903), 1, 24. On James Dempsey, a Michigan lumberman (Manistee Lumber Company, Manistee) with interests in Arizona (Saginaw & Manistee Lumber Company, Williams) and in Oregon and Washington (Dempsey Lumber Company).

A216 *American Lumberman.* "A Notable Northern-Southern Lumber Enterprise." *American Lumberman* 69 (Mar. 14, 1903), 43-48. Wisconsin and Arkansas Lumber Company, Malvern, Arkansas.

A217 *American Lumberman.* "Cypress and a Cypress Specialist." *American Lumberman* 69 (Mar. 28, 1903), 1, 27. Frank B. Williams and the F.B. Williams Cypress Lumber Company, Patterson, Louisiana.

A218 *American Lumberman.* "The Pine Lumber of Montana." *American Lumberman* 70 (Apr. 4, 1903), 36-38. Big Blackfoot Milling Company, Montana.

A219 *American Lumberman.* "Versatility a Factor in Success." *American Lumberman* 70 (Apr. 11, 1903), 1, 29. On John Paul, Wisconsin lumberman (John Paul Lumber Company, LaCrosse) and inventor of sawmill equipment since 1860s. He also organized the East Coast Lumber Company, Watertown, Florida.

A220 *American Lumberman.* "Lumber Manufacturer's Executive." *American Lumberman* 70 (Apr. 25, 1903), 1, 71. Edgar Charles Fosburgh of Fosburgh Lumber Company, Norfolk, Virginia, and president of the National Lumber Manufacturers Association.

A221 *American Lumberman.* "Story of a Great Enterprise." *American Lumberman* 70 (May 9, 1903), 43-90. Includes some history of the following firms in Missouri and Louisiana: Cordz-Fisher Lumber Company, Louisiana Central Lumber Company, Louisiana Long Leaf Lumber Company, Missouri Lumber and Land Exchange, Missouri Lumber and Mining Company, and Ozark Land and Lumber Company.

A222 *American Lumberman.* "Changes and Ramifications of a Great Southern Company." *American Lumberman* 70 (May 23, 1903), 20-21. Allen-Wadley Lumber Company, Allentown, Louisiana.

A223 *American Lumberman.* "Forty Years of Faithful Toil." *American Lumberman* 70 (June 13, 1903), 1, 37-38. Seth Thomas Foresman of Bowman-Foresman Company, Williamsport, Pennsylvania, since 1877, and his timber interests in West Virginia.

A224 *American Lumberman.* "Self Sufficiency at Its Best." *American Lumberman* 70 (June 27, 1903), 1, 56. David Lindsay Gillespie, wholesale lumberman of Pittsburgh, Pennsylvania, since 1886.

A225 *American Lumberman.* "A Commercial Invasion." *American Lumberman* 71 (July 18, 1903), 1, 35-36. John M. Hastings, wholesale lumberman of Pittsburgh, Pennsylvania, with lumber interests in Nova Scotia, since 1875.

A226 *American Lumberman.* "Use of Inherited Opportunity." *American Lumberman* 71 (Aug. 8, 1903), 1, 48. Henry C. Atkins, president of a saw manufacturing firm, E.C. Atkins & Company, Indianapolis, Indiana.

A227 *American Lumberman.* "From Riverman to Lumberman." *American Lumberman* 71 (Sept. 12, 1903), 1, 66. John Henry Berkshire, associated with the Ozark Land & Lumber Company and other Missouri firms.

A228 *American Lumberman.* "The Business Man in Politics." *American Lumberman* 71 (Sept. 19, 1903), 1, 52. Joseph Warren Fordney, lumberman of Saginaw, Michigan, and U.S. representative, 1899-1923.

A229 *American Lumberman.* "Saw Milling in Mississippi." *American Lumberman* 72 (Oct. 3, 1903), 43-48. Camp and Hinton Company, Lumberton, Mississippi.

A230 *American Lumberman.* "Of Pioneer Ancestry." *American Lumberman* 72 (Oct. 10, 1903), 1, 46. William Garrett Wadley, Arkansas lumberman with interests in Texas, Louisiana, and Missouri.

A231 *American Lumberman.* "Mill Work Manufacture in Georgia." *American Lumberman* 72 (Oct. 24, 1903), 34-35. O'Neill Manufacturing Company, Rome and Tifton, Georgia.

A232 *American Lumberman.* "The Science of Wood Preservation." *American Lumberman* 72 (Oct. 31, 1903), 44-46. International Creosoting and Construction Company, Galveston, Texas.

A233 *American Lumberman.* "A Producer of Powerful Machinery." *American Lumberman* 72 (Nov. 28, 1903), 50b-50c. Brown Corliss Engine Company, Corliss, Wisconsin, manufacturers of sawmill machinery.

A234 *American Lumberman.* "The Manufacture of Fine Veneers." *American Lumberman* 72 (Dec. 19, 1903), 52-53. Coe Manufacturing Company, Painesville, Ohio, pioneers in the making of veneering machinery.

A235 *American Lumberman.* "Appreciation of Opportunity." *American Lumberman* 72 (Dec. 26, 1903), 1, 19. Delos A. Blodgett, Michigan lumberman since 1848, and his investments in southern timber.

A236 *American Lumberman.* "A Young-Old Man." *American Lumberman* 73 (Jan. 2, 1904), 1, 48. James Edwin Lindsay, owner of Lindsay & Phelps Company, Davenport, Iowa, Lindsay Land & Lumber Company, Arkansas, and other interests in Washington and Louisiana, since 1861.

A237 *American Lumberman.* "The Warrant for Use of a Royal Name." *American Lumberman* 73 (Jan. 9, 1904), 43-46. Monarch Lumber Company, St. Louis, Missouri.

A238 *American Lumberman.* "Business Experience in Banking." *American Lumberman* 73 (Jan. 16, 1904), 1, 63. Charles Fremont Latimer, Wisconsin lumberman (Ashland Lumber Company, Ashland) with interests in Oregon (Peninsula Lumber Company, Portland), Louisiana (Lyon Cypress Company), and California (West Coast Timber Company).

A239 *American Lumberman.* "Utilizing a Great Discovery." *American Lumberman* 73 (Jan. 24, 1904), 1, 82. Thomas J. Shrycock, a wholesale lumber dealer of Baltimore with operations in West Virginia.

A240 *American Lumberman.* "Possibilities in Small Beginnings." *American Lumberman* 73 (Feb. 13, 1904), 1, 59. Peter Musser, owner of Musser Lumber Company, Muscatine, Iowa, and his interests in Wisconsin, Minnesota, Washington, Idaho, and the South, since 1870.

A241 *American Lumberman.* "In the Van of Young Lumberman." *American Lumberman* 73 (Mar. 5, 1904), 1, 81. John B. Flint, wholesale lumberman with Flint, Erving & Stoner Company, Pittsburgh, Pennsylvania.

A242 *American Lumberman.* "The Story of a Yellow Pine Sextet." *American Lumberman* 73 (Mar. 5, 1904), 43-74. Includes some history of the lumber interests of the Gates brothers in Arkansas, Louisiana, and Alabama.

A243 *American Lumberman.* "Greatest of Wholesale Lumbermen." *American Lumberman* 73 (Mar. 19, 1904), 1, 53. Edward Hines of Edward Hines Lumber Company, Chicago, with operations in the Great Lakes states and the South.

A244 *American Lumberman.* "Southern Pine from Forest to Retailer." *American Lumberman* 74 (Apr. 16, 1904), 35-42. Southern Pine Lumber Company, Texarkana, Arkansas, and Diboll, Texas.

A245 *American Lumberman.* "Founder of Retailers' Associations." *American Lumberman* 74 (Apr. 23, 1904), 1, 43. Franklin Darwin Arnold, Iowa lumberman and president of the Iowa Retail Lumber Dealers' Association, which in 1877 became the National Retail Lumber Dealers' Association.

A246 *American Lumberman.* "The Financial End in Lumbering." *American Lumberman* 74 (June 11, 1904), 1, 57. Harvey J. Hollister of Fosburgh Lumber Company, Norfolk, Virginia.

A247 *American Lumberman.* "A Retail Association President." *American Lumberman* 74 (June 25, 1904), 1, 47. John W. Barry, Iowa retail lumberman and president of the Northwestern Lumbermen's Association, since the 1880s.

A248 *American Lumberman.* "Wisconsin Hemlock and Hardwoods." *American Lumberman* 74 (June 11, 1904), 36-37. Foster-Latimer Company, Mellen, Wisconsin.

A249 *American Lumberman.* "From Tree to Trade in Yellow Pine." *American Lumberman* 75 (July 2, 1904), 47-116. Includes some history of the Long-Bell Lumber Company of Kansas City, Missouri, and its various mills.

A250 *American Lumberman.* "A Type of the Vigorous American." *American Lumberman* 75 (July 9, 1904), 1, 47-48. Samuel "Diaz" Park, lumberman with interests in Texas, Louisiana, and Mexico, since the 1870s.

A251 *American Lumberman.* "Cloquet, Home of White Pine." *American Lumberman* 75 (July 9, 1904), 35-46. Includes some history of the lumber industry and sawmills at Cloquet, Minnesota.

A252 *American Lumberman.* "The Systematized and Expert Handling of Fir Lumber." *American Lumberman* 75 (July 30, 1904), 30-32. On the Lumber Manufacturer's Agency of Centralia, Washington, a marketing agency for eleven Douglas fir lumber companies.

A253 *American Lumberman.* "Manufacture of Sash and Doors at St. Louis." *American Lumberman* 75 (Aug. 6, 1904),

48-53. History and operations of Hafner Manufacturing Company, St. Louis, Missouri.

A254 *American Lumberman.* "The Uses of Prosperity." *American Lumberman* 75 (Aug. 20, 1904), 1, 49. Peter Miller Musser, nephew of Peter Musser, affiliated with Musser Lumber Company, Muscatine, Iowa, and other companies in Minnesota, Wisconsin, and Washington, since the 1870s.

A255 *American Lumberman.* "Profiting by Experience." *American Lumberman* 75 (Sept. 24, 1904), 1, 55. On Wellington R. Burt, Michigan lumberman (W.R. Burt & Co.) in Saginaw since 1869. Burt owned a lumber yard in Buffalo, New York, and operated Burt & Brabb Lumber Company of Ford, Kentucky, manufacturer of poplar lumber.

A256 *American Lumberman.* "How Rich Harvests of California Pine Are Secured." *American Lumberman* 76 (Oct. 1, 1904), 24-25. Sierra Lumber Company, Red Bluff, California.

A257 *American Lumberman.* "A Doer of Big Things." *American Lumberman* 76 (Oct. 8, 1904), 1, 53. On Frank William Gilchrist, Michigan lumberman and Great Lakes lumber trader (Gilchrist Transportation Company), since the 1860s.

A258 *American Lumberman.* "Some History and Suggestions Relative to Maple Flooring." *American Lumberman* 76 (Oct. 22, 1904), 16-17. Especially in Michigan.

A259 *American Lumberman.* "Canada's Foremost Lumberman." *American Lumberman* 76 (Oct. 29, 1904), 1, 51. John R. Booth, a prominent lumberman in the Ottawa Valley of Ontario and Quebec in the late 19th century.

A260 *American Lumberman.* "Of Sturdy Stock." *American Lumberman* 76 (Nov. 5, 1904), 1, 53. Lucius Kellogg Baker, Wisconsin lumberman with interests in Arkansas and Virginia during the 1890s.

A261 *American Lumberman.* "Ambition Backed by Forcefulness." *American Lumberman* 76 (Nov. 12, 1904), 1, 52. Elbert Milton Wiley, North Carolina lumberman with retail outlets in New York and other northern cities.

A262 *American Lumberman.* "A Man and a Mart." *American Lumberman* 76 (Nov. 19, 1904), 1, 51. Orson Ellsworth Yeager, hardwood wholesaler of Buffalo, New York, and president of the Buffalo Hardwood Exchange.

A263 *American Lumberman.* "In Forefront of Hemlock." *American Lumberman* 76 (Dec. 3, 1904), 1, 57. Charles Sumner Horton of the Central Pennsylvania Lumber Company, Williamsport, Pennsylvania, and his previous interests in the tanning business.

A264 *American Lumberman.* "Faith and Confidence." *American Lumberman* 76 (Dec. 10, 1904), 1, 46. On Matthew George Norton, Minnesota lumberman (Laird, Norton Company, Winona; Winona Lumber Company, Winona; Mississippi River Logging Company) since the 1850s.

A265 *American Lumberman.* "A Lumberman from Youth." *American Lumberman* 76 (Dec. 24, 1904), 1, 53. John Jacob Rumbarger, a lumberman of Dobbin, West Virginia, with operations in Coketon, West Virginia, and Philadelphia.

A266 *American Lumberman.* "The Heart of Arkansas' Shortleaf Pine Belt and Phases of Its Growth." *American Lumberman* 76 (Dec. 31, 1904), 35-40. A.J. Neimeyer Lumber Company, Little Rock, Arkansas.

A267 *American Lumberman.* "Genesis and Personnel of the Fourche River Lumber Company, of Esau, Arkansas." *American Lumberman* 77 (Jan. 21, 1905), 45-46.

A268 *American Lumberman.* "Light in a Dark Corner." *American Lumberman* 77 (Jan. 28, 1905), 51-82. C.D. Johnson and his lumber enterprises in Arkansas and Louisiana, including the Union Saw Mill Company and the Little Rock & Monroe Railway Company.

A269 *American Lumberman.* "Lumberman and Legislator." *American Lumberman* 77 (Feb. 18, 1905), 1, 71. Elias Deemer, lumberman of Williamsport, Pennsylvania, since 1868, and U.S. representative after 1900.

A270 *American Lumberman.* "An Honor to the Far West." *American Lumberman* 77 (Mar. 4, 1905), 1, 51. Samuel S. Johnson engaged in the lumber business in Michigan and Minnesota, 1870s-1903. He became manager and president of Scott & Van Arsdale Lumber Company (later McCloud River Lumber Company) in California after 1904.

A271 *American Lumberman.* "Honored Through Service." *American Lumberman* 77 (Mar. 11, 1905), 1, 29. Lewis Dill, Baltimore wholesale lumber dealer and president of the National Wholesale Lumber Dealers' Association.

A272 *American Lumberman.* "True to His Traditions." *American Lumberman* 77 (Mar. 18, 1905), 1, 35. On Selwyn Eddy, a Maine-born lumberman operating in Michigan near Bay City (Eddy Bros. & Co.), 1870s and 1880s. He also had interests in Canada and California (Pacific Lumber Company, Scotia).

A273 *American Lumberman.* "Sugar and White Pine Manufacture in California." *American Lumberman* 78 (Apr. 8, 1905), 35-38. McCloud River Lumber Company, McCloud, California.

A274 *American Lumberman.* "A Long Life without Blemish." *American Lumberman* 78 (May 13, 1905), 1, 65. Addison P. Brewer, Michigan surveyor and lumberman since the 1840s, member of A.P. Brewer & Sons, Saginaw, Michigan, and Brewer, Knapp & Company of Ashland, Wisconsin, and Portland, Oregon.

A275 *American Lumberman.* "An Embodiment of Perseverance." *American Lumberman* 79 (July 29, 1905), Samuel Evans Slaymaker, a West Virginia lumberman with operations in Pennsylvania and New York.

A276 *American Lumberman.* "Profiting by Advantages." *American Lumberman* 79 (July 15, 1905), 1, 63. John W. Blodgett, Michigan lumberman who took over his father's business (Blodgett & Byrne Lumber Company) in Muskegon in 1882 and became president of the Muskegon Boom Company in 1886. Later he invested in timberlands in the South and Pacific Northwest.

A277 *American Lumberman.* "A Journey Through The Vast Dowman Cypress Interests with Camera and Pen." *American Lumberman* 79 (Aug. 5, 1905), 1, 42-82. R.H. Dowman inherited the operations of William Cameron to become the nation's largest manufacturer of cypress lumber. Includes some history of the five Louisiana companies presided over by Dowman: Whitecastle Lumber and Shingle Company, Whitecastle; Bowie Lumber Company, Bowie; Jeanerette Lumber & Shingle Company, Jeanerette; Iberia Cypress Company, New Iberia; and Des Allemands Lumber Company, Allemands.

A278 *American Lumberman.* "Success Through Wise Deviation." *American Lumberman* 79 (Aug. 12, 1905), 1, 41. Frank Ellsworth Sheldon, a lumberman with extensive operations throughout the South.

A279 *American Lumberman.* "A Story of Two Brothers." *American Lumberman* 79 (Sept. 16, 1905), 1, 63. Alexander Wilson and Frank Elliott Wilson, lumbermen of Pittsburgh, Pennsylvania, with extensive interests in North Carolina, South Carolina, and West Virginia.

A280 *American Lumberman.* "Alfred Gustavus Hauenstein." *American Lumberman* 80 (Oct. 7, 1905), 1, 61. Hauenstein, a wholesale lumber dealer in Buffalo, New York, since 1876.

A281 *American Lumberman.* "James Jennings Mead." *American Lumberman* 80 (Nov. 4, 1905), 1, 59. And the wholesale lumber business in St. Louis, Missouri; Toledo, Ohio; and Pittsburgh, Pennsylvania.

A282 *American Lumberman.* "The Evolution of a Great Manufacturer of Woodworking Machinery." *American Lumberman* 80 (Nov. 4, 1905), 44-45. Stearns Manufacturing Company, Erie, Pennsylvania, manufacturers of sawmill machinery.

A283 *American Lumberman.* "Henry S. Thayer." *American Lumberman* 80 (Nov. 11, 1905), 1, 59. Pennsylvania lumberman.

A284 *American Lumberman.* "Influence of a Railroad on the Lumber Trade of Mississippi." *American Lumberman* 80 (Nov. 11, 1905), 29-30. The Gulf and Ship Island Railroad.

A285 *American Lumberman.* "Jacob Louis Kendall." *American Lumberman* 80 (Nov. 18, 1905), 1, 57. Associated with the H.C. Huston Lumber Company, Pennsylvania, 1881-1902.

A286 *American Lumberman.* "D. Clint Prescott." *American Lumberman* 80 (Nov. 25, 1905), 1, 50b-50d. And the Prescott Company, Menominee, Michigan, manufacturer of band sawmills.

A287 *American Lumberman.* "George M. Hinckley." *American Lumberman* 80 (Dec. 23, 1905), 1, 37. Supervisor of the sawmill machinery department, Allis-Chalmers Company.

A288 *American Lumberman.* "John L. Kaul." *American Lumberman* (Feb. 3, 1906), 1, 27. A former Pennsylvanian who organized the Kaul Lumber Company in Alabama and became president of the Yellow Pine Manufacturers' Association.

A289 *American Lumberman.* "Russell Alexander Alger." *American Lumberman* (Mar. 3, 1906), 1, 53. Michigan lumberman with interests also in Minnesota, Florida, and Alabama; Alger was governor of Michigan, 1885-1887, Secretary of War in the McKinley administration, 1897-1899, and U.S. senator from Michigan, 1902-1907.

A290 *American Lumberman.* "Wood Beal." *American Lumberman* (Mar. 31, 1906), 1, 63. Affiliated since 1889 with James D. Lacey Company, timberland brokers.

A291 *American Lumberman.* "Edwin Ambrose Frost." *American Lumberman* (Apr. 14, 1906), 1, 61. A lumberman with operations in Texas and Louisiana.

A292 *American Lumberman.* "Twelve Years in Yellow Pine: Being an Intimate Story of the Rise and Progress of the W.R. Pickering Lumber Company of Missouri, Louisiana, and Texas." *American Lumberman* (Apr. 21, 1906), 47-94.

A293 *American Lumberman.* "Rowland Harold Ewing." *American Lumberman* (Apr. 28, 1906), 1, 65. Wholesale lumberman of Pittsburgh, Pennsylvania.

A294 *American Lumberman.* "Daniel Simonds." *American Lumberman* (May 19, 1906), 1, 42A. And the Simonds Manufacturing Company, makers of circular saws since 1876.

A295 *American Lumberman.* "Robert Fullerton." *American Lumberman* (May 26, 1906), 1, 33. President of Chicago Lumber and Coal Company of Chicago, with sawmills and retail lumber yards in Louisiana, Mississippi, Arkansas, and Wisconsin.

A296 *American Lumberman.* "Perfected Manufacture of Hardwood Flooring." *American Lumberman* (July 14, 1906), 60-61. Kerry and Hanson Company, Grayling, Michigan.

A297 *American Lumberman.* "William Edmund Ramsay." *American Lumberman* (June 23, 1906), 1, 59. And the Bradley-Ramsay Lumber Company, Lake Charles, Louisiana, since 1887.

A298 *American Lumberman.* "Clement E. Lloyd, Jr." *American Lumberman* (June 30, 1906), 1, 38. Associated with the Cherry River Boom and Lumber Company in Pennsylvania.

A299 *American Lumberman.* "Arthur Hill." *American Lumberman* (Aug. 11, 1906), 1, 31. Michigan civil engineer and lumberman with operations in Minnesota, Ontario, and the Pacific states.

A300 *American Lumberman.* "Arthur Clark Ramsay." *American Lumberman* (Sept. 15, 1906), 1, 41. A lumberman with interests in Missouri and Arkansas.

A301 *American Lumberman.* "Victor Thrane." *American Lumberman* (Sept. 29, 1906), 1, 47. Affiliated since 1900 with James D. Lacey Company, timberland brokers.

A302 *American Lumberman.* "Nathan B. Bradley." *American Lumberman* (Nov. 17, 1906), 1, 61. Michigan lumberman with sawmills at Bay City and elsewhere in the state since the 1840s; he was also a partner in the Bradley-Ramsay Lumber Company, Lake Charles, Louisiana.

A303 *American Lumberman.* "The Wonderful Resources of the Arkansas Lumber Company." *American Lumberman* (Nov. 24, 1906), 59-86. On its operations at Warren, Bradley County, Arkansas, including biographical sketches of lumbermen M.F. Rittenhouse and J.W. Embree.

A304 *American Lumberman.* "Striking Changes in the Geography of the Lumber Traffic." *American Lumberman* (Dec. 22, 1906), 27.

A305 *American Lumberman.* "Pioneering in the Michigan-Chicago Lumber Trade—First Deck Load from St. Joe." *American Lumberman* (Dec. 22, 1906), 28-29. Alfred L. Driggs and logging on Michigan's St. Joseph River, 1832.

A306 *American Lumberman.* "Rufus Farrington Sprague." *American Lumberman* (Dec. 29, 1906), 1, 55. Manufacturer of sawmill machinery since the 1860s.

A307 *American Lumberman. American Lumbermen: The Personal History and Public and Business Achievements of One Hundred Eminent Lumbermen of the United States.* Three Series (Volumes). Chicago, 1905, 1906. Illus. Three hundred biographical sketches of lumbermen. Many were featured in early issues of the *American Lumberman.*

A308 *American Lumberman.* "Andrew McBride Turner." *American Lumberman* (Jan. 12, 1907), 1, 39. A Pennsylvania lumberman and president of the American Lumber & Manufacturing Company.

A309 *American Lumberman.* "The Old and the New in the Retail Lumber Business." *American Lumberman* (Jan. 19, 1907), 33. Changes since 1890 in the Great Lakes states.

A310 *American Lumberman.* "Edwin Stanton Nail." *American Lumberman* (Jan. 26, 1907), 1, 62. President and manager of the Lumbermen's Mutual Insurance Company.

A311 *American Lumberman.* "Frank Raymond Whiting." *American Lumberman* (Feb. 2, 1907), 1, 58. A Philadelphia hardwood manufacturer and distributor.

A312 *American Lumberman.* "Merritt Ashmun Potter." *American Lumberman* (Feb. 16, 1907), 1, 62. In the saw manufacturing business since 1878 and affiliated with E.C. Atkins & Company, Indianapolis, Indiana.

A313 *American Lumberman.* "John Leighton Alcock." *American Lumberman* (Mar. 9, 1907), 1, 35. Baltimore hardwood exporter and president of the National Exporters' Association.

A314 *American Lumberman.* "John M. Hastings." *American Lumberman* (Mar. 16, 1907), 50. Lumberman, exporter, and president of the National Wholesale Lumber Dealers' Association.

A315 *American Lumberman.* "A Graphic Story of the Frost-Trigg Interests in Louisiana, Arkansas, and Texas." *American Lumberman* (Mar. 30, 1907), 51-114. Includes some history and biographical sketches of lumbermen with the following companies: Red River Lumber Company, Noble Lumber Company, Union Saw Mill Company, DeSoto Land and Lumber Company, Black Lake Lumber Company, and Star and Crescent Lumber Company.

A316 *American Lumberman.* "Joseph Harry Foresman." *American Lumberman* (Jan. 11, 1908), 1, 63. Associated with Long-Bell Lumber Company, Kansas City, Missouri.

A317 *American Lumberman.* "Neches Valley Pine." *American Lumberman* (Jan. 18, 1908), 67-106. Southern Pine Company, Diboll, Texas.

A318 *American Lumberman.* "William A. Wimbish." *American Lumberman* (Feb. 1, 1908), 1, 46. A lumbermen's attorney in many railroad rate cases.

A319 *American Lumberman.* "Machinery Supplies for the Southwest." *American Lumberman* (Mar. 7, 1908), 101-12. On the Henderson Iron Works and Supply Company, Shreveport, Louisiana, manufacturers of sawmill machinery since 1892.

A320 *American Lumberman.* "Fred J. Bannister." *American Lumberman* (Mar. 14, 1908), 1, 109. Associated with Long-Bell Lumber Company, Kansas City, Missouri.

A321 *American Lumberman.* "Yellow Poplar, as Scientifically Produced by the Yellow Poplar Lumber Co., Coal Grove, Ohio." *American Lumberman* (Mar. 21, 1908), 63-110. Includes incidental historical references to the company and its logging operations in the Cumberland Mountains.

A322 *American Lumberman.* "The Manufacture of Mississippi White Oak by the Carrier Lumber & Mfg. Company, at Sardis, Mississippi." *American Lumberman* (Apr. 18, 1908), 71-87.

A323 *American Lumberman.* "Yellow Poplar of the Ohio River Produced and Distributed by the W.H. Dawking Lumber Company, Ashland, Kentucky." *American Lumberman* (May 23, 1908), 63-79.

A324 *American Lumberman.* "Perfection in Shingle Machinery Manufacture." *American Lumberman* (June 13, 1908), 91-94, 116-21. Challoner Company, Oshkosh, Wisconsin.

A325 *American Lumberman.* "'Bliss-Cook Oak': Its Conversion into Lumber by the Bliss-Cook Oak Company, Blissville, Arkansas." *American Lumberman* (June 20, 1908), 63-81.

A326 *American Lumberman.* "The Largest Lumber Manufacturing Proposition in the World." *American Lumberman* (July 4, 1908), 53-68. Great Southern Lumber Company, Bogalusa, Louisiana.

A327 *American Lumberman.* "Mack Barnabas Nelson." *American Lumberman* (July 18, 1908), 3, 59. A former Texas lumberman associated with the Long-Bell Lumber Company, Kansas City, Missouri.

A328 *American Lumberman.* "Uriah Spray Epperson." *American Lumberman* (July 25, 1908), 1, 63. Organized Lumbermen's Underwriting Alliance.

A329 *American Lumberman.* "C. Fred Yegge." *American Lumberman* (Aug. 22, 1908), 1, 51. President of the National Association of Box Manufacturers.

A330 *American Lumberman.* "Charles Holden Prescott, Jr." *American Lumberman* (Aug. 29, 1908), 1, 54. Of C.H. Prescott & Son, Michigan, and president of the National Wholesale Lumber Dealers' Association.

A331 *American Lumberman.* "Frederic S. Underhill." *American Lumberman* (Sept. 5, 1908), 1, 61. A Philadelphia lumberman.

A332 *American Lumberman.* "George Franklin Hawley." *American Lumberman* (Sept. 19, 1908), 1, 57. Eastern Lumber Company, Tonawanda, New York.

A333 *American Lumberman.* "Lone Star Pine." *American Lumberman* (Sept. 26, 1908), 67-150. John Martin Thompson and family and some history of the Thompson lumber interests in Texas since the 1850s.

A334 *American Lumberman.* "Revolutionary Idea in Lumber Manufacture." *American Lumberman* (Oct. 31, 1908), 43-46. Concerns the aggregate unit system of lumber handling with monorail, developed by the Grayson-McLeod Lumber Company, Graysonia, Arkansas.

A335 *American Lumberman.* "James H. De Veuve." *American Lumberman* (Nov. 7, 1908), 1, 43. Organized mutual insurance for lumbermen in the Pacific Coast states.

A336 *American Lumberman.* "Eugene Franklin Perry." *American Lumberman* (Dec. 12, 1908), 1, 86. A Connecticut and New York lumberman who served as secretary of the National Wholesale Lumber Dealers' Association.

A337 *American Lumberman.* "A Big Factor in the Yellow Pine Trade: A Portrayal of the Hogg-Harris Lumber Co., St. Louis, and Its Associated Manufacturing Institutions." *American Lumberman* (Jan. 30, 1909), 59-74. Includes some history of the following Louisiana lumber-producing firms: J.F. Ball & Brothers, White Sulphur Company, T.U. Norton Lumber Company, and Magnolia Manufacturing Company.

A338 *American Lumberman.* "Raymond Thomas Jones." *American Lumberman* (Feb. 27, 1909), 1, 62. A wholesale lumberman of North Tonawanda, New York, since 1883.

A339 *American Lumberman.* "E. Pemberton Gebhard and E.L. Polk." *American Lumberman* (Mar. 20, 1909), 1, 61. Both were associated with Strasburg Manufacturing Company, a hardwood flooring firm of Strasburg, Virginia.

A340 *American Lumberman.* "Willapa Harbor, A Great Saw Mill Port on the Pacific Coast." *American Lumberman* (Mar. 27, 1909), 58-59. Includes historical statistics on lumber production in Pacific County, Washington.

A341 *American Lumberman.* "Hon. Joseph Warren Fordney." *American Lumberman* (Apr. 3, 1909), 1, 40. Lumberman of Saginaw, Michigan, and U.S. representative, 1899-1923.

A342 *American Lumberman.* "Huie-Hodge Lumber Company, Ltd., Hodge, La." *American Lumberman* (May 8, 1909), 67-82.

A343 *American Lumberman.* "Nelson Platt Wheeler." *American Lumberman* (July 10, 1909), 31. A third-generation Pennsylvania lumberman associated with Wheeler and Dusenbury Lumber Company and a U.S. representative, 1907-1911.

A344 *American Lumberman.* "Edward Hines." *American Lumberman* (July 24, 1909), 42-43. A Chicago lumber wholesaler who in the 1890s moved into production by buying timberlands and mills in the Great Lakes states and the South.

A345 *American Lumberman.* "The Breaking of a Rifle River Jam—A Tale of a Drive." *American Lumberman* (Oct. 2, 1909), 95-96. Michigan, 1882.

A346 *American Lumberman.* "Thomas B. Walker." *American Lumberman* (Oct. 16, 1909), 1, 33. A Minnesota lumberman since the 1860s (Red River Lumber Company), with interests in California.

A347 *American Lumberman.* "From Forest to Consumer: A Story of the Paepcke-Leicht Companies, of Chicago." *American Lumberman* (Nov. 20, 1909), 51-146. A largely pictorial feature; incidental history.

A348 *American Lumberman.* "George Robert Hogg." *American Lumberman* (Dec. 4, 1909), 1. Lumberman of St. Louis, Missouri.

A349 *American Lumberman.* "Leonard Bronson." *American Lumberman* (Jan. 1, 1910), 1. Manager of the National Lumber Manufacturers' Association.

A350 *American Lumberman.* "George Wing Sisson." *American Lumberman* (Jan. 15, 1910), 1, 139. A New York lumberman and timberlands owner since 1867.

A351 *American Lumberman.* "J. Lewis Thompson and N.H. Clapp." *American Lumberman* (Jan. 29, 1910), 1, 35. Of the Yellow Pine Manufacturers' Association. Clapp was manager of the Southern Lumber Company, Warren, Arkansas, and was associated with the Weyerhaeuser interests.

A352 *American Lumberman.* "Raising the Efficiency of a Great Lumber Producing Plant." *American Lumberman* (Feb. 19, 1910), 59-66. Grayson-McLeod Lumber Company, Graysonia, Arkansas.

A353 *American Lumberman.* "Robert Winfield Higbie." *American Lumberman* (Mar. 5, 1910), 1. A wholesale lumber dealer and hardwood manufacturer of Newton Falls, New York.

A354 *American Lumberman.* "Marvelous Growth and Development of City of Smokestacks." *American Lumberman* (Mar. 5, 1910), 56-57. The lumber industry in Everett, Washington.

A355 *American Lumberman.* "Joseph T. Burlingame." *American Lumberman* (Mar. 12, 1910), 1. President of the Alabama-West Florida Lumber Manufacturers Association.

A356 *American Lumberman.* "Tennessee Hardwoods of the New River Lumber Co., Norma, Tenn." *American Lumberman* (Apr. 2, 1910), 51-74. Incidental history.

A357 *American Lumberman.* "Daniel Page Simons." *American Lumberman* (May 21, 1910), 1. A timber buyer for Sage Land & Improvement Company in Wisconsin and California since the 1860s.

A358 *American Lumberman.* "E.L. Davidson." *American Lumberman* (May 28, 1910), 1. Parkersburg Mill Company, Parkersburg, West Virginia.

A359 *American Lumberman.* "Bay City—Saginaw." *American Lumberman* (Sept. 3, 1910), 51-86. A pictorial feature focusing on its hardwood resources and forest industries; incidental history.

A360 *American Lumberman.* "Peter Musser." *American Lumberman* (Oct. 1, 1910), 1, 77. An Iowa lumberman since the 1860s (Musser Lumber Company, Muscatine), with interests in Wisconsin, Minnesota, Washington, Idaho, and the South.

A361 *American Lumberman.* " 'Peach River Pine' and the 'Peach River Lines'." *American Lumberman* (Oct. 8, 1910), 51-102. Miller and Vidor Lumber Company, Galveston, Texas.

A362 *American Lumberman.* "Chauncey Wright Griggs." *American Lumberman* (Nov. 5, 1910), 1, 73. Cofounder (with Addison G. Foster) of the St. Paul & Tacoma Lumber Company, Tacoma, Washington.

A363 *American Lumberman.* "Lumbering at Wausau: Brief Histories of Men and Institutions." *American Lumberman* (Dec. 3, 1910), 53-57. Wisconsin.

A364 *American Lumberman.* "The White Pine." *American Lumberman* (Dec. 10, 1910), 1, 85. On its role in the history of the lumber industry.

A365 *American Lumberman.* "History and Evolution of Log Scaling Methods." *American Lumberman* (Dec. 24, 1910), 29.

A366 *American Lumberman.* "Strength, Durability, Adaptability." *American Lumberman* (Jan. 14, 1911), 1, 83. On the role of white oak in the history of the lumber industry.

A367 *American Lumberman.* "Robert Laird McCormick." *American Lumberman* (Feb. 11, 1911), 1, 77. Of Laird, Norton Company in Wisconsin and Minnesota, 1860s-1880s; later affiliated with the Weyerhaeusers in Minnesota and Washington.

A368 *American Lumberman.* "Loren Leslie Ashley." *American Lumberman* (Feb. 25, 1911), 1. Associated with the Norwood Manufacturing Company, Utica, New York.

A369 *American Lumberman.* "Horton Corwin, Jr." *American Lumberman* (Apr. 1, 1911), 1, 68. Associated with Branning Manufacturing Company of North Carolina and president of the North Carolina Pine Association.

A370 *American Lumberman.* "Gen. Francis E. Waters." *American Lumberman* (Apr. 15, 1911), 1, 71. A Maryland and Virginia lumberman and incorporator of the Baltimore Lumber Exchange.

A371 *American Lumberman.* "Solidarity and Progress." *American Lumberman* (May 13, 1911), 1, 77. Eli B. Hallowell and Ralph Sounder, wholesale lumbermen of Philadelphia.

A372 *American Lumberman.* "Wisconsin Lumbering." *American Lumberman* (May 13, 1911), 45-50. Lumber companies and lumbermen in Rhinelander, Wisconsin.

A373 *American Lumberman.* "Jacob B. Conrad." *American Lumberman* (June 3, 1911), 1, 78. Of the Bond Lumber Company of Florida and president of the Georgia-Florida Sawmill Association.

A374 *American Lumberman.* "Edward Rutledge." *American Lumberman* (Aug. 5, 1911), 1. A Canadian-born lumberman with operations in Chippewa Falls, Wisconsin, in the late 19th century.

A375 *American Lumberman.* "John Alexander Humbird." *American Lumberman* (Aug. 12, 1911), 1, 55. A Wisconsin lumberman since the 1870s with operations in British Columbia and Idaho.

A376 *American Lumberman.* "A Vigorous Lumber Center of the Far West." *American Lumberman* (Aug. 24, 1912), 38-40. Klamath Falls, Oregon.

A377 *American Lumberman.* "William Andrew Brown." *American Lumberman* (Sept. 9, 1911), 1, 63. A retail lumber dealer with Kendrick and Brown Company, Glens Falls, New York.

A378 *American Lumberman.* "North Carolina Pine." *American Lumberman* (Oct. 14, 1911), 51-54. Includes some incidental history of the Camp Manufacturing Company, North Carolina.

A379 *American Lumberman.* "Selwyn Eddy." *American Lumberman* (Oct. 21, 1911), 1. A Maine-born lumberman operating in Michigan's Saginaw Valley, 1870s-1890s; formed the Pacific Lumber Company of Scotia, California, in 1904.

A380 *American Lumberman.* "Edward D. Wetmore." *American Lumberman* (Oct. 28, 1911), 1, 71. In business with his father, Pennsylvania lumberman Lansing D. Wetmore, since 1885.

A381 *American Lumberman.* "The Realization of a Great Commercial Dream." *American Lumberman* (Nov. 11, 1911), 43-142. Includes some history of the C.A. Smith Lumber Company operations in Coos County, Oregon.

A382 *American Lumberman.* "Seattle as a Lumber Center." *American Lumberman* (Dec. 2, 1911), 47-49. Includes sketches of individual lumber companies.

A383 *American Lumberman.* "Thomas H. Shevlin." *American Lumberman* (Jan. 20, 1912), 1, 69. A Minnesota lumberman with interests in Wisconsin and Ontario.

A384 *American Lumberman.* "Thomas A. Charshee." *American Lumberman* (Jan. 27, 1912), 1. A hardwood lumber manufacturer and wholesaler of Baltimore, Maryland.

A385 *American Lumberman.* "Louis Germain, Jr." *American Lumberman* (Mar. 23, 1912), 1, 75. A wholesale lumberman of Pittsburgh, Pennsylvania, with mill operations in Florida and Georgia.

A386 *American Lumberman.* "Thomas E. Coale." *American Lumberman* (Mar. 30, 1912), 1. A wholesale lumberman of Philadelphia.

A387 *American Lumberman.* "George W. Jones." *American Lumberman* (June 29, 1912), 1, 67. Associated with Camp Manufacturing Company, Virginia, and experienced as a sawmill manager in the South Atlantic states since the 1880s.

A388 *American Lumberman.* "Samuel Stewart Henderson." *American Lumberman* (Oct. 12, 1912), 1, 69. A lumberman of Jefferson County, Pennsylvania, with interests in West Virginia timber.

A389 *American Lumberman.* "40 Years of Retail Lumber Trade." *American Lumberman* (Dec. 7, 1912), 42-43; (Dec. 14, 1912), 38-39; (Dec. 21, 1912), 41-42; (Dec. 28, 1912), 36-37; (Jan. 4, 1913), 42. Especially in Chicago.

A390 *American Lumberman.* "Welfare Work in a Sawmill Town." *American Lumberman* (Jan. 18, 1913), 34-37. Crossett, Arkansas.

A391 *American Lumberman.* "Resources of the Klamath Falls Region." *American Lumberman* (Mar. 29, 1913), 44-45. Includes some history of the lumber industry and forest resources near Klamath Falls, Oregon.

A392 *American Lumberman.* "George Washington Wilson." *American Lumberman* (May 17, 1913), 1. Associated with the Wildell Lumber Company, Elkins, West Virginia.

A393 *American Lumberman.* "Herman Dierks." *American Lumberman* (May 24, 1913), 1, 73. And the Dierks Lumber and Coal Company, Kansas City, Missouri.

A394 *American Lumberman.* "Bloedel Donovan Lumber Mills." *American Lumberman* (July 19, 1913), 43-57. History and description of the company's operations in Bellingham, Washington.

A395 *American Lumberman.* "Richard Chaffey." *American Lumberman* (Aug. 9, 1913), 1. Lumberman of Elkins, West Virginia.

A396 *American Lumberman.* "S.E. Slaymaker." *American Lumberman* (Sept. 6, 1913), 1, 65. Samuel Evans Slaymaker, associated with the West Virginia Spruce Lumber Company, Cass, West Virginia.

A397 *American Lumberman.* "Woodsmen Old and New." *American Lumberman* (Nov. 22, 1913), 34. On the changing nationality of Maine woodsmen.

A398 *American Lumberman.* "Atlantic Coast Lumber Corporation, Georgetown, S.C." *American Lumberman* (Jan. 8, 1916), 43-67. Includes some history of this South Carolina firm.

A399 *American Lumberman.* "A History of the Cady Lumber Corporation (McNary, Arizona) and Its Wonderful Future in the Production of Arizona White Pine." *American Lumberman* (Apr. 10, 1926), 53-108.

A400 *American Lumberman.* "In Retrospect and in Prospect, 1883, 1908, 1933." *American Lumberman* (Dec. 9, 1933), 1 ff.

A401 *American Lumberman.* "300 Years of American Houses." *American Lumberman* (July 10, 1943), 49-70, 176-86.

A402 *American Lumberman.* "Progress in Lumber Production." *American Lumberman* (July 10, 1943), 73-84, 191-92. Survey of logging and milling developments since the 19th century.

A403 *American Lumberman.* "The Story of Mill Work." *American Lumberman* (July 10, 1943), 116-28.

A404 *American Lumberman.* "The Retail Dealer." *American Lumberman* (Sept. 11, 1948), 86-104. Includes some history of the retail lumber business.

A405 *American Lumberman.* "Story of the 'American Lumberman'." *American Lumberman* (Sept. 11, 1948), 144-47. Since 1899 an important weekly trade journal published in Chicago.

A406 *American Lumberman.* "Building Products That Build Profits." *American Lumberman* (Sept. 11, 1948), 148-207. Historical sketches of many building materials and products.

A407 American Paper and Pulp Association. *The Background and Present Status of the United States Paper Industry.* New York, 1940. 45 pp.

A408 American Paper and Pulp Association. *A Capital and Income Survey of the United States Pulp and Paper Industry, 1934-1946.* New York, 1947. 15 pp. Diags., tables. There are later editions.

A409 American Paper and Pulp Association. *New Horizons, a History of Seventy-Five Years of the American Paper and Pulp Association.* New York, 1952. 32 pp. Illus. History of the association since 1878, based largely on the proceedings of annual meetings.

A410 American Tree Association. *Forestry Almanac.* Washington, 1924. 225 pp. Illus., tables, index. A compendium of information, some historical, on such subjects as the USFS, national forests, forest legislation, forestry and conservation associations, lumber industry, demonstration forests, forestry education, forestry in other nations (including Canada), etc. Subsequent editions were published in 1926, 1929, and 1933, with even broader coverage. Editions in 1943 and 1949, under the title *Forestry Directory,* were compiled by Tom Gill and Ellen C. Dowling.

A411 American Tree Association. *Forestry Legislative Survey.* Washington, 1927. 67 pp. Tables. A state-by-state survey including some historical references. There are other editions.

American West, editors of. See Webster, Paul, #W130

A412 *American West,* editors of. *The Great Northwest: The Story of a Land and Its People.* Palo Alto, California: American West Publishing Company, 1973. 288 pp. Illus., maps, diags., apps., bib., index. Includes a chapter on the lumber industry and many historical references to conservation, the environment, and development of natural resources.

A413 *American Wildlife.* "John Bird Burnham, 1869-1939." *American Wildlife* 28 (Nov.-Dec. 1939), 244-46. Burnham, a noted wildlife conservationist, labored for legislation to protect migratory birds and served as chairman of the USFS Committee on Game in the National Forests.

A414 American Wood-Preservers' Association. *Handbook on Wood Preservation.* Baltimore, 1916. 73 pp. Includes some chronology of American wood preservation.

A415 Ames, Carleton C. "Paul Bunyan: Myth or Hoax?" *Minnesota History* 21 (Mar. 1940), 55-58. Author claims that the legend was not indigenous to lumber camps. See discussion in subsequent issues.

A416 Ames, Charles A. "A History of the Forest Service." *Smoke Signal* (Tucson Corral of the Westerners) (Fall 1967), 117-43. Brief history of the USFS with emphasis on Arizona's Coronado National Forest. Also included are reminiscences of the following USFS rangers: Fred W. Croxen, Roger Morris, Virgil Smith, Fred Knipe, and Gilbert Sykes.

A417 Ames, E.G. "Port Gamble, Washington." *Washington Historical Quarterly* 16 (Jan. 1925), 17-19. Origins of a lumber town, 1853.

A418 Amidon, George B. *The Development of Industrial Forestry in the Lake States.* Colonel William B. Greeley Lectures in Industrial Forestry, No. 5. Seattle: University of Washington, College of Forestry, 1961. 44 pp. Illus., map, tables.

A419 Amsden, Perham L. "A History of the Belfast and Moosehead Lake Railroad." Master's thesis, Univ. of Maine, 1949. Maine logging railroad.

A420 Andersen Corporation. *The Andersen Story.* Bayport, Minnesota, 1953. 43 pp. Illus., tables. Company history of a firm established in 1903 as the Andersen Lumber Company and later evolved into a manufacturer of window frames.

A421 Anderson, A.A., and Hayden, Margaret. "Introduction to a History of the Shoshone National Forest." *Annals of Wyoming* 4 (Apr. 1927), 373-88. Of Wyoming.

A422 Anderson, A.A. *The Yellowstone Forest Reserve: Its Foundation and Development.* New Rochelle, New York, 1927. 22 pp. Illus.

A423 Anderson, Antone A., and McDermott, Clare (Anderson). *The Hinckley Fire.* New York: Comet Press, 1954. 157 pp. Illus. Recollections by two survivors of a forest fire that destroyed the lumber town of Hinckley, Minnesota, in 1894, killing 418 persons. Includes narratives of more than thirty other survivors.

A424 Anderson, Bette Roda. "The Olympic Rain Forest." *American West* 10 (Sept. 1973), 30-35. Incidental history of the western Olympic Peninsula, Washington.

A425 Anderson, C.F. "Lumber Manufacturing and Wholesaling Fifty Years Ago and Today." *Southern Lumber Journal* 52 (Feb. 15, 1926), 35, 50, 55, 62.

A426 Anderson, C.R. "Ten Years' Forestry Extension Work in Pennsylvania." *Journal of Forestry* 29 (Jan. 1931), 100-04. Since 1919.

A427 Anderson, Clinton P. *Outsider in the Senate: Senator Clinton Anderson's Memoirs.* Ed. by Milton Viorst. New York: World Publishing Company, 1970. viii + 328 pp. Illus., index. Anderson was Secretary of Agriculture, 1945-1948. As U.S. senator from New Mexico, 1949-1973, Anderson was an advocate and sponsor of wilderness and conservation legislation. See chapter 8.

A428 Anderson, Darlene G., and Wellons, John C. "Upper Klamath River Issue." *Siskiyou Pioneer and Yearbook* 4 (1974), complete. Contains six articles on the lumber industry at Klamathon, California, and lumber rafting on the Klamath River, 1880s-1900s.

A429 Anderson, Donald F. *William Howard Taft: A Conservative's Conception of the Presidency.* Ithaca, New York: Cornell University Press, 1973. Includes references to the Ballinger-Pinchot controversy and other conservation issues ca. 1910.

A430 Anderson, Frederick R., with Robert H. Daniels. *NEPA in the Courts: A Legal Analysis of the National Environmental Policy Act.* Baltimore: Johns Hopkins University Press, for Resources for the Future, 1973. 424 pp.

A431 Anderson, George. "Economics of Site Preparation and Land Regeneration in the South: Example of an Industry Concept." *Journal of Forestry* 56 (Oct. 1958), 754-56. On forest management by the Brunswick Pulp and Paper Company, Brunswick, Georgia, since 1940.

A432 Anderson, George L. *Essays on the Public Lands: Problems, Legislation, and Administration.* Lawrence, Kansas: Coronado Press, 1971. 73 pp.

A433 Anderson, Henry W. "Flood Frequencies and Sedimentation from Forest Watersheds." *Transactions of the American Geophysical Union* 30 (Aug. 1949), 567-86. Flooding in southern California since 1860.

A434 Anderson, I.V. "The Forest Problem in Western Montana." *Journal of Forestry* 31 (Jan. 1933), 4-13. Case history of a section of forested land near Greenough, Montana, logged in 1887 and again in 1927.

A435 Anderson, I.V., and Rapraeger, E.F. *Forest Industries of the Inland Empire.* USFS, Northern Rocky Mountain Forest and Range Experiment Station, Division of Forest Products, Bulletin No. 2. Missoula, 1940. 14 pp. Tables, graph. Of eastern Washington, northern Idaho, and western Montana, since the mid-19th century.

A436 Anderson, John S. "West Side Lumber: History Born Again." *Trains* 18 (Oct. 1958), 16-26. California.

A437 Anderson, Julie. *I Married a Logger: Life in Michigan's Tall Timber.* New York: Exposition Press, 1951. 328 pp. Illus. Reminiscences of life in Michigan logging camps.

A438 Anderson, Sven A. "Trends in the Pulp and Paper Industry." *Economic Geography* 18 (Apr. 1942), 195-202.

A439 Anderson, Thomas D. "The Geography of Christmas Tree Production and Marketing in Anglo-America: With Special Attention to Twelve Counties in Western Pennsylvania." Ph.D. diss., Univ. of Nebraska, 1966. 293 pp.

A440 Anderson Lumber Company. *Seventy-Five Years of Service, Anderson Lumber Company, 1890-1965.* Salt Lake City, 1965. 26 pp. Illus.

A441 Andreas, A.T., comp. *History of Northern Wisconsin, Containing an Account of Its Settlement, Growth, Development and Resources . . . .* Chicago: Western Historical Company, 1881. 1218 pp. Illus., maps. Includes some history of the lumber industry.

A442 Andress, Joel Max. "The Role of Wood in the Economy of Maine." Master's thesis, Univ. of California, Berkeley, 1960.

Andrews, Alice E. See Andrews, Christopher, C., #A444

A443 Andrews, Christopher C. "Brief History of Itasca Park." *North Woods* 7 (No. 17, 1920), 25-32; 7 (No. 18, 1920), 5-15. Minnesota.

A444 Andrews, Christopher C. *Christopher C. Andrews, Pioneer in Forestry Conservation in the United States: For Sixty Years a Dominant Influence in the Public Affairs of Minnesota; Lawyer; Editor; Diplomat; General in the Civil War; Recollections: 1829-1922.* Ed. by Alice E. Andrews with intro. by William Watts Folwell. Cleveland: Arthur H. Clark Company, 1928. 327 pp. Illus. By Minnesota's first state forester and a leading conservationist.

A445 Andrews, Clarence L. "Russian Shipbuilding in the American Colonies." *Washington Historical Quarterly* 25 (Jan. 1934), 3-10. Russian shipbuilding in Alaska and at Fort Ross, California, 1792-1867.

A446 Andrews, H.J., and Cowlin, Robert W. *Forest Resources of the Douglas-Fir Region.* USDA, Miscellaneous Publication No. 389. Washington: GPO, 1940. 169 pp. Incidental history.

A447 Andrews, Henry N., Jr. "The King's Pines." *Historical New Hampshire* (Mar. 1947), 1-14. On British mast policy and the naval stores industry in colonial New Hampshire.

A448 Andrews, Henry N., Jr. "The Royal Pines of New Hampshire." *Appalachia* 27 (Dec. 1948), 186-98.

A449 Andrews, Ralph W. *This Was Logging!* Seattle: Superior Publishing Company, 1954. 157 pp. Illus. A biographical sketch of photographer Darius Kinsey (1871-1945) and a selection from some 8,000 photographs taken by him in the forests and lumber camps of Oregon, Washington, and British Columbia, 1892-1940.

A450 Andrews, Ralph W. *Glory Days of Logging.* Seattle: Superior Publishing Company, 1956. 176 pp. Illus. Anecdotes and sketches of logging in British Columbia. Washington, Oregon, Idaho, Montana, and northern California, heavily illustrated with reproductions of early photographs, 1852-1920.

A451 Andrews, Ralph W. *This Was Sawmilling.* Seattle: Superior Publishing Company, 1957. 176 pp. Illus. A heavily illustrated history of sawmilling in the Pacific Northwest and northern California, 1825-1925. Includes reminiscences and stories from a variety of sources.

A452 Andrews, Ralph W. *Redwood Classic.* Seattle: Superior Publishing Company, 1958. 174 pp. Illus. A largely pictorial account of redwood forests, logging, and the lumber industry in California, 1847-1929.

A453 Andrews, Ralph W. "Up Today and Down Tomorrow." *True West* 6 (May-June 1959), 20-22, 36. On water-powered sash mills in the lumber industry in Oregon and northern California, 1867-1910.

A454 Andrews, Ralph W. *Heroes of the Western Woods.* New York: E.P. Dutton and Company, 1960. 192 pp. Illus., bib. A popular or juvenile account of lumbering in the Pacific Northwest.

A455 Andrews, Ralph W. *Timber!: Toil and Trouble in the Big Woods.* Seattle: Superior Publishing Company, 1968. 182 pp. Illus., index. A heavily illustrated history of the lumber industry in the Pacific Northwest and British Columbia since the beginning of white settlement.

A456 Andrews, Russell P. *Wilderness Sanctuary.* Inter-University Case Program Case Series, No. 13. University, Alabama: University of Alabama Press, for the Inter-University Case Program, 1953. vi + 10 pp. On the antecedents of an executive order of 1949 prohibiting air traffic above the roadless area of the Superior National Forest, Minnesota.

A457 Angier, F.J. "Wood Preservation: Its Past, Present and Future." *Lumber World Review* (No. 9, 1915), 42-46.

A458 Angle, Grant Colfax. *Lumbering in the Northwest and the Logger at His Work.* Shelton, Washington: G.C. Angle and Journal Press, 1905. 24 pp. Includes some history of logging operations and pioneer sawmills in Mason County, Washington, since the 1850s.

A459 Angus, Henry Forbes, ed. *British Columbia and the United States: The North Pacific Slope from Fur Trade to Aviation.* By F.W. Howay, W.N. Sage, and Henry Forbes Angus. Toronto: Ryerson Press; New Haven: Yale University Press, for the Carnegie Endowment for International Peace, 1942. xv + 408 pp. Illus., maps, notes. Includes a chapter on the lumber industry in British Columbia, 1860-1913.

A460 Anonsen, Andrew E. *Autobiographical Sketches.* Kerkhoven, Minnesota: N.p., 1939. 23 pp. Includes an account of his years in a Washington logging camp.

A461 Anspach, Marshall R. "Log Marks as Used in Connection with Operation of the Susquehanna Boom." *Now and Then* 14 (No. 1, 1962), 15-17. Pennsylvania log marks.

A462 Anthony, Bruce. "Ninety Years Later." *Southern Lumberman* 223 (Dec. 15, 1971), 76-78. Some history of the southern lumber industry and its trade associations.

A463 Antrei, Albert C. "A Geographic Interpretation of Timber Production in Utah." Master's thesis, Univ. of Utah, 1951. 133 pp.

A464 Antrei, Albert C. "A Western Phenomenon: The Origin and Development of Watershed Research: Manti, Utah, 1889." *American West* 8 (Mar. 1971), 42-47, 59. Concerns overgrazing, flooding, and the beginnings of watershed research and management by the USFS on the La Sal National Forest, Utah, 1880s-1920s.

A465 Apgar, William B. "The Administration of Grazing on the National Forests." Master's thesis, Cornell Univ., 1922.

A466 Appalachian Mountain Club. "Appalachian Mountain Club, 1876-1951: Seventy-Fifth Anniversary." *Appalachia* 28 (May 1951), 273-368. Includes articles by many contributors on the history of the club and its members. Robert S. Monahan writes of the "Conservation Crusade," for example.

A467 Applegate, M. Richard. *Massachusetts Forest and Park Association: A History, 1898-1973.* Boston: Massachusetts Forest and Park Association, 1974. 76 pp. Illus. A brief account of its activities and leaders, especially Harris A. Reynolds. Until 1933 the organization was known as the Massachusetts Forestry Association.

A468 Appleman, Roy E. "Timber Empire from the Public Domain." *Mississippi Valley Historical Review* 26 (Sept. 1939), 193-208. A study of the lumber industry and federal land disposal in the Pacific Northwest, 1897-1910.

A469 Appleton, John. "A Study of White Birch and of the Spool and Novelty Industries in Maine." Master's thesis, Yale Univ., 1904.

A470 Arend, John L. "An Early Eastern Red Cedar Plantation in Arkansas." *Journal of Forestry* 45 (May 1947), 358-60. Established in Independence County, Arkansas, 1902.

A471 Argow, Keith A. "Social Action for Land Preservation." Ph.D. diss., North Carolina State Univ., 1970. 143 pp. Includes some history of preservation of wild and scenic lands by citizens groups.

A472 Argy, Michael J. "Reminiscences of a Ground Wood Man." *Paper Trade Journal* 90 (June 5, 1930), 79-80. Of groundwood mills and preparing wood pulp for papermaking, late 19th century.

A473 Arkell & Smiths. *Arkell & Smiths: 90 Years of Know How, the Oldest Name in Paper Bags.* Canajoharie, New York, 1949. 78 pp. Illus.

A474 Armstrong, Andrew K. *Wood-Using Industries of California.* California State Board of Forestry, Bulletin No. 3 Sacramento, 1912. 114 pp. Incidental history.

A475 Armstrong, Chester H., comp. *History of the Oregon State Parks, 1917-1963.* Salem: Oregon State Highway Department, Parks Division, 1965. x + 268 pp. Illus., maps, tables. A history of the state parks movement and system and individual sketches of Oregon's state parks and waysides.

A476 Armstrong, E.P. "History of the Band Saw Mill and Its Bearing on Present Day Mill Problems." *West Coast Lumberman* (1922-1923). Serialized, Nos. 489-527. Also appeared in *Southern Lumberman* (Mar. 3-Oct. 13, 1923).

A477 Armstrong, George R. *The Forests and Economy of Lewis County, New York.* State University of New York, College of Forestry, Bulletin No. 33. Syracuse, 1954. 49 pp. Map, diags., tables. Incidental history.

A478 Armstrong, George R., and Bjorkbom, John C. *The Timber Resources of New York.* Upper Darby, Pennsylvania: USFS, Northeastern Forest Experiment Station, 1956. 37 pp. Incidental history.

A479 Armstrong, George R. "The Forest Resources of New York State." *Northeastern Logger* 4 (May 1956), 24-25, 74-75.

A480 Armstrong, George R. "Economic Development in the Lumber and Plywood Industries of Western North America: A Study in Forecasting Methods." Master's thesis, State Univ. of New York, College of Forestry, 1959.

A481 Armstrong, George R., and Kranz, Marvin W., eds. *Forestry College: Essays on the Growth and Development of New York State's College of Forestry, 1911-1961.* Syracuse: Alumni Association, New York State University, College of Forestry, 1961. viii + 360 pp. Illus., tables, graphs, apps., index.

Armstrong, George R. See Guthrie, John A., #G385

A482 Armstrong, George R. "An Economic Study of New York's Pulp and Paper Industry." Ph.D. diss., State Univ. of New York, College of Environmental Science and Forestry, 1965. 142 pp.

A483 Armstrong, R.H. "The History and Mechanics of Forest Planting and Aerial Spraying." *Pulp and Paper Magazine of Canada* 64 (June 1963), 268-70. Reforestation since 1950 on the Ontario lands of the Spruce Falls Power and Paper Company.

A484 Armstrong Cork Company. *Cork: Being the Story of the Origin of Cork, the Processes Employed in Its Manufacture & Its Varied Uses in the World To-Day.* Pittsburgh, 1909. 46 pp. Illus. Includes some history of the Armstrong Cork Company, Pittsburgh, Pennsylvania, and of its products.

A485 Arneson, Ben A. "Federal Aid to the States." *American Political Science Review* 16 (Aug. 1922), 443-54. Includes reference to the National Forest Fund Act of 1907 and its subsequent amendments.

A486 Arnold, Charles I. "The History of Lumbering and Logging in Maine." Master's thesis, Yale Univ., 1941.

A487 Arnold, Frank A. "The Man with the Thousand-Year Tree Garden." *American Forestry* 24 (May 1918), 297-98. Charles Sprague Sargent and Arnold Arboretum, Massachusetts, since 1872.

A488 Arnot, Merl Jay. "Employment Effects of Log Exports to Japan from the Oregon South Coastal Region." Master's thesis, Univ. of Oregon, 1968.

A489 Arnst, Albert. "Planning a Public Information Program." *Journal of Forestry* 49 (June 1951), 427-30. Brief history of the public relations program of Weyerhaeuser Timber Company, Tacoma, Washington, 1947-1950.

A490 Arnst, Albert, ed. *Fifty Years of Forestry at Oregon State College.* Corvallis: Oregon State College, 1956. 24 pp.

A491 Artman, James O., and Dean, George W. "Forest Fire Control in Virginia." *Journal of Forestry* 43 (June 1945). 393-97. Since 1914.

A492 Artman, James O. *Twenty Years of Fire Records for State and Private Forest Lands in the Tennessee Valley.* Tennessee Valley Authority, Division of Forestry, Publication No. 138. Knoxville, 1954. 27 pp. Maps, diags., tables, notes.

A493 Artman, James O. "Trees for the Tennessee." *American Forests* 61 (Apr. 1955), 8-12. Reforestation by and for the Tennessee Valley Authority since 1933.

A494 Artman, James O. *Forest Fires and Area Burned, State and Private Lands, Tennessee Valley, 1934-1958: A 25-Year Record.* Tennessee Valley Authority, Division of Forestry Relations, Report No. 227-59. Norris, Tennessee, 1959. 27 pp.

A495 Arvola, T.F. "Forest Practice Regulation in California." *Journal of Forestry* 60 (Dec. 1962), 872-76. Includes a brief history of forest practice regulation since 1813, with emphasis on the California Forest Practice Act of 1945 and its effects.

A496 Ashby, Charlotte M., comp. *Record Group 95: Cartographic Records of the Forest Service.* National Archives, Preliminary Inventories, No. 167. Washington, 1967. 71 pp. App., index. Includes a brief history of the USFS.

A497 Ashe, William W. *The Forests, Forest Lands, and Forest Products of Eastern North Carolina.* North Carolina Geological Survey, Bulletin No. 5. Raleigh, 1894. 128 pp. Illus., tables, index. Includes some history of forest resources and industries, especially the lumber and naval stores industries.

Ashe, William W. See Pinchot, Gifford, #P197

A498 Ashe, William W., and Ayers, H.B. *The Southern Appalachian Forests.* U.S. Geological Survey, Professional Paper No. 37. Washington: GPO, 1905. 291 pp. Includes some history of resources and industries.

A499 Ashe, William W. "The Creation of the Eastern National Forests." *American Forestry* 28 (Sept. 1922), 521-25. On the work of the National Forest Reservation Commission since passage of the Weeks Law in 1911.

A500 Ashe, William W. "The Place of the Eastern National Forests in the National Economy." *Geographical Review* 13 (Oct. 1923), 532-39. Incidental history.

A501 Ashman, Robert I. "Fifty Years of Forestry at Maine." In *Papers Presented at Fiftieth Anniversary Celebration, Forestry Department, October 1, 2, 3, 1953.* Orono: University of Maine, Forestry Department, 1954. Pp. 3-9.

A502 Ashton, Bessie L. *The Geonomic Aspects of the Illinois Waterway.* Urbana: University of Illinois, 1926. 177 pp. Maps, tables, charts, bib. Includes some history of the lumber trade on the Illinois and other rivers since the mid-19th century.

A503 Association for the Protection of the Adirondacks. *The Adirondack Park: A Sketch of the Origin, Romantic Charms and Practical Uses of the Adirondack Park, and Some Reasons for the Acquisition of Land and Reforestation by the State of New York.* New York, 1903. 32 pp. Illus.

A504 Association for the Protection of the Adirondacks. *Brief Review of the Depredations upon the Adirondack Forests Accomplished or Attempted during the Past Few Years, with Reference to the Proposed Amendment to Section 7 of Article VII of the Constitution.* New York, 1907. 20 pp. Illus.

A505 Aston, G.F. "History of the Diesel Engine." *Timberman* 27 (Nov. 1925), 63-65.

Astorino, Samuel J. See Kehl, James A., #K45

A506 Athearn, Robert G. *High Country Empire: The High Plains and Rockies.* New York: McGraw-Hill, 1960. viii + 360 pp. Illus., maps, notes, bib. A general history of upper Missouri River region. Chapter 12, "Uncle Sam's West," contains some history of the USFS, the National Park Service, and the general effects of the conservation movement upon the region.

A507 Atkinson, J.B. "Planting Forests in Kentucky." *American Forestry* 16 (Aug. 1910), 449-56. On the lands of the St. Bernard Mining Company of Earlington, Kentucky, since 1890.

A508 Attaway, Herbert B. "Florida's Forest Economy." Master's thesis, Univ. of Florida, 1950.

A509 Attwood, Lloyd M. "Cheboygan as a Nineteenth-Century Lumber Area." Master's thesis, Wayne State Univ., 1947. Cheboygan, Michigan.

A510 Atwater, Montgomery M. *The Forest Rangers.* Philadelphia: Macrae Smith Company, 1969. 190 pp. Illus. A popular history of the USFS, including descriptions of jobs and career opportunities with the agency.

A511 Atwood, Ann. *The Kingdom of the Forest.* New York: Charles Scribner's Sons, 1972. 32 pp. Illus. Brief historical references to forest ecology.

A512 Audubon, John James. *Audubon and His Journals.* Ed. by Maria R. Audubon. 2 Volumes. New York: Charles Scribner's Sons, 1897. Reprint. New York: Dover Publications, 1960. In part an autobiographical account of Audubon's life to 1820.

Audubon, Maria R. See Audubon, John James, #A512

A513 Aughanbaugh, John. "Experimental Woodlands as a Means of Encouraging Improved Management of Small Tracts." *Journal of Forestry* 57 (June 1959), 409-12. In Ohio since 1945.

A514 Augspurger, Marie M. *Yellowstone National Park, Historical and Descriptive.* Middletown, Ohio: Naegele-Auer, 1948. xxxi + 247 pp. Illus., bib.

A515 Averill, Gerald. *Ridge Runner: The Story of a Maine Woodsman.* Philadelphia: J.B. Lippincott, 1948. 217 pp. Autobiographical account of pulpwood logging early in the 20th century.

Ayers, H.B. See Ashe, William W., #A498

A516 Ayres, Robert W. *History of Timber Management in the California National Forests, 1850-1937.* Washington: USFS, 1958. 86 pp.

B1 Baar, C.F. "A History of Reforestation in New York." *New York Forester* 13 (Feb. 1956), 2-6.

B2 Babcock, H.M., and Nicolaiff, J.E. *The Christmas Tree Industry in Canada.* Miscellaneous Publication No. 10. Ottawa: Forestry Branch, 1958. 16 pp.

B3 Babcock, Thorpe. *Broke at Forty Five: A Letter to His Grandchildren.* Los Angeles: Fashion Press, 1967. 223 + 15 pp. Illus. Includes reminiscences of the lumber industry in Hoquiam, Washington, and of the author's connections with the West Coast Lumbermen's Association, 1910s-1920s.

B4 Babcock, Willoughby M. "The St. Croix Valley as Viewed by Pioneer Editors." *Minnesota History* 17 (Sept. 1936), 276-87. Includes some history of logging and the lumber industry.

B5 Bach, Arthur L. "Administration of Indian Resources in the United States, 1933-1941." Ph.D. diss., Univ. of Iowa, 1943.

B6 Bachman, Earl E. *Recreation Facilities: A Personal History of Their Development in the National Forests of California.* San Francisco: USFS, California Region, 1967. 52 pp. Illus. Since ca. 1900.

B7 Bachmann, Elizabeth M. "Minnesota Log Marks." *Minnesota History* 26 (June 1945), 126-37. On the practice of branding or marking logs in Minnesota since the 1850s.

B8 Bachmann, Elizabeth M. "Fire!—An Underwriter of the History of Minnesota Forestry: A Review of the Development of State Forest Management." *Conservation Volunteer* 27 (Jan. 1964), 4-9.

B9 Bachmann, Elizabeth M. "Log Marks." *Northern Logger* 12 (Mar. 1964), 22-23, 35, 41, 56. In Minnesota, 1850s-1890s.

B10 Bachmann, Elizabeth. *A History of Forestry in Minnesota, with Particular Reference to Forestry Legislation.* N.p.: Association of Minnesota Division of Lands and Forestry Employees, 1969. iv + 65 pp. Illus., maps, apps., bib., index. Compilation of historical information with emphasis on state forestry since 1895. This printing is a slightly enlarged version of the account which first appeared in 1965. An abridged version also appeared in successive issues of *Conservation Volunteer,* July 1960-May 1961.

Badè, William Frederic. See Muir, John, #M612

B11 Badè, William Frederic. *The Life and Letters of John Muir.* 2 Volumes. Boston: Houghton-Mifflin, 1923, 1924. Illus. Biography and selected writings of Muir (1838-1914), California preservationist, wilderness philosopher, and founder of the Sierra Club.

B12 Badger, K. "Bradford Torrey: New England Nature Writer." *New England Quarterly* 18 (June 1945), 234-46.

B13 Badger Paper Mills. *A Word and Picture Story of 25 Years of Progress, 1929-1954.* Peshtigo, Wisconsin, 1954. 57 pp. Illus.

Baer, Donna Degen. See Russell, Curran N., #R379

Baggs, John T. See Billings, C.L., #B267

Bagley, W.T. See Stone, Robert N., #S669

B14 Bahr, Henry. "Henry Bahr Remembers." *Southern Lumberman* 223 (Dec. 15, 1971), 117-19. Reminiscences of the National Lumber Manufacturers Association and National Forest Products Association since 1928.

B15 Bailey, Irving Widmer, and Spoehr, H.A. *The Role of Research in the Development of Forestry in North America.* New York: Macmillan, 1929. xiii + 118 pp.

B16 Bailey, Reed W. "Watershed Management: Key to Resource Conservation." *Journal of Forestry* 48 (Sept. 1950), 393-96. Includes some history of improvements in watershed management in the 20th century.

Bailey, Reed W. See Croft, A. Russell, #C714

B17 Bailey, Richard R. *Pioneer Lumbering, Northeastern Minnesota, 1880-1930.* Ely, Minnesota: J. William Trygg, 1966. 19 pp.

B18 Bailey, William Francis, ed. *History of Eau Claire County, Wisconsin.* Chicago: C.F. Cooper, 1914. 920 pp. Illus. Includes account of the lumber industry.

B19 Baille, D. "Conservation in Canada: Part Three—Trees and Forests." *Canadian Banker* 64 (Winter 1957), 38-55.

B20 Baird, Dan W. "Early Sawmill Reminiscences." *Southern Lumberman* 29 (July 1, 1903). Tennessee.

B21 Baird, Dan W. "The History of the *Southern Lumberman.*" *Southern Lumberman* 193 (Dec. 15, 1956), 110-15. Reprint from a 1906 issue in which the editor recalls the history of the trade journal since 1881.

B22 Baird, Richard Edward. "The Politics of Echo Park and Other Water Development Projects in the Upper Colorado River Basin, 1946-1956." Ph.D. diss., Univ. of Illinois, 1960. 580 pp. Includes treatment of conservation controversies surrounding proposed and actual developments in Utah, Wyoming, Colorado, and New Mexico.

B23 Baker, A.E.M. "End Matched Lumber Suggested Some Twenty Years Ago." *Canada Lumberman* 49 (Jan. 1, 1929), 44. The author's firm manufactured spruce lumber in Manitoba, 1909-1912.

B24 Baker, Floyd P., and Furnas, Robert Wilkinson. *Preliminary Report on the Forestry of the Mississippi Valley and Tree Planting on the Plains.* USDA, Report No. 28. Washington: GPO, 1883. 45 pp.

B25 Baker, G. "The Primary Wood-Using Industries of Maine." *Maine Agricultural Experiment Station Bulletin.* No. 448 (Apr. 1947), 143-260. Includes some history.

B26 Baker, G. "A Century of Logging in the Pine Tree State." *Northeastern Logger* 4 (Apr. 1956), 24-25, 49. Maine.

B27 Baker, Gladys L.; Rasmussen, Wayne D.; Wiser, Vivian; and Porter, Jane M. *Century of Service: The First 100 Years of the United States Department of Agriculture.* Washington: GPO, 1963. 575 pp. Bib. Emphasizes the department's organization, development, and response to changing conditions; includes material on the USFS.

Baker, Gladys L. See Rasmussen, Wayne D., #R55

Baker, H.L. See Poli, Adon, #P256

B28 Baker, Harry Lee. *Forest Fires in Florida.* Tallahassee: Florida Forestry Association, 1926. 37 pp. Illus.

B29 Baker, Hugh P. "Paper History of New England." *Paper Mill and Wood Pulp News* 49 (No. 42, 1926), 20-28.

B30 Baker, James H., and Hafen, LeRoy R., eds. *History of Colorado.* 5 Volumes. Denver: Linderman Company, 1927. Illus., maps. Volume 2, chapter 14, written by Walter J. Morrill, contains history of forestry and forest conservation in Colorado.

Baker, W.J. See Panshin, Alexis John, #P26

B31 Baker, Willis M. "Reminiscing About the TVA." *American Forests* 75 (May 1969), 30-31, 56-60. By a Tennessee Valley Authority forester, since 1933.

Balaam, Dorothy Ross. See Ross, John Simpson II., #R314

B32 Balch, R.E. "John William B. Sisam." *Journal of Forestry* 48 (May 1950), 372-73. Sketch of Sisam's career, 1937-1950, especially with the Commonwealth Forestry Bureau and as dean of the Faculty of Forestry, University of Toronto.

B33 Baldridge, Kenneth W. "Nine Years of Achievement: The Civilian Conservation Corps in Utah." Ph.D. diss., Brigham Young Univ., 1971. 403 pp. Covers 1933-1942.

B34 Baldridge, Kenneth W. "Reclamation Work of the Civilian Conservation Corps, 1933-1942." *Utah Historical Quarterly* 39 (Summer 1971), 265-85. Including its work on the national forests and national parks of Utah.

B35 Baldwin, Donald Nicholas. "A Historical Study of the Western Origin, Application and Development of the Wilderness Concept, 1919 to 1933." Ph.D. diss., Univ. of Denver, 1965. 387 pp.

B36 Baldwin, Donald Nicholas. "Wilderness: Concept and Challenge." *Colorado Magazine* 44 (Summer 1967), 224-40. On the history of the wilderness concept and wilderness preservation, especially the leading role of Arthur Hawthorne Carhart of the USFS and his work in Colorado since 1919.

B37 Baldwin, Donald Nicholas. *The Quiet Revolution: Grass Roots of Today's Wilderness Preservation Movement.* Boulder, Colorado: Pruett, 1972. xxii + 295 pp. Illus., maps, notes, bib., apps., index. Emphasizes Arthur Hawthorne Carhart's seminal efforts to advance a wilderness concept and to preserve areas of scenic beauty within the national forests. The author asserts that Carhart's work for the USFS, beginning in Colorado in 1919, deserves recognition equal to that given to Aldo Leopold.

B38 Baldwin, Henry I. "The Trees of Nantucket." *American Forests and Forest Life* 34 (Nov. 1928), 664-65, 684. On forests and tree planting experiments on Nantucket Island, Massachusetts, since the 18th century.

B39 Baldwin, Henry I. *Forestry in New England.* Preliminary edition. U.S. National Resources Planning Board, Publication No. 70. Washington, 1942. 57 pp. Maps. Includes some history of forestry.

B40 Baldwin, Henry I., and Heermance, Edgar L. *Wooden Dollars: A Report on the Forest Resources of New England, Their Condition, Economic Significance and Potentialities.* Boston: Federal Reserve Bank of Boston, 1949. 127 pp. Illus., maps, tables, graphs. Includes some historical references.

B41 Baldwin, Henry I. "John H. Foster." *Journal of Forestry* 49 (June 1951), 451-52. Sketch of Foster's career with the USFS, in forestry education, and as state forester of New Hampshire, 1907-1951.

B42 Baldwin, Henry I. *Annals of the Class of 1922, Yale Forest School.* New Haven: Yale Forest School, Class of 1922, 1972. 86 pp. Illus.

B43 Ball, Howard E., and Clepper, Henry. "H.A. Smith: Southern Forester from the North." *Journal of Forestry* 47 (Oct. 1949), 835-36. On Homer Arthur Smith's career in state forestry (Pennsylvania, Florida, and South Carolina) and with the Tennessee Valley Authority, 1916-1940s.

B44 Ballaine, Wesley Charles. "The Revested Oregon and California Railroad Grant Lands: A Problem in Land Management." *Land Economics* 29 (Aug. 1953), 219-32. Includes some history of these forested lands in western Oregon since 1866, with special attention to federal management after 1916.

B45 Ballantyne, John N. "The Prairie States Shelterbelt Project." Master's thesis, Yale Univ., 1949.

B46 Bamford, Paul Walden. "French Naval Timber: A Study of the Relation of Forests to French Sea Power, 1660-1789." Ph.D. diss., Columbia Univ., 1951. 409 pp.

B47 Bamford, Paul Walden. "France and the American Market in Naval Timber and Masts, 1776-1786." *Journal of Economic History* 12 (Winter 1952), 21-34.

B48 Bamford, Paul Walden. *Forests and French Sea Power, 1660-1789.* Toronto: University of Toronto Press, 1956. ix + 240 pp. Notes, bib., index. Includes some history of French attempts to acquire naval masts and timbers from North American forests.

Bancroft, Frederic. See Schurz, Carl, #S129

B49 Bancroft, Hubert Howe. *History of the Pacific States of North America.* 34 Volumes. San Francisco: A.L. Bancroft & Company, 1882-1890. Illus., maps. Many volumes contain useful historical references to 19th-century use of forest resources in the Pacific states and British Columbia.

B50 Banfield, Edward C. "Organization for Policy Planning in the U.S. Department of Agriculture." *Journal of Farm Economics* 34 (Feb. 1952), 14-33.

B51 Banger, George S. *History of the Susquehanna Boom Co. from 1846 to 1876.* Williamsport, Pennsylvania, 1876. 11 pp.

B52 Banker, Harry J. "Arbor Day: The First 100 Years." *American Forests* 78 (Apr. 1972), 8-11, 60-61.

B53 Banks, Laura Stockton Voorhees. "John Aston Warder: First President of the American Forestry Association." *American Forests* 73 (Nov. 1967), 10-13, 66-68. From 1875 to 1882.

B54 Banner, Gilbert. "A Study of the Role of Citizen Organizations in the Conservation Movement." Master's thesis, Univ. of Michigan, 1951.

Baptie, Sue. See Whittaker, Jack, #W265

B55 Barbeau, Marius. "The Modern Growth of the Totem Pole on the Northwest Coast." *Journal of the Washington Academy of Science* 28 (Sept. 15, 1938), 385-93.

B56 Barber, Olive. *The Lady and the Lumberjack.* New York: Crowell, 1952. vi + 250 pp. On the author's life as the wife of a lumberjack near Coos Bay, Oregon, ca. 1940.

B57 Barbour, Ian G., ed. *Western Man and Environmental Ethics: Attitudes toward Nature and Technology.* Reading, Massachusetts: Addison-Wesley, 1973. 276 pp. An anthology of writings containing some history of the conservation movement.

B58 Barclay, Frank H. "The Natural Vegetation of Johnson County, Tennessee, Past and Present." Ph.D. diss., Univ. of Tennessee, 1957.

B59 Barclay, George. "Chicago—The Lumber Hub." *Southern Lumberman* 193 (Dec. 15, 1956), 177-178B. Since 1834.

B60 Barcus, Frank. "The Oak in Michigan's History." *Michigan History Magazine* 29 (Spring 1945), 180-88.

B61 Bardin, P.C. "Some Veneer and Panel History." *Veneers* 16 (No. 6, 1922), 22-24.

B62 Bardin, P.C. "The Outline of History of the Sawmill." *Hardwood Record* 55 (Aug. 10, 1923), 35-38, 46.

B63 Bardon, John A. "Early Logging Methods." *Minnesota History* 15 (June 1934), 203-06. In 19th-century Minnesota.

B64 Barfield, Claude E. "'Our Share of the Booty': The Democratic Party, Cannonism, and the Payne-Aldrich Tariff." *Journal of American History* 57 (Sept. 1970), 308-23. Includes brief treatment of the tariff on lumber, 1909.

B65 Barker, Elliott S. "Fifty Years of Conservation Progress." *New Mexico Magazine* 31 (May 1953), 27. On the New Mexico Department of Game and Fish since 1903 and the author's services as state game warden since 1931.

B66 Barker, Elliott S. *Beatty's Cabin: Adventures in the Pecos High Country.* Albuquerque: University of New Mexico Press, 1953. x + 220 pp. Illus. Reminiscence and history of the upper Pecos River watershed and the Santa Fe National Forest, New Mexico, since 1896.

Barker, Elliott S. See Myers, John M., #M668

B67 Barker, Ernest F. "History of Papermaking in Ohio." *Paper Trade Journal* 141 (Sept. 2, 1957), 36-40.

B68 Barker, Ernest F. "History of Papermaking in California." *Paper Trade Journal* 142 (June 9, 1958), 25-26, 28-31. Dates of establishment and location of paper mills.

B69 Barker, Fred C. *Lake and Forest As I Have Known Them.* Boston: Lee and Shepard, 1903. 230 pp. Account of a Maine steamboat captain, logger, and resort operator.

B70 Barker, Shirley Frances. "The Fire of '47." *New Hampshire Profiles* 6 (Oct. 1957), 26-28. Forest fires in New Hampshire, October 1947.

B71 Barland, Lois. *Sawdust City, A History of Eau Claire, Wisconsin From Earliest Times to 1910.* Stevens Point, Wisconsin: Worzalla Publishing Company, 1960. 147 pp. Illus., diags., maps, tables, bib. History of Eau Claire, Wisconsin, a center of the lumber industry in the late 19th century.

B72 Barlowe, Raleigh. "Forest Policy in Wisconsin." *Wisconsin Magazine of History* 26 (Mar. 1943), 261-79. On the origins and development of forest policy in Wisconsin, with emphasis on the forest conservation movement, forest fire protection, the Forest Crop Taxation Law of 1927, and county forestry programs, since the 19th century.

Barlowe, Raleigh. See Wehrwein, George S., #W138

B73 Barlowe, Raleigh. "The Wisconsin Forest Crop Law: An Appraisal and Evaluation." Ph.D. diss., Univ. of Wisconsin, 1946. 259 pp. Since its enactment in 1927.

B74 Barlowe, Raleigh. *Administration of Tax-Reverted Lands in the Lake States.* Michigan Agricultural Experiment Station, Bulletin No. 225. East Lansing, 1954. 77 pp. Land policy and legislation for Michigan, Wisconsin, and Minnesota, especially for forested and cutover lands, 1869-1950.

Barlowe, Raleigh. See Johnson, Vernon W., #J126

B75 Barlowe, Raleigh. *Land Resource Economics: The Political Economy of Rural and Urban Land Resource Use.* Englewood Cliffs, New Jersey: Prentice-Hall, 1958. 585 pp. Illus., bib. Includes some history of forest land use.

B76 Barnard, Joseph E., and Meyer, Carl E. "Penn's Sylvania—The Current Forest Resources of Pennsylvania." *Pennsylvania Forests* 63 (Fall 1972), 60-64. Includes some history of forest land use and resources inventories in Pennsylvania.

Barnes, A.S.L. See Richardson, Arthur Herbert, #R174

Barnes, Burton V. See Spurr, Stephen H., #S530

B77 Barnes, Irston Robert. "The Quest for a Conservation Ethic." *Atlantic Naturalist* 14 (July-Sept. 1959), 179-81. Since 1908.

B78 Barnes, Will Croft. "Gifford Pinchot, Forester." *McClure's Magazine* 31 (July 1908), 319-27. Includes some early history of the USFS.

B79 Barnes, Will Croft. *Western Grazing Grounds and Forest Ranges; A History of the Livestock Industry as Conducted on the Open Ranges of the Arid West, with Particular Reference to the Use Now Being Made of the Ranges in the National Forests.* Chicago: Breeder's Gazette, 1913. 390 pp. Illus., bib.

B80 Barnes, Will Croft. "Retirement of Albert F. Potter." *Journal of Forestry* 18 (Mar. 1920), 211-13. Potter (1859-1944) is regarded as the architect of the USFS grazing policy. He was chief of grazing in the USFS, 1905-1910, and associate forester, 1910-1920.

B81 Barnes, Will Croft. "Ranger Shinn: The Story of a Man Who Shaped His Life to Get the Greatest Happiness." *Sunset Magazine* 53 (June 1924), 32, 60, 62. Charles Howard Shinn was forest supervisor of California's Sierra National Forest.

B82 Barnes, Will Croft. *The Story of the Range: An Account of the Occupation of the Public Domain Ranges by the Pioneer Stockmen, the Effect on the Forage and the Land of Unrestricted Grazing, and the Attempts That Have Been Made to Regulate Grazing Practice and Perpetuate the Great Natural Forage Resources of the Open Ranges.* Washington: GPO, 1926. iii + 60 pp. Maps, tables, diags., bib. Reprinted from Senate hearings.

B83 Barnes, Will Croft. "George Bishop Sudworth." *Science* 66 (1927), 6-8. A forester.

B84 Barnes, Will Croft. "Winning the Forest Range." *American Forests and Forest Life* 36 (July 1930), 398-400, 466. Reminiscences of the introduction of grazing regulations to the forest reserves and national forests of Arizona and New Mexico, 1897-1905, and of the early years of the USFS Branch of Grazing.

B85 Barnes, Will Croft. *Apaches and Longhorns: The Reminiscences of Will C. Barnes.* Ed. and with an intro. by Frank C. Lockwood. Los Angeles: Ward Ritchie Press, 1941. xxiii + 210 pp. Illus. One chapter details the reaction of Barnes and other Arizona cattlemen to the arrival in 1897 of federal foresters and the subsequent restrictions on grazing in national forests. Barnes later became an inspector and eventually chief of the Branch of Grazing, USFS, 1907-1928.

Barnes, Will Croft. See Raines, William McLeod, #R10

B86 Barnett, H.G. "The Southern Extent of Totem Pole Carving." *Pacific Northwest Quarterly* 33 (Oct. 1942), 379-89. Indian totem pole carving in coastal Alaska, British Columbia, and Washington in the 19th century.

B87 Barnett, Harold J. *Malthusianism and Conservation—Their Role as Origins of the Doctrine of Increasing Economic Scarcity of Natural Resources.* Washington: Resources for the Future, 1959. 32 pp. Graphs. Includes some history of the conservation movement and the economics of resource conservation.

B88 Barnett, Harold J., and Morse, Chandler. *Scarcity and Growth: The Economics of Natural Resource Availability.* Baltimore: Johns Hopkins Press, for Resources for the Future, 1963. 288 pp. Notes. Includes an economic analysis and some history of theories concerning the scarcity of resources and a brief review of the conservation movement.

B89 Barney, Daniel R. *The Last Stand: Ralph Nader's Study Group Report on the National Forests.* New York: Grossman Publishers, 1974. 185 pp. Illus., map, app., notes, index. An argumentative work containing some history of the USFS, the national forests, and the "timber lobby" which influences federal forest policy.

B90 Barney, Richard Joshel. "A Study of Educational Forest Fire Prevention Media in Western Montana." Master's thesis, Univ. of Montana, 1961.

B91 Barnjum, Frank John Dixie. *Startling Facts and Fallacies Regarding Canada's Forests.* Montreal: By the author, 1930. 89 pp. New Brunswick lumberman views the timber supply question; includes some history.

B92 Barnum, Charles R. "Redwood Cruisers' Marks." *Timberman* 45 (Aug. 1944), 18-19. Brief sketches of redwood timber cruisers and collection of cruisers' marks used in northern California.

B93 Barr, P.M. "The Aleza Lake Forest Experiment Station." *Forestry Chronicle* 4 (Sept. 1928), 3-8. British Columbia.

Barr, P.M. See Garman, E.H., #G30

B94 Barraclough, K.E., and Walker, C.E. "A Study of Land Utilization in Grafton County." *Journal of Forestry* 32 (Oct. 1934), 695-700. Includes some history of forest uses since 1772 near Dorchester, New Hampshire.

B95 Barraclough, D.E., and Herr, Clarence S. "Forest Products Association, Inc." *Journal of Forestry* 34 (May 1936), 498-502. On the work of a cooperative marketing association in New Hampshire since 1929.

B96 Barraclough, Solon Lovett. *Forest Land Ownership in New England, with Special Reference to Forest Holdings of Less Than Five Thousand Acres.* Cambridge: Harvard University Press, 1949. 269 pp.

B97 Barrett, John W.; Ketchledge, Edwin H.; and Satterlund, Donald R., eds. *Forestry in the Adirondacks.* Syracuse: New York State University, College of Forestry, 1961. 139 pp. Maps, tables, charts.

B98 Barrett, John W., ed. *Regional Silviculture in the United States.* New York: Ronald Press, 1962. 610 pp. There are eleven regional chapters by specialists which include some history of forest use.

B99 Barrett, Louis A. *A Record of Forest and Field Fires in California from the Days of the Early Explorers to the Creation of the Forest Reserves.* San Francisco: USFS, California Region, 1935. 171 pp. Illus., bib. Processed.

B100 Barrett, Louis A. "Two Thousand Miles With a Pack Train." *American Forests* 47 (May 1941), 234-35, 249. Reminiscences by a forest inspector for the General Land Office of a pack trip in 1903 through the Yellowstone Forest Reserve of Wyoming and Idaho.

B101 Barrow, Susan H.L., and Evans, J. Allan. *Green Gold Harvest: A History of Logging and Its Products; An Exhibition of the Whatcom Museum of History and Art, Bellingham, Washington.* Bellingham: Whatcom Museum of History and Art, 1970. 78 pp. Illus. Primarily a catalog of art work, photographs, artifacts, and dioramas at the museum illustrating the history of logging and the lumber industry in northwestern Washington.

B102 Barrows, J.S. *Forest Fires in the Rocky Mountains.* Missoula, Montana: USFS, Northern Rocky Mountain Forest and Range Experiment Station, 1951. 252 pp. Analysis of 36,000 forest fires in the region since 1908.

B103 Barrus, Clara. *Our Friend, John Burroughs.* Boston: Houghton Mifflin, 1914. Includes Burroughs's "Autobiographical Sketches" on his early life in New York's Catskill Mountains and how he became a nature writer.

B104 Barrus, Clara, ed. *The Life and Letters of John Burroughs.* 2 Volumes. Boston: Houghton Mifflin, 1925. Burroughs (1837-1921), a naturalist and nature writer, influenced the conservation movement in part through his associations with John Muir, Theodore Roosevelt, and others.

B105 Barry, Phillips, ed. *The Maine Woods Songster.* Cambridge, Massachusetts: Powell Printing Company, 1939. 102 pp. Texts, tunes, analytic notes. Includes some history of lumberjack songs.

B106 Bartelle, J.P. *Forty Years on the Road; or the Reminiscences of a Lumber Salesman.* Cedar Rapids, Iowa: Torch Press, 1925. 161 pp. In the Midwest since 1878.

B107 Bartells, E.J. "History of Wood Pipe and Some Data on Its Use." *Proceedings of the American Wood Preservers' Association* 17 (1921), 369-81.

B108 Barth, N. "A Study of the Conflicts of Interest Arising from Competing Resource Uses in the Allagash Regions of Northwestern Maine." Master's thesis, Univ. of Michigan, 1962.

B109 Bartlett, Richard A. "Clarence King's Fortieth Parallel Survey." *Utah Historical Quarterly* 24 (Apr. 1956), 131-47. On the U.S. Geological Survey of the 40th parallel from the 120th to the 105th meridian, 1867-1879.

B110 Bartlett, Richard A. *Great Surveys of the American West.* Norman: University of Oklahoma Press, 1962. xxiv + 408 pp. Illus., maps, bib., index. Includes the Hayden, King, Powell, and Wheeler surveys of the 19th century.

Bartlett, Richard A. See Chittenden, Hiram Martin, #C304

B111 Bartlett, Richard A. "'Will Anyone Come Here for Pleasure?'" *American West* 6 (Sept. 1969), 10-16. Early history of Yellowstone National Park with emphasis on tourism and the 1883 visit of President Chester A. Arthur.

B112 Bartlett, Richard A. "Those Infernal Machines in Yellowstone." *Montana, Magazine of Western History* 20 (Summer 1970), 16-29. On the successful campaign to open Wyoming's Yellowstone National Park to automobiles in 1915.

B113 Bartlett, Richard A. *Nature's Yellowstone.* Albuquerque: University of New Mexico Press, 1974. xiii + 250 pp. Illus., map, notes, bib., index. Natural and human history of the Yellowstone area prior to 1872, including its establishment as a national park.

B114 Bartlett, Stanley Foss. *Beyond the Sowdyhunk.* Portland, Maine: Falmouth Book House, 1937. 164 pp. Illus. Short stories dealing with the life and customs of Maine lumberjacks.

B115 Bartlett, William W. *History, Tradition and Adventure in the Chippewa Valley.* Chippewa Falls, Wisconsin: Chippewa Printery, 1929. 244 pp. Illus. Includes history of the lumber industry.

Barton, James H. See Seigworth, Kenneth J., #S163

Bartoo, Ronald A. See Meyer, Hans Arthur, #M426

B116 Bartram, J.C. "How About Lumber Prices in the 80's? Records of Sales That Make Interesting Comparisons." *Canada Lumberman* 60 (Aug. 1, 1940), 64. On Ontario white pine prices from mill to wholesaler, 1880s to 1890s.

B117 Bartram, William. *Travels to North and South Carolina, Georgia, East and West Florida.* 1792. Reprint. Intro. by Gordon DeWolf. Savannah: Beehive Press, 1973. xx + xxiv + 535 pp. Illus. Bartram, a naturalist and one of the earliest interpreters of American wilderness, gives his account of botanical explorations through the Southeast in the 1770s. There are other editions of the 1791 version of this work under the title *The Travels of William Bartram.* Mark Van Doren edited and John Livingston Lowes introduced the Barnes and Noble reprint (New York, 1940). Francis Harper edited the "naturalist's edition" reprinted by Yale University Press (New Haven, 1958).

B118 Bartz, Melvin E. "Origin and Development of the Paper Industry in the Fox River Valley (Wisconsin)." Master's thesis, Univ. of Iowa, 1940.

Baser, Nort. See Say, Harold Bradley, #S53

B119 Bass, Robert P. "The Progress of Forestry." *American Forestry* 18 (Feb. 1912), 75-81. Brief sketch by a governor of New Hampshire and president of the American Forestry Association.

B120 Bates, Carlos G., and Peirce, G.R. *Forestation of the Sand Hills of Nebraska and Kansas.* USFS, Bulletin No. 121. Washington: GPO, 1913. 49 pp.

B121 Bates, Carlos G., and Zon, Raphael. *Research Methods in the Study of Forest Environment.* USFS, Bulletin No. 1059. Washington: GPO, 1922. 209 pp. Includes some history of forest research.

B122 Bates, Carlos G., and Henry, Alfred Judson. "Forest and Stream-Flow Experiment at Wagon Wheel Gap, Colorado: Final Report on Completion of the Second Phase of the Experiment." *U.S. Monthly Weather Review,* Supplement No. 30 (1928), 79 pp. Illus. The stream-flow controversy divided the ranks of conservationists beginning in 1907. The authors wrote several earlier accounts of this USFS experiment in forest influences.

B123 Bates, James Leonard. "Senator Walsh of Montana, 1918-1924: A Liberal Under Pressure." Ph.D. diss., Univ. of North Carolina, 1952. 247 pp. Thomas J. Walsh, a Democratic senator from Montana, was involved in conservation politics, especially as the chairman of the Senate committee investigating the Teapot Dome scandals in the mid-1920s.

B124 Bates, James Leonard. "The Teapot Dome Scandal and the Election of 1924." *American Historical Review* 60 (Jan. 1955), 303-22. Democrats were unable to capitalize politically on the Teapot Dome scandal.

B125 Bates, James Leonard. "Fulfilling American Democracy: The Conservation Movement, 1907-1921." *Mississippi Valley Historical Review* 44 (June 1957), 29-57. Argues that "organized conservationists were concerned more with economic justice and democracy in the handling of resources than with mere prevention of waste." Includes sketches of conservationist leaders, an analysis of the Ballinger-Pinchot controversy, and an explanation of World War I's impact on the conservation movement.

B126 Bates, James Leonard. "The Midwest Decision, 1915: A Landmark in Conservation History." *Pacific Northwest Quarterly* 51 (Jan. 1960), 26-34. On *United States v. Midwest Oil Company,* in which the Supreme Court "upheld the executive power to withdraw and protect public lands of the United States," thus preserving the naval petroleum reserve policies of President William H. Taft, and significantly influencing natural resource conservation efforts.

B127 Bates, James Leonard. *The Origins of Teapot Dome: Progressives, Parties, and Petroleum, 1909-1921.* Urbana: University of Illinois Press, 1963. x + 278 pp. Illus., notes, bib., index. Although primarily concerned with federal oil policy, this book contains much history of the conservation movement in general.

B128 Bates, John S. "Reminiscences of Technical Section Early Days." *Pulp and Paper Magazine of Canada* 75 (Jan. 1974), 20-25. On the work of the Technical Section of the Canadian Pulp and Paper Association since 1913.

B129 Bates, Marston. *The Forest and the Sea: A Look at the Economy of Nature and the Ecology of Man.* New York: Random House, 1960. 216 pp. Incidental references to the history of forest ecology and of man's attitudes and philosophies toward nature.

B130 Bauer, Clyde Max. *Yellowstone—Its Underworld: Geology and Historical Anecdotes of Our Oldest National Park.* Albuquerque: University of New Mexico Press, 1948. x + 122 pp. Illus., maps, bib. Includes some history of the park area since 1830.

B131 Bauer, Patricia M. "History of Lumbering and Tanning in Sonoma County, California, since 1812." Master's thesis, Univ. of California, Berkeley, 1951.

B132 Bauer, Patricia M. "The Beginnings of Tanning in California." *California Historical Society Quarterly* 33 (Mar. 1954), 59-72. From 1814 to 1862.

B133 Baumgartner, David Carroll. "An Analysis of Forest and Associated Land Taxation in Illinois." Ph.D. diss., Southern Illinois Univ., 1973. 191 pp.

B134 Baxter, Henry. "Rafting on the Allegheny and Ohio, 1844." *Pennsylvania Magazine of History and Biography* 51 (Jan. 1927), 27-78; 51 (Apr. 1927), 143-71; 51 (July 1927), 207-43. A diary account of rafting lumber from Friendship, New York, downstream on the Allegheny and Ohio rivers and marketing it along the way.

B135 Baxter, Maurice G. "Orville H. Browning: Conservative in American Politics." Ph.D. diss., Univ. of Illinois, 1948.

B136 Baxter, Maurice G. *Orville H. Browning, Lincoln's Friend and Critic.* Indiana University Publications, Social Science Series, No. 16. Bloomington: Indiana University Press, 1957. vii + 351 pp. Notes, bib. Browning was Secretary of the Interior in the Johnson administration, 1866-1869.

B137 Bayard, Charles Judah. "The Development of the Public Land Policy, 1783-1820, with Special Reference to Indiana." Ph.D. diss., Indiana Univ., 1956. 331 pp.

B138 Bazeley, W.A.L. "State Forests in New England." *Journal of Forestry* 24 (May 1926), 559-61. Includes historical references to their acquisition in New Hampshire, Vermont, Connecticut, and Massachusetts.

B139 Beadel, H.L. "Fire Impressions." *Tall Timbers Fire Ecology Conference Proceedings* 1 (1962), 1-6. Includes some history of forest fire ecology.

B140 Beal, Junius E. "Michigan Forests: Historical Notes." *Michigan Alumnus* 35 (1929), 590-91. Brief sketch of logging and forestry in Michigan, 1890s-1920s.

B141 Beal, Merrill D. "A History of Yellowstone National Park." Ph.D. diss., Washington State Univ., 1945. 341 pp.

B142 Beal, Merrill D. *The Story of Man in Yellowstone.* Caldwell, Idaho: Caxton Printers, 1949. 320 pp. Illus., maps, notes, bib., index. A history of Yellowstone National Park emphasizing efforts to preserve its wildlife and scenic beauty. The Yellowstone Library and Museum Association published revised editions in 1956 and 1960.

B143 Beall, H.W. *Forest Fire Research.* Canadian Pulp and Paper Association, Woodlands Section, Index No. 1928. Montreal, 1947. 5 pp. Some history of forest fire research in Canada.

B144 Beall, H.W. "Highlights in the Development of Forest Fire Protection in Canada." *Forestry Chronicle* 31 (Dec. 1955), 332-337.

B145 Beals, Charles E. *Passaconaway in the White Mountains.* Boston: R.G. Badger, 1916. 343 pp. Includes some account of lumbering in New Hampshire's White Mountains since 1900.

B146 Beals, Ralph L. *History of Glacier National Park, with Particular Emphasis on the Northern Developments.* Berkeley, California: U.S. National Park Service, Field Division of Education, 1935. 31 pp. Processed.

B147 Beard, Daniel B.; Semingsen, Earl M.; and Vinten, C.R. "Florida's Royal Palm Forest." *National Parks Magazine* 22 (Oct.-Dec. 1948), 32-34. On Fahkahatchee Slough, 1906-1948, containing the nation's largest stand of royal palms.

B148 Beard, Daniel B. "Let 'er Burn?" *Everglades Natural History* 2 (Mar. 1954), 2-8. On prevention of fire in Florida's Everglades National Park since 1928.

B149 Beard, F. "Regional Development of the Pulp and Paper Industry." Master's thesis, Yale Univ., 1933.

B150 Beasley, Norman. *Freighters of Fortune: The Story of the Great Lakes.* New York: Harper, 1930. ix + 311 pp. Bib. Commerce on the Great Lakes, including reference to the lumber trade.

B151 Beatley, Janice Carson. "The Primary Forests of Vinton and Jackson Counties, Ohio." Ph.D. diss., Ohio State Univ., 1953. 297 pp. On the forests shortly before settlement began, as indicated in surveyors' field notes and other records, 1798-1805.

B152 Beattie, George William, and Beattie, Helen Pruitt. *Heritage of the Valley: San Bernardino's First Century.* Oakland: Biobooks, 1951. xxix + 459 pp. Illus., maps, notes, bib., index. Includes some history of logging in the San Bernardino Mountains of California in the 1850s.

Beattie, Helen Pruitt. See Beattie, George William, #B152

B153 Beattie, Rolla Kent, and Dillard, Jesse D. "Fifty Years of Chestnut Blight in America." *Journal of Forestry* 52 (May 1954), 323-29. Since 1905.

B154 Beatty, Jeanne K. *Lookout Wife.* New York: Random House, 1953. 311 pp. Experiences at a lookout station on the Salmon National Forest of Idaho.

B155 Beatty, Leslie R. "A Forest Ranger's Diary." Ed. by Julius F. Wolff, Jr. *Conservation Volunteer* 25-31 (Mar. 1962 through July 1968). A thirty-two-part series. From 1911 to 1958 in Minnesota.

B156 Beatty, Robert O. "The Conservation Movement." *Annals of the American Academy of Political and Social Science* 281 (May 1952), 10-19.

B157 Beck, Earl C. *Lore of the Lumber Camps.* Ann Arbor: University of Michigan Press, 1948. xii + 348 pp. Illus., music, bib., index. A revised and enlarged edition of the author's *Songs of the Michigan Lumberjacks* (1941); includes texts and music for over 100 ballads, with a historical introduction.

B158 Beck, Earl C. *They Know Paul Bunyan.* Ann Arbor: University of Michigan Press, 1956. 255 pp. Illus., notes. A collection of lumberjack songs and stories, mostly from the Great Lakes area and the Pacific Northwest, with some historical commentary on the Paul Bunyan legend since the 1880s.

B159 Becker, F.D., and Bellows, S.B., eds. *History of the Pacific Coast Shipper's Association, and Organization of Wholesalers and Manufacturers of Pacific Coast Forest Products.* Seattle: Pacific Coast Shipper's Association, 1914. 80 pp.

B160 Becker, Folke. "Trees for Tomorrow." *Wisconsin Magazine of History* 36 (Autumn 1952), 43-47. An organization established by nine paper mills in the Wisconsin River Valley of Wisconsin, and its interest in reforestation since 1944.

B161 Beckham, Stephen Dow. "Asa Mead Simpson: Lumberman and Shipbuilder." *Oregon Historical Quarterly* 68 (Sept. 1967), 259-73. Simpson (1826-1915) operated lumber and shipping enterprises in Coos County, Oregon, from the 1850s to 1915.

B162 Beckham, Stephen Dow. *The Simpsons of Shore Acres.* Coos Bay, Oregon: Arago Books, 1971. 37 pp. Illus. On the careers of Asa Mead Simpson and his son, Louis Jerome Simpson, lumbermen of Coos Bay, Oregon, 1850s-1920s.

B163 Beckham, Stephen Dow. *Coos Bay: The Pioneer Period, 1851-1890.* Coos Bay, Oregon: Arago Books, 1973.

Coos Bay became a leading Oregon lumber port during this period.

B164 Bédard, Avila. "History of the Quebec Forestry School." *Forestry Chronicle* 4 (Mar. 1928), 3-6. Laval Université Forestry School, Quebec, Quebec, since 1910.

B165 Bédard, Avila, and Boutin, Fernand. "Administration of Crown Timberlands in Quebec." *Journal of Forestry* 36 (Oct. 1938), 933-37. Includes some history of the Quebec Forest Service since 1910.

B166 Bédard, Avila. "Forestry in Quebec, Past—Present—Future." *Canadian Geographic Journal* 28 (June 1944), 258-80.

B167 Bédard, Avila. "Forestry in Quebec: Past, Present and Future." *Canadian Geographic Journal* 57 (Aug. 1958), 36-49.

B168 Bédard, Paul W., and Ylvisaker, Paul N. *The Flagstaff Federal Sustained Yield Unit.* Inter-University Case Program, ICP Case No. 37. University: University of Alabama Press, 1957. 24 pp. Explores policy and political problems encountered by the USFS in establishing a sustained-yield program to restrict cutting of timber near Flagstaff, Arizona, on the Coconino National Forest, 1947-1949.

B169 Beebe, Lucius M., and Clegg, Charles. *Narrow Gauge in the Rockies.* Berkeley: Howell-North, 1958. 224 pp. Illus., maps, diags. Includes logging railroads.

B170 Beer, George L. *The Commercial Policy of England toward the American Colonies.* Studies in History, Economics and Public Law, Volume 3, Number 2. New York: Columbia University, 1893. 167 pp. Includes some history of English naval stores policy in the colonial period.

B171 Beers, Howard W., and Heflin, Catherine P. *People and Resources in Eastern Kentucky.* Kentucky Agricultural Experiment Station, Bulletin No. 500. Lexington, 1947. 59 pp. Includes forest resources and industries.

B172 Beers, J.H., and Company. *History of the Great Lakes.* 2 Volumes. Chicago: J.H. Beers and Company, 1899. Includes mention of the lumber trade.

Beeson, Lewis. See Lyon, Alanson Forman, #L332

B173 Begg, Hugh M. "Resource Endowment and Economic Development: The Case of the Gulf Islands of British Columbia." *Scottish Geographical Magazine* 89 (Sept. 1973), 119-30.

Beh, John L. See Highsmith, Richard Morgan, Jr., #H360

B174 Behan, Richard W. "The Succotash Syndrome, or Multiple Use: A Heartfelt Approach to Forest Management." *Natural Resources Journal* 7 (Oct. 1967), 473-84. Some history and critique of the multiple use concept of forest management.

B175 Behan, Richard W. "Wilderness Decisions in Region I, United States Forest Service: A Case Study of Professional Bureau Policy Making." Ph.D. diss., Univ. of California, Berkeley, 1972.

B176 Behan, Richard W. "Forestry and the End of Innocence." *American Forests* 81 (May 1975), 16-19, 38-49. Includes some general history of forestry and its basic concepts, especially the influence of German forestry through Bernhard E. Fernow in the late 19th century.

Behre, C. Edward. See Yale University, School of Forestry, #Y2

B177 Behre, C. Edward, and Lockard, C.R. *Centralized Management and Utilization Adapted to Farm Woodlands in the Northeast.* Syracuse: Charles Lathrop Pack Forestry Foundation and New York State College of Forestry, 1937. Includes some history of the Cooperstown Forest Unit in New York since the late 18th century.

B178 Beilmann, August P., and Brenner, Louise G. "The Recent Intrusion of Forests in the Ozarks." *Annals of the Missouri Botanical Garden* 38 (Sept. 1951), 261-82. Evidence since 1541 that the Ozark Mountains of Missouri were formerly open and parklike.

B179 Beilmann, August P., and Brenner, Louise G. "The Changing Forest Flora of the Ozarks." *Annals of the Missouri Botanical Garden* 38 (Sept. 1951), 283-91. Study of a forested area in Franklin County, Missouri, 1938-1950.

B180 Belcher, C. Francis. "The Logging Railroads of the White Mountains." *Appalachia* 32 (Dec. 1959), 516-28; 33 (June 1960), 37-49; 33 (Dec. 1960), 200-16; 33 (June 1961), 353-74; 33 (Dec. 1961), 501-25; 34 (Dec. 1962), 279-95; 34 (Dec. 1963), 689-715; 35 (Dec. 1964), 325-39; 35 (Dec. 1965), 715-31. Subtitles for this series on New Hampshire logging railroads include: "Upper Ammonoosuc Railroads," "The Sawyer River Railroad," "The Zealand Valley Railroad," "East Branch and Lincoln Railroad," "Upper Saco Valley," "The Conway Lumber Co. Operations," and "Woodstock and Thornton Gore Railroad."

B181 Belding, George Angus. "Thunder in the Forest." *Michigan History* 33 (Mar. 1949), 30-42. Reminiscences of the lumber industry of Michigan in the early 20th century and a historical review of the lumberjack from colonial times.

B182 Belknap, Jeremy. *The History of New Hampshire.* 3 Volumes. 1784-1792. There are several later editions and facsimile reprints of the various volumes. Included is a detailed account of the British Broad Arrow policy and other forest-related disputes of the colonial period.

B183 Bell, Frank Carter. "Federal Legislation Concerning the Disposition of Grazing Lands (1862-1900)." Ph.D. diss., Indiana Univ., 1959. 292 pp.

B184 Bell, Keller J. "History and Growth of the Treated Wooden Silo." *Proceedings of the American Wood Preservers' Association* 17 (1921), 352-57.

B185 Bell, L.C., ed. *The Golden Anniversary Dinner, February 24, 1940, in Honor of William McClellan Ritter, Founder of W.M. Ritter Lumber Company: Fifty Years of Service.* Richmond, Virginia, 1940. 96 pp.

B186 Bell, Laird. *The Mid-West Lumber Cycle.* Princeton, New Jersey: Newcomen Society, American Branch, and Princeton University Press, 1940. 36 pp. On the lumber industry of the upper Midwest and Great Lakes states, particularly in the late 19th century.

B187 Bell, Laird. *William H. Laird: A Sort of Profile by His Grandson.* Chicago: Privately printed, 1963. 73 pp. Owner of the Laird, Norton Company, Winona, Minnesota, 1850s-1890s.

B188 Bellaire, John I. "The Greatest 'Jack' Battle of the Ages." *Michigan History Magazine* 24 (Summer 1940), 339-44. A fist fight between Tim Kaine and John Dugan, lumberjacks in Seney, Michigan.

B189 Bellaire, John I. "Silver Jack Driscoll; with Text of Ballad Silver Jack." *Michigan History Magazine* 25 (Winter 1941), 14-22. On a Michigan lumberjack who achieved legendary status.

B190 Bellaire, John I. "Greater Michigan's Lumberjacks." *Michigan History Magazine* 26 (Spring 1942), 173-87. Reminiscences of the lumber industry in Michigan's Upper Peninsula, including log driving on the Fox and Indian rivers.

Bellaire, John I. See Cookson, Edwin, #C582

B191 Bellomy, M.D. "The Unconquered Gypsy." *American Forests* 61 (June 1955), 26-27, 56-59. On efforts to control the destructive gypsy moth since its accidental introduction into the United States in 1869.

Bellows, S.B. See Becker, F.D., #B159

B192 Bellush, Bernard E. *Franklin D. Roosevelt as Governor of New York.* Columbia Studies in the Social Sciences, No. 585. New York: Columbia University Press, 1955. xiii + 338 pp. Notes. Includes an account of his conservation activities, 1929-1933, especially those programs that foreshadowed the Civilian Conservation Corps.

B193 Belt, Charles Banks. *History of the Committee on Conservation of Forests and Wildlife of the Camp Fire Club of America, 1909-1956.* New York: Camp Fire Club of America, 1956. 76 pp.

B194 Belthuis, Lyda C. "The Geography of Lumbering in the Mississippi River Section of Eastern Iowa." Ph.D. diss., Univ. of Michigan, 1947. 250 pp.

B195 Belthuis, Lyda C. "The Lumber Industry in Eastern Iowa." *Iowa Journal of History and Politics* 46 (Apr. 1948), 115-55. Since 1833.

B196 Belyea, Harold Cahill. "Log Rules, with Special Reference to the Scribner and the Doyle Diameter." *Southern Lumberman* 187 (Dec. 15, 1953), 276-86. History of rules for the measurement of logs, published by Edward Doyle, 1825-1854, John Marston Scribner, 1846-1872, and George W. Fisher and heirs, 1872-1944.

B197 Belyea, Harold Cahill. "A Postscript on the Lost Identity of Doyle and Scribner." *Journal of Forestry* 51 (May 1953), 326-29. On the rules for the measurement of logs, published by Edward Doyle and John Marston Scribner, 1825-1846, with biographical sketches, and an explanation of the subsequent confusion of the roles.

B198 Benedict, Darwin. "The New York Forest Preserve: Formative Years, 1872-1895." Master's thesis, Syracuse Univ., 1953.

B199 Benedict, M.A. "Twenty-One Years of Fire Protection in the National Forests of California." *Journal of Forestry* 28 (May 1930), 707-10. Since 1908.

B200 Benedict, Murray R. *Farm Policies of the United States: A Study of Their Origins and Development, 1790-1950.* New York: Twentieth Century Fund, 1953. 548 pp. Notes.

B201 Benedict, Warren V. "The Fight Against Blister Rust: A Personal Memoir." *Forest History* 17 (Oct. 1973), 21-28. The author, former director of the Division of Forest Pest Control, USFS, recalls his involvement in the Blister Rust Control program in Oregon, Washington, Idaho, and Montana, 1924-1953.

B202 Benincasa, Frederick Albert. "An Analysis of the Historical Development of the Tennessee Valley Authority from 1933 to 1961." Ph.D. diss., St. John's Univ., 1961.

B203 Bennett, A.L. "Profile of a Year: 1910." *American Forests* 66 (Sept. 1960), 30-33, 40-45. Exercise in nostalgia; some remarks on the politics of conservation, especially the Ballinger-Pinchot controversy.

Bennett, Edwin. See Myers, John M., #M668

B204 Bennett, F.W. "Some Developments of the Pine Lumber Industry of the South." Master's thesis, Yale Univ., 1931.

B205 Bennett, Guy Vernon. *Grant to Eisenhower: Political Giveaways Unlimited.* New York: Comet Press Books, 1956. xviii + 134 pp. Bib. On the "corrupt giving away" of publicly owned natural resources to wealthy individuals or firms, primarily by Republican administrations since the 1870s.

B206 Bennett, H.D. "Appalachian Hardwood Manufacturers, Inc.: The Story of 'America's Finest' Hardwoods." *Southern Lumberman* 193 (Dec. 15, 1956), 148-50. Since 1926.

B207 Bennett, Hugh Hammond. *Soil Conservation.* New York: McGraw-Hill, 1939. Reprint. New York: Arno Press, 1973. xvii + 993 pp. Illus., index. Includes history of soil erosion, flood control, soil conservation, and other topics pertaining to forest influences.

B208 Bennett, Hugh Hammond. *Elements of Soil Conservation.* 1947. Second edition. New York: McGraw-Hill, 1955. 358 pp. Includes some history of soil conservation.

B209 Bennett, Hugh Hammond. "Soil and Water Relationships in Western Political Economy." *Western Political Quarterly* 1 (Dec. 1948), 404-12. Offers historical examples of western soil and water conservation problems and explains use of the soil conservation district as a political device to ensure appropriate relationships of soil and water.

B210 Bennett, Hugh Hammond. *Our American Land: The Story of Its Abuse and Its Conservation.* USDA, Miscellaneous Publication No. 596. Washington, 1950. 32 pp.

B211 Bennett, John David. "Economics and the Folklore of Forestry." Ph.D. diss., Syracuse Univ., 1968. 253 pp. A history of the development of forestry economics, 1902-1939.

B212 Bennett, Russell W. "An Historical Outline of the Lumbering Industries." *Southern Lumber Journal* 125 (Dec. 15, 1926), 58-60, 169.

B213 Bennett, W.T.; Malkin, B.; and Jones, H.M. "Development of High Speed Paper Machines." *Paper Trade Journal* 103 (Aug. 13, 1936), 91-97. Also published in *Pulp and Paper of Canada* 37 (Aug. 1936), 505-11.

B214 Benns, F.L. *The American Struggle for the British West India Carrying Trade, 1815-1830.* Indiana University Studies, Volume 10, No. 56. Bloomington, 1923. 207 pp. Includes some history of the Atlantic lumber and timber trades.

B215 Benson, Elmer A. "Conservation and the Lumberjack." *American Forests* 43 (Aug. 1937), 381-83, 418-19. Some history of the effects of forest devastation and the conservation movement upon the lumber industry and its labor force, especially in Minnesota.

B216 Benson, Ezra Taft. *Cross Fire: The Eight Years with Eisenhower.* Garden City, New York: Doubleday, 1962. 627 pp. Benson was Dwight D. Eisenhower's Secretary of Agriculture, 1953-1961.

B217 Benson, H.K. *The Pulp and Paper Industry of the Pacific Northwest.* University of Washington, Engineering Experiment Station, Report No. 1. Seattle, 1929. 89 pp. Illus. Includes some history.

B218 Benson, Ivan. *Paul Bunyan and His Men, Being Exploits of the Men in the Logging Camps of Paul Bunyan, Lumberjack Hero of the North.* Rutland, Vermont: C.E. Tuttle Company, 1955. 231 pp. Paul Bunyan lore collected by the author during his days as a lumberjack and forest ranger in northern Minnesota, 1920s.

B219 Bentley, A.W. "Forestry in Canada's Tenth Province: Comprehensive Outline of History, Development, Forest Stands and Forest Activities of Newfoundland." *Canada Lumberman* 69 (Dec. 1949), 51-53; 70 (Jan. 1950), 39-41; 71 (Feb. 1950), 82 ff.

Bentley, John, Jr. See Recknagel, Arthur B. #R69

B220 Berberet, William Gerald. "The Evolution of a New Deal Agricultural Program: Soil Conservation Districts and Comprehensive Land and Water Development in Nebraska." Ph.D. diss., Univ. of Nebraska, 1970. 576 pp.

B221 Bercaw, T.E. "Management of Southern Pine Plantations for Pulpwood Production." In *Management of Young Even-Aged Stands of Southern Pines.* Louisiana State University and Agricultural and Mechanical College, School of Forestry, 1st Forestry Symposium. Baton Rouge, 1952. Pp. 9-15. On the Gaylord Container Corporation, Bogalusa, Louisiana, 1920-1952.

B222 Berdahl, Erick J. *Autobiographies of Andrew J. and Erick J. Berdahl.* N.p., n.d. 29 + 46 pp. Includes some account of logging in Wisconsin, 1870s and later.

B223 Berglund, Abraham; Starnes, George T.; and DeVyver, Frank T. *Labor in the Industrial South.* Charlottesville: University of Virginia, Institute for Research in the Social Sciences, 1930. 176 pp. Illus., tables, index. Includes chapters on labor in the lumber and furniture industries.

B224 Bergland, George. "I Remember the Fire: Lest We Forget . . . This Reminder of the Power and Devastation of a Forest Fire." *Conservation Volunteer* 33 (Mar.-Apr. 1970), 59-62. Reminiscence of a fire in northern Minnesota, 1931.

B225 Bergman, Hermas J. "The Reluctant Dissenter: Governor Hay of Washington and the Conservation Problem." *Pacific Northwest Quarterly* 62 (Jan. 1971), 27-33. Marion E. Hay favored conservation measures but opposed federal control of natural resources during his term, 1909-1913.

B226 Bergman, M.M. "Forest Fire Control under Clarke-McNary Act." *Michigan Conservation* 18 (July-Aug. 1949), 18-20. Concerns fire control, 1924-1948, and includes mention of forest fires in Michigan between 1871 and 1911.

B227 Bergman, M.M. "Science, Skill, and Forest Fire." *Michigan Conservation* 25 (May-June 1956), 22-26. On the Forest Fire Experiment Station of the Michigan Department of Conservation at Roscommon since 1928.

B228 Bergman, M.M. "Lest We Forget." *Michigan Conservation* 29 (Sept.-Oct. 1960), 49-51. On forest fires in Michigan since 1871.

B229 Bergquist, James M. "The Oregon Donation Act and the National Land Policy." *Oregon Historical Quarterly* 58 (Mar. 1957), 17-35. Operation of the law in Oregon Territory, 1850-1855.

B230 Bergren, Myrtle. *Tough Timber: The Loggers of B.C. — Their Story.* Toronto: Progress Books, 1966. 254 pp. Illus. History of efforts to organize a woodworkers union (International Woodworkers of America) in the Lake Cowichan area of British Columbia in the 1930s.

B231 Berkes, H.C. "Trend of Southern Pine Production and Supply." *Southern Lumberman* 117 (Dec. 20, 1924), 154-60.

B232 Berland, Oscar. "Giant Forest's Reservation: The Legend and the Myth." *Sierra Club Bulletin* 47 (Dec. 1962), 68-82. On the role of the Southern Pacific Railroad in preserving watersheds to foster agricultural growth.

B233 Berner, Richard C. "Source Materials for Pacific Northwest History: The Port Blakely Mill Company Records." *Pacific Northwest Quarterly* 49 (Apr. 1958), 82-83. Information on the business records of a lumber and land company covering the period 1876 to 1923. The major operations of the Port Blakely Mill Company were in Washington, and the records are held at the University of Washington.

B234 Berner, Richard C. "Sources for Research in Forest History: The University of Washington Manuscript Collection." *Business History Review* 35 (Autumn 1961), 420-425. Briefly describes history and papers of the following lumber industries and lumbermen of western Washington: Yesler, Denny and Company; Washington Mill Company; Port Blakely Mill Company; Puget Mill Company; Stimson Mill Company; Merrill and Ring Lumber Company; Northwestern Lumber Company; Will C. Ruegnitz, and Mark E. Reed.

B235 Berner, Richard C. "The Port Blakely Mill Company, 1876-89." *Pacific Northwest Quarterly* 57 (Oct. 1966), 158-71. On Puget Sound, Washington; one of the largest sawmills in the world during this era.

B236 Berner, Richard C. "Business Archives in Perspective." *Journal of Forest History* 18 (Apr. 1974), 32-34. On continuing efforts to solicit business records and make them available for researchers in archival institutions, with special reference to the business records of the forest products industries.

B237 Berry, Edward Wilber. *Tree Ancestors: A Glimpse into the Past.* Baltimore: Williams & Wilkins Company, 1923. vi + 270 pp. Illus., maps, index. Includes some forest history of North America based on research in paleontology.

B238 Berry, Fern. "Unchanging Land: The Jack-Pine Plains of Michigan." *Michigan History* 47 (Mar. 1963), 15-28. Reminiscence and history of pulpwood logging, the Civilian Conservation Corps, reforestation, and tree farming in Michigan since ca. 1900.

B239 Berry, James B. "A Conservative Lumbering Operation." *American Forestry* 19 (Jan. 1913), 1-6. Includes some history of the logging operations and sawmills of Nelson P. Wheeler in Forest and Warren counties of Pennsylvania since the mid-19th century.

B240 Berry, James B. *Farm Woodlands: A Textbook for Students of Agriculture in Schools and Colleges and a Handbook for Practical Farmers and Estate Managers.* Yonkers-on-Hudson, New York: World Book Company, 1923. vi + 425 pp. Illus., maps, index. Contains many incidental references to forest history, especially involving farm woodlands and forest products industries.

B241 Berry, Swift. *Lumbering in the Sugar and Yellow Pine Region of California.* USDA, Bulletin No. 440. Washington: GPO, 1917. 99 pp. Incidental history.

B242 Bertholf, John Rossman. *Men and Mutuality.* Seattle: Metropolitan Press, 1951. 223 pp. Illus. A history of the Northwestern Insurance Company, including a chapter on writing insurance for Pacific Northwest sawmills in the early 20th century.

B243 Berton, John Louis. "An Evaluation of the Marketing Program of Arkansas Lumber Manufacturers in Selling to the Residential Construction Industry." Ph.D. diss., Univ. of Arkansas, 1965. 273 pp. Since 1945.

B244 Besley, Fred Wilson. "Forest Mapping and Timber Estimating as Developed in Maryland." *Proceedings of the Society of American Foresters* 4 (1909), 196-206. Includes some history and description of methods of early forest surveys.

Besley, Fred Wilson. See Maxwell, Hu, #M322

B245 Besley, Fred Wilson. "State Forest Problems in Maryland." *American Forestry* 18 (July 1912), 446-52. Includes some history of state forestry in Maryland since 1906, by the state forester.

B246 Besley, Fred Wilson. *The Forests of Maryland.* Baltimore: Maryland Board of Forestry, 1916. 152 pp. Illus., maps, tables, index. The results of a seven-year survey of forest resources, including incidental historical information.

B247 Besley, Fred Wilson, and Dorrance, John G. *The Wood-Using Industries of Maryland.* Baltimore: Maryland Board of Forestry, 1919. 122 pp. Illus.

B248 Besley, Lowell. *Taxation of Crown-Granted Timberlands in British Columbia.* Vancouver: University of British Columbia, Faculty of Forestry, 1950. 86 pp. Includes history of forest taxation.

B249 Besley, Lowell. "Western Logging Engineering Schools: University of British Columbia." *Loggers Handbook* 11 (1951), 79-84. Brief history of forestry education since 1920.

B250 Bessey, Roy F. "The Political Issues of the Hells Canyon Controversy." *Western Political Quarterly* 9 (Sept. 1956), 676-90. Primarily concerning dams and hydroelectric power in the Hells Canyon of Oregon and Idaho, but with broad implications for natural resource policy and the conservation movement since 1946.

B251 Best, Gerald M.; Hogan, C.S.; and Richter, D.S. "McCloud River Railroad Company." *Western Railroader* 11 (July 1948), 4-5. A logging railroad between McCloud and Mount Shasta City, California, 1901-1948.

B252 Best, Jerome W. "Lumberjack Legacy." *Wisconsin Conservation Bulletin* 35 (May-June 1970), 23-25. Logging by the Mellon Lumber Company in Ashland County, Wisconsin, 1918-1927.

B253 Beswick, R. "Regreening the Yacolt Burn." *Western Conservation Journal* 11 (Jan.-Feb. 1954), 18, 27. An area of southwestern Washington burned in 1902.

B254 Bethea, John M. "Florida's Forest Resources." *Forest Farmer* 33 (Apr. 1974), 8-9, 40-42. Includes brief history of the forestry movement in Florida since the 1920s. The issue contains other articles on Florida forestry and foresters with incidental historical references.

Bethel, James Samuel. See Brown, Nelson Courtlandt, #B502

B255 Bethune, J.E., and Le Grande, W.P. "On-the-Ground Example in the Coastal Plain of South Carolina: Profitable Forest Management—A Case History." *Forest Farmer* 20 (Oct. 1960), 12-13, 38.

B256 Bethune, W.C., ed. *Canada's Western Northland: Its History, Resources, Population and Administration.* Ottawa: Department of Mines and Resources, Lands, Parks, and Forests Branch, 1937. 162 pp.

B257 Bettendorf, Harry J. *Paperboard and Paper Board Boxes—A History.* Chicago: Board Products Publishing Co., 1946. 135 pp. Also serialized in *Fibre Containers*, Jan., Apr., and July 1946.

Betts, H.S. See Schorger, Arlie W., #S111

B258 Beuter, Arthur J. "Lumber Railroads of the Weyerhaeuser Timber Company." *Pacific Railway Journal* 2 (Mar. 1957), 3-16. Washington and Oregon.

B259 Bevan, Arthur. *Foreign Trade Study of the Forest Products Industries.* NRA Work Materials, No. 32. Washington: GPO, 1936. 69 pp. Map, tables, diag., bib. National Recovery Administration.

B260 Bidwell, Percy Wells, and Falconer, John I. *History of Agriculture in the Northern United States, 1620-1860.* Washington: Carnegie Institution of Washington, 1925. Reprint. New York: Peter Smith, 1941. xii + 512 pp. Illus., maps, diags., charts, bib., index. Includes many references to the history of forest use and settlement.

B261 Bien, Morris. "The Public Lands of the United States." *North American Review* 192 (Sept. 1910), 387-402. Sketches the history of public land policy.

B262 Bien, Morris. "Le Domain public des États-Unis." *Journal des Économistes*, Sixth Series, 63 (Aug. 1919), 243-60. History of American public land policy.

B263 Bigfork Commercial Club. *On the Banks of the Bigfork: The Story of the Bigfork River Valley*. Bigfork, Minnesota, 1956. 56 pp. Illus., map. Includes much history of the lumber industry and forest lands near Bigfork, Minnesota.

B264 Bigelow, Martha Mitchell. "Isle Royale National Park Movement or a Study in Frustrations." *Michigan History* 41 (Mar. 1957), 35-44. The movement to create Isle Royale National Park, Michigan, 1921-1946.

B265 Billeb, Emil W. "Bodie's Railroad That Was." *Pony Express* 24 (June 1957), 3-12. On the Bodie Railroad & Lumber Company and its narrow-gauge line between Bodie and Mono Lake, California, 1881-1918.

B266 Billings, C.E. "The Great Fires of 1870." *Transactions of the Women's Canadian Historical Society of Ottawa* 3 (1910), 27-35.

B267 Billings, C.L.; Rettig, E.C.; Baggs, John T.; and Rapraeger, E.F. "Forest Management by Potlatch Forests, Inc., with Special Reference to Clearwater County, Idaho." *Journal of Forestry* 40 (May 1942), 364-70. Contains incidental history of the Clearwater Unit of Potlatch Forests, Inc., located in Lewiston, Idaho, since 1927.

B268 Billings, Robert W. "Diamond City: Timber Down the Hill (1920-1928)." *The Big Smoke 1972* (1972), 3-10. Reminiscences of logging, the lumber industry, and especially Diamond Match Company operations in Pend Oreille County, Washington.

B269 Billington, Ray Allen. "The Origin of the Land Speculator as a Frontier Type." *Agricultural History* 19 (Oct. 1945), 204-12.

B270 Bingaman, John W. *Guardians of Yosemite: Story of the First Rangers.* Washington: U.S. National Park Service, 1961. 123 pp.

B271 Bining, Arthur Cecil. *Pennsylvania Iron Manufacture in the Eighteenth Century.* Harrisburg: Pennsylvania Historical Commission, 1938. 227 pp. Illus., map, tables, diags., apps., bib. Including the charcoal iron industry.

B272 Bining, Arthur Cecil. *The Rise of American Economic Life.* New York: Scribner's, 1943. xii + 732 pp. Bibs.

B273 Binns, Archie. *The Roaring Land.* New York: Robert M. McBride and Company, 1942. 284 pp. Illus. Contains several chapters with history of the lumber industry in the Pacific Northwest.

B274 Binns, Archie. *Sea in the Forest.* Garden City, New York: Doubleday and Company, 1953. 256 pp. Index. Includes two chapters on the Industrial Workers of the World and the lumber industry of Washington.

B275 Birch, Brian. "The Environment and Settlement of the Prairie-Woodland Transition Belt—A Case Study of Edwards County, Illinois." *Southampton Research Series in Geography* 6 (1971), 3-31.

B276 Birch, John Worth. "The Changing Location of the North American Wood Pulp Industry, 1880 to 1955." Ph.D. diss., Johns Hopkins Univ., 1962.

Birch, Robert W. See Harmston, Floyd K., #H128

B277 Bird, Annie Laurie. *Boise, The Peace Valley.* Caldwell, Idaho: Caxton Printers, 1934. 408 pp. Illus. Includes some history of the lumber industry in Idaho's Boise Valley.

B278 Bird, Barbara (Kephart). *Calked Shoes: Life in Adirondack Lumber Camps.* Prospect, New York: Prospect Books, 1952. x + 141 pp. Illus. On the life of the author and her husband, Royal Gould Bird, in pulpwood logging for the Gould Paper Company, Lyons Falls, New York, 1922-1925. Parts were printed in *North Country Life* (Spring 1951-Summer 1952) as "I Married a Forester."

B279 Bird, Ralph Durham. *Ecology of the Aspen Parkland of Western Canada in Relation to Land Use.* Ottawa: Department of Agriculture, Research Branch, 1961. 155 pp. Illus., bib.

B280 Bird & Son, Inc. *One Hundred Twenty-Fifth Anniversary of Bird and Son, Inc.: A Graphic Record of a Manufacturing Enterprise Extending throughout a Century and a Quarter under the Guidance of Three Generations of Massachusetts Paper Makers.* East Walpole, Massachusetts, 1919. 36 pp. Illus.

B281 Birkeland, Torger. *Echoes of Puget Sound: Fifty Years of Logging and Steamboating.* Caldwell, Idaho: Caxton Printers, 1960. 251 pp. Illus., map. Includes an account of the author's work in Washington logging camps, 1903 ff.

B282 Bishop, G.N., and Grant, B.F. "Our First Fifty Years." University of Georgia Forestry Club, *Cypress Knee* 33 (1957), 10-15, 104-09. History of the University of Georgia School of Forestry since 1906.

B283 Bishop, J. Leander. *A History of American Manufactures from 1608 to 1860: Exhibiting the Origin and Growth of the Principal Mechanic Arts and Manufactures, from the Earliest Colonial Period to the Adoption of the Constitution; and Comprising Annals of the Industry of the United States in Machinery, Manufactures, and Useful Arts, with a Notice of the Important Inventions, Tariffs, and the Result of Decennial Census. To Which Are Added Statistics of the Principal Manufacturing Centres, and Descriptions of Remarkable Manufactories at the Present Time.* 2 Volumes. Philadelphia: Edward Young and Company, 1864. Includes some account of the lumber and wood-using industries. A revised edition in three volumes appeared in 1868, and there is a recent reprint by Augustus M. Kelley.

B284 Bishop, L.L. "Texas National Forests." *Texas Geographic Magazine* 1 (Nov. 1937), 1-15.

B285 Bissell, Lewis P. "The Forest Industries of Maine." *Northern Logger* 13 (Apr. 1965), 12-13, 42-43, 46-47. Since the early 19th century.

B286 Biswell, Harold Hubert. "Prescribed Burning in Georgia and California Compared." *Journal of Range Man-*

agement 11 (Nov. 1958), 293-97. Studies of prescribed burning, 1942-1958.

Bjorkbom, John C. See Armstrong, George R., #A478

B287 Black, John D. *The Rural Economy of New England, a Regional Study.* Cambridge: Harvard University Press, 1950. 820 pp. Maps, tables. Includes a chapter with some history of woodlands and forest land use.

B288 Black, Robson. "Canada's Deadly Forest Fires." *American Forestry* 22 (Sept. 1916), 521-24. Contemporary account of a forest fire in the "Clay Belt" of northern Ontario, with some mention of previous forest fires in Canada. See also the following article (Clyde Leavitt, "The Cause of the Fire—And Future Prevention," *ibid.*, 524-28) for incidental references to Canadian forest fires.

B289 Black, Robson. "The Canadian Forestry Association." *Journal of Forestry* 44 (Sept. 1946), 660-61. A prominent conservation organization since its founding in 1900.

B290 Black, S.R. "Forestry by Private Enterprise on Michigan-California Holdings." *Timberman* 27 (Aug. 1926), 113-14. An early experiment in industrial forestry in California.

B291 Blackburn, Benjamin Coleman. "Mettler's Woods in New Jersey." *Bulletin of the Garden Club of America* 42 (Jan. 1954), 44-47. On a 65-acre tract of primeval forest surviving in Somerset County in an area settled ca. 1703.

B292 Blackburn, George M., and Ricards, Sherman L., Jr. "A Demographic History of the West: Manistee County, Michigan, 1860." *Journal of American History* 57 (Dec. 1970), 600-18. A socioeconomic interpretation of life in 1860 on Michigan's "timber frontier."

B293 Blackburn, George M., and Ricards, Sherman L., Jr. "The Timber Industry in Manistee County, Michigan: A Case History in Local Control." *Journal of Forest History* 18 (Apr. 1974), 14-21. During the period from 1841 to the 1870s, the lumber industry was locally owned, locally financed, and resulted in considerable local wealth (although unevenly distributed).

B294 Blackhurst, W.E. *Riders of the Flood.* New York: Vanguard Press, 1954. Reprint. Parsons, West Virginia: McClain Printing Company, 1974. Fictional account of logging, river driving, and the white pine industry along West Virginia's Greenbrier River, 1884-1900. This and other works by Blackhurst combine fact and fiction.

B295 Blackhurst, W.E. *Of Men and a Mighty Mountain.* Parsons, West Virginia: McClain Printing Company, 1965. Includes some history of lumbering near Cheat Mountain, West Virginia.

B296 Blackhurst, W.E. *Your Train Ride Through History.* Parsons, West Virginia: McClain Printing Company, 1968. Illus. History of the Cass Scenic Railroad and of logging operations near Cass, West Virginia.

B297 Blackorby, Edward C. "Theodore Roosevelt's Conservation Policies and Their Impact upon America and the American West." *North Dakota History* 25 (Oct. 1958), 107-17. On Roosevelt's conservation policies, 1902-1908, and his belief that the presidency and the federal government should act as stewards for future generations.

B298 Blackwell, Lloyd P., and Burns, Paul Y. "The Beginning of Forestry Education in Louisiana." *Forests & People* 13 (First quarter, 1963), 70-71, 106-07, 114-15. Beginning in 1902, with emphasis on Louisiana State University, Louisiana Technical University, and McNeese State College.

B299 Blaess, A.F. "Twenty-Five Years of Timber Treatment on the Illinois Central System." *Proceedings of the American Wood Preservers' Association* 25 (1929), 162-69.

B300 Blair, Calvin Hobson. "Pacific Northwest Shingle Industry and the Tariff Act of 1930." Master's thesis, Washington State Univ., 1952.

Blair, J.C. See Simmons, Roger E., #S287

B301 Blair, T.B. "How the CCC Has Paid Off." *American Forests* 60 (Feb. 1954), 28-30, 44-45. Some history of tree planting by the Civilian Conservation Corps in southern states, 1930s.

B302 Blair, Walter Acheson. *A Raft Pilot's Log: A History of the Great Rafting Industry on the Upper Mississippi, 1840-1915.* Cleveland: Arthur H. Clark Company, 1930. 328 pp. Illus., map, bib., index. An account by a rafting captain.

B303 Blake, Oscar W. *Timber Down the Hill.* St. Maries, Idaho: By the author, 1967. 133 pp. Illus. Reminiscences of logging in Oregon and northern Idaho since ca. 1900.

B304 Blake, Peter, *God's Own Junkyard: The Planned Deterioration of America's Landscape.* New York: Holt, Rinehart and Winston, 1964. 144 pp. Incidental historical references to man's ruination of the environment.

B305 Blakelock, Chester R. "Four of Our State Parks." *Long Island Forum* 12 (June 1949), 113-16. On Montauk Point, Hither Hills, Wildwood, and Sunken Meadow state parks on Long Island, New York, since 1924.

B306 Blakelock, Chester R. "The Story of Bethpage State Park." *Long Island Forum* 14 (Mar. 1951), 43-44, 53-54. Long Island, New York, since 1931. See a similar article by the same title, *Long Island Forum* 21 (June 1958), 105-06, 112-14.

B307 Blakey, Roy G., and Associates. *Taxation in Minnesota.* University of Minnesota Studies in Economics and Business, No. 4. Minneapolis: University of Minnesota Press, 1932. xii + 627 pp. Illus., tables, diags. Includes two chapters with some history on forest taxation.

B308 Blanchard, Louis. *The Lumberjack Frontier; the Life of a Logger in the Early Days of the Chippeway.* Retold from the Recollections of Louis Blanchard by Walker D. Wyman, with the assistance of Lee Prentice. Lincoln: University of Nebraska Press, 1969. 88 pp. Illus., map. Based on oral reminiscences of a lumberjack who worked along the Chippewa River, Wisconsin, 1888-1912.

B309 Blancke, Harold. *Celanese Corporation of America: The Founders and the Early Years.* New York: Newcomen Society in North America, 1952. 24 pp. On a company established by Dr. Camille Dreyfus at Cumberland, Maryland, 1918, for the manufacture of cellulose acetate yarn and acetate compounds for plastics.

B310 Blanks, Carolyn. "Industry in the New South: A Case History." *Arkansas Historical Quarterly* 11 (Autumn 1952),

164-75. On the Crossett Lumber Company, Ashley County, Arkansas, since 1899.

B311 Blatnik, John A. "Voyageurs—The Wilderness Park." *National Parks & Conservation Magazine* 48 (Sept. 1974), 4-7. Includes some incidental history of Voyageurs National Park, Minnesota.

B312 Blegen, Theodore C. *Minnesota: A History of the State.* Minneapolis: University of Minnesota Press, 1963. 688 pp. Illus., maps. Includes a chapter on the lumber industry.

B313 Blegen, Theodore C. "With Ax and Saw: A History of Lumbering in Minnesota." *Forest History* 7 (Fall 1963), 2-13. From *Minnesota: A History of the State* (1963).

B314 Blewett, Marilyn Bowman, and Potzger, John Ernest. "The Forest Primeval of Marion and Johnson Counties, Indiana, in 1819." *Butler University Botanical Studies* 10 (Aug. 1951), 40-52. Based on land-survey records.

B315 Bloch, Don. "The Tie Hacks' Last Stand." *American Forests* 54 (Feb. 1948), 72-74. On the "Wind River tie drive," 1914-1946, in which hand-hewn and sawed crossties were floated on the Wind River from the Shoshone National Forest to Riverton, Wyoming.

B316 Blohm, Ernest V. "Albert E. Sleeper State Park." *Michigan Conservation* 23 (Sept.-Oct. 1953), 27-29. On the lumber industry in the area of Caseville, Michigan, 1830-1953, and the subsequent establishment of a state park.

B317 Bloom, John Porter, ed. *The American Territorial System.* Athens: Ohio University Press, 1973. xvi + 248 pp. Notes, bib. Includes an essay by Thomas G. Alexander on the operation of the federal land survey in the West in the late 19th century.

B318 Bloomer, P.A. *Economic Conditions in Southern Pine Industry.* New Orleans: Southern Pine Association, 1935. A statement and brief in behalf of the southern pine industry before the National Industrial Recovery Board, containing some history of economic conditions.

Bluestone, David W. See Van Tassel, Alfred J., #V31

B319 Blyth and Company. *The Georgia-Pacific Story.* New York, 1957. 39 pp. Illus., maps.

B320 Boardman, Leona. "Geologic Mapping in the United States." *Bulletin of the Geological Society of America* 60 (July 1949), 1125-31. Since 1879.

B321 Boardman, Samuel H. "Oregon State Park System: A Brief History." *Oregon Historical Quarterly* 55 (Sept. 1954), 179-233. Boardman, Oregon's state parks superintendent (1929-1951), gives historical and descriptive sketches of fifteen parks in the system.

B322 Boehler, Charles F. "The Pines of Interlochen." *Michigan Conservation* 23 (Sept.-Oct. 1954), 23-25. On 200 acres of original pine forest purchased by Michigan in 1917 and established as Interlochen State Park.

B323 Boerker, Richard H. "A Historical Study of Forest Ecology: Its Development in the Fields of Botany and Forestry." *Forestry Quarterly* 14 (Sept. 1916), 380-432. From roots in the 14th century to developments of the late 19th and early 20th centuries.

B324 Boerker, Richard H. *Our National Forests: A Short Popular Account of the Work of the United States Forest Service on the National Forests.* New York: Macmillan, 1919. lxix + 238 pp. Illus., tables, app.

B325 Boerker, Richard H. *Behold Our Green Mansions.* Chapel Hill: University of North Carolina Press, 1945. xv + 313 pp. Illus., maps, tables, index. Largely descriptive work on the extensive role of forests in national life; includes incidental historical references.

Bohn, David. See Farquhar, Francis P. #F28

B326 Bohn, David. *Backcountry Journal: Reminiscences of a Wilderness Photographer.* Santa Barbara, California: Capra Press, 1974. By the author of other books advocating scenic preservation and wilderness appreciation.

B327 Bohn, Frank P. "This Was the Forest Primeval." *Michigan History Magazine* 21 (Winter 1937), 21-38; 21 (Spring 1937), 178-96. Descriptions of the forest and forest industries near Seney, Michigan, 1890s.

B328 Boisen, Anton Theophilus. *Out of the Depths: An Autobiographical Study of Mental Disorder and Religious Experience.* New York: Harper, 1960. 216 pp. Includes an account of the author's employment with the USFS, early in the 20th century.

B329 Boisfontaine, A.S. "The Southern Pine Association in Retrospect: Seventeen Years of Trail Blazing in the Trade Association Field." *Southern Lumberman* 144 (Dec. 15, 1931), 109-14.

B330 Boisfontaine, A.S. "Grading: Lumber's Quality Insurance." *American Lumberman* (Sept. 11, 1948), 236-40. Short history of lumber grading standards and practices.

B331 Boles, Donald E. "Administrative Rule Making in Wisconsin Conservation." Ph.D. diss., Univ. of Wisconsin, 1956. 335 pp. On the administrative function of the Wisconsin Conservation Commission; includes some history of the shift from regulation by legislation to regulation by commission.

B332 Bolin, L.A. *The National Parks of the United States.* New York: Alfred A. Knopf, 1962. 105 pp. Illus.

B333 Bolinger, Margaret Ann. *S.H. Bolinger & Co., Ltd.* Shreveport, Louisiana: By the author, 1973. 40 pp. Illus., genealogical charts. Concerns a wood products enterprise in Bossier City, Louisiana, since 1898.

Bollaert, R. See Crafts, Edward C., #C672

B334 Bolles, Albert Sydney. *Industrial History of the United States from the Earliest Settlements to the Present Time: Being a Complete Survey of American Industries. . . .* Norwich, Connecticut: Henry Bill Publishing Company, 1879. x + 936 pp. Illus. Includes some material on the forest products industries, shipbuilding, and the lumber trade with Canada. There is a recent reprint by Augustus M. Kelley.

B335 Bolsinger, Charles L. *Changes in Commercial Forest Area in Oregon and Washington, 1945-1970.* USFS, Pacific Northwest Forest and Range Experiment Station, Resource Bulletin No. 46. Portland, 1973. 16 pp. Illus., tables. Examines conversion of forested lands to nonforest uses such as roads, urban and industrial expansion, agricultural clearing, power-line clearing, reservoir construction, etc. Reservation of park lands and changes in forest ownership are also noted.

Bond, Courtney C.J. See Hughson, John W., #H610

B336 Bond, Jay F. "Notes on the Girard Estate Forest Plantations." *Forest Quarterly* 6 (Mar. 1908), 34-39. Brief history of forest plantations on the lands of the Girard Water Company near Shenandoah, Pennsylvania, since 1881.

B337 Bond, Walter Edwin, and Campbell, Robert Samuel. *Planted Pines and Cattle Grazing: A Profitable Use of Southwest Louisiana's Cut-Over Pine Land.* Louisiana Forestry Commission, Bulletin No. 4. Baton Rouge, 1951. 28 pp. Illus., map, diag., tables, bib. Since 1925.

B338 Bones, James T. *The Timber Industries of New Jersey and Delaware.* Resource Bulletin, NE-28. Upper Darby, Pennsylvania: USFS, Northeast Forest Experiment Station, 1973. 17 pp. Illus. Includes some recent trends and statistics on the forest industries of the two states.

B339 Bonner, E., and Sexsmith, E.R. "Ten Years of Industrial Planning." *Pulp and Paper Magazine of Canada* 61 (Feb. 1960), 118-19, 122-23. On a reforestation program conducted by Spruce Falls Power and Paper and Kimberly-Clark Corporation in Ontario, since 1950.

Bonney, Lorraine. See Bonney, Orrin H., #B340, B341, B342

B340 Bonney, Orrin H., and Bonney, Lorraine. *Bonney's Guide: Grand Teton National Park and Jackson's Hole.* Houston: By the authors, 1966. 144 pp. Illus. Includes some history of the region and the movement to establish the park. The role of John D. Rockefeller, Jr., is emphasized.

B341 Bonney, Orrin H., and Bonney, Lorraine. "Lieutenant G.C. Doane: His Yellowstone Exploration Journal." *Journal of the West* 9 (Apr. 1970), 222-39. Gustavus C. Doane was part of the Washburn-Langford expedition which explored the Yellowstone region in 1870.

B342 Bonney, Orrin H., and Bonney, Lorraine. *Battle Drums and Geysers: The Life and Journals of Lt. Gustavus Cheyney Doane, Soldier and Explorer of the Yellowstone and Snake River Regions.* Chicago: Swallow Press, 1970. 622 pp. Illus.

B343 Booth, J.R. "Ground Wood Actually Discovered in Canada." *Pulp and Paper Magazine of Canada* 18 (Nov. 18, 1920), 1179-80. On the manufacture of pulp and paper from spruce in Halifax, Nova Scotia, 1838-1839.

B344 Booth, John. "Forestry in North America: The Pertinent Laws and Regulations and the Future of North American Forests." *Gardener's Monthly* 22 (1880), 277-78, 305-07, 341-44.

B345 Booth, John Derek. "Changing Forest Utilization Patterns in the Eastern Townships of Quebec, 1800 to 1930." Ph.D. diss., McGill Univ., 1972.

B346 Borden, Stanley T. "The Pacific Lumber Co." *Western Railroader* 12 (June 1949), 7-10. On the Humboldt & Eel River Railroad (later the Humboldt & Eureka Railway), Southport to Eureka, California, 1882-1935.

B347 Borden, Stanley T. "Arcata and Mad River, 100 Years of Railroading in the Redwood Empire." *Western Railroader* 17 (June 1954), 1-38. A short line logging railroad in the Humboldt Bay region of California since 1854.

B348 Borden, Stanley T. *The California Western Railroad.* San Mateo, California: Western Railroader, 1957. 40 pp. Illus., maps. A logging railroad between Ft. Bragg and Willits, California, since ca. 1900. A reprint from *Western Railroader* 20 (1957).

B349 Borden, Stanley T. "E.J. Dodge Lumber Co." *Western Railroader* 21 (June 1958), 3-8. A logging railroad (and its predecessors) in Humboldt County, California, since the 1880s.

B350 Borden, Stanley T. "Oregon & Eureka Railroad." *Western Railroader* 22 (Nov. 1958), 2-18. A logging railroad in Humboldt County, California, since the 1870s.

B351 Borden, Stanley T. "Redwood Trio." *Western Railroader* 22 (Apr. 1959), 1-14. Brief articles on the logging railroads of three lumber companies of Humboldt County, California: Elk River Mill & Lumber Company, Metropolitan Redwood Lumber Company, and McKay & Company, since the 1880s.

B352 Borden, Stanley T. "Minarets & Western Railway and Sugar Pine Lumber Co." *Western Railroader* 22 (Oct. 1959), 2-9. A logging railroad in Madera County, California, since the 1920s.

B353 Borden, Stanley T. "Bucksport and Elk River Railroad." *Western Railroader* 24 (Jan. 1961), 2-8. A logging railroad in Humboldt County, California, since the 1880s.

B354 Borden, Stanley T. "The Albion Branch." *Western Railroader* 24 (Dec. 1961), 2-32. A logging railroad and branch of the Northwestern Pacific Railroad in Mendocino County, California, since 1885.

B355 Borden, Stanley T. "San Francisco & Northwestern Ry." *Western Railroader* 26 (Jan. 1963), 2-16. A logging railroad in Humboldt County, California, since the late 19th century.

B356 Borden, Stanley T. *Caspar Lumber Company: Caspar, South Fork & Eastern Railroad.* San Mateo, California: Western Railroader, 1966. 48 pp. Illus., maps. Mendocino County, California, since the 1860s.

B357 Bordin, Ruth B. "A Michigan Lumbering Family." *Business History Review* 34 (Spring 1960), 64-76. On the business activities of Gideon Olin Whittemore, his sons James, Charles, and William, and his son-in-law A.B. Mathews, all of Pontiac, in the white pine lumber industry in the vicinity of Tawas City, Michigan, 1853-1866.

B358 Borland, Hal G., comp. and ed. *Our Natural World: The Land and Wildlife of America as Seen and Described by Writers Since the Country's Discovery.* Philadelphia: J.B. Lippincott Company, 1969. 849 pp. Illus. An anthology of writings, some treating forested lands descriptively and historically.

B359 Born, Wolfgang. "The Panoramic Landscape as an American Art Form." *Art in America* 36 (Jan. 1948), 3-10. On the period after 1840.

B360 Born, Wolfgang. "With Fresh Eyes (American Primitive Landscape Painting)." *American Collector* 17 (May 1948), 12-14. From 1820 to 1880.

B361 Born, Wolfgang. *American Landscape Painting: An Interpretation.* New Haven: Yale University Press, 1948. 228 pp. Gives attention to the Hudson River School (beginning in the 1820s) and subsequent artistic interpreters of scenic beauty.

B362 Bothwell, George E. *Co-operative Forest Fire Protection: A Brief History of the Movement in Canada and the United States, with Special Reference to the St. Maurice Fire Protective Association.* Forestry Branch, Bulletin No. 42. Ottawa: Department of the Interior, 1914. 28 pp. St. Maurice Fire Protective Association, Quebec.

B363 Botting, David C., Jr. "Bloody Sunday." *Pacific Northwest Quarterly* 49 (Oct. 1958), 162-72. On the armed clash between a shipload of Industrial Workers of the World (including lumber workers) and a sheriff's posse at Everett, Washington, November 5, 1916.

B364 Botsford, Robert Joe. *The Curran Story: An Account of the Life and Times of John Curran, Rhinelander Pioneer, and of the Early Development of the Wisconsin River Valley.* Rhinelander, Wisconsin: N.p., 1953. 52 pp. Illus. John C. Curran (1838-1931) was a pioneer lumberman near Rhinelander, Wisconsin.

B365 Bouffard, Jean. "Du Régime légal forestier, ou de la tenure des limites forestières en notre province." *Société de Géographie de Québec Bulletin* 6 (Mar. 1912), 84-91. A study of forestry regulation in Quebec, under both French and English rule.

B366 Boult-bee, Horace. "Organizing the Retailers for Service to the Lumber Industry: History of the O.R.L.D.A." *Canada Lumberman* 60 (Aug. 1, 1940), 88-91. Ontario Retail Lumber Dealers Association.

B367 Boulter, E.L. "Nova Scotia's Small Tree Conservation Act." *Journal of Forestry* 46 (Nov. 1948), 827-28. On the Small Tree Conservation Act of 1942 and its application in Nova Scotia; includes brief history of forest conservation in the province since 1909.

B368 Boulton, Harold, ed. *Century of Wood Preserving.* London: Philip Allan, 1930. 150 pp. Bib.

B369 Bounds, Harvey. "Wilmington Match Companies." *Delaware History* 10 (Apr. 1962), 3-32. The match industry in Wilmington, Delaware.

B370 Bourdo, Eric A., Jr. "A Validation of Methods Used in Analyzing Original Forest Cover." Ph.D. diss., Univ. of Washington, 1955. 224 pp. Concerns the presettlement vegetation of the western part of Michigan's Upper Peninsula; based in part on records of early land surveys.

B371 Bourdo, Eric A., Jr. "A Review of the General Land Office Survey and of Its Use in Quantitative Studies of Former Forests." *Ecology* 37 (Oct. 1956), 754-68. Since 1785.

B372 Bourgeois, Euclid J.; Morrison, John G., Jr.; and Wight, Charles L. *Mainly Logging.* Collected by Charles Vandersluis. Minneota, Minnesota: Minneota Clinic, 1974. 372 pp. Illus., bib., index. Reminiscences of logging and the lumber industry in Beltrami County, Minnesota. Individual contributions are as follows: Bourgeois, "Thoughts While Strolling," Morrison, "Never a Dull Moment," and Wight, "Reminiscences of a Cruiser."

Boutin, Fernand. See Bédard, Avila, #B165

B373 Bower, Dick Harry. "Conservation of Natural Resources: A Study of Governmental and Educational Activity in Santa Clara County, California, 1950-1964." Ed.D. diss., Stanford Univ., 1965. 160 pp.

B374 Bower, Ray F. "The President's Forests." *American Forests* 40 (Jan. 1934), 7-9, 46. On the practice of forestry at Franklin D. Roosevelt's estate, Hyde Park, New York.

B375 Bowers, Maynard O. *Through the Years in Glacier National Park: An Administrative History.* West Glacier, Montana: Glacier Natural History Association, 1960. v + 111 pp. Illus., map, notes, bib. Based in part upon materials assembled by Donald H. Robinson. Includes some history of the area since the first white visitors, ca. 1792.

B376 Bowers, William L. *The Country Life Movement in America, 1900-1920.* Port Washington, New York: National University Publications of Kennikat Press, 1974. 189 pp. Notes, app., bib., index. Contains much of interest to environmental and conservation history.

B377 Bowler, Frank Colburn. *It Began with the Wasps.* New York: Newcomen Society of England, American Branch, 1949. 24 pp. On the pulp and paper industry in the United States and Canada, 1900-1940s, especially the International Paper Company, a New York-based firm.

B378 Bowman, Francis F., Jr. *Paper in Wisconsin: Ninety-Two Years of Industrial Progress.* 1940. 30 pp.

B379 Bowman, James Cloyd. "Lumberjack Ballads." *Michigan History Magazine* 20 (Spring-Summer 1936), 231-45.

B380 Bowman, James Cloyd. "Life in the Michigan Woods." *Michigan History Magazine* 21 (Summer-Autumn 1937), 267-83. General account of 19th-century settlers, sawmills, surveyors, lumberjacks, and river drivers of Michigan.

B381 Bowman, James Cloyd. "Paul Bunyan Yarns and Other Frontier Legends." *Michigan History Magazine* 25 (Winter 1941), 25-28. Their place in American folklore.

Boyce, W.G.H. See Lewis, R.G., #L177, L178

B382 Boyd, James. "Fifty Years in the Southern Pine Industry." *Southern Lumberman* 144 (Dec. 15, 1931), 59-67; 145 (Jan. 1, 1932), 23-34. Includes a state-by-state survey which chronicles the establishment of individual lumber companies. Reprinted in pamphlet form by the Southern Pine Association in 1961.

B383 Boyd, James. "Lumber—World's First Industry." *Southern Lumberman* 177 (Dec. 15, 1948), 256, 258, 260-62, 264-66. On the lumber industry in the South since the colonial period.

B384 Boyd, Robert K. "Up and Down the Chippewa River." *Wisconsin Magazine of History* 14 (Mar. 1931), 243-61. Rafting on this Wisconsin river.

B385 Boyer, Charles S. *Early Forges and Furnaces in New Jersey.* Philadelphia: University of Pennsylvania Press, 1931. xvi + 287 pp. Illus., maps, bib. Includes some account of the charcoal iron industry. There is a modern reprint.

B386 Boyle, Clarence. "I Remember: Interesting Recollections of a Hardwoods Trail-Blazer." *Southern Lumberman* 144 (Dec. 15, 1931), 77-79. Tennessee, 1880s.

B387 Boyle, L.C. "Lumber Production and Prices." *Lumber World Review* 41 (Sept. 10, 1921), 27-33. Includes some history of the World War I period.

B388 Bracklin, John L. "A Forest Fire in Northern Wisconsin." *Wisconsin Magazine of History* 1 (Sept. 1917), 16-24. Reminiscence of a fire near Rice Lake, Wisconsin, 1898.

B389 Bracklin, James. "A Tragedy of the Wisconsin Pinery." *Wisconsin Magazine of History* 3 (Sept. 1919), 42-51. Logging boss of Knapp, Stout and Company recalls clash between Indians and loggers in Wisconsin, 1864.

B390 Braden, Leo. "Old Rafting Days on the Clarion River." *Northeastern Logger* 7 (Apr. 1959), 10-11, 29. Clarion River of northwestern Pennsylvania.

B391 Braden, Leo, and Reed, Frank A. "Early Lumbering in Northern New York." *Northern Logger* 12 (Apr. 1964), 14-15, 62-63.

B392 Bradford, Samuel Sydney. "The Ante-Bellum Charcoal Iron Industry of Virginia." Ph.D. diss., Columbia Univ., 1958. 210 pp.

B393 Bradley, C.B. "A Reference List of John Muir's Newspaper Articles." *Sierra Club Bulletin* 10 (Jan. 1916), 55-59.

B394 Bradley, Douglas H. "Canadian Newsprint Industry and Its Importance to the United States Market." Master's thesis, Univ. of Pennsylvania, 1948.

B395 Bradley, Harold C. "Colonel Benson—Rover." *Sierra Club Bulletin* 34 (June 1949), 15-16. Reminiscences of Colonel Benson, superintendent of Yosemite National Park, California, 1905.

B396 Bradley, James G. "When Smoke Blotted Out the Sun." *American West* 11 (Sept. 1974), 4-9. On the forest fire which burned much of northern Idaho in 1910.

B397 Bradshaw, James S. "Grand Rapids Furniture Beginnings." *Michigan History* 52 (Fall 1968), 279-98. Traces origin of the industry, 1836-1870, and attributes location to abundance of wood, waterpower, and New England immigrant technology.

B398 Bradshaw, James S. "Grand Rapids, 1870-1880: Furniture City Emerges." *Michigan History* 55 (Winter 1971), 321-42. Abundance of wood a factor in location of furniture industry.

B399 Bradwin, Edmund W. *The Bunkhouse Man: A Study of Work and Pay in the Camps of Canada, 1903-1914.* New York: Columbia University Press, 1928. Reprint. Toronto: University of Toronto Press, 1972. 306 pp. Illus., maps, tables, index. Although primarily concerned with working conditions among railroad construction workers, there are references to the lumber industry and lumberjacks.

B400 Brady, Eugene Allen. "The Role of Government Land Policy in Shaping the Development of the Lumber Industry in the State of Washington." Master's thesis, Univ. of Washington, 1959.

B401 Bramble, William Clark. "Landmarks in Forestry Research at Penn State." *Pennsylvania Forests* 38 (Fall 1953), 100-01.

B402 Bramble, William Clark. "Occupations of Penn State Forestry Graduates." *Journal of Forestry* 55 (Nov. 1957), 848-50. Since 1906.

B403 Bramble, William Clark, ed. *Forestry and Conservation in Indiana.* Lafayette, Indiana: Purdue University, Department of Forestry and Conservation, 1965. viii + 237 pp. Illus., bibs. Includes essays on the history of forestry, wildlife management, forestry education, forestry extension, industrial forestry, and other aspects of forestry and conservation in Indiana.

B404 Bramble, William Clark. "Is Conservation Education Failing?" *Transactions of the North American Wildlife Conference* 24 (1959), 52-58. Primarily concerned with forestry education in North America since 1886.

B405 Bramble, William Clark. "The Story of McCormick Woods." *American Forests* 81 (Jan. 1975), 16-21. Mostly natural history of a forty-acre plot of Indiana hardwoods since the early 19th century.

Bramlett, G.A. See Rose, B.B., #R295

B406 Branch, Maurice Lloyd. "The Paper Industry in the Lake States Region, 1834-1947." Ph.D. diss., Univ. of Wisconsin, 1955. 105 pp. Emphasis on Wisconsin.

B407 Brander, John Robert Gordon. "The Economic Importance of the Pulp and Paper Industry to the Maritime Provinces." Master's thesis, Queen's Univ., 1962.

B408 Brandes, Hans-Gunther. "An Investigation of the Canada-United States Trade in Newsprint." Ph.D. diss., New York State College of Forestry, 1955. 105 pp.

B409 Brandis, Dietrich. "The Late Franklin B. Hough." *Indian Forester* 11 (Sept. 1885), 426-31. Sketch of Hough (1822-1885), director of federal forestry activities, 1876-1883.

B410 Brandstrom, Axel J.F. *Analysis of Logging Costs and Operating Methods in the Douglas Fir Region.* Seattle: Charles Lathrop Pack Forestry Foundation, 1933. 117 pp. Incidental history.

B411 Brandstrom, Axel J.F. *Development of Industrial Forestry in the Pacific Northwest.* Colonel William B. Greeley Lectures in Industrial Forestry, No. 1. Seattle: University of Washington, College of Forestry, 1957. 33 pp. A general historical sketch of the lumber industry and forestry movement with emphasis on the development of industrial forestry and tree farming in the Pacific Northwest since 1920.

B412 Brandt, Ray. "The History of the Kisatchie National Forest." *Forests & People* 13 (First quarter, 1963), 52-53, 92. Reforestation on cutover lands of Louisiana since the 1920s—designated as Kisatchie National Forest in 1930.

B413 Braun, E. Lucy. *Deciduous Forests of Eastern North America.* Philadelphia: Blakiston, 1950. 596 pp. Illus., maps, bib. Includes some history.

B414 Braune, Yvonne O. "An Attempt to Place the IWW in Proper Historical Perspective as It Relates to the Development of Organized Labor in the Lumber Industry in the Northwest." Master's thesis, Univ. of Puget Sound, 1965.

B415 Bray, William L. *Forest Resources of Texas.* USDA, Bureau of Forestry, Bulletin 47. Washington: GPO, 1904. 71 pp.

B416 Bray, William L. *History of Forest Development on an Undrained Sand Plain in the Adirondacks.* New York State College of Forestry, Technical Publication No. 13. Syracuse, 1921. 47 pp. Illus., maps.

Brayer, Herbert Oliver. See Hamp, Sidford, #H71

B417 Breckenridge, W.J. "A Century of Minnesota Wild Life." *Minnesota History* 30 (June 1949), 123-34; 30 (Sept. 1949), 220-31. Its study and conservation since 1823.

B418 Brender, Ernest V. "From Forest to Farm to Forest Again." *American Forests* 58 (Jan. 1952), 24-25, 40-41, 43. On Georgia's lower Piedmont and the progression from hardwoods to homesteads to loblolly pine, 1773-1952.

B419 Brender, Ernest V. "Impact of Past Land Use on the Lower Piedmont Forest." *Journal of Forestry* 72 (Jan. 1974), 34-36. Historical survey of land use (by Indians as well as white and black settlers) in the lower Piedmont forests of Georgia, Alabama, and South Carolina.

Brenner, Louise G. See Beilmann, August P., #B179, B180

B420 Breton, J.A. "The Quebec Forestry Association, Inc." *Journal of Forestry* 45 (Feb. 1947), 92-93. Its activities since 1939.

B421 Brewer, George S. "Timber, Arkansas' Leading Resource." *Southern Lumberman* 193 (Dec. 15, 1956), 153-56. Some history since 1826.

B422 Brewer, Virginia W. *Forestry Activities of the Federal Government.* U.S. Library of Congress, Legislative Reference Service, Public Affairs Bulletin, No. 47. Washington, 1946. 185 pp. Includes some history.

B423 Bridgham, Lawrence Donald. "Maine Public Lands, 1781-1795; Claims, Trespassers and Sales." Ph.D. diss., Boston Univ., 1959. 395 pp.

Briegleb, Philip A. See Mason, David T. #M276

Briegleb, Philip A. See Demmon, Elwood L., #D131

Briegleb, Philip A. See Cowlin, Robert W., #C645

B424 Brier, Howard Maxwell. *Sawdust Empire: The Pacific Northwest.* New York: Alfred A. Knopf, 1958. xiv + 269 + xi pp. Illus., bib. Includes some history of logging and the lumber industry in Oregon and Washington.

B425 Briggs, Lloyd Vernon. *History of Shipbuilding on the North River, Plymouth County, Massachusetts.* Boston: Coburn Brothers, 1889. xv + 420 pp. Illus., maps. Wooden sailing vessels.

B426 Brigham, Johnson. *James Harlan.* Iowa City: State Historical Society of Iowa, 1913. xvi + 308 pp. Harlan (1820-1899) served Andrew Johnson as Secretary of the Interior, 1865-1866.

B427 Bright, Pascal A. "The Making of Pine Tar in Hocking County." *Ohio Archaeological and Historical Quarterly* 41 (Apr. 1932), 151-60. History and description of a primitive forest industry of southern Ohio.

B428 Brilhart, John K. "Gifford Pinchot as a Conservation Crusader in 1909." Master's thesis, Pennsylvania State Univ., 1957.

B429 Brimlow, George Francis. *Harney County, Oregon, and Its Range Land.* Portland: Binfords & Mort, for Harney County Historical Society, 1951. 332 pp. Illus., map, notes. Includes sections on the history of logging and the forest products industries.

B430 Brink, Wellington. *Big Hugh: The Father of Soil Conservation.* New York: Macmillan, 1951. xii + 167 pp. Illus., bib. Biography of Hugh Hammond Bennett (1881-1960), soils specialist in the USDA and chief of the Soil Conservation Service, 1935-1951.

B431 Brinks, Herbert. "The Effect of the Civil War in 1861 on Michigan Lumbering and Mining Industries." *Michigan History* 44 (Mar. 1960), 101-07.

B432 Brisbin, Bryce James. "Marketing Problems of the Arizona Lumber Industry." D.B.A. diss., Univ. of Southern California, 1967. 264 pp. Since 1959.

B433 Briscoe, Mark W. "Damariscotta-Newcastle Ships and Shipbuilding." Master's thesis, Univ. of Maine, 1967.

B434 Briscoe, Vera; Martin, James W.; and Reeves, J.E. *Safeguarding Kentucky's Natural Resources.* Bulletin No. 14. Lexington: Bureau of Business Research, College of Commerce, University of Kentucky, 1948. x + 224 pp. Diags. Examines nature, scope, and history of natural resources management in Kentucky.

B435 Brissenden, Paul Frederick. *The I. W. W.: A Study of American Syndicalism.* Columbia University Studies in History, Economics and Public Law, Volume 83, No. 193. New York: Columbia University, 1919. Reprint. New York: Russell & Russell, 1957. 432 pp. The Industrial Workers of the World organized many lumberjacks and millworkers in the Pacific Coast states early in the 20th century.

B436 Bristow, Arch. *Old Time Tales of Warren County.* Meadville, Pennsylvania: Tribune Publishing Company, 1932. 389 pp. Illus. Includes some history of logging and the lumber industry in Warren County, Pennsylvania.

B437 British Columbia. Department of Lands. *Forests and Forestry in British Columbia.* Victoria, 1920. 35 pp. Illus. Incidental history.

B438 British Columbia. Forest Service. *The Cowichan Lake Forest Experiment Station: A Brief Description of Its Purpose, Development, and Environment.* Victoria, 1947. 24 pp. On Vancouver Island, British Columbia.

B439 British Columbia. Forest Service. *Forestry in British Columbia.* Victoria, 1956. 15 pp. Incidental history.

B440 British Columbia. Forest Service. *Provincial Inventory and Forest Surveys in British Columbia.* Forest Survey Notes, No. 1. Victoria, 1957. 50 pp.

B441 British Columbia. Forest Service. *Crown Charges for Early Timber Rights; Royalties and Other Levies for Harvesting Rights on Timber Leases, Licences and Berths in British Columbia: First Report of the Task Force on Crown Timber Disposal.* Victoria, 1974. 67 pp. Tables, graphs. Includes some history of timber rights and related policies on crown lands in British Columbia since 1865.

B442 British Columbia. Forest Service. *Forest Tenures in British Columbia: Policy Background Paper Prepared by the Task Force on Crown Timber Disposal.* Victoria, 1974. 127 pp. Tables, notes, map. Includes some history of tenure policies and systems since the mid-19th century.

B443 British Columbia, University of. Faculty of Forestry. *The First Decade of Management Research on the U.B.C. Forests, 1949-1958.* Vancouver, 1958. 82 pp. Maps.

B444 *British Columbia Lumberman.* "Lumber Industry of Vancouver Island Rooted Deep in Coast History." *British Columbia Lumberman* 27 (Apr. 1943), 82-86.

B445 *British Columbia Lumberman.* "From Oxen to Rail Grade." *British Columbia Lumberman* 50 (Aug. 1966), 40-41. Transportation improvements in the British Columbia lumber industry.

B446 *British Columbia Lumberman.* "100 Years After the Birth of B.C.s Glory Industry." *British Columbia Lumberman* 50 (Aug. 1966), 32-35. Advancements in British Columbia logging technology.

B447 *British Columbia Lumberman Greenbook, 1973.* "Historical Review: A Chronology of the B.C. Forest Industries from 1778 to 1972." *British Columbia Lumberman Greenbook, 1973.* Vancouver: Journal of Commerce Limited, 1973. Pp. 97-106.

B448 Britton, Blaine S. *History of Paper Merchandising in New York City.* Chicago: Howard Publishing Company, 1939. 80 pp. Illus. Includes some history of paper manufacturing firms.

B449 Brockman, Christian Frank. "Coordination of Multiple Recreation Administration on Forest Lands." *Journal of Forestry* 48 (Jan. 1950), 8-10. Problems and trends, 1940s.

B450 Brockman, Christian Frank. *The Story of Mount Rainier National Park.* 1940. Second revision. Longmire, Washington: Mount Rainier National Park Natural History Association, 1952. 63 pp. Illus., map, bib. Includes some history of the area since 1833.

B451 Brockman, Christian Frank, and Merriam, Lawrence C., Jr. *Recreational Use of Wild Lands.* American Forestry Series. 1959. Second edition. New York: McGraw-Hill, 1973. 329 pp. Illus., bib., index. A text which contains a historical account of outdoor recreation in relation to the conservation movement; the emphasis is on recreational use of wild lands in state parks, national parks, and national forests.

B452 Brockway, Chauncey. "Frontier Days: An Autobiographical Sketch of Chauncey Brockway." Ed. by James W. Silver. *Pennsylvania History* 25 (Apr. 1958), 137-61. Brockway (1793-1886) was unsuccessful in the lumber industry and in rafting in Elk County, Pennsylvania, 1817-1854.

B453 Broderick, R.E. "321 Years Later." *Southern Lumberman* 185 (Dec. 15, 1962), 159-60. On the lumber industry in the Northeast since 1631.

B454 Bromley, Edward A. "The Old Government Mills at the Falls of St. Anthony." *Collections of the Minnesota Historical Society* 10, part 2 (1905), 635-43. A sawmill was built at St. Anthony (Minneapolis) in 1822 to supply lumber for Fort Snelling.

B455 Bromley, Stanley W. "The Original Forest Types of Southern New England." *Ecological Monographs* 5 (1935), 61-89.

B456 Brookfield, Charles M., and Griswold, Oliver T. *They All Called It Tropical: True Tales of the Romantic Everglades National Park, Cape Sable, and the Florida Keys.* Miami: Data Press, 1949. 77 pp. Illus., maps. Since 1513.

B457 Brooks, A.B. *Forestry and Wood Industries.* West Virginia Geological Survey, Volume 5. Morgantown, 1911. xvi + 481 pp. Illus., index. Includes much county-by-county history of the lumber industry and other wood-using industries.

B458 Brooks, Bryant Butler. *Memoirs of Bryant Butler Brooks, Cowboy, Trapper, Lumberman, Stockman, Oilman, Banker, and Governor of Wyoming.* Glendale, California: Arthur H. Clark, 1939. 370 pp.

B459 Brooks, Maud D. "Rafting on the Allegheny." *New York Folklore Quarterly* 1 (1945), 224-30.

B460 Brooks, Paul. *Roadless Area.* New York: Alfred A. Knopf, 1964. xiii + 259 pp. Illus., map. Collection of essays and memoirs of visits to wilderness areas of North America, including some history of man's attitude toward wilderness.

B461 Brooks, Paul. *The Pursuit of Wilderness.* Boston: Houghton Mifflin, 1971. xiii + 220 pp. Illus., maps. Includes some history of recent wilderness controversies over Everglades and North Cascades national parks.

B462 Brooks, Paul. "A Roadless Area Revisited." *Audubon* 77 (Mar. 1975), 28-37. Includes some history of the efforts to preserve the Quetico-Superior area of northern Minnesota and western Ontario in the 20th century.

B463 Brooks, Robert H. "The Origin of the Wooden Floor." *Southern Lumberman* 125 (Dec. 18, 1926), 204-06.

B464 Brookshire, Douglas C. "Carolina's Lumber Industry." *Southern Lumberman* 193 (Dec. 15, 1956), 161-62. On the history of hardwood lumber operations in western North Carolina since the 1880s.

B465 Broome, Harvey. "Origins of the Wilderness Society." *Living Wilderness* 5 (July 1940), 10-14. Author was a founder in 1935.

B466 Broome, Harvey. "Thirty Years—Decades of the Wilderness Society." *Living Wilderness* 29 (Winter 1965-1966), 15-26. On the history of the Wilderness Society and its leaders since 1935.

B467 Broome, Harvey. *Faces of the Wilderness.* Missoula, Montana: Mountain Press Publishing Company, in cooperation with the Wilderness Society. Washington, D.C., 1972. xiii + 271 pp. Illus. Descriptive and reminiscent essays of national parks and wilderness areas by a founder and former president of the Wilderness Society.

B468 Brotslaw, Irving. "Trade Unionism in the Pulp and Paper Industry." Ph.D. diss., Univ. of Wisconsin, 1964. 376 pp.

B469 Brouillet, Benoît. "L'Industrie des pâtes et du papier." In *La Forêt, étude prepareé avec la collaboration de l'ecole de génie forestier de Quebec*. Montreal, 1944. Pp. 171-231. Includes some history of the pulp and paper industry and of individual companies in Quebec.

B470 Brouillet, Benoît. "Les Courants commerciaux entre l'ile du Prince-Edouard et l'exterieur." *Transactions of the Royal Society of Canada* 2 (1964), 63-68. Includes reference to Prince Edward Island's trade in forest products.

B471 Brower, David, ed. *The Sierra Club, a Handbook*. 1951. Second edition. San Francisco: Sierra Club, 1957. viii + 110 pp. Illus., tables. Includes some history of the Sierra Club, a California-based preservationist organization founded in 1892. There are several editions.

Brower, David. See Muir, John, #M619

B472 Brower, David, ed. *Wilderness: America's Living Heritage*. Totowa, New Jersey: Sierra Club Books, 1972.

B473 · Brower, J.V. *Itasca State Park: An Illustrated History*. Minnesota Historical Collections, Volume 11. St. Paul: Minnesota Historical Society, 1904. xxiii + 285 pp. Illus., maps, index.

Brower, Philip P. See Yoshpe, Harry B., #Y22

B474 Browin, Frances Williams. "When Forests Floated Away." *American Forests* 63 (Jan. 1957), 30-31, 62-64. History and description of log rafting on the Delaware River of New York-Pennsylvania-New Jersey, 1870s.

Brown, A.A. See Folweiler, Alfred D., #F140

B475 Brown, Alan K. *Sawpits in the Spanish Redwoods, 1787-1849*. San Mateo, California: San Mateo County Historical Association, 1966. 27 pp. Illus., maps, notes. Of the Santa Cruz Mountains and San Francisco Peninsula of California.

B476 Brown, Alan K., and Stanger, Frank M. "October 9, 1769; Discovery of the Redwoods." *Forest History* 13 (Oct. 1969), 6-11. The account of Father Juan Crespi, Spanish explorer in California. Excerpted from Frank M. Stanger and Alan K. Brown, *Who Discovered the Golden Gate?* (San Mateo, California: San Mateo County Historical Association, 1969).

B477 Brown, B.S. "Naval Stores Past, Present and Future." *Naval Stores Review* (Oct. 18, 1930), 12ff.

B478 Brown, Babette I. "The History and Development of the Nature Trail Idea in the U.S." Master's thesis, Cornell Univ., 1940.

B479 Brown, Carroll T. "The Pulp and Paper Industry of the Pacific Coast States." Master's thesis, Yale Univ., 1940.

B480 Brown, Clair A. "Historical Commentary of the Distribution of Vegetation in Louisiana and Some Recent Observations." *Proceedings of the Louisiana Academy of Sciences* 8 (1943), 34-47.

B481 Brown, D.C. "Lumbering in Randolph County." *Northeastern Logger* 4 (Sept. 1955), 14-15, 30-31. History of logging and the lumber industry in this West Virginia county.

B482 Brown, Eben. "Cattle, Christmas Trees, & Conservation: Riley Bostwick Proves the Small Mountain Farm Still Has A Place in Vermont." *Vermont Life* 13 (Winter 1959), 54-59. On Bostwick's application of sustained-yield forestry to his Mountain Meadows farm, near Rochester, Vermont, assembled since 1930 from parcels of abandoned farm land.

B483 Brown, Elton T. *The History of the Great Minnesota Forest Fires, Sandstone, Mission Creek, Hinckley, Pokegama, Skunk Lake*. St. Paul: Brown Brothers, 1894. 238 pp. Illus., maps. "Instant" history of the fires of 1894.

B484 Brown, James H. "The Role of Fire in Altering the Species Composition of Forests in Rhode Island." *Ecology* 41 (Apr. 1960), 310-16. Since the 1930s.

B485 Brown, James W. "The Administration of Law in Yellowstone National Park." *Wyoming Law Journal* 14 (Fall 1959), 9-16. On the exclusive jurisdiction of the U.S. government over the administration of justice in the Wyoming park since 1872.

B486 Brown, John Howard. "Arbor Day and Its Founders: Who Morton, Northrup and Peaslee Were and What They Did to Inaugurate, Propagate and Popularize the Day." *Americana* 8 (Apr. 1913), 313-20. J. Sterling Morton, B.G. Northrup, and John B. Peaslee.

B487 Brown, John P. *Practical Arboriculture: How Forests Influence Climate, Control the Winds, Prevent Floods, Sustain National Prosperity*. Connersville, Indiana: Privately published, 1906. 460 pp. Illus., index. Contains some historical references to forest influences, tree planting, and forest conservation.

B488 Brown, John Willcox. "Forest History of Mount Moosilauke." Master's thesis, Yale Univ., 1941.

B489 Brown, John Willcox. "Recent Forest Tax Legislation in New Hampshire." *Papers of the Michigan Academy of Science, Arts and Letters* 36 (1953), 85-92. Compares forest taxation in New Hampshire and Michigan since 1913.

B490 Brown, John Willcox. "Forest History of Mount Moosilauke." *Appalachia* 32 (June 1958), 23-32; 32 (Dec. 1958), 221-33. Mount Moosilauke, New Hampshire, 1712-1917.

B491 Brown, Leahmae. "The Development of National Policy with Respect to Water Resources." Ph.D. diss., Univ. of Illinois, 1937.

Brown, Nelson Courtlandt. See Moon, Frederick F. #M524

B492 Brown, Nelson Courtlandt. *Forest Products, Their Manufacture and Use; Embracing the Principal Commercial Features in the Production, Manufacture, and Utilization of the Most Important Forest Products Other Than Lumber, in the United States*. New York: John Wiley, 1919. xix + 471 pp.

Illus., tables, diag., bib., index. Includes some history of the following industries: pulp and paper, tanning, plywood, cooperage, naval stores, charcoal, box manufacturing, crosstie, poles and piling, mining timbers, fuelwood, shingle and shake, maple sugar, dyewoods, and cork. A revised edition appeared in 1927.

B493 Brown, Nelson Courtlandt. *The American Lumber Industry, Embracing the Principal Features of the Resources, Production, Distribution, and Utilization of Lumber in the United States.* New York: John Wiley and Sons, 1923. xviii + 279 pp. Illus., maps, tables, graphs, bib., index. Contains historical treatment of forested lands and the lumber industry, including the seasoning, grading, inspecting, sizing, marketing, transportation, preservation, exporting, and importing of lumber, as well as sketches of trade associations and journals.

B494 Brown, Nelson Courtlandt. "Recent Developments in Lumber Distribution." *Journal of Forestry* 22 (Jan. 1924), 62-64. Factors affecting the lumber trade since World War I.

B495 Brown, Nelson Courtlandt. *Logging—Principles and Practices in the United States and Canada.* New York: John Wiley and Sons, 1934. xvii + 284 pp. Illus., maps, tables, bib., index. Includes much history of logging technology. A revised version is titled *Logging: The Principles and Methods of Harvesting Timber in the United States and Canada* (1949), xix + 418 pp.

B496 Brown, Nelson Courtlandt. *A General Introduction to Forestry in the United States, with Special Reference to Recent Forest Conservation Practices.* New York: John Wiley and Sons, 1935. xix + 293 pp. Illus. Contains a historical section on New Deal forestry programs, as well as many other historical references to forestry.

B497 Brown, Nelson Courtlandt. *Logging—Transportation: The Principles and Methods of Log Transportation in the United States and Canada.* New York: John Wiley and Sons, 1936. xv + 327 pp. Illus., maps, tables, bib. Includes many historical references to the changing technology of log transportation, including tractors, cable hauling systems, trucks, logging railroads, river driving, boom, rafts, and flumes.

B498 Brown, Nelson Courtlandt. *Timber Products and Industries; the Harvesting, Conversion, and Marketing of Materials Other Than Lumber, Including the Principal Derivatives and Extractives.* New York: John Wiley, 1937. xviii + 316 pp. Illus., tables, graphs, bib., index.

B499 Brown, Nelson Courtlandt. "Progress in Community Forests." *Journal of Forestry* 37 (Jan. 1939), 25-28. Includes some history of community or municipal forests throughout the United States since 1710.

B500 Brown, Nelson Courtlandt. "The President Practices Forestry." *Journal of Forestry* 41 (Feb. 1943), 92-93. On the practice of forestry since about 1912 at the estate of Franklin D. Roosevelt, Hyde Park, New York.

B501 Brown, Nelson Courtlandt. "America's Eternal Wealth: Growing Forests." *American Lumberman* (July 10, 1943), 35-45. A brief sketch of forest use, especially the lumber industry, since the colonial period.

B502 Brown, Nelson Courtlandt. *Lumber: Manufacture, Conditioning, Grading, Distribution, and Use.* New York: John Wiley and Sons, 1947. xvi + 344 pp. Illus., maps, tables, index. Includes historical treatment of the lumber industry in much the same fashion as in the author's earlier books. A second edition with James Samuel Bethel appeared in 1958.

B503 Brown, Nelson Courtlandt. *Forest Products: The Harvesting, Processing, and Marketing of Materials Other Than Lumber, Including the Principal Derivatives, Extractives, and Incidental Products in the United States and Canada.* New York: John Wiley and Sons, 1950. xv + 399 pp. Illus., tables, diags., bib., index. A much revised version of the author's works of 1919 and 1937, containing history of the many forest products industries.

B504 Brown, Nelson Courtlandt. "The King's Arrow Pine." *American Forests* 58 (July 1952), 22-23, 40. On the protection and near extinction of white pine in New England since 1691, and its recent rediscovery in northern Maine.

B505 Brown, Nelson Courtlandt, and Recknagel, Arthur B. "Fifty Years of Industrial Forestry: The Empire State Forest Products Association." *Society of American Foresters Meeting, Proceedings, 1957* (1958), 86-89.

B506 Brown, Nelson Courtlandt. "The Cradle of American Forestry Education." *Northeastern Logger* 8 (Oct. 1959), 14-15, 42. On the summer camp of the Yale Forest School at Milford, Pennsylvania.

Brown, R.M. See Allison, John H., #A98

B507 Brown, Ralph Adams. "The Lumber Industry in the State of New York, 1790-1830." Master's thesis, Columbia Univ., 1933.

Brown, Stuart E., Jr. See Pollock, George Freeman, #P260

Brown, Virginia Holmes. See Greene, Lee Seifert, #G320

Brown, Walter L. See Twiford, Ormond H., #T298

B508 Brown, William H., ed. *History of Warren County, New York.* Glens Falls, New York: Board of Supervisors of Warren County, 1963. 302 pp. Illus., maps. Includes a chapter on the Adirondack Forest Preserve and Adirondack State Park.

B509 Brown, William John. "Forestry and the Growth of Kimberly-Clark Corporation." *Journal of Forestry* 51 (Nov. 1953), 792-94. On the manufacture of paper and wood pulp in the Great Lakes states and Canada, with emphasis on the operations of Kimberly-Clark Corporation, headquarters in Neenah, Wisconsin.

B510 Brown, William Robinson. *Our Forest Heritage: A History of Forestry and Recreation in New Hampshire.* Concord: New Hampshire Historical Society, 1958. xv + 341 pp. Illus., index. Includes reminiscences of the author's work as a forester and executive of his family's lumber, pulp, and paper mill, the Brown Company of Berlin, New Hampshire.

B511 Brown, William S., and Show, Stuart B. *California Rural Land Use and Management: A History of the Use and Occupancy of Rural Lands in California.* 2 Volumes. San Francisco: USFS, California Region, 1944. Processed. Includes several chapters on national forests and the conservation movement.

B512 Brown, William S. *History of the Los Padres National Forest.* 1945. Reprint. Santa Barbara, California: USFS, Los Padres National Forest, 1972. 175 pp. App., tables, bib. Processed.

B513 Browning, Bryce C. "Watershed Management in the Muskingum Watershed Conservancy District." *Journal of Forestry* 58 (Apr. 1960), 296-98. In the Muskingum Valley of Ohio since 1913.

B514 Browning, Orville Hickman. *The Diary of Orville Hickman Browning, 1850-1881.* 2 Volumes. Ed. with intro. and notes by Theodore Calvin Pease and James G. Randall. Springfield: Illinois State Historical Library, 1925, 1931. Browning was Secretary of the Interior in the Andrew Johnson administration, 1866-1869.

B515 Brubaker, Sterling. *To Live on Earth: Man and His Environment in Perspective.* Baltimore: Johns Hopkins University Press, for Resources for the Future, 1972. 218 pp.

B516 Bruce, David, and Court, Arnold. "Trees for the Aleutians." *Geographical Review* 35 (July 1945), 418-24. On tree planting activities in the Aleutian Islands of Alaska during World War II.

B517 Bruce, Donald, and Schumacher, Francis X. *Forest Mensuration.* American Forestry Series. 1935. Third edition. New York: McGraw-Hill, 1950. 483 pp. Illus. Incidental history.

B518 Bruce, Kathleen. *Virginia Iron Manufacture in the Slave Era.* New York: Century Company, 1931. xiii + 482 pp. Illus., map. Includes charcoal iron industry.

B519 Bruce, Mason. "National Forests in Alaska." *Journal of Forestry* 58 (June 1960), 437-42. Since 1892.

B520 Bruce, Richard Wilburt. "Intra-regional Competition in Lumber Markets of the Eleven Western States." Ph.D. diss., Washington State Univ., 1968. 113 pp.

B521 Bruce, Robert K. "History of the Medicine Bow National Forest, 1902-1910." Master's thesis, Univ. of Wyoming, 1959. Concerns grazing and timber policies.

B522 Bruère, Martha Bensley. *Your Forests.* Philadelphia: Junior Literary Guild and J.B. Lippincott Company, 1945. x + 159 pp. Illus., maps, index. Contains some history (juvenile) of forests, forestry, and the forest products industries. A revised edition was published in 1957.

B523 Bruins, Elton J. "Holocaust in Holland, 1871." *Michigan History* 55 (Winter 1971), 289-304. On the forest fire which destroyed Holland, Michigan.

B524 Bruncken, Ernest. *North American Forests and Forestry: Their Relations to the National Life of the American People.* New York: G.P. Putnam's Sons, 1899. vi + 265 pp. Index. An early survey of forests and their relation to the young science of forestry; includes many historical references.

B525 Brundage, Roy C. "Utilization Trends in the Central Hardwood Region." *Journal of Forestry* 50 (Mar. 1952), 211-13. Incidental historical references to the hardwood timber supply since ca. 1900 in the Midwest.

B526 Bruner, M.H. "Recent Developments in Cooperative Farm Forestry between the Extension Service and the S.C. State Commission of Forestry." *Journal of Forestry* 41 (Mar. 1943), 186-89. Includes a brief history of the South Carolina State Commission of Forestry and of the Cooperative Extension Service of Clemson University.

B527 Bruns, Paul E. *A New Hampshire Everlasting and Unfallen.* Concord: Society for the Protection of New Hampshire Forests, 1969. 95 pp. Illus. On the Society for the Protection of New Hampshire Forests and especially the work of Philip Wheelock Ayres, the society's forester from 1901 to 1935.

Brush, Warren D. See Sparhawk, William N., #S494

B528 Bryan, Charles W., Jr. "The Racquette—River of the Forest." *North Country Life* 5 (Winter 1951), 24-29; 5 (Spring 1951), 20-25. Includes some history of log driving since 1860.

B529 Bryan, Charles W., Jr. *The Racquette: River of the Forest.* Blue Mountain Lake, New York: Adirondack Museum, 1964. xii + 122 pp. Illus., map. Includes some history of the lumber industry and log driving along the Racquette River of northern New York.

B530 Bryan, J.Y. "Mountain Monarchs." *National Parks & Conservation Magazine* 48 (Apr. 1974), 4-9. On efforts to protect the giant sequoia trees of Sequoia and Kings Canyon national parks, California, since the 1860s.

B531 Bryan, L.W. "Twenty-Five Years of Forestry Work on the Island of Hawaii." *Hawaiian Planters' Record* 51 (1947), 1-80. Since 1921.

B532 Bryant, Claude W. *Lumbering along in Texas.* San Antonio: Naylor Company, 1960. ix + 202 pp. Illus. On the author's early life and his activities since 1897 in the lumber industry and retail lumber trade in Texas.

B533 Bryant, David G., ed. *The Fiftieth Anniversary of the Faculty of Forestry at the University of New Brunswick, 1908-1958.* Fredericton: University of New Brunswick Forestry Association, 1958. 155 pp. Illus. A compilation of short articles, some historical, and lists of forestry graduates of the university.

B534 Bryant, Ralph C. "The Economic Feasibility of a Permanent Pulp and Paper Industry in Central Colorado." Ph.D. diss., Duke Univ., 1953.

B535 Bryant, Ralph Clement. *Logging: The Principles and General Methods of Operation in the United States.* New York: John Wiley and Sons, 1913. xviii + 590 pp. Illus., maps, tables, bib., glossary, index. Contains many incidental historical references to logging technology and log transportation. Second edition by same publisher in 1923.

B536 Bryant, Ralph Clement. "The Panama Canal and the Lumber Trade." *American Forestry* 20 (Feb. 1914), 81-91. Discussion of world financial conditions since 1907 and the probable effect of the Panama Canal on the lumber trade.

B537 Bryant, Ralph Clement. "The European War and the Lumber Trade." *American Forestry* 20 (Dec. 1914), 881-86. The effects of world financial conditions and World War I on the lumber trade, 1906-1914.

B538 Bryant, Ralph Clement. "The War and the Lumber Industry." *Journal of Forestry* 17 (Feb. 1919), 125-34. On the weaknesses of industrial organization evidenced by World War I.

B539 Bryant, Ralph Clement. *Prices of Lumber.* U.S. War Industries Board, Price Bulletin No. 43. Washington: GPO, 1919. 112 pp. Includes some history of the lumber industry and of lumber price movements since the 1860s.

B540 Bryant, Ralph Clement. *Lumber, Its Manufacture and Distribution.* New York: John Wiley and Sons, 1922. xxi + 539 pp. Illus., apps., bib., index. Includes historical references to many aspects of the lumber industry. Second edition by same publisher in 1938.

B541 Bryden, Clifford M. "Federal Forestry Legislation, 1930-1945." Master's thesis, Yale Univ., 1946.

B542 Brydon, Norman. "New Jersey Wildlife Conservation and the Law." *New Jersey History* 86 (Winter 1968), 215-35.

B543 Buchanan, Iva Luella. "An Economic History of Kitsap County, Washington, to 1889." Ph.D. diss., Univ. of Washington, 1930. 334 pp. An early Puget Sound logging region.

B544 Buchanan, Iva Luella. "Lumbering and Logging in the Puget Sound Region in Territorial Days." *Pacific Northwest Quarterly* 27 (Jan. 1936), 34-53. From the 1850s to 1880s.

B545 Buchanan, Millard. "Logging." OHI by Wendell Culpepper, Doug James, Jim Renfro, Steve Smith, Peter Reddick, and Matt Young. *Foxfire* 9 (Summer-Fall 1975), 177-208. Reminiscences of primitive logging and sawmilling in North Carolina's Great Smoky Mountains since the 1930s.

B546 Buchen, Gustave W. "Sheboygan County—Out of a Wilderness." *Wisconsin Magazine of History* 25 (June 1942), 425-43. Some information on logging in relation to clearing land for agriculture since 1834.

B547 Buchheister, Carl W., and Graham, Frank, Jr. "From the Swamps and Back: A Concise and Candid History of the Audubon Movement." *Audubon* 75 (Jan. 1973), 4-45. On the origins and history of the National Audubon Society since its establishment in 1896, especially in the field of wildlife conservation.

B548 Buchholtz, Curtis W. "The Historical Dichotomy of Use and Preservation in Glacier National Park." Master's thesis, Univ. of Montana, 1969.

B549 Buchholtz, Curtis W. "W.R. Logan and Glacier National Park." *Montana, Magazine of Western History* 19 (Summer 1969), 2-17. On William R. Logan, superintendent of Glacier National Park, Montana, 1910-1912, and a variety of problems and achievements under his administration.

Buchwalter, Nichelsen E. See Moore, Robert M., #M543

B550 Buck, Paul H. "The Evolution of the National Park System in the United States." Master's thesis, Ohio State Univ., 1922.

B551 Buck, Paul H. *The Evolution of the National Park System of the U.S.* Washington: GPO, 1946. 74 pp.

B552 Buehler, J. Marshall. "Birthplace of the Wisconsin River Paper Industry." *Paper Maker* 35 (Oct. 1966), 12-25. On the Centralia Pulp and Water Power Company, Wisconsin Rapids, Wisconsin, established in 1891 and now owned by the Nekoosa-Edwards Paper Company.

B553 Buell, Guy A. "Logging, Past and Present." *Pioneer Western Lumberman* 61 (May 15, 1914), 15, 23-26. Redwood logging in northern California.

B554 Buell, Guy A. "Operating a Planning Mill in the Yosemite Valley." *Pioneer Western Lumberman* 61 (June 15, 1914), 15, 19. Reminiscences of a mill operated by Ike Noble in 1886.

B555 Buell, Walter. "The Michigan Lumber Industry as Told in Sketches of Some of Its Leading Men." *Magazine of Western History* 4 (Sept. 1886), 712-17; 5 (Nov. 1886), 126-39.

B556 Buie, Thomas Stephen. "From Pines to Pines." *American Forests* 62 (June 1956), 20-23, 54-55. On woodland salvage and reforestation in the Sand Hills of South Carolina, 1905-1956; with the author's memories of part of Chesterfield County, stripped of timber between 1899 and 1904, farmed until the 1930s, and later replanted to pine.

B557 Buley, R. Carlyle. *The Old Northwest: Pioneer Period, 1815-1840.* 2 Volumes. Bloomington: Indiana University Press, 1950. Illus. Some history of pioneer settlement in relation to the forested lands.

B558 Bullis, Raymond S. "Spencer S. Bullis by C.F.H. Allen." *Bulletin of the Railway & Locomotive Historical Society* 100 (Apr. 1959), 85-91. Bullis (1849-1929) was an operator of lumber mills near Buffalo, New York, owner of railroads in New York, Pennsylvania, and Mississippi, and lumber and mine operator in eastern Oregon, 1849-1929.

B559 Bullock, Warren B. *The Romance of Paper.* Boston: Richard G. Badger, The Gorham Press, 1933. 88 pp. Includes some history of papermaking and the pulp and paper industry.

B560 Bullock, Warren B. "Paper and Pulp Mills' Industrial Forestry Program—History and Enormous Progress." *Southern Pulp and Paper Manufacturer* 23 (Jan. 11, 1960), 46-52. Concerns the contributions of H.P. Baker, Royal S. Kellogg, and G.W. Sisson to formulating a forestry program for the pulp and paper industry.

Bulmer, R.M. See Hawboldt, Lloyd S., #H209

B561 Bunce, Frank H. "Dreams from a Pack: Isaac Wolfe Bernheim and Bernheim Forest." *Filson Club History Quarterly* 47 (Oct. 1973), 323-32. Bernheim (1848-1945) established a private forest of 10,000 acres in Bullitt and Nelson counties, Kentucky, in 1929. It is managed by the Isaac W. Bernheim Foundation as a wilderness sanctuary.

B562 Bundy, C.S. *Early Days in the Chippewa Valley.* Menomonie, Wisconsin: Flint-Douglas, 1916. 16 pp. Includes some history of logging.

B563 Burcalow, Donald W., and Marshall, William H. "Deer Numbers, Kill, and Recreational Use on an Intensively Managed Forest." *Journal of Wildlife Management* 22 (Apr. 1958), 141-48. On the relationship of forest management and wildlife management programs in the University of Minnesota's Cloquet Experimental Forest, Carlton County, Minnesota, 1930-1956.

B564 Burch, William Richard, Jr. "Nature as Symbol and Expression in American Social Life: A Sociological Exploration." Ph.D. diss., Univ. of Minnesota, 1964. 525 pp. Interprets the social meaning of "Nature" by examining its rhetorical use in American history and literature; also examines attitudes of contemporary Oregon campers.

B565 Burch, William Richard, Jr. *Daydreams and Nightmares: A Sociological Essay on the American Environment.* New York: Harper and Row, 1971. 175 pp. Index. Includes an examination of man's relationship with nature as expressed in history and literature.

B566 Burcham, L.T. *California Rangeland, an Historico-Ecological Study of the Range Resource of California.* Sacramento: California Department of Natural Resources, Division of Forestry, 1957. 261 pp. Illus., maps, tables, graphs, apps., bib. Includes much history of forested lands in California.

B567 Burchill, J.P. "Lumbering on the Miramichi." *Report of the Canadian Forestry Association* 11 (1910), 46-52. A contemporary description with some historical references to lumbering in New Brunswick.

B568 Burger, William H. "Some Conservation Men." *North Country Life* 4 (Summer 1950), 4-9. Biographical sketches of Clinton West, Walter Rice, William E. Petty, and Lucius Russell, employees of the New York Conservation Department stationed in the Adirondack Mountains.

B569 Burgess, Sherwood D. "The Forgotten Redwoods of the East Bay." *California Historical Society Quarterly* 30 (Mar. 1951), 1-14. On the redwood forest of the East Bay hills near Oakland, California, completely logged off between 1776 and 1860.

B570 Burgess, Sherwood D. "Lumbering in Hispanic California." *California Historical Society Quarterly* 41 (Sept. 1962), 237-48. Logging and the lumber industry in the redwood forests of the Santa Cruz Mountains and the Monterey Bay area of California, 1777-1847, with emphasis on the lumber business of Thomas O. Larkin.

B571 Burgh, Robert. *The Region of Three Oaks.* Ed. by Albert J. Chapman. Three Oaks, Michigan: Edward K. Warren Foundation, 1939. xx + 234 pp. Illus., maps, index. Includes some history of the lumber industry in this southwestern Michigan community during the mid-19th century.

B572 Burke, Fred C. *Logs on the Menominee: The History of the Menominee River Boom Company.* Menasha, Wisconsin: George Banta Publishing Company, 1946. xiv + 98 pp. Illus., maps. On logging in the Menominee River Basin of Wisconsin and Michigan, and the Menominee River Boom Company, Marinette, Wisconsin, 1867-1920s.

B573 Burkett, Charles William. *History of Ohio Agriculture: A Treatise on the Development of the Various Lines and Phases of Farm Life in Ohio.* Concord, New Hampshire: Rumford Press, 1900. 211 pp. Illus. Includes chapters on forestry and horticulture.

B574 Burleigh, Charles Abner. "The Development and Use of Power Saws in the Woods." Master's thesis, Yale Univ., 1947.

B575 Burnett, Edmund Cody. "Shingle Making on the Lesser Waters of the Big Creek of the French Broad River." *Agricultural History* 20 (Oct. 1946), 225-35. Making shingles from white pine in North Carolina, 1868-1880.

B576 Burns, Anna Maria Cannaday. *A History of the Louisiana Forestry Commission.* Louisiana Studies Institute, Monograph Series, No. 1. Natchitoches: Northwestern State College, 1968. xviii + 137 pp. Illus., maps, tables, graphs, notes, bib., index. On the Louisiana Forestry Commission, its predecessor agencies, and other aspects of the forest products industries and forest conservation in the state since the 19th century.

B577 Burns, Findley. *The Crater National Forest: Its Resources and Their Conservation.* USFS, Bulletin No. 100. Washington: GPO, 1911. 20 pp. Oregon.

B578 Burns, Francis P. "The Spanish Land Laws of Louisiana." *Louisiana Historical Quarterly* 11 (Oct. 1928), 557-81.

B579 Burns, Matthew J. "54 Years of Union Progress." *Paper Trade Journal* 139 (Nov. 7, 1955), 39-44, 56. In the pulp and paper industry.

B580 Burns, Paul Y., and Cole, John F. "History of the Gulf States Section, Society of American Foresters." *Gulf States Section Newsletter* 5 (Spring 1963), 17-30.

Burns, Paul Y. See Blackwell, Lloyd P., #B298

B581 Burr, Samuel Engle, III. "The Rise of the Furniture Industry in the South." *Southern Lumberman* 183 (Dec. 15, 1951), 199-202. Since 1888.

B582 Burrell, Jim. "The 3500-Year History of Hardwood Plywood." *Plywood and Panel* 12 (June 1971), 30-36.

B583 Burrier, Tom. "The Park Misnamed Deception." *American Forests* 66 (June 1960), 36-37, 43-44. History of forested Deception Pass State Park, Puget Sound, Washington, since 1792.

B584 Burris, Martin. *True Sketches of the Life and Travels of Martin Burris on the Western Plains, the Rocky Mountains and the Pacific Coast, U.S.A.* Salina, Kansas: Padgett, 1910. 67 pp. Burris was a logger during part of his career.

B585 Burroughs, John. *John James Audubon.* Boston: Small, Maynard & Company, 1902. 144 pp. On Audubon (1785-1851) as a naturalist.

B586 Burroughs, John. *Camping and Tramping with Roosevelt.* Boston: Houghton Mifflin, 1907. 110 pp. Burroughs, a naturalist and nature writer, influenced the conservation movement through his association with Theodore Roosevelt. The book describes a trip to Yellowstone National Park with Roosevelt in 1903 and subsequent visits to his Oyster Bay home.

B587 Burroughs, John. *The Heart of Burroughs' Journals.* Boston: Houghton Mifflin, 1928. xvii + 361 pp.

B588 Burroughs, John. *John Burroughs' America: Selections from the Writings of the Hudson River Naturalist.* Ed. with an intro. by Farida A. Wiley. New York: Devin-Adair, 1951. xv + 304 pp. Illus. Essays about forests, meadows, wildlife, and birds, especially in New York, 1871-1921.

B589 Burroughs, Raymond Darwin. "Conservation is Big Business." *Michigan Conservation* 20 (Mar.-Apr. 1951), 6-8, 22-23. On the Michigan Department of Conservation since 1921.

B590 Burroughs, Raymond Darwin. "The Big Wheels." *American Forests* 59 (Feb. 1953), 16-18, 43. On logging wheels manufactured by Silas C. Overpack and his son of Manistee, Michigan, 1870-1936.

B591 Burroughs, Raymond Darwin. "Michigan's Training Center for Conservation." *Michigan Conservation* 26 (Jan. 1957), 15-18. On the Michigan Department of Conservation Training School, Higgins Lake, since 1941.

B592 Burroughs, Raymond Darwin, ed. *The Natural History of the Lewis and Clark Expedition.* East Lansing: Michigan State University Press, 1961. 340 pp. Their expedition to the Pacific Northwest, 1804-1806.

B593 Burton, Ian, and Kates, Robert W., eds. *Readings in Resource Management and Conservation.* Chicago: University of Chicago Press, 1965. 609 pp. Maps, tables, graphs, index. An anthology of articles, some with historical information on the conservation movement.

B594 Burton, Larry. "No Prison Riots Here." *American Forests* 62 (Sept. 1956), 28-29, 44-47. On the use of convicts on road crews and fire crews in California forests since 1915.

B595 Burton, T.L. *Natural Resource Policy in Canada: Issues and Perspectives.* Toronto: McClelland & Stewart, 1972. 174 pp. Includes some history of forest policy.

B596 Business Executives Research Committee. *The Forest Products Industry of Oregon: A Report.* Portland, 1954. 36 pp. Map, diag., tables. Some history since 1947.

B597 Buskirk, Richard Hobart. "A Description and Critical Analysis of the Marketing of Douglas Fir Plywood." D.B.A. diss., Univ. of Washington, 1955. 391 pp. Incidental history.

B598 Butcher, Devereux. "National Parks Association." *Journal of Forestry* 44 (Mar. 1946), 184-85. On the history and activities of this conservation organization, founded in 1919 for the purpose of defending the existing national parks and promoting the park concept.

B599 Butcher, Devereux. *Exploring Our National Parks and Monuments.* Prepared under the auspices of the National Parks Association. New York: Oxford University Press, 1947. 106 pp. Illus., maps, bib. Subsequent editions, revised and much enlarged, were published by Houghton Mifflin. Contains description and history of the national parks and monuments and of the policies governing their administration by the U.S. National Park Service.

B600 Butcher, Devereux. "Going, Going—Florida's Royal Palm-Big Cypress Forest." *National Parks Magazine* 22 (Apr.-June 1948), 3-7. On a forest in Fahkahatchee Slough, near the Florida Everglades, since 1906.

B601 Butcher, Devereux. *Exploring the National Parks of Canada.* Washington: National Parks Association, 1951. 84 pp. Illus., map, bib. Includes some history.

B602 Butcher, Devereux. "Do Our Historic Areas Deserve the Dignity of a Separate Bureau?" *National Parks Magazine* 30 (Oct.-Dec. 1956), 152-57. On the historical preservation work of the National Park Service since 1916.

B603 Butcher, Edward B. "An Analysis of Timber Depredations in Montana to 1900." Master's thesis, Univ. of Montana, 1967.

Butcher, Edward B. See Toole, K. Ross, #T236

Butler, Edwin R. See Hanrahan, Frank J., #H87

B604 Butler, Frank O. *The Story of Paper-Making: An Account of Paper-Making from Its Earliest Known Record down to the Present Time.* Chicago: J.W. Butler, 1901. 136 pp. Illus.

B605 Butler, June Rainsford. "America—A Hunting Ground for Eighteenth-Century Naturalists, with Special Reference to Their Publications about Trees." *Papers of the Bibliographical Society of America* 32 (1938), 1-16.

B606 Butler, Ovid M. *The Distribution of Softwood Lumber in the Middle West: Wholesale Distribution.* USDA, Report No. 115. Washington: GPO, 1917. 96 pp. Maps, diags. Includes history of wholesaling conditions since 1880.

B607 Butler, Ovid M. *The Distribution of Softwood Lumber in the Middle West: Retail Distribution.* USDA, Report No. 116. Washington: GPO, 1918. 100 pp.

B608 Butler, Ovid M. "What Forestry Means to Southern Commerce." *American Forestry* 28 (July 1922), 433-35. Includes some historical references to the influence of forest industries upon the southern economy.

B609 Butler, Ovid M. "Henry Ford's Forest." *American Forestry* 28 (Dec. 1922), 725-31. On the timbered lands and sawmill near Iron Mountain, Michigan, belonging to the Ford Motor Company.

B610 Butler, Ovid M., ed. *Rangers of the Shield: A Collection of Stories Written by Men of the National Forests of the West.* Washington: American Forestry Association, 1934. 270 pp. Illus. The 29 stories, some historical in nature, were first published in *American Forests.*

B611 Butler, Ovid M., comp. and ed. *American Conservation in Picture and in Story.* Washington: American Forestry Association, 1935. 144 pp. Illus., bib. A collection of essays, many on the history of the conservation movement, reprinted from the "Anniversary Number" of *American Forests* 41 (Sept. 1935), marking the 60th year of the American Forestry Association. A revised edition appeared in 1941.

B612 Butler, Ovid M. "The Oregon Checkmate: How the Federal Government is Blocking the Conservation of the Nation's Greatest Remaining Forest." *American Forests* 42 (Apr. 1936), 156-62, 196-97. Includes some history of Oregon and California Railroad grant lands in Oregon, which were revested to the federal government in 1916, and the legislative attempts to bring them under modern forest management.

B613 Butler, Ovid M. "The American Forestry Association." *Journal of Forestry* 44 (Jan. 1946), 17-18. On its history since 1875.

B614 Butler, Ovid M. "70 Years of Campaigning for American Forestry." *American Forests* 52 (Oct. 1946), 456-59, 512. On the origins of the forestry movement, and the work of the American Forestry Association in the 1870s-1880s, and especially on the first two American forest congresses held in Cincinnati, 1882, and Washington, 1905.

B615 Butler, Ovid M. "A Foundation for the Forest." *American Forests* 54 (Mar. 1948), 105-07, 126, 144. History of the Charles Lathrop Pack Forestry Foundation, Washington, D.C., since 1930.

Butler, Ovid M. See Schenck, Carl Alwin, #S80

B616 Butler, R.H. "Early Logging on the South Platte District." *Colorado Magazine* 13 (Sept. 1936), 180-83. Brief account of logging and sawmills near Buffalo and South Platte, 1874-1890, on what later became the South Platte District of the Pike National Forest, Colorado.

B617 Butt, Archibald Willingham. *Taft and Roosevelt: The Intimate Letters of Archie Butt, Military Aide.* 2 Volumes. Garden City, New York: Doubleday, Doran & Company, 1930. Includes some account of the rift between Theodore Roosevelt and William H. Taft over conservation and other issues, 1908-1912.

B618 Butterfield, George E. *Bay County Past and Present.* Bay City, Michigan: C. & J. Gregory, 1918. 212 pp. Illus. Includes some history of the lumber industry in this Michigan county.

B619 Butterfield, Roy L. "The Great Days of Maple Sugar." *New York History* 39 (Apr. 1958), 151-64. On projects for the large-scale manufacture of maple sugar in New York, 1765-1801.

B620 Buttrick, Philip Laurence. "Commercial Use of the Longleaf Pine." *American Forestry* 25 (Sept. 1915), 896-908. Contains some history of the many uses of this species since the colonial period, including its importance in the naval stores industry.

B621 Buttrick, Philip Laurence. "Backgrounds, Methods, and Problems of Public Regulation of Private Forests." *Journal of Forestry* 39 (Mar. 1941), 283-87.

B622 Buttrick, Philip Laurence. *Forest Economics and Finance.* New York: John Wiley and Sons, 1943. xviii + 484 pp. Illus., maps, tables, notes, apps., index. Contains many scattered historical references to forest conservation and forest economics.

B623 Byers, Archie M. "The Timber Industry and Industrial Forestry in Alaska." *Journal of Forestry* 58 (June 1960), 474-77. Since 1867.

B624 Byers, W.L. "Sidelights of History in the Buchanan Forest District." *Forest Leaves* 22 (Oct. 1929), 72-73. Pennsylvania.

Byrne, A.R. See Nelson, J.G., #N60

B625 Byrne, J.J., ed. "Engineering in the Forest Service: Six Memoirs." *Forest History* 14 (Jan. 1971), 6-17. Individual contributors include: Fleming K. Stewart, "Communicating the Hard Way"; Hartley A. Calkins, "Bitterroot to Big Hole Project"; Henry M. Shank, "Mapping the West"; L.H. LaFaver, "Engineering and the Depression"; Verne V. Church, "CCCs and Fire Fighting"; Jack Hamblet, "Chrome Mines Road Project."

B626 Byshe, F.H. "Origin and Development of the Forestry Branch, Department of Interior, Canada." *Empire Forestry Journal* 4 (1925), 76.

C1 Cadman, A.E. "The Canadian Pulp and Paper Industry, 1917-1928." *Pulp and Paper Magazine of Canada,* International number, (1929), 86-88.

C2 Cadman, A.E. "Development of the Pulp and Paper Industry, 1919-1929." *Pulp and Paper Magazine of Canada* 29 (Feb. 6, 1930), 201-05.

C3 Caffey, Francis G. "A Brief History of the United States Department of Agriculture." *Case and Comment* 22 (Feb. 1916), 723-33; 22 (Mar. 1916), 850-56.

C4 Caffey, Francis G. *A Brief Statutory History of the United States Department of Agriculture.* Washington: GPO, 1916. 26 pp.

C5 Cahn, Robert. *Will Success Spoil the National Parks?* Boston: Christian Science Publishing Company, 1968. 55 pp. Illus.

C6 Cahn, Robert. "Alaska: A Matter of 80,000,000 Acres." *Audubon* 76 (July 1974), 2-13, 66-81. A report on the public lands of Alaska, with recommendations for extensive additions to the national forest, park, wildlife refuge, and wild and scenic river systems. There are many incidental historical references. See also "Wildlands for Tomorrow: An Album," pp. 14-65.

C7 Cail, Robert Edgar. "Disposal of Crown Lands in British Columbia, 1871-1913." Master's thesis, Univ. of British Columbia, 1956.

C8 Cail, Robert Edgar. *Land, Man, and the Law: The Disposal of Crown Lands in British Columbia, 1871-1913.* Vancouver: University of British Columbia Press, 1974. xv + 333 pp. Illus., maps, tables, apps., bib., index. A study of British Columbia land policy; includes a chapter on timber legislation which permitted the province to avoid alienation of forest land in the sense that occurred in the United States.

C9 Cain, Cyril Edward. *Four Centuries on the Pascagoula.* Volume 1. *History, Story, and Legend of the Pascagoula River Country.* State College, Mississippi: N.p., 1953. xii + 216 pp. Maps, notes. Includes history of the lumber and naval stores industries and reminiscences of log rafting on Mississippi's Pascagoula River.

C10 Cain, W.J. "Pioneer North American Loggers." *Timberman* 31 (Apr. 1930), 186ff.

C11 Cairney, Daniel W. "The Effect of Some Economic Disturbances on the Lumber Trade of Washington and British Columbia." Master's thesis, Univ. of Washington, 1935.

C12 Caldwell, Lynton K., ed. *Environmental Studies: Papers on the Politics and Public Administration of Man-Environment Relationships.* 4 Volumes. Bloomington, Indiana: Institute of Public Administration, Indiana University, 1967.

C13 Caldwell, Lynton K. *Environment: A Challenge for Modern Society.* Garden City, New York: Natural History Press, for the American Museum of Natural History, 1970. xvi + 292 pp. Notes, index. Includes some environmental history.

C14 Calef, Wesley. *Private Grazing and Public Lands: Studies of Local Management of the Taylor Grazing Act.* Chicago: University of Chicago Press, 1960. xviii + 292 pp. Illus., notes, maps, bib. On the Bureau of Land Management and its administration of grazing lands, especially in Wyoming, under the Taylor Grazing Act of 1934. The author finds the BLM to be an ineffective agency, primarily due to its weak political position and lack of money for research and operations.

C15 Calhoun, Charles E. "Financing the Pulp and Paper Industry on the Pacific Coast." Master's thesis, Univ. of Washington, 1930.

C16 *California Monthly.* "The Brothers Drury." *California Monthly* 52 (May 1944), 14-16, 37-39. Biographical sketches of Newton and Aubrey Drury, including an account of their work for the Save-the-Redwoods League.

C17 Calkins, E.A. "Michigan Railroads since 1850." *Michigan History Magazine* 13 (Jan. 1929), 5-25. Includes information on logging railroads.

Calkins, Hartley A. See Byrne, J.J., #B625

C18 Call, H.M. "A Case History: Forestry in Farm Management." *Northeastern Logger* 6 (June 1958), 24-25, 44.

C19 Callahan, J.M., ed. *Semi-Centennial History of West Virginia.* Charleston: Semi-Centennial Commission of West Virginia, 1913. ix + 594 pp. Illus., maps, tables. Includes a chapter on "forest and timber industries," pp. 322-28.

C20 Callahan, James D. "Crossett—Monument to Planned Forestry." *American Forests* 54 (Apr. 1948), 152-54. On the Crossett Lumber Company, Crossett, Arkansas, an industrial leader in modern forestry methods since 1901.

C21 Callison, Charles. *Man and Wildlife in Missouri: The History of One State's Treatment of Its Natural Resources.* Harrisburg, Pennsylvania: Stackpole Company, by arrangement with the Edward K. Love Conservation Foundation, 1953. 136 pp. Illus., bib. Deals especially with the Missouri Conservation Commission, established in 1936.

C22 Callison, Charles, ed. *America's Natural Resources.* New York: Ronald Press, for the Natural Resources Council of America, 1957. v + 211 pp. Includes chapters on forests, parks, and wilderness.

C23 Calvin, Delano Dexter. *A Saga of the St. Lawrence: Timber and Shipping through Three Generations.* Toronto: Ryerson Press, 1945. x + 176 pp. Illus. Lumbermen Delano Dexter Calvin, Hiram Augustus Calvin, and the family lumber and shipping firm in the St. Lawrence Valley, New York, Ontario, and Quebec, since 1825.

C24 Calvin, Delano Dexter. "Rafting on the St. Lawrence." *Canadian Geographical Journal* 67 (No. 5, 1963), 158-65. On the use of timber rafts, until 1911, on the St. Lawrence River from Kingston, Ontario, to Montreal.

C25 Cameron, D. Roy. *Report of Timber Conditions around Lesser Slave Lake.* Bulletin No. 29. Ottawa: Forestry Branch, 1912. 54 pp. Illus. Northern Alberta.

C26 Cameron, D. Roy. "Dominion Forest Experiment Stations." *Journal of Forestry* 36 (Oct. 1938), 1086-91. The Dominion Forester's account, largely descriptive, of forest experiment stations in New Brunswick, Quebec, Ontario, Manitoba, and Alberta.

C27 Cameron, D. Roy. "Canada's Forests." *Canadian Geographical Journal* 18 (May 1939), 248-74. Incidental history.

C28 Cameron, George M. "The Cruise of the St. Eugene: How a Canadian Vessel Captain Stole a Cargo of Cork Pine for the Liverpool Market—An Early Incident in Michigan Lumbering." *Lumber World Review* (Nov. 10, 1922), 58-61.

C29 Cameron, J.O. "The Genesis of the Wooden Shipbuilding Industry in British Columbia." *Timberman* 18 (Sept. 1917), 38-40.

C30 Cameron, Jenks. *The National Park Service; Its History, Activities and Organization.* Institute for Government Research, Service Monographs of the United States Government, No. 11. New York: Appleton, 1922. Reprint. New York: AMS Press, 1973. xii + 172 pp. Map. General history and analysis of the agency and some history of the national parks movement.

C31 Cameron, Jenks. "President Adams' Acorns and How They Came to Be Planted at Santa Rosa." *American Forests and Forest Life* 34 (Mar. 1928), 131-34. On Henry Marie Brackenridge and the live oak plantation at Santa Rosa Island, Florida, 1820s.

C32 Cameron, Jenks. "An Anchor to Forestward: How America Tried to Grow Trees for Sail of the Line at Santa Rosa." *American Forests and Forest Life* 34 (Apr. 1928), 199-201, 235. An attempt to provide timber for American naval vessels, 1828-1830.

C33 Cameron, Jenks. "Who Killed Santa Rosa? Wherein America's First Tree Planting Experiment Is Abandoned—Through Sheer Cussedness, Some Say, While Others Lay It to Politics." *American Forests and Forest Life* 34 (May 1928), 263-66, 312. On the failure of the experiment to raise live oak trees for naval purposes at Santa Rosa Island, Florida, 1829-1831.

C34 Cameron, Jenks. *The Development of Governmental Forest Control in the United States.* Institute for Government Research, Studies in Administration. Baltimore: Johns Hopkins Press, 1928. Reprint. New York: DaCapo Press, 1972. x + 471 pp. Notes, bib., index. A major study of American forestry and forest policy at the level of national government since the colonial period.

C35 Cameron, Jenks. *The Bureau of Biological Survey—Its History, Activities and Organization.* Institute for Government Research, Service Monographs of the United States Government, No. 54. Baltimore: Johns Hopkins University Press, 1929. Reprint. New York: AMS Press, 1973. x + 339 pp. Illus., map, tables, bib.

C36 Camp Manufacturing Company. *Sixty Years of Progress.* Franklin, Virginia, 1948. 80 pp. Illus. On a lumber, paper, and chemical producer headquartered since 1887 in Franklin, Virginia.

C37 Campana, Richard J. "A History of Introduced Forest Tree Diseases in the U.S." Master's thesis, Yale Univ., 1947.

C38 Campbell, Archer Stuart. *Studies in Forestry Resources in Florida.* 3 Volumes. Gainesville: University of Florida, 1932-1934. The subjects of the three volumes are timber conservation, the lumber industry, and the naval stores industry. Also issued as University of Florida Publications, Economic Series, Volume 1, Nos. 3-5.

C39 Campbell, Carlos Clinton. *Birth of a National Park in the Great Smoky Mountains: An Unprecedented Crusade Which Created, As Gift of the People, the Nation's Most Popular Park.* Knoxville: University of Tennessee Press, 1960.

xii + 155 pp. Illus., maps, tables, notes. On the work of the Great Smoky Mountains Conservation Association (founded in 1923), the establishment of the Great Smoky Mountains National Park in North Carolina and Tennessee in 1934, and its subsequent development and expansion. A new edition appeared in 1970.

C40 Campbell, G. Murray. "The Rich Lumber Company and Its Manchester, Vt. Railroad, 1912-'19." *Northern Logger and Timber Processor* 15 (Feb. 1967), 10-11, 26-27.

C41 Campbell, J.L. "'Merchandising' Lumber on the Prairies Back in the Eighties and Nineties." *Canada Lumberman* 50 (Aug. 1, 1930), 111-12.

C42 Campbell, Johnson A. "Timber Resources of Northern Manitoba." *Canadian Forestry Journal* 13 (Sept. 1917), 1305-06.

C43 Campbell, R.S. "Milestones in Range Management." *Journal of Range Management* 1 (Oct. 1948), 4-8.

C44 Campbell, Robert Henry. "Rocky Mountain Forest Reserve." *Report of the Canada Commission of Conservation* 3 (1912), 64-75. Some history of the Alberta reserve.

C45 Campbell, Robert Henry. *Manitoba, A Forest Province.* Forestry Branch, Circular No. 7. Ottawa, 1914. 16 pp.

C46 Campbell, Robert Samuel. "Vegetational Changes and Management in the Cutover Longleaf Pine—Slash Pine Area of the Gulf Coast." *Ecology* 36 (Jan. 1955), 29-34. Review of studies of "secondary plant succession" in this area, 1926-1951.

Campbell, Robert Samuel. See Bond, Walter Edwin, #B337.

C47 Campbell, Roy L. *Turtle Mountain Forest Reserve.* Forestry Branch, Bulletin No. 32. Ottawa, 1912. 20 pp. In Manitoba.

C48 Canada. Forest Service. *The Forests of Canada: Their Extent, Character, Ownership, Management, Products and Probable Future.* Revised edition. Ottawa, 1928. 56 pp.

C49 Canada. Forest Service. *Success in Prairie Tree Planting.* Bulletin No. 72, Ottawa, 1930. 45 pp.

C50 Canada. Forest Service. *Statistical Record to 1940 of the Forests and Forest Industries of Canada.* Ottawa, 1943. 41 pp. Graphs, tables. Resources and production figures since 1908.

C51 Canada. Forestry Branch. *Wood is Wealth: Canada's Forest Economy, 1938 to 1949.* Bulletin No. 105. Ottawa, 1952. 52 pp.

C52 Canada. Forestry Branch. *Forest and Forest Product Statistics.* Bulletin No. 106. Ottawa, 1952. 65 pp.

C53 Canada. Forestry Branch. *Canada's Forests, 1946-1950: Report to the Sixth British Commonwealth Forestry Conference, Held in Canada, 1952.* Ottawa, 1952. 49 pp.

C54 Canada. Forestry Branch. *Canada's Forests, 1951-1955: Report to the Seventh British Commonwealth Forestry Conference, 1957.* Ottawa, 1957. 75 pp.

C55 Canada. Forestry Branch. *Forestry Inventory and Reforestation under the Canada Forestry Act, 1952-1956.* Miscellaneous Publication No. 9. Ottawa, 1957. 87 pp.

C56 Canada. National Parks Bureau. *The National Parks of Canada: A Brief Description of Their Scenic and Recreational Aspects.* Third edition. Ottawa, 1938. 64 pp. Includes some history.

C57 Canada. National Research Council. *Farm Woodlots in Eastern Canada.* Ottawa, 1940. 120 pp.

C58 *Canada Lumberman.* "Early Days of Canadian Shipbuilding." *Canada Lumberman* 37 (Nov. 15, 1917), 48-49. Issue contains other articles on shipbuilding.

C59 *Canada Lumberman.* "Lack of Raw Material Closes Historic Mill: Georgian Bay Lumber Company's Plant at Waubaushene, Ontario, Has Enjoyed Unique Record for Sixty Years—Now Idle for First Time in Its History." *Canada Lumberman* 41 (Jan. 1, 1921), 45-46.

C60 *Canada Lumberman.* "Capt. Dollar Recalls Days of Early Bush Life." *Canada Lumberman* 43 (Feb. 15, 1923), 50-51. Life in Ontario logging camps.

C61 *Canada Lumberman.* "Historic Lumber Firm is Still Going Strong." *Canada Lumberman* 43 (May 15, 1923), 46. D. Aitchison Company, Ltd., Hamilton, Ontario.

C62 *Canada Lumberman.* "The Final Whitewood Operation in Ontario." *Canada Lumberman* 43 (June 15, 1923), 35-36, 59. On logging yellow poplar in southwestern Ontario and the demise of this hardwood.

C63 *Canada Lumberman.* "The Only Living Commissioned Deal Culler in Quebec." *Canada Lumberman* 43 (July 1, 1923), 39, 48. Concerns Thomas Malone of Trois Rivières, Quebec, and the third generation of log graders under the Culler's Act of 1845.

C64 *Canada Lumberman.* "Historic Canadian Firm is Celebrating Its Golden Jubilee in Business of Making Saws." *Canada Lumberman* 44 (Jan. 1, 1924), 41-44. Shurly-Dietrich Company, Ltd., Galt, Ontario, 1870s-1920s.

C65 *Canada Lumberman.* "In Old Square Timber Days." *Canada Lumberman* 44 (Dec. 15, 1924), 66-68. Reminiscences of P.J. Loughrin of rafting on the Ottawa River, Ontario-Quebec, 1870s.

C66 *Canada Lumberman.* "The Early Days of Log Driving on St. John River." *Canada Lumberman* 45 (Sept. 15, 1925), 128-29. St. John River Log Driving Company, New Brunswick, 1885-1924.

C67 *Canada Lumberman.* "The Evolution of the Log Hauler: A Description of Its Development during Twenty-Six Years." *Canada Lumberman* 46 (June 15, 1926), 42-43. The Lombard log hauler since 1900.

C68 *Canada Lumberman.* "Early Lumber Trade Also Had Its Trade Problems to Solve." *Canada Lumberman* 47 (Aug. 1, 1927), 160. Montreal lumberman reminisces about trade conditions of the 1880s and 1890s.

C69 *Canada Lumberman.* "Progress and Worth of Canadian Lumbermen's Association." *Canada Lumberman* 48 (Feb. 1, 1928), 57-58.

C70 *Canada Lumberman.* "Historic Sawmill at Port Rowan Started 130 Years Ago: John C. Backus, Port Rowan, Ontario, Is the Great-Grandson of the Founder." *Canada Lumberman* 48 (Sept. 15, 1928), 33-34.

C71 *Canada Lumberman.* "Old Woodworking Plant Has Unique History." *Canada Lumberman* 49 (Aug. 1, 1929), 141-42. J.C. Risteen, Fredericton, New Brunswick, since 1872.

C72 *Canada Lumberman.* Golden Jubilee, 1880-1930. *Canada Lumberman* 50 (Aug. 1, 1930). Anniversary issue contains articles on many aspects of Canadian forest history, including sketches of individuals and companies. See following entries.

C73 *Canada Lumberman.* "A Saga of the Lumbering Industry: A Chronological Review of the Highlights of Canadian Lumbering Development as Recorded in Past Issues of *Canada Lumberman.*" *Canada Lumberman* 50 (Aug. 1, 1930), 60-62, 76. Selected events, 1880-1920.

C74 *Canada Lumberman.* "Launched When Canada Was Young, the Name of W.C. Edwards Has Been a Familiar One in the Ottawa Valley for Over Sixty Years." *Canada Lumberman* 50 (Aug. 1, 1930), 89-91. An Ontario lumber merchant and his firm.

C75 *Canada Lumberman.* "Remembers John Waldie as Storekeeper: Old Timers in the Trade Will Recollect Many Incidents Recalled by Josh Collins." *Canada Lumberman* 50 (Aug. 1, 1930), 93-94. Collins, a Montreal wholesaler, brought in the first fir timbers from the West via the Canadian Pacific Railway in 1891.

C76 *Canada Lumberman.* "Pioneers on the Ottawa: The Name of Mason Has Been Associated With Lumbering for More Than Sixty Years." *Canada Lumberman* 50 (Aug. 1, 1930), 95-96. Mason, Gordon Lumber Company of Montreal; sawmillers in Ontario and wholesalers of British Columbia fir, 1900-1925.

C77 *Canada Lumberman.* "Pioneer Firm Specializes in Timbers." *Canada Lumberman* 50 (Aug. 1, 1930), 96. James Sheppard & Son, Sorel, Quebec, early importer of longleaf pine timbers, 1870s.

C78 *Canada Lumberman.* "Rutherfords of Montreal Among the Pioneers." *Canada Lumberman* 50 (Aug. 1, 1930), 102-03. Suppliers of wood and lumber for construction since 1865.

C79 *Canada Lumberman.* "When New Brunswick Lumbermen Dominated the Sport of Kings." *Canada Lumberman* 50 (Aug. 1, 1930), 104.

C80 *Canada Lumberman.* "In Picturesque Argenteuil: Strong's Mills in the Municipality of Mille Isles Is a Spot Where Lumbering Is an Old Industry but with Modern Ideas and Equipment." *Canada Lumberman* 50 (Aug. 1, 1930), 106. Mille Isles, Quebec.

C81 *Canada Lumberman.* "Recollections of Old Time Auctions: Memories of the Late Peter Ryan." *Canada Lumberman* 50 (Aug. 1, 1930), 109-10. Concerns a timber sale in Toronto, 1904.

C82 *Canada Lumberman.* "The Frasers Are Old Timers in the Maritimes: An Industrial Organization Which Is Placing New Brunswick on Foreground of Forest Products World." *Canada Lumberman* 50 (Aug. 1, 1930), 113-14. Concerning the Fraser Companies, Ltd., producers of lumber and paper in New Brunswick.

C83 *Canada Lumberman.* "Wholesaling Fifty Years Ago: Here's an Interesting Description of Toronto Half a Century Ago, in the Days When the First Wholesale Lumberman Started Operations." *Canada Lumberman* 50 (Aug. 1, 1930), 115-16.

C84 *Canada Lumberman.* "Waterous Organization of Brantford, Ontario Has Served Lumber Industry for 86 Years." *Canada Lumberman* 50 (Aug. 1, 1930), 118-24. Sawmill machinery by Waterous, Ltd., since 1844.

C85 *Canada Lumberman.* "Backus Operations Started in 1798 Near Port Rowan: Sawmill Business on Lake Erie Thought to Be Oldest in Canada." *Canada Lumberman* 50 (Aug. 1, 1930), 125-26. Port Rowan, Ontario.

C86 *Canada Lumberman.* "Johnston Mill Has Cut for Over Century: Old York County Sawmill at Pefferlaw, Ontario, Was First Operated by Capt. William Johnston." *Canada Lumberman* 50 (Aug. 1, 1930), 127-28.

C87 *Canada Lumberman.* "Old Hardwood Firm is Link with Past: Gall Lumber Co., Toronto, Which Was Founded More Than Fifty Years Ago, Specializes in Custom Drying." *Canada Lumberman* 50 (Aug. 1, 1930), 129.

C88 *Canada Lumberman.* "D'You Mind the Time! When John B. Reid Supplied Lumber for Old Massey Hall, Toronto." *Canada Lumberman* 50 (Aug. 1, 1930), 130. Reid & Company, Toronto retail lumber yard since 1880.

C89 *Canada Lumberman.* "Old Nassau Mill Near Peterborough Was Challenger in Early Days." *Canada Lumberman* 50 (Aug. 1, 1930), 134-35. Peterborough, Ontario, 1858.

C90 *Canada Lumberman.* "Fifty-Nine Years in Business and Going Strong." *Canada Lumberman* 50 (Aug. 1, 1930), 136. G.A. Grier & Sons, lumber wholesalers of Montreal, Quebec, 1870s-1920s.

C91 *Canada Lumberman.* "The Story of Gillies Bros. of Braeside: This Old Established Mill Operation Has Many Years of Sawing Ahead of It." *Canada Lumberman* 50 (Aug. 1, 1930), 141-45. Braeside, Ontario.

C92 *Canada Lumberman.* "Has Served Lumber Trade for Half Century; Kerr Engine Co., Limited, Dates Its Activities from 1872 When Small Machine Shop Was Started at Walkerville, Ontario." *Canada Lumberman* 50 (Aug. 1, 1930), 146.

C93 *Canada Lumberman.* "An Industrial Romance: Price Brothers & Co., Limited, of Old Quebec, 1817-1930." *Canada Lumberman* 50 (Aug. 1, 1930), 151-58.

C94 *Canada Lumberman.* "Truax Has Made Doors for Half Century: R. Truax, Son & Co. Was First to Start the Wholesale Manufacturing of Doors." *Canada Lumberman* 50 (Aug. 1, 1930), 165, 169-70. Walkerton, Ontario.

C95 *Canada Lumberman.* "Ottawa Company Comes Ahead: MacDonell-Conyers Lumber Co. of Ottawa Has Made Substantial Progress Since 1918—Operate Mill on Gatineau." *Canada Lumberman* 50 (Aug. 1, 1930), 171. Sawmill in Quebec and retail yard in Ottawa, Ontario.

C96 *Canada Lumberman.* "Pulpwood Man at Eighty-Nine Goes to Work Every Day." *Canada Lumberman* 50 (Aug. 1, 1930), 186. Includes reminiscences of Quebec lumber trade, 1860s.

C97 *Canada Lumberman.* "Montreal Retail Firm Busy as Ever: L. Villeneuve & Co. Limited, Has Been Serving the Building Trade of Montreal for Fifty-Five Years." *Canada Lumberman* 50 (Aug. 1, 1930), 191-92.

C98 *Canada Lumberman.* "Montreal Wholesale Concern Operating Fifty Years." *Canada Lumberman* 50 (Aug. 1, 1930), 193-94. E.H. Lemay, Ltd., Montreal, Quebec.

C99 *Canada Lumberman.* "In Beautiful Oromocto Valley: The Picturesque Oromocto River in New Brunswick Has Been a Lumbering Stream for More Than a Hundred Years." *Canada Lumberman* 50 (Aug. 1, 1930), 201-02.

C100 *Canada Lumberman.* "Evolution of the Lumber Business." *Canada Lumberman* 51 (Sept. 15, 1931), 29-30. In the Midland District of Ontario—counties of Peterborough and Victoria, 1830s-1850s.

C101 *Canada Lumberman.* "Early Lumbering Activities in Muskoka District." *Canada Lumberman* 53 (June 1, 1933), 17. Near Lake Muskoka, Ontario.

C102 *Canada Lumberman.* "When Rafts of Square Timber Were Taken from Hull to Quebec over Century Ago." *Canada Lumberman* 54 (Apr. 1, 1934), 11-12. Philemon Wright and rafting on the Ottawa River, Ontario-Quebec, 1830s.

C103 *Canada Lumberman.* "Export of Lumber, Timber, Lath and Shingle from Canada to United Kingdom." *Canada Lumberman* 54 (May 1, 1934), 33ff.

C104 *Canada Lumberman.* "Early History of Lumbering on Pacific Coast." *Canada Lumberman* 54 (June 15, 1934), 38. British Columbia.

C105 *Canada Lumberman.* "Review of Pulpwood Production, Domestic Consumption, Exportation and Importation from 1908-1933." *Canada Lumberman* 54 (Nov. 15, 1934), 26.

C106 *Canada Lumberman.* "Early Days of Lumbering at Fort Coulonge." *Canada Lumberman* 55 (May 1, 1935), 43. Lumberjack Larry Frost, logging camps, and log rafts at Fort Coulonge, Quebec.

C107 *Canada Lumberman.* "Veteran Recalls Early Lumbering Days in Georgian Bay Activities." *Canada Lumberman* 56 (June 1, 1936), 21-22. Concerns the activities of D.L. White, an American who rafted logs from Ontario to Michigan.

C108 *Canada Lumberman.* "Historic Timber Days in Quebec." *Canada Lumberman* 56 (July 1, 1936), 13-14. Lumber companies near Sillery Cove, Quebec.

C109 *Canada Lumberman.* "Eighty-Six Year Old Lumberman Recalls Bush Life of Seventy Years Ago." *Canada Lumberman* 56 (July 15, 1936), 17-18. Allen McPherson of Ovillia, Ontario.

C110 *Canada Lumberman.* "Historic Firm of Wm. Milne and Sons." *Canada Lumberman* 56 (Aug. 15, 1936), 38-39. Timagami, Ontario.

C111 *Canada Lumberman.* "In Old Square Timber Days: Interesting Reminiscences of Background of Lumber Industry in Canada." *Canada Lumberman* 57 (Sept. 1, 1937), 17-19.

C112 *Canada Lumberman.* "Canada's First Sawmill at Port Rowan Is 139 Years Old." *Canada Lumberman* 57 (Sept. 1, 1937), 20. John C. Backus, fourth-generation operator of the sawmill in Port Rowan, Ontario.

C113 *Canada Lumberman.* "Century Old Waterloo County Sawmill." *Canada Lumberman* 58 (Apr. 15, 1938), 9-10; 58 (July 15, 1938), 23. The Hallman Lumber Plant, New Dundee, Ontario, founded in 1828.

C114 *Canada Lumberman.* "Several Century Old Sawmills Still Going." *Canada Lumberman* 58 (Nov. 15, 1938), 24. Ontario.

C115 *Canada Lumberman.* "Ancient Documents Throw Light on Log Driving and Lumber Deliveries in the Ottawa Valley." *Canada Lumberman* 58 (Dec. 15, 1938), 22-23. Contracts for procuring logs on the Gatineau River, Quebec, by the W.C. Edwards & Co., 1877.

C116 *Canada Lumberman.* "A Lumbering Background of 135 Years Reviewed by Bill Blair of Montreal." *Canada Lumberman* 60 (May 15, 1940), 11-12. Blair Brothers of Montreal, founded in 1805.

C117 *Canada Lumberman.* Diamond Jubilee Issue. *Canada Lumberman* 60 (Aug. 1, 1940). Similar to Golden Jubilee issue of Aug. 1, 1930. See following entries.

C118 *Canada Lumberman.* "Sixty Years of Lumbering and Publishing." *Canada Lumberman* 60 (Aug. 1, 1940), 53. The role of *Canada Lumberman.*

C119 *Canada Lumberman.* "In the Days When Muscle Counted: Savoie Bros. of Manseau, Quebec, Trace Their Present Transit Operations Back to 1904." *Canada Lumberman* 60 (Aug. 1, 1940), 59.

C120 *Canada Lumberman.* "Thirty-Two Years of Service to the Lumbermen of Canada: History and Achievements of the Canadian Lumbermen's Association." *Canada Lumberman* 60 (Aug. 1, 1940), 69.

C121 *Canada Lumberman.* "Canada's Export Lumber Trade Grew in the Coves of Quebec: Old Time Lumber Firms Used Quebec Coves Extensively and Found Them Ideal for Booming, Loading and Piling." *Canada Lumberman* 60 (Aug. 1, 1940), 72-73.

C122 *Canada Lumberman.* "Booth Interests Have Been in the Forefront of Progress for 85 Years." *Canada Lumberman* 60 (Aug. 1, 1940), 76. J.R. Booth, Ltd., Ontario.

C123 *Canada Lumberman.* "Lumber Firm Keeps Pace with Growth of Montreal's Largest Suburb: J.P. Dupuis, Limited." *Canada Lumberman* 60 (Aug. 1, 1940), 77-78. Retail firm established in 1908.

C124 *Canada Lumberman.* "Progress Evidenced by Steady Growth of 86-Year-Old Guelph Lumber Firm." *Canada Lumberman* 60 (Aug. 1, 1940), 79-80. Robert Stewart, Ltd., Guelph, Ontario.

C125 *Canada Lumberman.* "Montreal Concern Carries Huge Variety of Rare and Quality Goods: E.J. Maxwell, Ltd." *Canada Lumberman* 60 (Aug. 1, 1940), 81. Established in 1862.

C126 *Canada Lumberman.* "Retail Lumber Concern Traces Back History for Nearly Seventy Years: S.F. Stinson and Son Continue to Render Reliable Service to Their Many Customers." *Canada Lumberman* 60 ®Aug. 1, 1940), 86-87. Toronto, since 1872.

C127 *Canada Lumberman.* "Sons Continue Business Established by Frank A. Bowden 60 Years Ago: Bowden Lumber & Coal Co. Is Active in Retail Lumber Trade in Toronto and Suburbs." *Canada Lumberman* 60 (Aug. 1, 1940), 98.

C128 *Canada Lumberman.* "Sixty-Six Years of Progress for James Davidson's Sons." *Canada Lumberman* 60 (Aug. 1, 1940), 100. James Davidson's Sons, Ontario.

C129 *Canada Lumberman.* "R.R. Campbell Active with Company He Founded Fifty-Six Years Ago: Present Firm of Chappells, Ltd., of Sydney, N.S., Traces Its History Back to 1884." *Canada Lumberman* 60 (Aug. 1, 1940), 106, 119. Nova Scotia.

C130 *Canada Lumberman.* "Haley & Son Ltd. Have Witnessed Many Changes in Lumber Industry in 52 Years of Business, St. Stephen, N.B." *Canada Lumberman* 60 (Aug. 1, 1940), 114. New Brunswick.

C131 *Canada Lumberman.* "When Pine Timber Was King of the Ottawa Valley." *Canada Lumberman* 60 (Aug. 15, 1940), 19-20. Rafting and driving of square timber on the Ottawa River, Ontario-Quebec.

C132 *Canada Lumberman.* "Historic New Brunswick Lumbering Centres." *Canada Lumberman* 61 (May 15, 1941), 44-46, 51. History of the lumber industry in New Brunswick, including a list of firms operating at mouth of St. John River in 1880.

C133 *Canada Lumberman.* "Evolution of Tractor Logging: Invention of Power Winch an Outstanding Factor in Its Success." *Canada Lumberman* 61 (Nov. 15, 1941), 17-18.

C134 *Canada Lumberman.* "Forestry Policies and Progress in Four Western Provinces." *Canada Lumberman* 67 (Sept. 1, 1947), 113-16.

C135 *Canadian Geographical Journal.* "Forest Protection in Ontario." *Canadian Geographical Journal* 46 (Feb. 1953), 42-59.

C136 Canadian Lumbermen's Association. *The Story of the Canadian Lumbermen's Association.* Ottawa, 1952. 15 pp. Illus. Since 1907.

C137 Canadian Pulp and Paper Association, Western Branch. "Historical Notes on British Columbia's Pulp and Paper Industry." *Pulp & Paper Magazine of Canada* 49 (July 1948), 67-82.

C138 Cancell, Benton. "A Study of the Paper Industry in the Northeastern U.S." Master's thesis, Univ. of Michigan, 1943.

C139 Candee, Richard M. "Merchant and Millwright: The Water Powered Sawmills of the Piscataqua." *Old-Time New England* 60 (Spring 1970), 131-49. Development of sawmills in 17th-century Maine and New Hampshire.

C140 Candy, R.H. *Reproduction on Cut-Over and Burned-Over Land in Canada.* Toronto: Ontario Department of Resources and Development, 1951. 224 pp.

C141 Cannon, Robert Walter. "A Comparison of the Trends in Prices of Lumber, Logs, and Stumpage with the General Price Level of All Commodities." Master's thesis, Univ. of Georgia, 1950.

C142 Cantwell, Robert. *Alexander Wilson: Naturalist and Pioneer.* Philadelphia: J.B. Lippincott, 1961. 318 pp. Illus. Wilson (1766-1813) glorified nature and wilderness in his writings.

C143 Cantwell, Robert. *The Hidden Northwest.* Philadelphia: J.B. Lippincott, 1972. 335 pp. Maps, bib., app., index. Includes an essay on George Weyerhaeuser, Washington lumberman.

C144 Carbone, Mario G. *Economic Difficulties of the Lumber Industry of the United States, 1850-1932.* New York: N.p., 1937. viii + 80 pp. Tables, diags., bib. Concerns forest depletion, the capital structure of the industry, and a national forestry plan.

C145 Carder, David Ross. "Unified Planning and Decision Making: A Conceptual Framework for U.S. Forest Service Management." Ph.D. diss., Stanford Univ., 1974. 263 pp. Includes some incidental history of the USFS and its administrative organization.

C146 Carey, Daniel. "Michigan's Foremost Unique Logger." *Michigan History* 32 (Sept. 1948), 301-02. On Scott Gerrish of Evart, the first Michigan lumberman to log by rail, ca. 1873.

C147 Carhart, Arthur H. "Recreation in the Forests." *American Forestry* 26 (May 1920), 268-72. Incidental historical references to USFS interest in managing national forests for outdoor recreation purposes.

C148 Carhart, Arthur H. "Our Public Lands in Jeopardy." *Journal of Forestry* 46 (June 1948), 408-16. On the attacks against the national forests and other public lands by congressional representatives of the livestock interests and other opponents of conservation, 1940s.

C149 Carhart, Arthur H. "Forest in the Rockies." *American Forests* 54 (July 1948), 296-99, 335-36. White River National Forest, Colorado, since 1891.

C150 Carhart, Arthur H. "Golden Anniversary." *American Forests* 54 (Sept. 1948), 400-02, 420. Black Hills National Forest, South Dakota, since 1898.

C151 Carhart, Arthur H. "Mass Murder in the Spruce Belt." *American Forests* 55 (Mar. 1949), 14-15, 41-42. On the destruction of Engelmann spruce in Colorado by *Dendroctonus engelmanni,* since 1939.

C152 Carhart, Arthur H. *Timber in Your Life.* Intro. by Bernard De Voto. Philadelphia: J.B. Lippincott, 1954. 317 pp. Index. Includes some history of the lumber industry, the development of national forest policies, the profession of forestry, and other aspects of forest conservation.

C153 Carhart, Arthur H. "Forest Beacon in Michigan." *American Forests* 61 (Apr. 1955), 16-17, 53-55. On the Au Sable Forest Products Association, a cooperative selling organization of small farmers in the vicinity of East Tawas, Michigan, since 1946.

C154 Carhart, Arthur H. "The First Ranger." *American Forests* 62 (Feb. 1956), 26-27, 55-56. On the service of William R. Kreutzer as a USFS ranger in Colorado, 1898-1939.

C155 Carhart, Arthur H. *Trees and Game—Twin Crops.* Washington: American Forest Products Industries, 1958. Illus. 32 pp. Includes some history of game and wildlife management on forested lands.

C156 Carhart, Arthur H. *The National Forests.* New York: Alfred A. Knopf, 1959. 289 pp. Illus. On the establishment of national forests, including history and description of each USFS region.

C157 Carhart, Arthur H. "Shelterbelts, a 'Failure' That Didn't Happen." *Harper's* 221 (Oct. 1960), 75-76. Corrects a widespread misconception that the New Deal tree planting and shelterbelt programs on the Great Plains were failures.

C158 Carlisle, George T., Jr. "The Maine Wilderness is Not Passing." *American Forests and Forest Life* 34 (June 1928), 323-26, 358. Incidental historical references to forest uses in Maine since the early 19th century.

C159 Carlson, Elmer J. "The Cherokee Indian Forest of the Appalachian Region." *Journal of Forestry* 51 (Sept. 1953), 628-30. Cherokee Indian Reservation, North Carolina.

C160 Carlson, Norman K. "Honaunau Forest." *American Forests* 66 (Apr. 1960), 16-18, 53, 55-58. Forestry practiced on the Honaunau Forest, owned by the Bernice P. Bishop Estate, on the island of Hawaii. Includes some history of the area since 1778.

C161 Carlson, Paul H. "Forest Conservation on the South Dakota Prairies." *South Dakota History* 2 (Winter 1971), 23-45. A general history of tree planting, shelterbelts, and forest conservation in South Dakota since 1862, with special attention to federal and state legislation and programs.

Carlson, Reynold E. See Raup, Hugh M., #R58

C162 Carlson, Valdemar. "Associations and Combinations in the American Paper Industry." Ph.D. diss., Harvard Univ., 1931. Historical survey of the industry since 1819, with emphasis on period from 1900. Treats International Paper and Power Company and American Paper and Pulp Association.

C163 Carlton, William R. "New England Masts and the King's Navy." *New England Quarterly* 12 (Mar. 1939), 4-18. On the logging of New England pines and their use as masts for English naval vessels during the colonial period.

C164 Carmer, Carl L. *The Susquehanna.* New York: Rinehart & Company, 1955. 493 pp. Illus., bib. Includes some history of lumbering and rafting in Pennsylvania.

C165 Carmichael, Herbert. "Pioneer Days in Pulp and Paper." *British Columbia Historical Quarterly* 9 (July 1945), 201-12. Reminiscences of the British Columbia Paper Manufacturing Company, established in 1894 and predecessor of the Powell River Paper Company.

C166 Carothers, June E. *Estes Park, Past and Present.* Denver: University of Denver Press, 1951. 89 pp. Illus., map, bib. Estes Park and Rocky Mountain National Park, Colorado, since 1859.

C167 Carper, Edith T. *Illinois Goes to Congress for Army Land.* Inter-University Case Program, ICP Case No. 71. University: University of Alabama Press, 1962. 32 pp. Describes struggle between conservationists and industrial and real estate developers over the disposition of army lands near Des Plaines, Illinois, 1950s.

C168 Carr, Archie. *The Everglades.* New York: Time-Life, 1973. 184 pp. Illus., bib. Includes some history of Everglades National Park, Florida.

C169 Carranco, Lynwood F. "Logger Lingo in the Redwood Region." *American Speech* 31 (May 1956), 149-52; 34 (Feb. 1959), 76-80.

C170 Carranco, Lynwood F. "Logging Railroad Language in the Redwood Country." *American Speech* 37 (May 1962), 130-36.

C171 Carranco, Lynwood F. "Americanisms in the Redwood Country." *Western Folklore* 22 (No. 4, 1963), 263-67. On the influence of the redwood lumber industry on speech and vocabulary in California's redwood region.

C172 Carranco, Lynwood F., and Fountain, Mrs. Eugene. "California's First Railroad: The Union Plank Walk, Rail Track, and Wharf Company Railroad." *Journal of the West* 3 (Apr. 1964), 243-56. Logging railroads in Humboldt County since 1854.

C173 Carranco, Lynwood F. "A Miscellany of Folk Beliefs from the Redwood Country." *Western Folklore* 26 (July 1967), 169-76. Some concerning California lumberjacks.

C174 Carranco, Lynwood F. "Logger Language in Redwood Country." *Journal of Forest History* 18 (July 1974), 52-59. On the development of specialized logging terminology and profanity in northern California since the 1880s.

C175 Carranco, Lynwood F., and Labbe, John T. *Logging the Redwoods.* Caldwell, Idaho: Caxton Printers, 1975. Illus. Since the mid-19th century.

C176 Carroll, Charles Francis. "The Forest Civilization of New England: Timber, Trade, and Society in the Age of Wood, 1600-1688." Ph.D. diss., Brown Univ., 1970. 659 pp.

C177 Carroll, Charles Francis. *The Timber Economy of Puritan New England.* Providence: Brown University Press, 1973. xiii + 221 pp. Maps, tables, apps., notes, index. An account of the 17th-century settlers' adjustment to and exploitation of the New England forest environment. Topics of interest include the primeval forest, pioneer uses of wood, logging and the lumber industry, shipbuilding, the timber trade, and "timber imperialism"—the political struggles over timber-rich New Hampshire and Maine.

C178 Carroll, Peter Neil. "Puritanism and the Wilderness: The Intellectual Significance of the New England Frontier, 1629-1675." Ph.D. diss., Northwestern Univ., 1968. 249 pp.

C179 Carroll, Peter Neil. *Puritanism and the Wilderness: The Intellectual Significance of the New England Frontier, 1629-1700.* New York: Columbia University Press, 1969. xi +

243 pp. Notes, bib. An examination of Puritan attitudes toward the New England forest and the influence of the wilderness upon Puritan social thought as revealed in 17th-century literature.

C180 Carrothers, W.A. "Forest Industries of British Columbia." In Lower, Arthur R.M.; Carrothers, W.A.; and Saunders, S.A. *The North American Assault on the Canadian Forest.* Ed. by Harold A. Innis. Relations of Canada and the United States Series. Toronto: Ryerson Press; New Haven: Yale University Press, 1938. Pp. 225-344. Map, tables, notes, bib., index. On the provincial timber policy, lumber industry, lumber trade, shingle industry, pulp and paper industry, and other aspects of British Columbia forest history since the 1870s.

C181 Carrott, M. Browning. "The Supreme Court and American Trade Associations, 1921-1925." *Business History Review* 44 (Autumn 1970), 320-38. Especially concerned with the American Hardwood Manufacturers Association and the Maple Flooring Manufacturers Association.

C182 Carruth, Viola. "Would Fisher Die?" *Forests & People* 24 (Second quarter, 1974), 31-35. On recent efforts of the Boise Southern Company to preserve Fisher, a Louisiana sawmill town founded in the 1890s by the Louisiana Long Leaf Lumber Company.

C183 Carruthers, George. *Paper-Making. I. First Hundred Years of Paper-Making by Machine. II. First Century of Paper-Making in Canada.* Toronto: Garden City Press Co-operative, 1947. 712 pp. Illus., maps.

C184 Carruthers, Guy. "Our Strange Debt to the 'Gasoline Tree'." *Westways* 48 (Jan. 1956), 14-15. On the preparation of heptane, a hydrocarbon compound similar to gasoline, from Jeffrey pine trees in California since 1860s.

C185 Carson, Joan. *Tall Timber and the Tide.* 1971. Revised edition. Poulsbo, Washington: Kitsap Weeklies, 1972. 113 pp. Illus., maps, bib. Includes some history of the lumber industry and lumber towns of north Kitsap County, Washington.

C186 Carson, Rachel. *Silent Spring.* New York: Houghton Mifflin, 1962. 368 pp. Bib. On the environmental effects of the use of man-made chemicals, including some historical references to forests.

C187 Carson, Russell Mack Little. "The Adirondack Mountain Club." *High Spots* 14 (Dec. 1937), 17-20. History of the Adirondack Forest Preserve, New York.

C188 Carson, W.J. "Developments in Mechanization of Woods Operations." *Canada Lumberman* 60 (Aug. 1, 1940), 54-55, 95-96.

C189 Carstenson, Vernon R. *Farms or Forests: Evolution of a State Land Policy for Northern Wisconsin, 1850-1932.* Madison: College of Agriculture, University of Wisconsin, 1958. 130 pp. Maps, tables, notes. Concerns the era of logging, the movement for farms on the cutover lands, and the revival of interest in non farm land use that culminated in the Wisconsin Rural Zoning Law of 1929.

Carstenson, Vernon R. See Rasmussen, Wayne D., #R54

C190 Carstenson, Vernon R., ed. *The Public Lands: Studies in the History of the Public Domain.* Madison: University of Wisconsin Press, 1963. xxvi + 522 pp. Maps, charts, notes, app., index. An anthology of historical articles, some of which pertain to forested lands.

C191 Cart, Theodore W. "The Struggle for Wildlife Protection in the United States, 1870-1900: Attitudes and Events Leading to the Lacey Act." Ph.D. diss., Univ. of North Carolina, 1971. 221 pp.

*C192 Cart, Theodore W. "'New Deal' for Wildlife: A Perspective on Federal Conservation Policy, 1933-40." *Pacific Northwest Quarterly* 63 (July 1972), 113-20. Surveys the growth of wildlife conservation prior to the Franklin D. Roosevelt administration and includes some history of the conservation movement in general throughout the 1930s.

C193 Cart, Theodore W. "The Lacey Act: America's First Nationwide Wildlife Statute." *Forest History* 17 (Oct. 1973), 4-13. Background to the passage of the Lacey Act of 1900, sponsored by Iowa congressman John Fletcher Lacey (1841-1913).

C194 Carter, Jane. "Man's Place in the Sun." *Agricultural History* 22 (Oct. 1948), 209-20. Survey of land ownership and exploitation from ancient times, with special reference to the United States.

C195 Carter, Jane. "Old Rail Fences." *American Forests* 54 (Dec. 1948), 547-48, 569-71. Includes some history of European and American fence making, with personal reminiscences of post and rail fence making in Pennsylvania.

C196 Carter, Jane. "Old Arboretum Lives Again." *American Forests* 55 (Sept. 1949), 18-20, 45. On the John J. Tyler Arboretum, Delaware County, Pennsylvania, since 1825.

C197 Carter, Lawrence. "Mr. Holcombe Comes to Town." *Great Lakelands* 9 (Dec. 1959), 13-15, 24. Concerns James Holcombe, a lumberman and farmer in the Au Sable River area and later on the Saginaw River of Michigan, 1880s-1915.

C198 Carter, Luther J. *The Florida Experience: Land and Water Policy in a Growth State.* Baltimore: Johns Hopkins University Press, for Resources for the Future, 1974. xvi + 355 pp. Illus., maps, index. Includes some history of Everglades National Park, Big Cypress Swamp, and other endangered forested areas of south Florida.

C199 Cartwright, Walter J. "The Cedar Chopper." *Southwestern Historical Quarterly* 70 (Oct. 1966), 247-55. Some historical observations of the men who cut fence posts from the cedar brakes of central Texas.

C200 Caruso, John Anthony. *The Appalachian Frontier: America's First Surge Westward.* Indianapolis: Bobbs-Merrill, 1959. 408 pp. Maps, notes, bib. On the settlement of North Carolina, Kentucky, and Tennessee, 1750-1800, with many references to the forest environment.

Carvell, K.L. See Clarkson, Roy B., #C377, C378

Carvell, K.L. See Percival, W.C., #P113

C201 Carver, Clifford N. *John Carver: Builder of Wooden Ships Upon the Penobscot Bay.* New York: Newcomen Society in North America, 1957. 32 pp. Illus.

C202 Cary, Austin. "An Appreciation of Dr. Schenck." *Forestry Quarterly* 12 (Dec. 1914), 562-66. German-born Carl Alwin Schenck (1868-1955) directed the Biltmore Forest School near Asheville, North Carolina, from 1898 to 1909.

C203 Cary, Austin. "Forest People: 'Get a Living and Let Your Forest Grow'." *American Forests and Forest Life* 30 (Feb. 1924), 94-95. On W.B. Deering of Hollis, Maine, and his farm forestry since 1886.

C204 Cary, Austin. "Forty Years of Forest Use in Maine." *Journal of Forestry* 33 (Apr. 1935), 366-72. On the forestry activities of paper companies in Maine.

C205 Cary, Austin. "Austin Cary Speaks Out." *Journal of Forestry* 33 (Nov. 1935), 916-22. A letter to Franklin D. Roosevelt which contains much forest history of Maine.

C206 Cary, Austin. "White Pine and Fire." *Journal of Forestry* 34 (Jan. 1936), 62-65. Observations since the 1890s on the effects of fires upon the white pine forests of the New England states.

C207 Casamajor, Paul; Teeguarden, Dennis; and Zivnuska, John A. *Timber Marketing and Land Ownership in Mendocino County.* University of California, Agriculture Experiment Station, Bulletin No. 772. Berkeley, 1960. 56 pp.

C208 Casamajor, Paul, ed. *Forestry Education at The University of California: The First Fifty Years.* Berkeley: California Alumni Foresters, 1965. 422 pp. Illus., maps, tables, apps., index. Biographies of staff and alumni and a general account of forestry education in California since 1873.

Casamajor, Paul. See Teeguarden, Dennis, #T30

Casler, Walter. See Taber, Thomas T., #T2

C209 Casson, Henry. *"Uncle Jerry." Life of General Jeremiah M. Rusk. . . .* Madison, Wisconsin: J.W. Hill, 1895. xi + 490 pp. Rusk was Benjamin Harrison's Secretary of Agriculture, 1889-1893.

C210 Castor, Marilyn. "Timber!" *Michigan Conservation* 21 (May-June 1952), 11-14. On the Edith E. Pettee Forest, Roscommon County, Michigan, established in 1930 with money contributed by Detroit high school students.

C211 Cate, Donald. "Recreation and the U.S. Forest Service: A Study of Organizational Response to Changing Demands." Ph.D. diss., Stanford Univ., 1963. 661 pp. Includes some history of administration of national forests for recreational use.

C212 Cate, Wirt A. *Lucius Q.C. Lamar: Secession and Reunion.* Chapel Hill: University of North Carolina Press, 1935. xii + 594 pp. Bib. Lamar was Grover Cleveland's Secretary of the Interior, 1885-1888. There is a modern reprint.

C213 Caterpillar Tractor Company. *Harvesting the Timber of the West.* Peoria, 1946. 15 pp. Illus.

C214 Caterpillar Tractor Company. *Fifty Years on Tracks.* Peoria, 1954. 102 pp. Illus., maps. A manufacturer of heavy equipment used in logging.

C215 Caterpillar Tractor Company. *Men of Timber.* Peoria, 1955. 88 pp. Illus. Biographical sketches of presidents of the Pacific Logging Congress, 1909-1955, and the Intermountain Conference, 1939-1956. There are other editions.

C216 Catesby, Mark. *Natural History of Carolina, Florida, and Bahama Islands.* 2 Volumes. London, 1754. There are subsequent editions and reprints.

C217 Caudill, Harry M. *Night Comes to the Cumberlands, a Biography of a Depressed Area.* Boston: Little, Brown, 1963. 394 pp. Illus. Includes historical references to the lumber industry and other forest uses of the Cumberland and Appalachian plateaus.

C218 Caughey, John W. "The Californian and His Environment." *California Historical Quarterly* 51 (Fall 1972), 195-204. General statement of the problem in a historical context.

C219 Cauvin, Dennis M. "Measurement of a Forest's Contribution to the Economy of Alberta." Ph.D. diss., Univ. of Washington, 1972. 218 pp. An economic analysis of Alberta's forest resources and of their development; incidental history.

C220 Caverhill, P.Z. "Forest Policy in New Brunswick." Master's thesis, Univ. of New Brunswick, 1917.

C221 Caverhill, P.Z. "The Development of the Forest Policy in British Columbia." *Empire Forestry Journal* 4 (1925), 66ff.

C222 Caverhill, P.Z. "Forestry and Lumbering in British Columbia." *Journal of Forestry* 29 (Nov. 1931), 1067-74. Contains incidental references to the lumber industry since 1860s and provincial forestry programs in the 20th century.

C223 Cazden, Norman. "Regional and Occupational Orientations of American Traditional Song." *Journal of American Folklore* 72 (Oct. 1959), 310-44. Concerns in part the contributions of loggers to the preservation of American folk songs and traditions.

C224 Chaffe, John. "New Orleans, the Logical Lumber Port." *Southern Lumberman* 36 (Dec. 24, 1910), 51-56. Incidental history.

C225 Challenger, J.W. "Power Saws: Experiments in British Columbia." *Timberman* 38 (Aug. 1937), 18-22.

C226 Challenger, J.W. "Operate Fifty-Four Power Saws." *Canada Lumberman* 63 (Mar. 15, 1943), 9-12. Report on five years of experimental chain saw use by Bloedel, Stewart & Welch, Ltd., in British Columbia. Reprint in *Timberman* 44 (Nov. 1943), 10-13.

C227 Chamberlain, Arthur Henry. *Thrift and Conservation.* Philadelphia: J.B. Lippincott, 1919. 272 pp. Illus.

C228 Champ, F.P. "National Parks and National Forests in Relation to the Development of the Western States." *Proceedings of the American Forestry Association* (1941), 11-22.

C229 Champion, F.J. *Forest Products of American Forests.* USFS, Miscellaneous Publication No. 861. Washington: GPO, 1961. v + 30 pp. Illus.

C230 Champion Paper and Fibre Company. *This is Champion.* Hamilton, Ohio, 1959. 61 pp. Illus. Includes some history of the company since its founding in 1893, with attention to operations at Hamilton, Ohio; Canton, North Carolina; and Pasadena, Texas.

C231 Chandler, Charles Lyon. *Early Shipbuilding in Pennsylvania, 1683-1812.* Philadelphia: Colonial Press, 1932. 43 pp. As a continuously subsidized colonial industry, especially in Philadelphia.

Chandler, Robert F., Jr. See Lutz, Harold J., #L324

C232 Chandler, Robert W. "Robert W. Sawyer: 'He Thought in Terms of Forever.'" *American Forests* 65 (Dec. 1959), 16-17, 44, 46-49. The Bend, Oregon, newspaper editor was a leader in the state parks movement and championed many other conservation causes, 1910s-1950s.

C233 Chaney, Donald E. "Constitutional Validity of a Federal Reforestation Program for Upper Tributaries of Navigable Rivers." *Missouri Law Review* 25 (June 1960), 317-23. On the whole application of the commerce clause of the Constitution to streams and waterways since 1829.

C234 Chaney, Esket B. *The Story of Portage.* Onekama, Michigan: By the author, 1960. 75 pp. Illus., maps, bib. Includes some history of the 19th-century lumber industry near Portage Lake and Onekama, Michigan.

C235 Chaney, Ralph W. "John Campbell Merriam." *Yearbook of the American Philosophical Society* (1945), 381-87. A biographical sketch including his role in the conservation and preservation movements, especially the Save-the-Redwoods League.

C236 Chaney, Ralph W. *The Ancient Forests of Oregon.* Eugene: Oregon State System of Higher Education, 1948. xiv + 56 pp. Illus., bib.

C237 Chapelle, Howard I. *The History of American Sailing Ships.* New York: W.W. Norton & Company, 1935. xvi + 400 pp. Illus. Includes some history of ships used in the lumber trade.

C238 Chaplin, Ralph H. *The Centralia Conspiracy: The Truth about the Armistice Day Tragedy.* 1920. Third edition. Chicago: General Defense Committee, Industrial Workers of the World, 1924. 143 pp. Illus. On the violent conflict between Wobblies and American Legionnaires in the lumber town of Centralia, Washington, 1919.

C239 Chaplin, Ralph H. *Wobbly: The Rough and Tumble Story of an American Radical.* Chicago: University of Chicago Press, 1948. vi + 435 pp. Illus. On the author's career as a socialist and labor organizer, 1911-1943, especially among lumberjacks and millworkers of the Pacific Northwest.

C240 Chapline, William R. "Douglas C. Ingram." *Journal of Forestry* 28 (Mar. 1930), 403-05. With the USFS in the Pacific Northwest since 1909.

C241 Chapline, William R. "Range Management History and Philosophy." *Journal of Forestry* 49 (Sept. 1951), 634-38. Range research and public range policy from the early 1900s.

Chapman, Albert J. See Burgh, Robert, #B571

C242 Chapman, Berlin Basil. "Federal Management and Disposition of the Lands of Oklahoma Territory, 1866-1907." Ph.D. diss., Univ. of Wisconsin, 1931.

Chapman, Herman Haupt. See Woolsey, Theodore S., Jr., #W463

C243 Chapman, Herman Haupt. "Recreation as a Federal Land Use." *American Forests* 31 (June 1925), 349-51, 378-80. Includes incidental history of recreation in national parks and national forests.

C244 Chapman, Herman Haupt. "The Origin of the Minnesota National Forest." University of Minnesota Forestry Club, *Gopher Peavey* (1928), 46-51. Now the Chippewa National Forest.

C245 Chapman, Herman Haupt. "The Chippewa National Forest." *American Forests and Forest Life* 35 (Sept. 1929), 561-64. Brief history of the Chippewa or Minnesota National Forest, Minnesota, since the 1880s.

C246 Chapman, Herman Haupt. "Conservation, and the Department of the Interior." *Journal of Forestry* 30 (May 1932), 544-53. Historical account of conservation (and anti-conservation) in the USDI since late 19th century. This article appeared as a critical review of Ray Lyman Wilbur and William A. Du Puy's *Conservation in the Department of the Interior* (1931).

C247 Chapman, Herman Haupt. "Education in Forestry in Minnesota: A Historical Sketch." *Journal of Forestry* 33 (July 1935), 695-96. Since the 1880s at University of Minnesota.

C248 Chapman, Herman Haupt. "The Pennsylvania Forest Service: Past, Present, and Future." *Journal of Forestry* 34 (Apr. 1936), 409-13. Since its establishment in 1903.

C249 Chapman, Herman Haupt. "Reorganization of the Forest Service." *Journal of Forestry* 35 (May 1937), 427-34. Includes some history of forest conservation and the USFS under the USDA, as part of an argument against transferring the USFS to the USDI or proposed Department of Conservation.

C250 Chapman, Herman Haupt. "Some Important Trends in Forestry in the United States." *Journal of Forestry* 36 (July 1938), 653-58. Reviews achievements of the USFS and the Society of American Foresters.

C251 Chapman, Herman Haupt. "A Factual Analysis of the Quetico-Superior Controversy." *Journal of Forestry* 43 (Feb. 1945), 97-103. On the effort to preserve a memorial wilderness area along the Minnesota-Ontario boundary within the Quetico Provincial Park and the Superior National Forest.

C252 Chapman, Herman Haupt. "The Cure for the O. and C. Situation." *Journal of Forestry* 43 (Aug. 1945), 569-74. Includes some history of the revested Oregon and California Railroad grant lands since 1916, with particular attention to the formula by which eighteen western Oregon counties receive federal payments in lieu of taxes.

C253 Chapman, Herman Haupt. "Origin and Results of the Seed-Tree Experiment with Norway Pine on the Chippewa National Forest." *Journal of Forestry* 44 (Mar. 1946), 178-83. On the origins of the Morris Act of 1902, which provided for logging on northern Minnesota Indian reservations later to become the Chippewa National Forest. It was the government's first attempt to legislate silviculture by requiring loggers to leave seed trees and dispose of slash.

C254 Chapman, Herman Haupt, and Meyer, Walter H. *Forest Valuation, with Special Emphasis on Basic Economic*

Principles. American Forestry Series. New York: McGraw-Hill, 1947. xii + 521 pp. A successor to Chapman's *Forest Valuation* (1914, 1925) and *Forest Finance* (1926, 1935). Contains some incidental historical references to forest finance, forest valuation, and other aspects of forest economics.

C255 Chapman, Herman Haupt. "The Case of the Public Range." *American Forests* 54 (Feb. 1948), 56-60, 92-93; 54 (Mar. 1948), 116-18, 136, 138-39. Examines long and often bitter conflict between conservationists and western livestock interests since 1891.

C256 Chapman, Herman Haupt. "Recreational Interests as Affecting Professional Forestry Activities." *Journal of Forestry* 46 (Apr. 1948), 290-93. On the influence of recreationists and wilderness enthusiasts in determining forest policy since ca. 1900.

C257 Chapman, Herman Haupt. "The Initiation and Early Stages of Research on Natural Reforestation of Longleaf Pine." *Journal of Forestry* 46 (July 1948), 505-10. Cooperative research program of the USFS, Yale School of Forestry, and Urania Lumber Company, Urania, Louisiana, 1915-1944.

C258 Chapman, Herman Haupt. *Forest Management.* Bristol, Connecticut: Hildreth Press, 1950. xix + 582 pp. Illus., maps, bibs., index. Includes a chapter on the "Evolution of Forest Organization" and many incidental historical references to forest management. The original version of this text appeared in 1931.

C259 Chapman, Herman Haupt. "An Ancient and Original Transportation System for Logs in Southern Alabama." *Journal of Forestry* 49 (Mar. 1951), 209-10. On the use of ditches or sluiceways with timbered sides as a means of transporting logs in the 1870s.

C260 Chapman, Herman Haupt. "The Development of the Profession of Forestry in the United States." *Forestry Chronicle* 27 (June 1951), 129-35.

C261 Chapman, Herman Haupt. "Bernhard Eduard Fernow." *American-German Review* 19 (Feb. 1953), 13-14. On his contributions to forestry and forestry education in the United States and Canada. Fernow (1851-1923) was chief of the Division of Forestry, 1886-1898.

C262 Chapman, Herman Haupt. *The Menominee Indian Timber Case History: Proposals for Settlement.* N.p., 1957. 70 pp. On legislation and administration of the timberlands belonging to the Menominee Indians of Wisconsin since 1890.

C263 Chapman, Herman Haupt. "Evolution of the Society of American Foresters, 1934-1937, as Seen in the Memoirs of H.H. Chapman." Ed. by David Montgomery. *Forest History* 6 (Fall 1962), 2-9. During Chapman's term as president of the SAF.

C264 Chapman, Leonard B. "Mast Industry of Old Falmouth." *Collections and Proceedings of the Maine Historical Society*, Second Series 7 (1896), 390-405.

C265 Chapman, Oscar L. *A Century of Conservation, 1849-1949.* USDI, Conservation Bulletin No. 39. Washington: GPO, 1950. 35 pp. Secretary of the Interior's account of conservation in his department.

C266 Chapman, Oscar L. *Geography and Natural Resources Development in the American Way of Life.* Madison: University of Wisconsin, 1953. 16 pp.

C267 Chappell, Gordon S. *Logging Along the Denver & Rio Grande: Narrow Gauge Logging Railroads of Southwestern Colorado and Northern New Mexico.* Golden: Colorado Railroad Museum, 1971. 190 pp. Illus., maps, app., bib., index. Since the 1880s.

C268 Charters, W.W. "Paul Bunyan in 1910." *Journal of American Folklore* 57 (July-Sept. 1944), 188-89. On the written origins of the Paul Bunyan legend.

Chase, C.D. See Warner, John R., #W76

C269 Chase, Doris Harter. *They Pushed Back the Forest.* Sacramento: By the author, 1959. 78 pp. Illus., maps, bib. Logging in the redwoods, Del Norte County, California, since the mid-19th century.

C270 Chase, Edward Everett. *Maine Railroads: A History of the Development of the Maine Railroad System.* Portland: Beyer and Small, 1926. 145 pp. Diags. Includes logging railroads.

C271 Chase, Lew W. "Michigan's Upper Peninsula." *Michigan History Magazine* 20 (Autumn 1936), 313-49. Historical sketch since white settlement, including mention of lumber industry.

C272 Chase, Stuart. *Rich Land, Poor Land: A Study of Waste in the Natural Resources of America.* New York: Whittlesey House of McGraw-Hill, 1936. x + 361 pp. Illus., maps, tables, bib., index. Includes a chapter on forest devastation, with some history of the lumber industry.

C273 Chase, Warren W. "Recent Advances in Forest Game Management." *Journal of Forestry* 47 (Nov. 1949), 882-85.

C274 Chatelain, Verne E. "The Public Land Officer on the Northwestern Frontier." *Minnesota History* 12 (Dec. 1931), 379-89. From 1847 to 1862.

C275 Chatelain, Verne E. "The Federal Land Policy and Minnesota Politics, 1854-60." *Minnesota History* 22 (Sept. 1941), 227-48.

Chatham, John H. See Shoemaker, Henry W., #S241

Chatham, John H. See Walker, James Herbert, #W28

C276 Cheatham, Owen R., and Pamplin, Robert B. *The Georgia-Pacific Story.* New York: Newcomen Society of North America, 1966. 28 pp. Illus. Goergia-Pacific Corporation of Portland, Oregon, a forest products firm with operations throughout the United States and abroad.

C277 Cheatham, Owen R. "Of Timber, Men, and Money: The Georgia Pacific Story." *Growth* 8 (Dec. 1967), 4-11. The founder of Georgia-Pacific Corporation recalls its history since 1927.

C278 Cherry, Edgar. *Redwood and Lumbering in California Forests.* San Francisco: Privately printed, 1884. 107 pp. Contemporary description but includes some historical references.

C279 Chessman, G. Wallace. *Governor Theodore Roosevelt: The Albany Apprenticeship, 1898-1900.* Cambridge: Harvard University Press, 1965. ix + 335 pp. Illus., notes, bib., index. Includes some history of Roosevelt's conservation efforts as governor of New York.

C280 Chevalier, Louis Jacques George. *Les Bois d'oeuvre pendant la guerre.* Paris: Les Presses Universitaires de France; New Haven: Yale University Press, 1927. 196 pp. Includes a chapter on the forestry regiments in France.

C281 Chew, Arthur P. *The Response of Government to Agriculture: An Account of the Origin and Development of the United States Department of Agriculture on the Occasion of Its 75th Anniversary.* Washington: GPO, 1937. 108 pp.

C282 Chew, Arthur P. *The United States Department of Agriculture; Its Structure and Functions.* USDA, Miscellaneous Publication No. 88. Washington: GPO, 1940. 242 pp.

C283 Cheyney, Edward G., and Wentling, J.P. *The Farm Woodlot: A Handbook of Forestry for the Farmer and the Student in Agriculture.* 1914. Second edition. New York: Macmillan, 1926. xii + 349 pp. Illus., tables, index. A chapter on the history of forests and forestry in the United States and Canada.

C284 Cheyney, Edward G. "The Development of the Lumber Industry in Minnesota." *Journal of Geography* 14 (Feb. 1916), 189-95.

C285 Cheyney, Edward G. "The Passing of an Industry—An Epic of the Great American Forest." *American Forestry* 28 (June 1922), 323-28. An illustrated historical sketch of the lumber industry since the colonial period.

C286 Cheyney, Edward G. "History of the Minnesota Forest School." University of Minnesota Forestry Club, *Gopher Peavey* (1923), 7-9.

C287 Cheyney, Edward G. "A Phoenix of the Lumber Industry." *American Forests and Forest Life* 30 (Apr. 1924), 239-42. On the lumber town of Cloquet, Minnesota, since 1880, the forest fire that destroyed it in 1918, and its rebuilding.

C288 Cheyney, Edward G., and Levin, Oscar R. *Forestry in Minnesota.* St. Paul: Minnesota Forest Service, 1929. 55 pp. Includes some history of forested lands, forestry, and the lumber industry.

C289 Cheyney, Edward G., and Schantz-Hansen, Thorvald. *This Is Our Land: The Story of Conservation in the United States.* St. Paul: Webb, 1946. xii + 345 pp. Illus., map, bib.

C290 Cheyney, Edward G. "History of Forestry in Minnesota." *Lake States Timber Digest* 1 (July 31, 1947), 3-4, 6; 1 (Aug. 14, 1947), 9-10, 15.

C291 Cheyney, Robert K. "Industries Allied to Shipbuilding in Newburyport." *American Neptune* 17 (Apr. 1957), 114-27. Some references to wood-using crafts and industries, Newburyport, Massachusetts.

C292 *Chicago Hardwood Record.* "History of the National Hardwood Lumber Association." *Chicago Hardwood Record* 7 (Nov. 19, 1898), 26-27. On the origins of this trade association.

C293 Chicanot, E.L. "Garden Cities in Forest Wilds." *Journal of Forestry* 24 (Jan. 1926), 52-59. Description and history of company towns built by pulp and paper companies in Ontario and Quebec, 1910s-1920s.

C294 Chicanot, E.L. "The Transformation of the Canadian Prairie." *American Forests and Forest Life* 34 (Sept. 1928), 549-52. Tree planting and shelterbelt projects in the Prairie provinces encouraged by the federal government and the Canadian Pacific Railway since 1901.

C295 Chicanot, E.L. "Reforestation in Canada." *American Forests and Forest Life* 36 (May 1930), 285-88. Some history of public and private reforestation since 1900.

C296 Chinard, Gilbert. "The American Philosophical Society and the Early History of Forestry in America." *Proceedings of the American Philosophical Society* 89 (July 18, 1945), 444-48. Concerns the commercial, cultural, and scientific ideas about forests held by Americans, and especially the American Philosophical Society, in the late 18th and early 19th centuries.

C297 Chinard, Gilbert. *L'Homme contre la nature: essais d'histoire de l'Amérique.* Paris: Hermann, 1949. 179 pp. Illus., bib. Includes "La Forête américaine: l'homme et la forêt ameri caine." Previously published in English in the *Proceedings of the American Philosophical Society,* 1945 and 1947.

Chinard, Gilbert. See Academie des Sciences, Paris, #A13

C298 Chisholm, Hugh J. "History of Paper Making in Maine and the Future of the Industry." *Annual Report of the Maine Bureau of Industrial and Labor Statistics* 20 (1906), 161-69. By the founder of the Oxford Paper Company, Rumford, Maine.

C299 Chisholm, Hugh J., Jr. *A Man and the Paper Industry: Hugh J. Chisholm.* New York: Newcomen Society in North America, 1952. 28 pp. Illus. On the Oxford Paper Company, Rumford, Maine, the International Paper Company, and the hydroelectric and railroad industries in New England and New York established by Hugh Joseph Chisholm between 1887 and 1912.

C300 Chisholm, Hugh J., Jr. "A Man and the Paper Industry: Hugh J. Chisholm (1847-1912)." *Tappi* 38 (Oct. 1955), 28A-40A. Biographical sketch of the founder of the Oxford Paper Company, Rumford, Maine, the International Paper Company, and related enterprises.

C301 Chittenden, Alfred K. "The Development of Forestry in the U.S." Master's thesis, Yale Univ., 1902.

C302 Chittenden, Alfred K. *Forest Conditions of Northern New Hampshire.* USDA, Bureau of Forestry, Bulletin No. 55. Washington: GPO, 1905. 100 pp.

C303 Chittenden, Hiram Martin. *Yellowstone National Park, Historical and Descriptive.* Cincinnati: Robert Clarke Company, 1895. Fifth edition, revised by Eleanor Chittenden Cress and Isabelle F. Story. Stanford, California: Stanford University Press, 1949. xi + 286 pp. Illus., maps. On the history of the Yellowstone region, the movement to create a national park, and its history to 1895. Subsequent editions and revisions in 1915, 1924, 1933, and 1949 brought the history up to date. A descriptive section of the book made it popular also as a guide to the Wyoming park.

C304 Chittenden, Hiram Martin. *The Yellowstone National Park.* Ed. with an intro. by Richard A. Bartlett. Norman: University of Oklahoma Press, 1964. xxi + 208 pp. Illus., maps, notes, app., bib., index. This edition contains the historical section of the 1895 edition plus a biographical sketch of the author.

C305 Choate, Grover A. *The Forests of Wyoming.* Resource Bulletin, INT-2. Fort Collins, Colorado: USFS, Rocky Mountain Forest and Range Experiment Station, 1963. 47 pp.

Choate, Grover A. See Miller, Robert L., #M468

C306 Choate, Grover A. *New Mexico's Forest Resource.* Resource Bulletin, INT-5. Fort Collins, Colorado: USFS, Rocky Mountain Forest and Range Experiment Station, 1966. 58 pp.

C307 Choate, Grover A., and Spencer, John S., Jr. *Forests in South Dakota.* Resource Bulletin, INT-8. Fort Collins, Colorado: USFS, Rocky Mountain Forest and Range Experiment Station, 1969. 40 pp.

C308 Christ, J.H. "Soil Conservation Society of America." *Journal of Forestry* 44 (Dec. 1946), 1074-75. On its origins in 1943 and subsequent conservation activities.

Christen, H.E. See Devall, W.B., #D147

C309 Christensen, Thomas P. "The State Parks of Iowa." *Iowa Journal of History and Politics* 26 (July 1928), 331-414. On the movement for conservation of forests and scenery in Iowa, with a detailed account of individual park areas.

C310 Christi, Aldis J. "Forestry at Manchester Water Works—Fact or Fancy." *Journal of the New England Water Works Association* 74 (Mar. 1960), 46-56. On the tree planting and sustained-yield forestry program of the water board of Manchester, New Hampshire, to protect its water supply in Massabesic Lake, since 1913.

C311 Christie, H.R. "The Forests of Central British Columbia." *Forestry Quarterly* 13 (Dec. 1915), 495-503. Incidental history.

C312 Christie, Jean. "The Mississippi Valley Committee: Conservation and Planning in the Early New Deal." *Historian* 32 (May 1970), 449-69. On the committee's failure to achieve natural resource planning in the 1930s.

C313 Christie, Robert A. "Empire in Wood: A History of the United Brotherhood of Carpenters and Joiners of America." Ph.D. diss., Cornell Univ., 1954. 612 pp.

C314 Christie, Robert A. *Empire in Wood: A History of the Carpenter's Union.* Cornell Studies in Industrial and Labor Relations. Volume 7. Ithaca, New York: Cornell University, 1956. xvii + 356 pp. Notes, bib., index. General history of an AFL union, including its effort to organize lumber workers in the Pacific Northwest, 1881-1941.

C315 Christner, Margaret Louise. "The Ballinger Controversy: An Analysis of Political Intrigue." Master's thesis, Univ. of Kansas, 1942.

Christy, Francis T., Jr. See Potter, Neal, #P287

C316 Chriswell, F.W. "History of Railroad Logging: Early Times and Early Names Recalled." *Loggers Handbook* 9 (1949), 64-69. Emphasis on the Pacific states.

Church, Verne V. See Byrne, J.J., #B625

C317 Churchill, Edwin A. "Merchants and Commerce in Falmouth (1740-1775)." *Maine Historical Society Newsletter* 9 (May 1970), 93-104. On the lumber and mast trade centered in Falmouth, Maine.

C318 Churchill, Howard L. "An Example of Industrial Forestry in the Adirondacks." *Journal of Forestry* 27 (Jan. 1929), 23-26. As practiced since 1908 by Finch, Pruyn & Company of Glens Falls, New York.

C319 Churchill, Sam. "That Wonderful Old Fashioned Shay." *American Forests* 68 (July 1962), 16-19. On a Shay engine operated by the Klickitat Log and Lumber Company, Klickitat, Washington, since 1942, and the Klickitat Log and Lumber Railroad since 1915.

C320 Churchill, Samuel. *Big Sam.* Garden City, New York: Doubleday and Company, 1965. 184 pp. Illus. Biography of Sam Churchill (author's father), a logger with the Western Cooperage Company in Oregon's Clatsop County, 1902-1941.

C321 Ciriacy-Wantrup, Siegfried von. "Taxation and the Conservation of Resources." *Quarterly Journal of Economics* 58 (Feb. 1944), 157-95.

C322 Ciriacy-Wantrup, Siegfried von. "Administrative Coordination of Conservation Policy." *Journal of Land and Public Utility Economics* 22 (Feb. 1946), 48-58.

C323 Ciriacy-Wantrup, Siegfried von. "Resource Conservation and Economic Stability." *Quarterly Journal of Economics* 60 (May 1946), 412-52.

C324 Ciriacy-Wantrup, Siegfried von. *Resource Conservation: Economics and Policies.* Berkeley: University of California Press, 1952. Third edition. Berkeley: University of California, Division of Agricultural Sciences, Agricultural Experiment Station, 1968. 395 pp. Notes, index. Includes many historical references to the conservation movement and to individual elements of natural resource conservation.

C325 Ciriacy-Wantrup, Siegfried von. "Social Objectives of Conservation of Natural Resources with Particular Reference to Taxation of Forests." In *Taxation and Conservation of Privately Owned Timber: Proceedings of a Conference Held at the University of Oregon, January 27, 28, 1959.* Eugene: University of Oregon, Bureau of Business Research, 1959. Pp. 1-9.

C326 Ciriacy-Wantrup, Siegfried von, and Parsons, James J., eds. *Natural Resources: Quality and Quantity.* Berkeley: University of California Press, 1967.

C327 Clapp, Earle H. "The Long Haul from the Woods." *American Forestry* 29 (May 1923), 259-64, 320. Historical survey of the effect of transportation distances and costs on lumber prices.

C328 Clapp, Earle H. "The Decennial of the McSweeney-McNary Act." *Journal of Forestry* 36 (Sept. 1938), 832-36. The lead article of an issue devoted to commemorating the accomplishments in forest research made under the McSweeney-McNary Act of 1928.

C329 Clapp, Earle H. "Education and Demonstration in American Forestry." *Journal of Politics* 13 (Aug. 1951), 345-68. Since 1900.

C330 Clapp, Earle H. "Zon." *American Forests* 62 (Dec. 1956), 6, 44-46. Biographical sketch of Raphael Zon (1874-1956), who served the USFS from 1901 to 1944. He was director of the Lake States Forest Experiment Station for twenty-one of those years. Zon was also managing editor and editor-in-chief of the *Journal of Forestry* from 1917 to 1928.

C331 Clapp, Earle H. "Ramifications of Reorganization: The Land, The Sea, and The Air." *American Forests* 74 (Sept. 1968), 16-19, 44-47; 74 (Oct. 1968), 24-27, 56. Includes some history and reminiscence of efforts to reorganize federal natural resource agencies since the 1930s.

C332 Clapp, Gordon R. *The TVA: An Approach to the Development of a Region.* Charles R. Walgreen Foundation Lectures. Chicago: University of Chicago Press, 1955. xiii + 206 pp. Maps, notes, bib. A former chairman presents an inside view of administration and policy formation in the Tennessee Valley Authority since 1933.

C333 Clapperton, Robert H. *The Paper-Making Machine: Its Invention, Evolution, and Development.* Elmsford, New York: Pergamon Press, 1967. 365 pp. Illus.

C334 Clar, C. Raymond. *Forest Use in Spanish-Mexican California: A Brief History of the Economic Interest in the Forest and the Actions of Government Prior to American Statehood.* Sacramento: Division of Forestry, California Department of Natural Resources, 1957. 39 pp. Notes. Processed.

C335 Clar, C. Raymond. *Brief History of the California Division of Forestry.* Second edition. Sacramento: California State Board of Forestry, 1957. 52 pp. And its predecessor agencies since the 1860s.

C336 Clar, C. Raymond. "John Sutter, Lumberman." *Journal of Forestry* 56 (Apr. 1958), 259-65. On John August Sutter (1803-1880), emigrant to California in 1839, and logging near Coloma.

C337 Clar, C. Raymond. *California Government and Forestry from Spanish Days until the Creation of the Department of Natural Resources in 1927.* Sacramento: California Division of Forestry, 1959. xv + 623 pp. Illus., map, tables, notes. A detailed history of forestry and forest administration in California.

C338 Clar, C. Raymond. "One Thing Led to Another." *Forest History* 7 (Spring/Summer 1963), 2-9. On the author's research for his *California Government and Forestry from Spanish Days until the Creation of the Department of Natural Resources in 1927* (1959).

C339 Clar, C. Raymond. *California Government and Forestry-II: During the Young and Rolph Administrations.* Sacramento: California Division of Forestry, 1969. x + 319 pp. Illus., map, graphs, charts, notes, index. On state forestry in California from 1927 through the mid-1930s, with some topics discussed through the 1940s.

C340 Clar, C. Raymond. *Evolution of California's Wildland Fire Protection System.* Sacramento: California State Board of Forestry, 1969. 35 pp.

C341 Clar, C. Raymond. *Harvesting and Use of Lumber in Hispanic California.* Sacramento: Sacramento Corral of Westerners, 1971. 12 pp.

C342 Clar, C. Raymond. *Out of the River Mist.* 1973. Reprint. Santa Cruz, California: Forest History Society, 1974. vi + 135 pp. Illus., maps. History and reminiscences of the Russian River area and of Guerneville, an important lumber town of the California redwood region, 1860s-1930s.

C343 Clare, Warren L. "James Stevens: The Laborer and Literature." *Research Studies* 4 (Dec. 1964), 355-67. Stevens, a Washingtonian, was the author of Paul Bunyan tales and other works concerning lumberjacks and the forest industries.

C344 Clare, Warren L. "Big Jim Stevens: A Study in Pacific Northwest Literature." Ph.D. diss., Washington State Univ., 1967. 174 pp.

C345 Clare, Warren L. "The Slide-Rock Bolter, Splinter Cats and Paulski Bunyanovitch." *Idaho Yesterdays* 15 (Fall 1971), 2-8. On the career of James Stevens, writer and collector of Paul Bunyan stories.

C346 Clark, Charles E. *The Eastern Frontier: The Settlement of Northern New England, 1610-1763.* New York: Alfred A. Knopf, 1970. xxiv + 419 + xvi pp. Maps, notes, bib., index. Includes many references to the forest environment and economy.

C347 Clark, Charles E. "Beyond the Frontier: An Environmental Approach to the Early History of Northern New England." *Maine Historical Society Newsletter* 11 (Summer 1971), 5-21. Considers the forest, landscape, appreciation of nature, and "sense of place" as factors in the interpretation of northern New England's history, ca. 1763-1840.

C348 Clark, Dan E. *Samuel Jordan Kirkwood.* Iowa City: State Historical Society of Iowa, 1917. 464 pp. Notes, bib. Kirkwood was Secretary of the Interior under James A. Garfield and Chester A. Arthur, 1881-1882.

C349 Clark, Donald H. *18 Men and a Horse.* Seattle: Metropolitan Press, 1949. Reprint. Bellingham, Washington: Whatcom Museum of History and Art, 1969. xxi + 217 pp. Illus., maps. On the Bloedel-Donovan Lumber Mills and the lumber industry around Puget Sound, Washington, since 1898.

C350 Clark, Donald H. "Shake Maker." *American Forests* 56 (Apr. 1950), 18-19, 42. Includes some incidental history of cedar shake making in the Pacific Northwest.

C351 Clark, Donald H. "Sawmill on the Columbia." *Beaver* 281 (June 1950), 42-44. Operated by the Hudson's Bay Company near Ft. Vancouver (Vancouver, Washington), 1828-1847.

C352 Clark, Donald H. "Bror L. Grondal: Forest Products Specialist." *Journal of Forestry* 49 (Mar. 1951), 211. Sketch of Grondal's career as a forestry educator at the University of Washington College of Forestry, 1910s-1950.

C353 Clark, Donald H. "An Analysis of Forest Utilization as a Factor in Colonizing the Pacific Northwest and in Subsequent Population Transitions." Ph.D. diss., Univ. of Washington, 1952.

C354 Clark, Donald H. "The Yacolt Burn: Forest Graveyard." *American Forests* 60 (Sept. 1954), 18-20. On forest fires in Clark, Skamania, and Cowlitz counties of southwestern Washington in 1902, 1917, 1918, 1922, 1929, and 1949, and efforts to rehabilitate the burned area.

C355 Clark, Donald H. "First There Were Skidroads—Now Oregon Forests Lead the Nation." *Oldtimer* (Josephine County Historical Society) (1962), 3-8. Brief history of the lumber industry in southwestern Oregon.

C356 Clark, Ella E. "Forest Lookout." *National Geographic Magazine* 90 (July 1946), 73-96. Includes some history of USFS fire spotting in the West.

C357 Clark, Ella E. "Smokejumpers—The Quickest Way to the Fire." *Pacific Discovery* 4 (July-Aug. 1951), 4-14. On the use of USFS smokejumpers to fight forest fires in the Pacific Northwest and Rocky Mountain states since 1939.

C358 Clark, Fay G. "Trends in the Recreational Policy of the U.S. Forest Service." *Northwest Science* 14 (1940), 33-38.

C359 Clark, Galen. *The Big Trees of California, Their History and Characteristics.* Redondo, California: Press of Reflex Publishing Company, 1907. 104 pp. Illus.

Clark, Giles. See Smith, Mowry, Jr., #S393

C360 Clark, J.W. "Lumberjack Lingo." *American Speech* 7 (Oct. 1931), 47-53.

C361 Clark, Jack Dale. "Special Interests, Public Policy, and the Bureau of Land Management in Western Oregon: A Case History of the 1963 Cruise-Log Scale Controversy." Master's thesis, Univ. of Oregon, 1969.

C362 Clark, James I. *The Wisconsin Pineries: Logging on the Chippewa.* Chronicles of Wisconsin Series, No. 9. Madison: State Historical Society of Wisconsin, 1956. 20 pp. Bib. From 1840 to 1910.

C363 Clark, James I. *Farming the Cutover: The Settlement of Northern Wisconsin.* Chronicles of Wisconsin Series, No. 10. Madison: State Historical Society of Wisconsin, 1956. 20 pp. Bib. From 1873 to 1915.

C364 Clark, James I. *Cutover Problems: Colonization, Depression, Reforestation.* Chronicles of Wisconsin Series, No. 13. Madison: State Historical Society of Wisconsin, 1956. 20 pp. Bib.

C365 Clark, James I. *The Wisconsin Pulp and Paper Industry.* Chronicles of Wisconsin Series, No. 15. Madison: State Historical Society of Wisconsin, 1956. 20 pp. Illus., bib. A short historical survey.

C366 Clark, Norman H. "Everett, 1916, and After." *Pacific Northwest Quarterly* 57 (Apr. 1966), 57-64. On the Everett Massacre, which pitted members of the Industrial Workers of the World against a sheriff's posse of middle-class citizens, including many managers of Everett sawmills, 1916.

C367 Clark, Norman H. *Mill Town: A Social History of Everett, Washington, from Its Earliest Beginnings on the Shores of Puget Sound to the Tragic and Infamous Event Known as the Everett Massacre.* Seattle: University of Washington Press, 1970. x + 267 pp. Illus., notes, index. Everett, a major center of forest products industries, was the site of industrial violence involving Wobblies, millworkers, and mill owners in 1916.

C368 Clark, Paul Odell. "Stephen Sears Smith, Hardwood Lumber Dealer of San Francisco." *California Historical Society Quarterly* 33 (Dec. 1954), 321-28; 34 (Mar. 1955), 65-82. Includes transcripts of letters sent by Smith (1819-1907) to his relatives and friends in San Francisco and also a biographical sketch by Clark.

C369 Clark, Thomas D. "Early Lumbering Activities in Kentucky." *Northern Logger* 13 (Mar. 1965), 14-15, 42-43. A social and economic interpretation of lumbering, ranging from the earliest pioneers to the commercial era from 1870 to 1920.

C370 Clark, Thomas D. *Three Paths to Modern South: Education, Agriculture, and Conservation.* Athens: University of Georgia Press, 1965. xiii + 103 pp. Notes, bib., index. One essay includes some history of resource conservation in the South.

C371 Clark, Thomas D. "The Impact of the Timber Industry on the South." *Mississippi Quarterly* 25 (Spring 1972), 141-64. A broad survey of the social, cultural, and economic impact of the forest and forest products industries on the South since the colonial period.

C372 Clark, Victor S. "Manufacturers in the South from 1865 to 1880" and "Modern Manufacturing Development in the South, 1880-1905." In *South in the Building of the Nation,* Volume 6, ed. by J.C. Ballagh. Richmond, 1910. Pp. 253-304. Includes some account of the lumber and naval stores industries.

C373 Clark, Victor S. *History of Manufactures in the United States, 1607-1914.* 3 Volumes. New York: McGraw-Hill, for the Carnegie Institution of Washington, 1929. Reprint. Gloucester, Massachusetts: Peter Smith, 1967. Illus., maps, diags., bib. Based on the author's *History of Manufactures in the United States, 1607-1860,* first published in 1916. Includes some history of the lumber and forest product industries.

C374 Clark, William F. "National Parks Survey: The Interpretive Programs of the National Parks, Their Development, Present Status, and Reception by the Public." Ph.D. diss., Cornell Univ., 1949. 80 pp.

C375 Clark Equipment Company. *First Fifty Years, 1903-1953 of Clark Equipment; a Story of an American Manufacturing Enterprise.* Buchanan, Michigan, 1953. 75 pp. Illus. Manufacturers of lift trucks and other equipment used in the lumber industry.

C376 Clarke, Carrel Eugene. "The History and Development of the Pulp and Paper Industry in the Maritime Provinces and Newfoundland." Master's thesis, Acadia Univ., 1951. 106 pp.

C377 Clarkson, Roy B., and Carvell, K.L. "West Virginia's Logging Railroads—Its Past and Present." *Northeastern Logger* 10 (Dec. 1961), 20-21, 63.

C378 Clarkson, Roy B., and Carvell, K.L. "Logging Locomotives in the Mountain State." *Northeastern Logger* 10 (June 1962), 10-11. West Virginia.

C379 Clarkson, Roy B. *Tumult on the Mountains: Lumbering in West Virginia, 1770-1920.* Parsons, West Virginia: McClain Printing Company, 1964. 410 pp. Illus., index.

Heavily illustrated work on the lumber industry, railroad logging, and changing logging technology.

C380 Clary, David A. *'The Place Where Hell Bubbled Up': A History of the First National Park*. Washington: U.S. National Park Service, 1972. 68 pp. Illus., map. Yellowstone National Park, Wyoming.

C381 Clawson, Marion. "The Administration of Federal Range Lands." *Quarterly Journal of Economics* 53 (1939), 435-53. Incidental history.

C382 Clawson, Marion. *The Western Range Livestock Industry*. American Forestry Series. New York: McGraw-Hill, 1950. xii + 401 pp. Illus., maps, notes. Includes some history of range and grazing policies on forested lands.

C383 Clawson, Marion. "Administration of Federal Lands in the Public Interest." *Journal of Politics* 13 (Aug. 1951), 441-60. Director of the Bureau of Land Management reviews the history, uses, and problems of the public lands, especially in their administration.

C384 Clawson, Marion. *Uncle Sam's Acres*. New York: Dodd, Mead, and Company, 1951. xvi + 414 pp. Illus., maps, bib. History of land acquisition and disposal, federal land management agencies, and the policies and politics of public lands since 1781.

Clawson, Marion. See Penny, J. Russell, #P110

C385 Clawson, Marion, and Held, R. Burnell. *The Federal Lands: Their Use and Management*. Baltimore: Johns Hopkins Press, for Resources for the Future, 1957. Reprint. Lincoln: University of Nebraska Press, 1965. xxi + 501 pp. Illus., diags., maps, tables, notes, index. The history, uses, policies, and management of federal lands since 1781.

C386 Clawson, Marion. "Reminiscences of the Bureau of Land Management, 1947-1948." *Agricultural History* 33 (Jan. 1959), 22-28. By the director of the BLM (1948-1953), with some history of the General Land Office and the administration of public lands by the USDI since the 1890s.

C387 Clawson, Marion. "The Crisis in Outdoor Recreation." *American Forests* 65 (Mar. 1959), 22-31, 40-41; (Apr. 1959), 28-35, 61-62. Includes some history of outdoor recreation in the national forests and national parks in the 20th century.

C388 Clawson, Marion; Held, R. Burnell; and Stoddard, Charles H. *Land for the Future*. Baltimore: Johns Hopkins Press, for Resources for the Future, 1960. xix + 570 pp. Maps, diags., tables, notes. Includes some historical account of changes in land use in the United States.

C389 Clawson, Marion. *Land for Americans: Trends, Prospects, and Problems*. Chicago: Rand McNally Company, for Resources for the Future, 1963. 141 pp. Illus., maps. Incidental history of public land ownership. This is a summary of *Land for the Future*.

C390 Clawson, Marion. *Land and Water for Recreation: Opportunities, Problems, and Policies*. Chicago: Rand McNally Company, for Resources for the Future, 1963. ix + 144 pp. Illus.

C391 Clawson, Marion. *Man and Land in the United States*. Lincoln: University of Nebraska Press, 1964. ix + 178 pp. Illus., maps, diags., graphs, index. A brief history of land use in the United States, including sections on the use of forested lands.

Clawson, Marion. See Held, R. Burnell, #H282

C392 Clawson, Marion, and Stewart, Charles L. *Land Use Information: A Critical Survey of U.S. Statistics, Including Possibilities for Greater Uniformity*. Baltimore: Johns Hopkins Press, for Resources for the Future, 1966. xvii + 402 pp. Apps., bib., index. Includes some history of land use information and of its use by government agencies in establishing policy.

C393 Clawson, Marion, and Knetsch, Jack L. *Economics of Outdoor Recreation*. Baltimore: Johns Hopkins Press, for Resources for the Future, 1966. xx + 328 pp. Illus., notes. Incidental history.

C394 Clawson, Marion. *The Federal Lands Since 1956: Recent Trends in Use and Management*. Washington: Resources for the Future, 1967. xi + 113 pp. Tables, graphs. A supplement to *The Federal Lands: Their Use and Management* (1957). Indicates that uses of federal lands increased much more rapidly than receipts, especially in the area of outdoor recreation.

C395 Clawson, Marion. *The Land System of the United States: An Introduction to the History and Practice of Land Use and Land Tenure*. Lincoln: University of Nebraska Press, 1968. ix + 145 pp. Illus., maps, tables, graphs. Based on *Man and Land in the United States* (1964), this work is intended as an introduction to the American land system for foreign readers. Includes history of the public domain and of federal land disposal.

C396 Clawson, Marion. *The Bureau of Land Management*. Praeger Library of U.S. Departments and Agencies, No. 27. New York: Praeger Publishers, 1971. xiii + 209 pp. Illus., maps, tables, apps., bib., index. Includes some history of the federal lands and especially of the Bureau of Land Management and its predecessor agencies within the USDI, the General Land Office and Grazing Service. There is a chapter on forestry.

C397 Clawson, Marion. *America's Land and Its Uses*. Baltimore: Johns Hopkins University Press, for Resources for the Future, 1972. x + 166 pp. Illus., bib. Based largely on *Land for the Future*. Includes some history of land policy.

Clawson, Marion. See Seaton, Fred A., #S157

C398 Clayson, Edward. *Historical Narrative of Puget Sound, Hoods Canal, 1865-1885: The Experience of an Only Free Man in a Penal Colony*. Seattle: R.L. Davis Printing Company, 1911. 'On Seabeck, Washington, a lumber town operated by the Washington Mill Company.

C399 Clayton, A.G. "A Brief History of the Washakie National Forest and the Duties and Some Experiences of a Ranger." *Annals of Wyoming* 4 (Oct. 1926), 277-95. Includes some history of tie operations and tie drives on the Wind River since 1914. The Washakie is now part of the Shoshone National Forest, Wyoming.

C400 Cleaveland, Frederic Neill. "Federal Reclamation Policy and Administration: A Case Study in the Development of Natural Resources." Ph.D. diss., Princeton Univ., 1951. 316 pp.

Clegg, Charles. See Beebe, Lucius M., #B169

C401 Cleland, Robert G. *A History of Phelps-Dodge, 1834-1950.* New York: Alfred A. Knopf, 1952. xiv + 307 + xxii pp. Illus., map, tables, notes. Known primarily as a copper mining firm, the Phelps-Dodge Corporation also had interest in the lumber industry.

C402 Clement, G.E. "The Manufacture of Slack Cooperage Staves in the U.S." Master's thesis, Yale Univ., 1902.

C403 Clemente, Ricardo Ama. "Interregional Competition in the Pulpwood Industry of the United States and Canada." Ph.D. diss., Pennsylvania State Univ., 1970. 160 pp.

C404 Clements, F.E. *The Life History of Lodgepole Burn Forests.* USFS, Bulletin No. 79. Washington: GPO, 1910. 56 pp. On forest fires at Estes Park, Colorado, since the 18th century.

C405 Clendening, Carl H. "Early Days in the Southern Appalachians." *Southern Lumberman* 144 (Dec. 15, 1931), 101-05. Reminiscences of logging, log driving, and sawmill operations in West Virginia and elsewhere in the southern Appalachian Mountains, 1880s and 1890s.

C406 Clendening, Carl. "Looking Back: An Appalachian Veteran Recalls Early Days in West Virginia and the State's First Band Sawmill." *Southern Lumberman* 195 (Dec. 15, 1957), 232-37. Reminiscences of James C. West.

C407 Clepper, Henry. "In Penn's Woods: Where State Forestry Serves the Public." *American Forests* 41 (June 1935), 269-71, 301-02. Includes some history of state forests in Pennsylvania since the 1890s.

C408 Clepper, Henry. "The Mont Alto State Forest." *Journal of Forestry* 34 (Jan. 1936), 30-35. Of Pennsylvania, since 1902.

C409 Clepper, Henry. "The Genesis of the *Journal of Forestry.*" *Journal of Forestry* 39 (Dec. 1941), 971-72. On the predecessors of *Journal of Forestry—Forestry Quarterly,* founded in 1902, and the *Proceedings of the Society of American Foresters,* first issued in 1905. The two were amalgamated in 1917.

C410 Clepper, Henry. "A Quarter-Century of Service." *Journal of Forestry* 40 (Jan. 1942), 1-2. Editors and editorial policies of the *Journal of Forestry* since 1917.

C411 Clepper, Henry. "Herbert A. Smith, 1866-1944." *Journal of Forestry* 42 (Sept. 1944), 625-27. Sketch of Smith's career in the USFS, where from 1901 to the 1930s he was an advisor to successive chiefs, and as editor-in-chief of the *Journal of Forestry*, 1935-1937.

C412 Clepper, Henry. "Rise of the Forest Conservation Movement in Pennsylvania." *Pennsylvania History* 12 (July 1945), 200-16. With emphasis on early forest legislation, Joseph T. Rothrock and the Pennsylvania Forestry Association, state forestry commissions, state forests, and forestry education.

C413 Clepper, Henry. "George H. Wirt, First Forester of Pennsylvania." *Journal of Forestry* 43 (Sept. 1945), 687-88. Biographical sketch of Wirt, appointed state forester in 1901.

C414 Clepper, Henry. "A Man Named Smith." *Journal of Forestry* 46 (July 1948), 531-32. On Harry Frederick Smith (1895-1947), inspector for the Alabama Commission of Forestry, 1925-1931, and later with the statistical staff of the USFS, Southern Forest Experiment Station, New Orleans, Louisiana.

Clepper, Henry. See Ball, Howard E., #B43

C415 Clepper, Henry. "The 10 Most Influential Men in American Forestry." *American Forests* 56 (May 1950), 10-11, 30, 37-39. Brief evaluations of J. Sterling Morton, Bernhard E. Fernow, Joseph T. Rothrock, Filibert Roth, Gifford Pinchot, Theodore Roosevelt, Henry S. Graves, William B. Greeley, Ovid Butler, and Franklin D. Roosevelt.

C416 Clepper, Henry. "Forestry's First Fifty Years." *Scientific Monthly* 71 (Dec. 1950), 387-92.

C417 Clepper, Henry. "T. Edward Shaw." *Journal of Forestry* 49 (May 1951), 365-66. On Shaw's career in state forestry, extension forestry, and forestry education, Pennsylvania and Indiana, 1921-1950.

C418 Clepper, Henry. "Professional Education in Forestry." *Higher Education* 8 (Oct. 15, 1951), 37-41. Since 1898.

C419 Clepper, Henry. "The *Journal of Forestry*: An Historical Summary of the First Fifty Years." *Journal of Forestry* 50 (Dec. 1952), 899-912. On *Forestry Quarterly, Journal of Forestry*, their editors, and their relations with the Society of American Foresters since 1902.

C420 Clepper, Henry. "The Conservation Association." *American Forests* 59 (Jan. 1953), 16-17, 36-38. Brief sketches of many organizations formed to conserve natural resources since the 1870s.

C421 Clepper, Henry. "William S. Swingler." *Journal of Forestry* 51 (Mar. 1953), 207-08. On his work as a forester for Pennsylvania and with the USFS, 1921-1950s.

C422 Clepper, Henry. "The Pennsylvania State Forest School." *Pennsylvania Forests* 38 (Spring 1953), 28-40. Established as the Pennsylvania State Forest Academy in 1903 at Mont Alto. The article covers the period from 1903 to 1929.

C423 Clepper, Henry. "Our State Forestry Associations." *American Forests* 60 (Jan. 1954), 30-34. Brief historical sketches of forestry associations in the following states: Colorado, Connecticut, Florida, Georgia, Louisiana, Massachusetts, Mississippi, New Hampshire, North Carolina, Ohio, Pennsylvania, Texas, Virginia, and West Virginia.

C424 Clepper, Henry. "The Ten Most Important Events in American Forestry." *American Forests* 61 (Oct. 1955), 52-53, 100-03. Since 1875.

C425 Clepper, Henry. "Early Forestry Instruction at Penn State." *Pennsylvania Forests* 36 (Winter 1956), 4-7. From 1889 to 1907.

C426 Clepper, Henry. "Forestry Education in America." *Journal of Forestry* 54 (July 1956), 455-57. Since 1898.

C427 Clepper, Henry. "Clarence S. Herr, Industrial Forester." *Journal of Forestry* 54 (Dec. 1956), 848-49. On Herr's career with the USFS and in state and industrial forestry in the Northeast.

C428 Clepper, Henry. "Women in Conservation." *American Forests* 62 (Dec. 1956), 20-22, 52-53. Since 1875.

C429 Clepper, Henry, ed. *Forestry Education in Pennsylvania*. University Park, Pennsylvania: Penn State—Mont Alto Forestry Alumni Association, 1957. 269 pp. Illus., map, tables. Forestry education in Pennsylvania since the 1870s, with particular attention to the Pennsylvania State Forest Academy, Mont Alto, 1903-1929, and the School of Forestry of Pennsylvania State University. Includes a directory of alumni.

C430 Clepper, Henry. "The Cooperative Role of the Society of American Foresters." *Journal of Forestry* 56 (Sept. 1958), 666-70, 674-76. On the cordial relations of the Society with schools, governmental agencies, international agencies, and other professional organizations, since 1913.

C431 Clepper, Henry. "Foresters in Uniform." *American Forests* 64 (Nov. 1958), 20-23, 47-48. On the origins and development of federal and state foresters' uniforms since 1908.

C432 Clepper, Henry. "Pennsylvania's Forestry Heritage." *American Forests* 65 (Oct. 1959), 15, 78-82. On the forestry movement in Pennsylvania since the 1870s, with emphasis on the contributions of Joseph Trimble Rothrock.

C433 Clepper, Henry. "Ovid Butler (1880-1960)." *Journal of Forestry* 58 (Apr. 1960), 317-18. Butler began his career with the USFS in 1907 and served the American Forestry Association after 1923 as forester, executive secretary, editor, and executive director emeritus. He wrote several books and was a major publicist of the forestry and conservation movements.

Clepper, Henry. See Hosmer, Ralph S., #H531

C434 Clepper, Henry, and Meyer, Arthur B., eds. *American Forestry: Six Decades of Growth.* Washington: Society of American Foresters, 1960. x + 319 pp. Apps., index. A collection of historical essays by specialists in various aspects of professional forestry, including education, research, literature, silviculture, forest management, forest utilization, range management, wildlife management, watershed management, recreation, and trade and professional associations. Forestry in the USDA, USDI, and other federal agencies is surveyed, as well as developments in state, industrial, and farm forestry.

C435 Clepper, Henry. "Chiefs of the Forest Service." *Journal of Forestry* 59 (Nov. 1961), 795-803. Brief sketches of Franklin B. Hough, Nathaniel H. Egleston, Bernhard E. Fernow, Gifford Pinchot, Henry S. Graves, William B. Greeley, Robert Y. Stuart, Ferdinand A. Silcox, Earle H. Clapp, Lyle F. Watts, and Richard E. McArdle.

C436 Clepper, Henry. "The Allagash: Wilderness in Controversy." *Journal of Forestry* 60 (Nov. 1962), 777-81. On the movement to preserve the Allagash region of Maine as a national recreation area, with incidental historical references to the region from the 1850s.

C437 Clepper, Henry, ed. *Careers in Conservation: Opportunities in Natural Resources.* New York: Ronald Press, for the Natural Resources Council of America. 1963. v + 141 pp. Illus., apps., index. Essays by various contributors on educational and career opportunities in resource fields such as forestry, wildlife management, range management, parks and recreational development, etc.; includes many historical references.

C438 Clepper, Henry. "Pennsylvania Forestry's Ten Most Influential Men." *Pennsylvania Forests* 54 (Fall 1964), 56-58. Brief evaluations of Joseph T. Rothrock, George H. Wirt, Robert S. Conklin, John A. Ferguson, Edwin A. Ziegler, Joseph S. Illick, Gifford Pinchot, R. Lynn Emerick, Arthur C. McIntyre, and Maurice K. Goddard.

C439 Clepper, Henry. "Gifford Pinchot and the SAF." *Journal of Forestry* 63 (Aug. 1965), 590-92. On Pinchot as a founder of the Society of American Foresters in 1900 and his subsequent services and relations with the organization.

C440 Clepper, Henry, and Meyer, Arthur B. *The World of the Forest.* Prepared in cooperation with the Society of American Foresters. Boston: D.C. Heath, 1965. vi + 122 pp. Illus., index. Contains some incidental historical references to forested lands and forestry in Canada and the United States.

C441 Clepper, Henry. "World Forestry Under FAO." *Journal of Forestry* 64 (Jan. 1966), 5-9. Some history of forestry activities in the Food and Agriculture Organization of the United Nations since 1945.

C442 Clepper, Henry, ed. *Origins of American Conservation.* New York: Ronald Press, for the Natural Resources Council of America. x + 193 pp. Illus., bib., index. Collection of essays by several contributors surveys the sweep of organized conservation, including forestry, parks, wildlife, ranges, soil, etc.

C443 Clepper, Henry. "Plunder of the Pineries." *Journal of Forestry* 65 (Feb. 1967), 114-19. On efforts to halt timber trespass in Michigan, especially by U.S. Timber Agent Isaac W. Willard, 1850s.

C444 Clepper, Henry. "Tree Farming in America." *Unasylva* 21 (No. 2, 1967), 2-8. Includes some history of tree farming since ca. 1940, with earlier references to forest conservation.

C445 Clepper, Henry. "Forestry Education in Pennsylvania: Its Origins and Historical Highlights." *Pennsylvania Forests* 57 (Spring 1967), 28-30.

C446 Clepper, Henry. "Conservation's Grand Lodge." *American Forests* 73 (Oct. 1967), 22-27, 58-61. On the origins and history of the Natural Resources Council of America, a federation of conservation organizations, since 1944.

C447 Clepper, Henry. "The Forest Service Backlashed." *Forest History* 11 (Jan. 1968), 6-15. On USFS Chief Henry S. Graves and problems resulting from the Ballinger-Pinchot controversy, 1910s.

C448 Clepper, Henry. "The First White House Conference on Natural Resources: May 13-15, 1908." *American Forests* 74 (May 1968), 28-31, 50-51.

C449 Clepper, Henry. "Industrial Forestry in the Northeastern States." *Northern Logger and Timber Processor* 17 (Nov. 1968), 9-10, 32-36; 17 (Dec. 1968), 16-17, 66-67, 90; 17 (Jan. 1969), 14-15, 34; 17 (Feb. 1969), 10-11, 32. Industrial foresters and forestry since the 1890s.

C450 Clepper, Henry. "Forestry's Uncertain Beginning." *Journal of Forestry* 67 (Apr. 1969), 218-21. Origins of federal forestry and forest reserves, 1870s-1890s.

C451 Clepper, Henry. "Industrial Forestry in the South." *Forest Farmer* 28 (June 1969), 12-14; 28 (Aug. 1969), 13-16. In the 20th century.

C452 Clepper, Henry. "First Wings Over the Forest." *American Forests* 75 (June 1969), 24-27, 53-55; 75 (July 1969), 20-23, 47-48. On aerial forest fire patrol flights by the USFS and by state forestry agencies since the 1910s. The second part of the article is subtitled "Smokejumpers: The Corps d'Elite," and deals with the origins of parachute dropping forest fire fighters in the 1930s.

C453 Clepper, Henry. "The 75th Anniversary of State Forestry in Pennsylvania." *Pennsylvania Forests* 60 (Spring 1970), 10-12.

C454 Clepper, Henry. "A Century of Fish Conservation." *American Forests* 76 (Nov. 1970), 16-19, 54-56. On the American Fisheries Society and its predecessor organization (American Fish Culturists' Association) since 1870.

C455 Clepper, Henry. *Professional Forestry in the United States.* Baltimore: Johns Hopkins University Press, for Resources for the Future, 1971. ix + 337 pp. App., notes, bib., index. A topical history of forestry in the United States at the federal, state, and industrial levels.

C456 Clepper, Henry, ed. *Leaders of American Conservation.* New York: Ronald Press, 1971. vii + 353 pp. Index of contributors. Alphabetically arranged biographical sketches of ca. 300 leading conservationists in American history.

C457 Clepper, Henry. "History and Forest Farming." *Forest Farmer* 31 (May 1972), 12-14, 18. On private and industrial forestry, forest farming, and the work of the Forest History Society.

C458 Clepper, Henry. "The Romance of Forest History." *American Forests* 78 (Dec. 1972), 20-23. On the work of the Forest History Society and its publication, *Forest History.*

C459 Clepper, Henry. "The Birth of the C.C.C." *American Forests* 79 (Mar. 1973), 8-11. The origins of the Civilian Conservation Corps and the roles of Franklin D. Roosevelt and Ovid M. Butler, 1932-1933.

C460 Clepper, Henry. "A Salute to Samuel T. Dana at 90." *Journal of Forestry* 71 (Apr. 1973), 200-02. Biographical sketch and appreciation of Dana, a versatile forestry educator and frequent contributor of forest history.

C461 Clepper, Henry. "The Genesis of Federal Forestry." *Journal of Forestry* 71 (Aug. 1973), 486-87. Its origins with Franklin B. Hough in the 1870s.

C462 Clepper, Henry. "Homage to 'Pennsylvania Trees.'" *Pennsylvania Forests* (Dec. 1973), 92-93. Career sketch of Joseph S. Illick (1884-1967), author of *Pennsylvania Trees,* Pennsylvania state forester, 1927-1931, and professor and dean at the New York State College of Forestry, 1932-1952.

C463 Clepper, Henry. "*American Forests:* Magazine of Record in Conservation." *American Forests* 80 (Apr. 1974), 25-56. History of the oldest continuously published conservation magazine in the United States, including much material on prominent issues, contributors, editors, and relations with the American Forestry Association, since 1897.

C464 Clepper, Henry. "A Foundering Journal Rescued." *Journal of Forestry* 72 (May 1974), 291. Brief history of *Forestry Quarterly,* 1902-1917, predecessor to the *Journal of Forestry.*

C465 Clepper, Henry. "Who Was Robert Douglas?" *Journal of Forest History* 19 (Jan. 1975), 22-23. Douglas (1813-1897), an Illinois nurseryman and horticulturist, was a founder of the American Forestry Association and served briefly (one hour) as president of the conservation organization in 1875.

C466 Cleveland, George A. "Pine Tree Fringed Penobscot." *Sprague's Journal of Maine History* 9 (Jan.-Mar. 1921), 13-14. Maine.

C467 Cleveland, Treadwell. "Forest Law in the United States." *Forester* 6 (July 1900), 153-60; 6 (Aug. 1900), 183-86; 6 (Sept. 1900), 210-12; 6 (Oct. 1900), 238-40. A general history of forestry legislation.

C468 Cleveland, Treadwell. "The Forest Laws of New York." *Forester* 7 (Apr. 1901), 81-85. A historical summary.

C469 Cleveland, Treadwell. *What Forestry Has Done.* USFS, Circular No. 140. Washington: GPO, 1908. 31 pp. Brief history.

C470 Cleveland, Treadwell. *The Status of Forestry in the U.S.* USFS, Circular No. 167. Washington: GPO, 1909. 39 pp. Includes some history.

C471 Clevinger, Woodrow R. "The Appalachian Mountaineers in the Upper Cowlitz Basin." *Pacific Northwest Quarterly* 29 (Apr. 1938), 115-34. Many immigrants from Appalachia sought work in Washington's lumber industry.

C472 Clevinger, Woodrow R. "Southern Appalachian Highlanders in Western Washington." *Pacific Northwest Quarterly* 33 (Jan. 1942), 3-25. Many came to work in the lumber industry and/or found other familiar forest occupations.

C473 Clevinger, Woodrow R. "The Western Washington Cascades: A Study of Migration and Mountain Settlement." Ph.D. diss., Univ. of Washington, 1955. 380 pp.

C474 Cliff, Edward P. "Changes in the Status of Wildlife and Its Habitat in the Northwest." *University of Washington Forest Club Quarterly* 9 (Spring 1936), 25-30. Historical sketch of wildlife based on diaries and other records.

C475 Cliff, Edward P. "The National Forests Serve." *Journal of Forestry* 53 (Feb. 1955), 112-15. On the benefits of national forests since 1905, especially those of watershed, timber, range forage, and recreation.

C476 Cliff, Edward P. *Timber: The Renewable Material.* Prepared for the National Commission on Materials Policy. Washington: GPO, 1973. 151 pp. Illus. A report on the nation's wood supply, including some history of forest resources and USFS forest policies.

Clifford, J. Nelson. See Huffington, Paul, #H603

C477 Clifton, Francis H. "The Importance of the Forests to Arkansas Economy." Master's thesis, Yale Univ., 1948.

C478 Cline, Albert C. *The Virgin Upland Forest of Central New England.* Harvard Forest Bulletin No. 21. Petersham, Massachusetts: Harvard Forest, 1942. 58 pp.

Cline, Albert C. See Lutz, Harold J., #L325

C479 Cline, Bernard G. "Wood in the Automobile Industry." Master's thesis, Yale Univ., 1948.

C480 Cline, McGarvey. "An Interview with McGarvey Cline." OHI by Donald G. Coleman. *American Forests* 68 (May 1962), 18-19, 40-44. On Cline's career in forestry and as the first director of the USFS's Forest Products Laboratory, Madison, Wisconsin, since 1900.

C481 Clinton, Ruth H. "Evolution of a Lumberman." *Niagara Frontier* 14 (Autumn 1967). On Henry P. Smith, 1836-1846.

C482 Clise, James William. *Personal Memoirs, 1855-1932.* N.p., 1935. 56 pp. Includes some of the author's early experiences in the lumber business in Colorado.

C483 Clowse, Converse D. *Economic Beginnings in Colonial South Carolina: 1670-1730.* Columbia: University of South Carolina Press, 1971. ix + 283 pp. Map, tables, notes, app., bib., index. Includes some account of the naval stores industry.

C484 Clyne, J.V. *'What's Past is Prologue': The History of MacMillan, Bloedel and Powell River, Limited.* New York: Newcomen Society in North America, 1965. 28 pp. Illus. A forest products firm based in Vancouver, British Columbia.

C485 Coate, Charles E. "Water, Power, and Politics in the Central Valley Project, 1933-1967." Ph.D. diss., Univ. of California, Berkeley, 1969. 242 pp. Resource development in California's Central Valley.

C486 Cobb, B.F. "Evolution of the Lumber Delivery Wagon in Boston." *Lumber World Review* (Jan. 10, 1912), 30-31.

Cobb, Francis E. See Wilson, Robert, #W360

C487 Cobb, Francis E. "A Study of Afforestation in the Great Plains Region from Its Early Settlement to the Present Time." Master's thesis, Cornell Univ., 1925.

C488 Cobb, S.H. *The Story of the Palatines.* London: G.P. Putnam, 1897. ix + 319 pp. Maps. Includes some history of an English naval stores experiment in New York involving German immigrants, ca. 1710.

C489 Cobb, Samuel S. "Development of Aircraft Use for Fire Control in Pennsylvania." *Journal of Forestry* 65 (June 1967), 394-97. Since the 1930s.

C490 Cobb, Samuel S. "The Development of Forest Conservation in Pennsylvania." *Journal of Forestry* 66 (Sept. 1968), 662-65. Especially since the 1880s.

C491 Cobbe, Thomas James. "Secondary Forest Successions of Clermont, Brown, and Adams Counties in Southwestern Ohio." Ph.D. diss., Univ. of Michigan, 1953. 88 pp. Includes review of the history of soils and forests in this area since ca. 1800.

C492 Coberly, A.C. *Life in a Lumber Yard.* Kansas City, Missouri: Retail Lumberman, 1918. 144 pp. Illus., diags. Anecdotal reminiscences of a retail lumberman.

C493 Cochrell, Albert N. *A History of the Nezperce National Forest.* 1960. Third edition, revised by Gayle Hauger and Morris Reynolds. Missoula, Montana: USFS, Northern Region, 1970. iii + 129 pp. Illus., map, tables, notes. Idaho.

C494 Coffin, David Linwood. *The History of the Dexter Corporation, 1767-1967.* New York: Newcomen Society in North America, 1967. 24 pp. Manufacturer of paper products, Windsor Locks, Connecticut.

Coffin, Tris. See Kerr, Robert S., #K116

C495 Coffman, John D. "Forestry in the Department of the Interior." *Journal of Forestry* 39 (Feb. 1941), 84-91. A historical account of forestry and forest conservation in the USDI and its agencies since the mid-19th century.

C496 Coffman, Lotus D., and Associates. *Land Utilization in Minnesota—A State Program for Cut-Over Lands.* Minneapolis: University of Minnesota Press, 1934. 289 pp. Includes some history of land utilization in northeastern Minnesota leading to the critical problems of cutover lands.

C497 Coggeshall, Robert W. "A Survey of the Efforts to Reorganize the Federal Wildlife and Forestry Agencies." Master's thesis, George Washington Univ., 1964. 132 pp.

C498 Cohen, Michael Peter. "The Pathless Way: Style and Rhetoric in the Writings of John Muir." Ph.D. diss., Univ. of California, Irvine, 1973. Muir (1838-1914), the California preservationist, was a prolific writer and wilderness philosopher.

C499 Cohn, Edwin J., Jr. *Industry in the Pacific Northwest and the Location Theory.* New York: King's Crown Press, 1954. x + 214 pp. Includes a chapter on the forest products industries.

C500 Colburn, William H. "7 Million Acres and All Yours." *Michigan Conservation* 25 (Jan. 1956), 19-23. On state and federal lands in Michigan since 1893.

C501 Colby, William E. "William Frederic Badè." *Sierra Club Bulletin* 22 (Feb. 1937), 19-28. A California preservationist and biographer of John Muir.

C502 Colby, William E. "Yosemite and the Sierra Club." *Sierra Club Bulletin* 23 (Apr. 1938), 11-19. On John Muir and the recession of Yosemite Valley from California to the United States, to be made part of Yosemite National Park, ca. 1890-1906.

Colcord, Mahlon J. See Walker, James Herbert, #W28

C503 Cole, Arthur H. *Wholesale Commodity Prices in the United States, 1700-1861.* Cambridge: Harvard University, for the International Scientific Committee on Price History, 1938. Reprint. New York: Johnson Reprint Corporation, 1969. xxiii + 187 pp. Tables, diags. Largely statistical; includes monthly prices for lumber, naval stores, and other forest products in Boston, Philadelphia, Charleston, New Orleans, and Cincinnati.

C504 Cole, Arthur H. "The Mystery of Fuel Wood Marketing in the United States." *Business History Review* 44 (Autumn 1970), 339-59. An exploratory essay on the declining use of wood for fuel.

C505 Cole, Douglas, and Tippett, Maria. "Pleasing Diversity and Sublime Desolation: The 18th-Century British Perception of the Northwest Coast." *Pacific Northwest Quarterly* 65 (Jan. 1974), 1-7. On the preference of maritime explorers for the pastoral scenery of the Puget Sound area as opposed to the forested wilderness of the British Columbia coast.

C506 Cole, Douglas. "Early Artistic Perceptions of the British Columbia Forest." *Journal of Forest History* 18 (Oct. 1974), 128-31. On the esthetic prejudices against evergreen forests by 18th- and 19th-century artists.

C507 Cole, Fred C. "The Texas Career of Thomas Affleck." Ph.D. diss., Louisiana State Univ., 1942. 488 pp. Affleck, a scientific agriculturist and horticulturalist, operated a sawmill and woodworking mill in Washington County, Texas, 1859-1876.

Cole, John F. See Burns, Paul Y., #B580

C508 Cole, Lela. "The Early Tie Industry along the Niangua River." *Missouri Historical Review* 48 (Apr. 1954), 264-72. On the raft transportation of railroad crossties in Camden County, Missouri, 1870s-1930.

C509 Cole, R.O. "A Chronological History of the Soil Conservation Service and Related Events." *Proceedings of the Indiana Academy of Science* 66 (1957), 291-96. Since 1907.

C510 Cole, W.E. "Impact of the T.V.A. on the Southeast." *Social Forces* 28 (May 1950), 435-40. Including wildlife and forestry development of the Tennessee Valley Authority.

C511 Coleman, Bevley R. "A History of State Parks in Tennessee." Ph.D. diss., George Peabody College for Teachers, 1963. 430 pp. The Tennessee Valley Authority played a large role in the development of a state park system.

Coleman, Donald G. See Gerry, Eloise, #G94

Coleman, Donald G. See Cline, McGarvey, #C480

Coleman, Donald G. See Tiemann, Harry D., #T105, T106

C512 Coles, Harry L., Jr. "A History of the Administration of Federal Land Policies and Land Tenure in Louisiana, 1803-1860." Ph.D. diss., Vanderbilt Univ., 1949. 407 pp.

C513 Coles, Harry L. "Applicability of the Public Land System to Louisiana." *Mississippi Valley Historical Review* 43 (June 1956), 39-58.

Colles, George Wetmore. See Deckert, E., #D112

C514 Colleta, Paolo E. *The Presidency of William Howard Taft.* American Presidency Series. Lawrence: University Press of Kansas, 1973. ix + 306 pp. Notes, essay on sources, index. Includes a chapter on conservation policies and conflicts, including the Ballinger-Pinchot affair of 1910 and an interpretation of its background.

C515 Colley, Reginald H., technical director. *'Economy in Action': Postwar Trends in the Production and Consumption of Treated Wood in the United States, 1945-1953.* New York: Bernuth, Lembcke Company, 1955. 42 pp. Illus., tables.

C516 Collier, George W. *Soil Conservation During the War.* USDA, War Records Monograph No. 2. Washington: USDA, 1946. 25 pp.

C517 Collier, Gerald L. "The Evolving East Texas Woodland." Ph.D. diss., Univ. of Nebraska, 1964. 444 pp. On the development of logging and the forest products industry in the Southern pine region—and its cultural impact.

C518 Collier, H.L. "Historical Review of the Growing Use of Creosoted Wood Block Paving." *Pacific Lumber Trade Journal* 19 (No. 1, 1913), 22-23, 51.

C519 Collier, John M. "The Southern Pine Story." *Forests & People* 13 (First quarter, 1963), 42-45, 116-17. On the lumber industry in Louisiana since the early 19th century, with reference to the work of the Southern Pine Association since 1914.

C520 Collier, John M. *The First Fifty Years of the Southern Pine Association, 1915-1965.* New Orleans: Southern Pine Association, 1965. 173 pp. Illus. Includes much history of the lumber industry in the South, as well as of the trade association.

C521 Collingwood, George Harris. *Farm Forestry Extension: Early Development, and Status in 1923.* USDA, Circular No. 345. Washington: GPO, 1925. 14 pp.

C522 Collingwood, George Harris. "Filibert Roth—An Appreciation." *American Forests and Forest Life* 32 (Jan. 1926), 43-44. Roth (1858-1925) was a prominent forestry educator at the University of Michigan and elsewhere.

C523 Collingwood, George Harris. "The Elements of the National Forestry Program." *Journal of Forestry* 37 (Feb. 1939), 83-87. Includes some review of forest resources and forestry since the 19th century.

C524 Collingwood, George Harris. "Sincerely Yours, Harris: Being the Selected Letters of George Harris Collingwood to Miss Jean Cummings, written by the Young Ranger in 1914 and 1915 while stationed on the Apache National Forest in Arizona." Ed. by Joseph A. Miller and Judith C. Rudnicki. *Forest History* 12 (Jan. 1969), 10-29.

C525 Collingwood, George Harris. "The Lost Identity of Doyle and Scribner." *Journal of Forestry* 50 (Dec. 1952), 943-44. On Edward Doyle and J.M. Scribner, 19th-century authors of formulae for measuring the board foot content of logs.

C526 Collins, A.E. "Forest Resources of Northern British Columbia: A Preliminary Reconnaissance." *Empire Forestry Review* 32 (Mar. 1953), 21-27. Some historical references.

C527 Collins, B.C. "Land Use Conflict in the North Cascades Wilderness of Washington State." Master's thesis, Univ. of Michigan, 1962.

C528 Collins, Chapin. "The First Ten Years." *American Forests* 64 (May 1958), 24-27, 49-51. On the social and economic impact since 1946 to residents of Shelton and McCleary, Washington, of the sustained-yield contract between the USFS and Simpson Logging Company under Public Law 273.

C529 Collins, Chapin. "25 Years Later: The Circle That Works." *American Forests* 77 (Oct. 1971), 15-19, 58-60. History of Simpson Timber Company logging operations near Shelton and McCleary, Washington, especially since passage of Public Law 273 of 1944, providing for sustained-yield forestry by Simpson on their own lands and Olympic National Forest lands.

C530 Collins, George L., and Sumner, Lowell. "Northeast Arctic: The Last Great Wilderness." *Sierra Club Bulletin* 38 (Oct. 1953), 13-26. On a partly forested area in northeastern Alaska and northwestern Yukon Territory since 1826, proposed for "perpetual preservation" as a wilderness.

C531 Collins, Hubert E. "Edwin Williams, Engineer: An Account of the First Steam Saw-Mill Installed and Operated in Western Oklahoma." *Chronicles of Oklahoma* 10 (Sept. 1932), 331-347. Established in 1870 at the Cheyenne and Arapaho Agency at Darlington; includes some account of subsequent logging operations in nearby stands of cottonwood.

C532 Collins, James. *Life in a Lumber Camp.* Alpena, Michigan: Alpena News Publishing Company, 1914. 20 pp.

C533 Collins, Robert F. "Daniel Boone National Forest: Historic Sites." *Filson Club History Quarterly* 42 (Jan. 1968), 26-48. Includes references to the charcoal iron industry of 19th-century Kentucky.

C534 Collins, Truman W. "Evolution of Logging Trucks." *Timberman* 40 (Aug. 1939), 15-16, 18. In the Pacific Northwest since 1926.

C535 Collins, Walter G. "How Safety Came to Lumbering." *American Forests* 61 (May 1955), 20-23, 58-59. On provisions for the protection of working loggers since 1910.

C536 Collins, Walter G. "Steam Schooners of the Redwood Coast." *American Forests* 62 (Aug. 1956), 20-21, 46-48. On shipment of redwood lumber by steamer from the northern California coast, 1852-1912.

C537 Collins, S.W., Company. *Lumbering Then and Now: S. W. Collins Co. 1844-1959.* Caribou, Maine, 1960.

C538 Collister, L.C. "Fiftieth Anniversary of the Santa Fe's Experiment with Eucalyptus." *Cross Tie Bulletin* 39 (Feb. 1958), 9-10. On tree planting experiments intended to supply the railroad with timber, since 1908.

C539 Colton, L.J. "Early Day Timber Cutting along the Upper Bear River." *Utah Historical Quarterly* 25 (Summer 1967), 202-08. On the construction of a flume near Evanston, Wyoming, 1872-1885. Includes general remarks on the history of the lumber industry in Utah, Wyoming, and Colorado.

C540 Colton, Louis A. "Paper Manufacturing on the Pacific Coast: A Bit of History." *Pacific Pulp and Paper Industry* 6 (Apr. 1932), 42-45, 83.

C541 Coman, Edwin Truman, and Gibbs, Helen M. *Time, Tide, and Timber: A Century of Pope & Talbot.* Stanford Business Series, No. 7. Stanford, California: Stanford University Press, 1949. xvi + 480 pp. Illus., diags., notes, bib., index. Business history of the lumber and shipping firm founded by Andrew Jackson Pope and Frederic Talbot at San Francisco, 1849-1850, with principal operations on Puget Sound, Washington.

C542 Comer, F.G. "The Role of the Pulp and Paper Industry in Alabama's Piedmont." *Journal of the Alabama Academy of Science* 31 (Apr. 1960), 253-57.

C543 Commons, John R.; Saposs, David J.; Sumner, Helen L.; Mittelman, E.B.; Hoagland, H.E.; Andrews, John B.; Perlman, Selig; and Taft, Philip. *History of Labour in the United States.* 4 Volumes. New York: Macmillan, 1918-1935. Volume 4, by Perlman and Taft, covers labor movements since 1896, including mention of workers in the lumber and forest products industries.

C544 Comp, Frances, ed. "Deer Park, Maryland." *Glades Star* 2 (Sept. 1951), 97-112. Includes a biographical sketch of West Virginia Senator Henry G. Davis (1823-1902) and some account of his development of timber resources in the upper Potomac Valley of Maryland and West Virginia.

C545 Compton, Wilson M. "Recent Tendencies in the Reform of Forest Taxation." *Journal of Political Economics* 23 (Dec. 1915), 971-79.

C546 Compton, Wilson M. *The Organization of the Lumber Industry, with Special Reference to the Influences Determining the Prices of Lumber in the United States.* Chicago: American Lumberman, 1916. x + 153 pp. Tables, diags., notes, apps. A pioneering economic analysis, including much history of the lumber industry and prices since the 1860s. Issued as a doctoral dissertation at Princeton University, 1915.

C547 Compton, Wilson M. "The Price Problem in the Lumber Industry." *American Economic Review* 7 (Sept. 1917), 582-97. Includes some history.

C548 Compton, Wilson M. "Lumber—An Old Industry—and the New Competition." *Harvard Business Review* 10 (1932), 161-69. History of the development of lumber mills and industrial organizations as a background to study of contemporary problems.

C549 Compton, Wilson M. "Recent Developments in the Lumber Industry." *Journal of Forestry* 30 (Apr. 1932), 440-50.

C550 Compton, Wilson M. "Forestry Under a Free Enterprise System." *American Forests* 66 (Aug. 1960), 26-27, 50-54. A review of the laws and policies regulating forestry in the United States since 1920.

C551 Compton, Wilson M. "Looking Ahead from Behind at American Forestry." *Southern Lumberman* (Dec. 15, 1960), 123-27. Landmark events in the history of American forestry.

C552 Condit, Carl W. *American Building: Materials and Techniques from the First Colonial Settlements to the Present.* Chicago: University of Chicago Press, 1968. xiv + 329 pp. Illus., bib., index. Includes some history of the earliest log, timber, and frame construction.

C553 Condrell, William K. "How Has Taxation Affected the Growth of the Forest Products Industries?" In *First National Colloquium on the History of the Forest Products Industries, Proceedings,* ed. by Elwood R. Maunder and Margaret G.

Davidson. New Haven: Forest History Society, 1967. Pp. 144-63. Since 1909.

C554 Condron, H. David. "The Knapheide Wagon Company, 1848-1943." *Journal of Economic History* 3 (May 1943), 32-41. This firm manufactured logging vehicles, especially trucks, in Quincy, Illinois.

C555 Condry, William M. *Thoreau.* Great Naturalists Series. American edition. New York: Philosophical Library, 1954. 114 pp. Illus., bib. On the lifelong interest of Henry David Thoreau (1817-1862) in natural history.

C556 Congressional Quarterly Service. "Natural Resources." In *Congress and the Nation, 1945-64.* Washington: Congressional Quarterly Service, 1965. Pp. 771-1109. This chapter considers natural resources in detail, including topics such as soil conservation, public lands, forestry programs, fish and wildlife conservation, and national parks and recreation. Each topic includes a chronology of legislation, statistics, program activities, political participants, and controversies.

C557 Conklin, Charles Franklin. "A Study of the Economic Development of the Northern Panhandle of West Virginia." Ph.D. diss., Univ. of Pittsburgh, 1959. 304 pp. From 1749 to 1940; includes the lumber industry.

C558 Conklin, Edwin P. "Logging on Puget Sound, as Illustrated in the Lives of Sol Simpson and Mark E. Reed." *Americana* 29 (1935), 256-83. The Simpson Logging Company, Shelton, Washington, since the late 19th century.

C559 Conklin, Edwin P. "A Saga of Pulp and Paper Making." *Americana* 35 (Apr. 1941), 351-80. A general history of papermaking since 1840, the development of the wood pulp industry, and the westward movement of the industry.

C560 Conlin, Joseph R. "The Wobblies: A Study of the Industrial Workers of the World before World War I." Ph.D. diss., Univ. of Wisconsin, 1966. 435 pp. A general history of the radical union which had a significant following among lumberjacks and millworkers in the Pacific Northwest.

C561 Conlin, Joseph R. *Big Bill Haywood and the Radical Union Movement.* Men and Movements Series. Syracuse: Syracuse University Press, 1969. xii + 244 pp. Illus., notes, bib., note, index. One of the organizers and leaders of the Industrial Workers of the World in the 1910s, Haywood was associated with lumberjacks and millworkers in the South and the Pacific Northwest.

C562 Conlin, Joseph R. *Bread and Roses Too: Studies of the Wobblies.* Contributions in American History, No. 1. Westport, Connecticut: Greenwood Press, 1969. xvi + 166 pp. Notes, note on sources, index. Essays on the Industrial Workers of the World.

C563 Connaughton, Charles A. *Forestry in Mid-Century.* Portland: By the author, 1973. 120 pp. Processed. Autobiography of a career forester with the USFS. Connaughton (b. 1908) served as president of the American Forestry Association and the Society of American Foresters.

C564 Connecticut. State Forester. *Forest Fires in Connecticut, 1910-1922 (Inclusive).* Hartford, 1923. 28 pp. Illus.

C565 Connecticut. State Park and Forest Commission. *Connecticut State Parks: A Report on Their Growth, Cost, and Use, 1914-1931.* Hartford, 1932. 23 pp.

C566 Connery, Robert H. *Governmental Problems in Wildlife Conservation.* Columbia University Studies in History, Economics and Public Law, No. 411. New York: Columbia University Press, 1935. 251 pp. Includes a legislative history of wildlife conservation in the United States.

C567 Connolly, Frank A. "Lumber Organization Activity in the Half-Century." *Southern Lumberman* 144 (Dec. 15, 1931), 107-08. General history of trade associations.

C568 Connor, Mary Roddis. *A Century with Connor Timber: Connor Forest Industries, 1872-1972.* Wausau, Wisconsin: Privately published, 1972. 158 pp. Illus., app., bib. Company history of a firm with timberlands in Michigan and Wisconsin and sawmills and other wood-using plants in Wisconsin.

C569 Connor, Seymour V. "Log Cabins in Texas." *Southwestern Historical Quarterly* 53 (Oct. 1949), 105-16. On construction practices after 1835.

C570 Connor Lumber and Land Company. *Connor Forest Products Since 1872.* Marshfield, Wisconsin: Connor Lumber and Land Company, 1947. 36 pp. Illus.

C571 Conover, Milton. *The General Land Office: Its History, Activities and Organization.* Institute for Government Research, Service Monographs of the United States Government, No. 13. Baltimore: Johns Hopkins Press, 1923. Reprint. New York: AMS Press, 1973. xii + 224 pp. Bib. Since 1812.

C572 Conroy, William Brown. "The Changing Recreational Geography of the Adirondack Mountain Area." D.S.S. diss., Syracuse Univ., 1963. 290 pp. A history of the recreational use of Adirondack State Park, New York.

C573 *Conservation Volunteer.* "A History of Natural Resources: The Centennial Story." *Conservation Volunteer* 20 (Jan. through Dec. 1957). A six-part series on Minnesota.

C574 Cook, A.B. "The Evolution of the Maple Sugar Industry." *Report of the Michigan Forestry Commission, 1902* (1903), 70-75. In Shiawassee County.

C575 Cook, Charles W.; Folsom, David E.; and Peterson, William. *The Valley of the Upper Yellowstone.* Ed. and with an intro. by Aubrey L. Haines. American Exploration and Travel Series, No. 47. Norman: University of Oklahoma Press, 1965. xxxii + 79 pp. Illus., maps, bib. An account of exploration in the Yellowstone region, 1869.

C576 Cook, Harold Oatman. *The Forests of Worcester County.* Boston: Office of Massachusetts State Forester, 1917. 88 pp. Results of a forest survey and some historical references to the local lumber industry.

C577 Cook, Harold Oatman. *Fifty Years a Forester.* Boston: Massachusetts Forest and Park Association, 1961. 63 pp. Illus. Autobiographical account including history of state forestry and the state forest system of Massachusetts since 1907.

C578 Cook, Richard C. "The Forgotten Industry in the Tar Heel State." *Naval Stores Review* 77 (June 1967), 8-9; 77 (July 1967), 8-9. History of the naval stores industry in North Carolina since the 18th century.

C579 Cook, Richard C. "Wood Naval Stores—Early Industry Accounts Through 1920." *Naval Stores Review* 77 (Aug. 1967), 6-8. On the development of distillation processes in the naval stores industry.

C580 Cook, Rufus George. "A Study of the Political Career of Weldon Brinton Heyburn through His First Term in the U.S. Senate, 1852-1909." Master's thesis, Univ. of Idaho, 1964.

C581 Cook, Rufus George. "Senator Heyburn's War Against the Forest Service." *Idaho Yesterdays* 14 (Winter 1970), 12-15. Weldon B. Heyburn of Idaho was a spokesman for the opponents of Gifford Pinchot and the USFS, 1902-1912.

C582 Cookson, Edwin. "One Camp That John Barleycorn Didn't Rule." Ed. by John I. Bellaire. *Michigan History Magazine* 27 (Winter 1943), 51-57. Recollections of a logging foreman about the operations of the Chicago Lumbering Company near Seney on Michigan's Upper Peninsula.

C583 Cool, Robert A.; Kielbaso, J. James; and Myers, Wayne L. "A Survey of Forestry Activities of Michigan Cities." *Michigan Academician* 6 (Fall 1973), 223-32.

C584 Cooley, Richard A. "State Land Policy in Alaska: Progress and Prospects." *Natural Resources Journal* 4 (Jan. 1965), 455-67. Problems attendant to Alaska's legal right, since statehood, to transfer large amounts of federal lands to state ownership for use or resale.

C585 Cooley, Richard A. *Alaska, a Challenge in Conservation.* Madison: University of Wisconsin Press, 1966. xv + 170 pp. Illus., maps, tables, notes, bib., index. Includes some history of land disposition and policy and other aspects of resource conservation.

C586 Cooley, Richard A., and Wandesforde-Smith, Geoffrey, eds. *Congress and the Environment.* Seattle: University of Washington Press, 1970. xix + 277 pp. Maps, notes, bib., index. A collection of essays dealing with specific environmental problems and attempts to solve them by congressional legislation. Legislation treated includes the Wild and Scenic Rivers Act of 1968, the Highway Beautification Act of 1965, the Land and Water Conservation Act of 1965, the Wilderness Act of 1964, and the acts to create the Redwoods and North Cascades national parks.

C587 Coolidge, Edwin H. "An Old-Time Sawmill at Sterling, Massachusetts." *Old-Time New England* 29 (1938), 5-8. Reminiscences since 1867.

C588 Coolidge, Philip T. "Beginnings of Mechanical Traction in the Northeast." *Northeastern Logger* 4 (Apr. 1956).

C589 Coolidge, Philip T. "Consulting Forestry in the Old Days." *Northeastern Logger* 7 (Apr. 1959), 22-23, 38. Experiences since 1916 in cruising, mapping, and surveying Maine timberlands.

C590 Coolidge, Philip T. "Colorado Forestry Fifty Years Ago." *Colorado Magazine* 38 (July 1961), 188-94. Reminiscences of work for the USFS on national forests of Wyoming and Colorado, 1906-1912, and of forestry education at Colorado College.

C591 Coolidge, Philip T. *History of the Maine Woods.* Bangor, Maine: Furbush-Roberts Printing Company, 1963. xvi + 806 pp. Illus., maps, tables, apps., bib., index. On the history since colonial times of forests, forestry, and forest industries.

C592 Coolidge, Philip T. *Park Holland: Revolutionary Soldier, Maine Surveyor.* Bangor, Maine: Furbush-Roberts Printing Company, 1967. 32 pp. Illus., map. Brief biographical sketch of Holland (1752-1844), Maine surveyor and probable originator of the Holland log rule.

C593 Coombs, Charles Ira. *High Timber, the Story of American Forestry.* Cleveland: World Publishing Company, 1960. 223 pp. Illus. A general history covering many aspects of forestry in the United States.

C594 Coombs, Whitney. *Wages of Unskilled Labor in Manufacturing Industries in the United States, 1890-1924.* New York: Columbia University Press, 1926. 163 pp. Tables, diags. Includes forest industries.

C595 Coon, Shirley Jay. "The Economic Development of Missoula, Montana." Ph.D. diss., Univ. of Chicago, 1926. 420 pp. A center of the lumber industry and of USFS regional administration.

C596 Cooper, Charles F. "Multiple Land Use on the Salt River Watershed, Arizona." *Journal of Forestry* 57 (Oct. 1959), 729-34. Since 1900.

C597 Cooper, Charles F. "Vegetation Changes in Southwestern Pine Forests Since White Settlement." Ph.D. diss., Duke Univ., 1959. Arizona and New Mexico.

C598 Cooper, Charles F. "The Ecology of Fire." *Scientific American* 204 (1961), 150-56. A historical treatment.

Cooper, E.N. See Wollenberg, R.P., #W422

C599 Cooper, Ellwood. *Forest Culture and Eucalyptus Trees.* San Francisco: Cubery and Company, 1876. 237 pp. Includes some history of the 19th-century crusade to plant eucalyptus trees in California.

C600 Cooper, William E. "Virginia Stands Out in Practice of Good Forestry." In *The Story of Virginia: A Quarter Century of Progress, 1924-1949.* Richmond: Virginia State Chamber of Commerce, 1949. 112 pp. Illus. Also appeared in *Commonwealth* 16 (June 1949).

C601 Cooper, William Skinner. *A Contribution to the History of the Glacier Bay National Monument.* Minneapolis: Department of Botany, University of Minnesota, 1956. 36 pp. Illus., bib. On the author's work as an ecologist and his participation in the establishment of Glacier Bay National Monument, Alaska, 1914-1937.

C602 Cope, Joshua A. "Ten Years of Farm Forestry on a New York Farm." *Journal of Forestry* 41 (Mar. 1943), 169-73.

C603 Cope, Joshua A. *Farm Forestry in Eastern United States.* Washington: Charles Lathrop Pack Forestry Foundation, 1943. 40 pp. Illus., tables, maps. Incidental history.

C604 Copes, Parzival. "The Place of Forestry in the Economy of Newfoundland." *Forestry Chronicle* 36 (Dec. 1960), 330-41.

C605 Cordes, O.C. "Development of the Electric Log Carriage Drive." *Timberman* 28 (July 1927), 146-47.

C606 Corey, A.B. *Crisis of 1832-1842 in Canadian-American Relations.* Relations of Canada and the United States Series. Toronto: Ryerson Press; New Haven: Yale University Press, 1941. xi + 203 pp. Illus., maps. Includes an account of the Maine-New Brunswick boundary dispute, which involved in part rivalries among lumbermen over forest resources.

C607 Corle, Edwin. *The Story of the Grand Canyon.* New York: Duell, Sloan, and Pearce, 1951. viii + 312 pp. Illus., maps. Published in 1946 as *Listen, Bright Angel.* Includes some history of Arizona's Grand Canyon National Park.

Cornell, William B. See Glover, John G., #G158

C608 Corning, Howard McKinley. "The Lookout's Mechanical Eye." *American Forests* 58 (Mar. 1952), 18-19, 51. The Osborne fire finder, invented by William Bushnell Osborne, and its use in Oregon since 1910.

C609 Cornwall, George F. "Deep River Camp: Pioneer Columbia River Logging Operation." *Timberman* 35 (May 1934), 13-14. In Washington, since 1897.

C610 Cornwall, George F. "History of Lumbering in Western Nevada." *Timberman* 42 (June 1941), 11-14, 50-62. On the production of mine timbers between 1853 and 1914; also some description of log driving on rivers and the use of flumes.

C611 Cornwall, George F. "Dealers in Billions: Pacific Lumber Inspection Bureau." *Timberman* 43 (May 1942), 18-32. A historical sketch of the bureau, including biographical material on its officers and inspectors.

C612 Cornwall, George F. "History Research Uncovers Forgotten Log Rule." *Timberman* 46 (June 1945), 59-60.

C613 Cornwall, George F. "Lumber and Gold." *Timberman* 49 (Feb. 1948). On John Sutter's sawmill and the discovery of gold in California, 1848.

C614 Cornwall, George F. "The First Half Century." *Timberman* 50 (Oct. 1949), 50-53, 78. Historical sketch of this Portland trade journal since 1899, by the editor and son of founder.

C615 Cornwall, George M. "The Logging Industry in Retrospect and Prospect." *Timberman* 27 (Nov. 1925), 50-53. Logging technology in the Pacific Northwest since the 1880s.

C616 Cornwall, George M. "The Lumber Industry and Port Development." *Timberman* 27 (Oct. 1926), 42-43. Relationship between two on Pacific Coast.

C617 Cornwall, George M. "Christening Pacific Coast Woods." *Timberman* 28 (Nov. 1926), 87, 90-92. Historical circumstances in the naming of five species.

C618 Cornwall, George M. "The Men In Our Industry." *Timberman* 34 (Nov. 1932), 8, 41-42, 44. Includes biographical material on prominent lumbermen of the Pacific Coast states.

C619 Cornwall, George M. "Pioneer Loggers of the Lower Columbia River." *Timberman* 35 (Nov. 1933), 36-43. In Oregon and Washington since the 1880s.

C620 Cornwall, George M. "The Passing of a Stalwart Lumberman." *Timberman* 35 (Jan. 1934), 62. Biographical sketch of Pacific Northwest lumberman Andrew B. Hammond (1848-1934).

C621 Cornwall, George M. "Logging Clatsop County: Oregon Coast Region Illustrated Evolution in Logging Methods." *Timberman* 37 (Sept. 1936), 18-24.

C622 Cornwall, George M. "Fifty Years: Three Generations. History of St. Paul and Tacoma Lumber Company." *Timberman* 39 (May 1938), 16-30. Of Tacoma, Washington.

C623 Cornwall, George M. *Founding of the Pacific Logging Congress.* Portland: Ivy Press, 1939. 21 pp. On its origins in Seattle in 1909, by one of the founders.

C624 Cornwall, George M. "Colorful Career of Simon Benson." *Timberman* (Dec. 1940), 46-48. Sketch of Portland lumberman whose logging operations were located on the lower Columbia River of Washington and Oregon.

C625 Cortese, Jim. "Memphis: The Hardwood Capital." *Southern Lumberman* 193 (Dec. 15, 1956), 137-39. Since 1859.

C626 Cory, Floyd W. "The Influence of Canadian Competition on the Pulp-Wood Industry of the Pacific Coast." Master's thesis, Univ. of Washington, 1927.

C627 Cosbey, Robert C. "John Muir." Ph.D. diss., Ohio State Univ., 1949. 307 pp. On the California preservationist, wilderness philosopher, and founder of the Sierra Club.

C628 Cosgrove, George F. "On Sawmills, Past, Present, Future." *Hardwood Record* 61 (Nov. 1926), 46-50, 56.

C629 Cote, P. Emile. "Reminiscences of Lumbering in Quebec: Ever Since the First Sawmill Was Established in 1646, the Development of Quebec's Forests Has Become the Province's Major Industry." *Canada Lumberman* 60 (Aug. 1, 1940), 66-67.

C630 Cotroneo, Ross R. "The History of the Northern Pacific Land Grant, 1900-1952." Ph.D. diss., Univ. of Idaho, 1967. 482 pp. Included vast stands of timber in Montana, Idaho, and Washington, much of which was sold to Frederick Weyerhaeuser and other lumbermen.

C631 Cotroneo, Ross R. "Western Land Marketing by the Northern Pacific Railway." *Pacific Historical Review* 27 (Aug. 1968), 299-320. Includes account of the sale of timberlands in Oregon, Washington, and Idaho, especially to lumberman Frederick Weyerhaeuser, ca. 1900-1920.

C632 Cottam, Walter P., and Stewart, George. "Plant Succession as a Result of Grazing and of Meadow Desiccation by Erosion Since Settlement in 1862." *Journal of Forestry* 38 (Aug. 1940), 613-26. On ecological changes at Mountain Meadows, Washington County, Utah, since 1862.

C633 Cottell, Philip L. *Occupational Choice and Employment Stability Among Forest Workers.* Bulletin No. 82. New Haven: Yale University School of Forestry and Environmental Studies, 1974. xiv + 161 pp. A study of forest industry workers of British Columbia's northern interior region.

C634 Cotterill, Ralph S. "The National Land System in the South: 1803-1812." *Mississippi Valley Historical Review* 16 (Mar. 1930), 495-506.

C635 Coufal, James E. "The School the Students Built." *American Forests* 68 (May 1962), 34-35, 53-54. Short history of the New York Ranger School at Wanakena in the Adirondack Mountains, founded in 1912.

C636 Coulter, E. Merton. "The Okefenokee Swamp, Its History and Legends." *Georgia Historical Quarterly* 48 (No. 2,

1964), 166-92; 48 (No. 3, 1964), 291-312. Includes some history of lumbering in the Georgia swamp, ca. 1900, and of later conservation efforts.

C637 Cour, Robert M. *The Plywood Age: A History of the Fir Plywood Industry's First Fifty Years.* Portland: Binfords and Mort, for the Douglas Fir Plywood Association, 1955. 171 pp. Illus. On the manufacture of plywood from Douglas fir in the Pacific Northwest, especially since 1905.

Court, Arnold. See Bruce, David, #B516

C638 Courtenay, John. "Texas National Forests Are 40 Years Old." *Texas Forestry* 14 (Apr. 1974), 18-19. Some history of Sam Houston, Davy Crockett, Angelina, and Sabine national forests in Texas, since the 1930s.

C639 Coville, Frederick V. *Forest Growth and Sheep Grazing in the Cascade Mountains of Oregon.* USDA, Division of Forestry, Bulletin No. 15. Washington: GPO, 1898. 54 pp. Some history of controversy between sheepmen and forest conservationists.

C640 Cowan, Charles S. "An Interview with Charles S. Cowan: Forest Protection Comes under the Microscope." OHI by Elwood R. Maunder. *Forest History* 2 (Winter 1959), 3-14. On Cowan's career in forest protection and fire prevention in Washington and British Columbia, 1913-1950s.

C641 Cowan, Charles S. *The Enemy Is Fire.* Seattle: Superior Publishing Company, 1961. 135 + [24] pp. Illus., app., graphs, index. A history of 20th-century forest fires and forest protection in Washington, especially the role of the Washington Forest Fire Association.

C642 Cowan, Fred W. "Logging the Connecticut Headwaters." *Northeastern Logger* 3 (Apr. 1955), 8-9, 50. Some history of logging in northern Vermont and New Hampshire lands by the Connecticut Valley Lumber Company and the St. Regis Paper Company.

C643 Cowen, William F. "Forestry Operations on the Scituate Reservoir Watershed." *Journal of the New England Water Works Association* 67 (Mar. 1953), 1-8. Providence County, Rhode Island, since 1925.

C644 Cowlin, Robert W. "The Wholesale Middleman in the Lumber Industry." Master's thesis, Univ. of California, Berkeley, 1928.

C645 Cowlin, Robert W.; Briegleb, Philip A.; and Moravets, F.L. *Forest Resources of the Ponderosa Pine Region of Washington and Oregon.* USDA, Miscellaneous Publication No. 490. Washington: GPO, 1942. 99 pp.

Cowlin, Robert W. See Andrews, H.J., #A446

C646 Cox, Herbert J. "On the Other Side of Certain Public Lands." *Journal of Forestry* 43 (May 1945), 315-21. Includes some administrative and legal history of the revested Oregon and California Railroad grant lands in western Oregon since 1916. See rejoinder by Samuel Trask Dana, pp. 321-23.

C647 Cox, Herbert J. *Random Lengths: Forty Years with 'Timber Beasts' and 'Sawdust Savages.'* Eugene, Oregon: Shelton-Turnbull-Fuller Company, 1949. 310 pp. Illus. On the author's life in Oregon since 1901 and his activities as a lumberman.

C648 Cox, John H. "Organizations of the Lumber Industry in the Pacific Northwest, 1889-1914." Ph.D. diss., Univ. of California, Berkeley, 1937. 265 pp. Especially trade associations.

C649 Cox, John H. "Trade Associations in the Lumber Industry of the Pacific Northwest, 1899-1914." *Pacific Northwest Quarterly* 41 (Oct. 1950), 285-311.

Cox, Laurie D. See Hoyle, Raymond J., #H580

C650 Cox, Thomas R. "Lower Columbia Lumber Industry, 1880-93." *Oregon Historical Quarterly* 67 (June 1966), 160-78. On lumbermen and the lumber industry on the lower Columbia River of Oregon and Washington; the author argues that the arrival of transcontinental railroads had relatively little impact on marketing patterns.

C651 Cox, Thomas R. "The Crusade to Save Oregon's Scenery." *Pacific Historical Review* 37 (May 1968), 179-99. On the efforts of Robert W. Sawyer and other preservationists to halt despoliation of Oregon's natural beauty, especially along highways, 1919-1922. These efforts laid the groundwork for the state park system.

C652 Cox, Thomas R. "Sails and Sawmills: The Pacific Lumber Trade to 1900." Ph.D. diss., Univ. of Oregon, 1969. 536 pp.

C653 Cox, Thomas R. "The Passage to India Revisited: Asian Trade and the Development of the Far West, 1850-1900." In *Reflections of Western Historians,* ed. by John A. Carroll. Tucson: University of Arizona Press, 1969. Pp. 85-103. Includes some history of the Pacific lumber trade.

C654 Cox, Thomas R. "Lumber and Ships: The Business Empire of Asa Mead Simpson." *Forest History* 14 (July 1970), 16-26. His lumber and shipping operations in California, Oregon, and Washington, 1850-1915.

C655 Cox, Thomas R. "Pacific Log Rafts in Economic Perspective." *Forest History* 15 (July 1971), 18-19. Case study of the Benson Lumber Company's rafting operations, Columbia River to San Diego, California, 1906-1940s.

C656 Cox, Thomas R. "Conservation by Subterfuge: Robert W. Sawyer and the Birth of the Oregon State Parks." *Pacific Northwest Quarterly* 64 (Jan. 1973), 21-29. State parks and the scenery preservation movement in Oregon developed from the administrative initiative of Sawyer and other members of the Oregon State Highway Commission in the 1920s.

C657 Cox, Thomas R. *Mills and Markets: A History of the Pacific Coast Lumber Industry to 1900.* Seattle: University of Washington Press, 1974. xx + 332 pp. Illus., maps, charts, tables, notes, bib., index. On the 19th-century development of the lumber industry in British Columbia, Washington, Oregon, and California, with attention to exports, markets, transportation, logging and mill technology, and efforts to organize the industry.

C658 Cox, Thomas R. "William Kyle & the Pacific Lumber Trade: A Study in Marginality." *Journal of Forest History* 19 (Jan. 1975), 4-14. Kyle owned a sawmill at Florence, Oregon, and operated a coastal lumber schooner. His relative lack of success in the 1890s illustrates the business problems of small-time operators, especially on the inferior harbors of the Pacific Coast.

C659 Cox, W.K. "Forty Years of Tractor Logging." *Southern Lumberman* 144 (Dec. 15, 1931), 118-21. The development and use of Holt and Best tractors since 1893.

C660 Cox, William T. "When Forest Fires Began." *American Forests and Forest Life* 34 (Aug. 1928), 477-79. Argues that the era of large forest fires began with the extermination of the beaver.

C661 Coy, Owen C. "The Settlement and Development of the Humboldt Bay Region, 1850-1875." Ph.D. diss., Univ. of California, Berkeley, 1918.

C662 Coy, Owen C. *The Humboldt Bay Region, 1850-1875: A Study in the American Colonization of California.* Los Angeles: California State Historical Association, 1929. xiii + 346 pp. Illus., maps, notes, bib., index. Includes three chapters on the lumber and other forest industries in a region comprising Humboldt, Del Norte, Trinity, and parts of Siskiyou and Mendocino counties in northwestern California.

C663 Coyle, David Cushman. *Our Forests.* Washington: National Home Library Foundation, 1940. 150 pp. Concerned with federal forest conservation efforts, including some history of the movement.

C664 Coyle, David Cushman. "The Attack on Soil Conservation." *Colorado Quarterly* 3 (Autumn 1954), 202-14. On the Soil Conservation Service since 1933.

C665 Coyle, David Cushman. *Conservation: An American Story of Conflict and Accomplishment.* New Brunswick, New Jersey: Rutgers University Press, 1957. xii + 284 pp. Illus., tables, index. A general history of the conservation movement in the United States in the 20th century. Much of the book concerns the forestry movement and the politics of conservation, but there are chapters on the conservation of soil, water, minerals, wildlife, and other natural resources.

C666 Coyne, Franklin E. *The Development of the Cooperage Industry in the United States, 1620-1940.* Chicago: Lumber Buyers Publishing, 1940. 112 pp. Illus.

C667 Cozzens, A.B. "Conservation in German Settlements of the Missouri Ozarks." *Geographical Review* 33 (Apr. 1943), 286-98. Includes some history of forest and wildlife conservation practices.

C668 Crabtree, Kay. "Remember . . . Reminiscing on Progress in Fine Paper Manufacturing." *Pulp and Paper Magazine of Canada* 43 (Sept. 1942), 746-47.

C669 Craddock, George Washington. "Floods Controlled on Davis County Watersheds." *Journal of Forestry* 58 (Apr. 1960), 291-93. USFS efforts to prevent floods by controlling fire in forested watershed, regulating grazing, cutting contour trenches, and seeding perennial grasses, Davis County, Utah, 1923-1958.

C670 Crafts, Edward C. "A Forest Industry and Its Dependent Population." Master's thesis, Univ. of Michigan, 1933.

C671 Crafts, Edward C. "Some Effects of Defense on Wood Utilization in California." *Journal of Forestry* 49 (Apr. 1942), 285-90. Summarizes trends within the forest products industries since 1920, with emphasis on the influence of defense industries.

C672 Crafts, Edward C., and Bollaert, R. *Some Social and Economic Effects of Timber Utilization and Management in Modoc County, California.* Berkeley: USFS, California Forest and Range Experiment Station, 1942. 41 pp.

C673 Crafts, Edward C. "Public Forest Policy in a National Emergency." *Journal of Forestry* 50 (Apr. 1952), 266-70. Concerns the World War II and post-war years, 1941-1951.

C674 Crafts, Edward C. "Congressional Liaison in the Forest Service." OHI by Susan R. Schrepfer. *Forest History* 16 (Oct. 1972), 12-17. As assistant chief for the USFS Division of Program Planning and Legislation, Crafts campaigned for regulation of private logging and for multiple-use forestry, 1950-1962.

C675 Crafts, Edward C. *Forest Service Researcher and Congressional Liaison: An Eye to Multiple Use.* OHI by Susan R. Schrepfer. Santa Cruz, California: Forest History Society, 1972. xiii + 188 pp. Illus., apps., bib., index. Processed. On his career in the USFS, 1932-1962, especially in research in the Southwest, in the Division of Forest Economics, as assistant chief, and his role in securing passage of the Multiple Use-Sustained Yield Act of 1960. Crafts was also director of the U.S. Bureau of Outdoor Recreation, 1962-1969.

Crafts, Eleanor Bait. See Wiswall, Clarence A., #W400

C676 Craig, Douglass A. "The Federal Assistance Role in Private Forestry." *Southern Lumberman* 227 (Dec. 15, 1973), 131-33. Includes some account of cooperative forestry since 1911.

C677 Craig, James B. "Blueprint for Public Service—the Story of Grandfather Mountain." *American Forests* 54 (May 1948), 200-04, 236-37. On the activities of the USFS and the National Park Service in the Grandfather Mountain-Linville Gorge Area of western North Carolina, since 1918.

C678 Craig, James B. "The Miracle of Muskingum." *American Forests* 55 (July 1949), 18-23, 37, 46. On the Muskingum Watershed Conservancy District of Ohio, since 1933.

C679 Craig, James B. "Seaboard's Green Gold." *Railroad Magazine* 57 (Apr. 1952), 62-69. On development of the woodpulp industry by the Seaboard Air Line Railroad since 1936.

C680 Craig, James B. "Muskingum Revisited." *American Forests* 60 (June 1954), 7-13, 36-39. On the Muskingum Watershed Conservancy District, Ohio, since 1933.

C681 Craig, James B. "Forestry's Ambassador without Portfolio." *American Forests* 66 (May 1960), 20-21, 41-42. On Thomas Harvey Gill (1891-1972), his studies at Yale School of Forestry, his work in the USFS, 1915-1925, his service to the Lathrop Pack Forestry Fund since 1926, and his promotion of forestry and conservation education throughout the world.

C682 Craig, James T. "Muskegon and the Great Chicago Fire." *Michigan History Magazine* 28 (Oct.-Dec. 1944), 610-33. Muskegon provided lumber for the rebuilding of Chicago.

C683 Craig, Mary Eleanor. "Recent History of the North Carolina Furniture Manufacturing Industry with Special Attention to Locational Factors." Ph.D. diss., Duke Univ., 1959. 286 pp. The abundance of hardwood timber was a factor in the development of the industry since the 1880s.

Craig, Roland D. See Whitford, Harry N., #W256

C684 Craig, Roland D. "Softwood Resources of Canada." *Empire Forestry Journal* 2 (1923), 198-207.

C685 Craig, Roland D. "The Forest Resources of Canada." *Economic Geography* 2 (Mar. 1926), 394-413.

C686 Craig, Roland D. "Canada's Ups and Downs in the Lumber Industry in Quarter Century, 1908-1936." *Canada Lumberman* 57 (Aug. 15, 1937), 24-25.

C687 Craig, Roland D. "The Lumber Industry in Canada." *Canadian Geographical Journal* 15 (Nov. 1937), 225-47.

C688 Craig, Roland D. "Canadian Forest Resources: Their Relation to the War of 1914-1918 and to the Present Effort." *Canada Yearbook, 1940* (1940), 251-58.

C689 Craig, Ronald B. "The Forest Tax Delinquency Problem in the South." *Southern Economic Journal* (Oct. 1939), 145-64. Includes some history.

C690 Craig, Ronald B. "The Past and Future of Forest Taxation in Mississippi." *Conservationist* 7 (Jan. 1941), 5, 11.

C691 Craig, Ronald B. *Forestry in the Economic Life of Knott County, Kentucky.* Kentucky Agricultural Experiment Station, Bulletin No. 236. Lexington, 1932. 39 pp. Illus.

C692 Craig, Ronald B. *Virginia Forest Resources and Industries.* USDA, Miscellaneous Publication No. 681. Washington: GPO, 1949. iv + 64 pp. Illus., maps, diags., tables, bib. Incidental history.

C693 Craighead, Frank Cooper; Miller, J.M.; Evenden, J.C.; and Keen, Frederick Paul. "Control Work Against Bark Beetles in Western Forests and an Appraisal of Its Results." *Journal of Forestry* 29 (Nov. 1931), 1001-18. Includes sections on the history of bark beetle control work in various regions and national forests of the West, since 1906.

C694 Craine, Lyle E. "The Muskingum Watershed Conservancy District: An Appraisal of a Watershed Management Agency." Ph.D. diss., Univ. of Michigan, 1956. 345 pp. A study of watershed management in an Ohio district since 1933.

C695 Craine, Lyle E. "The Muskingum Watershed Conservancy District: A Study of Local Control." *Law and Contemporary Problems* 22 (Summer 1957), 378-404.

C696 Crampton, C. Gregory. *Standing-Up Country: Canyonlands of Utah and Arizona.* New York and Salt Lake City: Alfred A. Knopf and University of Utah Press, 1964. xv + 191 + iv pp. Illus., maps. Includes some history of national parks.

C697 Cramton, Louis C. *Early History of Yellowstone National Park and Its Relation to National Park Policies.* Washington: GPO, 1932. iii + 148 pp. Bib. Includes a chronological legislative history of the park, 1871-1897.

C698 Crandall, Vine. "A Study of the Cotton and Hanlon Lumber Company." Master's thesis, Cornell Univ., 1936. New York.

C699 Crane, J.R. "Monument to a Conservationist." *American Forests* 63 (Feb. 1957), 28-30, 62-65. On George Buckman Dorr's efforts to save land on Mount Desert Island, Maine, from real estate speculators, and his part in the movement to create Acadia National Park, 1901-1935.

C700 Crane, J.R. "Ships and Sawmills." *American Forests* 63 (Aug. 1957), 24-26, 54-56. On the lumber industry and lumber export from Maine since 1691.

C701 Cranston, Robert Brooks. "The Forests and Forest Industries of British Columbia." Master's thesis, Univ. of Washington, 1952.

C702 Crapo, Henry H. *The Story of Henry Howland Crapo, 1804-1869.* Boston: Thomas Todd Company, 1933. 272 pp. Illus. On the career of a Michigan lumberman and governor.

C703 Crasse, J.A. "Forest Industries Productivity in British Columbia." *British Columbia Lumberman* 46 (Sept. 1962), 18-21. Concerns productivity or output-per-man indexes for the period 1949-1960.

C704 Crawford, Blaine G. "Activities of the CIO and AFL in the Pacific Northwest Lumber Industry, 1935-1940." Master's thesis, Univ. of Idaho, 1942.

C705 Crawford, Charles W. "A History of the R.F. Learned Lumber Company, 1865-1900." Ph.D. diss., Univ. of Mississippi, 1968. 329 pp. Natchez, Mississippi.

C706 Crawford, Finla Goff. "The Paper Industry, 1860-1870." Ph.D. diss., Univ. of Wisconsin, 1922.

C707 Crawford, Finla Goff. "The Paper Industry 1860-70." *Paper Industry* 7 (Apr. 1925), 53-58; 7 (May 1925), 223-28. Charts industrial changes during a critical period when wood pulp was being used increasingly in papermaking.

C708 Creel, George. "Feudal Towns of Texas." *Harper's Weekly* 60 (Jan. 23, 1915), 76-78. Lumber towns.

C709 Creighton, Donald Grant. *The Commercial Empire of the St. Lawrence (1760-1850).* Toronto: Ryerson Press; New Haven, Connecticut: Yale University Press, 1937. 441 pp. Illus. Reprinted as *The Empire of the St. Lawrence* (Toronto: Macmillan, 1956). Contains history of the lumber industry and trade.

C710 Creighton, G.W.I. "Forestry in Nova Scotia." *Journal of Forestry* 35 (July 1937), 671-73. Includes some historical references to the Nova Scotia Department of Lands and Forests.

C711 Crerar, T.A. "Dominion Forest Service Protects Forests." *Pulp and Paper Magazine of Canada* 37 (Apr. 1936), 266-68.

Cress, Eleanor Chittenden. See Chittenden, Hiram Martin, #C303

C712 Crittenden, Henry Temple. *The Maine Scenic Route: A History of the Sandy River and Rangeley Lakes Railroad.* Parsons, West Virginia: McClain Printing Company, 1966. 229 pp. Illus., maps, app. A logging railroad in Franklin County, Maine, 1879-1935.

C713 Crittenden, Henry Temple. *The Comp'ny; the Story of the Surry, Sussex & Southampton Railway and the Surry Lumber Company.* Parsons, West Virginia: McClain Printing Company, 1967. 246 pp. Illus., maps. In southeastern Virginia, 1880-1930.

Croffut, W.A. See Hitchcock, Ethan Allen, #H387

C714 Croft, A. Russell, and Bailey, Reed W. *Mountain Water.* Ogden, Utah: USFS, Intermountain Region, 1964. 64 pp. Illus., maps, charts, tables, bib. Includes some history of flooding and watershed management in Utah, Nevada, western Wyoming, and southern Idaho, mainly in the 20th century.

C715 Cronau, Rudolf. *Our Wasteful Nation: The Story of American Prodigality and the Abuse of Our Natural Resources.* New York, 1908. 134 pp. Illus.

C716 Cronemiller, L.F. "State Forestry in Oregon." Master's thesis, Oregon State College, 1936.

C717 Cronin, James E. *Hermann von Schrenk, A Biography: Botanist, Plant Pathologist, Wood Preserving Scientist, Pioneer in American Wood Preservation, Forest Scientist, Forester, Timber Engineer.* Chicago: Kuehn, 1959. xiii + 257 pp. Illus., apps., index. On the career of Schrenk (1873-1953), including his establishment of the firm Von Schrenk, Fulks, and Kranmer, Timber Engineers, and his investigations of methods for preserving wood, particularly railroad ties.

C718 Cronk, C.P. *Forest Industries of New Hampshire and Their Trend of Development.* Concord: New Hampshire Forestry and Recreation Commission, 1936. 237 pp.

Cronquist, Arthur. See Gleason, Henry A., #G153

C719 Crosby, Charles P. "Memories of Lumbering on the Black River." *La Crosse County Historical Sketches,* Series 3 (1937), 43-56. Black River, Wisconsin.

C720 Crosby, Lucius Osmond, Jr. *Crosby, a Story of Men and Trees.* New York: Newcomen Society in North America, 1960. 32 pp. Illus. On the life of Lucius Olen Crosby, Sr. (1869-1948) in Mississippi as a farmer, wholesale lumber dealer, organizer of the Goodyear Yellow Pine Company at Picayune in 1916, and his success in rehabilitating the dwindling forests of Pearl River County and in developing a tung-tree industry there.

C721 Crosby, William. "Forestry in 49th State." *American Forests* 59 (July 1953), 20-22, 43-45. Includes some history of forestry in Hawaii.

C722 Cross, Clark Irwin. "Factors Influencing the Abandonment of Lumber Mill Towns in the Puget Sound Area." Master's thesis, Univ. of Washington, 1946.

C723 Cross, John K. "Tar Burning, a Forgotten Art?" *Forests & People* 23 (Second quarter, 1973), 21-23. Brief history of pine tar production in the South.

C724 Cross, Michael Sean. "The Dark Druidical Groves: The Lumber Community and the Commercial Frontier in British North America, to 1854." Ph.D. diss., Univ. of Toronto, 1968. Emphasis on Ontario and Quebec.

C725 Cross, Michael Sean. "The Lumber Community of Upper Canada, 1815-1867." *Ontario Historical Society Papers and Records* 52 (Autumn 1960), 213-33. Considers the social impact of the lumber industry and lumbermen in Ontario.

C726 Cross, Michael Sean. "The Age of Gentility: The Formation of an Aristocracy in the Ottawa Valley." Canadian Historical Association, *Report, 1967* (1967), 105-17. Including reference to 19th-century lumbermen in Ontario and Quebec.

C727 Cross, Michael Sean. "The Shiners' War: Social Violence in the Ottawa Valley in the 1830s." *Canadian Historical Review* 54 (Mar. 1973), 1-26. Concerns ethnic conflict (Irish-English-French) in the lumber industry.

C728 Cross, Whitney R. "Road to Conservation." *Antioch Review* 8 (Dec. 1948), 432-46.

C729 Cross, Whitney R. "W J McGee and the Idea of Conservation." *Historian* 15 (Spring 1953), 148-62. On his work as a geologist and anthropologist and his association with John Wesley Powell, Lester Frank Ward, and Gifford Pinchot, 1878-1912, all leading to his program for the conservation of natural resources.

C730 Cross, Whitney R. "Ideas in Politics: The Conservation Policies of the Two Roosevelts." *Journal of the History of Ideas* 14 (June 1953), 421-38. Theodore Roosevelt and Franklin D. Roosevelt. See reply by Richard Hofstadter, *ibid.* 15 (Apr. 1954), 328-29.

C731 Crowe, W.S. "Historical Notes on Michigan's Lumber Industry." *Timber Producers Bulletin* 119 (Feb. 1953), 6-7.

C732 Crowell, Benedict, and Wilson, Robert F. *The Giant Hand: Our Mobilization and Control of Industry and Natural Resources, 1917-1918.* New Haven: Yale University Press, 1921. 333 pp. Illus.

C733 Crown Zellerbach Corporation. *The Years of Paper: Isadore Zellerbach, 1866-1941.* San Francisco, 1941. 19 pp. Illus. Biographical sketch of a company founder.

C734 Crown Zellerbach Corporation. *The Years of Paper, in Memory of Louis Bloch.* San Francisco, 1951. Tribute and biographical sketch of a founder and longtime chairman of the board.

C735 Crowther, Simeon John. "The Shipbuilding Industry and the Economic Development of the Delaware Valley, 1681-1776." Ph.D. diss., Univ. of Pennsylvania, 1970. 242 pp.

C736 Croxton, Frederick C. "Wage and Salary Payments in Manufacture of Lumber and Lumber Products in Ohio, 1916-1932." *Monthly Labor Review* 39 (Aug. 1934), 423-30.

C737 Cruikshank, Helen G. *John & William Bartram's America: Selections from the Writings of the Philadelphia Naturalists.* New York: Devin-Adair, 1957. xxii + 418 pp. Illus., notes. The Bartrams were 18th-century natural scientists whose commentary on the wilderness of the Southeast helped promote a more romantic appreciation of forests.

C738 Cruikshank, J.W., and Eldredge, Inman F. *Forest Resources of Southeastern Texas.* USDA, Miscellaneous Publication No. 326. Washington: GPO, 1939. 37 pp.

C739 Cruikshank, J.W. *North Carolina Forest Resources and Industries.* USDA, Miscellaneous Publication No. 533. Washington: GPO, 1943. 76 pp.

C740 Crumley, J.J. *Constructive Forestry for the Private Owner.* New York: Macmillan, 1926. xviii + 322 pp. Illus., index. Contains incidental references to the history of Arbor Day and other aspects of the forestry movement.

C741 Crump, John. "The History of the Peoples Lumber Company." *Ventura County Historical Society Quarterly* 3 (Nov. 1957), 12-13. Ventura, California, since 1890.

C742 Crump, Spencer. *Redwoods, Iron Horses, and the Pacific: The Story of the California Western 'Skunk' Railroad.* Los Angeles: Trans-Anglo Books, 1974. 176 pp. Illus. On the line between Fort Bragg and Willits, California, since 1885 an important logging railroad of the Union Lumber Company, but now primarily a tourist attraction.

C743 Cubby, Edwin A. "Timbering Operations in the Tug and Guyandot Valleys in the 1890's." *West Virginia History* 26 (Jan. 1965), 110-20. On the lumber industry in southwestern West Virginia, including some account of lumberjacks and log drives.

C744 Cummings, H.R. *Early Days in Haliburton.* Toronto: Ontario Department of Lands and Forests, 1963. ix + 180 pp. Illus., maps. Includes some history of the lumber industry near Haliburton, Ontario.

Cummings, William H. See Kaufert, Frank H., #K20

C745 Cummins, John Gaylord. "Concentration and Mergers in the Pulp and Paper Industries of the United States and Canada, 1895-1955." Ph.D. diss., Johns Hopkins Univ., 1961.

C746 Cunningham, Russell N. "The Fire Protective Associations of Idaho and Montana." *Journal of Forestry* 21 (Nov. 1923), 736-41. Contains some incidental historical references to forest fires and forest fire protection since 1905.

C747 Cunningham, Russell N. *Forest Resources of the Lake States Region.* USFS, Forest Resource Report No. 1. Washington: GPO, 1950. vi + 57 pp. Illus., diags., tables, notes, bib. Concerns the forest resources of Minnesota, Wisconsin, and Michigan, since 1869.

C748 Cunningham, Russell N. *Changes in Forest Conditions, 1936-1949, North Central Minnesota and Upper Peninsula of Michigan.* Paper No. 25. St. Paul: USFS, Lakes States Forest Experiment Station, 1951. 20 pp.

C749 Cunningham, Russell N.; Horn, Arthur G.; and Quinney, Dean N. *Minnesota's Forest Resources.* USFS, Forest Resource Report No. 13. Washington: GPO, 1958. 52 pp. Summary of forest survey data collected from 1947 to 1954 and comparison of it with earlier surveys.

Cunningham, Russell N. See Dana, Samuel Trask, #D29

C750 Cunningham, William P. "Magruder Corridor Controversy: A Case History." Master's thesis, Univ. of Montana, 1968. Part of the Bitterroot National Forest, Idaho-Montana, considered for inclusion in the wilderness system.

C751 Curl, Melvin James. "Fact and Fable of the Life of the Northern Woodsmen." *Stone & Webster Journal* 24 (1919), 200-10. In the New England states.

C752 Curran, C.E. "Broadening the Basis of America's Pulpwood Supply." *Journal of Forestry* 36 (Sept. 1938), 879-81. A historical survey of pulpwood species and pulpwood supply since ca. 1900.

C753 Current, Richard N. *Pine Logs and Politics: A Life of Philetus Sawyer, 1816-1900.* Madison: State Historical Society of Wisconsin, 1950. xii + 330 pp. Illus., map, notes, bib., essay, index. On Sawyer's career as a lumberman in the Fox Valley and as a U.S. representative (1865-1875) and U.S. senator (1881-1893) from Wisconsin.

C754 Curry, Corliss C. "Early Timber Operations in Southeast Arkansas." *Arkansas Historical Quarterly* 19 (Summer 1960), 111-18. On the lumber industry in Ashley, Bradley, and Drew counties, 1830s-1890s.

C755 Curry, John R. "The Morse Sawmill at Cooptown." *American Forests* 38 (June 1932), 354-56, 382. Constructed in 1840 at Cooptown, Maryland, and believed to be the longest-lived sawmill in the country.

C756 Curry, John R. "Early Lumbering in the Adirondacks." *Northeastern Logger* 1 (May 1953), 13, 22. Of New York.

C757 Curry, John R. "The Management of Whitney Park, 1897-1957." *Northeastern Logger* 6 (July 1957), 20-22, 48-49, 71. Private forestry in the Adirondack Mountains of New York.

C758 Curry, W.H. "Some Highlights in the History of the Texas Association." *Gulf Coast Lumberman* 50 (Apr. 1, 1962), 22, 30, 32. By a former president of the Lumbermen's Association of Texas.

C759 Curry-Lindahl, Kai, and Harroy, Jean-Paul. *National Parks of the World.* 2 Volumes. New York: Golden Press, for the International Union for Conservation of Nature and Natural Resources, 1972. Illus., bibs., index. Includes some history of the international expansion of the national park "idea" and incidental historical references to individual parks (over 200) in 73 countries. North American national parks are covered in Volume 1.

C760 Curti, Merle. *The Making of an American Community: A Case Study of Democracy in a Frontier County.* Stanford, California: Stanford University Press, 1959. vii + 483 pp. Maps, diags. Includes brief mention of the lumber industry in Trempealeau County, Wisconsin, in the mid-19th century.

C761 Curtis, Charles T., ed. *Stories of the Raftsmen.* Callicoon, New York: Sullivan County Democrat, 1922. 35 pp. Reminiscences of 19th-century lumber rafting on the Delaware River, New York-Pennsylvania-New Jersey.

C762 Curtis, Elisha B. "The Old Ship-Building Days." *Medford Historical Register* 15 (Oct. 1912), 77-80. On the shipbuilding industry of Medford, Massachusetts, ca. 1850.

C763 Curtis, James D. "Historical Review of Artificial Forest Pruning." *Forestry Chronicle* 13 (June 1937), 390-95.

C764 Curtis, Michael. "Early Development and Operations of the Great Southern Lumber Company." *Louisiana History* 14 (Fall 1973), 347-68. In Bogalusa, Louisiana, 1870-1940.

C765 Custer, Dale H. "Obituary of a Boom Town." *Northern Logger* 13 (June 1965), 12-13, 30. On Cross Fork, Pennsylvania, a company town operated by the Lackawanna Lumber Company, 1890s-1910s.

C766 Custis, V. "Lumber Grading in the Pacific Northwest." *Quarterly Journal of Economics* 26 (May 1912), 538-44. Brief history of grading.

C767 Cutler, Malcolm R. "A Study of Litigation Related to Management of Forest Service Administered Lands and Its Effect on Policy Decisions. Part Two: A Comparison of Four Cases." Ph.D. diss., Michigan State Univ., 1972. 538 pp. On four conflicts between USFS and wilderness user or citizens' groups: Sylvania Recreation Area, Michigan; East Meadow Creek drainage, Colorado; Mineral King Valley, California; and Boundary Waters Canoe Area, Minnesota. Part One of the study is the author's master's thesis, "A Study of Litigation Related to Management of Forest Service Administered Lands and Its Effect on Policy Decisions: The Gandt v. Hardin Case," Michigan State University, 1971.

C768 Cutright, Paul Russell. *Theodore Roosevelt, the Naturalist.* New York: Harper, 1956. xiv + 297 pp. Illus., notes, bib. On Roosevelt's amateur studies of natural history in the United States and abroad, his attacks upon "nature fakers," and his measures to promote conservation of natural resources, 1871-1919.

C769 Cutright, Paul Russell. "Lewis and Clark and Cottonwood." *Missouri Historical Society Bulletin* 22 (Oct. 1965), 35-44. On their observations of cottonwood groves along the banks of the Missouri River, 1804-1806.

C770 Cutright, Paul Russell. *Lewis and Clark: Pioneering Naturalists.* Urbana: University of Illinois Press, 1969. xiii + 506 pp. Illus., maps, notes, apps., bib., index. On the Lewis and Clark Expedition of 1804-1806, and the influence of their work in natural history and science.

C771 Cutter, Emma E. "Lumbering." *Vermont Quarterly* 21 (Jan. 1953), 47-48. Reminiscence of logging and the lumber industry in Centerville, Vermont, ca. 1880.

D1 Dadisman, Andrew Jackson. "Lumbering in the North Woods." *Proceedings of the West Virginia Academy of Science* 31 (1960), 110-13. Logging in the Great Lakes states, Idaho, Washington, and California, 1830s-1910.

D2 Dahl, Jerome. "Progress and Development of the Prairie States Forestry Project." *Journal of Forestry* 38 (Apr. 1940), 301-06. Includes some history and evaluation of the Shelterbelt Project and Prairie States Forestry Project, as well as reference to earlier tree planting projects.

D3 Dahllöf, Tell. "Pehr Kalm's Concern About Forests in America, Sweden and Finland Two Centuries Ago." *Swedish Pioneer Historical Quarterly* 17 (No. 3, 1966), 123-45. Includes his observations of American forests in the 1750s.

D4 Dale, Edward Everett. "Wood and Water: Twin Problems of the Prairie Plains." *Nebraska History* 29 (June 1948), 87-104. On the difficulties of obtaining wood on the prairies.

D5 Dall, William H. *Spencer Fullerton Baird, a Biography, Including Selections from His Correspondence with Audubon, Agassiz, Dana, and Others.* Philadelphia: Lippincott, 1915. xvi + 462 pp. Baird (1823-1887), a leading natural scientist and conservationist, was the first commissioner of the U.S. Commission of Fish and Fisheries.

Daly, Dorothy. See Riley, R.D., #R200

D6 Dalzell, K.E. *The Queen Charlotte Islands, 1774-1966.* Terrace, British Columbia: C.M. Adam, 1968. 340 pp. Illus., maps. Includes some history of forests and forest industries.

D7 Dambach, Charles A. "On the Consolidation of Resource Departments." *American Forests* 66 (Apr. 1960), 32, 60-66. Natural resource agencies in Ohio since 1900.

D8 Damtoft, Walter J. "Fifty Years of Industrial Forestry Progress." *Southern Pulp and Paper Manufacturer* 14 (Oct. 1, 1951), 175-76, 183. In the Southern pulp and paper industry since 1900.

D9 Dana, Edward B. "Muskegon Fifty Years Ago." *Michigan History Magazine* 16 (Autumn 1932), 413-21. On the lumber industry and working conditions in Muskegon, Michigan, 1880s.

D10 Dana, Julian. *Sutter of California: A Biography.* New York: Press of the Pioneers, 1934. xi + 423 pp. Illus., bib. John Augustus Sutter (1803-1880) had lumber interests in California. His sawmill was the site of a gold discovery in 1848.

D11 Dana, Marshall N. "Reclamation, Its Influence and Impact on the History of the West." *Utah Historical Quarterly* 27 (Jan. 1959), 38-49. Since 1902.

D12 Dana, Samuel Trask. "Pennsylvania, a Forest Tragedy: The Rise and Fall of a Lumber Town." *Munsey's Magazine* 60 (1917), 353-63.

D13 Dana, Samuel Trask. *Forest Fires in Maine, 1916-1925.* Maine Forest Service, Bulletin No. 6. Augusta, 1926. 73 pp. Diags.

D14 Dana, Samuel Trask. "The Editor's Silver Jubilee." *Journal of Forestry* 24 (Dec. 1926), 845-46. A brief tribute to Raphael Zon.

D15 Dana, Samuel Trask. "Forest Fires in New Hampshire, 1921-1925." New Hampshire Forestry Commission, *Biennial Report, 1927-28* (1928), 68-112.

D16 Dana, Samuel Trask. "Donald Maxwell Matthews, 1886-1948." *Journal of Forestry* 46 (Dec. 1948), 922-24. Biographical sketch of Matthews, especially in forestry education at the University of Michigan, and as a consultant and writer, since 1927.

D17 Dana, Samuel Trask. "The Growth of Forestry in the Past Half Century." *Journal of Forestry* 49 (Feb. 1951), 86-92. On developments in federal, state, and private forestry, advances in education and research, and improvement in the status of the profession, since 1900.

D18 Dana, Samuel Trask. "Private Forestry Transition." *Annals of the American Academy of Political and Social Science* 281 (May 1952), 84-92. Includes some history of private and industrial forestry.

D19 Dana, Samuel Trask, and Watson, Russell. "Willett Forrest Ramsdell (1890-1951)." *Journal of Forestry* 50 (May 1952), 398-99. On his career with the USFS, 1914-1930, and as a forestry educator at the University of Michigan.

D20 Dana, Samuel Trask. *Forest Policy in the United States.* University of British Columbia Lecture Series, No. 21. Vancouver, 1953. 26 pp. A historical sketch.

D21 Dana, Samuel Trask, ed. *History of Activities in the Field of Natural Resources, University of Michigan.* Ann Arbor: University of Michigan Press, 1953. xii + 353 pp.

Illus., tables, apps., bib. Forestry and conservation education since 1881; includes list of faculty, alumni, and publications.

D22 Dana, Samuel Trask. "Filibert Roth—Master Teacher." *Michigan Alumnus Quarterly Review* 61 (Winter 1955), 100-10. Roth (1858-1925) was head of the Department of Forestry at the University of Michigan, 1903-1923, following service with the USDA's Division of Forestry and the Forestry Division of the General Land Office.

D23 Dana, Samuel Trask. "The American Forestry Association's First Eighty Years." *Journal of Forestry* 54 (Mar. 1956), 163-71. Since its founding in 1875.

D24 Dana, Samuel Trask. "The First Eighty Years." *American Forests* 62 (Apr. 1956), 13-19, 42-48. The American Forestry Association since its founding in 1875.

D25 Dana, Samuel Trask. *Forest and Range Policy: Its Development in the United States.* American Forestry Series. New York: McGraw-Hill, 1956. xi + 455 pp. Charts, apps., bib., index. An encyclopedic history of American forestry and forest legislation from the colonial period to the 1950s. Appendix 2 (pp. 372-425) is a useful "Chronological Summary of Important Events in the Development of Colonial and Federal Policies Relating to Natural Resources."

D26 Dana, Samuel Trask, and Krueger, Myron Edward. *California Lands: Ownership, Use, and Management.* Washington: American Forestry Association, 1958. xx + 308 pp. Maps, diags., tables, notes, bib., index. Includes a chronological summary of state and federal legislation relating to wildland ownership, as well as other historical references to forested lands.

D27 Dana, Samuel Trask. "Half Century of Progress." *American Forests* 66 (July 1960), 32-35, 90. On the USFS, Forest Products Laboratory, Madison, Wisconsin, since 1910.

D28 Dana, Samuel Trask. "Forest Ownership in Minnesota—Problems and Prospects." *American Forests* 66 (Oct. 1960), 32-38, 48-60. Since 1858.

D29 Dana, Samuel Trask; Allison, John H.; and Cunningham, Russell N. *Minnesota Lands: Ownership, Use, and Management of Forest and Related Lands.* Washington: American Forestry Association, 1960. xxi + 463 pp. Maps, tables, apps., bib., index. Includes much history.

D30 Dana, Samuel Trask. "Chapman of Yale." *American Forests* 69 (Sept. 1963), 1, 59, 61. On Herman Haupt Chapman (1874-1963), prolific writer, educator, and forestry consultant.

D31 Dana, Samuel Trask, and Johnson, Evert W. *Forestry Education in America, Today and Tomorrow.* Washington: Society of American Foresters, 1963. ix + 402 pp. Map, tables, graphs, apps., bib., index. Includes a chapter and many incidental references to the history of forestry education.

D32 Dana, Samuel Trask, and Pomeroy, Kenneth B. "Redwoods and Parks." *American Forests* 71 (May 1965), 1-32. The American Forestry Association's report concerning preservation of California coastal redwoods. Included is some early history of preservation and exploitation of the redwoods.

D33 Dana, Samuel Trask. "Gifford Pinchot, Forester." *Journal of Forestry* 63 (Aug. 1965), 603-07. History and evaluation of Pinchot as a forester and conservationist.

D34 Dana, Samuel Trask. "AFA's Birthday: 90 Years of Service." *American Forests* 71 (Sept. 1965), 9-14.

D35 Dana, Samuel Trask. "Samuel Trask Dana: The Early Years." OHI by Elwood R. Maunder and Amelia Fry. *Forest History* 10 (July 1966), 2-13. Recollections of his career in the USFS, as Maine's forest commissioner, and controversies within the Society of American Foresters, 1905-1920s.

D36 Dana, Samuel Trask. "The Dana Years." OHI by Elwood R. Maunder and Amelia Fry. *American Forests* 72 (Nov. 1966), 32-35, 62-66; 72 (Dec. 1966), 26-29, 50-55.

D37 Danford, Ormund S. "The Social and Economic Effects of Lumbering on Michigan, 1835-1890." *Michigan History Magazine* 26 (Summer 1942), 346-64.

D38 Danhof, Clarence H. *Change in Agriculture; The Northern United States, 1820-1870.* Cambridge: Harvard University Press, 1969. x + 322 pp. Illus., tables, bib., index. Incidental history of forested lands in relation to agriculture.

D39 Daniels, Jonathan Worth. *The Forest is the Future: A Southerner Looks at the Revolution Which Has Been Taking Place All Over the South As the Tall Chimneys of Pulp and Paper Mills Have Risen High Above the Nation's Fastest-Growing Trees.* New York: International Paper Company, 1957. 66 pp. Illus. The Southern pulp and paper industry since 1909.

Daniels, Robert H. See Anderson, Frederick R., #A430

D40 Dannenbaum, Jed. "John Muir and Alaska." *Alaska Journal* 2 (Autumn 1972), 14-20. On Muir's exploration of the Glacier Bay area in 1879 and his promotion of Alaska's scenic beauty, 1879-1900s.

D41 Danziger, Edmund J., Jr. "They Would Not Be Moved: The Chippewa Treaty of 1854." *Minnesota History* 43 (Spring 1973), 175-85. Concerns in part the logging and sawmilling activities of the Lake Superior Chippewa Indians in northern Wisconsin and northeastern Minnesota, 1870s-1890s.

D42 Darley, Albert D., Jr. "He Made Legal History." *American Forests* 62 (Oct. 1956), 27, 87-90. On litigation concerning the validity of the Maryland Forest Conservancy Districts Act of 1943, and a ruling by Chief Judge George Henderson affirming its constitutionality in 1947.

D43 Darling, Birt. *City in the Forest: The Story of Lansing.* New York: Stratford House, 1950. viii + 280 pp. Illus. Includes some history of the Michigan city's forest industries.

D44 Darling, F. Fraser, and Eichhorn, Noel D. *Man and Nature in the National Parks: Reflections on Policy.* Washington: Conservation Foundation, 1967. 80 pp. Illus. Incidental history of the impact of tourism and increasing visitor use.

D45 Darling, Jay Norwood. "The Story of the Wildlife Refuge Program." *National Parks Magazine* 28 (Jan.-June 1954), 6-10, 43-46, 53-56, 86-91. Since 1949.

D46 Darling, Sid L. "In Retrospect." *Southern Lumberman* 193 (Dec. 15, 1956), 250-51. Historical sketch of the National-American Wholesale Lumber Association.

D47 Darr, David R. "The Comparative Advantage of Minnesota's Wood Pulp Industry: Some Indirect Evidence."

Ph.D. diss., Univ. of Minnesota, 1971. 191 pp. Incidental history of pulp and paper industry.

D48 Darrah, William Culp. *Powell of the Colorado.* Princeton, New Jersey: Princeton University Press, 1951. ix + 426 pp. Illus., notes, bib., index. The biography of John Wesley Powell treats his Western explorations and other scientific work for the federal government, including his proposals for reform of public land policy.

D49 Darrah, William Culp. "Powell of the Colorado." *Utah Historical Quarterly* 28 (July 1960), 222-31. On explorations of the Colorado River under the direction of John Wesley Powell, 1868-1877, and his later proposals for surveying the lands and conserving the waters, soils, and forests of the Colorado Valley.

D50 Dasmann, Raymond F. *Environmental Conservation.* Second edition. New York: John Wiley and Sons, 1959. xiii + 375 pp. Illus., maps, bib., index. A textbook including some broadly historical accounts of man's use of natural resources and some specific accounts of conservation problems; includes a chapter on "Forests and Man."

D51 Dasmann, Raymond F. *The Last Horizon.* New York: Macmillan, 1963. vi + 279 pp. Illus., maps, graphs, notes, index. A general treatment of ecological problems, including some history of man's use and misuse of forests.

D52 Dasmann, Raymond F. *The Destruction of California.* New York: Macmillan, 1965. vii + 247 pp. Illus., maps, notes, bib., index. Contains much history of environmental deterioration in California, including incursions on its once vast forests.

D53 Dasmann, Raymond F. *A Different Kind of Country.* New York: Macmillan, 1968. viii + 276 pp. Illus., maps, notes, index. Includes some history of the conservation and wilderness preservation movements in the United States.

D54 Dasmann, Raymond F. *No Further Retreat: The Fight to Save Florida.* New York: Macmillan, 1971. xii + 244 pp. Illus., maps, bib., index. Includes some history of efforts to conserve the state's natural resources.

D55 Dasmann, William P. "Conservation: Can We Hold the Western Range?" *Pacific Discovery* 3 (July-Aug. 1950), 16-23. On land destruction and conservation measures in the West since the 1860s.

Dassow, Ethel. See Jackson, William H., #J9

D56 Davenport, Eugene. *Timberland Times.* Urbana: University of Illinois Press, 1950. 274 pp. Illus. Reminiscences of the lumbering era in Michigan's Grand River Valley.

D57 Davenport, F. Garvin. "Robert Ridgway: Illinois Naturalist." *Journal of the Illinois State Historical Society* 63 (Autumn 1970), 271-89. Ridgway (1850-1929) operated a private bird sanctuary that was one of the most important wildlife conservation areas in the United States. He anticipated the dangers to wildlife of indiscriminate hunting and insecticide use.

Davenport, William A. See Johannessen, Carl L., #J82

Davidson, Margaret G. See Maunder, Elwood R., #M312

D58 Davidson, Marshall B. "The 'American Woodsman'." *American Heritage* 11 (Dec. 1959), 12-23, 94-99. On John James Audubon (1785-1851) and his work as an ornithologist, painter, and precursor of the conservation movement.

D59 Davidson, R.R. "Comparisons of Iowa Forest Resources in 1832 and 1954." *Iowa State Journal of Science* 36 (No. 2, 1961), 133-36.

Davies, Alfred H. See Simmons, Perez, #S284

D60 Davies, Rosemary R. "The Rosenbluth Affair." *Forest History* 14 (Oct. 1970), 17-26. On the involvement of Yale Forest School faculty and alumni in a celebrated murder case, 1918-1924.

D61 Davies, Rosemary R. *The Rosenbluth Case: Federal Justice on Trial.* Ames: Iowa State University Press, 1971. xxvii + 252 pp. Illus., notes, app., bib., index. On the murder trial and exoneration of Robert Rosenbluth, a Yale Forest School graduate falsely charged with the death of a fellow army officer at Camp Lewis, Washington, 1918-1924.

D62 Davies, William A. "Western Logging Engineering Schools: Oregon State College." *Loggers Handbook* 11 (1951), 87-89. Education in logging engineering since 1913.

D63 Davis, Charles M. "The Development of Settlements in Northern Michigan." *Michigan Alumnus Quarterly Review* 42 (Summer 1936), 268-74. Including lumber towns.

D64 Davis, Charles M. "The Cities and Towns of the High Plains of Michigan." *Geographical Review* 28 (Oct. 1938), 664-73. On the rise and decline of the lumber towns since ca. 1900.

D65 Davis, Darrell H. "Return of the Forest in Northeastern Minnesota." *Economic Geography* 16 (Apr. 1940), 171-87.

D66 Davis, David. "Fifty Years at Bonner: Anaconda Copper Mining Company's Lumber Department Outstanding Montana Operation." *Timberman* 37 (Oct. 1936), 48-52, 54-55.

D67 Davis, Donald W. "Logging Canals: A Distinct Pattern of the Swamp Landscape of South Louisiana." *Forests & People* 25 (First quarter, 1975), 14-17, 33-35. On their construction and use for the transportation of cypress logs, early 19th century to ca. 1930.

D68 Davis, Douglas F. "Fort Humboldt Logging Museum: A Brief History." *Forest History* 17 (July 1973), 26-30. Originated by Stanwood S. Schmidt in the late 1950s, the museum is now part of Fort Humboldt State Park, Eureka, California.

D69 Davis, Douglas F. "Logging, Lumbering, & Forestry Museums: A Review." *Forest History* 17 (Jan. 1974), 28-31. A brief report on the origins of several museums and recent efforts to create an association of such organizations. Museums included are: Adirondack Museum, Blue Mountain Lake, New York; Lumbertown, U.S.A., Brainerd, Minnesota; Cowichan Valley Forest Museum, British Columbia; Collier State Park, Oregon; Camp Six, Tacoma Lumber Museum, Tacoma, Washington; Texas Forestry Museum, Lufkin, Texas; and Georgia-Pacific Historical Museum, Portland, Oregon.

D70 Davis, Douglas F. "First Forest Service Wireless: Primary Sources in Forest History." *Journal of Forest History* 18

(Apr. 1974), 22. On the USFS career of John A. Adams (1883-1968), in whose files was found L.V. Slonaker's account of the USFS's first wireless station (q.v.).

D71 Davis, Douglas F. "Port Gamble: Unique Historical Restoration Project." *Journal of Forest History* 19 (July 1975), 137-39. Port Gamble, Washington, a 19th-century lumber town restored by Pope & Talbot, Inc.

D72 Davis, Earle. "Sixty Years of Lima Locomotives." *Railroad Magazine* 27 (Dec. 1939), 6-27. Widely used in railroad logging.

D73 Davis, Edward Manning. "Development of American Lumber Standards." *American Architect* 130 (1926), 67-72.

D74 Davis, Elrick B. "Paul Bunyan Talk." *American Speech* 17 (Dec. 1942), 217-25. Lumberjack terms.

D75 Davis, Elrick B. "Some New Words by War Out of Wood." *American Speech* 19 (Apr. 1944), 91-96. On the origins of new words used in the forest products industries, as a consequence of wartime research.

D76 Davis, Harold A. *An International Community on the St. Croix (1604-1930)*. University of Maine Studies, Second Series, No. 64. Orono, Maine: University Press, 1950. xi + 412 pp. Maps, tables, app., notes, bib. Includes several chapters on the history of the lumber and shipbuilding industries in the St. Croix Valley of Maine and New Brunswick. Originally a doctoral dissertation at Columbia University (1950), this work was issued also as *The Maine Bulletin,* Volume 52, No. 12 (1950), complete.

D77 Davis, Harold A. "Shipbuilding on the St. Croix." *American Neptune* 15 (July 1955), 173-90. From the 1780s to 1870s.

D78 Davis, Harwell G. *The Life and Achievements of Joseph Linyer Bedsole.* New York: Newcomen Society in North America, 1962. 28 pp. Illus. Includes some account of the S.B. Adams Lumber Company.

D79 Davis, Hugh Cuthbert. "Demographic Changes and Resource Use in the Western Counties of Michigan's Upper Peninsula, 1860-1950." Ph.D. diss., Univ. of Michigan, 1962. 388 pp. The forest products industries were a major factor in region's growth and decline.

D80 Davis, Kenneth P. "Development of Forest Practice Controls in the United States." *Journal of Forestry* 44 (Nov. 1946), 934-39. Some history of forest practice controls and legislation since the 1920s, including the issue of public regulation.

D81 Davis, Kenneth P. "The Montana Conservation Council: A Significant Experiment in Conservation Organization." *Papers of the Michigan Academy of Science, Arts, and Letters* 36 (1953), 93-99. Established in 1949.

D82 Davis, Kenneth P. *American Forest Management.* American Forestry Series. New York: McGraw-Hill, 1954. xiii + 482 pp. Diags., bib., index. Incidental history.

D83 Davis, Kenneth P. *Forest Fire: Control and Use.* American Forestry Series. New York: McGraw-Hill, 1959. 584 pp. Illus., bib. Some history.

D84 Davis, Richard Carter. "Wilderness, Politics, and Bureaucracy: Federal and State Policies in the Administration of San Jacinto Mountain, Southern California, 1920-1968." Ph.D. diss., Univ. of California, Riverside, 1973. 527 pp. On efforts to preserve the wilderness characteristics of the mountain since ca. 1920. Controversial commercial developments have included the Palm Springs Aerial Tramway.

D85 Davis, Wilbur A., comp. "Logger and Splinterpicker Talk." *Western Folklore* 9 (Apr. 1950), 111-23. A glossary of terms collected in northwestern Oregon woods and sawmills.

D86 Davis Brothers Lumber Company. *The Story of Davis Brothers Lumber Company.* Ansley, Louisiana, 1952. 24 pp. Illus. Since 1902.

D87 Davison, Stanley R. "The Leadership of the Reclamation Movement, 1875-1902." Ph.D. diss., Univ. of California, Berkeley, 1951.

D88 Daw, T.E. "Remember the CCC? Time Stamps Approval." As told to Russell McKee. *Michigan Conservation* 24 (July-Aug. 1960), 12-14. On the Civilian Conservation Corps in Michigan, 1930s.

D89 Dawson, Hazel, comp. *Gifford Pinchot, A Bio-Bibliography.* Washington: USDI, Information Services Division, Office of Library Services, 1971. 37 pp. A list of publications by and about Gifford Pinchot.

D90 Dawson, J.W. "On the Destruction and Partial Reproduction of Forests in British North America." *American Journal of Science* 2 (1847), 161-70.

D91 Dawson, J.W. "The Removal and Restoration of Forests." *Canada Naturalist,* Second series, 3 (1868), 405-17.

D92 Day, Clarence A. *A History of Maine Agriculture, 1604-1860.* University of Maine Studies, Second Series, No. 68. Orono, Maine: University Press, 1954. ix + 318 pp. Illus., notes, bib. Includes some history of forested lands in relation to agriculture. Also published as *University of Maine Bulletin,* Volume 56, No. 11 (1954), complete.

D93 Day, Gordon M. "The Indian As an Ecological Factor in the Northeastern Forest." *Ecology* 34 (Apr. 1953), 329-46. Largely concerned with fires set by the Indians in New England and New York, 1580-1800.

D94 Day, Ralph K. *The Wood Pallet Industry: Its Development and Progress toward Standardization.* USFS, Forest Products Laboratory, Report No. R1957. Madison, 1953.

D95 Dayton, William A. *William Willard Ashe (1872-1932).* Washington: By the author, 1936. 22 pp. Bib. Ashe was employed as forester of the North Carolina Geological Survey, 1892-1905, and was with the USFS, 1905-1932, mostly in the South.

D96 Dayton, William A. "William Willard Ashe." *Journal of Forestry* 44 (Mar. 1946), 213-14. Career sketch.

D97 Dayton, William A. "Historical Sketch of U.S. Forest Service Botanical Activity, 1905-1954." *Journal of Forestry* 53 (July 1955), 505-07.

D98 Dean, George W. "Virginia's Great Wealth in Forest Dollars." *Commonwealth* 26 (Mar. 1959), 22-25, 48. On commercial forestry in Virginia since 1940.

Dean, George W. See Artman, James O. #A491

D99 Dean, H.A. "The Lumber Manufacturer's Trade Association: A History of Its Development and Economic Aspects in the U.S." Master's thesis, Yale Univ., 1929.

D100 Dean, Leon. *I Became a Ranger.* New York: Farrar & Rinehart, 1938. 240 pp. In New York's Adirondack Mountains.

D101 Dean, T.N. "History of Lumber Industry with Compensation Board." *Canada Lumberman* 59 (Nov. 15, 1939), 10.

D102 Deas, Stanley P. "The Southern Pine Association Story." *Forest Farmer* 26 (Apr. 1967), 22, 41. Historical sketch of the New Orleans-based trade association.

D103 Deatherage, Charles P. *The Early History of the Lumber Trade of Kansas City.* Kansas City, Missouri: Retail Lumberman, 1924.

D104 De Berti, M.J. "Controlling Blister Rust in Pennsylvania." *Pennsylvania Forests and Waters* 3 (Mar.-Apr. 1951), 41-46. Fighting white pine blister rust since 1916.

D105 De Boer, M.J. "Sunken Log Salvage." *Michigan Conservation* 19 (July-Aug. 1950), 23-26. On logs sunk in Michigan rivers and litigations about their ownership since 1855.

D106 De Camp, David, and Newman, Thomas Stell, comps. "Smokejumping Words." *American Speech* 33 (Oct. 1958), 180-84. Words and phrases used by smokejumpers.

D107 DeCew, T.H. "For Sixty-Five Years: Reminiscences of Early Lumbering Days in Ontario and the West." *Canada Lumberman* 50 (Aug. 1, 1930), 105-06.

D108 Dechêne, Louise. "Les Entreprises de William Price." *Histoire Sociale* 1 (Apr. 1968), 16-52. Price (1790-1867) was a Canadian timber exporter and founder of Price Brothers and Company.

D109 Decker, Arlie D. "Lumbering Moves West." *Pacific Northwesterner* 3 (Spring 1959), 17-24. On the operations of Frederick Weyerhaeuser and other lumbermen of Minnesota and Wisconsin and the gradual transfer of the industry to Idaho, Washington, and Oregon, 1839-1911.

D110 Decker, Leslie E. "The Railroad and the Land Office: Administration Policy and the Land Patent Controversy, 1864-1896." *Mississippi Valley Historical Review* 46 (Mar. 1960), 679-99. On the administration of railroad land grants by the General Land Office.

D111 Decker, Leslie E. *Railroads, Land and Politics: The Taxation of the Railroad Land Grants, 1864-1897.* Providence: Brown University Press, 1964. xi + 435 pp. Illus., maps. Especially in Kansas and Nebraska.

D112 Deckert, E. "Forest Fires in North America: A German View." Translated by George Wetmore Colles. *American Forestry* 17 (May 1911), 273-79. Forest fire study and prevention and forest protection since 1880.

Deckert, Russell C. See Hoyle, Raymond J., #H582

D113 Deering, Ferdie. *USDA, Manager of American Agriculture.* Norman: University of Oklahoma Press, 1945. xvi + 213 pp. Illus., tables, app.

D114 Defebaugh, James E. "Lumber Michigan—Its Past and Future." *American Lumberman* 64 (Nov. 9, 1901), 15-16.

D115 Defebaugh, James E. *History of the Lumber Industry of America.* 2 Volumes. Chicago: American Lumberman, 1906-1907. Illus., map, tables, notes. The first half of Volume 1 concerns the history of the lumber industry and forestry in Canada; the second half focuses on lumber production, forest resources, land policy, forestry, and tariff policy in the United States. Volume 2 is a history of the lumber industry in New York, New Jersey, Pennsylvania, and the New England states.

D117 Deibler, Frederick S. *The Amalgamated Wood Workers' International Union of America: A Historical Study of Trade Unionism in Its Relation to the Development of an Industry.* Bulletin of the University of Wisconsin, No. 511, Economics and Political Science Series, Volume 7, No. 3. Madison: University of Wisconsin, 1912. 211 pp. Bib. Also issued as a Ph.D. dissertation at the University of Wisconsin in 1909.

D118 Deitz, W. "A Study in Michigan Forest Land Taxation." Master's thesis, Michigan State Univ., 1954.

DeLaittre, Calvin L. See DeLaittre, Joseph A., #D119

D119 DeLaittre, Joseph A. *A Story of Early Lumbering in Minnesota.* Ed. by Calvin L. DeLaittre. Minneapolis: DeLaittre-Dixon Company, 1959. 43 pp. Illus., maps. Reminiscences of the lumber interests of the DeLaittre family in Minnesota, 1870s-1910.

D120 Delaney, William E. "Early Hardwood Days." *Southern Lumberman* 144 (Dec. 15, 1931), 79. Kentucky Lumber Company, Williamsburg and Burnside, Kentucky, 1880s.

Delavan, C.C. See Larsen, J.A., #L51

D121 DeLong, Thomas S. "Logging Railroads and Their History in the Coastal Plain of North Carolina." Master's thesis, Duke Univ., 1947.

D122 De Lotbiniere, A. Joly. "Lumbering in Lotbiniere under the Old Seigniorial Regime." *Canada Lumberman* 60 (Aug. 1, 1940), 83-84. Lotbiniere, Quebec.

D123 Demarest, Doug. "Son of the Sierra Nevada." *American Forests* 64 (Apr. 1958), 32-34, 61-63. On John Muir (1838-1914) as a naturalist and preservationist in the Sierra Nevada, California, 1869-1914.

D124 Demars, Stanford E. "The Triumph of Tradition: A Study of Tourism in Yosemite National Park, California." Ph.D. diss., Univ. of Oregon, 1970. 248 pp. Evaluates changing patterns of visitor use since 1860s.

D125 Demmon, Elwood L. "Henry E. Hardtner." *Journal of Forestry* 33 (Oct. 1935), 885-86. Biographical sketch of Hardtner (1870-1935), owner of the Urania Lumber Company, Urania, Louisiana, first chairman of the Louisiana Conservation Department, and sometimes called "the father of forestry in the South."

D126 Demmon, Elwood L. "Forests in the Economy of the South." *Southern Economic Journal* 3 (Apr. 1937), 369-80. Incidental history.

D127 Demmon, Elwood L. "20 Years of Forest Research in the South, 1921-1941." *Southern Lumberman* 163 (Aug. 15, 1941), 29-31. At the USFS, Southern Forest Experiment Station, New Orleans.

D128 Demmon, Elwood L. "Southern Forestry, Past, Present, and Future." *Southern Lumberman* 163 (Dec. 15, 1941), 213-18.

D129 Demmon, Elwood L. "Twenty Years of Forest Research in the Lower South, 1921-1941." *Journal of Forestry* 40 (Jan. 1942), 33-36. At the USFS's Southern Forest Experiment Station.

D130 Demmon, Elwood L. "30 Years of Forest Research in the Southeast." *Southern Lumberman* 185 (Dec. 15, 1952), 196-98. By the USFS.

D131 Demmon, Elwood L., and Briegleb, Philip A. "Progress in Forest and Related Research in the South." *Journal of Forestry* 54 (Oct. 1956), 674-82, 687-92. By the USFS since 1921.

Demmon, Elwood L. See Robertson, Reuben B., #R238

Dence, W.A. See King, Ralph Terence, #K156

D132 Dennett, Fred. "The Story of the Public Lands: The Development of the General Land Office, the Oldest Government Bureau." *Americana* 5 (June 1910), 558-70.

D133 Denny, Arthur. *Pioneer Days on Puget Sound.* Seattle: C.B. Bagley, 1888. 83 pp. Includes some account of the author's experiences in the lumber industry in Washington.

D134 Den Uyl, Daniel. "From Field to Forest—A 50 Year Record." *Journal of Forestry* 49 (Oct. 1951), 698-704. On Clark State Forest, Indiana, since 1903.

D135 Den Uyl, Daniel. "Indiana's Old Growth Forests." *Proceedings of the Indiana Academy of Science* 63 (1954), 73-79. On the composition of Indiana forests at the beginning of settlement, ca. 1800.

D136 Den Uyl, Daniel. "Charles C. Deam." *Proceedings of the Indiana Academy of Science* 63 (1954), 232-39. Deam (1865-1953) was a botanist and state forester of Indiana.

D137 Den Uyl, Daniel. "History of Forest Conservation in Indiana." *Proceedings of the Indiana Academy of Science* 66 (1957), 261-67. Since 1870.

D138 Den Uyl, Daniel. *A Twenty Year Record of the Growth and Development of Indiana Woodlands.* Indiana Agricultural Experiment Station, Research Bulletin No. 661. Lafayette: Purdue University, 1958. 52 pp. Illus., diags., tables. From 1931 to 1951.

D139 Den Uyl, Daniel. "Forests of the Lower Wabash Bottomlands during the Period, 1870-1890." *Proceedings of the Indiana Academy of Science* 67 (1958), 244-48.

Depew, Chauncey M. See Fernow, Bernhard E., #F68

D140 Derlath, August. *The Wisconsin: River of a Thousand Isles.* Rivers of America Series. New York: Rinehart & Company, 1942. 366 pp. Maps, index. Includes a section on the history of lumbering.

D141 DeSormo, Maitland C. *The Heydays of the Adirondacks.* Saranac Lake, New York: Adirondack Yesteryears, 1974. 265 pp. Illus. A general anecdotal history of the northeastern Adirondack Mountains of New York, including several chapters on forest industries and "River-Driving Days."

D142 Despres, George C. "Entire 100 Year History of Manistee Based on Timber." *Michigan History Magazine* 26 (Winter 1942), 92-99. Chronological list of lumbermen and sawmills in Manistee, Michigan, 1841-1925; reprinted from *Manistee News-Advocate,* June 19, 1940.

D143 Despres, George C. "Railroad Logging." *Michigan History* 38 (June 1954), 182-84. Reminiscences of railroad logging in Michigan.

D144 Detweiler, Robert; Sutherland, Jon N.; and Werthman, Michael S., eds. *Environmental Decay in Its Historical Context.* Glenview, Illinois: Scott, Foresman, 1973. 142 pp. An anthology of writings, some of which pertain to forested lands.

D145 Detwiler, S.B. "The History of Shipmast Locust." *Journal of Forestry* 35 (Aug. 1937), 709-12. On the introduction of shipmast locust on Long Island, New York, presumably ca. 1700.

D146 Deutsch, Herman J. "Geographic Setting for the Recent History of the Inland Empire." *Pacific Northwest Quarterly* 49 (Oct. 1958), 150-61; 50 (Jan. 1959), 14-25. Contains some history of logging and the lumber industry in this region comprising eastern Washington, northern Idaho, and western Montana.

D147 Devall, W.B., and Christen, H.E. "Alabama Hardwoods, Past, Present, and Future." *Journal of Alabama Academy of Science* 23 (Feb. 1953), 34-38.

D148 Devall, William B. "The Governing of a Voluntary Organization: Oligarchy and Democracy in the Sierra Club." Ph.D. diss., Univ. of Oregon, 1970. Includes some history of the preservationist organization.

D149 Devenish, F.H. "Reminiscences over Twenty Years: Great Changes Have Revolutionized Eastern Lumber Trade during Past Two Decades—Some for the Better." *Canada Lumberman* 50 (Aug. 1, 1930), 139-40.

D150 Devore, Roy W. "The River Drivers." *Alberta Historical Review* 8 (Winter 1960), 21-23. Reminiscence of log driving on the Red Deer River, Alberta, 1909.

D151 De Voto, Bernard. "The West: A Plundered Province." *Harper's* 169 (Aug. 1934), 355-63. Includes some history of natural resource exploitation in the West.

D152 De Voto, Bernard. "The West Against Itself." *Harper's* 194 (Jan. 1947), 1-13. A general historical account of natural resource exploitation in the West, including public lands and grazing controversies of the 1940s. This and the following three essays, as well as others pertaining to historic and contemporary problems of the USFS and the conservation movement generally, are reprinted in De Voto's *The Easy Chair* (Boston: Houghton Mifflin, 1955).

D153 De Voto, Bernard. "Sacred Cows and Public Lands." *Harper's* 196 (July 1948). On contemporary grazing controversies involving the USFS, including some history of the problem.

D154 De Voto, Bernard. "The Smokejumpers." *Harper's* 203 (Nov. 1951), 54-61. History and description of USFS smokejumping in the West, especially from the major base in Missoula, Montana.

D155 De Voto, Bernard. "Conservation: Down and on the Way Out." *Harper's* 209 (Aug. 1954), 66-74. A brief account of conservation failures in the 1940s and 1950s, with emphasis on problems faced by the USFS and Soil Conservation Service.

D156 De Voto, Bernard. *The Letters of Bernard De Voto.* Ed. by Wallace E. Stegner. Garden City, New York: Doubleday and Company, 1975. xiv + 393 pp. Includes a chapter of correspondence on "Conservation and the Public Domain," illustrating De Voto's involvement in the conservation movement, 1948-1955.

D157 DeVries, Wade, and Mcdaniels, E.H. "Forest Taxation in Oregon and Washington." *West Coast Lumberman* 69 (Apr. 1942), 32-33. Since 1931.

De Vyver, Frank T. See Berglund, Abraham, #B223

D158 Dewers, Robert S. "A Problem in the Tight Cooperage Industry: Concerning the Diminishing Supply of White Oak." Master's thesis, Colorado Agricultural and Mechanical College, 1948.

D159 Dewhurst, J. Frederic, and Associates. *America's Needs and Resources: A New Survey.* New York: Twentieth Century Fund, 1955. xxix + 1148 pp. Notes, tables, apps., index. A chapter on "Land and Water Conservation and Development" by Robert W. Hartley contains some historical references to forest resources and conservation. The volume is a revision of a 1947 survey.

DeWolf, Gordon. See Bartram, William, #B117

D160 Dick, Everett. *The Lure of the Land: A Social History of the Public Lands from the Articles of Confederation to the New Deal.* Lincoln: University of Nebraska Press, 1970. xii + 413 pp. Illus., notes, bib., index. Focuses on the human dimension of the frontier experience and the unrealistic nature of the public land system. Includes chapters on the conservation movement and "Timber and the Public Domain."

D161 Dick, Roger S. "History of Lumbering in Cowlitz County." Master's thesis, Univ. of Washington, 1941.

D162 Dick, W. Bruce. "A Study of the Original Vegetation of Wayne County, Michigan." *Papers of the Michigan Academy of Science, Arts, and Letters* 22 (1936), 329-34.

D163 Dickerman, Murlyn B., and Hutchison, S. Blair. "Montana's Timber Base for Industrial Growth." *Proceedings of the Montana Academy of Science* 11 (1952), 43-46. Incidental history.

D164 Dickerman, Murlyn B. *The Changing Forests of the Lake and Central States Region.* USFS, Lake States Forest Experiment Station, Miscellaneous Report No. 31. St. Paul, 1954. 10 pp. Changes in forest area, forest resources, and forest ownership.

Dickerman, Murlyn B. See Morgan, James T., #M560

D165 Dickerman, Murlyn B. "Research: A Guide to Progress in Minnesota Forestry." *Northern Logger* 12 (Mar. 1964), 18-19, 52-53, 56. Includes some history of the USFS's Lake States Forest Experiment Station, St. Paul, Minnesota, since the 1920s.

D166 Dickie, David W. "The Pacific Coast Steam Schooner." In *Historical Transactions, 1893-1943.* New York: Society of Naval Architects and Marine Engineers, 1945. Pp. 39-48. Some account of ships constructed for the Pacific lumber trade.

D167 Dickie, Francis. "David Douglas: Forgotten, Heroic Explorer." *Southern Lumberman* 189 (Dec. 15, 1954), 107-08. Douglas (1799-1834) made expeditions to the Pacific Northwest to collect botanical specimens for the London Horticultural Society, 1824-1834.

D168 Dickinson, Fred E. "The Influence of Transportation Methods on Wood Distribution of Southern and Western Softwood Lumber, with Special Reference to the Effect on the Price Structure of Lumber." Ph.D. diss., Yale Univ., 1951.

D169 Dickinson, Fred E. "Development of Southern and Western Freight Rates on Lumber." *Southern Lumberman* 185 (Dec. 15, 1952), 229-45. History of rates since the 1860s; based largely on cases before the Interstate Commerce Commission.

D170 Dick-Paddie, William A. "Primeval Forest Types in Iowa." *Proceedings of the Iowa Academy of Science* 60 (1953), 112-16. Data from surveyors' notes on three eastern counties, 1836-1859.

D171 Dick-Paddie, William A. "Presettlement Forest Types in Iowa." Ph.D. diss., Iowa State Univ., 1955. 75 pp.

D172 Diebold, A.J. "Logging 90 Years Back on the Ottawa River Valley." *Timber of Canada* 14 (Aug. 1954), 20-23; 15 (Oct. 1954), 21-23, 33. Log driving in Quebec, 1862.

D173 Dierks, F. McD., Jr. *The Legacy of Peter Henry Dierks, 1824-1972.* Tacoma: By the author, 1972. xv + 133 pp. Illus., maps, charts, index. On the Dierks family genealogy and its lumber and business interests in Kansas City, Missouri, and elsewhere in the Midwest and South.

D174 Dietrich, A.C. "Historical Southern Lumber Production Facts." *Southern Lumberman* 179 (Dec. 15, 1949), 189-92. Statistics on production since 1869.

D175 Dietrich, Irvine T., and Hove, John, eds. *Conservation of Natural Resources in North Dakota.* Fargo: North Dakota Institute for Regional Studies, North Dakota State University, 1962. 327 pp. Illus., bib. Includes some history.

D176 Dietz, Martha A. "A Review of the Estimates of the Sawtimber Stand in the United States, 1880-1946." *Journal of Forestry* 45 (Dec. 1947), 865-74.

D177 Dill, Alonzo Thomas. "Chesapeake Corporation of Virginia." *Commonwealth* 25 (Dec. 1958), 33-43. A paper mill established at West Point, Virginia, in 1914.

D178 Dill, Alonzo Thomas. *Chesapeake, Pioneer Papermaker: A History of the Company and Its Community.* Charlottesville: University Press of Virginia, 1968. xvi + 356 pp. Illus., tables, notes, apps., bib., index. History of the Chesapeake Corporation of Virginia and its predecessor, Chesapeake Pulp and Paper Company, West Point, Virginia, since 1914.

D179 Dill, Dorothy. "Lumberjack Stories." *Michigan History* 41 (Sept. 1957), 327-34. On the daily life of Michigan lumberjacks in the 1880s.

Dillard, Jesse D. See Beattie, Rolla Kent, #B153

D180 Dillon, John Brown. *Oddities of Colonial Legislation in America as Applied to the Public Lands. . . .* Ed. by Benjamin Douglass. Indianapolis: Robert Douglass, 1879. 784 pp.

D181 Dinh, T.T. "A Historical Sketch of the Lumber Industry in the United States." Master's thesis, State Univ. of New York, College of Forestry, 1962.

D182 Dinsdale, Evelyn M. "The Lumber Industry of Northern New York: A Geographical Examination of Its History and Technology." Ph.D. diss., Syracuse Univ., 1963. 212 pp. Since 1750.

D183 Dinsdale, Evelyn M. "Spatial Patterns of Technological Change: The Lumber Industry of Northern New York." *Economic Geography* 41 (July 1965), 252-74. Includes some history of lumbering since the early 19th century.

Dinsdale, Evelyn M. See Stokes, Evelyn, #S656

D184 Dionne, Jack. *A Brief Story of the Life of John Henry Kirby.* Houston: Kirby Lumber Company, 1940. Texas lumberman.

D185 Disston, Henry, & Sons. *The Saw in History: A Comprehensive Description of the Development of This Most Useful of Tools from the Earliest Times to the Present Day.* Philadelphia: Henry Disston & Sons, 1915. 63 pp. Illus. In part a company history.

D186 Disston, Jacob S., Jr. *Henry Disston (1819-1878: Pioneer Industrialist, Inventor & Good Citizen.* New York: Newcomen Society in North America, 1950. 32 pp. Illus. On the Philadelphia saw manufacturer, Henry Disston & Sons, Inc., 1840-1878.

D187 Dixon, Albert. "The Conservation of Wilderness: A Study in Politics." Ph.D. diss., Univ. of California, Berkeley, 1968. 220 pp. On the growth of a wilderness preservation policy since the 1920s, with emphasis on political developments in Congress and the interests of the USFS and the National Park Service.

D188 Dixon, Les B. "Birth of the Lumber Industry in British Columbia." *British Columbia Lumberman* 39 (Nov.-Dec., 1955); 40 (Jan.-Sept. 1956). A serialized history of eleven installments, each three or four pages in length, on the lumber industry before 1870. The series was reprinted by *British Columbia Lumberman* in 1956, 24 pp. Illus., tables.

D189 Dixon, R.M. *The Forest Resources of Ontario, 1963.* Toronto: Ontario Department of Lands and Forests, 1971. 108 pp. Illus. Incidental history.

D190 Dobbin, F.H. "How Lumber Was Produced in the Late Fifties." *Canada Lumberman* 41 (Jan. 1, 1921), 44-45. Sawmills on the Otonabee River, Ontario, 1850s.

D191 Dobbin, F.H. "Romantic Days Recalled in Lumbering Line: Interesting Anecdotes and Historical Record of Many Ventures in Stony and Rice Lake Districts." *Canada Lumberman* 41 (Jan. 15, 1921), 53-55. Woods operations and sawmills near Peterborough, Ontario, in 1858.

D192 Dobbin, F.H. "Harwood's Palmy Days as Lumbering Town: Eastern Ontario Hamlet is Now 'Deserted Village'." *Canada Lumberman* 41 (Apr. 1, 1921), 51. The lumber industry at Harwood in the Rice Lake District of Ontario.

D193 Dobbin, F.H. "When Timber on Trent Waterway Was Cheap and Sawmills Flourished on Every Side." *Canada Lumberman* 44 (Mar. 15, 1924), 71-74. Ontario, 1850s.

D194 Dobbin, F.H. "Sawmilling Equipment of the Late Fifties: Yankee Gangs, English Gates, Water Wheels and Other Old Time Equipment Are Described in This Reference." *Canada Lumberman* 50 (Aug. 1, 1930), 163-64. Ontario.

D195 Dobbin, F.H. "Some Difficulties Publishing *Canada Lumberman* Back in the Eighties: An Interesting Reference to Early Struggles of the Publishers in Producing Canada's National Lumber Journal." *Canada Lumberman* 50 (Aug. 1, 1930), 187.

D196 Dobbin, F.H. "In Good Old Days When Trees Were Real Giants." *Canada Lumberman* 51 (Sept. 15, 1931), 26. Reminiscence of logging in Ontario's Midland District.

D197 Dobie, John Gilmore. *The Itasca Story.* Minneapolis: Ross & Haines, 1959. ix + 202 pp. Illus., maps, tables, bib. On Lake Itasca and Itasca State Park, Minnesota, since 1803.

Doctor, Joseph E. See Griggs, Monroe Christopher, #G347

D198 Dodds, Gordon B. "The Fight to Close the Rogue." *Oregon Historical Quarterly* 60 (Dec. 1959), 461-74. On the political and legislative efforts of conservationists, ultimately successful, to prohibit commercial fishing on Oregon's Rogue River, 1910-1935.

Dodds, Gordon B. See Hume, Robert D., #H620

D199 Dodds, Gordon B. *The Salmon King of Oregon: R.D. Hume and the Pacific Fisheries.* Chapel Hill: University of North Carolina Press, for the American Association for State and Local History, 1962. xiv + 257 pp. Illus., notes, app., bib., index. Includes a brief account of his timberlands and sawmill interests in Curry County, Oregon, 1882-1894.

D200 Dodds, Gordon B. "The Historiography of American Conservation: Past and Prospects." *Pacific Northwest Quarterly* 56 (Apr. 1965), 75-81. A historiographical essay listing and evaluating 117 major books and articles about the organized conservation movement.

D201 Dodds, Gordon B., ed. "H.M. Chittenden's 'Notes on Forestry Paper'." *Pacific Northwest Quarterly* 57 (Apr. 1966), 73-81. Concerning Hiram M. Chittenden's memoir (1916) about his essay, "Forests and Reservoirs in Their Relation to Stream Flow, with Particular Reference to Navigable Rivers," (1908), and its impact on the conservation movement.

D202 Dodds, Gordon B. "The Stream-Flow Controversy: A Conservation Turning Point." *Journal of American History* 56 (June 1969), 59-69. Controversy over the influence of forests in regulating stream flow began in 1907 and revealed differing theories of conservation among the federal agencies, especially between the USFS and Hiram H. Chittenden of the U.S. Army Corps of Engineers.

D203 Dodds, Gordon B., ed. "Conservation & Reclamation in the Trans-Mississippi West: A Critical Bibliography." *Arizona and the West* 13 (Summer 1971), 143-71. Evaluates major books and articles on the subject published between 1917 and 1970.

D204 Dodds, Gordon B. *Hiram Martin Chittenden: His Public Career.* Lexington: University Press of Kentucky, 1973. xi + 220 pp. Illus., map, notes, bib., essay, index. Chittenden (1858-1917), a noted civil engineer and historian of the American West, was an early advocate of scientific accuracy as a base for conservation policy. In particular, he quarreled with Gifford Pinchot and the USFS over the relationship between forest cover and stream flow.

D205 Dodge, Alex M. "New Versus Old Methods of Logging in the Lake States." *Lumber World Review* (Nov. 10, 1920).

D206 Doerflinger, William Main, ed. *Shantymen and Shantyboys: Songs of the Sailor and Lumberman.* New York: Macmillan, 1951. xxiii + 374 pp. Illus., music, notes, bib., index. Contains history and criticism of the songs of Canadian and American lumberjacks.

D207 Doherty, Paul C. "The Columbia-Providence Plank Road." *Missouri Historical Review* 57 (Oct. 1962), 53-59. On the promotion, construction, and operation of a plank road between Columbia and Providence, Missouri, 1850s-1860s.

Doherty, William T., Jr. See Smith, Frank E., #S366

D208 Doig, Ivan C. "John J. McGilvra: The Life and Times of an Urban Frontiersman, 1827-1903." Ph.D. diss., Univ. of Washington, 1969. 298 pp.

D209 Doig, Ivan C. "John J. McGilvra and Timber Trespass: Seeking a Puget Sound Timber Policy, 1861-1865." *Forest History* 13 (Jan. 1970), 6-17. McGilvra, a U.S. district attorney in Washington Territory, prosecuted timber trespassers.

D210 Doig, Ivan C. "Timber and the Law: A Civil War Chapter." *Pacific Search* 8 (July 1974), 18-20. John J. McGilvra and timber trespass in Washington Territory.

D211 Doig, Ivan C. "When Forests Went to Sea." *Oceans* 7 (July-Aug. 1974), 24-29. On the use of tall pines for masts on American and British sailing vessels until the mid-19th century.

D212 Dolan, T.J. *Twenty Years of Conservation on the Upper Thames Watershed, 1947-1967.* London, Ontario: Upper Thames River Conservation Authority, 1969. 110 pp. Illus. Includes a chapter on forestry and land use.

D213 Dollar, Robert. *Memoirs of Robert Dollar.* 4 Volumes. San Francisco: Robert Dollar Company, 1918-1928. Illus. Includes an account of his career in the lumber industry in California, Michigan, and Canada, although mostly devoted to his shipping interests.

D214 Donaldson, Alfred Lee. *A History of the Adirondacks.* 2 Volumes. New York: Century Company, 1921. Illus., maps, charts, bib. A comprehensive history of the Adirondack Mountains of New York.

D215 Donaldson, Thomas Corwin. *The Public Domain: Its History, with Statistics.* 1880. Third edition. Washington: GPO, 1884. xi + 1343 pp. Maps, tables, diags. Several early editions of this work were issued as government documents. There are also modern reprints.

D216 Donery, J.A. "Forest Resources of Minnesota." *American Forests* 54 (Jan. 1948), 26-28, 44. Since 1835.

D217 Donia, Robert. "From Swampland to City: The Settlement of Onaway." *Michigan History* 53 (Winter 1969), 292-306. A lumber town in Michigan, 1880s-1900s.

D218 Donnelly, Florence. "Pioneer Paper Mill of the West." *Paper Maker* 18 (Sept. 1949), 14-19. The Pioneer Paper Mill of Paperville, California, made paper and bags from rags beginning in 1856. The owner, Samuel Penfield Taylor, located his mill in a redwood forest to obtain the necessary fuelwood for his operation.

D219 Donnelly, Florence. "First in the West: The Story of Peter and James Brown and How They Successfully Overcame Obstacles to Produce Strawboard on the Pacific Coast." *Paper Maker* 19 (Sept. 1950), 1-9. The Caledonia Paper Mills of Saratoga, California, produced paperboard made from straw in the 1870s. Its proximity to nearby sawmills guaranteed a source of wood for fuel.

D220 Donnelly, Florence. "The Beautiful Mill: With Interior Finish of Finest Inlaid Mahogany It Was 'The Wonder of the Time'." *Paper Maker* 20 (Feb. 1951), 23-32. Concerns the mill of James Lick near San Jose, California, which milled flour and then paper from woodpulp until ca. 1900.

D221 Donnelly, Florence. "The Pride of the San Joaquin." *Paper Maker* 20 (Sept. 1951), 32-41. Rufus B. Lane's California Paper Company mill in Stockton, California, first produced paper from straw and rags in 1878 and gradually converted to pulp wood in the late 1880s.

D222 Donnelly, Florence. "The Paper Mill at Floriston in the Heart of the Sierras." *Paper Maker* 21 (Feb. 1952), 59-71. On the Floriston Pulp and Paper Company, Nevada County, California; includes a career sketch of Louis Bloch, who became president of the Crown-Willamette Paper Company and chairman of the board of the Crown Zellerbach Corporation.

D223 Donnelly, Florence. "The San Lorenzo Paper Mill." *Paper Maker* 22 (Feb. 1953), 11-23. Santa Cruz, California, 1860-1872.

D224 Donnelly, Florence. "Saratoga Paper Mill." *Paper Maker* 23 (Feb. 1954), 29-39. This mill at Saratoga, California, manufactured paper from straw, 1869-1883. It was located near redwood sawmills at the base of the Santa Cruz Mountains, where building materials and fuel wood were abundant and cheap.

D225 Donnelly, Florence. "Trail of Ventures." *Paper Maker* 25 (Feb. 1956), 17-27. Papermaking in Mendocino County, California, 1870s-1880s.

D226 Donnelly, Florence. "Oregon's Second Venture in Papermaking, the Clackamas Mill." *Paper Maker* 27 (Sept.

1958), 21-28. On the H.L. Pittock and Company, established at Oregon City, Oregon, 1867, to provide paper for the *Portland Oregonian.*

D227 Donnelly, Florence. "Camas Paper Mill, First in Washington." *Paper Maker* 29 (Sept. 1960), 14-28. On the Columbia River Paper Company, established by William Lewthwaite, Henry L. Pittock, and J.K. Gill in 1884, was the first Pacific Northwest mill to make paper from wood pulp. It is now the Camas Division of the Crown Zellerbach Corporation.

D228 Donovan, J.J. "Lumber Industry in Foreign Commerce." *Export and Shipping Journal* 3 (Mar. 1922), 39-40. Incidental history of the Pacific lumber trade.

D229 Dopp, Mary. "Geographical Influences in the Development of Wisconsin. V. The Lumber Industry." *Bulletin of the American Geographical Society* 45 (1913), 490-99, 585-609, 736-49.

D230 Doran, Jennie Elliot. "A Bibliography of John Muir." *Sierra Club Bulletin* 10 (Jan. 1916), 41-54. A list of his writings, mostly on natural history and preservationist themes.

D231 Doree, Bill. "Texas' First Steam Powered Sawmill." *Gulf Coast Lumberman* (Apr. 1963), 13, 26, 28. On a mill built by William Plunkett Harris in 1830.

D232 Dorenfelt, L.J. "History of Sulphate Pulp Manufacture: Extracts from the Private Papers of C.F. Dahl, Original Inventor of the Sulphate Process." *Paper Industry* 10 (Aug. 1928), 809-13.

D233 Dorf, Philip. *Liberty Hyde Bailey, an Informal Biography.* Ithaca, New York: Cornell University Press, 1956. 259 pp. Illus., bib. Bailey (1858-1954), a distinguished botanist and horticulturist, also made contributions to forestry and originated the Cornell University arboretum.

Doriot, Georges F. See Fraser, Cecil E., #F214

D234 Dornberger, Suzette. "The Struggle for Hetch Hetchy, 1900-1913." Master's thesis, Univ. of California, Berkeley, 1935.

D235 Dorr, George Bucknam. *Acadia National Park: Its Origins and Background.* Bangor, Maine: Burr Printing Company, 1942. Maine.

D236 Dorr, George Bucknam. *Acadia National Park: Its Growth and Development. . . . Book II.* Bangor, Maine: Burr Printing Co., 1948. 46 pp. Illus. Maine. Reminiscences of a chief promoter of the park.

Dorrance, J.G. See Besley, Fred Wilson, #B247

D237 Dorson, Richard M. "The Lumberjack Code." *Western Folklore* 8 (Oct. 1949), 358-65. Reminiscences collected in Michigan's Upper Peninsula, 1946.

D238 Dorson, Richard M. "Paul Bunyan in the News, 1939-1941." *Western Folklore* 15 (Jan. 1956), 26-39; 15 (July 1956), 179-93; 15 (Oct. 1956), 247-61. Newspaper anecdotes and editorial comment.

D239 Doty, H.L. "Development of Wood Veneers." *Hardwood Record* 48 (June 25, 1920), 29-30, 34.

D240 Doucet, J. Andre. *Timber Conditions in Little Smoky River Valley, Alberta.* Forestry Branch, Bulletin No. 41. Ottawa, 1914. 52 pp. Illus.

D241 Doucet, J. Andre. *Timber Conditions in the Smoky River Valley and Grande-Prairie Country.* Forestry Branch, Bulletin No. 53. Ottawa, 1915. 55 pp. Illus. Northern Alberta.

Doucet, J. Andre. See Lewis, R.G., #L79

D242 Doughty, Robin W. "Concern for Fashionable Feathers." *Forest History* 16 (July 1972), 4-11. On the plume trade, the antiplumage movement, and their relation to the wildlife preservation movement, ca. 1880s-1900s.

D243 Doughty, Robin W. *Feather Fashions and Bird Preservation: A Study in Nature Protection.* Berkeley: University of California Press, 1975. ix + 184 pp. Illus., notes, tables, bib., apps., index. On feather wearing, the plumage trade, and the movement against it, as part of a world wide wildlife conservation movement, late 19th and early 20th century.

D244 Douglas, Byrd. *Steamboatin' on the Cumberland.* Nashville: Tennessee Book Company, 1961. 407 pp. Illus., bib. Includes some history of lumber rafts and the hardwood lumber industry of eastern Tennessee and eastern Kentucky in the late 19th century.

D245 Douglas, Richard. "Logging in the Big Hatchie Bottoms." *Tennessee Historical Quarterly* 25 (Spring 1966), 32-49. Reminiscences of a former sawmill owner about lumbering in western Tennessee in the early 20th century.

D246 Douglas, William O. *My Wilderness: The Pacific West.* Garden City, New York: Doubleday and Company, 1960. 206 pp. Illus., table, maps. Reminiscences of outings in Western wilderness areas, especially in Oregon and Washington.

D247 Douglas, William O. *My Wilderness: East to Katahdin.* Garden City, New York: Doubleday and Company, 1961. 290 pp. Reminiscences of outings in Maine and other Eastern wilderness areas.

D248 Douglas, William O. *A Wilderness Bill of Rights.* Boston: Little, Brown and Company, 1965. 192 pp. Illus., app., tables, index. Reviews some history of conservation and federal natural resource agencies as part of a plea for wilderness preservation and acceptance of a conservation ethic.

Douglass, Benjamin. See Dillon, John Brown, #D180

D249 Douglass, William. *A Summary, Historical and Political, of the First Planting, Progressive Improvements and Present State of the British Settlements in North America.* 2 Volumes. London: R. and J. Dodsley, 1760. Includes some history of the manufacture of lumber and naval stores in colonial America.

D250 Dovell, Junius E. "A History of the Everglades in Florida." Ph.D. diss., Univ. of North Carolina, 1947. 324 pp.

D251 Dovell, Junius E. "The Railroads and the Public Lands of Florida, 1879-1905." *Florida Historical Quarterly* 34 (Jan. 1956), 236-58.

D252 Dowell, Eldridge F. *A History of Criminal Syndicalism Legislation in the United States.* Johns Hopkins University Studies in Historical and Political Science, Series 57, No.

1. Baltimore: Johns Hopkins Press, 1939. 176 pp. Directed in part against the Industrial Workers of the World, a radical union which organized many lumberjacks and millworkers.

D253 Dowling, D.B. "The Pioneers of Jasper Park." *Transactions of the Royal Society of Canada,* Third series, 11 (Mar. 1918), 241-52. Jasper National Park, Alberta.

Dowling, Ellen C. See American Tree Association, #A410

D254 Downs, Norton. "A Tribute to James Lippincott Goodwin." *Connecticut Woodlands* 36 (Winter 1971-1972), 15-17. A Connecticut forester and conservationist.

D255 Drabek, Stanley. "Headquarters-Field Relationships in the Ontario Department of Lands and Forests." Ph.D. diss., Univ. of Toronto, 1972. Examines the administrative organization of the department from 1941 to 1967.

D256 Drake, George L. "The U.S. Forest Service, 1905-1955: An Industry Viewpoint." *Journal of Forestry* 53 (Feb. 1955), 116-20. From the point of view of an industrial forester of the Pacific Northwest.

D257 Drake, George L. "Pacific Logging Congress: An Inside View." OHI by Elwood R. Maunder. *Forest History* 16 (Oct. 1972), 24-29. A lumbermen's association of the Pacific states, 1919-1930s.

D258 Draper, A.S. "Reminiscences of the Lumber Camp." *Michigan History Magazine* 14 (Summer 1930), 438-54. Northern Michigan, 1890s.

D259 Drefahl, L.C. "Zinc Chloride as a Wood Preservative: Its Past, Present, and Future." *Proceedings of the American Wood Preservers' Association* 26 (1930), 78-95.

D260 Drexel Enterprises, Inc. *Reflections: A History of Drexel Enterprises, Inc., 1903-1963.* Drexel, North Carolina, 1963. 92 pp. Illus., maps. A North Carolina furniture manufacturer.

D261 Drinker, Henry S. "The Spread of the Forestry Movement." *American Forestry* 19 (Mar. 1913), 175-90. On the growth of forestry and forest conservation since the 1870s; includes sketches of state forestry organizations, conservation organizations, and forestry education.

D262 Drinker, Henry S.; Pinchot, Gifford; and Stuart, Robert Y. "Dr. Joseph Trimble Rothrock." *Journal of Forestry* 21 (Nov. 1923), 669-76. Biographical sketches and tributes to Rothrock (1839-1922), Pennsylvania forester and forestry educator.

D263 Droze, Wilmon H. *High Dams and Slack Waters: TVA Rebuilds a River.* Baton Rouge: Louisiana State University Press, 1965. vii + 174 pp. Illus., bib. A history of the navigation and flood control program of the Tennessee Valley Authority, with references to other aspects of conservation.

D264 Droze, Wilmon H.; Wolfskill, George; and Leuchtenburg, William E. *Essays on the New Deal.* Ed. by Harold M. Hollingsworth and William F. Holmes. Walter Prescott Webb Memorial Lectures, Volume 2. Austin: University of Texas Press, for the University of Texas, Arlington, 1969. 115 pp. Notes. Droze's essay is entitled "The New Deal's Shelterbelt Project, 1934-1942."

D265 Druce, Eric. "The Canadian Society of Forest Engineers." *Journal of Forestry* 44 (Aug. 1946), 573-74. Brief history of the organization, precursor of the Canadian Institute of Forestry, since 1908.

D266 Drury, Aubrey. "Saving the Redwoods." *Pacific Discovery* 2 (Sept.-Oct. 1949), 25-30. On the work of the Save-the-Redwoods League in California since 1918.

D267 Drury, Aubrey. *John A. Hooper and California's Robust Youth.* San Francisco: N.p., 1952. 91 pp. Illus. Hooper, a San Franciscan, was involved in the lumber industry, shipping, and banking, 1853-1925.

D268 Drury, Aubrey. "John Albert Hooper." *California Historical Society Quarterly* 31 (Dec. 1952), 289-305.

Drury, Newton B. See Tilden, Freeman, #T107

D269 Dryfhout, John H. "The Saratoga of the West." *Inland Seas* 20 (No. 3, 1964), 185-95. History of Grand Haven, Michigan, a lumber town in the mid-19th century and later a resort town.

D270 Dubay, Robert W. "The Civilian Conservation Corps: A Study of Opposition, 1933-1935." *Southern Quarterly* 6 (Apr. 1968).

D271 Dubofsky, Melvyn. *We Shall Be All: A History of the Industrial Workers of the World.* Chicago: Quadrangle Books, 1969. xvi + 557 pp. Illus., notes, bib., index. A general history of the radical union, including its efforts to organize lumberjacks and millworkers.

D272 duBois, Coert. *Trail Blazers.* Stonington, Connecticut: Stonington Publishing Company, 1957. x + 85 pp. Illus., map. On the author's forestry education at the Biltmore Forest School near Asheville, North Carolina, 1890s, and his experiences in the USFS in California and the Rocky Mountain states, 1900-1919.

D273 duBois Coert. "Trail Blazers." *Forest History* 9 (Oct. 1965), 14-22. Recollections of a USFS forest inspector in the West, 1903-1906; an excerpt from duBois' autobiography, *Trail Blazers* (1957).

D274 Dubos, Rene Jules. "The Genius of the Place." *American Forests* 76 (Sept. 1970), 16-19, 61-62. Brief history of ecological thought.

D275 Duchaine, William J. "Charcoal." *American Forests* 63 (Feb. 1957), 50-51, 66-67. On the history of the charcoal industry in the United States, especially in Michigan, since the 1850s.

Dudley, Harold M. See Oliver, A.C., #O29

D276 Dudley, Susan, and Goddard, David R. "Joseph T. Rothrock and Forest Conservation." *Proceedings of the American Philosophical Society* 117 (Feb. 16, 1973), 37-50. Rothrock (1839-1922) long worked for forest conservation and forestry education in Pennsylvania and nationally.

D277 Dudley, William Russell. "Forestry Notes: The Redwood Reservation Act." *Sierra Club Bulletin* 3 (June 1901), 337-39. A brief history of the legislation which provided for the purchase of land creating Big Basin Redwood Park, California's first state park.

D278 Due, John F. "The City of Prineville Railway and the Economic Development of Crook County." *Economic Geography* 43 (No. 2, 1967), 170-81. Includes some history of the Oregon town's lumber industry and of its dependence on the municipal railroad since the 1930s.

D279 Due, John F., and Juris, Frances. *Rails to the Ochoco Country: The City of Prineville Railway.* San Marino, California: Golden West Books, 1968. 236 pp. Illus., maps, references, index. History of the municipally owned railway which has serviced the central Oregon lumber town of Prineville since 1918.

D280 Due, John F. "Dangers in the Use of the Subsidization Technique: The Central Oregon Wagon Road Grants." *Land Economics* 46 (May 1970), 105-17.

Duerr, William A. See Zon, Raphael, #Z29

D281 Duerr, William A. "The Economic Problems of Forestry in the Appalachian Region." Ph.D. diss., Harvard Univ., 1945. 223 pp.

D282 Duerr, William A., and Gustafson, R.O. *Management of Forests in an Eastern Kentucky Area.* Kentucky Agricultural Experiment Station, Bulletin No. 518. Lexington, 1948. 122 pp.

D283 Duerr, William A. *The Economic Problems of Forestry in the Appalachian Region.* Harvard Economic Studies, Volume 84. Cambridge: Harvard University Press, 1949. xi + 317 pp. Notes, bib. Includes incidental historical references.

D284 Duerr, William A., and Vaux, Henry J. "Research in the Economics of Forestry, 1940-1947." *Journal of Forestry* 47 (Apr. 1949), 265-70. A report by the Society of American Foresters' Committee on Scope and Method of Research in the Economics of Forestry.

D285 Duerr, William A. "The Southern Hardwood Problem: Some Resource Facts." *Southern Lumberman* 179 (Dec. 15, 1949), 147-52. On the replacement of pine by low-quality hardwoods in Mississippi since 1932.

D286 Duerr, William A., and Vaux, Henry J., eds. *Research in the Economics of Forestry.* Washington: Charles Lathrop Pack Forestry Foundation, 1953. xi + 475 pp. Notes, index. Contains many historical references to problems in forest economics.

D287 Duerr, William A. *Fundamentals of Forestry Economics.* American Forestry Series. New York: McGraw-Hill, 1960. 579 pp. Illus., bib.

D288 Duerr, William A., ed. *Timber! Problems, Prospects, Policies.* Ames: Iowa State University Press, 1973. xvi + 260 pp. An anthology of essays on environmental problems and national policies relating to forests, forestry, and the forest industries, including some historical references.

D289 Duffield, George. "Frontier Mills." *Annals of Iowa* 6 (July 1904), 425-36. Includes some sawmills in 19th-century Iowa.

D290 Duffus, Robert L. *The Valley and Its People: A Portrait of TVA.* New York: Alfred A. Knopf, 1944. 167 pp. Illus. Tennessee Valley Authority.

D291 Dulles, Foster Rhea. *A History of Recreation: America Learns to Play.* 1940. Second edition. New York: Appleton-Century-Crofts, 1965. xvii + 446 pp. Illus., notes, index. Includes some references to outdoor recreation and other pastimes associated with forested lands.

D292 Dunbar, Robert G. "The Economic Development of the Gallatin Valley." *Pacific Northwest Quarterly* 47 (Oct. 1956), 117-23. Includes a brief account of logging and tie operations in Montana's Gallatin Valley since 1872.

D293 Dunbar, Willis F. *Michigan Through the Centuries.* 4 Volumes. New York: Lewis Historical Publishing Company, 1955. Illus., maps, bibs., index. Chapter 18 of Volume 1 treats the history of the lumber industry; there are many other scattered references to forest history.

D294 Dunbar, Willis F. *All Aboard! A History of Railroads in Michigan.* Grand Rapids: W.B. Eerdmans Publishing Company, 1969. 308 pp. Illus., maps, notes, index. Includes some history of logging railroads.

D295 Duncan, J.S. "The Effect of the N.R.A. Lumber Code on Forest Policy." *Journal of Political Economy* 49 (Feb. 1941), 91-102. National Recovery Administration, 1930s.

D296 Duncan, Richard. "The Paper Bag, a History of Its Manufacture." *Paper Mill News* 69 (Mar. 23, 1946), 10-14.

D297 Dunford, Earl Gerard. "Watershed Management Research in the Lake States, Intermountain, and Pacific Northwest Regions." *Journal of Forestry* 58 (Apr. 1960), 288-90. By the USFS since 1912.

D298 Dunham, Harold Hathaway. "Some Crucial Years of the General Land Office, 1875-1890." *Agricultural History* 11 (Apr. 1937), 117-41. Emphasizes the lack of appropriations and staff during this period of land and timber frauds.

D300 Dunham, Harold Hathaway. *Government Handout: A Study in the Administration of the Public Lands, 1875-1891.* New York: By the author, 1941. Reprint. New York: DaCapo Press, 1970. vii + 364 pp. Bib. Includes chapters on "timber laws" and "timber companies." A Ph.D. dissertation at Columbia University, 1942.

D301 Dunn, B.C. "The American Forest Stand in Relation to the Economic Progress of the Country." *Vassar Journal of Undergraduate Studies* 7 (1933).

Dunn, Edgar S., Jr. See Perloff, Harvey S., #P119

D302 Dunn, James T. *Marine Mills: Lumber Village, 1838-1888.* Marine on St. Croix, Minnesota: By the author, 1963. 55 pp. Illus. A chronology of a Minnesota lumber town.

D303 Dunn, James T. *The St. Croix: Midwest Border River.* Rivers of America Series. New York: Holt, Rinehart and Winston, 1965. 309 pp. Illus., maps, bib. Includes some history of logging and the lumber industry in the St. Croix Valley of Wisconsin and Minnesota.

D304 Dunn, James T. *Marine on St. Croix; From Lumber Village to Summer Haven, 1838-1968.* Marine on St. Croix, Minnesota: Marine Historical Society, 1968. 105 pp. Illus., map, index. An expanded version of the author's *Marine Mills: Lumber Village, 1838-1888* (1963).

D305 Dunne, D. Michael. "Conservation on the American Frontier." *American West* 9 (Sept. 1972), 48, 59. Examples of leading conservationists from frontier regions, 19th century.

D306 Dunning, Carroll W. *Wood-Using Industries of Ohio.* Wooster: Ohio Agricultural Experiment Station, 1912. 133 pp. Illus.

D307 Dunning, Carroll W. "Timber Resources and Wood-Using Industries of the State of Idaho." *Pacific Lumber Trade Journal* 18 (July 1912), 23-26. Incidental history.

D308 Dupree, Anderson Hunter. *Science in the Federal Government: A History of Policies and Activities to 1940.* Cambridge: Belknap Press of Harvard University Press, 1959. x + 460 pp. Bib., index. Includes some history of forest conservation, the conservation movement generally, particularly in terms of federal policies and agencies.

D309 Dupree, Anderson Hunter. *Asa Gray, 1810-1888.* Cambridge: Belknap Press of Harvard University Press, 1959. x + 505 pp. Illus., bib., index. Biography of an eminent botanist and taxonomist who helped promote forestry as a member of the American Association for the Advancement of Science in the 1870s.

DuPuy, William A. See Wilbur, Ray Lyman, #W282

D310 DuPuy, William A. *The Nation's Forests.* New York: Macmillan, 1938. vii + 264 pp. Illus.

D311 DuPuy, William A. *Green Kingdom: The Way of Life of a Forest Ranger.* Evanston, Illinois: Row, Peterson and Company, 1940. 64 pp. Illus. Incidental historical references.

D312 Durant, Edward W. "Lumbering and Steamboating on the St. Croix River." *Collections of the Minnesota Historical Society* 10, pt. 2 (1905), 645-75. Of Minnesota and Wisconsin.

D313 Durham, George. "Canoes from Cedar Logs: A Study of Early Types and Designs." *Pacific Northwest Quarterly* 46 (Apr. 1955), 33-39. On the Pacific Northwest coast since 1778.

D314 Durisch, Lawrence L., and Macon, Hershal L. *Upon Its Own Resources: Conservation and State Administration.* University: Cooperatively published by the University of Alabama Press, University of Georgia Press, University of Mississippi, Bureau of Public Administration, University of South Carolina Press, and University of Tennessee Press, 1951. ix + 136 pp. Illus., maps, notes, bib. Summarizes separate studies of natural resource administration in six Southeastern states, including some history of each administrative agency.

D315 Durisch, Lawrence L., and Lowry, Robert E. "State Watershed Policy and Administration in Tennessee." *Public Administration Review* 15 (Winter 1955), 17-20. Describes the federal-state pilot cooperative program between TVA and Tennessee.

D316 Durland, William Davies. "An Inquiry into Forest Adequacy as It Concerns the United States." Ph.D. diss., Univ. of Texas, 1957. 321 pp. Includes some history of forest depletion and changes in forest industries.

D317 Durst, Ross C. "The Mills of New Germany." *Glades Star* 2 (Sept. 1955), 300-02. On grist mills and sawmills in what is now the New Germany Recreation Center of the Savage River Forest, Maryland, 1859-1927.

D318 Dutcher, William. "In the Beginning—An Early History of Our Origin and Growth." *Audubon* 57-58 (Mar. 1955-July 1956). A series of short articles on the National Audubon Society from 1883 to 1895, reprinted from *Audubon's* predecessor, *Bird-Lore,* 1905.

D319 Dutton, W.L. "History of Forest Service Grazing Fees." *Journal of Range Management* 6 (Nov. 1953), 393-98. Reviews the USFS grazing fee policy since 1905.

D320 Dwight, T.W. *Forest Conditions in the Rocky Mountains Forest Reserve.* Forestry Branch, Bulletin No. 33. Ottawa, 1913. 62 pp. Illus.

D321 Dwight, T.W. "The Contribution of R.H. Campbell to the Forestry Profession." *Canadian Forest and Outdoors* 22 (No. 1, 1926), 32-33.

Dworsky, Leonard B. See Smith, Frank E., #S366

D322 Dyar, Ralph E. *News for an Empire: The Story of the Spokesman-Review of Spokane, Washington, and of the Field It Serves.* Caldwell, Idaho: Caxton Printers, 1952. xlix + 494 pp. Illus., bib. Primarily a history of the newspaper, but includes some history of the lumber industry in eastern Washington and northern Idaho.

D323 Dyche, William K. "Log Drive on the Clearwater." *Forest History* 5 (Fall 1961), 2-12. Memoir of a logging operation on Idaho's Clearwater River, 1913-1914.

D324 Dyche, William K. "Tongue River Experience." *Forest History* 8 (Spring/Summer 1964), 2-16. Memoir of a logging and tie operation on the Big Horn National Forest, Wyoming, 1906-1909.

D325 Dyer, C. Dorsey. "History of the Gum Naval Stores Industry." *AT-FA Journal* 25 (Jan. 1963), 5-8. Since the colonial period.

D326 Dykstra, Daniel J. "Corporations in the Day of the Special Charter." *Wisconsin Law Review* (Mar. 1949), 310-35; (May 1949), 468-93. On the charters granted to "lumber, log driving, booming, and river improvement corporations," Wisconsin, 1858-1872.

D327 Dykstra, Daniel J. "Legislation and Change." *Wisconsin Law Review* (May 1950), 523-40. On Wisconsin legislation relating to the construction and operation of milldams since 1840.

D328 Dykstra, Daniel J. "Law and the Lumber Industry, 1861-1888." S.J.D. diss., Univ. of Wisconsin, 1950.

D329 Dykstra, Daniel J. "Legislative Efforts to Prevent Timber Conversion: The History of a Failure." *Wisconsin Law Review* (May 1952), 461-79. On efforts to protect timbered school and university lands in Wisconsin, 1860-1880.

E1 Eames, Ninetta. "Staging in the Mendocino Redwoods." *Overland Monthly* 20 (Aug. 1892), 113-31; 20 (Sept. 1892), 265-84. Includes some history of the lumber industry in the redwood region of Mendocino County, California.

E2 Earle, G. Harold. "Reminiscence of Excellence." *Chronicles of the Historical Society of Michigan* (Sept. 1966). On C.J.L. Meyer and the IXL Lumber Company of Michigan.

E3 Earnest, Ernest P. *John and William Bartram, Botanists and Explorers, 1699-1777, 1739-1823.* Philadelphia: University of Pennsylvania Press, 1940. vii + 187 pp. Illus., bib. The writings of the Bartrams advanced knowledge of American natural history and fostered a more appreciative attitude toward wilderness. This work focuses on the elder Bartram.

Earnshaw, Deanne. See Lowry, Alexander, #L285

E4 East, Dennis. "Water Power and Forestry in Wisconsin: Issues of Conservation, 1890-1915." Ph.D. diss., Univ. of Wisconsin, 1971. 526 pp.

E5 *Eastern Building Materials and Lumber Trade Journal.* "Eighty Eventful Years." *Eastern Building Materials and Lumber Trade Journal* 80 (Oct. 1966), 4-5. Brief history of the journal and its predecessor, the *New York Lumber Trade Journal,* as well as the New York Lumber Trade Association, since 1886.

E6 *Eastern Building Materials and Lumber Trade Journal.* "The Retail Lumber Business in Retrospect." *Eastern Building Materials and Lumber Trade Journal* 80 (Oct. 1966), 8-10, 29-30. Includes brief company histories of New England retail lumber firms which were founded in the 19th century.

E7 Eastman, H.B. "History and Industrial Uses of the Second Growth White Pine of New England." Master's thesis, Yale Univ., 1904.

E8 Eastom, Frank Amos. "Enterprise at Fraser." *Westerners Brand Book* 10 (1955), 117-38. On the founding of Fraser, Colorado, by George Eastom, who also organized the Middle Park Lumber Company.

E9 Easton, Augustus B., ed. *History of the St. Croix Valley.* 2 Volumes. Chicago: H.C. Cooper, Jr., 1909. Includes some history of the forest industries in the Minnesota-Wisconsin valley.

E10 Easton, Hamilton Pratt. "The History of the Texas Lumbering Industry." Ph.D. diss., Univ. of Texas, 1947. 271 pp.

E11 Eastwood, Alice. "Early Botanical Explorers on the Pacific Coast and the Trees They Found There." *California Historical Society Quarterly* 18 (Dec. 1939), 335-46.

E12 Eaton, Ida. *As It Looked to Me.* New York: Exposition Press, 1949. 38 pp. Girlhood memories of "Sawdust Town" in Michigan, 1880s.

E13 Eaton, Leonard K. *Landscape Artist in America: The Life and Work of Jens Jensen.* Chicago: University of Chicago Press, 1964. 336 pp. Illus. Danish-born Jensen (1860-1951) was a noted landscape architect and conservationist.

E14 Eaton, Louis Woodbury. *Pork, Molasses, and Timber; Stories of Bygone Days in the Logging Camps of Maine.* New York: Exposition Press, 1954. 75 pp. Illus. Includes reminiscences of the author's experience as a lumberman in Washington County, Maine, ca. 1909.

E15 Ebert, Charles H.V. "Furniture Making in High Point." *North Carolina Historical Review* 36 (July 1959), 330-39. On the furniture industry and its relation to forest resources near High Point, North Carolina, since 1889.

Ebert, Isabel J. See Sorden, Leland George, #S436

E16 Eccles, W.J. *The Canadian Frontier, 1534-1760.* New York: Holt, Rinehart and Winston, 1969. xv + 234 pp. Illus., maps. Includes references to a variety of forest uses.

E17 Eckardt, Hugo William. *Accounting in the Lumber Industry.* New York: Harper, 1929. xii + 291 pp. Incidental history.

E18 Eckstorm, Fannie H. *The Penobscot Man.* Boston: Houghton Mifflin, 1904. 351 pp. Stories of log driving on Maine's Penobscot River; later editions appeared in 1924 and 1931.

E19 Eckstorm, Fannie H. *David Libbey: Penobscot Woodsman and River Driver.* Boston: American Unitarian Association, 1907. 109 pp. Of Maine; there is a 1922 edition.

E20 Eckstorm, Fannie H., and Smyth, Mary Winslow. *Minstrelsy of Maine. Folk-Songs and Ballads of the Woods and the Coast.* Boston: Houghton Mifflin, 1927. 390 pp.

E21 Ecusta Paper Corporation. *10 Years, Ecusta Service.* Pisgah Forest, North Carolina, 1949. 96 pp. Illus. Producers of cigarette paper and paper products. Also issued as the anniversary edition of the company publication, *The Echo.*

Edwards, E.E. See Folsom, William H.C., #F135

E22 Edwards, George. *Public Domain or Government by Law.* San Francisco: National Publishing Company, 1934. 369 pp. General history of state and federal land policies.

E23 Edwards, J.D. *Paper and Paper Products: Development of Restrictive Regulations, 1917-1918.* U.S. Bureau of Labor Statistics, Historical Study No. 20. Washington: GPO, 1941. 19 pp.

E24 Edwards, N.L. "The Establishment of Paper Making in Upper Canada." *Ontario Historical Society Papers and Records* 39 (1947), 63-74.

Edwards, Paul Carroll. See Wilbur, Ray Lyman, #W284

E25 Edwards, Paul E. "Charles L. McNary: Promoter of Water Resource Development Programs in the United States, 1917-1944." Master's thesis, American Univ., 1963.

E26 Edwards, William Grimm. "The Hardwood-Using Industries of the San Francisco Bay Region." Master's thesis, Univ. of California, Berkeley, 1926.

E27 Eddy, Gerald Ernest. "Who Foots the Bill?" *Michigan Conservation* 29 (Sept.-Oct. 1960), 2-6. On the Michigan Department of Conservation since 1921.

E28 Eddy, John Mathewson, comp. *In the Redwood's Realm: By-ways of Wild Nature and Highways of Industry as Found in Humboldt Co., Calif.* San Francisco: D.S. Stanley & Company, 1893. 112 pp. Illus., maps, tables. Primarily an advertising brochure for the Humboldt Chamber of Commerce, it also includes some history of the lumber industry in the redwood region.

E29 Edgerton, Daisy Priscilla. *First Steps in Southern Forest Study.* New York: Rand McNally, 1930. xii + 308 pp. Illus., diags., maps.

Edgerton, Daisy Priscilla. See Randall, Charles E., #R27

E30 Egan, James A. "Forestry Above the Mogollon Rim." *Journal of Forestry* 47 (Jan. 1949), 14-17. Forest management on Arizona's Coconino National Forest since 1905.

E31 Edleston, Melville E. *The Land Systems of the New England Colonies.* Johns Hopkins Studies in Historical and Political Science, Series 4, Parts 11-12. Baltimore: Johns Hopkins University, 1886. 56 pp.

E32 Egleston, Nathaniel H. *Report on Forestry.* Volume 4. Washington: GPO, 1884. 421 pp. Tables, charts, index. The last of a series begun by Franklin B. Hough in 1876, much of this volume was written by Hough rather than Egleston. It includes information on forest conditions, the lumber trade, the use of timber by railroads, the maple sugar industry, and the results of tree planting experiments on the Great Plains.

E33 Egleston, Nathaniel H. "Summary of Legislation for the Preservation of Timber or Forests on the Public Domain." *Division of Forestry Bulletin* 2 (1889), 212-20. Since 1831.

E34 Egleston, Nathaniel H. *Arbor Day, Its History and Observance.* USDA, Report No. 56. Washington: GPO, 1896. 80 pp. Illus.

Eichhorn, Noel D. See Darling, F. Fraser, #D44

E35 Eichstedt, John H. *Payments in Lieu of Taxes on Public Lands under the Jurisdiction of the Michigan Department of Conservation.* Institute of Public Administration, Bureau of Government, Papers in Public Administration, No. 16. Ann Arbor: University of Michigan, 1956. iii + 48 pp. Diags., map, tables, notes, bib. Since 1869.

E36 Eifert, Virginia S. *Tall Trees and Far Horizons: Adventures and Discoveries of Early Botanists in America.* New York: Dodd, Mead & Company, 1965. 301 pp. Illus., index. Popular history of botanical explorers such as Mark Catesby, John Bartram, David Douglas, John Muir, and others.

E37 Eisenhower, Milton S., and Chew, Arthur P. *The United States Department of Agriculture: Its Growth, Structure and Functions.* USDA, Miscellaneous Publication No. 88. Washington: GPO, 1930. iv + 147 pp. Diags. Revised by Chew in 1940.

E38 Eisterhold, John A. "Lumber and Trade in the Seaboard Cities of the Old South, 1607-1860." Ph.D. diss., Univ. of Mississippi, 1970. 254 pp.

E39 Eisterhold, John A. "Colonial Beginnings in the South's Lumber Industry: 1607-1800." *Southern Lumberman* 223 (Dec. 15, 1971), 150-53.

E40 Eisterhold, John A. "Lumber and Trade in the Lower Mississippi Valley and New Orleans, 1800-1860." *Louisiana History* 13 (Winter 1972), 71-91.

E41 Eisterhold, John A. "Savannah: Lumber Center of the South Atlantic." *Georgia Historical Quarterly* 57 (Winter 1973), 526-43. In the 19th century.

E42 Eisterhold, John A. "Lumber and Trade in Pensacola and West Florida: 1800-1860." *Florida Historical Quarterly* 51 (Jan. 1973), 267-80.

E43 Eisterhold, John A. "Patrick Parker of Virginia, 1788-1795." *Forest History* 16 (Jan. 1973), 18-19. On the financial failure of a Norfolk lumber merchant.

E44 Eisterhold, John A. "Charleston: Lumber and Trade in a Declining Southern Port." *South Carolina Historical Magazine* 74 (Apr. 1973), 61-73. In the 19th century.

E45 Eisterhold, John A. "Mobile: Lumber Center of the Gulf Coast." *Alabama Review* 26 (Apr. 1973), 83-104. In the 19th century.

E46 Ekirch, Arthur A. *Man and Nature in America.* New York. Columbia University Press, 1963. Reprint. Lincoln: University of Nebraska Press, 1973. xii + 331 pp. Notes, index. On the historic impact of science and technology on the Romantic, agrarian, and Transcendental ideal of a harmonious relationship between man and the natural world. Henry David Thoreau, George Perkins Marsh, and other environmental and conservation philosophers are treated.

E47 Elazar, Daniel J. "Land Space and Civil Society in America." *Western Historical Quarterly* 5 (July 1974), 262-84. A historical survey of the influence of land on American political ideas and institutions.

E48 Elchibegoff, Ivan M. *United States International Timber Trade in the Pacific Area.* Stanford, California: Stanford University Press, 1949. xvii + 302 pp. Diags., maps, notes, bib. On the lumber industry and lumber trade in countries bordering the Pacific Ocean, with incidental historical references.

E49 Elchibegoff, Ivan M. "United States International Timber Trade with the Pacific Countries (Land Utilization and Forest Resources in the Pacific)." Ph.D. diss., New School for Social Research,1950.

E50 Eldredge, Inman F. *The Four Forests and the Future* W377

Eldredge, Inman F. See Cruikshank, J.W., #C738

Eldredge, Inman F. See Spillers, A.R., #S511

Eldredge, Inman F. *The Four Forests and the Future of the South.* Washington: Charles Lathrop Pack Forestry Foundation, 1947. 65 pp. Illus. Includes some historical references to the forest industries.

E51 Eldredge, Inman F. "Tom Gill." *Journal of Forestry* 46 (Nov. 1948), 846-47. Sketch of Gill's career in the USFS, 1915-1925, with the Charles Lathrop Pack Forestry Foundation after 1926, and his travels and writings in connection with tropical forestry.

E52 Eldredge, Inman F. "Southern Forests, Then and Now." *Journal of Forestry* 50 (Mar. 1952), 182-85. On the changes in Southern forests and forestry since ca. 1900.

E53 Eldredge, Inman F. "Forty Years of Forestry on Private Lands in the South." *Southern Lumberman* 184 (Apr. 15, 1952), 88-90.

E54 Eldredge, Inman F. "Ride the White Horse—Memories of a Southern Forester." OHI by Elwood R. Maunder. *Forest History* 3 (Winter 1960), 3-14; 4 (Spring 1960), 3-12. On Eldredge's attendance at the Biltmore Forest School, 1904-1905; his employment as forst-manager of the Superior Pine Products Company, Fargo, Georgia; his career in the USFS in the South and Far West; and his work as a forest consultant to 1956.

E55 Eldredge, Inman F. "Ride the White Horse . . . An Interview with 'Mr. Southern Forestry'." OHI by Elwood R. Maunder. *American Forests* 66 (Oct. 1960), 16-19, 60-89. Slightly edited version of the above.

E56 Eldredge, Inman F. "Backwoodsmen I Have Known." *American Forests* 67 (May 1961), 4-5, 44-47. Reminiscences about Hiram Hancock, Jim Hawks, W.H.B. Kent, Manuel Brown, and a Mr. Graddy, interesting men known by the author during his forestry career between 1906 and 1926.

E57 Eldridge, Albert G. "The Problem of Our Wood Pulp and Paper." *Journal of Geography* 29 (Sept. 1930), 240-58.

E58 Elgin, Max. "Girard, a Timberfaring Man of the Mountains." *American Forests and Forest Life* 30 (Jan. 1924), 34-35, 51-52. On the work of James W. Girard for the USFS in Montana and Idaho since 1907.

E59 Eliel, Paul. "Industrial Peace and Conflict: A Study of Two Pacific Coast Industries." *Industrial and Labor Relations Review* 2 (July 1949), 477-501. Concerns the relatively peaceful labor relations in the pulp and paper industry, in contrast to violent relations in the stevedoring industry, since 1934.

E60 Eliot, Thomas H. *Reorganizing the Massachusetts Department of Conservation.* Inter-University Case Program, ICP Case No. 14. University: University of Alabama Press, 1953. 48 pp. Account of an unsuccessful effort to reorganize the department.

E61 Ellarson, Robert H. "The Vegetation of Dane County, Wisconsin in 1835." *Transactions of the Wisconsin Academy of Sciences, Arts, and Letters* 39 (1949), 21-46.

Eller, Geraldine Crill. See Whisler, Ezra Leroy, #W229

E62 Ellickson, Donald Lien. "A History of Land Taxation Theory." Ph.D. diss., Univ. of Wisconsin, 1966. 260 pp.

E63 Elliott, Charles N. *Conservation of American Resources.* Atlanta: Turner E. Smith and Company, 1940. 672 pp. Illus., maps, diags. Incidental history.

E64 Elliott, Lloyd H. *Unique Partners in Progress: The University of Maine and the Pulp and Paper Industry.* New York: Newcomen Society in North America, 1964. 24 pp. Illus.

E65 Ellis, Albert G. "Upper Wisconsin Country." *Collections of the Wisconsin State Historical Society* 3 (1857), 435-52. Reminiscences of pioneer logging and rafting activities on the Wisconsin River.

E66 Ellis, Benjamin R. "The Evolution of Cypress Manufacturing." *Southern Lumberman* (Dec. 15, 1929), 225-26.

E67 Ellis, David M. "Railroad Land Grant Rates, 1850-1945." *Journal of Land and Public Utility Economics* 21 (Aug. 1945), 207-22.

E68 Ellis, David M. "The Forfeiture of Railroad Land Grants, 1867-1894." *Mississippi Valley Historical Review* 33 (June 1946), 27-60. Including forested lands.

E69 Ellis, David M. "The Oregon and California Railroad Land Grant, 1866-1945." *Pacific Northwest Quarterly* 39 (Oct. 1948), 253-83. These Oregon lands, largely forested, were revested to the federal government in 1916 and have often been in dispute since.

E70 Ellis, David M., ed. *The Frontier in American De-velopment: Essays in Honor of Paul Wallace Gates.* Ithaca, New York: Cornell University Press, 1969. xxx + 425 pp. Illus., map, tables, notes, app., index. Includes essays by David C. Smith (q.v.) and others on the history of public lands, land disposal, and other land topics.

E71 Ellis, Elmer. "The Public Career of Henry M. Teller." Ph.D. diss., Univ. of Iowa, 1930.

E72 Ellis, Elmer. *Henry Moore Teller, Defender of theWest.* Caldwell, Idaho: Caxton Printers, 1941. 409 pp. Illus., notes, bib., index. Teller was Chester A. Arthur's Secretary ofthe Interior, 1882-1885. As U.S. senator from Colorado in lateryears, Teller was often in the thick of political battles involving the forest reserves and national forests.

E73 Ellis, Everett L. *Education in Wood Science and Technology.* Madison: Society of Wood Science and Technology, 1964. vi + 187 pp. Tables, apps., bib., index. Includes a section on the "Historical Development of Programs in Wood Science and Technology."

E74 Ellis, H.D. "History of Industrial Forestry in Pennsylvania." *Pennsylvania Forests* 54 (Fall 1964), 82, 87.

E75 Ellis, Havelock. "The Love of Wild Nature." *Contemporary Review* 95 (1909), 180-99. Includes some history of man's conception of wilderness, primarily the Christian view.

E76 Ellis, J.S. "Trail of the Cedar Rails." *Sylva* 13 (July-Aug. 1957), 15-18. On the use of cedar for fencing in southern Ontario.

E77 omitted.

E78 Ellis, James Fernando. *The Influence of Environment on the Settlement of Missouri.* St. Louis: Webster PublishingCompany, 1929. 180 pp. Bib. Also issued as a Ph.D. dissertation at St. Louis University, 1929.

E79 Ellis, Lewis Ethan. *Reciprocity, 1911.* New Haven: Yale University Press, 1939. x + 207 pp. Includes some account of the role of newsprint in tariff reciprocity negotiations with Canada during the William H. Taft administration.

E80 Ellis, Lewis Ethan. *Print Paper Pendulum: Group Pressures and the Price of Newsprint.* New Brunswick, New Jersey: Rutgers University Press, 1948. ix + 215 pp. Notes, bib. On the newsprint industry in the United States and Canada, 1878-1936.

E81 Ellis, Lewis Ethan. *Newsprint: Producers, Publishers, Political Pressures.* New Brunswick, New Jersey: Rutgers University Press, 1960. 305 pp. App. The history of the newsprint industry shows a close relationship between economic conditions and political events in the development and use of American and Canadian forests since the 1870s. The appendix is a reprint of *Print Paper Pendulum* (1948).

E82 Ellison, Lincoln. "Trends of Forest Recreation in the U.S." *Journal of Forestry* 40 (Aug. 1942), 630-38. On thedevelopment of outdoor recreation in forested areas since the 1860s.

E83 Ellsworth, Rodney S. "Discovery of the Big Trees of California, 1833-1852." Master's thesis, Univ. of California, Berkeley, 1923.

E84 Ellsworth, Rodney S. *The Giant Sequoia: An Account of the History and Characterostocs pf Big Trees of California.* Oakland: J.D. Berger, 1924. 167 pp. Illus., bib.

E85 Elmer, Manuel Conrad. *Timber, America's Magic Resource.* Boston: Christopher Publishing House, 1961. 208 pp. Illus. Incidental history.

E86 Ely, Richard T. *The Foundations of National Prosperity: Studies in the Conservation of Permanent National Resources.* New York: Macmillan, 1917. 378 pp. Diags.

E87 Ely, Richard T., and Wehrwein, George S. *Land Economics.* New York: Macmillan, 1940. Reprint. Madison: University of Wisconsin Press, 1964. xi + 496 pp. Notes, tables, index. Includes chapters on forest and recreational land containing some history.

E88 Emerson, F.V. "The Southern Long-Leaf Pine Belt." *Geographical Review* 7 (Feb. 1919), 81-90. Incidental history of the lumber industry.

E89 Emery, K.M. "Tractor Logging, a Resume of Progress since 1893." *Southern Lumber Journal* 49 (July 1945), 19-20, 22.

E90 Emmerling, A.N. "Forests of Oklahoma." *American Forests* 54 (Dec. 1948), 554-55, 574. Forests and forestry since 1909.

E91 Emmerson, Donald W. "History of the Pulp and Paper Industry in Canada." *Indian Pulp and Paper* 2 (July 1947), 48-52, 55.

E92 Emmerson, Irma L., and Muir, Jean. *The Woods Were Full of Men.* New York: David McKay Company, 1963. 242 pp. Autobiographical account of a logging camp cook in Coos County, Oregon.

E93 Emmons, David M. "American Myth: Desert to Eden: Theories of Increased Rainfall and the Timber Culture Act of 1873." *Forest History* 15 (Oct. 1971), 6-14. Tree planting experiments on the Great Plains, 1860s-1890s.

E94 Emmons, David M. *Garden in the Grasslands: Boomer Literature of the Central Great Plains.* Lincoln: University of Nebraska Press, 1971. xi + 220 pp. Illus., notes, bib., index. Contains some incidental historical references to forests and tree planting in relation to settlement and promotion of the Great Plains region.

E95 Empire State Forest Products Association. *50 Years of Forest Management, 1906-1956.* Albany, 1956. 12 pp. Illus.

E96 Engbeck, Joseph H., Jr. *The Enduring Giants: The Giant Sequoias, Their Place in Evolution and in the Sierra Nevada Forest Community; History of the Calaveras Big Trees; The Story of Calaveras Big Trees State Park.* Berkeley: University Extension, University of California, in cooperation with the California Department of Parks and Recreation, Save-the-Redwoods League, and the Calaveras Grove Association, 1973. 120 pp. Illus., maps, tables. Includes some history of the discovery, exploitation, and preservation of sequoias since the mid-19th century.

E97 Engberg, George B. "The Rise of Organized Labor in Minnesota." *Minnesota History* 21 (Dec. 1940), 372-94. Includes incidental mention of organized labor and unionizing efforts among the workers in several forest industries in the late 19th century.

E98 Engberg, George B. "The Knights of Labor in Minnesota." *Minnesota History* 22 (Dec. 1941), 367-90. The Knights of Labor made some efforts to organize workers in the forest industries, 1880s-1890s.

E99 Engberg, George B. "Who Were the Lumberjacks?" *Michigan History* 32 (Sept. 1948), 238-46. Of the Great Lakes region, since 1828.

E100 Engberg, George B. "Labor in the Lake States Lumber Industry—1830-1930." Ph.D. diss., Univ. of Minnesota, 1950. 476 pp.

E101 Engberg, George B. "Collective Bargaining in the Lumber Industry of the Upper Great Lake States." *Agricultural History* 24 (Oct. 1950), 205-11. Since 1858.

E102 Engberg, George B. "Lumber and Labor in the Lake States." *Minnesota History* 36 (Mar. 1959), 153-66. On laborers in the lumber camps and sawmills of Minnesota, Wisconsin, and Michigan, 1830s-1930s.

E103 Engelbert, Ernest A. "American Policy for Natural Resources: A Historical Survey to 1862." Ph.D. diss., Harvard Univ., 1950. 227 pp.

E104 Engelbert, Ernest A. "Political Parties and Natural Resource Policies: An Historical Evaluation, 1790-1950." *Natural Resources Journal* 1 (Nov. 1961), 224-56. Credits Democrats with stronger support for conservation measures than Republicans.

E105 English, P.F. "The Wildlife Society." *Journal of Forestry* 44 (May 1946), 345-46. Includes some history of this wildlife conservation organization since its establishment in 1936.

E106 Engstrom, Emil. *The Vanishing Logger.* New York: Vantage Press, 1956. 135 pp. An autobiographical account of logging in Oregon, Washington, and British Columbia from 1903 to 1946, including reminiscences of the Industrial Workers of the World and strikes in 1919.

E107 Engstrom, Emil. *John Engstrom, the Last Frontiersman.* New York: Vantage Press, 1957. 156 pp. In part concerned with the careers of Emil, John, and Erick Engstrom as loggers in Oregon, Washington, and Alaska in the early 20th century.

E108 Erickson, Kenneth A. "Morphology of Lumber Settlements in Western Oregon and Washington." Ph.D. diss., Univ. of California, Berkeley, 1965. 461 pp. Lumber and sawmill towns since the 1850s.

E109 Erickson, Russell F. *The Story of Rayonier Incorporated.* New York: Newcomen Society in North America, 1963. 24 pp. On Rayonier, Inc., a pulp and paper firm of New York City, with operations in Washington, Florida, and Georgia.

E110 Eriksson, H.C., comp. *Historical Account of the Activities of the 800th Engineer Forestry Company during World War II, 1942-1945.* Hot Springs, Arkansas: N.p., 1954. 77 pp. Illus.

E111 Erisman, Fred. "The Environmental Crisis and Present-Day Romanticism: The Persistence of an Idea." *Rocky Mountain Social Science Journal* 10 (Jan. 1973), 7-14. On the 19th-century Romantic antecedents of contemporary environmentalism.

E112 Ernst, Dorothy J. "Daniel Wells, Jr., Wisconsin Commissioner to the Crystal Palace Exhibition of 1851." *Wisconsin Magazine of History* 42 (Summer 1959), 243-56. Wells, a Milwaukee businessman and lumberman, tried to market his "stave-dressing machine" at the exhibition.

E113 Ernst, Joseph W. "With Compass and Chain: Federal Land Surveyors in the Old Northwest, 1785-1816." Ph.D. diss., Columbia Univ., 1958. 328 pp.

E114 Errington, Paul L. "In Appreciation of Aldo Leopold." *Journal of Wildlife Management* 12 (Oct. 1948), 341-50. Leopold (1886-1948) was an eminent ecologist, wildlife conservationist, and wilderness advocate.

Escola, Nannie. See Moungovan, Thomas O., #M595

E115 Espeth, Edmund C. "Lodestar in the Northland." *Wisconsin Magazine of History* 36 (Autumn 1952), 23-27, 56. On the lumber industry in Vilas County, Wisconsin, since 1855.

E116 Espeth, Edmund C. "Early Vilas County—Cradle of an Industry." *Wisconsin Magazine of History* 37 (Autumn 1953), 27-34, 51-54. On the lumber industry and the later resort and recreational industries of Vilas County, Wisconsin, since 1852.

E117 Estall, R.C. *New England: A Study of Industrial Adjustment.* New York: Praeger Publishers, 1966. xv + 296 pp. Illus., maps, bibs. Includes a chapter on "Development in Paper and Paper Products Manufacture."

E118 Eubank, F.D. "Early Lumbering in East Tennessee." *Southern Lumberman* (Dec. 15, 1935), 100.

E119 Eureka Redwood Lumber Company. *The Titans, Story of the West's Oldest Redwood Lumber Mill.* Portland: M & M Wood Working Company, 1954. 9 pp. Illus. History of the Eureka Redwood Lumber Company, Eureka, California, since 1853.

E120 Evans, C.F. "A Saga of Southern Pine." *American Forests* 48 (Sept. 1942), 403-06, 428. On the pulp and paper industry in the South since 1909.

E121 Evans, Cerinda W. *Some Notes on Shipbuilding and Shipping in Colonial Virginia.* Williamsburg, Virginia: Anniversary Celebration Corporation, 1957. 77 pp.

E122 Evans, Charles F. "A Half Century of Service." *American Forests* 56 (Dec. 1950), 10-13, 33. On the Society of American Foresters since its founding in 1900.

Evans, J. Allan. See Barrow, Susan H.L., #B101

Evenden, J.C. See Craighead, Frank Cooper, #C693

E123 Evenson, W.T. "Ocean Log Rafts." *Timberman* 27 (July 1926), 37-38. Developments in the Pacific coastal trade since the 1880s.

E124 Everard, William P. "Economic Aspects of Cross Tie Preservation." Master's thesis, Yale Univ., 1934.

E125 Everard, William P. "William A. Dayton." *Journal of Forestry* 49 (Apr. 1951), 287-88. Sketch of Dayton's career as a forest ecologist and dendrologist with the USFS, 1910-1950.

E126 Everest, D.C. "Regional Shift in Production in the Paper Industry." *Paper Trade Journal* 90 (Mar. 13, 1930), 34-38.

E127 Everest, D.C. "Progress in Paper, 1900-1950." *Paper Trade Journal* 130 (Feb. 23, 1950), 65-66.

E128 Everest, D.C. "A Reappraisal of the Lumber Barons." *Wisconsin Magazine of History* 36 (Autumn 1952), 17-22. A favorable view by the president of the Marathon Corporation.

E129 Everhart, William C. *The National Park Service.* Praeger Library of U.S. Government Departments and Agencies, No. 31. New York: Praeger Publishers, 1972. xii + 276 pp. Illus., map, apps., bib., index. Includes some history of national parks since 1872 and of the National Park Service since 1916.

E130 Evers, Alf. *The Catskills: From Wilderness to Woodstock.* Garden City, New York: Doubleday & Company, 1972. xiv + 821 pp. Illus., bib. History of New York's forested Catskill Mountains.

E131 Everts, Truman C. "Thirty-Seven Days of Peril, or Lost in the Wilderness." *Montana, Magazine of Western History* 7 (Autumn 1957), 29-52. On the author's experiences while lost from the Washburn-Doane expedition at Yellowstone Lake, Wyoming, 1870.

E132 Ewan, Joseph A. *Rocky Mountain Naturalists.* Denver: University of Denver Press, 1950. xiv + 358 pp. Illus., notes, bib. There are chapters devoted to prominent naturalists and a roster of the less famed, including foresters, from several Rocky Mountain states, 1682-1932.

E133 Ewan, Joseph A., and Ewan, Nesta. *John Banister and His Natural History of Virginia, 1678-1692.* Urbana: University of Illinois Press, 1970. xxx + 485 pp. Illus., map, chart, notes, bib., indexes. An early botanist and student of natural history.

Ewan, Nesta. See Ewan, Joseph A., #E133

E134 Ewing, Cortez A.M. "The Impeachment of Colonel W.L. McGaughey (1893)." *Southwestern Social Science Quarterly* 15 (1934), 52-63. McGaughey, commissioner of the General Land Office, was impeached for an alleged violation of the Dawes Act of 1887.

E135 Ewing, James R. *Public Services of Jacob Dolson Cox, Governor of Ohio and Secretary of the Interior.* Washington: Neale Publishing Company, 1902. 31 pp. Cox was Ulysses S. Grant's Secretary of the Interior, 1869-1870. Also issued as a Ph.D. dissertation at Johns Hopkins University, 1899.

E136 Ewing, Thomas. "The Autobiography of Thomas Ewing." Ed. by Clement L. Martzolff. *Ohio Archaeological and Historical Quarterly* 22 (Jan. 1913), 126-204. Ewing was Secretary of the Interior, 1849-1850.

F1 Fabos, Julius Gy.; Milde, Gorden T.; and Weinmayr, V. Michael. *Frederick Law Olmsted, Sr., Founder of Landscape Architecture in America.* Amherst: University of Massachusetts Press, 1968. 114 pp. Illus., app., bib. Olmsted (1822-

1903) was a major figure of the 19th-century conservation movement. This heavily illustrated work emphasizes landscape architecture.

F2 Fackenthal, B.F., Jr. "Improving Navigation on the Delaware River. . . ." *Papers of the Bucks County Historical Society* 6 (1932). Includes some history of log rafting.

F3 Fadely, Marian E. "Isabelle's Story." *American Forests* 61 (Apr. 1955), 37, 51-52. Career sketch of Isabelle F. Story, an information officer and editor for the National Park Service from 1916 to the 1950s. Her writings publicized the national parks and conservation history.

F4 Fagerlund, Gunnar O. *Olympic National Park, Washington.* U.S. National Park Service, Natural History Handbook Series, No. 1. Washington, 1954. 67 pp. Illus., map, bib. Includes some history of the park and surrounding area.

F5 Fagin, Nathan Bryllion. *William Bartram: Interpreter of the American Landscape.* Baltimore: Johns Hopkins Press, 1933. ix + 229 pp. Bartram (1739-1823) was a botanist and naturalist who traveled widely in the American South and described the wilderness in appreciative terms.

F6 Fahl, Ronald J. "S.C. Lancaster and the Columbia River Highway: Engineer as Conservationist." *Oregon Historical Quarterly* 74 (June 1973), 101-44. Biographical sketch of Samuel Christopher Lancaster (1864-1941), builder of the scenic Columbia River Highway in Oregon, 1913-1916, and promoter and protector of scenic resources.

Fahnestock, G.R. See Olson, D.S., #O39

F7 Fairbanks, W.W. "Historical Sketch of the Redwood Industry." *Timberman* 27 (June 1926), 38-41, 168. On the beginnings (1850s-1860s) of the lumber industry in California's redwood region.

F8 Fairchild, Fred Rogers, and Associates. *Forest Taxation in the United States.* USDA, Miscellaneous Publication No. 218. Washington: GPO, 1935. 681 pp. Tables, graphs, bib. Report on studies of forest taxation (Forest Taxation Inquiry) authorized by the Clarke-McNary Act of 1924. Many incidental historical references to forest taxation are included.

F9 Falco, Joseph A. "Political Background and First Gubernatorial Administration of Gifford Pinchot, 1923-1927." Ph.D. diss., Univ. of Pittsburgh, 1956. 341 pp. Pinchot (1865-1946), a leading forester and conservationist, twice served as governor of Pennsylvania.

Falconer, John I. See Bidwell, Percy Wells, #B260

F10 Falconer, Robert. "Tribute to the Late Dr. B.E. Fernow." *Canadian Forestry Magazine* 19 (No. 5, 1923), 328-29. Fernow (1851-1923) was chief of the U.S. Division of Forestry, 1886-1898, and a leading forestry educator at several schools, including the University of Toronto.

F11 Falk, Harry W., Jr. *Timber and Forest Product Law.* Berkeley: Howell-North Books, 1958. xviii + 365 pp. Notes, table of cases, index. Incidental history.

F12 Falkenau, G.E. "A Study of the Wood-Preserving Industry." Master's thesis, State Univ. of New York, College of Forestry, 1936.

F13 Fall, Albert Bacon. "The Memoirs of Albert Bacon Fall." Ed. with annotations by David H. Stratton. *Southwestern Studies* 4 (No. 3, 1966), 1-63. Fall was Warren G. Harding's Secretary of the Interior, 1921-1923. As U.S. senator from New Mexico, 1912-1921, Fall was interested in resource development and regarded as an anticonservationist.

F14 Fannon, R.W. "The Growth of the Pulp and Paper Industry in the South." *Proceedings of the Southern Forestry Congress* 7 (1925), 95-101.

F15 Farb, Peter. "Messiah of the Soil." *American Forests* 66 (Jan. 1960), 18-19, 40-42. Biographical sketch of Hugh H. Bennett (1881-1960), chief of the Soil Conservation Service, 1935-1951.

F16 Farb, Peter. "Money Can Grow on Trees." *American Forests* 66 (Mar. 1960), 10-11, 48-52. On tree farming since ca. 1920.

F17 Farb, Peter. *Face of North America: The Natural History of a Continent.* New York: Harper, 1963. 316 pp. Illus. Includes some history of forested lands.

F18 Faris, John T. *Roaming the Rockies: Through National Parks and National Forests of the Rocky Mountain Wonderland.* New York: Farrar and Rinehart, 1930. xiv + 333 pp. Illus., map, bib. Incidental history.

Farley, Alan W. See Richardson, Elmo R., #R179

F19 Farnan, William T. "Land Claims Problems and the Federal Land System in the Louisiana-Missouri Territory." Ph.D. diss., St. Louis Univ., 1971. 279 pp. From 1803 to 1820.

F20 Farney, Dennis. "In the Sea of Grass a Forest Grows." *National Wildlife* 13 (Dec.-Jan. 1975), 18-21. On tree planting in the Sand Hills of northwestern Nebraska and establishment of the Nebraska National Forest, since ca. 1900.

F21 Farquhar, Francis P. "Stephen T. Mather: 1867-1930." *Sierra Club Bulletin* 16 (1931), 55-59. Mather was the first director of the U.S. National Park Service, 1917-1928, and a leader in the conservation movement.

F22 Farquhar, Francis P. "Colonel George W. Stewart, Founder of Sequoia National Park." *Sierra Club Bulletin* 17 (1932), 49-52. Stewart, a Visalia newspaperman, agitated for the establishment of the California park. It was protected by an act of Congress in 1890.

F23 Farquhar, Francis P. "Legislative History of Sequoia and Kings Canyon National Parks." *Sierra Club Bulletin* 26 (Feb. 1941), 42-58. Since 1880.

F24 Farquhar, Francis P. "Walker's Discovery of Yosemite." *Sierra Club Bulletin* 27 (Aug. 1942), 35-49. On the expedition of Joseph Reddeford Walker, which in 1833 discovered California's Yosemite Valley and the sequoias of the Sierra Nevada.

F25 Farquhar, Francis P. *Yosemite, the Big Trees and the High Sierra; a Selective Bibliography.* Berkeley: University of California Press, 1948. xii + 104 pp. Illus. Annotated guide to literature revealing the history of the Sierra Nevada and its national parks.

F26 Farquhar, Francis P. "California's Big Trees." *American West* 2 (May 1965), 58-64. On the discovery, promotion, and preservation of sequoia groves.

F27 Farquhar, Francis P. *History of the Sierra Nevada.* Berkeley: University of California Press, 1965. xiv + 245 pp. Maps, illus., bib., index. Includes the national parks and forests of the California range, with emphasis on their discovery, exploration, and efforts to preserve their natural and scenic features.

F28 Farquhar, Francis P. "Francis Farquhar at 84 Speaks of the Sierra Club—Then and Now." OHI by Dave Bohn. *Sierra Club Bulletin* 57 (June 1972), 8-14. Reminiscences of Sierra Club activities and achievements since the 1890s, by a former editor of *Sierra Club Bulletin.*

F29 Farquhar, Samuel T. "John Muir and Ralph Waldo Emerson in Yosemite, Gathered from Their Writings and Correspondence." *Sierra Club Bulletin* 19 (1934), 48-55.

F30 Farrar, Broadus F. "John Burroughs, Theodore Roosevelt, and the Nature Fakers." *Tennessee Studies in Literature* 4 (1959), 121-30. On a literary battle between writers who sentimentalized about natural history and wildlife preservation, and Burroughs and Roosevelt, who opposed their methods, 1898-1907.

F31 Fastabend, John A. "Gradual Evolution and Development of Ocean Log Rafting." *Loggers Handbook* 32 (1972), 30-32, 98. Especially on the Pacific Coast since the 1890s; originally a paper presented to the Pacific Logging Congress in 1909.

F32 Faubel, Arthur L. *Cork and the American Cork Industry.* 1938. Revised edition. New York: Cork Institute of America, 1941. 151 pp. Illus., notes. Incidental history.

F33 Faucher, Albert. "The Decline of Shipbuilding at Quebec in the Nineteenth Century." *Canadian Journal of Economics and Political Science* 33 (May 1957), 195-215. Includes some history of the timber trade from Quebec City.

F34 Fausold, Martin L. "Gifford Pinchot and the Progressive Movement: An Analysis of the Pinchot Papers, 1910-1917." Ph.D. diss., Syracuse Univ., 1953. 318 pp.

F35 Fausold, Martin L. "Gifford Pinchot." *Social Studies* 46 (Oct. 1955), 210-15. On his political activities in the Progressive or Bull Moose party, 1910-1912.

F36 Fausold, Martin L. "Gifford Pinchot and the Decline of Pennsylvania Progressivism." *Pennsylvania History* 25 (Jan. 1958), 25-38. From 1912 to 1917.

F37 Fausold, Martin L. *Gifford Pinchot, Bull Moose Progressive.* Syracuse: Syracuse University Press, 1961. viii + 270 pp. Illus., notes, bib., index. On Pinchot's political career from 1910 to 1917, relating progressivism and the conservation movement.

F38 Fauteux, Joseph-Noel. *Essai sur l'industrie au Canada sous le régime français.* 2 Volumes. Quebec: Ls-A. Proulx, 1927. Includes some history of naval stores policy and industry in French Canada.

F39 Fay, C.R. "Mearns and the Miramichi: An Episode in Canadian Economic History." *Canadian Historical Review* 4 (Dec. 1923), 316-20. On the activities of the Canadian branch of an English shipping firm, Pollock, Gilmour and Company, which in 1812 opened as timber merchants on the Miramichi River of New Brunswick.

F40 Fay, C.R. "Forest and Mining Frontiers." *Economic History* 4 (Feb. 1938), 98-107. Includes a lengthy review of Arthur R.M. Lower's *Settlement and the Forest Frontier in Eastern Canada* (1936).

F41 Fay, John Henry. "Masting the Fleets of Britannia." *American Forests* 80 (Oct. 1974), 16-19. With New Hampshire white pine during the colonial period.

F42 Fazio, James R. *Men on the Mountain: A Historical Look at the Vernal Ranger District.* Vernal, Utah: USFS, Ashley National Forest, Vernal Ranger District, 1967. Processed.

F43 Fedkiw, John. *Preliminary Review of 60 Years of Reforestation in New York State.* Syracuse: State University of New York, College of Forestry, 1959. vi + 80 pp. Map, diags., tables, bib.

F44 Fehren, R.B. "The Intercoastal Lumber Business; A Survey of the Economic Distribution of West Coast Woods on the Eastern States." Master's thesis, Yale Univ., 1928.

Fein, Albert. See Olmsted, Frederick Law, #O33

F45 Fein, Albert. *Frederick Law Olmsted and the American Environmental Tradition.* New York: George Braziller, 1972. xi + 180 pp. Illus., maps, notes, bib., index. As a landscape architect, planner, and social scientist, Olmsted strongly influenced the parks and conservation movements.

F46 Fellman, Evan L. "Fifty Years Development of Oak Flooring." *Southern Lumberman* 144 (Dec. 15, 1931), 125-26.

F47 Fellmeth, Robert C., et al. *Politics of Land: Ralph Nader's Study Group Report on Land Use in California.* New York: Grosman Publishers, 1973. 715 pp. Bib. Includes a section on logging and the forest products industries.

F48 Fellows, E.S. "The Lumber Industry of Eastern Canada." Master's thesis, Univ. of New Brunswick, 1935.

F49 Fellows, E.S. "Some Changes in the Lumber Industry of the Maritimes." *Forestry Chronicle* 22 (Sept. 1946), 172-81.

F50 Fellows, E.S. "The Impact of Technology on the Use of Forest Products." *Forestry Chronicle* 39 (Dec. 1963), 460-65.

F51 Felt, Margaret E. *Gyppo Logger.* Caldwell, Idaho: Caxton Printers, 1963. 315 pp. Illus. An autobiography with some incidental history of independent loggers in the Pacific Northwest.

F52 Feltner, George. *A Look Back: A 65-Year History of the Colorado Game and Fish Department.* Annual Report of the Colorado Game and Fish Department, 1961. Denver, 1962. 65 pp. Illus.

F53 Felton, Harold W., comp. and ed. *Legends of Paul Bunyan.* Foreword by James F. Stevens. New York: Alfred A. Knopf, 1947. xxi + 418 pp. Illus., bib. Anthology of 108 Paul Bunyan tales from the works of forty authors; contains a brief historical foreword.

F54 Fendall, Gary K. "Historical Aspects of the Willamette River Park System." Master's thesis, Oregon College of Education, 1968. Oregon.

F55 Fensom, Kenneth G. *Expanding Forestry Horizons: A History of the Canadian Institute of Forestry-Institut Forestier du Canada, 1908-1969.* Macdonald College, Quebec: Canadian Institute of Forestry, 1972. x + 547 pp. Illus., tables, apps., index.

F56 Fensom, Kenneth G. "Reminiscences of an Editor." *Forestry Chronicle* 50 (Feb. 1974), 2-4. Former editor examines the origins of *Forestry Chronicle* and sketches its editorial history since the 1920s.

F57 Ferguson, John Arden. "Some Recent Developments in Forestry Education." *Journal of Forestry* 21 (Mar. 1923), 278-83. Since 1896.

F58 Ferguson, John L. "No. 150 West: A New Era Was Born as an Old One Ended." *Forests & People* 17 (Third quarter, 1967), 16-17, 29, 34-35, 39, 44. On the migration of the W.M. Cady Lumber Company from McNary, Louisiana, to McNary, Arizona, in 1924. It is now Southwestern Forest Industries.

F59 Ferguson, Roland H., and Howard, M.C. *The Timber Resource in Massachusetts.* Upper Darby, Pennsylvania: USFS, Northeastern Forest Experiment Station, 1956. 45 pp. Incidental history.

F60 Ferguson, Roland H., and McGuire, John R. *The Timber Resources of Rhode Island.* Upper Darby, Pennsylvania: USFS, Northeastern Forest Experiment Station, 1957. 36 pp. Incidental history.

F61 Ferguson, Roland H. *The Timber Resources of Pennsylvania.* Upper Darby, Pennsylvania: USFS, Northeastern Forest Experiment Station, 1958. 46 pp. Incidental history.

F62 Ferguson, Roland H. *The Timber Resources of Delaware: A Report on the Forest Survey Made by the U.S. Forest Service.* Upper Darby, Pennsylvania: USFS, Northeastern Forest Experiment Station, 1959. 30 pp. Illus., map. Incidental historical references to forestry and the lumber industry.

F63 Ferguson, Roland H., and Longwood, Franklin R. *The Timber Resources of Maine: A Report on the Forest Survey Made by the U.S. Forest Service.* Upper Darby, Pennsylvania: USFS, Northeastern Forest Experiment Station, 1960. 75 pp. Illus., tables. Includes some history of forest industries.

F64 Ferguson, Roland H., and Kingsley, Neal P. *The Timber Resources of Maine.* Resource Bulletin NE-26. Upper Darby, Pennsylvania: USFS, Northeastern Forest Experiment Station, 1972. 129 pp. Illus., maps, tables. A statistical and analytical report including some history of forested lands and forest industries.

F65 Ferguson, Roland H., and Mayer, Carl E. *The Timber Resources of New Jersey.* Resource Bulletin NE-34. Upper Darby, Pennsylvania: USFS, Northeastern Forest Experiment Station, 1974. 58 pp. Illus., maps, tables. Incidental history of forestry and forest industries.

F66 Ferguson, Roland H., and Mayer, Carl E. *The Timber Resources of Delaware.* Resource Bulletin NE-32. Upper Darby, Pennsylvania: USFS, Northeastern Forest Experiment Station, 1974. 42 pp. Illus., maps, tables. Incidental history of forestry and the forest industries.

F67 Fernald, Merritt L. "Some Early Botanists of the American Philosophical Society." *Proceedings of the American Philosophical Society* 86 (1942), 63-71. Some were involved in the forestry movement.

F68 Fernow, Bernhard E. "American Lumber." In *One Hundred Years of American Commerce, 1795-1895,* ed. by Chauncey M. Depew. New York: D.O. Haynes, 1895. Volume 1, pp. 196-203.

F69 Fernow, Bernhard E. *Report Upon Forestry Investigations of the United States Department of Agriculture, 1877-1898.* Washington: GPO, 1899. 401 pp.

F70 Fernow, Bernhard E. *Forestry in the U.S. Department of Agriculture during the Period 1877-1898.* Washington: GPO, 1899. 44 pp. Summary of research by the former chief of the Division of Forestry; reprinted from House Document 181, 55th Congress, 3rd Session (1899).

F71 Fernow, Bernhard E. "Beginnings of Professional Forestry in the Adirondacks." In *Fifth Annual Report of the New York Fisheries, Game and Forest Commission* (1899), 401-44.

F72 Fernow, Bernhard E. *Beginnings of Professional Forestry in the Adirondacks, Being the 1st and 2nd Annual Reports of the New York State College of Forestry.* Bulletin No. 2. Ithaca, New York: New York State College of Forestry, 1900. 152 pp. Illus.

F73 Fernow, Bernhard E. *Economics of Forestry: A Reference Book for Students of Political Economy and Professional and Lay Students of Forestry.* New York: Thomas Y. Crowell, 1902. xii + 520 pp. Notes, tables, bib., index. Includes a chapter on the history of the forestry movement of the United States, as well as other historical references throughout.

F74 Fernow, Bernhard E. "The Movement of Wood Prices and Its Influence on Forest Treatment." *Forestry Quarterly* 3 (Feb. 1905), 18-31.

F75 Fernow, Bernhard E. *A Brief History of Forestry in Europe, the United States and Other Countries.* Toronto: University Press, 1907. x + 438 pp. Notes. Includes chapters on Canada and the United States. Subsequent editions (1911, 1913) are revised, enlarged, and contain indexes.

F76 Ferrell, John Robert. "Water Resource Development in the Arkansas Valley: A History of Public Policy to 1950." Ph.D. diss., Univ. of Oklahoma, 1968. 238 pp. Conservation as pork-barrel politics.

F77 Ferrell, Mallory H. *Rails, Sagebrush, and Pine; a Garland of Railroad and Logging Days in Oregon's Sumpter Valley.* San Marino, California: Golden West Books, 1967. 128 pp. Illus., index. On the Sumpter Valley Railway Company of eastern Oregon, which served many lumber companies in the Blue Mountains from 1891 to the 1960s.

Ferriday, Virginia Guest. See Vaughan, Thomas, #V41

F78 Feuerlicht, Roberta S. *The Legends of Paul Bunyan.* New York: Collier Books, 1966. 128 pp. Illus.

Fichter, Edson. See Mohler, Levi L., #M512

F79 Fickes, Clyde P. *Recollections by Clyde P. Fickes, Forest Ranger Emeritus.* Missoula, Montana: USFS, Northern Region, 1972. iii + 126 pp. Illus., map. Processed. Reminiscences of his career with the USFS, 1907-1944, mostly in the Northern Region and on the Lewis and Clark National Forest of Montana.

F80 Fickle, James E. "History of the Southern Pine Association, 1914-1920." Master's thesis, Louisiana State Univ., 1963.

F81 Fickle, James E. "The Origins and Development of the Southern Pine Association, 1883-1954." Ph.D. diss., Louisiana State Univ., 1970. 629 pp.

F82 Fickle, James E. "Management Looks at the 'Labor Problem': The Southern Pine Industry During World War I and the Postwar Era." *Journal of Southern History* 40 (Feb. 1974), 61-76. Labor-management relations and labor conditions for white and black workers in the Southern lumber industry from 1914 to 1926.

F83 Field, Earle. "The New York State College of Forestry at Syracuse University: The History, Founding and Early Growth, 1911-1922." Ph.D. diss., Syracuse Univ., 1954.

F84 Findell, Virgil E., and others. *Michigan's Forest Resources.* USFS, Lake States Forest Experiment Station, Paper No. 82. St. Paul, 1960. 46 pp. Illus., map, tables, diags. Incidental history.

F85 Fine, Homer. "Nebraska's Man-Made Oasis." *American Forests* 58 (Sept. 1952), 12-13, 30. On the Nebraska National Forest, a product of tree planting, since 1902.

F86 Finer, Herman. *The TVA: Lessons for International Application.* International Labour Office, Studies and Reports, Series B (Economic Conditions), No. 37. Montreal, 1944. viii + 289 pp. Diags., tables. Includes some incidental history of the Tennessee Valley Authority and its work in the area of natural resources.

F87 Finger, John R. "Henry Yesler's Seattle Years, 1852-1892." Ph.D. diss., Univ. of Washington, 1969. 416 pp.

F88 Finger, John R. "Seattle's First Sawmill, 1853-1869: A Study of Frontier Enterprise." *Forest History* 15 (Jan. 1972), 24-31. Operated by Henry L. Yesler, Washington lumberman.

F89 Finger, John R. "The Seattle Spirit, 1851-1893." *Journal of the West* 13 (July 1974), 28-45. Deals in part with the lumber industry as a factor in the growth of Seattle, Washington.

F90 Finkbeiner, Daniel T. "Pennsylvania Legislation on Water Supply Resources." *Pennsylvania Forests and Water* 3 (May-June 1951), 56-57. Including watershed management, 1905-1939.

F91 Finley, Robert W. "The Original Vegetation Cover of Wisconsin." Ph.D. diss., Univ. of Wisconsin, 1951.

F92 Finn, Terence T. "Conflict and Compromise: Congress Makes a Law, the Passage of the National Environmental Policy Act." Ph.D. diss., Georgetown Univ., 1973. 702 pp. Case study of the legislative process in passing this act of 1969.

F93 Firey, Walter I. *Man, Mind and Land: A Theory of Resource Use.* Glencoe, Illinois: Free Press, 1960. 256 pp. Illus., bib. Incidental history.

F94 Fischer, Duane D. "The Disposal of Federal Lands Within the Eau Claire Land District of Wisconsin, 1848-1925." Master's thesis, Univ. of Wisconsin, 1961.

F95 Fischer, Duane D. "The John S. Owen Enterprizes." Ph.D. diss., Univ. of Wisconsin, 1964. 606 pp. Business history of a Wisconsin-based entrepreneur whose logging and lumber operations stretched to the South and the Pacific Coast, 1874-1931.

F96 Fischer, Duane D. "The Short, Unhappy Story of the Del Norte Company." *Forest History* 11 (Apr. 1967), 12-25. History of a Wisconsin-owned lumber company's operation in the California-Oregon redwoods, 1902-1950.

F97 Fish, Carl R. "Phases of the Economic History of Wisconsin, 1860-1870." *Proceedings of the Wisconsin Historical Society*, 1907 (1908), 204-16. Includes the lumber industry.

F98 Fish, Gretchen Houghton. "Indian Lake." *North Country Life* 6 (Winter 1952), 55-59. On the lumber and resort industries at Indian Lake, New York, since the 19th century.

F99 Fishburn, Jesse J. "Ben Hershey, Lumber Baron." *Palimpsest* 28 (Oct. 1947), 289-99. A 19th-century sawmill operator in Muscatine, Iowa.

F100 Fisher, Aileen L. *Timber! Logging in Michigan.* New York: Aladdin Books, 1955. 191 pp. Illus.

F101 Fisher, Albert K. "In Memoriam: George Bird Grinnell." *Auk: A Quarterly Journal of Ornithology* 56 (Jan. 1939), 1-12. Biographical sketch and tribute to Grinnell (1849-1938), a leader in the conservation movement.

F102 Fisher, Arthur M. *Slivers, Knots, Selects, Clears: A Retail Lumberman's Story.* New York: Vantage Press, 1965. 231 pp. Illus. The author rose from a lumberyard apprentice in Indiana to president of Home Lumber and Supply Company, a retail lumber firm of Rockford, Illinois, since 1905.

F103 Fisher, Joseph L. "Resources Policies and Administration for the Future." *Public Administration Review* 21 (Spring 1961), 74-80. Includes some history of the degelopment of resource policies.

F104 Fisher, Joseph L. *Conservation as Research, Policy, and Action.* Horace M. Albright Conservation Lectures, No. 11. Berkeley: University of California, School of Forestry and Conservation, 1971. 22 pp. Includes some history of conservation in the United States since the late 19th century.

F105 Fisher, Richard T. *The Management of the Harvard Forest, 1909-1919.* Bulletin No. 1. Petersham, Massachusetts: Harvard Forest, 1921. 27 pp. History of an intensively managed forest.

F106 Fisher, W.R. "First Fire Protective Organization." *American Forestry* 22 (Apr. 1916), 234-35. On the Pocono Protective Fire Association, Monroe County, Pennsylvania, established in 1902.

F107 Fishwick, Marshall W. "Paul Bunyan: The Folk Hero as Tycoon." *Yale Review* 41 (Winter 1952), 264-74. On the origin of this mythical character in lumber advertising.

F108 Fitch, Edwin M. "The Lumber Industry and the Tariff." Ph.D. diss., Univ. of Wisconsin, 1933.

F109 Fitch, Edwin M. *The Tariff on Lumber.* Madison, Wisconsin: Tariff Research Committee, 1936. xv + 140 pp. Maps, diags.

F110 Fitch, Edwin M., and Shanklin, John F. *The Bureau of Outdoor Recreation.* Praeger Library of U.S. Government Departments and Agencies, No. 24. New York: Praeger, 1970. xii + 227 pp. Illus., tables, apps., bib., index. On the history of outdoor recreation and especially of the U.S. Bureau of Outdoor Recreation since its founding in 1963.

F111 Fite, Emerson D. *Social and Industrial Conditions in the North during the Civil War.* New York: Macmillan, 1910. vii + 318 pp. Includes a chapter on the lumber industry.

F112 Fitzgerald, Denis Patrick. "Pioneer Settlement in Northern Saskatchewan." Ph.D. diss., Univ. of Minnesota, 1966. 889 pp. And its relation to forested lands.

F113 Fitzgerald, O.A. "Last of the River Pigs." *American Forests* 58 (Apr. 1952), 12-16. Log drives on the Clearwater River, Idaho, since ca. 1910.

F114 Fitzmaurice, John W. *Shanty Boy, or Life in a Lumber Camp; Being Descriptions, Tales, Songs and Adventures in the Lumbering Shanties of Michigan and Wisconsin.* Cheboygan, Michigan: Democrat Steam Print, 1889. Reprint. Mount Pleasant, Michigan: Central Michigan University Press, 1963. 246 pp.

Fitzpatrick, George. See Tucker, Edwin A., #T280

F115 Flader, Susan L. "Aldo Leopold and the Evolution of an Ecological Attitude." Ph.D. diss., Stanford Univ., 1971. 358 pp.

F116 Flader, Susan L. "Thinking Like a Mountain: A Biographical Study of Aldo Leopold." *Forest History* 17 (Apr. 1973), 14-28. Leopold (1887-1948) was a forester, ecologist, naturalist, wildlife management specialist, and wilderness enthusiast. His advocacy of a "land ethic," as expressed in his sensitive essays, found an appreciative audience beyond his own lifetime.

Flader, Susan L. See Steinhacker, Charles, #S579

F117 Flader, Susan L. *Thinking Like a Mountain: Aldo Leopold and the Evolution of an Ecological Attitude Toward Deer, Wolves, and Forests.* Columbia: University of Missouri Press, 1974. xxv + 284 pp. Illus., maps, notes, bib., index. Leopold (1887-1948), a major figure in the conservation and preservation movements, worked for the USFS in Arizona, New Mexico, and Wisconsin in the 1910s and 1920s. This work explores his studies and philosophies regarding wildlife, wilderness, and an ecologically based "land ethic."

F118 Flatt, William D. "Lumber Veteran Tells of Days That Were." *Canada Lumberman* 48 (Dec. 15, 1928), 53-56. In Ontario.

F119 Flavelle, W. Guy. *A Cedar Saga: And the Man Who Made It Possible.* Port Moody, British Columbia: Flavelle Cedar, Ltd., 1966. ix + 130 pp. Illus. Company history of Flavelle Cedar Ltd. and its predecessors at Port Moody, British Columbia, and especially the role of Aird Flavelle, since ca. 1900.

F120 Fleishel, Marc Leonard. "The First Forty-Two Years." *Southern Lumberman* 193 (Dec. 15, 1956), 173-76. Reminiscences of the work of the Southern Pine Association by its treasurer, a Florida lumberman.

F121 Fleming, Donald. "Roots of the New Conservation Movement." *Perspectives in American History* 6 (1972), 7-91. On the conservation philosophies and proposals of Rachel Carson, Barry Commoner, Lewis Mumford, Aldo Leopold, and other modern environmentalists. The origins of conservation and anticonservation thought are treated within a framework of intellectual history.

F122 Fleming, Guy L. "Should We Cherish and Maintain the San Jacinto Wild Area?" *Sierra Club Bulletin* 34 (Oct. 1949), 4-10. Efforts to preserve the California state park and adjacent San Bernardino National Forest from commercial incursions since the 1920s.

F123 Fleming, Robben Wright, and Witte, E.E. *Marathon Corporation and Seven Labor Unions: A Case Study.* National Planning Association, Case Studies on the Causes of Industrial Peace under Collective Bargaining, No. 8. Washington, 1950. 13 + 65 pp.

F124 Fletcher, E.D. *The Use of Lumber and Wood in Connecticut.* Hartford: N.p., 1928. 51 pp. Illus., diags., app. Contains some incidental historical references to wood utilization and lumber marketing.

F125 Fletcher, Ed. *Memoirs of Ed Fletcher.* San Diego: Pioneer Printers, 1952. 751 pp. Illus. Fletcher, a California state senator, 1935-1947, was involved in the passage of forestry legislation. He also had timber interests in Del Norte County.

F126 Fletcher, Marvin. "Army Fire Fighters." *Idaho Yesterdays* 16 (Summer 1972), 12-15. A contemporary account of a Negro regiment of forest fire fighters in Idaho, 1910, with an editorial introduction.

F127 Flexner, James. *That Wilder Image: Paintings of the American Native School from Thomas Cole to Winslow Homer.* Boston: Little, Brown and Company, 1962. xxii + 407 pp. Illus., notes, bibs., index. Emphasis on 19th-century landscape painting.

Flock, Warren L. See Huberty, Martin R., #H590

Florek, Larry. See Place, Marian T., #P237

F128 Flynn, Ted P. "From Bulls to Bulldozers: A Memoir on the Development of Machines in the Western Woods from Letters of Ted P. Flynn." Ed. by Joseph A. Miller. *Forest History* 7 (Fall 1963), 14-17. The author was a USFS construction engineer who advanced use of the tractor bulldozer in the Pacific Northwest.

F129 Fobes, Charles B. "Lightning Fires in the Forests of Northern Maine, 1926-1940." *Journal of Forestry* 42 (Apr. 1944), 291-93.

F130 Fobes, Charles B. "Historic Forest Fires in Maine." *Economic Geography* 24 (Oct. 1948), 269-73. Since 1762.

F131 Fobes, Charles B. "Barren Mountain Tops in Maine and New Hampshire." *Appalachia* 29 (July 1953), 315-22. On evidence that many mountain summits were forested when seen by the first settlers and have been denuded by fire since 1800.

F132 Foehl, Harold M., and Hargreaves, Irene M. *The Story of Logging the White Pine in the Saginaw Valley.* Bay City, Michigan: Red Keg Press, 1964. vii + 70 pp. Illus., maps, glossary, notes, bib. During the mid and late 19th century.

F133 Foerster, Norman. *Nature in American Literature: Studies in the Modern View of Nature.* 1923. Second edition. New York: Russell & Russell, 1958. xiii + 324 pp. Important for its essays on John Burroughs, Henry David Thoreau, John Muir, and other American nature writers.

F134 Foley, John. "The Work of the Foresters of the Pennsylvania Railroad System." *Journal of Forestry* 22 (Feb. 1924), 162-70. Includes a brief historical account since the 19th century.

Folsom, David E. See Cook, Charles W., #C575

F135 Folsom, William H.C. *Fifty Years in the Northwest.* Ed. by E.E. Edwards. St. Paul: Pioneer Press Company, 1888. xliii + 763 pp. Includes reminiscences of mid-19th-century logging in Wisconsin and Minnesota.

F136 Folsom, William H.C. "History of Lumbering in the St. Croix Valley, with Biographic Sketches." *Collections of the Minnesota Historical Society* 9 (1901), 291-324.

F137 Folweiler, Alfred D. *The Theory and Practice of Forest Fire Protection in the United States.* Baton Rouge: Louisiana State University, 1937. 163 pp. Illus., bib., tables, diags. Incidental history.

F138 Folweiler, Alfred D. "Ownership of Forest Land in Selected Parishes in Louisiana and Its Effect on Forest Conservation." Ph.D. diss., Univ. of Wisconsin, 1943.

F139 Folweiler, Alfred D. "The Political Economy of Forest Conservation in the United States." *Journal of Land & Public Utility Economics* 20 (Aug. 1944), 202-16. Outlines four periods of federal policy affecting forest conservation.

F140 Folweiler, Alfred D., and Brown, A.A. *Fire in the Forests of the United States.* 1946. Second edition. St. Louis: John S. Swift Company, 1953. 223 pp. Illus., maps, diag., bibs. Incidental history.

F141 Folweiler, Alfred D. "The Place of Fire in Southern Silviculture." *Journal of Forestry* 50 (Mar. 1952), 187-90. On the changing concepts and uses of light burning as a silvicultural practice in the 20th century.

F142 Folwell, William Watts. *A History of Minnesota.* 4 Volumes. St. Paul: Minnesota Historical Society, 1921-1930. Corrected editions, 1956-1969. Contains much history on the lumber and forest products industries, forestry, and the conservation movement in Minnesota.

Folwell, William Watts. See Andrews, Christopher C., #A444

F143 Foner, Philip S. *History of the Labor Movement in the United States.* 4 Volumes. New York: International Publishers, 1947-1965. Volume 4 treats the Industrial Workers of the World, 1905-1917, a radical labor union that organized many lumberjacks and millworkers.

F144 Foner, Philip S. "The IWW and the Black Worker." *Journal of Negro History* 55 (Jan. 1970), 52-57. Includes the role of the Industrial Workers of the World in labor struggles in the Southern lumber industry, early 20th century.

F145 Forbes, John Douglas. *Stettinius, Sr.: Portrait of a Morgan Partner.* Charlottesville: University Press of Virginia, 1974. xii + 244 pp. Edward Reilly Stettinius (1864-1925) was president of Diamond Match Company, 1909-1915, an episode treated briefly here.

F146 Forbes, Reginald D. "Progress of Forestry in Louisiana." *Lumber World Review* 39 (Nov. 10, 1920), 119-23.

F147 Forbes, Reginald D. "Where the South Stands in Forestry: The Story of the Inception and Progress of Forestry in the South." *Lumber World Review* (Nov. 10, 1921), 64-67.

F148 Forbes, Reginald D. "The Passing of the Piney Woods." *American Forestry* 29 (Mar. 1923), 131-36, 185. Devastation of Southern pine forests by logging.

F149 Forbes, Reginald D. "G.P.—His Public Career." *Journal of Forestry* 63 (Aug. 1965), 593-95. A resume of Gifford Pinchot's political career, 1912-1938.

F150 Force, Lee E. *Exploits of Lee.* Boston: Christopher Publishing House, 1952. 99 pp. On the author's activities in the lumber industry of Washington since 1915 and his travels through the United States and abroad as a representative of the industry.

F151 Ford, Alice. *John James Audubon.* Norman: University of Oklahoma Press, 1964. xiv + 488 pp. Illus., notes, bib., index. Biography of Audubon (1785-1851), naturalist and "patron saint" of wildlife conservationists.

F152 Ford, Amelia C. *Colonial Precedents of Our National Land System as It Existed in 1800.* University of Wisconsin Bulletin, No. 352, History Series, Volume 2, No. 2. Madison: University of Wisconsin, 1910. 157 pp. Bib.

F153 Ford, Thomas A. "Wildlife in Alabama." *Alabama Historical Quarterly* 10 (1948), 68-76. On legal protection and conservation since 1803, with emphasis on the 20th century.

F154 Foreman, Carolyn (Thomas). *The Cross Timbers.* Muskogee, Oklahoma: By the author, 1947. 123 pp. Map, bib. Includes some history of forestry in Texas.

F155 *Forest History.* "Cowichan Forest Museum." *Forest History* 10 (Jan. 1967), 36-38. Origins of the Cowichan Valley Forest Museum near Duncan, British Columbia, 1960s.

F156 *Forest History.* "The Forest in Canadian Life: A Brief Historical Anthology." *Forest History* 11 (Oct. 1967), 14-27. Eleven selections from three centuries of writing about Canadian forests.

F157 *Forest Leaves.* "Early Lumber Industry in Potter County, Pa." *Forest Leaves* (Dec. 1930), 177ff.

F158 *Forest Log.* "Clatsop State Forest: Managed by the Oregon Department of Forestry." *Forest Log* 44 (Oct. 1974), 3-6. Some history of a previously cut and burned-over forest area in northwestern Oregon, now managed as Clatsop State Forest.

F159 *Forests & People*. Golden Anniversary of Louisiana Forestry Issue. *Forests & People* 13 (First quarter, 1963). Contains many articles on the history of forestry and the forest industries.

F160 *Forests & People*. "The Louisiana Forestry Commission Is Formed." *Forests & People* 13 (First quarter, 1963), 75, 120-23. Its work since 1944.

F161 *Forests & People*. "Formation of the Louisiana Forestry Association." *Forests & People* 13 (First quarter, 1963), 76, 112-13, 118-19. Its work since formed in 1947.

F162 Forman, Benno M. "Mill Sawing in Seventeenth-Century Massachusetts." *Old-Time New England* 60 (Spring 1970), 110-30. Mill sawing replaced pitsawing within several decades after initial English settlement in Massachusetts.

F163 Forness, Norman Olaf. "The Origins and Early History of the United States Department of the Interior." Ph.D. diss., Pennsylvania State Univ., 1964. 263 pp.

F164 Forrester, George, ed. *Historical and Biographical Album of the Chippewa Valley, Wisconsin*. Chicago: A. Warner, 1891-1892. 950 pp. Illus., maps, tables, index. Includes some history of the lumber industry and sketches of prominent lumbermen.

F165 Forsey, E.A. "The Pulp and Paper Industry." *Canadian Journal of Economics* 1 (Aug. 1935), 501-09. Some history.

F166 Forsythe, James L. "Clinton P. Anderson: Politician and Businessman as Truman's Secretary of Agriculture." Ph.D. diss., Univ. of New Mexico, 1970. 678 pp.

Fortt, Inez Long. See Young, Carl Henry, #Y24

F167 *Fortune*. "Bunyan in Broadcloth: The House of Weyerhaeuser." *Fortune* 9 (Apr. 1934), 63-76, 155-90. A broad survey of Weyerhaeuser Timber Company operations and timberlands, including some history of the family and firm since 1858.

F168 *Fortune*. "Crown Zellerbach Second Growth: How a Progressive California Corporation Changed Itself from a Paper Company to a Great Forest-Products Company." *Fortune* 49 (Jan. 1954), 88-93.

F169 Fort Wayne Corrugated Paper Company. *Forty Years of Container Making: Fort Wayne Corrugated Paper Company*. Fort Wayne, Indiana, 1949. 24 pp. Illus.

F170 Fosburgh, Pieter W. "The Big Boom." *New York State Conservationist* 1 (Apr.-May 1947), 16-17. Reminiscences of logging in the Adirondack Mountains of New York. Reprinted in *North Country Life* 1 (Summer 1947), 52-55.

F171 Fosburgh, Pieter W. *The Natural Thing: The Land and Its Citizens*. New York: Macmillan, 1959. 255 pp. On the problem of abandoned farm lands and a general economic history of New York's Adirondack Forest Preserve.

F172 Fosburgh, Pieter W. *New York State's Forest Preserve*. Albany: New York State Conservation Department, Division of Conservation Education, 1965. 16 pp. Illus., maps. A reprint of historical articles first appearing in *Conservationist*, 1963-1964.

F173 Foscue, Edwin J., and Quam, Louis Otto. *Estes Park: Resort in the Rockies*. Dallas: University Press, 1949. vi + 98 pp. Illus., maps, bib. Estes Park, near Rocky Mountain National Park, Colorado, since 1859.

F174 Foscue, Edwin J. "East Texas: A Timber Empire." *Journal of the Graduate Research Center* (Southern Methodist University) 28 (Apr. 1960), 1-60.

F175 Fosdick, Raymond B. *John D. Rockefeller, Jr., a Portrait*. New York: Harper & Row, 1956. ix + 477 pp. Illus. Rockefeller (1874-1960) did much to preserve America's natural and historical heritage through philanthropy; he gave millions of dollars to expand the national parks and to improve their administration.

F176 Foss, Phillip O. *Politics and Grass: The Administration of Grazing on the Public Domain*. Seattle: University of Washington Press, 1960. ix + 236 pp. Tables, notes, bib., index. Deals generally with the subject since 1780 and particularly with the Taylor Grazing Act and its applications since 1934.

F177 Foss, Phillip O. *The Grazing Fee Dilemma*. Inter-University Case Program, ICP Case No. 57. University: University of Alabama Press, 1960. 12 pp. Notes. On the Grazing Service of the USDI, established by the Taylor Grazing Act in 1934 to collect fees for grazing on public lands.

Foss, Phillip O. See Smith, Frank E., #S366

F178 Foss, Phillip O., comp. *Politics and Ecology*. Belmont, California: Duxbury Press, 1972. 298 pp. Bib. Collection of essays on environmental policy, some with historical references.

F179 Foster, Chapin D. "Collins Almanor Forest." *American Forests* 49 (Oct. 1943), 480-82. On a forest in Plumas County, California, operated on a sustained-yield basis by the Collins Pine Company, Portland, Oregon, including some incidental history of the firm and the Collins family.

F180 Foster, Clifford H., and Kirkland, Burt P. *The Charles Lathrop Pack Demonstration Forest, Warrensburg, N.Y.: Results of Twenty Years of Intensive Forest Management*. Washington: Pack Forestry Foundation, 1949. 36 pp. Illus., maps.

F181 Foster, Edward Halsey. "Picturesque America: A Study of the Popular Use of the Picturesque in Consideration of the American Landscape, 1835-1860." Ph.D. diss., Columbia Univ., 1971.

F182 Foster, Helen Laura. "Today's Forests—He Helped Them Grow." *Michigan Conservation* 18 (May-June 1949), 9-11. On Marcus Schaaf as state forester of Michigan, 1910-1949.

F183 Foster, James C. "The Deer of Kaibab: Federal-State Conflict in Arizona." *Arizona and the West* 12 (Autumn 1970), 255-68. The issue of whether the USFS or Arizona could control deer hunting in the Kaibab National Forest arose in 1924. It was resolved in 1928 by a Supreme Court decision which was an important precedent for later actions in the field of wildlife-game management.

F184 Foster, Laura. "Honeymoon in the Hetch Hetchy." *American West* 8 (May 1971), 10-15. Based on a personal narrative (1914) of engineer Robert Duryea. Dammed to provide a water supply for San Francisco, the Hetch Hetchy Valley was the focal point of a dispute between preservationists and utilitarian conservationists.

F185 Foster, Samuel Lynde. *In the Canyons of Yosemite National Park of California.* Boston: Humphries, 1949. 121 pp. Illus. Reminiscences, 1901-1939.

F186 Foster Lumber Company. *75th Anniversary, Foster Lumber Company, 1879-1954.* Kansas City, Missouri, 1954. 69 pp. Illus. A retail firm with lumber yards throughout the Midwest and woods operations in Texas.

Fountain, Mrs. Eugene. See Carranco, Lynwood, #C172

F187 Fountain, Paul. *The Great North-West and the Great Lake Region of North America.* London: Longmans, Green, 1904. viii + 355 pp. One chapter concerns "A Winter with the Lumberers," 1860s.

F188 Fowells, H.A. "Cork Oak Planting Tests in California." *Journal of Forestry* 47 (May 1949), 357-65. On USFS experiments, 1942-1945, and planting from cork oak acorns as early as 1858.

F189 Fowke, Edith. *Lumbering Songs from the Northern Woods.* Austin: University of Texas Press, for the American Folklore Society, 1970. 232 pp. Map, musical scores, bib., index. Especially of Ontario.

F190 Fowke, Edith. "Songs of the Northern Shantyboys." *Forest History* 14 (Jan. 1971), 22-28. On 19th-century Canadian lumberjack songs.

F191 Fowler, Albert, ed. *Cranberry Lake, from Wilderness to Adirondack Park.* Blue Mountain Lake, New York: Adirondack Museum, 1968. 256 pp. Illus., maps.

F192 Fowler, Barnett. "He Set the Stage for Better Conservation." *Ad-i-ron-dac* 25 (July-Aug. 1961), 70-71. John Samuel Apperson (1878-1963).

F193 Fowler, Dorothy C. *John Coit Spooner: Defender of Presidents.* New York: University Publishers, 1961. xii + 436 pp. Illus., notes, bib., index. Spooner (1843-1919), best known as a U.S. senator from Wisconsin, was legal counsel for Philetus Sawyer and other Wisconsin lumbermen during the latter half of the 19th century.

F194 Fowler, R.M. "The Merchandising of Idaho White Pine." Master's thesis, State Univ. of New York, College of Forestry, 1937.

F195 Fowler, William A., and Robnett, Ronald H. *Oregon Hardwood Industries.* University of Oregon Studies in Business, Volume 1, No. 4. Eugene, 1929. 96 pp. Bib., illus., maps, diags. Incidental history.

F196 Fox, Charles E. *Know Your National Forests in California: A Story of Conservation through Wise Use.* Washington: GPO, 1949. 42 pp.

F197 Fox, George R. "The Ark in Michigan." *Michigan History* 42 (Mar. 1958), 75-80. On the floating wooden boxes or chests used to ship produce down the St. Joseph River, 1832-1853.

F198 Fox, Lester. "Pleasure and Profit." *American Forests* 66 (June 1960), 17-19. Forest conservation practices of Henry U. Webster on his small farm and woodland in Cayuga County, New York, 1932-1957.

F199 Fox River Paper Company. *Nearly Half a Century: Fox River Papers.* Appleton, Wisconsin, 1930. 76 pp. Since 1883.

F200 Fox, Truman B. *History of Saginaw County, from the Year 1819 Down to the Present Time.* East Saginaw, Michigan: Enterprise Print, 1858. Reprint. Mount Pleasant, Michigan: Central Michigan University, 1963. 80 pp. Tables. A center of logging and the lumber industry during the 19th century.

F201 Fox, William F. *A History of the Lumber Industry in the State of New York.* USDA, Bureau of Forestry, Bulletin No. 34. Washington: GPO, 1902. 59 pp. Illus., apps. From the colonial period to ca. 1900, including a list of historic sawmills in each county.

F202 Fox, William S. *'T Ain't Runnin' No More: The Story of Grand Bend, the Pinery and the Old River Bed.* London, Ontario: Holmes, 1946. 55 pp. Includes some history of logging in southern Ontario.

F203 Fox, William S. *The Bruce Beckons: The Story of Lake Huron's Great Peninsula.* Toronto: University of Toronto Press, 1952. xviii + 235 pp. Includes some history of lumbering on Ontario's Bruce Peninsula.

F204 Fox, William S. *'T Ain't Runnin' No More—Twenty Years After: The Story of Grand Bend, the Pinery and the Watershed of the Aux Sables River.* London, Ontario: Oxford Book Shop, 1958. xii + 89 pp. Illus. Informal history of a southern Ontario logging region.

F205 Fox, William S. *When Sir John A. MacDonald Put His Foot Down . . .* Western Ontario Historical Nuggets, No. 29. London, Ontario: Lawson Memorial Library, University of Western Ontario, 1961. 13 pp. On his intervention in a lumber dispute on Ontario's Bruce Peninsula.

F206 France. Centre National de la Recherche Scientifique. *Les Botanistes français en Amérique du Nord avant 1850.* Paris, 1957. 360 pp. A collection of essays, including several concerned with forestry and botanical studies conducted by Frenchmen in North America before 1850.

F207 Francis, Robert John. "An Analysis of British Columbia Lumber Shipments, 1947-1957." Master's thesis, Univ. of British Columbia, 1961.

F208 Frank, Bernard, and Munns, Edward N. "Watershed Flood Control: Performance and Possibilities." *Journal of Forestry* 43 (Apr. 1945), 236-51. On the work of the USDA and the USFS since passage of the Omnibus Flood Control Act in 1936.

F209 Frank, Bernard. "Benton MacKaye." *Journal of Forestry* 45 (Apr. 1947), 295-96. MacKaye (b. 1879) was a forester for several agencies of the federal government, a forestry educator, an organizer of the Wilderness Society, and widely known as "Father of the Appalachian Trail."

F210 Frank, Bernard, and Netboy, Anthony. *Water, Land and People.* New York: Alfred A. Knopf, 1950. xviii + 331 + xi pp. Illus., maps, notes, bib., index. Includes many historical references to American forests and forestry, especially in relation to soil and water conservation.

F211 Frank, Bernard. *Our National Forests.* Norman: University of Oklahoma Press, 1955. xx + 238 pp. Illus., map, diags., apps., index. On the administration of forests by the federal government since the establishment of the first national forest reserves under legislation of 1891. Includes much general history of the USFS, professional forestry, and the forest conservation movement.

F212 Frantz, Harvey R. "A Forest with Colonial Roots." *American Forests* 57 (Dec. 1951), 26, 43. On a forest in Pennsylvania preserved by the Nazareth Moravian Church since 1741.

F213 Frantz, Joe B. "The Meaning of Yellowstone: A Commentary." *Montana, Magazine of Western History* 22 (Summer 1972), 5-11. On the history of the Wyoming area, the growth of the national park concept, and the problems faced by the National Park Service due to increasing tourism.

F214 Fraser, Cecil E., and Doriot, Georges F. *Analyzing Our Industries.* New York: McGraw-Hill, 1932. x + 458 pp. Includes some information on the pulp and paper industry since ca. 1900.

F215 Fraser, James A. *A History of the W.S. Loggie Co. Ltd., 1873-1973.* Fredericton: New Brunswick Provincial Archives, 1973. 116 pp. A New Brunswick lumber company.

F216 Fraser, Joshua. *Shanty, Forest and River Life in the Backwoods of Canada.* Montreal: J. Lovell and Son, 1883. 361 pp. Illus. Lumberjacks and lumber camps in Quebec.

F217 Fraunberger, R.C. "Lumber Trade Associations, Their Economic and Social Significance." Master's thesis, Temple Univ., 1951. Includes some general history of trade associations and brief sketches of individual organizations since the mid-19th century. Also distributed in processed form.

F218 Frayer, Hume C. "Max Rothkugel: West Virginia's Own Johnny Appleseed." *Northeastern Logger* 6 (Nov. 1957), 30-31, 51.

F219 Frazer, Percy Warner. "The Austin Cary Memorial—20 Years." *Journal of Forestry* 57 (Jan. 1959), 35-37. On the University of Florida School of Forestry, 1939-1959, including some account of Austin Cary (1865-1936), forestry educator and specialist in studies of the Southern pine.

F220 Frazer, William R. "Sustained Yield-Multiple Use Management Concepts, Evolution, and Prospects: A Case Study of Weyerhaeuser Company's Tree Farm Program in Washington." Master's thesis, Pacific Lutheran Univ., 1968.

F221 Frazier, George David. "The Relationship Between Forest Service Timber Sales Behavior and the Structure of the California Pine Lumber Industry." Ph.D. diss., Yale Univ., 1967. 137 pp. From 1951 to 1962.

F222 Frear, Dana W. "Early Sawmills of Hennepin County." *Hennepin County History* (Fall 1963), 3-5; (Winter 1964), 7-8, 31. Minnesota.

F223 Frederick, Duke; Howenstine, William L.; and Sochen, June, eds. *Destroy to Create: Interaction with the Natural Environment in the Building of America.* Hinsdale, Illinois: Dryden Press, 1972. 323 pp. Illus. A collection of readings featuring aspects of resource development and conservation in American history.

F224 Frederick, Robert A. "Colonel Richard Lieber, Conservationist and Park Builder: The Indiana Years." Ph.D. diss., Indiana Univ., 1960. 468 pp. Lieber was director of the Indiana Department of Conservation, 1919-1933, and a national leader in the states parks movement.

F225 Freedgood, Seymour. "Weyerhaeuser Timber: Out of the Woods." *Fortune* 60 (July 1959), 93-105, 234-44.

F226 Freedman, L.J., and Nutting, A.D. "Dwight D. Demeritt." *Journal of Forestry* 48 (Mar. 1950), 206-07. On Demeritt as a forestry educator at the University of Maine, 1923-1950.

F227 Freeman, A.M. *The Economics of Environmental Policy.* New York: John Wiley and Sons, 1973. xii + 184 pp. Illus. Incidental history.

F228 Freeman, O.W., and Raup, H.F. "Industrial Trends in the Pacific Northwest." *Journal of Geography* 43 (May 1944), 175-84. Including the forest industries.

F229 Freeman, Orville L., and Frome, Michael. *The National Forests of America.* New York: G.P. Putnam's Sons, in association with Country Beautiful Foundation, 1968, 194 pp. Illus., maps.

French, John C. See Walker, James Herbert, #W28

French, John C. See Shoemaker, Henry W., #S241

F230 French, Lewis C. "Wisconsin's Rebirth of Pine." *American Forests* 57 (Dec. 1951), 6-9, 44-47. On the revival of the forest products industries in Marinette County, Wisconsin, since 1928.

F231 French, Lewis C. "Coon Valley Saga." *American Forests* 63 (Oct. 1957), 20-23, 50-60. On experiments in soil conservation, flood control, and reforestation in a cutover area along the Wisconsin shore of the Mississippi River, 1933-1957.

F232 Freund, Rudolf. "Military Bounty Lands and the Origins of Public Domain." *Agricultural History* 20 (Jan. 1946), 8-18.

F233 Frick, George F. "Mark Catesby, Naturalist, 1683-1749." Ph.D. diss., Univ. of Illinois, 1957. 247 pp.

F234 Frick, George F., and Stearns, Raymond Phineas. *Mark Catesby: The Colonial Audubon.* Urbana: University of Illinois Press, 1961. x + 137 pp. Illus., notes, bib., app., index. Catesby (1679-1749) was a naturalist.

F235 Fricke, Edgar Frederick. "Soviet Lumber Exports and Their Effect upon the Export of Lumber from the United States." Master's thesis, Univ. of Washington, 1931.

F236 Fridley, Russell W. "Preserving Our Green Legacy." *Minnesota History* 40 (Fall 1966), 131-36. A plea for preserving natural areas; includes some incidental historical references.

F237 Fridley, Russell W. "Yellowstone to Voyageurs: The Evolution of an Idea." *Minnesota History* 43 (Summer 1972), 70-71. On the national park concept with special reference to Voyageurs National Park, Minnesota, established in 1971.

F238 Friedman, Bernard Hertz. "Economic Aspects of the Conservation of Soil, Water, and Timber Resources." Ph.D. diss., Univ. of Pittsburgh, 1956. 319 pp. Incidental history.

F239 Friedman, Robert P. "Arthur M. Hyde: Articulate Antagonist." *Missouri Historical Review* 55 (Apr. 1961), 226-34. On the rhetorical style of Hyde, Secretary of Agriculture in the Hoover administration, 1929-1933.

F240 Fries, Robert F. "Some Economic Aspects of the Lumber Industry at La Crosse." *La Crosse County Historical Sketches,* Series 3 (1937), 14-17. Wisconsin.

F241 Fries, Robert F. "A History of the Lumber Industry in Wisconsin." Ph.D. diss., Univ. of Wisconsin, 1940. 190 pp.

F242 Fries, Robert F. "The Founding of the Lumber Industry in Wisconsin." *Wisconsin Magazine of History* 26 (Sept. 1942), 23-35.

F243 Fries, Robert F. "The Mississippi River Logging Company and the Struggle for the Free Navigation of Logs, 1865-1900." *Mississippi Valley Historical Review* 35 (Dec. 1948), 429-48. Frederick Weyerhaeuser and other Mississippi River lumbermen organized a boom operation near the mouth of Wisconsin's Chippewa River.

F244 Fries, Robert F. *Empire in Pine: The Story of Lumbering in Wisconsin, 1830-1900.* Madison: State Historical Society of Wisconsin, 1951. 285 pp. Illus., map, bib. The standard and authoritative work which considers all aspects of the lumber industry.

F245 Friis, Herman Ralph. "Highlights in the First Hundred Years of Surveying and Mapping and Geographical Explorations of the United States by the Federal Government, 1775-1880." *Surveying and Mapping* 18 (Apr.-June 1958), 186-206.

F246 Fritz, Emanuel. "Willis Linn Jepson, 1867-1946: A Eulogy and a Bit of California Forestry History." *California Forester* 14 (Apr. 1947), 6-8. University of California botanist and lifetime forest conservationist.

F247 Fritz, Emanuel. "From Bull Teams to Slacklines to Tree Farms." *Redwood Region Logging Conference Bulletin* 12 (1950), 2-3, 13-14. Changing redwood logging techniques in California.

F248 Fritz, Emanuel. *California Coast Redwood, an Annotated Bibliography.* San Francisco: Foundation for American Resource Management, 1957. 267 pp. Many entries to historical literature.

F249 Fritz, Emanuel. *Reminiscences of Fort Valley, 1916-1917: Presented at the 50th Anniversary Celebration of the Fort Valley Experimental Forest.* Flagstaff, Arizona: USFS, Fort Valley Experimental Forest, 1958. 12 pp. Processed.

F250 Fritz, Emanuel. *The Development of Industrial Forestry in California.* Colonel William B. Greeley Lectures in Industrial Forestry, No. 4. Seattle: University of Washington, College of Forestry, 1960. 40 pp. Illus., tables. Brief history of the lumber industry, forest conservation, and industrial forestry.

F251 Fritz, Emanuel. "Recollections of Forest Fire Detection of Fifty Years Ago." *Logger's Handbook* 22 (1962), 24-27, 115. Primitive fire lookout stations in the West.

F252 Fritz, Emanuel. "The Redwood Region Logging Conference: How It Came About and What Its Policies Have Been." *Loggers Handbook* 23 (1963), 24-25. A California regional logging conference since 1936.

F253 Fritz, Emanuel. "Recollections of Fort Valley, 1916-1917." *Forest History* 8 (Fall 1964), 2-6. Account of research and life at a USFS experimental forest near Flagstaff, Arizona.

F254 Fritz, Emanuel. *Teacher, Editor, and Forestry Consultant.* OHI by Elwood R. Maunder and Amelia Fry. Santa Cruz, California: Forest History Society, 1972. xiii + 336 pp. Illus., apps., index. Processed. On Fritz's career as a forestry educator at the University of California, forestry consultant, editor of the *Journal of Forestry,* and authority on redwoods, since 1919.

F255 Fritzell, Peter A. "Landscapes of Anglo-America During Exploration and Early Settlement." Ph.D. diss., Standord Univ., 1966.

F256 Fritzell, Peter A. "The Wilderness and the Garden: Metaphors for the American Landscape." *Forest History* 12 (Apr. 1968), 16-23. On literary and artistic reactions to landscape in 17th-century Virginia and New England.

F257 Fritzen, John. *History of North Shore Lumbering.* Duluth: St. Louis County Historical Society, 1968. 47 pp. Minnesota.

F258 Frome, Michael. "Let's Go Trail Riding." *American Forests* 66 (Jan. 1960), 20-23, 42-44. Includes a historical sketch of the ridger National Forest, Wyoming.

F259 Frome, Michael. *Whose Woods These Are: The Story of the National Forests.* Garden City, New York: Doubleday & Company, 1962. 360 pp. Illus., maps, bib., index. Concerns the mvement to eetablish the national forests, the physical characteristics of each forest area, and the modern economic and political problems in national forest administration.

F260 Frome, Michael. *Strangers in High Places: The Story of the Great Smoky Mountains.* Garden City, New York: Doubleday & Company, 1966. ix + 394 pp. Illus., maps, bib., notes, index. Includes a chapter on the history of lumbering in the Great Smoky Mountains, Tennessee-North Carolina, and another on the movement to establish Great Smoky Mountains National Park.

Frome, Michael. See Freeman, Orville L., #F229

F261 Frome, Michael. "Forest Lands and Wilderness." *Current History* 58 (June 1970), 343-48, 369. On federal forest and wilderness legislation, 1862-1964.

F262 Frome, Michael. *The Forest Service.* Praeger Library of U.S. Government Departments and Agencies, No. 30. New York: Praeger, 1971. xiii + 241 pp. Illus., maps, charts, apps., bib., index. Only the first chapter is devoted entirely to USFS history, but the rest of the book, which concerns USFS administration, policy, and issues, is marked by its historical perspective. Frome is somewhat critical of the modern Forest Service.

F263 Frome, Michael. *Battle for the Wilderness.* New York: Praeger Publishers in cooperation with the Wilderness Society, 1974. ix + 246 pp. Illus., notes, apps., index. Emphasizes the cultural and political backgrounds of the wilderness preservation movement, culminating in the passage of the Wilderness Act of 1964 and contemporary issues relating to wilderness areas and philosophy.

F264 Frome, Michael. "The Wilderness Act—Saving a Birthright." *Living Wilderness* 38 (Summer 1974), 9-14. On the origins of the Wilderness Act of 1964, an excerpt from Frome's *Battle for the Wilderness* (1974).

F265 Fromme, Rudo L. *Memoirs of Early Forest Service Years, 1906-1917.* Pomona, California, 1955. 30 pp. On his career with the USFS in the Pacific Northwest.

F266 Front, W.C. "The Evolution of Saw Mill Machinery." *St. Louis Lumberman* 35 (No. 2, 1905), 59-61.

F267 Frost, Sherman L. "The 'Plymouth Rock of Forestry'." *American Forests* 56 (July 1950), 15, 28. On a reunion of alumni of Biltmore Forest School near Asheville, North Carolina, gathered to honor forestry educator Carl A. Schenck.

F268 Frost, Sherman L. "Southwestern's Forest Empire." *American Forests* 57 (July 195o), 6-10. On the Southwestern Settlement and Development Corporation of southeastern Texas, including some history of its predecessor companies and of timber management on its 700,000 acre tract since 1947.

F269 Frost, Sherman L. "The Dexter Case." *Journal of Forestry* 52 (Aug. 1954), 579-83. On a suit of Avery Dexter against the state of Washington, involving the right of an owner to cut timber on his own land without a license and the right of a state to require such a license, 1949.

F270 Frost, Sherman L., comp. *Facts on Ohio's Watersheds.* Columbus: Ohio Forestry Association, 1955. 61 pp. Diags., maps, tables. Since 1921.

F271 Frothingham, Earl H. *The Status and Value of Farm Woodlots in the Eastern United States.* USDA, Bulletin No. 481. Washington: GPO, 1917. 44 pp. Maps, tables, diags. An analysis of census data to show the relation of woodlots to agricultural development in the East, 1880-1910.

F272 Frothingham, Earl H. *Timber Growing and Logging Practice in the Southern Appalachian Region.* Washington: GPO, 1931. 93 pp. Illus., tables. Includes some historical references.

F273 Frothingham, Earl H. "Biltmore—Fountain-Head of Forestry in America." *American Forests* 47 (Apr. 1941), 214-17. Concerns forestry experiments at the Biltmore Estate of George W. Vanderbilt near Asheville, North Carolina, since 1892.

F274 Frothingham, Earl H. *South Carolina Forest Resources and Industries.* USDA, Miscellaneous Publication No. 552. Washington: GPO, 1944. 72 pp. Incidental history.

Fry, Amelia. See Dana, Samuel Trask, #D35, D36

Fry, Amelia. See Fritz, Emanuel, #F254

Fry, Martha R. See Smith, T. Lynn, #S401

F275 Fry, Walter, and White, John R. *Big Trees.* 1930. Revised edition. Stanford, California: Stanford University Press, 1938. 126 pp. Illus., maps, bib. Contains some history of California's sequoia groves.

F276 Frye, Mary Virginia. "The Historical Development of Municipal Parks in the United States—Concepts and Their Application." Ph.D. diss., Univ. of Illinois, 1964. 304 pp.

F277 Fryxell, Fritiof M. *The Tetons: Interpretations of a Mountain Landscape.* Berkeley: University of California Press, 1938. xiv + 77 pp. Illus., maps.

F278 Fryxell, Fritiof M. "Thomas Moran's Journey to Tetons." *Annals of Wyoming* 15 (Jan. 1943), 71-84. Moran visited Wyoming's Teton Range in 1879. His landscape paintings aroused interest in the beauty spots of the West.

F279 Fryxell, Fritiof M., ed. *Thomas Moran (1837-1926), Explorer in Search of Beauty, a Biographical Sketch; an Account of the History and Nature of the Thomas Moran Biographical Art Collection, in the Pennypacker Long Island Collection at the East Hampton Free Library, New York; and Selected Articles and Illustrations Relating to the Life and Work of Thomas Moran.* East Hampton, New York: East Hampton Free Library, 1958. xii + 84 pp. Illus., notes. Moran was an important landscape painter who accompanied the F.V. Hayden expedition to the Yellowstone region (1871) and the John Wesley Powell expedition to the Colorado River (1873).

F280 Fuchs, James R. *A History of Williams, Arizona, 1876-1951.* Tucson: University of Arizona Press, 1953. 168 pp. Illus., notes, apps., bib. Also published in *University of Arizona Bulletin* 5 (Nov. 1953), Social Science Bulletin No. 23, pp. 1-168. Includes some history of the Saginaw Lumber Company and other lumber operations in the vicinity since the 1890s.

F281 Fuess, Claude M. *Carl Schurz, Reformer.* New York: Dodd, Mead & Company, 1932. xv + 421 pp. Illus., notes, bib., index. Schurz was Secretary of the Interior, 1877-1881, and showed particular concern for forest protection and conservation.

F282 Fugina, Frank J. *Lore and Lure of the Upper Mississippi River: A Book About the River by a River Man.* Winona, Minnesota: By the author, 1945. xii + 311 pp. Illus., map. Includes some account of lumber rafting.

F283 Fuller, George N. *Economic and Social Beginnings of Michigan: A Study of Settlement of the Lower Peninsula during the Territorial Period, 1805-1837.* Lansing: Wynkoop, Hallenbeck, Craford Company, 1916. lxxii + 630 pp. Illus., bib. Includes some history of logging and pioneer use of the forest.

F284 Fuller, George N., ed. *Michigan: A Centennial History of the State and Its People.* 4 Volumes. Chicago: Lewis Publishing Company, 1939. Volume 1, chapter 27, contains some history of logging and the lumber industry.

Fuller, Lucius E. See Johnson, Bolling A., #J93

F285 Fulling, Edmund Henry. "White Pine: Backbone of the Early American Lumber Industry." *Garden Journal of the New York Botanical Garden* 1 (Sept.-Oct. 1951), 132-34. From 1605 to 1892.

F286 Fulling, Edmund Henry. "Botanical Aspects of the Paper Pulp and Tanning Industries in the United States—An Economic and Historical Survey." *American Journal of Botany* 43 (Oct. 1956).

F287 Fulton, Ambrose C. *A Portion of a Life's Voyage.* Davenport, Iowa: Osborn-Skelly, 1902. 144 pp. Fulton owned an Illinois sawmill in the mid-19th century.

Fulton, H. Allan. See Harmston, Floyd K., #H128

F288 Funderburk, Robert S. *The History of Conservation Education in the United States.* George Peabody College for Teachers, Contribution to Education, No. 392. Nashville: 1948. vii + 151 pp. Notes, bib. History of conservation and conservation education, 1860s-1940s. Issued in the same year as a Ph.D. dissertation.

Furnas, Robert Wilkinson. See Baker, Floyd P., #B24

G1 Gabrielson, Ira N. *Wildlife Refuges.* New York: Macmillan, 1943. xiii + 257 pp. Illus., tables, bib., index. History of wildlife refuges administered by the U.S. Fish and Wildlife Service, including some treatment of state, private, and Canadian refuges.

G2 Gabrielson, Ira N. *Wildlife Conservation.* 1941. Third edition. New York: Macmillan, 1959. 244 pp. Illus. Includes some history of the administration of wildlife resources.

G3 Gabrielson, Ira N. "Ducks Can't Lay Eggs on Picket Fences." *American Forests* 68 (Sept. 1962), 16-21, 54-55. Career sketch and appreciation of Jay Norwood "Ding" Darling (1876-1962), journalist, cartoonist, and promoter of wildlife and resource conservation. Darling served as chief of the Biological Survey, 1934-1935.

G4 Gadway, Rita Mary. "Life in the South Woods Lumber Camps." *North Country Life* 12 (Fall 1958), 15-18. In the Adirondack Mountains of New York.

G5 Gaer, Joseph. *Men and Trees: The Problem of Forest Conservation and the Story of the United States Forest Service.* New York: Harcourt, Brace and Company, 1939. x + 118 pp. Illus., maps, diags., glossary, index.

G6 Gaines, Edward M., and Shaw, Elmer W. *Half a Century of Research—Fort Valley Experimental Forest, 1908-1958.* USFS, Rocky Mountain Forest and Range Experiment Station, Station Paper No. 38. Fort Collins, Colorado, 1958. 17 pp. Near Flagstaff, Arizona.

G7 Gaines, Edward M., and Shaw, Elmer W. "Half a Century of Research—Fort Valley Experimental Forest, 1908-1958." *Journal of Forestry* 57 (Sept. 1959), 629-33.

G8 Gale, George. "Quebec Ship-Building, Past and Present." *Canada Lumberman* 37 (Nov. 15, 1917), 61-63.

G9 Galenson, Walter. *The CIO Challenge to the AFL: A History of the American Labor Movement, 1935-1941.* Cambridge: Harvard University Press, 1960. xx + 732 pp. Illus., notes, index. Includes some history of AFL vs. CIO competition for workers in the lumber industry.

Galloway, Harry M. See Millar, Charles E., #M446

G10 Galloway, John A. "John Barber White: Lumberman." Ph.D. diss., Univ. of Missouri, 1961. 278 pp. White owned the Missouri Lumber and Mining Company and was an early industrial practitioner of forest conservation.

G11 Galloway, John A. "John Barber White and the Conservation Dilemma." *Forest History* 5 (Winter 1962), 9-16. On the Missouri lumberman's business interests and involvement in the conservation movement, 1880-1919.

G12 Galusha, Hugh D., Jr. "Yellowstone Years." *Montana, Magazine of Western History* 9 (July 1959), 2-21. On the work of the House of Haynes, official photographers in Yellowstone National Park since 1881.

G13 Gamble, Thomas. *Naval Stores: History, Production, Distribution and Consumption.* Savannah: Review Publishing and Printing Company, 1921. 286 pp. Illus. By the editor of *Naval Stores Review,* the prominent trade journal of the field.

G14 Gambs, John S. *The Decline of the I. W. W.* Columbia University Studies in History, Economics and Public Law, No. 361. New York: Columbia University Press, 1932. 268 pp. Diags., bib. A sequel to Paul Brissenden's *The I. W. W.: A Study of American Syndicalism* (1919), this work treats the Industrial Workers of the World during the period from 1917 to 1931, including its efforts to organize lumberjacks and millworkers in the Pacific Northwest.

G15 Gamertsfelder, Joseph W. "A Study of Glacier National Park." Master's thesis, Ohio Univ., 1947.

Gannett, Henry. See U.S. Congress, Senate, #U28

G16 Gannon, William R. "History Repeats at Old Winchester." *Michigan Conservation* 24 (Mar.-Apr. 1955), 27-29. On a dam built in 1890 by the Mecosta Lumber Company on a branch of the Chippewa River at Winchester, Michigan. The resulting lakes subsequently were used for recreational purposes.

G17 Ganoe, John T. "The History of the Oregon and California Railroad." *Quarterly of the Oregon Historical Society* 25 (Sept. 1924), 236-83; 25 (Dec. 1924), 330-52. Includes a brief account of the forested railroad grant lands in Oregon which were revested to the federal government in 1916.

G18 Ganoe, John T. "The Origin of a National Reclamation Policy." *Mississippi Valley Historical Review* 18 (June 1931), 34-52. On the development of federal irrigation policy from the Desert Land Act in 1877 to the Newlands Act of 1902.

G19 Ganoe, John T. "Some Constitutional and Political Aspects of the Ballinger-Pinchot Controversy." *Pacific Historical Review* 3 (Sept. 1934), 323-33. A major dispute in the conservation movement, 1909-1910.

G20 Ganoe, John T. "The Desert Land Act in Operation, 1877-1891." *Agricultural History* 11 (Apr. 1937), 142-57. On the operation of this reclamation act in the Rocky Mountain states.

G21 Ganoe, John T. "The Desert Land Act Since 1891." *Agricultural History* 11 (Oct. 1937), 266-77. On its operation and opposition to this reclamation act in the West, 1891-1930.

Ganser, William R. See Hanrahan, Frank J., #H87

G22 Gara, Larry. *Westernized Yankee: The Story of Cyrus Woodman.* Madison: State Historical Society of Wisconsin, 1956. x + 256 pp. Illus., notes, bib. Woodman (1814-1889) was a land speculator, sawmill owner, and dealer in Wisconsin and Michigan pine lands.

G23 Garber, C.Y. "Fire on Pine Creek." *Idaho Yesterdays* 11 (Summer 1967), 26-30. Reminiscence of a forest fire near Pine Creek in northern Idaho, 1924.

G24 Gardiner, Robert H. *Early Recollections of Robert Hallowell Gardiner.* Hallowell, Maine: By the author, 1936. 226 pp. Gardiner (1782-1864) was a Maine lumberman.

G25 Gardner, Albert F. "National Conservation Movement under Theodore Roosevelt." Master's thesis, Univ. of Nebraska, 1935.

Gardner, B. Delworth. See Roberts, N. Keith, #R230

G26 Gardner, Frank H. *The History & Literature of Wood Naval Stores.* Wilmington: Hercules Powder Company, 1963.

G27 Garey, Carl. "How Industry is Utilizing the Forest—Past, Present, and Future." *Tappi* 43 (Sept. 1960), 213a-16a. On the pulp and paper industry of the Pacific Northwest.

G28 Garfield, Charles W. "Forestry in Michigan." *Michigan Pioneer and Historical Collections* 35 (1907), 176-80.

G29 Garin, Alexis N. "Corporate Profits in the Forest Industries." Ph.D. diss., Yale Univ., 1936.

G30 Garman, E.H., and Barr, P.M. "A History Map Study in British Columbia." *Forestry Chronicle* 6 (Dec. 1930), 14-24. On a survey of natural regeneration after logging on ten operations covering 87,000 acres.

G31 Garratt, George A., and Hunt, George M. *Wood Preservation.* 1938. Second edition. New York: McGraw-Hill, 1953. 417 pp. Illus., tables, notes, app., index. Includes some history of wood preservation since the 18th century.

G32 Garratt, George A. "Wood in War and Peace." *Journal of Forestry* 42 (Sept. 1944), 636-44. Comparison of wood use in World War I and World War II and the development of new forest products.

G33 Garratt, George A. "Forestry for Connecticut." *Connecticut Woodlands* 14 (June 1949), 39-41. Since 1906.

G34 Garratt, George A. "Gifford Pinchot and Forestry Education." *Journal of Forestry* 63 (Aug. 1965), 597-600. On Pinchot as a student and teacher of forestry, 1885-1946.

G35 Garratt, George A. *Forestry Education in Canada.* Macdonald College, Quebec: Canadian Institute of Forestry, 1971. viii + 408 pp. Tables, charts, notes, apps., index. Includes some history of forestry and forestry education, especially at the University of Toronto, University of New Brunswick, Laval University, and University of British Columbia.

G36 Garratt, George A. "A Tribute to Harry McKusick." *Connecticut Woodlands* 38 (Fall 1973), 12-13. McKusick was long active in New England forestry affairs and served as Connecticut's state forester, 1965-1972.

G37 Garratt, George A., and Maunder, Elwood R. "David Mason: Architect of Forestry." *American Forests* 79 (Dec. 1973), 20-21. Career sketch of Mason (1883-1973), early USFS employee, forestry educator, forestry consultant, and leading advocate of sustained-yield forestry.

G38 Garrett, Ray. "In Less Than a Life Span." *American Forests* 58 (Oct. 1952), 23-25. Industrial forestry of the Champion Coated Paper Company, Canton, North Carolina, since 1906.

G39 Garrison, P.M. "Building an Industry on Cut-Over Land." *Journal of Forestry* 50 (Mar. 1952), 185-87. Concerns Gaylord Paper Company, Bogalusa, Louisiana, its predecessor companies, Great Southern Lumber and Bogalusa Paper, and planting experiments beginning in the 1920s.

G40 Gartenberg, Max. "W.B. Laughead's Great Advertisement." *Journal of American Folklore* 63 (Oct.-Dec. 1950), 444-49. Laughead's Paul Bunyan stories were used to promote the Red River Lumber Company of Westwood, California, beginning in 1914.

G41 Gary, George. "Hon. Philetus Sawyer." *Magazine of Western History* 10 (Aug. 1889), 457-83. On Sawyer's career as a Wisconsin lumberman, U.S. representative, and U.S. senator.

G42 Gaskill, Alfred. "The Progress of Forestry in the United States." *Forestry and Irrigation* 13 (Mar. 1907), 138-41. Includes some history of forestry since the 1870s.

G43 Gates, Charles M. "A Historical Sketch of the Economic Development of Washington Since Statehood." *Pacific Northwest Quarterly* 39 (July 1948), 214-32. Includes some account of the lumber industry.

Gates, Charles M. See Johansen, Dorothy O., #J83

G44 Gates, E.C. "Five Fruitful Years." *Southern Lumberman* 199 (Dec. 15, 1959), 82-83. On the Southern Pine Association, 1955-1959.

G45 Gates, Lillian F. "The Land Policies of Upper Canada." Ph.D. diss., Radcliffe College, 1956.

G46 Gates, Lillian F. *The Land Policies of Upper Canada.* Canadian Studies in History and Government, No. 9. Toronto: University of Toronto Press, 1968. ix + 378 pp. Maps, tables, notes, app., bib., index. A political and administrative history of land policy in Ontario, 1783-1867.

G47 Gates, Paul W. "Disposal of the Public Domain in Illinois, 1849-1856." *Journal of Economic and Business History* 3 (Feb. 1931), 216-40.

G48 Gates, Paul W. *The Illinois Central Railroad and Its Colonization Work.* Harvard Economic Studies, Volume 42. Cambridge: Harvard University Press, 1934. Reprint. New York: Johnson Reprint Corporation, 1966. xiii + 374 pp. Illus.' maps, notes, bib., index. Includes a brief account of the railroad's timberlands and the problem of timber thievery in southern Illinois, 1870s.

G49 Gates, Paul W. "American Land Policy and the Taylor Grazing Act." *Land Policy Circular* (Oct. 1935), 15-37.

G50 Gates, Paul W. "Homestead Law in an Incongruous Land System." *American Historical Review* 41 (July 1936), 652-81. Includes some history of legal and illegal acquisition of timberlands.

G51 Gates, Paul W. "Southern Investments in Northern Lands before the Civil War." *Journal of Southern History* 5 (May 1939), 155-85.

G52 Gates, Paul W. "Federal Land Policy in the South, 1866-1888." *Journal of Southern History* 6 (Aug. 1940), 303-30. Includes some history of the acquisition of timberlands.

G53 Gates, Paul W. "Land Policy and Tenancy in the Prairie States." *Journal of Economic History* 1 (May 1941), 60-82. In the 19th century.

G54 Gates, Paul W. "The Role of the Land Speculator in Western Development." *Pennsylvania Magazine of History and Biography* 66 (July 1942), 314-33.

G55 Gates, Paul W. *The Wisconsin Pine Lands of Cornell University: A Study in Land Policy and Absentee Ownership.* Ithaca, New York: Cornell University Press, 1943. Reprint. Madison: State Historical Society of Wisconsin, 1965. xi + 265 pp. Illus., maps, tables, notes, bib., index. On the acquisition, management, and sale of timberlands in northern Wisconsin by Cornell University, 1860s-1902.

G56 Gates, Paul W. "From Individualism to Collectivism in American Land Policy." In *Liberalism as a Force in History: Lectures on Aspects of the Liberal Tradition,* ed. by Chester McArthur Destler. New London, Connecticut, 1953. Pp. 14-35.

G57 Gates, Paul W. *Fifty Million Acres, Conflicts over Kansas Land Policy, 1854-1890.* Ithaca, New York: Cornell University Press, 1954. xiii + 311 pp. Illus., maps, tables, notes, bib., index.

G58 Gates, Paul W. "Weyerhaeuser and Chippewa Logging Industry." In *The John H. Hauberg Historical Essays,* ed. by O. Fritiof Ander. Augustana Library Publications, No. 26. Rock Island, Illinois: Denkmann Memorial Library, Augustana College, 1954. Pp. 50-64. On logging and log transportation in the Chippewa Valley of Wisconsin, particularly the activities of Frederick Weyerhaeuser and his Mississippi River Logging Company, 1869-1893.

G59 Gates, Paul W. "The Railroad Land-Grant Legend." *Journal of Economic History* 14 (Spring 1954), 143-46. Latter half of 19th century.

G60 Gates, Paul W. "Research in the History of American Land Tenure: A Review Article." *Agricultural History* 28 (July 1954), 121-26.

G61 Gates, Paul W. "Private Land Claims in the South." *Journal of Southern History* 22 (May 1956), 183-204.

G62 Gates, Paul W. *The Farmer's Age: Agriculture, 1815-1860.* Economic History of the United States, Volume 3. New York: Holt, Rinehart and Winston, 1960. xviii + 460 pp. Illus., map, tables, notes, bib., index. Includes many references to federal and state land policies and some on agricultural wood utilization.

Gates, Paul W. See Hibbard, Benjamin H., #H332

Gates, Paul W. See Hedrick, Ulysses P., #H265

G63 Gates, Paul W., with Robert W. Swenson. *History of Public Land Law Development.* Written for the Public Land Law Review Commission. Washington: GPO, 1968. xv + 828 pp. Notes, bib., apps., index. A comprehensive and authoritative history; includes chapters on "Early Efforts to Protect Public Timberlands," "Administration of the Public Forest Lands," and "Administration of Public Grazing Lands," as well as many scattered references to forested lands.

G64 Gates, Paul W. "Public Land Issues in the United States." *Western Historical Quarterly* 2 (Oct. 1971), 363-376. A survey of political issues deriving from public land management.

G65 Gates, Paul W. *Landlords and Tenants on the Prairie Frontier: Studies in American Land Policy.* Ithaca, New York: Cornell University Press, 1973. xii + 333 pp. A collection of essays (revised from their previously published form) concerning the history of public land policy in the Midwest, with emphasis on the role of land specualtors.

G66 Gates, Paul W. "Research in the History of the Public Lands." *Agricultural History* 48 (Jan. 1974), 31-50. A review of major research and literature on the history of the public lands.

G67 Gates, Paul W. "Public Land Disposal in California." *Agricultural History* 49 (Jan. 1975), 158-78. On the disposal of federal and state lands in the 19th century.

G68 Gates, Warren J. "The Broad Arrow Policy in Colonial America." Ph.D. diss., Univ. of Pennsylvania, 1951. 423 pp. British forest and mast policy.

G69 Gatewood, Willard B., Jr. "North Carolina's Role in the Establishment of the Great Smoky Mountains National Park." *North Carolina Historical Review* 37 (Apr. 1960), 165-84. On the activities of North Carolina in conjunction with Tennessee and the federal government, 1899-1940.

G70 Gatewood, Willard B., Jr. "Conservation and Politics in the South, 1899-1906." *Georgia Review* 16 (Spring 1962), 30-42. On the movement for a national park in the southern Appalachians of North Carolina, Tennessee, and Georgia.

G71 Gatewood, Willard B., Jr. "Theodore Roosevelt, Champion of Governmental Aesthetics." *Georgia Review* 21 (Summer 1967).

G72 Gaus, John M.; Wolcott, Leon O.; and Lewis, Verne B. *Public Administration and the United States Department of Agriculture.* Studies in Administration, Volume 10. Chicago: Public Administration Service, for the Committee on Public Administration of the Social Service Research Council, 1940. x + 534 pp. Illus., notes, tables, apps., index. Part I includes some history of the USDA and USFS.

G73 Gavin, Mortimer H. "Labor Union Policies in the U.S. Primary Pulp and Paper Industry." Ph.D. diss., St. Louis Univ., 1950.

G74 Gedosch, Thomas F. "Seabeck, 1857-1886: The History of a Company Town." Master's thesis, Univ. of Washington, 1967. Washington Mill Company, Seabeck, Washington.

G75 Gedosch, Thomas F. "A Note on the Dogfish Oil Industry of Washington Territory." *Pacific Northwest Quarterly* 59 (Apr. 1968), 100-02. Dogfish oil was used as a grease on logging skid roads in the 19th century.

G76 Gee, Willard S. "The Advantages to Society of Large Scale Timber Operations as Exemplified by the Weyerhaeuser Timber Company." Master's thesis, College of Puget Sound, 1948.

G77 Gehm, Harry W., and Lardieri, Nichols J. "Waste Treatment in the Pulp, Paper, and Paperboard Industries."

Sewage and Industrial Wastes 28 (Mar. 1956), 287-95. Since 1937.

G78 Gehrke, William H. "The Ante-Bellum Agriculture of the Germans in North Carolina." *Agricultural History* 9 (July 1935), 143-60. Includes mention of soil and timber conservation practices.

G79 Geibel, F.B. *Conservation Commissions in Worcester County, Massachusetts: A Survey of Progress, 1957-1971.* Amherst: Cooperative Extension Service, University of Massachusetts, 1972.

G80 Geiger, Harold W. "P.H. Glatfelter Co., a Brief History." *Northeastern Logger* 3 (Feb. 1955), 6-7, 24. A pulp and paper company of Spring Grove, Pennsylvania.

G81 Geiger, Robert L. *A Chronological History of the Soil Conservation Service and Related Events.* Washington: GPO, 1955. 27 pp.

G82 Gempel, E.P.H. "An Early Nebraska Sawmill." *Nebraska History* 31 (Dec. 1950), 283-91. Parts of a sawmill were transported from Missouri to Fort Kearny, Nebraska, by Meredith T. Moore, 1849. Includes excerpts from Moore's manuscript reminiscence.

G83 Gennett, A. "Timber Taxation Past and Present." *Southern Lumberman* 112 (Nov. 3, 1923), 39-40.

G84 Gentry, North Todd. "Plank Roads in Missouri." *Missouri Historical Review* 31 (Apr. 1937), 272-87. Mid-19th century.

G85 Genzoli, Andrew M., and Martin, Wallace E. *Redwood Bonanza, a Frontier's Reward: Lively Incidents in the Life of a New Empire.* Eureka, California: Schooner Features, 1967. 76 pp. Illus. Includes some history of the lumber industry in Humboldt County, California.

G86 Genzoli, Andrew M., and Martin, Wallace E. *Redwood Pioneer, a Frontier Remembered.* Eureka, California: Schooner Features, 1972. 112 pp. Illus. Includes mention of the lumber industry in Humboldt County, California.

G87 George, Dolores M. "The Plano Mill." *Wisconsin Magazine of History* 38 (Summer 1955), 237-38. A wind-driven sawmill in Taylor County, Wisconsin.

G88 George, Ernest J. *Growth and Survival of Deciduous Trees in Shelterbelt Experiments at Mandan, N.D., 1915-1934.* USDA, Technical Bulletin No. 496. Washington: GPO, 1936. 48 pp. North Dakota shelterbelts.

G89 George, Ernest J. *Cultural Practices for Growing Shelterbelt Trees on the Northern Great Plains.* USDA, Technical Bulletin No. 1138. Washington: USDA, Agricultural Research Service, 1956. 33 pp. Illus., diags., tables. Since 1918.

G90 George, Henry. *Our Land and Land Policy.* New York: Doubleday and McClure Company, 1901. iv + 345 pp. Map. First edition published in 1871. Some history and much criticism of federal land policy, along with George's single tax proposals.

G91 George, Mary Karl. "Zachariah Chandler: Radical Revisited." Ph.D. diss., St. Louis Univ., 1965. 392 pp.

G92 George, Mary Karl. *Zachariah Chandler: A Political Biography.* East Lansing: Michigan State University Press, 1969. x + 301 pp. Notes, bib., index. Chandler was Secretary of the Interior in the Grant administration, 1875-1877.

G93 Germantown Historical Society. "Mills of the Wissahickon." *Germantowne Crier* 4 (Mar. 1952), 14-15, 23; 4 (June 1952), 14-16, 18-22, 29. Rittenhouse Mill, 1690-1891, and Continental Mills, 1899-1952, paper mills along Wissahickon Creek, near Philadelphia, Pennsylvania.

G94 Gerry, Eloise. *An Oral History Interview with Dr. Eloise Gerry.* OHI by Donald G. Coleman. Madison: USFS, Forest Products Laboratory, 1961. 18 pp. Illus. Reminiscences of a botanist with the Forest Products Laboratory, Madison, Wisconsin, since 1910.

G95 Gessner, Lawrence K. *The Paper Industry Today.* New York: Smith, Barney, and Company, 1956. 139 pp. Illus., tables, diags., bib. Includes some history of the industry since 1939.

Getty, Russell. See Hartman, George B., #H181

G96 Gibbons, C.H. "Passing of Historic Hastings Mill at Vancouver." *Canada Lumberman* 50 (Aug. 1, 1930), 98. British Columbia Mills, Timber & Trading Company, producer of dock and bridge timbers since 1862.

G97 Gibbons, William H. *Logging in the Douglas Fir Region.* USDA, Bulletin No. 711. Washington: GPO, 1918. 256 pp. Incidental history of logging and the lumber industry in the Pacific Northwest.

G98 Gibbons, William H., and Johnson, Herman M. "Oregon-Washington Furniture Industry." *Timberman* 26 (Apr. 1925), 130-38. Brief history.

G99 Gibbs, Helen M. "Pope & Talbot's Tugboat Fleet." *Pacific Northwest Quarterly* 42 (Oct. 1951), 302-23. On the towing of sailing vessels and of log booms in the Puget Sound lumber trade, 1853-1925, including ocean-going tows to San Francisco and to Alaska.

Gibbs, Helen M. See Coman, Edwin Truman, #C541

G100 Gibbs, John M. "History of the North Carolina Pine Industry." *Southern Lumberman* 125 (Dec. 18, 1926), 187.

G101 Gibson, D.L. *Socio-Economic Evolution in a Timbered Area in Northern Michigan.* Michigan Agricultural Experiment Station, Technical Bulletin No. 193. Lansing, 1944. 76 pp. Illus.

G102 Gibson, Ethel, and Gibson, Harley. "The Ilwaco Mill and Lumber Company, 1903-1939." *Sou'wester* (Pacific County Historical Society) (Autumn 1971), 47-53. Ilwaco, Washington.

G103 Gibson, H.H. "A Remarkable Logging Railroad." *Hardwood Record* (Oct. 25, 1912), 25-28. Of the Little River Lumber Company, Sevier County, Tennessee.

Gibson, Harley. See Gibson, Ethel, #G102

G104 Gibson, J. Miles. *The History of Forest Management in New Brunswick.* H.R. Macmillan Lecture, No. 1. Vancouver: University of British Columbia, 1953. 14 pp.

G105 Giddens, Tandy Key. *Tandy Key Giddens.* Shreveport: Journal Printing Company, 1929. 63 pp. Giddens (b. 1868) was a dealer in Louisiana timberlands.

G106 Gifford, John C. *The Luquillo Forest Reserve, Puerto Rico.* USDA, Bureau of Forestry, Bulletin No. 54. Washington: GPO, 1905. 52 pp. Incidental history.

G107 Gifford, John C. *The Everglades and Other Essays Relating to Southern Florida.* 1911. Second edition. Miami: Everglade Land Sales Company, 1912. 226 pp. Illus., tables, index. Ironically, this eminent forester and conservationist published these essays (some with historical references to forested lands of southern Florida) to promote the Everglades Drainage Project—a stand he later reversed.

G108 Gifford, John C. *The Tropical Subsistence Homestead: Diversified Tree Crops in Forest Formation for the Antillian Areas.* Coral Gables, Florida: University of Miami, 1934. 158 pp. Some history of forestry in Florida.

G109 Gifford, John C. *Reclamation of the Everglades with Trees.* Coral Gables, Florida: University of Miami, 1935. 92 pp. Incidental history.

G110 Gifford, John C. "Recollections of a Faculty Member of the First College of Forestry." *Proceedings of the Society of American Foresters, 1948* (1949), 333-35. At the New York State College of Forestry at Cornell University, 1900-1903.

G111 Gifford, John C. *On Preserving Tropical Florida.* Ed. and with a biography by Elizabeth Ogren Rothra. Coral Gables, Florida: University of Miami Press, 1972. 222 pp. Illus., index. One of America's pioneer professional foresters, Gifford (1870-1949) was a specialist in tropical forestry and conservation at the University of Miami. This work includes excerpts of his published writings as well as a biographical sketch.

G112 Giles, H.F. *The Logged-Off Lands of Western Washington.* Olympia: Washington Bureau of Statistics and Immigration, 1911. 71 pp. Includes some incidental history of these lands; revised edition was issued in 1915.

G113 Gill, C.B. "The Development of Forestry in Manitoba." *Empire Forestry Journal* 18 (1939), 30-43.

G114 Gill, C.B. "Manitoba Looks Northward for Forest Development." *Pulp and Paper Magazine of Canada* 58 (Apr. 1957), 110-17. Some account of forest management by the Canadian Department of Mines and Natural Resources.

G115 Gill, C.B. *The Forests of Manitoba.* Manitoba Forest Service, Forest Resources Inventory, Report No. 10. Winnipeg, 1960. 43 pp. Incidental history.

Gill, Thomas H. See American Tree Association, #A410

Gill, Thomas H. See Pack, Charles Lathrop, #P7

Gill, Thomas H. See Yale University, School of Forestry, #Y2

G116 Gill, Thomas H. "Charles Lathrop Pack, 1857-1937." *Journal of Forestry* 35 (July 1937), 622-23. Pack established the forestry foundation which bears his name and promoted forest conservation, education, and research through it and other activities.

G117 Gill, Thomas H. "Tropical Forest Odyssey: Tom Gill's Diary, 1925." *Forest History* 16 (Jan. 1973), 13-17. Brief account of an eminent tropical forestry specialist in Venezuela and Trinidad.

G118 Gillett, Charles A. "Forests and Forestry in North Dakota." *Journal of Forestry* 25 (Jan. 1927), 38-43. A brief sketch of forest uses, forestry legislation, forest resources, forestry education, and tree planting experiments since the 19th century.

G119 Gilligan, James P. "The Development of Policy and Administration of Forest Service Primitive and Wilderness Areas in the Western United States." Lh.D. diss., Univ. of Michigan, 1954. 625 pp. Includes some history of USFS wilderness policy.

G120 Gilligan, James P. "Wildlife Values in Western Wilderness Area Management." *Journal of Wildlife Management* 18 (Oct. 1954), 425-32. Since 1924.

G121 Gillis, Robert Peter. "The Ottawa Lumber Barons and the Conservation Movement, 1880-1914." *Journal of Canadian Studies* 9 (Feb. 1974), 14-30. A historical analysis which finds Ottawa Valley lumbermen more interested in forest conservation than their "robber baron" image would suggest.

G122 Gillis, Robert Peter. " 'Great Britain's Woodyard': A Critical Appraisal." *Journal of Forest History* 18 (Oct. 1974), 110-12, 125-27. A review of Arthur R.M. Lower's *Great Britain's Woodyard: British America and the Timber Trade, 1763-1867* (1973).

G123 Gilman, Fred H. "History of the Development of Sawmill and Woodworking Machinery." *Mississippi Valley Lumberman* 26 (Feb. 1, 1895), 59-69.

G124 Gilman, Fred H. "History of Minneapolis as a Lumber Manufacturing Point." *Mississippi Valley Lumberman* 26 (Feb. 1, 1895), 71-76.

G125 Gilmour, James Frederick. "The Forest Industry as a Determinant of Settlement in British Columbia: The Case for Integration Through Regional Planning." Master's thesis, Univ. of British Columbia, 1965.

G126 Gilmour, John D. "Forestry Practice—Examples of, and Progress in Constructive Woods Operations." *Forestry Chronicle* 5 (June 1929), 7-9. Brief account of the beginnings of industrial forestry in Canada.

G127 Gilmour, John D. "Logging, Past and Future." *Pulp and Paper Magazine of Canada* 37 (Feb. 1936), 153-59, 166; 37 (Mar. 1936), 213-16. Pulpwood logging in Canada.

G128 Gilmour, John D. *The Forest Situation in the Province of Quebec.* H.R. Macmillan Lectureship Address, No. 2. Vancouver: University of British Columbia, 1951. 14 pp. Incidental history.

G129 Gilmour, Robert Scott. "Policy-Making for the National Forests." Ph.D. diss., Columbia Univ., 1968. 408 pp. On the issues, participants, and processes of federal legislative policy-making for management of the national forests.

G130 Gipson, Lawrence H. *Jared Ingersoll: A Study of American Loyalism in Relation to British Colonial Government.* New Haven: Yale University Press, 1920. 432 pp. Bib.

Ingersoll (1722-1781), a royal official in colonial Connecticut, had extensive timber interests in New England.

G131 Girard, James W. "Forest Service Stumpage Appraisals." *Journal of Forestry* 15 (Oct. 1917), 708-25. Includes some history of appraisals and the corresponding developments in logging engineering in the national forests of Montana, Idaho, and northeastern Washington, since ca. 1910.

G132 Girard, James W. "Logging the New Way." *American Forests* 43 (Apr. 1937), 165-72, 199. Illustrated essay on the evolution of logging techniques—from horse to power tractor logging.

G133 Girard, James W. *The Man Who Knew Trees: The Autobiography of James W. Girard.* Intro. by Rodney C. Loehr. Forest Products History Foundation Series, No. 4. St. Paul: Forest Products History Foundation, Minnesota Historical Society, 1949. 35 pp. Illus. On Girard's career as a timber cruiser in Tennessee, Idaho, Montana, and Alaska; as a USFS employee; as assistant director of the Forest Survey; as a consultant to the War Production Board; and as a consulting forester in Portland, 1892-1945.

G134 Giroux, Télesphore. *Anciens chantiers du St.-Maurice.* Trois-Rivières, Quebec: Les Éditions du Bien Public, 1935. 131 pp. Illus., map. Some history of logging and the lumber industry in the St. Maurice Valley of Quebec.

G135 Gisborne, Harry T. "Principles of Measuring Forest Fire Danger." *Journal of Forestry* 34 (Aug. 1936), 786-93. On the author's research since 1922 at the USFS's Northern Rocky Mountain Forest and Range Experiment Station, Missoula, Montana, leading to the invention of a fire danger meter.

G136 Gisborne, Harry T. "Mileposts of Progress in Fire Control and Fire Research." *Journal of Forestry* 40 (Aug. 1942), 597-608. A year-by-year chronicle since 1904.

G137 Gist, Evalyn Slack. "Forgotten Mill of the Joshuas." *Desert Magazine* 15 (Jan. 1952), 9-11. On a mill operated by the Atlantic and Pacific Fiber Company of London in Los Angeles County, California, to convert Joshua trees to pulp for the manufacture of paper, 1884-1886.

G138 Glaab, Charles N., and Larsen, Lawrence H. "Neenah-Menasha in the 1870s: The Development of Flour Milling and Paper-Making." *Wisconsin Magazine of History* 52 (Autumn 1968), 19-34.

G139 Glaab, Charles N., and Larsen, Lawrence H. *Factories in the Valley: Neenah-Menasha, 1870-1915.* Madison: State Historical Society of Wisconsin, 1969. xii + 293 pp. Illus., maps, tables, app., essay on sources, index. The Wisconsin cities of Neenah and Menasha were centers of the paper industry.

G140 Glacken, Clarence J. "The Origins of the Conservation Philosophy." *Journal of Soil and Water Conservation* 11 (No. 2, 1956), 63-66.

G141 Glacken, Clarence J. *Traces on the Rhodian Shore: Nature and Culture in Western Thought from Ancient Times to the End of the Eighteenth Century.* Berkeley: University of California Press, 1967. xxviii + 763 pp. Illus., notes, bib., index. Contains brief reference to North America in the 18th century.

G142 Gladding, Nelson A. "Recollections of a Pioneer Saw Man." *Southern Lumberman* (Dec. 15, 1931), 208-10. Gladding sold saws for E.C. Atkins & Company throughout the country since 1886.

G143 Glaser, Emma. "How Stillwater Came to Be." *Minnesota History* 24 (Sept. 1943), 195-206. An early Minnesota lumber town.

G144 Glasgow, James. *Muskegon, Michigan: The Evolution of a Lake Port.* Chicago: University of Chicago Libraries, 1939. x + 102 pp. Maps, diags., bib. This is a litho-printed version of a University of Chicago Ph.D. dissertation. Muskegon was a center of the lumber industry and an important 19th-century lumber port.

Glasier, Gilson G. See Marshall, Roujet DeLisle, #M235

G145 Glass, Remley J. "Early Transportation and the Plank Road." *Annals of Iowa* 21 (Jan. 1939), 502-34. Some history of the 19th-century plank roads. See *ibid.* 22 (1940), 77-81, for a supplement entitled "Plank Roads in Northeastern Iowa."

Glassie, Henry. See Kniffen, Fred, #K211

G146 Glasson, Philip S. "Chronology of Brown Company Technical Pioneering." *Pulp and Paper Industry* 21 (Feb. 1947), 26, 56. Brown Company, Berlin, New Hampshire.

G147 Glaster, Charles F. *The West Brancher.* New York, 1970. Reminiscences of river driving and a career with the Great Northern Paper Company, Maine.

G148 Glatfelter, P.H., and Company. *History of One of America's Oldest Paper Mills.* Spring Grove, Pennsylvania, 1922. 24 pp.

G149 Glaze, Olive A. "Rafting on the Susquehanna River." *Bulletin of the Snyder County Historical Society* 2 (1944). Pennsylvania.

G150 Glazer, Sidney. "Labor and Agrarian Movements in Michigan, 1876-1896." Ph.D. diss., Univ. of Michigan, 1932. Includes some history of the labor movement among lumberjacks and millworkers in Michigan.

G151 Glazer, Sidney. "The Lumber Frontier." *American Heritage* 2 (Summer 1951), 46-49. Of northern Michigan, Wisconsin, and Minnesota, 1865-1900.

G152 Gleason, Henry A. "The Vegetational History of the Middle West." *Annals of the Association of American Geographers* 12 (1922), 39-85.

G153 Gleason, Henry A., and Cronquist, Arthur. *The Natural Geography of Plants.* New York: Columbia University Press, 1964. viii + 420 pp. Illus., maps, index. Contains some incidental forest history of North America.

G154 Glenn, E.C. "A Few Facts Concerning the Cypress Industry." *Southern Lumberman* 144 (Dec. 15, 1931), 68-73. On the uses of cypress since the 16th century.

G155 Glesinger, Egon. *The Coming Age of Wood.* New York: Simon and Schuster, 1949. xv + 279 pp. Illus., maps, index. Includes some history of wood technology and utilization.

G156 Glickman, Freida F. "Forestry Legislation in the History of the Forest Reserves in Oregon." Master's thesis, Univ. of Oregon, 1940.

G157 Glock, Margaret S. *Collective Bargaining in the Pacific Northwest Lumber Industry.* Berkeley: Institute of Industrial Relations, University of California, 1955. viii + 62 pp. Notes, bib. From 1917 to 1952.

Glock, W.S. See Agerter, S.R., #A31

G158 Glover, John G., and Cornell, William B., eds. *The Development of American Industries: Their Economic Significance.* 1932. Third edition. New York: Prentice-Hall, 1951. xxvii + 1121 pp. Illus., maps, diags., tables. Contains historical chapters on the lumber and pulp and paper industries. The fourth edition (1959) is edited by Glover and Rudolph L. Lagai.

G159 Glover, Katherine. *America Begins Again.* New York: McGraw-Hill, 1939. 382 pp. Illus. Includes some history of American use and abuse of natural resources.

G160 Glover, Wilbur H. "Lumber Rafting on the Wisconsin River." *Wisconsin Magazine of History* 25 (Dec. 1941), 155-77; 25 (Mar. 1942), 308-24.

G161 Gober, William. "Lumbering in Florida." *Southern Lumberman* 193 (Dec. 15, 1956), 164-66. Since 1743.

Goddard, David R. See Dudley, Susan, #D276

G162 Goddard, Maurice K. "The Teaching of Forestry at Penn State." *Pennsylvania Forests* 38 (Fall 1953), 92-94, 114. Forestry education at Pennsylvania State College since 1889.

G163 Goddard, Robert H.I., Jr. "Anna R. Heidritter." *American Neptune* 21 (No. 1, 1961), 23-27. On a lumber schooner which operated between New York and Charleston, South Carolina, 1924-1934.

Goddell, B.C. See Love, Lawrence D., #L262

G164 Godwin, David P. "The Evolution of Fire-Fighting Equipment." *American Forests* 45 (Apr. 1939), 205-07, 235-37.

G165 Godwin, Gordon. *The Development of Portable Mechanical Saws for the Felling and Bucking of Timber.* Canadian Pulp and Paper Association, Woodlands Section Index, No. 733. Montreal: Canadian Pulp and Paper Association, 1944. 11 pp.

Goe, Vernon. See Labbe, John T., #L1

Goertzen, Dorine. See Rhodenbaugh, Beth, #R150

G166 Goetzmann, William H. *Army Exploration in the American West, 1803-1863.* Yale Publications in American Studies, No. 4. New Haven: Yale University Press, 1959. xx + 489 pp. Illus., map, bib., essay, index. Including the work of the U.S. Army Corps of Topographical Engineers.

G167 Goetzmann, William H. *Exploration and Empire: The Explorer and the Scientist in the Winning of the American West, 1805-1900.* New York: Alfred A. Knopf, 1966. xxii + 656 + xviii pp. Illus., maps, notes, bib., index. Includes some history of early efforts to catalogue the natural resources of the West.

G168 Goff, John H. "The Great Pine Barrens." *Emory University Quarterly* 5 (Mar. 1949), 20-31. On geography in relation to economic life in the sandy plains from Norfolk, Virginia, to New Orleans, Louisiana, since 1513.

Goforth, M.H. See Larson, Robert W., #L74

G169 Goldberg, H.M. "The Timber Cut of the Past in the U.S." Master's thesis, Yale Univ., 1929.

G170 Goldenberg, Joseph A. "The Shipbuilding Industry in Colonial America." Ph.D. diss., Univ. of North Carolina, 1969. 417 pp.

G171 Goltra, William F. "History of Wood Preservation." *Proceedings of the American Wood Preservers' Association* 9 (1913), 178-202.

G172 Golze, Alfred A. *Reclamation in the United States.* New York: McGraw-Hill, 1952. xiii + 451 pp. Illus., maps, diags., tables. A general history including some references to forests and forestry.

G173 Good, Thomas; Recknagel, Arthur B.; and Reed, Frank A. "The Growth of the Paper Industry in Northern New York." *Northern Logger* 12 (Apr. 1964), 18-19, 50-51, 56-59, 64-65.

G174 Goode, Robert D. "The Economic Growth of the Pulp and Paper Industry in Maine." Master's thesis, Univ. of Maine, 1934.

G175 Goodenough, Luman Webster. *Lumber, Lath, and Shingles: An Autobiographical Sketch Written for His Children during His Retirement Years, 1939-1946.* Detroit: By the author, 1954. 252 pp. Illus. Goodenough (1873-1947) writes of his early life in Ludington, Michigan, and his memories of the lumber industry of western Michigan.

G176 Goodlett, John Campbell. *The Development of Site Concepts at the Harvard Forest and Their Impact upon Management Policy.* Harvard Forest, Bulletin No. 28. Petersham, Massachusetts, 1960. 128 pp. Tables, bib. On the management of an area of woodland in central Massachusetts in the white pine era, 1908-1938, and the hardwood era, 1938-1960.

G177 Goodman, Robert B. "The Lumber Industry and the Income Tax." *Lumber World Review* (July 25, 1919), 25-27.

G178 Goodman, Robert B. "Forests and the Forest Industry." *Harvard Business Review* 17 (Winter 1939), 189-98. Includes some history of the forest products industries since the mid-19th century.

G179 Goodman Lumber Company. *The Goodman Forest.* Goodman, Wisconsin, 1952. 27 pp. Since 1907.

G180 Goodrum, Charles A. *History of the National Arboretum.* Washington: N.p., 1950. 15 pp. Map. Reprinted from the proceedings of the 75th annual convention of the American Association of Nurserymen, 1950.

G181 Goodspeed, Edgar Johnson. *History of the Great Fires in Chicago and the West.* New York: H.S. Goodspeed, 1871. xiv + 667 pp. See for accounts of the forest fires at Peshtigo, Wisconsin, and elsewhere.

G182 Goodstein, Anita Shafer. "Labor Relations in the Saginaw Valley Lumber Industry, 1865-1885." *Bulletin of the*

Business Historical Society 27 (Dec. 1953), 193-221. Saginaw Valley, Michigan.

G183 Goodstein, Anita Shafer. "Henry Williams Sage, 1814-1897: Biography of a Businessman." Ph.D. diss., Cornell Univ., 1958. 452 pp.

G184 Goodstein, Anita Shafer. *Biography of a Businessman: Henry W. Sage, 1814-1897.* Ithaca, New York: Cornell University Press, 1962. xiv + 279 pp. Illus., maps, notes, index. Sage owned sawmills in Canada and Michigan (Saginaw), lumber yards in Albany and New York City, and had a wide range of investments in forested lands in Wisconsin (Cornell University pinelands), California, and the South.

G185 Goodwin, James L. *A History of Pine Acres Farm, Hampton, Connecticut, 1941-1951.* Hartford: Privately printed, 1952. 68 pp. Illus., map.

G186 Goodwin, James L. "History of Great Pond Forest." *Connecticut Woodlands* 31 (Jan.-Feb. 1966), 6-7. Simsbury, Connecticut.

G187 Goodwin, O.C. *Eight Decades of Forestry Firsts: A History of Forestry in North Carolina, 1889-1969.* North Carolina Forest History Series, Volume 1, Number 3. Raleigh: School of Forest Resources, North Carolina State University, 1969. 30 pp. Bib.

G188 Goodwin, Rutherford. "The William Parks Paper Mill." *Southern Pulp and Paper Journal* 4 (Oct. 1941), 48-64. Also appeared in *Dyestuffs* 37 (Dec. 1941), 97-119.

G189 Goodyear, Charles W. *Bogalusa Story.* Buffalo: By the author, 1950. 208 pp. Illus., chart, map. Incidents and anecdotes of Bogalusa, Louisiana, 1890-1937, with particular reference to the lumber industry and the Goodyear family. The author founded the Great Southern Lumber Company.

G190 Goodyear, George F. "Goodyear Lumbering in Pennsylvania." *Niagara Frontier* 15 (Autumn 1968).

G191 Gooley, Walter R., Jr. "The Forest Industries of Southeastern Maine." *Northern Logger and Timber Processor* 18 (May 1970), 26-27, 54-57. Contains brief history since the 17th century.

G192 Gordon, David E. "Humboldt Lumber Mills." *Pacific Coast Wood and Iron* 42 (July-Dec. 1904), series of six articles. Historical sketches of individual redwood sawmills in Humboldt County, California, since the 1850s.

G193 Gordon, Robert B. *The Primeval Forest Types of Southwestern New York.* Albany: University of the State of New York, 1940. 102 pp. Illus., bib. Also issued as No. 321 of the *New York State Museum Bulletin,* April, 1940.

G194 Gorman, Lawrence. "Larry Gorman and 'Old Henry'." Ed. by Edward D. Ives. *Northeast Folklore* 2 (Fall 1959), 40-45. A lumberjack ballad composed by Gorman ca. 1892, concerning the habits of James E. Henry and his sons in their treatment of lumberjacks at their camp in Zealand Valley, New Hampshire. An introduction examines conflicting accounts of the Henrys and New Hampshire lumbermen.

G195 Gossett, Gretta. "Stock Grazing in Washington's Nile Valley: Receding Ranges in the Cascades." *Pacific Northwest Quarterly* 55 (July 1964), 119-27. Includes some account of grazing controversies involving the national forests.

G196 Gottlieb, Joel. "The Preservation of Wilderness Values: The Politics and Administration of Conservation Policy." Ph.D. diss., Univ. of California, Riverside, 1972. 300 pp. An examination of man-wilderness-policy relationships and study of conservation and environmental politics at the national level since 1872.

G197 Gottwald, Floyd D. *Albemarle, From Pines to Packaging: 75 Years of Papermaking Progress, 1887-1962.* New York: Newcomen Society in North America, 1962. 24 pp. Albemarle Paper Manufacturing Company, Richmond, Virginia.

G198 Gould, Alan B. "Secretary of the Interior Walter L. Fisher and the Return to Constructive Conservation: Problems and Policies of the Conservation Movement, 1909-1913." Ph.D. diss., West Virginia Univ., 1969. 605 pp.

G199 Gould, Alan B. " 'Trouble Portfolio' to Constructive Conservation: Secretary of the Interior Walter L. Fisher, 1911 1913." *Forest History* 16 (Jan. 1973), 4-12. On conservation problems and policies of the Taft administration subsequent to the Ballinger-Pinchot controversy.

G200 Gould, Clarence P. "Trade Between the Windward Islands and the Continental Colonies of the French Empire, 1638-1763." *Mississippi Valley Historical Review* 25 (Mar. 1939), 473-90. Includes information on the trade in lumber between the mainland and Caribbean possessions of France.

G201 Gould, Clark W., and Maxwell, Hu. "The Wood-Using Industries of Mississippi." *Lumber Trade Journal* 61 (Mar. 15, 1912), 19-29.

G202 Gould, Clark W., and Maxwell, Hu. "The Wood-Using Industries of Tennessee." *Southern Lumberman* 38 (May 25, 1912), 39-52.

G203 Gould, E. "Letter to the Editor." *Michigan History Magazine* 18 (Summer-Autumn 1934), 323-28. Reminiscences of logging near Muskegon, Michigan, 1860s-1870s.

G204 Gould, Ernest M., Jr. *Fifty Years of Management at the Harvard Forest.* Harvard Forest, Bulletin No. 29. Petersham, Massachusetts, 1960. 29 pp. Illus., tables, graphs, bib.

G205 Gould, Ernest M., Jr. "Changing Economics of the Forest Products Industries." In *First National Colloquium on the History of the Forest Products Industries, Proceedings,* ed. by Elwood R. Maunder and Margaret G. Davidson. New Haven: Forest History Society, 1967. Pp. 49-68. Since ca. 1900.

G206 Gould, Ernest M., Jr. "Whatever Became of the Invisible Hand?" *Forest History* 12 (Jan. 1969), 6-9. On the implications of planning for forest economics and democratic values; incidental historical references.

G207 Gould, Ernest M., Jr. "Scholars in the Forest." *Southern Lumberman* 219 (Dec. 15, 1969), 185-88. Brief history of the Harvard Forest, Petersham, Massachusetts.

G208 Gould, Ernest M., Jr. "The Future of Forests in Society." *Virginia Forests* 25 (Summer 1970), 10-15, 22. Includes some historical insights.

G209 Gould, Nathaniel. *Sketch of the Timber Trade of British North America.* London: Fisher, Fisher & Jackson, 1833. Includes some history of the timber trade between Canada and Great Britain.

G210 Gourlay, J.G. *History of the Ottawa Valley.* N.p., 1896. See for reference to lumbermen and the lumber industry.

G211 Gove, William. "The East Branch and Lincoln—A Logger's Railroad." *Northern Logger and Timber Processor* 16 (Apr. 1968), 16-17, 48-54. James E. Henry's logging operations and railroads near Lincoln, New Hampshire, 1890s-1930s.

G212 Gove, William. "Mountain Mills, Vermont, and the Deerfield River Railroad." *Northern Logger and Timber Processor* 17 (May 1969), 16-20, 36-38. The lumber industry, the pulp and paper industry, and the railroads which serviced them, 1890s-1920s.

G213 Gove, William. "William L. Sykes and the Emporium." *Northern Logger and Timber Processor* 18 (June 1970), 14-17, 34-37. The Emporium Lumber Company in eastern Pennsylvania, a major hardwood processor from the 1880s to 1918.

G214 Gove, William. "The Emporium Forestry Company of New York and the Grasse River R.R." *Northern Logger and Timber Processor* 19 (Sept. 1970), 10-13, 32-33. On William L. Sykes, his Emporium Forestry Company in northern New York, and the Grasse River Railroad, a logging line, 1905-1910s.

G215 Gove, William. "Latter Days—William L. Sykes and the Emporium—Part III." *Northern Logger and Timber Processor* 19 (Dec. 1970), 12-13, 32, 35, 38. Emporium Forestry Company of northern New York, 1920s-1950.

G216 Gove, William. "Sidewinders in the Great Smokies." *Northern Logger and Timber Processor* 19 (Feb. 1971), 14-15, 32-35. On the Bemis Hardwood Lumber Company, Robbinsville, North Carolina, and the logging railroads that have served it since the 1920s.

G217 Gove, William. "Burlington, the Former Lumber Capital." *Northern Logger and Timber Processor* 19 (May 1971), 18-19, 38-43. Lumbering around Lake Champlain and Burlington, Vermont, since 1786.

G218 Gove, William. "The Meadow River Lumber Company: Last of the West Virginia Pioneers." *Northern Logger and Timber Processor* 20 (Oct. 1971), 10-13, 26-29. Logging and logging railroads near Rainelle, West Virginia, since 1906.

G219 Gove, William. "Rough Logging on the Wild River Railroad." *Northern Logger and Timber Processor* 20 (Feb. 1972), 8-10, 23, 26, 37. In the Wild River Valley, New Hampshire-Maine, 1850s-1914, now part of the White Mountain National Forest. See *ibid.*, pp. 12-13, for the author's "Wrecks on the Wild River Railroad."

Gove, William. See Allen, Richard S., #A88

G220 Gove, William. "New England's First Logging Railroad: Brown's Lumber Company—Whitefield, New Hampshire." *Northern Logger and Timber Processor* 21 (Apr. 1973), 18-19, 40-45. From 1870 to 1903.

G221 Gove, William. "John Eaton." *Northern Logger and Timber Processor* 22 (Feb. 1974), 28-36, 42-45. A three-part biographical sketch and tribute to a prominent Vermont lumberman. ("Part I—The Eaton Lumber Company"; "Part II—Development of the Chipping Industry in the Northeast"; "Part III—The Man of Many Corporations.")

G222 Gove, William. "The Forest Industries of Lake Memphremagog." *Northern Logger and Timber Processor* 23 (Mar. 1975), 18-19, 31-34. History of the lumber and other forest products industries surrounding Lake Memphremagog in Vermont and Quebec, including some account of log transportation on the lake since the 1870s.

G223 Gowen, George M. "Progress in Forest Fire Control." *Scientific Monthly* 52 (June 1941), 523-29.

G224 Gower, Calvin W. "The Civilian Conservation Corps and American Education: Threat to Local Control?" *History of Education Quarterly* 7 (Spring 1967).

G225 Gower, Calvin W. "The CCC Indian Division: Aid for Depressed Americans, 1933-1942." *Minnesota History* 43 (Spring 1972), 3-13. Special attention to its work in Minnesota.

G226 Gower, Calvin W. "The CCC, The Forest Service, and Politics in Maine, 1933-1936." *New England Social Studies Bulletin* 30 (Spring 1973), 15-21.

G227 Gowland, John S. *Smoke over Sikanaska; the Story of a Forest Ranger.* New York: Ives Washburn, 1955. 224 pp. Illus. Reminiscence of a forest ranger in the Canadian Rockies of Alberta.

G228 Grady, Joseph F. *The Adirondacks: Fulton-Chain-Big Moose Region, the Story of a Wilderness.* Little Falls, New York: Press of the Journal & Courier Company, 1933. x + 320 pp. Illus., map.

G229 Graebner, Norman A. "The Public Land Policy of the Five Civilized Tribes." *Chronicles of Oklahoma* 23 (Summer 1945), 107-18. Including forest regulations and problems of timber trespass on lands in Indian Territory (Oklahoma), 19th century.

G230 Graff, Leo W. "The Senatorial Career of Fred T. Dubois of Idaho, 1890-1907." Ph.D. diss., Univ. of Idaho, 1968. 534 pp. Dubois was a defender of the forest reserves and the Roosevelt conservation program.

G231 Graham, Edward H., and Van Dersal, William R. *Water for America: The Story of Water Conservation.* New York: Oxford University Press, 1956. 111 pp. Illus.

Graham, Edward H. See Van Dersal, William R., #V17

G232 Graham, Frank. "Sawmill Evolution: Mixed Cars." *Timberman* 29 (Jan. 1928), 78. On the trend toward lumber distribution through retail outlets rather than through manufacturer, 1920s.

G233 Graham, Frank. "Sawmill Evolution." *Timberman* 29 (Feb. 1928), 156-58. On changes over fifty years at the Red River Lumber Company, Westwood, California.

G234 Graham, Frank, Jr. *Man's Dominion: The Story of Conservation in America.* New York: M. Evans and Company, distributed by J.B. Lippincott, 1971. xii + 339 pp. Illus., notes, index. A popular history of the conservation movement since the 1880s. Dramatic incidents and contributions of leaders such as Theodore Roosevelt, Gifford Pinchot, John Muir, Stephen T. Mather, and William T. Hornaday are emphasized.

G235 Graham, Gerald S. *British Policy and Canada, 1774-1791: A Study in 18th Century Trade Policy.* London: Longmans, Green and Company, 1930. xi + 161 pp. Maps. Includes some account of the timber trade.

G236 Graham, Gerald S. *Sea Power and British North America, 1783-1820: A Study in British Colonial Policy.* Harvard Historical Studies, Volume 46. Cambridge: Harvard University Press, 1941. xii + 302 pp. Maps, tables, diags., notes, index. Includes some history of British naval stores policy in Canada.

G237 Graham, Harry Edward. *The Paper Rebellion: Development and Upheaval in Pulp and Paper Unionism.* Iowa City: University of Iowa Press, 1970. xv + 170 pp. Tables, notes, bib., index. Includes some history of unionism in the pulp and paper industry, with emphasis on the Pacific Coast states and the intraunion rivalries and reorganization of the 1960s.

G238 Graham, Samuel A. *Forest Entomology.* 1929. Third edition. American Forestry Series. New York: McGraw-Hill, 1952. 351 pp. Illus. Includes some history of forest entomology.

G239 Graham, Samuel A. "The Larch Sawfly in the Lake States." *Forest Science* 2 (Jan. 1956), 132-60. On the defoliation of timber, particularly tamarack, in Michigan and Minnesota since 1906.

G240 Graham, Suzan, and Owsley, Cliff. "The Forest Service History Program at Age Three." *Southern Lumberman* 227 (Dec. 15, 1973), 117-18. On the development of the program from 1970 to 1973, including some of its achievements.

G241 Grahame, Arthur. "Father of Conservation." *Outdoor Life* 101 (Jan. 1948), 9-11, 95-97. Gifford Pinchot.

G242 Grainger, M. Allerdale. *Woodsmen of the West.* 1908. Reprint. Toronto: McClelland and Stewart, 1964. 152 pp. A novel about logging in British Columbia by the chief forester of the province; included here because of its factual detail.

G243 Grainger, M. Allerdale. "Forestry Progress in British Columbia." *Timberman* 21 (Oct. 1920), 99-101.

G244 Granberg, Wilbur J. "Tacoma: Lumber Port on an Inland Sea." *Ships and the Sea* 6 (Winter 1956), 22-25, 61-62. Tacoma, Washington, since 1852.

G245 Granger, Christopher M. "The National Forests at War." *American Forests* 49 (Mar. 1943), 112-15, 138. On the contributions of the national forests and the USFS to World War II efforts.

G246 Granger, Christopher M. "The Second 25 Years." *American Forests* 61 (Mar. 1955), 16-19, 82-90. On the USFS from 1930 to 1955, with special attention to the Great Depression, the Civilian Conservation Corps, and World War II.

G247 Grant, Adam, ed. "The Reminiscences of Joseph D. Grant." *California Historical Society Quarterly* 30 (Sept. 1951), 207-15; 30 (Dec. 1951), 339-51. Grant was a prominent leader in California's Save-the-Redwoods League.

Grant, B.F. See Bishop, G.N., #B282

G248 Grant, Joseph D. *Redwoods and Reminiscences: A Chronicle of Traffics and Excursions, of Work and Play, of Ups and Downs, During More Than Half a Century Happily Spent in California and Elsewhere.* San Francisco: Save-the-Redwoods League and the Menninger Foundation, 1973. xi + 216 + 14 pp. Illus., index. Autobiography of Grant (1858-1942), California conservationist and longtime official of the Save-the-Redwoods League.

G249 Grant, Madison. *Early History of Glacier National Park, Montana.* Washington: GPO, 1919. 12 pp.

G250 Grantham, Dewey W., Jr. "Hoke Smith: Secretary of the Interior, 1893-1896." *Georgia Historical Quarterly* 32 (Dec. 1948), 252-76.

G251 Grantham, Dewey W., Jr. "Hoke Smith: Representative of the New South." Ph.D. diss., Univ. of North Carolina, 1950.

G252 Grantham, Dewey W., Jr. *Hoke Smith and the Politics of the New South.* Baton Rouge: Louisiana State University Press, 1958. 396 pp. Illus., notes, bib., index. Smith was Grover Cleveland's Secretary of the Interior, 1893-1896, governor of Georgia, 1907-1909 and 1911, and U.S. senator from Georgia, 1911-1921. His secretaryship is covered here in two chapters.

G253 Grantham, John Bernard. "Wood-Using Industries of Virginia." *Virginia Polytechnic Institute Bulletin* 34 (Nov. 1940), 1-129.

G254 Grantham, John Bernard. "The Oregon Forest Products Research Center." *Journal of Forestry* 56 (Aug. 1958), 574-77. Corvallis, Oregon, since 1941.

G255 Grater, Russell K. *Grater's Guide to Mount Rainier National Park.* Portland: Binfords & Mort, 1949. 134 pp. Illus., maps. Includes a historical sketch of the area since its discovery in 1792.

G256 Graustein, Jeannette. *Thomas Nuttall, Naturalist: Explorations in America, 1808-1841.* Cambridge: Harvard University Press, 1967. xiii + 481 pp. Illus., maps, notes, index. Biography of Nuttall (1786-1859), an English naturalist whose scientific explorations and botanical studies in the South and Far West influenced similar work among Americans.

Graves, Alvin C. See Wurm, Theodore G., #W489

G257 Graves, Charles P. *John Muir.* New York: Thomas Y. Crowell, 1973. 33 pp. Illus. On the California preservationist, wilderness philosopher, and founder of the Sierra Club.

G258 Graves, Henry S. "Who Is Practicing Forestry?" *Lumber World Review* 39 (Nov. 10, 1920), 123-25. On forestry achievements among private owners of timberlands.

G259 Graves, Henry S. "Dr. Sargent's Contributions to Forestry in America." *American Forestry* 27 (Nov. 1921), 684-87. Charles Sprague Sargent (1841-1927), a noted student of trees and forestry, was the first director of the Arnold Arboretum.

G260 Graves, Henry S. "Education in Forestry." *Journal of Forestry* 23 (Feb. 1925), 108-25. Includes some history of forestry education since the 1890s.

Graves, Henry S. See Yale University, School of Forestry, #Y2

Graves, Henry S. See Havemeyer, Loomis, #H204

G261 Graves, Henry S. "A Look Ahead in Forestry." *Journal of Forestry* 29 (Feb. 1931), 166-74. Also reviews past advances in forestry.

G262 Graves, Henry S. "James William Toumey." *Journal of Forestry* 30 (Oct. 1932), 665-69. Toumey (1865-1932) served briefly in the Division of Forestry, 1899-1900, and was subsequently a forestry educator at Yale University and held the post of dean from 1910 to 1922.

G263 Graves, Henry S., and Guise, Cedric H. *Forest Education.* New Haven: Yale University Press, 1932. xvii + 421 pp. Tables, bib., index. Chapter 2 is entitled "History of Forest Education in the United States."

G264 Graves, Henry S. "Early Days with Gifford Pinchot." *Journal of Forestry* 43 (Aug. 1945), 550-53. Reminiscences of one ex-chief of the USFS about another, 1890s. An abridged version appears in the *Journal of Forestry* 63 (Aug. 1965), 585-86.

G265 Graves, Henry S. "Beginnings of Education in Forestry." *Proceedings of the Society of American Foresters,* 1948 (1949), 329-33.

G266 Graves, Henry S. "When Forestry Education Began." *American Forests* 55 (Feb. 1949), 18-19, 37, 40. On the establishment of the first schools of forestry in the United States, 1898-1914.

G267 Graves, John. "Redwood National Park: Controversy and Compromise." *National Parks and Conservation Magazine* 48 (Oct. 1974), 14-19. Includes some history of the park and of logging controversies near the California park's borders.

G268 Gray, Asa. "Sequoia and Its History." *American Naturalist* 6 (1872), 577-96.

G269 Gray, Asa. *Letters of Asa Gray.* Ed. by Jane Loring Gray. 2 Volumes. Boston: Houghton Mifflin, 1893. Volume 1, pp. 1-28, contains Gray's "Autobiography, 1810-1843." Gray (1810-1888) was a leading 19th-century botanist.

G270 Gray, James. *Pine, Stream and Prairie: Wisconsin and Minnesota in Profile.* New York: Alfred A. Knopf, 1945. xi + 312 + x pp. Illus., index. Chapter 8, "Timber Is a Crop," contains some history of the lumber industry.

Gray, Jane Loring. See Gray, Asa, #G269

G271 Gray, John Andrew. "Pricing and Sale of Public Timber: A Case Study of the Province of Manitoba." Ph.D. diss., Univ. of Michigan, 1971. 192 pp. Includes some history of the lumber industry, stumpage pricing, and timber disposal.

G272 Gray, John S. "Trails of a Trailblazer: P.W. Norris and Yellowstone." *Montana, Magazine of Western History* 22 (Summer 1972), 54-63. Biographical sketch of Philetus W. Norris, who attempted to explore the Yellowstone region in 1870 and who served as superintendent of Yellowstone National Park, 1877-1882.

G273 Gray, Lewis C. "The Economic Possibilities of Conservation." *Quarterly Journal of Economics* 27 (May 1913), 497-519. Relates the conservation movement to historic problems of economic theory.

G274 Gray, Lewis C., and Thompson, Esther K. *History of Agriculture in the Southern United States to 1860.* 2 Volumes. Carnegie Institute of Washington, Publication No. 430. Washington, 1933. Maps, tables, bib. A standard history containing some history of lumbering, naval stores, and the relation of agriculture to forested lands.

G275 Gray, Lewis C. "National Land Policies in Retrospect and Prospect." *Journal of Farm Economics* 13 (1931), 231-45.

G276 Gray, Norman H. "The Growth of a Wood Using Industry." *Northeastern Logger* 3 (June 1955), 8-9, 32. On C.B. Cummings and Sons, Norway, Maine, since 1860.

G277 Gray, Oscar S. *Cases and Materials on Environmental Law.* Second Edition. Washington: Bureau of National Affairs, 1973. 1442 pp. Table of cases, index. Statutes, administrative source materials, and court decisions on every facet of environmental protection and management.

G278 Gray, R.T. "History of Electric Logging." *Timberman* 28 (Sept. 1927), 184-88. In the Pacific Northwest since 1911.

G279 Gray, Roland P., ed. *Songs and Ballads of Maine Lumberjacks.* Cambridge: Harvard University Press, 1924. Reprint. Detroit: Singing Tree Press, 1969. xxi + 191 pp. Notes. Includes some history of lumberjack songs and ballads.

G280 Grayce, Robert L. "Cape Ann Forests: A Review." *Essex Institute Historical Collections* 88 (July 1952), 207-18. Cape Ann, Massachusetts.

G281 Greathouse, Charles H., comp. *Historical Sketch of the U.S. Department of Agriculture; Its Objects and Present Organization.* USDA, Division of Publications, Bulletin No. 3. Washington: GPO, 1898. 74 pp. Illus., tables, index. Includes some account of federal forestry activities under the USDA. A revised and enlarged edition was issued in 1907.

G282 Greeley, William B. "The Relation of Forest to Water Supply and Stream Flow." Master's thesis, Yale Univ., 1904.

G283 Greeley, William B. *Some Public and Economic Aspects of the Lumber Industry.* Studies of the Lumber Industry, Part I. USDA, Report No. 114. Washington: GPO, 1917. 100 pp. Maps, tables, graphs. Includes some history of the problems contributing to instability in the lumber industry.

G284 Greeley, William B. "The American Lumberjack in France." *American Forestry* 25 (June 1919), 1093-1108. On the Twentieth Engineers, the forestry regiment in World War I. The entire issue is given over to articles on the subject, many with historical references.

G285 Greeley, William B. "Westward Ho of the Sawmill." *Sunset* 50 (June 1923), 56-58, 96-99. Brief history of the westward movement of the lumber industry.

G286 Greeley, William B. "The Evolution of Forest Industries in the United States." *Western Society of Engineers Journal* 29 (No. 1, 1924), 1-12.

G287 Greeley, William B. "A Quarter Century's Achievement." *Journal of Forestry* 24 (Dec. 1926), 847-49. On Raphael Zon's career with the USFS since 1901.

G288 Greeley, William B. "The West Coast Problem of Stabilizing Lumber Production." *Journal of Forestry* 28 (Feb.

1930), 191-98. Includes some production figures and trends since 1909.

G289 Greeley, William B. "Forest Fire—The Red Paradox of Conservation." *American Forests* 45 (Apr. 1939), 153-57. Includes references to forest fires since the 1870s.

G290 Greeley, William B. "Forty Years of Forest Conservation." *Journal of Forestry* 38 (May 1940), 386-89. A tribute to Walter Mulford, forestry educator at the University of California; also includes general comments on the history of forestry.

G291 Greeley, William B. "The West Coast Lumber Industry, Past, Present and Postwar." *Mississippi Valley Lumberman* 76 (Jan. 12, 1945), 28-32.

G292 Greeley, William B. "Twenty-Five Years: Cut-Over Lands Come into Their Own." *Loggers Handbook* 5 (1945), 73-77. Pacific Northwest.

G293 Greeley, William B. "Forty Years of Southern Forestry." *Southern Lumberman* 176 (Apr. 15, 1948), 44, 46. Reminiscences, 1904-1940s.

G294 Greeley, William B. "George L. Drake." *Journal of Forestry* 46 (Aug. 1948), 610-11. Sketch of Drake's career in the USFS and with the Simpson Logging Company, Shelton, Washington, 1912-1948.

G295 Greeley, William B. *Industrial Forest Management in the Pacific Northwest as Influenced by Public Policies.* Duke University, School of Forestry Lectures, No. 7. Durham, North Carolina, 1948. 15 pp. On federal and state influences upon industrial forestry since ca. 1900.

G296 Greeley, William B. "Forestry Background of Pacific Northwest." *Journal of Forestry* 48 (Mar. 1950), 161-64. The development of forest conservation practices, programs, and legislation in Oregon and Washington, 1920s to 1950.

G297 Greeley, William B. "Tree Farms' 10th Anniversary." *American Forests* 57 (Aug. 1951), 12-14, 41-43. On tree farming in Grays Harbor County, Washington, since 1940.

G298 Greeley, William B. "Cooperative Forest Management in the Olympics." *Journal of Forestry* 49 (Sept. 1951), 627-29. On unified management of 270,000 acres of federal and private forest lands by agreement of the USFS and Simpson Logging Company, Shelton, Washington, since 1946.

G299 Greeley, William B. *Forests and Men.* Garden City, New York: Doubleday & Company, 1951. 255 pp. Bib., index. History of American forestry by former USFS chief (1920-1928), including many reminiscences since ca. 1905.

G300 Greeley, William B. "Oregon Restores a Green Tillamook." *American Forests* 59 (June 1953), 12-14, 30, 43. On the Tillamook Burn of 1933, subsequent forest fires, and subsequent reforestation efforts.

G301 Greeley, William B. "It Pays to Grow Trees." *Pacific Northwest Quarterly* 44 (Oct. 1953), 152-56. On the progress of forestry since 1900, with special reference to tree farming in Washington.

G302 Greeley, William B. *Forest Policy.* American Forestry Series. New York: McGraw-Hill, 1953. viii + 278 pp. Bib., index. Contains much history of forest policy in the United States; there are also chapters on Canada and other major forested nations.

G303 Greeley, William B. "Man-Made Fires." *Atlantic Monthly* (Aug. 1954), 78-80. Includes references to historic forest fires.

G304 Greeley, William B. "Eight Decades of Progress." *American Forests* 60 (Sept. 1954), 14-17. On the American Forestry Association, 1875-1954.

G305 Greeley, William B. "The First 25 Years." *American Forests* 61 (Mar. 1955), 12-15, 75-82. On the USFS, 1905-1930.

G306 Greeley, William B. "Henry Graves . . . The Great Conserver." *American Forests* 61 (Apr. 1955), 20-21, 40-48. On his work as USFS chief, 1910-1920, and his subsequent career in forestry education at Yale University.

G307 Greeley, William B. "What Price Competition?" *American Forests* 61 (Nov. 1955), 31, 48-49. Includes some incidental history of management of the revested Oregon and California grant lands in Oregon since 1937.

G308 Greeley, William B. "A Forester at War—Excerpts from the Diaries of Colonel William B. Greeley, 1917-1919." Ed. by George Thomas Morgan, Jr. *Forest History* 4 (Winter 1961), 3-15. Greeley was chief of the Forestry Section, Twentieth Engineers.

G309 Green, Charles L. "The Administration of the Public Domain in South Dakota." Ph.D. diss., Univ. of Iowa, 1940. 166 pp.

G310 Green, Charles L. *The Administration of the Public Domain in South Dakota.* Pierre, South Dakota: Hipple Printing Company, 1940. 280 pp. Notes, bib. Also issued as *South Dakota Historical Collections,* Volume 20, part 1 (1940).

Green, Charles Sylvester. See North Carolina, Department of Conservation and Development, #N136

G311 Green, Constance M. *Holyoke, Massachusetts: A Case History of the Industrial Revolution in America.* New Haven: Yale University Press, 1939. 425 pp. Illus. Includes some history of its paper mills.

G312 Green, Harold P. *The National Environmental Policy Act in the Courts.* Washington: Conservation Foundation, 1973. 31 pp. A review of court decisions under NEPA through April, 1972.

G313 Green, James R. "The Brotherhood of Timber Workers, 1910-1913: A Radical Response to Industrial Capitalism in the Southern U.S.A." *Past and Present* 60 (Aug. 1973), 161-200.

G314 Green, Samuel A. *Principles of Forest Entomology.* New York: McGraw-Hill, 1929. xiv + 339 pp. Illus., bib., index. Includes a chapter on the history of forest entomology.

G315 Green, Samuel B. *Forestry in Minnesota.* Delano, Minnesota: Minnesota State Forestry Association, 1898. 311 pp. Illus., tables, index. Incidental history. An enlarged edition was issued by the Geological and Natural History Survey of Minnesota in 1902.

G316 Green, Samuel B. *Principles of American Forestry.* New York: John Wiley & Sons, 1903. xiii + 334 pp. Illus., bib. Incidental history.

Green, Thornton A. See Johnson, Bolling A., #J93

G317 Greenamyre, Harold H. "Lumbering in Colorado." University of Nebraska, *Forest Club Annual* 1 (1909), 44-60.

G318 Greenberg, Irwin F. "Pinchot, Prohibition and Public Utilities: The Pennsylvania Election of 1930." *Pennsylvania History* 40 (Jan. 1973), 21-35. Gifford Pinchot and his candidacy for governor.

G319 Greene, A.C. "Mesquite and Mountain Cedar." *Southwest Review* 54 (No. 3, 1969), 314-20. Reflections of the history, use, and value of these trees native to western Texas.

G320 Greene, Lee Seifert; Brown, Virginia Holmes; and Iverson, Evan A. *Rescued Earth: A Study of the Public Administration of Natural Resources in Tennessee.* Knoxville: University of Tennessee Press, for the Bureau of Public Administration, 1948. x + 204 pp. Map, diags., tables. Since 1925.

G321 Greene, Lorenzo J., and Woodson, Carter G. *The Negro Wage Earner.* Washington: Association for the Study of Negro Life and History, 1930. xiii + 338 pp. Tables, diags., bib. Includes some history of labor conditions in the Southern lumber industry.

G322 Greening, William E. *Paper Makers in Canada: A History of the Paper Makers Union in Canada.* Cornwall, Ontario: International Brotherhood of Paper Makers, 1952. 96 pp.

G323 Greening, William E. *The Ottawa.* Toronto: McClelland and Stewart, 1961. 208 pp. Illus. Includes some history of the lumber industry in the Ottawa Valley of Ontario and Quebec.

G324 Greenwood, C.R. "The Crossett Story." *Forest Farmer* 17 (Mar. 1958), 16, 26, 28. Includes some history of industrial forestry and the Crossett Lumber Company, Crossett, Arkansas.

G325 Greenwood, Ned H. "An Outside View of Wilderness Preservation in British Columbia." *Living Wilderness* 32 (Autumn 1968), 30-42. Includes some history of provincial parks and other efforts to preserve wilderness since 1908.

G326 Greenwood, Ramon. "Apostle of the Forest Congress." *American Forests* 61 (June 1955), 20-22. On William Logan Hall (1873-1960) as a forester for the USFS and lumbermen since 1898, and promoter of the second American Forest Congress, Washington, 1905.

G327 Greenwood, Richard N. *The Five Heywood Brothers, 1826-1951: A Brief History of the Heywood-Wakefield Company during 125 Years.* New York: Newcomen Society in North America, 1951. 32 pp. On Levi Heywood and his brothers, furniture manufacturers of Gardner, Massachusetts.

G328 Greer, Cassie Conrad. "The Forest Frontier and the Development of Government Control in the Pacific Northwest, 1871-1891." Master's thesis, Univ. of Chicago, 1930.

G329 Greever, William S. "The Santa Fe Railway and its Western Land Grant." Ph.D. diss., Harvard Univ., 1949.

G330 Greever, William S. "Two Arizona Forest Lieu Land Exchanges." *Pacific Historical Review* 19 (May 1950), 137-49. On the disposition of forested land owned by the Santa Fe Railway in Arizona's San Francisco Mountains and the Grand Canyon, 1899-1905.

G331 Greever, William S. "A Comparison of Railroad Land-Grant Policies." *Agricultural History* 25 (Apr. 1951), 83-90. From 1850 to 1871.

G332 Greever, William S. *Arid Domain: The Santa Fe Railroad and Its Western Land Grant.* Stanford, California: Stanford University Press, 1954. x + 184 pp. Maps, notes, bib., index. On the Atlantic and Pacific Railroad's land grant, especially in Arizona and New Mexico, including the Santa Fe Railroad's disposition of its timberlands through the Forest Lieu Act of 1897 and other exchanges.

G333 Gregg, William Cephas. *Autobiography.* Hackensack, New Jersey: Privately printed, 1933. 364 pp. The author, a Minnesota lumberman, was interested in the preservation of national parks.

G334 Gregory, Annadora F. "Creating the Fruited Plains." *Nebraska History* 49 (Fall 1968), 299-321. On the early history of the Nebraska Horticultural Society and the work of nurseryman Ezra F. Stephens, who left a heritage of orchards and forest groves after forty years in Nebraska, 1871-1911.

Gregory, Edith. See Morris, John Milton, #M575

G335 Gregory, G. Robinson. *Forest Resource Economics.* New York: Ronald Press, 1972. viii + 548 pp. Illus., bib. Incidental history.

G336 Gregory, John G., ed. *West Central Wisconsin History.* 4 Volumes. Indianapolis: S.J. Clarke Publishing Company, 1933. Illus., index. Volumes 1 and 2 contain chapters on the lumber industry and forest conservation, especially in the Chippewa Valley.

G337 Gregory, L.F. "Lumbering Operations on the St. John." *Canada Lumberman* 33 (Sept. 1, 1913), 33-40. St. John River, New Brunswick.

G338 Gregory, U.S. *Recollections of U.S. Gregory, Old Time Sheriff, Resident of California Sixty Years.* San Francisco: Pernau-Walsh, n.d. 47 pp. Gregory was also a Nevada lumberman.

G339 Greller, Andrew M. "Observations on the Forests of Northern Queens County, Long Island, from Colonial Times to the Present." *Bulletin of the Torrey Botanical Club* 99 (July-Aug. 1972), 202-06. Queens County, Long Island, New York.

G340 Gressley, Gene M. "Arthur Powell Davis, Reclamation, and the West." *Agricultural History* 42 (July 1968), 241-57. Davis, an employee and director of the U.S. Reclamation Service, 1902-1923, was a major figure in the conservation movement.

G341 Grief, William G., and Hafner, Louis Albert. "Constitutional Aspects of Timber Conservation Legislation." *Notre Dame Lawyer* 25 (Summer 1950), 673-84. On legislation since 1908.

G342 Griffee, W.E. "Twenty-Five Years of Progress in the Western Pine Region of California." *California Lumber Merchant* 26 (July 1, 1947), 90-94.

G343 Griffith, Henry L. *Minneapolis, the New Sawdust Town.* Minneapolis: Privately published, 1968. 76 pp. Includes some reminiscences of Minneapolis, Minnesota, as a lumber and sawmilling center.

G344 Griffin, M.L. "History of the Manufacture of Soda Pulp." *Paper Trade Journal* 74 (Apr. 13, 1922), 55-59.

G345 Griffith, Walter. "The Age of Wood." *Michigan History* 32 (Dec. 1948), 374-77. On the use of wood in Detroit, 1880s.

G346 Grigg, David Henry. *From One to Seventy.* Vancouver: Mitchell Printing and Publishing Company, 1955. Reprint. New York: Vantage Press, 1957. 262 pp. An autobiography including reminiscences of logging and sawmilling in British Columbia and northern Washington, 1896-1953.

G347 Griggs, Monroe Christopher. *Wheelers, Pointers, and Leaders.* Ed. by Joseph E. Doctor and Annie R. Mitchell. Fresno, California: Academy Library Guild, for the Tulare County Historical Society, 1956. 60 pp. Illus. Reminiscences since 1876, including the author's work as a teamster in lumber camps in Tulare County, California.

G348 Griggs, Robert F. "The Timberlines of Northern America and Their Interpretation." *Ecology* 27 (Oct. 1946), 275-89.

G349 Grimshaw, Robert. *Saws: The History, Development, Action, Classification and Comparison of Saws of All Kinds. . . .* 1880. Second edition, with supplement. Philadelphia: E. Claxton and Company, 1882. iv + 279 pp. Illus., tables, diags.

G350 Grindle, Roger Lee. "The Maine Lime Industry: A Study in Business History, 1880-1900." Ph.D. diss., Univ. of Maine, 1971. 487 pp. Wood used in the burning process.

G351 Griner, James E.P. "The Growth of Manufactures in Arkansas, 1900-1950." Ph.D. diss., George Peabody College for Teachers, 1957. 245 pp. Includes some history of forest industries.

G352 Grinnell, George Bird, ed. *Brief History of the Boone and Crockett Club, with Officers, Constitution and List of Members for the Year 1910.* New York: Forest and Stream Publishing Company, 1910. 71 pp. A hunting and conservation organization.

G353 Grinnell, George Bird, and Sheldon, Charles, eds. *Hunting and Conservation; the Book of the Boone and Crockett Club.* New Haven: Yale University Press, 1925. Reprint. New York: Arno Press, 1970. xiv + 548 pp. Includes essays with some history of Mt. McKinley and Glacier national parks, the Save-the-Redwoods League, the National Recreation Conference, wildlife conservation, and wilderness preservation.

G354 Grinnell, George Bird. "The King of the Mountains." *American Forests and Forest Life* 35 (Aug. 1929), 489-93. On 19th-century explorations leading to the establishment of Glacier National Park, Montana, in 1910.

G355 Grinnell, George Bird. *The Passing of the Great West: Selected Papers of George Bird Grinnell.* Ed. by John F. Reiger. New York: Winchester Press, 1972. 182 pp. Illus., maps, notes, bib., index. From the journals (1870-1883) of a young conservationist, ornithologist, and advocate of national parks.

G356 Griswold, Oliver T. "Have We Saved the Everglades?" *Living Wilderness* 13 (Winter 1948-1949), 1-10. On the establishment and present problems of the Everglades National Park, Florida, 1928-1948.

Griswold, Oliver T. See Brookfield, Charles M., #B456

G357 Gritzner, Charles Frederick. "Spanish Log Construction in New Mexico." Ph.D. diss., Louisiana State Univ., 1969. 221 pp. In the forested part of northern New Mexico since the 17th century.

G358 Gritzner, Charles F. "Log Housing in New Mexico." *Pioneer America* 3 (July 1971), 54-62.

G359 Grobey, John H. "An Economic Analysis of the Hardwood Industry of Western Washington." Master's thesis, Univ. of Washington, 1964.

G360 Grondal, Bror L. "The Tanning Industry in the State of Washington." *West Coast Lumberman* 42 (1922), 34-41. Incidental history.

G361 Gross, James A. "The Making and Shaping of Unionism in the Pulp and Paper Industry." *Labor History* 5 (Spring 1964), 183-208. With emphasis on the International Brotherhood of Pulp, Sulphite and Paper Mill Workers, the United Papermakers and Paperworkers, and their predecessors, since 1884.

G362 Gross, H.H. "The Land Grant Legend." *Railroad Magazine* 55 (Aug. 1951), 28-49; 55 (Sept. 1951), 64-77; 56 (Oct. 1951), 68-79; 56 (Nov. 1951), 30-41; 56 (Dec. 1951), 40-53. On lands sold by the Illinois Central, Union Pacific, Northern Pacific, Burlington, and Santa Fe railroads, 1851-1903.

G363 Gross, Stuart D. *Indians, 'Jacks, and Pines; a History of Saginaw.* Saginaw, Michigan: N.p., 1962. 92 pp. Illus., bib. Popular history of Saginaw and its lumber industry.

G364 Grossman, Mary Louise; Grossman, Shelly; and Hamlet, John N. *Our Vanishing Wilderness.* New York: Grosset & Dunlap, 1969. 324 pp. Illus. Heavily illustrated work on American wilderness, wildlife, and ecological problems, including some historical references.

Grossman, Shelly. See Grossman, Mary Louise, #G364

G365 Grover, Ernest Thornton. "Major State Forest Legislation in the U.S." Master's thesis, Univ. of Idaho, 1951.

G366 Grover, Frederick W. *Multiple Use in U.S. Forest Service Land Planning.* OHI by Elwood R. Maunder. Santa Cruz, California: Forest History Society, 1972. x + 212 pp. Illus., app., bib., index. Processed. On Grover's career in the USFS, 1930-1970, mostly in California and in Washington, D.C., as director of the Division of Land and Division of Land Classification.

G367 Grozier, W.D. "Seventy Years of Paper Making in Canada." *Pulp and Paper Magazine of Canada* 28 (Jan. 31, 1929), 165ff.

G368 Grubbs, Frank H. "Frank Bond: Gentleman Sheepherder of Northern New Mexico, 1883-1915." *New Mexico Historical Review* 35 (July 1960), 169-199; (Oct. 1960), 293-308. Bond also was involved in the lumber business in northern New Mexico and southern Colorado, 1911-1915.

G369 Grünwoldt, Franz. "Die Geschictliche Entwicklung und der Heutige Stand der Verwaltung der Walder in Kanada." *Zeitschrift für Weltforstwirtschaft* 2 (Oct.-Dec. 1934), 1-67. Includes some history of forest administration in Canada.

Guise, Cedric H. See Recknagel, Arthur B., #R69

Guise, Cedric H. See Graves, Henry S., #G263

G370 Guise, Cedric H. *The Management of Farm Woodlands.* American Forestry Series. 1939. Second edition. New York: McGraw-Hill, 1950. xii + 356 pp. Illus., maps, bib. Incidental history of farm forestry.

Guise, Cedric H. See Gustafson, Axel Ferdinand, #G378

G371 Guittard, Francis G. "Roosevelt and Conservation." Ph.D. diss., Stanford Univ., 1931. 470 pp. Theodore Roosevelt.

G372 *Gulf Coast Lumberman.* "The Kirby Story, 50th Anniversary of the Founding of a Lumber Empire." *Gulf Coast Lumberman* (July 15, 1951).

G373 *Gulf Coast Lumberman.* Golden Anniversary Issue. *Gulf Coast Lumberman* 51 (Nov. 1963). Includes several articles on the history of the lumber industry in Texas.

G374 Gulick, Luther H. *American Forest Policy: A Study of Government Administration and Economic Control.* New York: Duell, Sloan and Pearce, for the Institute of Public Administration, 1951. 252 pp. Notes, tables, index. Includes some historical references to USFS policy.

G375 Gunns, Albert F. "Ray Becker, the Last Centralia Prisoner." *Pacific Northwest Quarterly* 59 (Apr. 1968), 88-99. Becker, a logger and member of the Industrial Workers of the World, was imprisoned following the Centralia, Washington, riot of 1919; this is a biographical sketch and account of legal efforts to release him, 1920-1939.

G376 Gunsky, Fred. "Trouble on the Tuolumne." *National Parks Magazine* 35 (Aug. 1960), 8-10. On the Hetch Hetchy Dam in Yosemite National Park, California, built to supply San Francisco with water and the focal point of many conservation issues, since ca. 1910.

G377 Gunter, Peter. *The Big Thicket: A Challenge to Conservation.* Austin: Jenkins Publishing Company, 1971. xviii + 172 pp. Illus., app., bib., index. Includes some history of the forested Big Thicket in eastern Texas and recent efforts to preserve it from further logging.

G378 Gustafson, Axel Ferdinand; Guise, Cedric H.; Hamilton, W.J.; and Ries, H. *Conservation in the United States.* 1939. Third edition. Ithaca, New York: Comstock Publishing Company, 1949. xi + 534 pp. Illus., maps. Includes a section on the history of conservation.

Gustafson, R.O. See Duerr, William A., #D282

G379 Gutermuth, Clinton Raymond. "Forest-Wildlife Ideologies." *Journal of Forestry* 47 (Nov. 1949), 886-89. Problems of wildlife management in the USFS, 1940s.

G380 Gutermuth, Clinton Raymond. "Origins of the Natural Resources Council of America: A Personal View." OHI by Elwood R. Maunder. *Forest History* 17 (Jan. 1974), 4-17. Gutermuth, a founder of the Natural Resources Council of America in 1946, has been involved in a wide range of conservation organizations and activities.

G381 Gutermuth, Clinton Raymond. *Pioneer Conservationist and the Natural Resources Council of America.* OHI by Elwood R. Maunder. Santa Cruz, California: Forest History Society, 1974. 166 pp. Illus., apps., index. Processed. On Gutermuth's career in conservation since the 1920s. He was director of the Indiana Department of Conservation and a founder of the Natural Resources Council of America.

G382 Guthrie, John A. "The Newsprint Industry." Ph.D. diss., Harvard Univ., 1939.

G383 Guthrie, John A. *The Newsprint Paper Industry: An Economic Analysis.* Harvard Economic Studies, Volume 68. Cambridge: Harvard University Press, 1941. xxiii + 274 pp. Illus., maps, tables, diags., bib. Contains some history of the pulp and paper industry and newsprint trade.

G384 Guthrie, John A. *The Economics of Pulp and Paper.* Pullman: State College of Washington Press, 1950. 194 pp. Maps, bib. Supplements his earlier study on the newsprint paper industry.

G385 Guthrie, John A., and Armstrong, George R. *Western Forest Industry; An Economic Outlook.* Baltimore: Johns Hopkins Press, for Resources for the Future, 1961. xxvi + 324 pp. Illus., notes, maps, graphs, tables, apps., index. Includes some history of the lumber and forest industries in Alaska, British Columbia, and eleven contiguous Western states since 1939.

G386 Guthrie, John A., and Iulo, William. *Some Economic Aspects of the Pulp and Paper Industry, with Particular Reference to Washington and Oregon.* Seattle: Pacific Northwest Pulp and Paper Association, 1963. 106 pp. Illus., maps, diags., tables, notes.

G387 Guthrie, John D. "John Muir: An Appreciation." *Forestry Quarterly* 13 (June 1915), 215-17. On Muir's role in alerting the public to threats on California forests.

G388 Guthrie, John D. "Alaska's Interior Forests." *Journal of Forestry* 20 (Apr. 1922), 363-73. Incidental history.

G389 Guthrie, John D. "Alaska's Interior Forests." *American Forestry* 28 (Aug. 1922), 451-55.

G390 Guthrie, John D. "Sterling Morton and American Forestry." *Union College Review* 11 (Aug. 1922), 270-73. J. Sterling Morton, a Union College graduate, was the "father" of Arbor Day and served as Grover Cleveland's Secretary of Agriculture, 1893-1897.

G391 Guthrie, John D. "Forestry on Arizona State Lands." *Journal of Forestry* 23 (Apr. 1925), 378-85. Since 1881.

G392 Guthrie, John D. "The History of Great Forest Fires of America." *Crow's Pacific Coast Lumber Digest* (Oct. 31, 1936).

G393 Guthrie, John D. *Great Forest Fires of America.* Washington: GPO, 1936. 11 pp. Illus.

G394 Guthrie, John D.; White, James A.; Steer, Henry B.; and Whitlock, Harry T. *'The Carpathians,' Tenth Engineers (Forestry), A.E.F., 1917-1919: Roster and Historical Sketch.* Washington: The Carpathians, 1940. 48 pp. Tables. On the foresters and lumberjacks who supplied timber for the American Expeditionary Force in Europe during World War I.

G395 Guthrie, John D. "Forestry in National Defense." *Journal of Forestry* 39 (Feb. 1941), 129-33. An account of forestry, foresters, and national forests in American defense efforts since 1826, with special attention given to war preparation and the Civilian Conservation Corps.

G396 Guthrie, John D. "Trees, People, and Foresters." *Journal of Forestry* 40 (June 1942), 477-80. Some history and evaluation of the USFS Shelterbelt Project and Prairie States Forestry Project since 1934.

G397 Guthrie, John D. "Historic Forest Fires of America." *American Forests* 49 (June 1943), 290-94, 316-17. Survey of destructive forest fires since 1825.

G398 Guthrie, John D. "W.H.B. Kent." *Journal of Forestry* 44 (Aug. 1946), 603-04. Kent worked for the USFS in the West, 1900-1910.

G399 Gutohrlein, Adolf. "Rayonier, Inc.: Railroading in the Northwest Pines." *Railway Historical Quarterly* 1 (Apr. 1964), 1-32. Pacific Northwest.

Haas, Robert B. See Hood, Mary V. Jessup, #H480

H1 Habeck, James R., and Curtis, John T. "Forest Cover and Deer Population Densities in Early Northern Wisconsin." *Transactions of the Wisconsin Academy of Sciences, Arts, and Letters* 48 (1959), 49-56. From 1736 to 1856.

H2 Habeck, James Robert. "The Original Vegetation of the Mid-Willamette Valley." *Northwest Science* 35 (May 1961), 65-77. Based on records of early land surveys in Oregon.

H3 Habib, Philip C. "Some Economic Aspects of the California Lumber Industry and Their Relation to Forest Use." Ph.D. diss., Univ. of California, Berkeley, 1952.

H4 Hacking, Norman. "Glory Days of Towboats." *British Columbia Lumberman* 50 (Aug. 1966), 44-45. Some history of log rafting and towboating in British Columbia waters.

H5 Hacking, Norman. "Schooners Pioneered Early Export Trade." *British Columbia Lumberman* 50 (Aug. 1966), 58-62. On the export of British Columbia forest products since the 1850s.

H6 Hacking, Norman. "Tugs and West Coast Logging." *British Columbia Lumberman* 47 (Apr. 1963), 16-27. On the evolution of log-towing tugs on the Pacific Coast since 1861.

H7 Haden-Guest, Stephen; Wright, John K.; and Teclaff, Eileen M., eds. *A World Geography of Forest Resources.* American Geographical Society, Special Publication No. 33. New York: Ronald Press, 1956. xviii + 736 pp. Illus., maps, diags., tables, bib., index. Includes some history of forests, forestry, and forest industries in the world economy. There are region-by-region descriptions of forest resources, including chapters on Alaska, Canada, and the United States.

H8 Hadley, Edith Jane. "John Muir's Views of Nature and Their Consequences." Ph.D. diss., Univ. of Wisconsin, 1956. 796 pp. Muir (1838-1914) was a leading naturalist, preservationist, and wilderness philosopher.

H9 Haefele, Edwin T. *Representative Government and Environmental Management.* Baltimore: Johns Hopkins University Press, for Resources for the Future, 1973. xii + 188 pp. On the role of bureaucrats in making decisions on environmental quality; incidental history.

H10 Haefner, Henry E. "Reminiscences of an Early Forester." *Oregon Historical Quarterly* 76 (Mar. 1975), 39-88. Reminiscences of forest rangers, forest fires, timber cruising, stage roads, logging, logging railroads, and settlers on the Siskiyou National Forest of southwestern Oregon and northwestern California, during the author's service with the USFS, 1909-1925.

Hafen, LeRoy R. See Baker, James H., #B30

Hafner, Louis Albert. See Grief, William G., #G341

H11 Hagen, Elizabeth. "Bridal Veil Joins the Past: Pioneer Columbia Gorge Lumber Enterprise Ends Colorful Career After Half a Century." *Timberman* 38 (Jan. 1937), 12-16, 58-60. Bridal Veil Lumber Company, Oregon.

H12 Hagenstein, William D. "Trees Grow: The Forest Economy of the Douglas Fir Region." *American Forests* 60 (Apr. 1954), 30-37. Incidental historical references to the lumber industry of Oregon and Washington, especially since 1933.

H13 Hagenstein, William D. *A Quarter Century of Industrial Forestry in the Douglas Fir Region.* Portland: Industrial Forestry Association, 1959. 8 pp. On the accomplishments of the Industrial Forestry Association in the Pacific Northwest.

H14 Hagg, Harold T. "Bemidji: A Pioneer Community of the 1890's." *Minnesota History* 23 (Mar. 1942), 24-34. A northern Minnesota lumber town.

H15 Hagg, Harold T. "The Beltrami County Logging Frontier." *Minnesota History* 29 (June 1948), 137-49. Logging near Bemidji, Minnesota, 1892-1905.

H16 Hagg, Harold T. "The Lumberjack's Sky Pilot." *Minnesota History* 31 (June 1950), 65-78. On Frank E. Higgins, a Presbyterian minister to loggers in northern Minnesota, 1890-1915.

H17 Hagg, Harold T. "Logging Line: A History of the Minneapolis, Red Lake and Manitoba." *Minnesota History* 43 (Winter 1972), 123-35. From Bemidji to Redby on Lower Red Lake, Minnesota, 1898-1938.

H18 Haig, Irvine T. "A Quarter Century of Silviculture in the Western White Pine Type." University of Montana, Forestry Club, *Forestry Kaimin* (1930), 36-41, 72-76.

H19 Haig-Brown, Roderick. *The Living Land: An Account of the Natural Resources of British Columbia.* New York: William Morrow and Company, 1961. 269 pp. Illus., maps. Primarily on the work of the British Columbia Natural Resources Conference, with some historical treatment of forest and other resources.

H20 Haight, Louis P. *The Life of Charles Henry Hackley, Drawn from Old Public and Family Records.* Volume I. Muskegon, Michigan: Dana Printing Co., 1948. vii + 150 pp. Illus. On Hackley as a Muskegon lumber dealer and philanthropist, 1855-1905.

Haines, Aubrey L. See Cook, Charles W., #C575

H21 Haines, Aubrey L. "Lost in the Yellowstone: An Epic of Survival in the Wilderness." *Montana, Magazine of Western History* 22 (Summer 1972), 31-41. On the search for Truman C. Everts, lost from the Washburn-Doane exploration party in 1870. Publicity given to the search brought support for the movement to create a Yellowstone National Park.

Haines, Aubrey L. See Langford, Nathaniel Pitt, #L45

H22 Haines, Aubrey L. *Yellowstone National Park: Its Exploration and Establishment.* Washington: USDI, National Park Service, 1974. xxiii + 218 pp. Illus., maps, biographical app., notes, bib., index.

H23 Hair, Dwight, comp. *Historical Forestry Statistics of the U.S.* USDA, Statistical Bulletin No. 228. Washington: GPO, 1958. 36 pp. Tables. Statistics on forestry and forest industries since the 1790s.

H24 Hair, Dwight. *The Economic Importance of Timber in the United States.* USFS, Miscellaneous Publication No. 941. Washington: GPO, 1963. 91 pp. Statistical report of the forest products industries in the 1950s.

H25 Hale, Peter M., comp. *The Woods and Timbers of North Carolina.* Raleigh: P.M. Hale, 1883. 272 pp. Map, index. Incidental history.

H26 Hale, Richard W., Jr. "The French Side of the 'Log Cabin Myth'." *Proceedings of the Massachusetts Historical Society* 72 (Oct. 1957-Dec. 1960), 118-25. On French-Canadian log cabin construction in the St. Lawrence Valley, 17th century.

H27 Hale, Warren F. "Town Forests in New Hampshire." *Journal of Forestry* 37 (July 1939), 525-28. Some history of town forests since 1710, especially those of Newington, Danville, Milton, Warner, Sunapee, and Grantham, New Hampshire.

H28 Hale, Warren F. "Town Forest." *New Hampshire Profiles* 1 (Nov. 1952), 63-65. On forest land owned by the town of Newington, New Hampshire, since 1714.

H29 Hale, Warren F. "Fighting Forest Fires in New Hampshire." *Appalachia* 32 (June 1958), 1-8. Forest fire fighting since 1820, especially reminiscences of 1941 and 1947.

H30 Haley, Jack D. "The Wichita Mountains: The Struggle to Preserve a Wilderness." *Great Plains Journal* 13 (Fall 1973), 70-99; 13 (Spring 1974), 149-86. On the Wichita Forest Reserve and Wichita National Forest and Game Preserve, especially since 1901, but including the 19th-century movement leading to federal preservation of this southwestern Oklahoma region.

H31 Hall, A. Dal. "Canadian Forestry Association: A Federation of Autonomous Provincial Forestry Associations." *Rod & Gun of Canada* (June 1965). Since 1900.

Hall, A.G. See Kylie, Harry R., #K285

H32 Hall, Albert George. "The Silent Saboteurs." *American Forests* 59 (Feb. 1953), 8-11, 46-49. On the progress of forest entomology and pathology since 1895.

H33 Hall, Ansel F., ed. *Handbook of Yosemite National Park.* New York: Putnam, 1921. xiii + 347 pp. Illus. Incidental history.

H34 Hall, C. Eleanor. "Joe Call, the Lewis Giant." *New York Folklore Quarterly* 9 (Spring 1953), 5-27. Call (1781-1834) of Essex County, New York, was a lumberjack of legendary strength.

H35 Hall, George Robert. "The Lumber Industry and Forest Policy: A Study in the Economics of Natural Resources." Ph.D. diss., Harvard Univ., 1960. Incidental history.

H36 Hall, Henry. *Report on the Ship-Building Industry of the United States.* Washington: GPO, 1884. 276 pp. Illus., tables, index. This government-sponsored study contains much history of shipbuilding in the United States; also issued as part of Volume 8 of the Tenth Census.

H37 Hall, J. Alfred. "The Battle for Wilderness: Another Skirmish in the Continuing Fight." *American Forests* 68 (Feb. 1962), 12-15, 42-44. Includes some account of wilderness preservation since the 1930s.

H38 Hall, J. Alfred. *The Pulp and Paper Industry and the Northwest.* Portland: USFS, Pacific Northwest Forest and Range Experiment Station, 1969. Contains some history of the industry and a list of pulp and paper mills established in Oregon and Washington since 1866.

H39 Hall, Jack. "Policies and Methods Used by the Weyerhaeuser Timber Company to Acquire Timber Land." Master's thesis, Pacific Lutheran Univ., 1953. In Washington.

H40 Hall, Leonard. "Wildlife in Missouri History." *Missouri Historical Review* 60 (Jan. 1966), 207-15.

H41 Hall, Otis F. "Sam Green Built a Forestry School." *American Forests* 59 (Oct. 1953), 24-25, 34, 36-37. On Samuel B. Green as a forestry educator in Minnesota and founder of the School of Forestry, University of Minnesota, 1888-1910.

H42 Hall, Otis F. "Manpower Shortage in Forestry." *American Forests* 63 (Mar. 1957), 10-12, 46-51. Since 1941.

H43 Hall, R. Clifford. "The Federal Income Tax and Forestry." *Journal of Forestry* 21 (Oct. 1923), 553-62. On its influence on the practice of forestry by private owners, since 1913.

H44 Hall, R. Clifford. "The Rise of Realism in Forest Taxation." *Journal of Forestry* 36 (Sept. 1938), 902-04. On the history of forest taxation since 1819, with emphasis on Fred R. Fairchild's study of 1935.

H45 Hall, R. Clifford. "The Pisgah Forest in 1912." *American Forests* 70 (Sept. 1964), 34-37, 48-50. A reminiscence of the North Carolina forest.

H46 Hall, Sherman. "Who Invented Mechanical Wood Pulp?" *Paper Trade Journal* 60 (Jan. 7, 1915), 28, 52-54.

H47 Hall, William L. "Building Up a Shortleaf-Loblolly Forest in Arkansas." *Journal of Forestry* 37 (July 1939), 538-40. Account of a consulting forester's own timberlands since 1928.

H48 Hall, William L. "Hail to the Chief." *Journal of Forestry* 43 (Aug. 1945), 553-57. Reminiscences of Gifford Pinchot during his period as chief of the USFS, 1898-1910.

H49 Hall, William L. "Is Pine Coming or Going in South Arkansas?" *Journal of Forestry* 43 (Sept. 1945), 634-37. Includes some history and reminiscences of the lumber industry and forestry in Arkansas since the 1890s.

H50 Hall, William L. "The Society of American Foresters—Its Contribution to Our National Economy." *Journal of Forestry* 49 (Mar. 1951), 165-68. Contributions of the SAF and the forestry profession since 1900.

H51 Hallberg, Gerald N. "Bellingham, Washington's Anti-Hindu Riot." *Journal of the West* 12 (Jan. 1973), 163-75. The riot of September 4, 1907, against East Indian sawmill workers, is interpreted as a phase of agitation against Oriental cheap labor on the Pacific Coast.

H52 Hallett, W.E.S. "Lumbering Cottonwood in Nebraska." University of Nebraska, *Forest Club Annual* 1 (1909), 35-38.

H53 Hallgren, A.R. "Minnesota's Forest Products Industries." *Northern Logger* 12 (Mar. 1964), 10-11, 36-37, 46, 51. History of the lumber and related forest industries since 1839.

H54 Hallin, Otis D. "Has Productivity of Forest Products Workers Kept Pace with Wages?" In *First National Colloquium on the History of the Forest Products Industries, Proceedings,* ed. by Elwood R. Maunder and Margaret G. Davidson. New Haven: Forest History Society, 1967. Pp. 164-75. On increasing productivity through advances in logging technology and transportation, based on Crown Zellerbach Corporation operations since the 1910s.

H55 Hallock, Charles. "Life among the Loggers." *Harper's* 20 (Mar. 1860), 437-54.

H56 Halm, Joe B. "The Great Fire of 1910." *American Forests and Forest Life* 36 (July 1930), 424-28, 479-80. On the forest fire which burned parts of northern Idaho and western Montana.

Halm, Joe B. See Reynolds, George W., #R133

Hamblet, Jack. See Byrne, J.J., #B625

H57 Hambleton, Jack. *Forest Ranger.* Toronto: Longmans, Green and Company, 1948. 226 pp. Illus.

H58 Hambleton, Jack. *Fire in the Valley.* Toronto: Longmans, Green and Company, 1960. 156 pp. Illus. Contains some account of forest fires and fire suppression in northern Ontario, 1950s.

H59 Hambridge, Gove. *The Story of F A O.* New York: D. Van Nostrand Company, 1955. xii + 303 pp. Illus., tables, app., bib., index. Includes some history of the Food and Agriculture Organization of the United Nations, under which international forestry activities are conducted.

H60 Hamill, Louis. "A Preliminary Study of the Status and Use of the Forest Resources of Western Oregon in Relation to Some Objectives of Public Policy." Ph.D. diss., Univ. of Washington, 1963. 159 pp. Incidental history.

H61 Hamilton, Edward P. "The New England Village Mill." *Old-Time New England* 42 (Oct. 1951), 29-38. On grist mills and sawmills, their construction, equipment, operation, and sources of power, 1632-1800.

H62 Hamilton, Edward P. *The Village Mill in Early New England.* Old Sturbridge Village Booklet Series, No. 18. Sturbridge, Massachusetts: Old Sturbridge Village, 1964. 23 pp. Illus. Description and history of colonial sawmills and gristmills.

H63 Hamilton, Eloise. *Forty Years of Western Forestry: A History of the Movement to Conserve Forest Resources by Cooperative Effort, 1909-1949.* Portland: Western Forestry and Conservation Association, 1949. 64 pp. Illus. Primarily on the work of the Western Forestry and Conservation Association.

H64 Hamilton, James Francis. "The Pulp and Paper Industry of New Brunswick, Canada." Master's thesis, Indiana Univ., 1950.

H65 Hamilton, Lawrence S. "An Analysis of New York State's Forest Practice Act." Ph.D. diss., Univ. of Michigan, 1963. 278 pp. Includes some history of federal and state forestry developments.

H66 Hamilton, Lawrence S. "The Federal Forest Regulation Issue: A Recapitulation." *Forest History* 9 (Apr. 1965), 2-11. On the threat of federal forest regulation as a factor in the adoption of state forest practice legislation in the 20th century.

Hamilton, W.J. See Gustafson, Axel Ferdinand, #G378

Hamlet, John N. See Grossman, Mary Louise, #G364

H67 Hamlin, Helen. *Nine Mile Bridge, Three Years in the Maine Woods.* New York: Norton, 1945. 233 pp. Illus., map. Account of a school teacher and wife of a game warden in the Allagash region of Maine, with some reference to logging.

H68 Hamlin, Helen. *Pine, Potatoes, and People.* New York: W.W. Norton Company, 1947. Reminiscences of life in the Maine woods, including reference to forest industries.

Hammath, R.F. See Kotok, E.I., #K245

H69 Hammerle, W.C. "Progress of Forestry in the South." *Southern Lumberman* 181 (Dec. 15, 1950), 251-52. Since 1900.

H70 Hammond, Kenneth A. "The Land and Water Conservation Act: Development and Impact." Ph.D. diss., Univ. of Michigan, 1969. 275 pp. Compares accomplishments under the act with the original intentions of congressional framers; special attention to the state of Washington.

H71 Hamp, Sidford. "Exploring the Yellowstone with Hayden, 1872: Diary of Sidford Hamp." Ed. by Herbert Oliver Brayer. *Annals of Wyoming* 14 (Oct. 1942), 253-98. An insight to an expedition which contributed to the creation of Yellowstone National Park.

H72 Hampton, H. Duane. "Conservation and Cavalry: A Study of the Role of the United States Army in the Development of a National Park System, 1886-1917." Ph.D. diss., Univ. of Colorado, 1965. 403 pp.

H73 Hampton, H. Duane. "The Army and the National Parks." *Forest History* 10 (Oct. 1966), 2-17. On the U.S. Army's administration of Yellowstone National Park, 1886-1916.

H74 Hampton, H. Duane. *How the U.S. Cavalry Saved Our National Parks.* Bloomington: Indiana University Press, 1971. 246 pp. Map, illus., notes, bib., index. On the U.S. Army's management of Yellowstone National Park, 1886-1918, with some attention to Yosemite, General Grant, and Sequoia national parks. Includes much general history of the national parks movement.

H75 Hampton, H. Duane, ed. "With Grinnell in North Park." *Colorado Magazine* 48 (Summer 1971), 273-98. Introduction includes a biographical sketch of George Bird Grinnell (1849-1938), scientist, conservationist, and enthusiast of western national parks. Grinnell's account of a visit in 1879 to North Park, Colorado, follows.

H76 Hampton, H. Duane. "The Army and the National Parks." *Montana, Magazine of Western History* 22 (Summer 1972), 64-79. On the U.S. Army's administration of Yellowstone National Park from 1886 to 1916. There is reference to the area's history since 1870.

H77 Haney, Gladys J. "Paul Bunyan Twenty-Five Years After." *Journal of American Folklore* 55 (July-Sept. 1942), 155-66. A bibliographical essay. See additional items in Herbert Halpert, "A Note on Haney's Bibliography of Paul Bunyan," *ibid.* 56 (Jan.-Mar. 1943), 57-59.

H78 Hanft, Robert M. *Pine Across the Mountain: California's McCloud River Railroad.* San Marino, California: Golden West Books, 1971. 224 pp. Illus., bib. A logging railroad in northern California.

H79 Hanks, Carlos C. "The Pulpwood Fleet." *Inland Seas* 10 (Summer 1954), 79-83. On a fleet of ships owned by Captain John Roen and operated on the Great Lakes since 1923.

H80 Hanlon, Howard. *The Bull-Hunchers: A Saga of the Three and a Half Centuries of Harvesting the Forest Crops of the Tidewater Low Country.* Parsons, West Virginia: McClain Printing Company, 1970. 352 pp. Illus. A novel but included here for its history of logging and the lumber industry in the coastal areas of the East.

H81 Hanlon, Howard A. "From the Forests They Felled—Cities Grew." *Northeastern Logger* 6 (June 1958), 14-15, 32-35, 46. On the management of state forests in New York.

H82 Hanlon, Howard A. *The Ball-Hooter: From the Forests They Felled—Cities Grew.* Prospect, New York: Prospect Books, 1960. xvi + 368 pp. Illus., maps, notes. A novel containing much factual information on logging on the Susquehanna and the upper Allegheny watersheds of Pennsylvania, and the upper reaches of the Greenbrier and Cheat rivers in West Virginia, 1881-1921.

H83 Hanlon, Howard A. *Delta Harvest: An Authentic Story of a Hardwood Harvest Interwoven with Intriguing Romance.* Watkins Glen, New York: Watkins Review, 1966. xv + 395 pp. Illus. A historical novel on the lumber industry in the Mississippi Delta and Memphis, Tennessee, since 1902; included here for its factual content.

H84 Hannay, James. *History of New Brunswick.* 2 Volumes. Saint John, New Brunswick, 1909. Includes much history of the mast trade and other forest industries.

H85 Hansell, Dorothy Ebel. "Desmond Arboretum." *Garden Journal of the New York Botanical Garden* 9 (Sept.-Oct. 1959), 170-72. Established by Thomas C. Desmond near Newburgh, New York, 1930.

H86 Hanna, Frances C. *Sand, Sawdust, and Saw Logs: Lumber Days in Ludington.* Ludington, Michigan, 1955. 73 pp. Illus. Ludington, Michigan, 1849-1917.

H87 Hanrahan, Frank J.; Ganser, William R.; and Butler, Edwin R., eds. *Proceedings of Wood Symposium: One Hundred Years of Engineering Progress with Wood, The Centennial of Engineering Convocation, September 3-13, 1952, Chicago, Illinois.* Washington: Timber Engineering Company, 1952. 111 pp. Illus., tables, references. Includes numerous articles on the history of wood technology, utilization, engineering, and construction, especially since 1852.

H88 Hansbrough, Thomas. "A Sociological Analysis of Man-Caused Forest Fires in Louisiana." Ph.D. diss., Louisiana State Univ., 1961. 317 pp. Incidental history.

H89 Hansbrough, Thomas, ed. *Southern Forests and Southern People.* Baton Rouge: Louisiana State University Press, 1963. vii + 115 pp. Diags., tables. Incidental history of incendiary forest fires.

H90 Hansen, Chris S. "Scattered Settlers: Resettlement Administration and Forest Service." *The Big Smoke 1972* (Pend Oreille County [Washington] Historical Society) (1972), 11-22. Reminiscences of work in the land acquisition and resettlement programs of the USFS and U.S. Resettlement Administration during the 1930s in northeastern Washington.

H91 Hanson, James Austin. "The Civilian Conservation Corps in the Northern Rocky Mountains." Ph.D. diss., Univ. of Wyoming, 1973. 404 pp. The CCC in Idaho, Montana, and Wyoming, 1930s.

H92 Haraszti, Zoltan. "The Invention of Wood Paper." *Paper and Printing Digest* 4 (Feb. 1938), 13-16.

H93 Harbaugh, William Henry. *Power and Responsibility: The Life and Times of Theodore Roosevelt.* New York: Farrar, Straus and Cudahy, 1961. 568 pp. Illus., bib. Includes some account of his interest in conservation.

H94 Harbeson, T.C. "Thirty-Six Years of Forestry in the Penn State Forest." *Pennsylvania Forests and Waters* 15 (Sept.-Oct. 1944), 71-74.

H95 Hardie, Robert E. *History of Conservation in the Missouri Valley.* Mitchell, South Dakota: By the author, 1964. 99 pp. Illus.

H96 Hardin, Charles M. "The Politics of Conservation: An Illustration." *Journal of Politics* 13 (Aug. 1951), 461-81. Making soil conservation policy during the 1930s.

H97 Hardin, Charles M. *The Politics of Agriculture: Soil Conservation and the Struggle for Power in Rural America.* Glencoe, Illinois: Free Press of Macmillan, 1952. 282 pp. Index. Includes some history of the Soil Conservation Service and other agencies and interest groups.

H98 Harding, T. Swann. "The Rise of the United States Department of Agriculture." *Scientific Monthly* 53 (Dec. 1941), 554-64.

H99 Harding, T. Swann. *Two Blades of Grass: A History of Scientific Development in the U.S. Department of Agri-*

culture. Norman: University of Oklahoma Press, 1947. xvi + 352 pp. Illus., app., notes, index. Chapter 8, "Man Can Help to Make a Tree," concerns the history of forestry research and is drawn from a manuscript written by Clark F. Hunn.

H100 Harding, T. Swann. *Some Landmarks in the History of the Department of Agriculture.* 1942. Agricultural History Series, No. 2, Revised. Washington: USDA, 1951. 116 pp. Bib. Since 1862.

H101 Hardtner, Henry E. "Pioneering in Reforestation." *Southern Lumberman* 121 (Dec. 19, 1925), 151-54. Urania Lumber Company, Urania, Louisiana, since ca. 1900.

H102 Hardtner, Q.T. "Southern Pine's 350th Anniversary." *Southern Lumberman* 195 (July 1, 1957), 32-34.

H103 Hardtner, Q.T. "American Southern Pine, est. 1608." *Southern Lumberman* 195 (Dec. 15, 1957), 87-91. On the commercial development of the Southern pine industry since colonial times.

H104 Hardwick, Walter G. "The Forest Industry of Coastal British Columbia: A Geographic Study of Place and Circulation." Ph.D. diss., Univ. of Minnesota, 1962. 197 pp.

H105 Hardwick, Walter G. *Geography of the Forest Industry of Coastal British Columbia.* Occasional Papers in Geography, No. 5. Vancouver: University of British Columbia, Department of Geography, for Canadian Association of Geographers, British Columbia Division, 1963. viii + 91 pp. Illus., maps, tables, graphs, notes. Historical geography of the industry since 1860.

H106 *Hardwood Record.* "Lumbering in the Adirondacks." *Hardwood Record* (Oct. 10, 1909), 19-22. Some history of hardwood logging and lumber operations of Robert W. Higbie Company, St. Lawrence County, New York.

H107 *Hardwood Record.* "The Making of a Trade Newspaper: Something of the History of 'Hardwood Record'." *Hardwood Record* (Aug. 10, 1909), 24-28. Published in Chicago since 1895.

H108 *Hardwood Record.* "Chicago: An Illustrated Review of the Men and Institutions of the Hardwood Business of Chicago." *Hardwood Record* (June 10, 1909), 21-43.

H109 *Hardwood Record.* "St. Louis As A Hardwood Market." *Hardwood Record* (Apr. 25, 1909), 36-41. Incidental history.

H110 *Hardwood Record.* "Cincinnati: The Importance of Its Hardwood Industry." *Hardwood Record* (Nov. 10, 1909), 37-60. Some history.

H111 *Hardwood Record.* "Memphis, the Hub of the Hardwood World." *Hardwood Record* (Aug. 25, 1910), 36-72. Incidental history.

H112 *Hardwood Record.* "Knoxville." *Hardwood Record* (Nov. 10, 1911), 27-32. Brief historical sketch of Knoxville, Tennessee, its lumber trade, companies, and leading lumbermen.

H113 *Hardwood Record.* "The Evolution of the Band Saw." *Hardwood Record* (July 10, 1913), 35-36.

H114 *Hardwood Record.* "World Markets for American Lumber." *Hardwood Record* (Aug. 25, 1914), 21-25; (Sept. 10, 1914), 18-21; (Nov. 25, 1914), 18-20; (Jan. 10, 1915), 16-19. Incidental history.

H115 *Hardwood Record.* "Louisiana's Oldest Sawmill." *Hardwood Record* (Feb. 10, 1915), 29.

H116 *Hardwood Record.* "James D. Lacey Timber Company." *Hardwood Record* (Aug. 25, 1915), 20-23. Nationally known as a broker in timberlands.

H117 *Hardwood Record.* "Early Sawmill and Its Builder." *Hardwood Record* (Nov. 10, 1917), 40-41. Built in Tucker County, West Virginia, 1776.

H118 *Hardwood Record.* "New Orleans, One of the Greatest Lumber Ports." *Hardwood Record* (Dec. 25, 1919), 27-30. Incidental history.

H119 *Hardwood Record.* "Stories of the Hardwoods: Making Mahogany Lumber." *Hardwood Record* (Jan. 10, 1920), 25-30. Freiberg Lumber Company, New Orleans, Louisiana.

H120 Hardy, Martha. *Tatoosh.* New York: Macmillan, 1946. 239 pp. Illus. Reminiscences of a woman forest fire lookout on the Columbia (Gifford Pinchot) National Forest, Washington.

Hargreaves, Irene M. See Foehl, Harold M., #F132

H121 Hargreaves, Leon A., Jr. "The Georgia Forestry Commission—Objectives, Organization, Policies and Procedures." Ph.D. diss., Univ. of Michigan, 1953. 252 pp. Incidental history.

Haring, Robert C. See Massie, Michael, #M285

H122 Harlan, W. Thomas. "K.O. Wilson: Firefighter." *American Forests* 80 (Mar. 1974), 8-11, 60. Career sketch of a USFS fire control specialist in the Pacific Northwest, since 1930.

H123 Harley, C.K. "On the Persistence of Old Techniques: The Case of North American Wooden Shipbuilding." *Journal of Economic History* 33 (June 1973), 372-98. A quantitative study of the wooden shipbuilding industry and its technology in the New England states and Maritime provinces, 1850s-1880s.

H124 Harley, John Richard. "Joseph Wood Krutch: Nature and Man." Ph.D. diss., Texas Christian Univ., 1971. 259 pp. On Krutch (1893-1970) as an environmentalist.

H125 Harley, Robert B. "The Land System in Colonial Maryland." Ph.D. diss., Univ. of Iowa, 1948.

H126 Harmon, Mont J. "Harold L. Ickes: A Study in New Deal Thought." Ph.D. diss., Univ. of Wisconsin, 1953. 222 pp. On Ickes's administrative theory and practice as Secretary of the Interior, 1933-1946.

H127 Harmon, Mont J. "Some Contributions of Harold L. Ickes." *Western Political Quarterly* 7 (June 1954), 238-52. On Ickes as Secretary of the Interior, 1933-1946, including his interest in and advocacy of natural resource conservation.

H128 Harmston, Floyd K.; Birch, Robert W.; and Fulton, H. Allan. *A Study of the Resources, People, and Economy of Southwestern Wyoming.* Prepared for the Wyoming Industrial Research Council. Laramie: University of Wyoming, Division of Economic Analysis, 1955. 164 pp. Maps, diags., tables, notes, bib. Since 1940.

H129 Harper, C. Armitage, and Henry, L.A. *Conservation in Arkansas.* Little Rock: Democrat Printing Company, 1939. 362 pp. Illus., bib. Incidental history of natural resources.

H130 Harper, Charles P. "The Administration of the Civilian Conservation Corps." Ph.D. diss., Johns Hopkins Univ., 1937.

H131 Harper, Charles P. *The Administration of the Civilian Conservation Corps.* Clarksburg, West Virginia: Clarksburg Publishers, 1939. 129 pp. Bib.

Harper, Francis. See Bartram, William, #B117

H132 Harper, Roland M. *Geographical Report on Forests.* Geological Survey of Alabama, Monograph 8, Economic Botany of Alabama, Part I. Montgomery, 1913. 222 pp. Includes incidental historical references to the forest resources and industries of Alabama.

H133 Harper, Roland M. "A Sketch of the Forest Geography of New Jersey." *Bulletin of the Philadelphia Geographical Society* 16 (1918), 107-25.

H134 Harper, Roland M. "Changes in the Forest Area of New England in Three Centuries." *Journal of Forestry* 16 (Apr. 1918), 442-52. Emphasis is on the increase in forest area since the mid-19th century.

H135 Harper, Roland M. *Resources of Southern Alabama: A Statistical Guide for Investors and Settlers, with an Exposition of Some of the Principles of Economic Geography.* Alabama Geological Survey, Special Report No. 11. University, Alabama, 1920. 152 pp. Includes some history of forests and forest industries.

H136 Harper, Roland M. *Natural Resources of Georgia.* University of Georgia, School of Commerce, Bureau of Business Research, Study No. 2. Athens, 1930. 105 pp. Incidental history.

H137 Harper, Roland M. *Forests of Alabama.* Alabama Geological Survey, Monograph 10. University, Alabama, 1943. 230 pp. Includes some history.

H138 Harper, Roland M. "Historical Notes on the Relation of Fire to Forest." *Proceedings of the Tall Timber Fire Ecology Conference* 1 (1962), 11-29.

H139 Harper, Verne L. "Some Highlights of Forest Research." *Journal of Forestry* 53 (Feb. 1955), 106-11. On forest research conducted by the USFS since 1905.

H140 Harper, Verne L. *A Forest Service Research Scientist and Administrator Views Multiple Use.* OHI by Elwood R. Maunder. Santa Cruz, California: Forest History Society, 1972. xii + 227 pp. Illus., apps., bib., index. Processed. On Harper's career in the USFS, 1927-1966, mostly in research at the Southern Forest Experiment Station, as director of the Northeastern Forest Experiment Station, and as deputy chief for research in Washington, D.C.

H141 Harper, Verne L. "The National Forest Multiple Use Act of 1960: Excerpts from an Oral History." OHI by Elwood R. Maunder. *Journal of Forestry* 71 (Apr. 1973), 203-05. Reminiscences of the act by a USFS deputy chief for research.

H142 Harr, J.L. "The Rise of the Wisconsin Timber Baronies." Washington State College, *Research Studies* 12 (Sept. 1944), 176-92.

Harrar, E.S. See Panshin, Alexis John, #P26

H143 Harrington, C.L. "The Story of the Kettle Moraine State Forest." *Wisconsin Magazine of History* 37 (Spring 1954), 143-45. Kettle Moraine State Forest, Wisconsin, since 1936.

H144 Harrington, Earl G. "Cadastral Surveys for the Public Lands of the United States." *Surveying and Mapping* 9 (Apr.-June 1949), 82-86. On official surveys or registers of lands conducted by the General Land Office since 1812.

H145 Harrington, Lyn. "Eastern Rockies Forest Conservation Project." *Canadian Geographical Journal* 48 (Apr. 1954), 129-43.

H146 Harrington, Robert F. "Prince George: Western White Spruce Capital of the World." *Canadian Geographical Journal* 77 (No. 3, 1968), 72-83. Includes some history of forest industries in and near the central British Columbia town.

H147 Harris, A.J. *State Lands of Alabama: A Brief History, Including Land Grants and Land Management.* Montgomery: Alabama Department of Conservation, 1951. iii + 47 pp. Since 1819.

Harris, Abram L. See Spero, Sterling, D., #S506

Harris, E.E. See Stamm, Alfred J., #S540

H148 Harris, John Tyre, and Kiefer, Francis. *Wood-Using Industries and National Forests of Arkansas.* USFS, Bulletin No. 106. Washington: GPO, 1912. 40 pp. Incidental history.

H149 Harris, John Tyre, and Maxwell, Hu. "The Wood-Using Industries of Alabama." *Lumber Trade Journal* 61 (May 1, 1912), 19-30. Incidental history.

Harris, John Tyre. See Maxwell, Hu, #M238

H150 Harris, John Tyre. *Wood-Using Industries of New York.* New York State College of Forestry, Series 14, No. 2. Syracuse, 1913. 213 pp. Illus.

Harris, John Tyre. See Nellis, Jesse Charles, #N46

H151 Harris, Marshall Dees. *Origin of the Land Tenure System in the United States.* Ames: Iowa State College Press, 1953. Reprint. Westport, Connecticut: Greenwood Press, 1970. xiv + 445 pp. Maps, bib. On landholding in the original colonies, the acquisition of Indian lands, and the activities of land companies, with incidental mention of forested lands, 1607-1790.

H152 Harris, Neil. *The Artist in American Society: The Formative Years, 1790-1860.* New York: George Braziller, 1966. xvi + 432 pp. Illus., bib. Discusses the esthetic and inspirational significance of nature and wilderness scenery to American artists.

H153 Harris, S.S. "Development of the Cedar Post and Pole Industry." University of Minnesota Forest Club, *Gopher Peavey* 2 (1922), 45-47.

H154 Harrison, Bettye Arnold. "A Journey Back Into Time." *Forests & People* 13 (First quarter, 1963), 46-49. Reminiscences of life in an unnamed lumber town in Louisiana, 1920s-1930s.

H155 Harrison, C. William. *Forest Fire Fighters and What They Do.* New York: Franklin Watts, 1962. ix + 142 pp. Illus. Largely descriptive but contains some mention of historic forest fires.

H156 Harrison, C. William. *Conservation: The Challenge of Reclaiming Our Plundered Land.* New York: Messner, 1963. 191 pp. Bib. Popular history of natural resource use and the conservation movement.

H157 Harrison, Frederick G. *Cinders and Timber: A Bird's-Eye View of Logging Railroads in Northeastern Minnesota, Yesterday and Today.* Cloquet, Minnesota: By the author, 1967. 49 pp.

H158 Harrison, Frederick G. *Steel Rails and Iron Men: A Chronicle of Logging Railroads in Northeastern Minnesota and Wisconsin during the Great Logging Era in the 1890's to 1935, and an Account of Logging Railroads up to 1969.* Indian Rocks Beach, Florida: Books Unlimited, 1970. 166 pp.

H159 Harrison, John D.B. *The Forest of Manitoba.* Forest Service, Bulletin No. 85. Ottawa, 1934. 80 pp. Maps, charts. Incidental history.

H160 Harrison, John D.B. *Forests and Forest Industries of the Prairie Provinces.* Forest Service, Bulletin No. 88. Ottawa, 1936. 69 pp. Illus., maps, diags.

H161 Harrison, John D.B. *Economic Aspects of the Forests and Forest Industries of Canada.* Forest Service, Bulletin No. 92. Ottawa, 1938. 53 pp. Map, tables, diags. Historical statistics.

H162 Harrison, John D.B. "60 Years of Progress in Canada's Lumber Industry." *Canada Lumberman* 60 (Aug. 1, 1940), 56-58.

H163 Harrison, John D.B. "American Forestry in a World Perspective." *Journal of Forestry* 49 (Feb. 1951), 172-76. Since 1900.

H164 Harrison, Lowell H. "Planted Forests—Will They Succeed on the Plains?" *Great Plains Journal* 8 (Spring 1969), 75-78. On the opinions of Leo Lesquereux, Alexander Winchell, and John Strong Newberry, 19th-century scientists, concerning forests on the Great Plains.

H165 Harrison, Robert W., and Kollmorgen, Walter M. "Socio-Economic History of Cypress Creek Drainage District and Related Districts of Southeast Arkansas." *Arkansas Historical Quarterly* 7 (Spring 1948), 20-52. From 1900 to 1930.

H166 Harrison, Robert W. *Swamp Land Reclamation in Louisiana, 1849-1879.* Baton Rouge, 1951.

H167 Harrison, Robert W., comp. "Unpublished Studies on Public Land Policies and Problems." *Land Economics* 28 (Nov. 1952), 391-400. A bibliography of theses, dissertations, and other manuscripts since 1890.

H168 Harrison, Robert W. "Public Land Records of the Federal Government." *Mississippi Valley Historical Review* 41 (Sept. 1954), 277-88. A description of such records and their possible use by historians.

H169 Harrison, Robert W. *Alluvial Empire.* Little Rock: Deta Fund and USDA, Economic Research Service, 1961. Maps, diags., tables. A study of state and local efforts toward land development, flood control, and reclamation in the lower Mississippi Valley, including some history of land clearing and the disposal of cypress and other timber.

Harroy, Jean-Paul. See Curry-Lindahl, Kai, #C759

H170 Harshberger, John W. "Nature and Man in the Pocono Mountain Region, Pennsylvania." *Bulletin of the Philadelphia Geographical Society* 13 (Apr. 1915), 64-71. A brief historical sketch of this forested region since the colonial period.

H171 omitted.

H172 Hart, Albert Bushnell. "The Disposition of Our Public Lands." *Quarterly Journal of Economics* 1 (Jan. 1887), 169-83; 1 (Apr. 1887), 251-54.

H173 Hart, John Fraser. "Loss and Abandonment of Cleared Farm Land in the Eastern United States." *Annals of the Association of American Geographers* 58 (Sept. 1968). Including reference to purchases by forest industries.

H174 Hartesveldt, Richard J. "Forest-Tree Distribution in Jackson County, Michigan, According to Original Land Survey Records." Master's thesis, Univ. of Michigan, 1951.

H175 Hartesveldt, Richard J. "The Effects of Human Impact upon Sequoia Gigantea and Its Environment in the Mariposa Grove, Yosemite National Park, California." Ph.D. diss., Univ. of Michigan, 1963. 327 pp. Including history of the National Park Service's sequoia management policy.

H176 Hartesveldt, Richard J. "The 'Discoveries' of the Giant Sequioas." *Journal of Forest History* 19 (Jan. 1975), 15-21. An analysis of rival theories concerning the "first" discovery of the sequoias of California's Sierra Nevada, 1833-1852.

H177 Hartley, Carl. "A Decade of Research in Forest Pathology." *Journal of Forestry* 36 (Sept. 1938), 908-12.

H178 Hartley, Edward N. *Ironworks on the Saugus: The Lynn and Braintree Ventures of the Company of Undertakers of the Ironworks in New England.* Norman: University of Oklahoma Press, 1957. xvi + 328 pp. Illus., notes, bib. Includes some history of the charcoal iron industry in Saugus, Massachusetts, 1652-1684.

Hartley, Robert W. See Dewhurst, J. Frederic, and Associates, #D159

H179 Hartman, Frank. "Life in a Lumber Camp." *La Crosse County Historical Sketches,* Series 3 (1937), 18-24. La Crosse County, Wisconsin.

H180 Hartman, George B. "The Iowa Sawmill Industry." *Iowa Journal of History and Politics* 40 (Jan. 1942), 52-93. A history of logging and the lumber industry in Iowa from 1829 through the 1930s. The emphasis is on the period from 1860 to 1910, when large sawmills in towns along the Mississippi River made Iowa a leader in lumber production. Includes accounts of log rafting, sawmill technology, lumber marketing, and labor conditions.

H181 Hartman, George B.; Larsen, J.A.; and Getty, Russell. "Gilmour Byers MacDonald." *Journal of Forestry* 47 (Mar. 1949), 219-20. Sketch of MacDonald's career as a nursery specialist with the USFS, 1907-1910, as head of the Forestry School at Iowa State College, 1910-1948, and as Iowa's state forester, 1935-1949.

H182 Hartshorn, Mellor. "Paul Bunyan: A Study in Folk Literature." Master's thesis, Occidental College, 1934.

H183 Hartshough, Mildred L. *From Canoe to Steel Barge on the Upper Mississippi.* Minneapolis: University of Minnesota Press, for the Upper Mississippi Waterway Association, 1934. xviii + 308 pp. Illus. Includes history of log and lumber rafting.

H184 Hartt, Rollin Lynde. "Notes on a Michigan Lumber Town." *Atlantic Monthly* 85 (Jan. 1900), 100-09.

H185 Hartzog, George B., Jr. "Over the Years With the National Park Service." *National Parks Magazine* 43 (May 1969), 13-14, 19-20. Brief history since 1916.

H186 Harvard Forest. *The Harvard Forest Models.* Petersham, Massachusetts: Harvard Forest, 1941. 48 pp. Illus. Includes a series of models revealing the history of central New England forests since the beginning of white settlement.

H187 *Harvard Law Review.* "Forest Taxes and Conservation." *Harvard Law Review* 53 (Apr. 1940), 1018-24.

H188 Harvey, Athelstan George. "John Jeffrey: Botanical Explorer." *British Columbia Historical Quarterly* 10 (Oct. 1946), 281-90. Sent by the Oregon Botanical Association of Scotland to study trees and plants of the Pacific Coast from British Columbia to California, 1850s.

H189 Harvey, Athelstan George. *Douglas of the Fir, a Biography of David Douglas, Botanist.* Cambridge: Harvard University Press, 1947. x + 290 pp. Illus., maps, bib. On the English botanist who visited the Pacific Northwest in the 1820s and 1830s, and after whom the Douglas fir is named.

H190 Harwood, W.S. "In the Minnesota Pines." *Harper's Weekly* 37 (Mar. 25, 1893), 279-82.

H191 Haskell, Elizabeth H., and Price, Victoria S. *State Environmental Management: Case Studies of Nine States.* New York: Praeger Publishers, 1973. xv + 283 pp. Studies of government reorganization at the state level to deal with environmental problems; incidental historical references.

H192 Haskell, William E. *The International Paper Co., 1898-1924: Its Origin and Growth in a Quarter of a Century with a Brief Description of the Manufacture of Paper from the Harvesting of Pulpwood to the Finished Roll.* New York: International Paper Company, 1924. 41 pp. Illus., map, diags.

H193 Hasse, Adelaide Rosalie. *Index of Economic Material in Documents of the States of the United States.* 15 Volumes. Washington: Carnegie Institution of Washington, 1907-1922. Arranged by states, these volumes describe printed reports of administrative officers, legislative committees, and special commissions since 1789. Many of the documents pertain to the history of forests and forest industries.

H194 Hastings, Alfred B., and Woods, John B. *A Survey of State Forestry Administration in Ohio, under the Direction of a Joint Committee of the Society of American Foresters and the Charles Lathrop Pack Forestry Foundation.* Washington: Society of American Foresters, 1947. 44 pp. Incidental history.

H195 Hastings, Philip R. "East Branch & Lincoln." *Railroad Magazine* 44 (Jan. 1948), 118-25. On a logging railroad in northern New Hampshire since 1893.

H196 Hatch, C.F., and Maxwell, Hu. "Wood-Using Industries of Missouri." *St. Louis Lumberman* 49 (Mar. 15, 1912), 68-83.

H197 Hatch, Melville H. *A Century of Entomology in the Pacific Northwest.* Seattle: University of Washington Press, 1949. v + 42 pp. Illus. Including forest entomology.

H198 Hatcher, Harlan H. *The Great Lakes.* New York: Oxford University Press, 1944. xi + 384 pp. Illus., maps, bib. Includes a chapter on the white pine era of the lumber industry in the Great Lakes states.

H199 Hatton, John H. "Nebraska National Forest Celebrates Its Silver Anniversary." *Producer* 9 (No. 2, 1927), 3-7.

H200 Hauberg, John Henry. *Weyerhaeuser & Denkmann: Ninety-Five Years of Manufacturing and Distribution of Lumber.* Rock Island, Illinois: Augustana Book Concern, 1957. 167 pp. Illus., tables. On a firm in Rock Island, Illinois, established in 1857 by Frederick Weyerhaeuser (1834-1914) and Frederick Carl August Denkmann (1821-1904), emigrants from Germany.

H201 Haueter, Lowell. "Westwood, California: The Life and Death of a Lumber Town." Master's thesis, Univ. of California, Berkeley, 1956.

H202 Haugen, Nils P. "Pioneer and Political Reminiscences." *Wisconsin Magazine of History* 11 (Dec. 1927), 121-52. A Wisconsin congressman reminisces about logging in the 19th century.

Hauger, Gayle. See Cochrell, Albert N., #C493

H203 Hautala, K. *European and American Tar in the British Market during the Eighteenth and Early Nineteenth Centuries.* Helsinki: Suomalainen Hedeakatemiia, 1963. Translated, 1963. 159 pp. Illus.

H204 Havemeyer, Loomis, ed. *Conservation of Our Natural Resources: Based on Van Hise's 'The Conservation of Natural Resources in the United States'.* New York: Macmillan, 1930. xvii + 551 pp. Illus., maps, tables, index. Charles R. Van Hise's pioneer work was published in 1910. There is some history of conservation, including sections on forests and forestry written by Henry S. Graves.

H205 Havighurst, Walter. *The Upper Mississippi: A Wilderness Saga.* Rivers of America Series. New York: Farrar & Rinehart, 1937. x + 258 pp. Illus., maps, index. Includes a section on the white pine lumber industry. A revised edition, omitting this section, was published in 1944.

H206 Havighurst, Walter. *The Long Ships Passing: The Story of the Great Lakes.* New York: Macmillan, 1942. viii + 291 pp. Illus., map, bib. Includes reference to the lumber industry and trade.

H207 Havighurst, Walter. *Wilderness for Sale: The Story of Our First Western Land Rush.* New York: Hastings House, 1956. 372 pp. Bib.

H208 Hawboldt, Lloyd S. "Forestry in Nova Scotia." *Canadian Geographical Journal* 51 (Aug. 1955), 66-83. Incidental history.

H209 Hawboldt, Lloyd S., and Bulmer, R.M. *The Forest Resources of Nova Scotia.* Halifax: Nova Scotia Department of Lands and Forests, 1958. 171 pp.

H210 Hawes, Austin F. "A Study of Beech and Yellow Pine as They Occur in Northern Maine, Together with an Account of the Veneer Industry." Master's thesis, Yale Univ., 1903.

H211 Hawes, Austin F. *Forest Fires In Vermont.* Vermont Forest Service, Publication No. 2. Burlington, 1909. 48 pp.

Hawes, Austin F. See Hawley, Ralph C., #H227

H212 Hawes, Austin F. "New England Forests in Retrospect." *Journal of Forestry* 21 (Mar. 1923), 209-24. General history of forest use in New England since the 1630s.

H213 Hawes, Austin F. "A Chapter in American Forestry History: The Association of Eastern Foresters." *Journal of Forestry* 25 (Mar. 1927), 325-37. The predecessor of the Association of State Foresters, 1908-1920.

H214 Hawes, Austin F. "Forty Years of State Forestry." *Journal of Forestry* 39 (Feb. 1941), 95-99. Historical sketch of state forestry since ca. 1900.

H215 Hawes, Austin F. "A Brief History of Forestry in Connecticut." *Connecticut Woodlands* 18 (May 1953), 45-48. Since 1866.

H216 Hawes, Austin F. "A Brief History of Forestry in Connecticut." *Connecticut Woodlands* 26 (Nov.-Dec. 1961), 93-96.

H217 Hawes, Harry B. "David Rowland Francis." *Missouri Historical Society Collections* 5 (Oct. 1927), 3-17. Francis was Grover Cleveland's Secretary of the Interior, 1896-1897, and urged the President to establish forest reserves.

H218 Hawk, Emory Q. *Economic History of the South.* New York: Prentice-Hall, 1934. xvii + 557 pp. Maps, tables, charts, bib. See for reference to forests and forest industries.

H219 Hawkes, Carl L. "Forestry in Guam." *American Forests* 79 (Dec. 1973), 46-50. On the history of Guam since its discovery in 1521, with special attention to its forest resources, lumber industry, and forestry programs in the 20th century.

H220 Hawkes, H. Bowman. *The Paradoxes of the Conservation Movement.* Annual Frederick William Reynolds Lecture, No. 24. Salt Lake City: Extension Division, University of Utah, 1960. 35 pp. Table, notes. On the ebb and flow of conflict between the "concept of harmony and balance" and the "concept of use," since 1901.

H221 Hawkins, Charles Franklin. "The Timber Severance Tax of Louisiana." Master's thesis, Louisiana State Univ., 1965.

H222 Hawkins, Guy C. "How One Wood-Using Industry Has Made Use of a Forester." *Journal of Forestry* 22 (Feb. 1924), 140-48. New England Box Company, New Hampshire, since 1913.

H223 Hawks, Graham P. "Increase A. Lapham, Wisconsin's First Scientist." Ph.D. diss., Univ. of Wisconsin, 1960. 314 pp. Lapham was an early advocate of forest conservation.

H224 Hawley, Curtis B. *Buckskin Mose.* New York: Worthington; n.d., 285 pp. Includes some account of the author's experiences as a forest ranger in the West.

H225 Hawley, Lee Fred. "Fifty Years of Wood Distillation." *Industrial and Engineering Chemistry* 18 (1926), 929-30.

H226 Hawley, Norm. "Naval Stores—America's First Widespread Forest Industry." *Southern Lumberman* 213 (Dec. 15, 1966), 162-64. Brief history of the industry in the South.

H227 Hawley, Ralph C., and Hawes, Austin F. *Forestry in New England, a Handbook of Eastern Forest Management.* New York: John Wiley & Sons, 1912. xv + 479 pp. Illus., maps, tables, bib., index. Contains a chapter on the "Original Forests and Their Early Development" and many historical references to state forestry in New England. A revised edition in two volumes is entitled *Manual of Forestry for the Northeastern States* (1918).

H228 Hawley, Ralph C. "Fifteen Years of Forestry." *Journal of Forestry* 21 (Mar. 1923), 225-30. By the New Haven Water Company on the watersheds of reservoirs furnishing water to New Haven, Connecticut, since 1907.

H229 Hawley, Ralph C., and Lutz, Harold J. *Establishment, Development and Management of Conifer Plantations in the Eli Whitney Forest, New Haven, Connecticut.* New Haven: Yale University, School of Forestry, 1946. 110 pp. Includes some history of this plantation since its establishment in 1901.

H230 Hawley, Robert Emmett. *Skqee Mus, or Pioneer Days on the Nooksack.* 1945. Reprint. Bellingham, Washington: Whatcom Museum of History and Art, 1971. xxiii + 189 pp. Illus., map, app. The author (b. 1862) owned a sawmill in Whatcom County, Washington.

H231 Hawley, Willis C. "Buying National Forests: A Personal Account of the Work of the National Forest Reservation Commission." *American Forests* 31 (May 1925), 293-96. Hawley, a congressman from Oregon, was a member of the commission.

H232 Haworth, Floyd B. *The Economic Development of the Wood Working Industry in Iowa.* Iowa Studies in Business, No. 13. Iowa City: State University of Iowa, College of Commerce, 1933. 128 pp. Map, tables, diags., bib. On the lumber, forest products, and wood-working industries since the 1830s.

H233 Hawthorne, Hildegarde, and Mills, Esther B. *Enos Mills of the Rockies.* Boston: Houghton Mifflin, 1935. 260 pp., plates. On the work of a Colorado conservationist (1870-1922) who promoted forest conservation and the Rocky Mountain National Park.

H234 Hayden, Elizabeth Wied. *From Trapper to Tourist in Jackson Hole.* N.p.: By the author, 1963. 48 pp. Illus., maps. A brief history of Jackson Hole, Wyoming, part of which became Grand Teton National Park.

Hayden, Margaret. See Anderson, A.A., #A421

H235 Hayden, Willard C. "The Hayden Survey." *Idaho Yesterdays* 16 (Spring 1972), 20-25. On Dr. Ferdinand Vandiveer Hayden, geologist and leader of the Hayden survey which explored and promoted the Yellowstone-Teton area of Wyoming in the 1870s.

H236 Hayes, E.P., and Heath, Charlotte. "History of the Dennison Manufacturing Company." *Journal of Economic and Business History* 1 (No. 4, 1929), 467-502; 2 (No. 1, 1929), 163-202. A paper manufacturing firm.

H237 Hayes, Marion. *Migration of the Lumber Industry.* New York: U.S. Works Progress Administration, Special Research Section, 1937. Bib. On reemployment opportunities and recent changes in industrial techniques.

Hayman, Donald B. See Wager, Paul W., #W11

H238 Haymond, Jay M. "History of the Manti Forest, Utah: A Case of Conservation in the West." Ph.D. diss., Univ. of Utah, 1972. 221 pp. Manti National Forest, Utah, 1903-1940.

H239 Hayner, Norman S. "Taming the Lumberjack." *American Sociological Review* 10 (Apr. 1945), 217-25. On the gradual "domestication" of Pacific Northwest lumberjacks from tough and lonely bachelors to family men, since 1895.

H240 Haynes, Jack Ellis. "The Expedition of President Chester A. Arthur to Yellowstone National Park in 1883." *Annals of Wyoming* 14 (Jan. 1942), 31-38. Based on the diary of Frank Jay Haynes, official photographer of the expedition.

H241 Haynes, John E. "Revolt of the 'Timber Beasts': IWW Lumber Strike in Minnesota." *Minnesota History* 42 (Spring 1971), 162-74. Industrial Workers of the World in northern Minnesota, 1917.

H242 Hays, Samuel P. "The First American Conservation Movement, 1891-1920." Ph.D. diss., Harvard Univ., 1953. 318 pp.

H243 Hays, Samuel P. *Conservation and The Gospel of Efficiency: The Progressive Conservation Movement, 1890-1920.* Harvard Historical Monographs, No. 40. Cambridge: Harvard University Press, 1959. 297 pp. Notes, bib., note, index. A standard work which argues that conservation did not develop as a mass movement but as a scientific movement led by specialists loyal to professional ideals. Early conservationists often believed that experts, not politicians, should formulate resource policies. All facets of conservation, including forests, are treated.

H244 Hayter, Roger. "An Examination of Growth Patterns and Locational Behaviour of Multi-Plant Forest Products Corporations in British Columbia." Ph.D. diss., Univ. of Washington, 1973. 290 pp. Trends in vertical and horizontal integration since 1950.

H245 Hayter, Roger. "Corporate Strategies in the Forest Product Industries of British Columbia." *Albertan Geographer* 10 (1974), 7-19. Since the 1950s.

H246 Hazard, Joseph P. "Shipbuilding in Narragansett." *Narragansett Historical Register* 2 (July 1883).

H247 Hazard, Joseph T. *Our Living Forests: The Story of Their Preservation and Multiple Use.* Seattle: Superior Publishing Company, 1948. xii + 302 pp. Illus. Broadly concerned with forests, forestry, and forest industries since colonial times, but with emphasis on the Pacific Northwest.

H248 Hazard, Joseph T. "Winter Sports in the Western Mountains." *Pacific Northwest Quarterly* 44 (Jan. 1953), 7-14. Includes incidental history of USFS involvement in ski areas of the Pacific Northwest.

H249 Hazard, Lucy Lockwood. *The Frontier in American Literature.* New York: Thomas Y. Crowell Company, 1927. xx + 308 pp. Bib. A Turnerian interpretation; many references to forested lands.

H250 Hazel, G.G. *Public Land Laws of Texas: An Examination of the History of the Public Domain of This State, with the Constitutional and Statutory Provisions, and Leading Cases, Governing Its Use and Disposition.* Austin: Gammels, 1938. viii + 121 pp.

H251 Hazeltine, Jean. "The Historical and Regional Geography of the Willapa Bay Area." Ph.D. diss., Ohio State Univ., 1956. 365 pp.

H252 Hazeltine, Jean. *The Historical and Regional Geography of the Willapa Bay Region, Washington.* South Bend, Washington: South Bend Journal, 1956. 308 pp. Illus., maps, tables, bib. An important lumbering region since the 1890s.

H253 Hazenberg, G. "An Analysis of New Brunswick Lumber Industry." Master's thesis, Univ. of New Brunswick, 1966.

H254 Headden, Harmon Clay. *Conservation of Wildlife and Forests in Tennessee.* Kingsport, Tennessee: Southern Publishers, 1936. 242 pp. Illus.

H255 Headley, Roy. "Budgets and Financial Control in the National Forest Service." *Annals of the American Academy of Political and Social Science* 113 (May 1924), 51-56.

H256 Heald, Weldon F. "Who Saved the Redwoods?" *Natural History* 62 (Jan. 1953), 24-31, 44. On the work of the Save-the-Redwoods League in California since 1918.

H257 Heald, Weldon F. "The Yellowstone Story: Genesis of the National Park Idea." *Utah Historical Quarterly* 28 (Apr. 1960), 98-110. History of the region prior to park status, 1807-1872.

H258 Heald, Weldon F. "Wanderer of the Wild Palms." *Pacific Discovery* 13 (Nov.-Dec. 1960), 14-16. On Randall Henderson's studies of fan palms in the southern California desert, 1920-1960.

H259 Heard, John P. "Records of the Bureau of Land Management in California and Nevada: Resource for Historians." *Forest History* 12 (July 1968), 20-26. Description of records at the Federal Record Center, San Francisco.

H260 Hearn, George, and Wilkie, David. *The Cordwood Limited: A History of the Victoria & Sidney Railway.* Victoria: British Columbia Railway Historical Association, 1966. 83 pp. Illus., maps. A logging railroad on Vancouver Island, British Columbia, 1892-1935.

Heath, Charlotte. See Hayes, E.P., #H236

H261 Heath, Virgil, and Hunt, John Clark. "Alaska CCC Days." *Alaska Journal* 2 (Spring 1972), 51-56. Civilian Conservation Corps, 1930s.

H262 Heatherly, L.J. "History of the Southern Hardwood Producers, Inc." *Southern Lumberman* 193 (Dec. 15, 1956), 131. On a trade association established in 1935.

H263 Hechler, Kenneth W. *Insurgency: Personalities and Politics of the Taft Era.* Columbia University Studies in History, Economics and Public Law, No. 470. New York: Columbia University Press, 1940. 252 pp. Bib. Includes a section on the Ballinger-Pinchot controversy and the conservation movement.

Hedgecock, George G. See Scheffer, Theodore C., #S72

H264 Hedges, James B. *The Federal Railway Land Subsidy Policy of Canada.* Harvard Historical Monographs, No. 3. Cambridge: Harvard University Press, 1934. viii + 151 pp. Forested lands comprised some of the subsidies.

H265 Hedrick, Ulysses P. *A History of Agriculture in the State of New York.* Albany: J.B. Lyon Company, for the New York State Agricultural Society, 1933. Reprint. With an intro. by Paul W. Gates. New York: Hill and Wang, for the New York State Historical Association, 1966. xxxiv + 465 pp. Illus., bib., index. Includes some history of forests in relation to agriculture.

H266 Hedrick, Ulysses P. *A History of Horticulture in America to 1860.* New York: Oxford University Press, 1950. xiii + 551 pp. Illus., notes, bib. Includes regional chapters and chapters on botanical explorers, botanical gardens, plant-breeding, horticultural literature, and horticultural societies.

H267 Hedstrom, Margaret. *Hedstrom Lumber Company: After 60 Years, 1914-1974.* Grand Marais, Minnesota: Hedstrom Lumber Company, 1974. 10 pp. Illus.

Heermance, Edgar L. See Baldwin, Henry I., #B40

Heflin, Catherine P. See Beers, Howard W., #B171

H268 Heidt, William. *History of Rafting on the Delaware.* Port Jervis, New York: Gazette Book Printing, 1922.

H269 Heilala, John J. "In An Upper Michigan Lumber Camp." *Michigan History* 36 (Mar. 1952), 55-79. Reminiscence of a camp near Michigamme Lake, Michigan, 1904.

H270 Heilala, John J. "With the Big Wheels." *Michigan History* 38 (Sept. 1954), 293-305. Recollections of the Lake and Independence Lumber Company near Big Bay in Michigan's Upper Peninsula, 1911.

H271 Heilman, John M. "The Historical Development of Grading Rules." Master's thesis, Yale Univ., 1930.

H272 Heilman, John M. "Northeast Municipal Watersheds." *Journal of Forestry* 58 (Apr. 1960), 305-07. Inventory of forested watersheds and forest-protected reservoirs owned and managed by ten cities of the Northeast, since 1890.

H273 Heim, A.L. *An Oral History Interview with A.L. Heim.* OHI by Donald G. Coleman. Madison: USFS, Forest Products Laboratory, 1961. 13 pp. Heim was a former chief of engineering at the Forest Products Laboratory, Madison, Wisconsin.

H274 Heim, Peggy. "Financing the Federal Reclamation Program, 1902-1919: The Development of Repayment Policy." Ph.D. diss., Columbia Univ., 1953. 386 pp.

H275 Heimart, Alan. "Puritanism, the Wilderness, and the Frontier." *New England Quarterly* 26 (Sept. 1953), 361-82. On Puritan New England's attitudes toward the westward movement and subjugation of the wilderness, 1620-1700.

Heinritz, Stuart F. See Thompson, Henry A., #T71

H276 Heintz, Emil. "Rafting on the Black River." *La Crosse County Historical Sketches,* Series 3 (1937). Black River, Wisconsin.

H277 Heintzleman, S.W. *The Cork Oak: Past, Present and Future on the Pacific Coast.* Corvallis, Oregon, 1940. 33 pp. Maps.

Heisley, Marie Foote. See Randall, Charles E., #R26

H278 Heitmann, John. "Julius B. Baumann: A Biographical Sketch." *Norwegian-American Studies and Record* 15 (1949), 140-75. On Baumann as a lumberman and poet in Wisconsin and Minnesota, 1891-1923.

H279 Helburn, N. "The Case for National Forest Roads." *Land Economics* 23 (Nov. 1947), 371-80. Incidental history.

H280 Helburn, Nicholas. "Geography of the Lumber Industry in Northwestern Montana." Ph.D. diss., Univ. of Wisconsin, 1950. 165 pp. Includes some history.

H281 Held, Jack. "Scotia, the Town of Concern." *Pacific Historian* 16 (Summer 1972), 76-92. A lumber town in Humboldt County, California, supported by the Pacific Lumber Company since the 1870s.

Held, R. Burnell. See Clawson, Marion, #C385, C388

H282 Held, R. Burnell, and Clawson, Marion. *Soil Conservation in Perspective.* Baltimore: Johns Hopkins Press, for Resources for the Future, 1965. 359 pp. Illus. Contains some history of the conservation movement and especially the development of soil conservation programs in the 20th century.

H283 Helfman, Elizabeth S. *Land, People, and History.* New York: McKay, 1962. 271 pp. Includes some history of the conservation movement and of man's attitude toward and use of the land since the colonial period.

H284 Helgeson, Arlan C. "The Promotion of Agricultural Settlement in Northern Wisconsin, 1880-1925." Ph.D. diss., Univ. of Wisconsin, 1952. 247 pp.

H285 Helgeson, Arlan C. "Nineteenth Century Land Colonization in Northern Wisconsin." *Wisconsin Magazine of History* 36 (Winter 1953), 115-21. On the "stump lands" or cutover lands, 1868-1900.

H286 Helgeson, Arlan C. *Farms in the Cutover: Agricultural Settlement in Northern Wisconsin.* Madison: State Historical Society of Wisconsin, for the Department of History, University of Wisconsin, 1962. viii + 184 pp. Maps, notes, bib., index. On the changing policies toward the cutover lands, including the influence of the forestry movement in discouraging settlement on lands proven marginal for farming.

H287 Helmers, Austin Edward. "Alaska Forestry—A Research Frontier." *Journal of Forestry* 58 (June 1960), 465-71. Since 1948.

H288 Helphenstine, R.K. *Wood-Using Industries of North Carolina.* North Carolina Geological and Economic Survey, Bulletin No. 30. Raleigh, 1923. 105 pp. Illus.

H289 Helphenstine, R.K. "Reducing the Drain on Our Forests." *American Forests and Forest Life* 34 (July 1928), 395-97, 430. Includes a historical sketch of the wood preserving industry.

H290 Hempstead, Alfred G. *The Penobscot Boom and the Development of the West Branch of the Penobscot River for Log Driving.* University of Maine Studies, Second Series, No. 18. Orono, Maine: University Press, 1931. 187 pp. Illus., maps, tables, bib. Also published as the *University of Maine Bulletin,* Volume 33, No. 11 (May 1931), complete. History of the boom under independent and cooperative effort and under corporate control, since 1828.

H291 Hench, Maynard. "The 'Up-and-Down Sawmill,' an Aid to the Early Development of Pennsylvania." *Pennsylvania Forests and Waters* 4 (Sept.-Oct. 1952), 106-07, 118. From 1662 to 1863.

H292 Hendee, Clare. *Organization and Management in the Forest Service: A Summary from the Manual and Handbook.* Washington: GPO, 1962. 84 pp. Includes some history.

H293 Henderson, G.L. *Yellowstone National Park, Past, Present and Future.* Washington: Gibson Brothers, 1891.

H294 Henderson, Patrick C. "The Public Domain in Arizona, 1863-1891." Ph.D. diss., Univ. of New Mexico, 1966. 276 pp.

H295 Henderson, S. "An Experiment in Forest-Farm Resettlement." *Land Economics* 22 (Feb. 1946), 10-21. Account of a family relocation project on Wisconsin's Chequamegon National Forest, 1935-1942.

Henry, A.J. See Bates, Carlos G., #B122

Henry, L.A. See Harper, C. Armitage, #H129

H296 Henry, Ralph Chester. [Eric Thane]. *The Majestic Land: Peaks, Parks & Prevaricators of the Rockies & Highlands of the Northwest.* Indianapolis: Bobbs-Merrill, 1950. 347 pp. Illus., maps. Description of parks, scenery, and outdoor industries in the mountains of Wyoming, Montana, Idaho, Washington, and Alberta, with incidental historical references.

H297 Henry, Ralph L. *St. Croix Boyhood.* St. Paul: North Central Publishing Company, 1972. 107 pp. Illus. Reminiscences of Point Douglas, Minnesota, a former lumber town, from 1900 to 1913.

H298 Henry, Ralph S. "The Railroad Land Grant Legend in American History Texts." *Mississippi Valley Historical Review* 32 (Sept. 1945), 171-94. Criticizes historians for exaggerating the extent of railroad land grants in the West.

Henze, Karl D. See Mason, David T., #M270, M276

H299 Hepting, George H. "Forest Pathology in the Southern Appalachians, 1900-1940." *Forest History* 8 (Fall 1964), 11-13.

H300 Hepting, George H. "Death of the American Chestnut." *Journal of Forest History* 18 (July 1974), 60-67. On the effects of chestnut blight in the United States since its identification early in the 20th century.

H301 Hepting, George H. "Quest of the Lonesome Pine." *Journal of Forest History* 19 (Jan. 1975), 30-33. Reminiscence of the author's search in 1935 for the tree immortalized by John Fox, Jr., in his *The Trail of the Lonesome Pine* (1908). Never found by the author, the tree allegedly stood near Norton, Virginia.

H302 Herbert, P.A. "Standing Timber Insurance." *Timberman* 24 (Nov. 1922), 152-58. Incidental history and comparison of American and European systems. See also *Southern Lumberman* (Dec. 23, 1922), 170-74

H303 Herbert, Paul Anthony. "Development of a Marginal Land County in Northern Michigan." Ph.D. diss., Univ. of Michigan, 1941.

H304 Heritage, William. "Forestry, Past and Future, on Indian Reservations in Minnesota." *Journal of Forestry* 34 (July 1936), 648-52. Since ca. 1900.

Heritage, William. See Kinney, Jay P., #K170

H305 Herndon, G. Melvin. "Naval Stores in Colonial Georgia." *Georgia Historical Quarterly* 52 (Dec. 1968), 426-33.

H306 Herndon, G. Melvin. "Timber Products of Colonial Georgia." *Georgia Historical Quarterly* 57 (Spring 1973), 56-62.

Herr, Clarence S. See Barraclough, K.E., #B95

H307 Herr, Clarence S. "Herman G. Schanche: Industrial Executive." *Journal of Forestry* 47 (Jan. 1949), 60-61. Sketch of Schanche's career, especially with Abitibi Power and Paper, Iroquois Falls, Ontario, 1920s, and with Brown Company, Berlin, New Hampshire, since 1943.

H308 Herr, Clarence S. *The Development of Industrial Forestry in the Northeast.* Colonel William B. Greeley Lectures in Industrial Forestry, No. 3. Seattle: University of Washington, College of Forestry, 1959. 50 pp. Illus., tables. History of industrial forestry in the lumber and pulp and paper industries.

H309 Herr, Clarence S. "Private Responsibility in Public Forestry: The New Hampshire Timberland Owners Association." *Journal of Forest History* 19 (Jan. 1975), 24-29. The organization, founded in 1910, has influenced forest fire control and forest policy in New Hampshire. The growth of the forestry movement since the 1880s is also treated.

H310 Herrick, Francis Hobart. *Audubon the Naturalist; A History of His Life and Time.* 2 Volumes. New York: Dover Publications, 1968. Illus., bib. John James Audubon (1785-1851) furthered American interest in conservation through his highly regarded work as a naturalist, ornithologist, and painter.

H311 Herrick, Rebecca B. "A Century of Shipbuilding in Blue Hill, Maine, 1792-1892." Master's thesis, Univ. of Maine, 1945.

H312 Herridge, A.J. "The Development of Forestry Policy." *Forestry Chronicle* 40 (Dec. 1964), 425-28. A survey of forestry legislation and policy in Canada since 1683. A reprint from *Ontario Economic Review* 2 (Aug. 1964).

H313 Herrity, George F. "The Tight Cooperage Business in the U.S." Master's thesis, Yale Univ., 1948.

H314 Herty, Charles H. "The Newsprint Gamble That Paid Off." *Pulp and Paper* 31 (Aug. 1957), 46-48. Southland Paper Mills, Lufkin, Texas.

H315 Hertzog, Dorothy. "Isaac Graham." Master's thesis, Univ. of California, Berkeley, 1942. A lumberman of the Santa Cruz Mountains, California, 1840s.

H316 Hess, Elmer B. "The Kalamazoo Valley Paper Industry." *Proceedings of the Indiana Academy of Science* 69 (1960), 224-35. On the dominant industry of the Kalamazoo Valley of Michigan and the reasons why the paper industry located there, since 1866.

H317 Hess, Roscoe R. "The Paper Industry in Its Relation to Conservation and the Tariff." *Quarterly Journal of Economics* 25 (Aug. 1911), 650-81. Some history of U.S. and Canadian forest products industries, although essentially an argument against the tariff on Canadian newsprint.

H318 Hession, Jack M. "The Legislative History of the Wilderness Act." Master's thesis, San Diego State College, 1967. A summary of wilderness legislation from the 1950s to the passage of the Wilderness Act in 1964; includes an analysis of the political tactics of the preservationists and their opponents.

H319 Hesterberg, Gene A., and Noblet, U.J. "River Drives on the Sturgeon." *Northeastern Logger* 5 (June 1957), 26-27, 42. Operations of the Sturgeon River Boom Company on the Sturgeon River of Michigan's Upper Peninsula, 1874-1917.

H320 Hetherington, Mary Elizabeth. "A Study of the Development of Journalism during the Lumbering Days of the Saginaws, 1853-1882." Master's thesis, Northwestern Univ., 1930.

H321 Hewes, Laurence I. "Some Features of Early Woodland and Prairie Settlement in a Central Iowa County." *Annals of the Association of American Geographers* 40 (Mar. 1950), 40-57.

H322 Hewes, Laurence I. "Western Forest Highways Come of Age." *American Forests* 66 (Feb. 1960), 42-44, 55-57; 66 (Mar. 1960), 34-40. On the effects of the Forest Highway Act of 1921 in the Far West,

H323 Hewett, Thomas. "Gifford Pinchot and His Fight for Our National Resources." *Review of Reviews* 39 (Jan. 1909), 88-89. As chief of the USFS.

H324 Hewlett, John D., and Metz, Louis J. "Watershed Management Research in the Southeast." *Journal of Forestry* 58 (Apr. 1960), 269-71. On the work of the USFS's Coweeta Hydrologic Laboratory in the mountains of North Carolina since the 1930s.

H325 Heydinger, Earl J. "Lumber and Its By-Products." *Bulletin of the Historical Society of Montgomery County, Pennsylvania* 10 (Oct. 1955), 16-30. Along the Schuylkill River, 1740-1826.

H326 Heyward, Frank. "The Trend of Forest Conservation in the Southern Pulpwood Industry." *Pulp and Paper Magazine of Canada* 41 (Feb. 1940), 254-56, 258.

H327 Heyward, Frank. "Austin Cary: Yankee Peddler in Forestry." *American Forests* 61 (May 1955), 29-30, 43-44; 61 (June 1955), 28-29, 52-53. Cary (1865-1936) was an industrial forester in New England, a forestry educator at Yale and Harvard, and a disciple of forestry in the South with the USFS.

H328 Heyward, Frank. *History of Industrial Forestry in the South.* Colonel William B. Greeley Lectures in Industrial Forestry, No. 2. Seattle: University of Washington, College of Forestry, 1958. 50 pp. Includes also a summary of forest use since colonial times.

H329 Heyward, Frank. "50 Years of Progress." *Forest Farmer* 17 (July 1958), 9, 22-24. Of the pulp and paper industry in the South.

H330 Heywood-Wakefield Company. *A Completed Century, 1826-1926: The Story of Heywood-Wakefield Company.* Boston, 1926. 111 pp. Illus. Company history of a Massachusetts furniture manufacturing firm.

H331 Hibbard, Benjamin H. "The Settlement of Public Lands in the United States." *International Review of Agricultural Economics* 61 (Jan. 1916), 97-117.

H332 Hibbard, Benjamin H. *A History of the Public Land Policies.* New York: Macmillan, 1924. Reprint. With a foreword by Paul W. Gates. Madison: University of Wisconsin Press, 1965. xix + 591 pp. Illus., maps, charts, tables, notes, bib., index. A standard history of American land policy; includes chapters on the Timber Culture Act of 1873, timberland disposal, grazing policy, national forests and parks, and the conservation movement.

Hibbard, John E. See Smith, David M., #S351

H333 Hibbard, John E. "A History of the Association Trails Program." *Connecticut Woodlands* 35 (Spring 1970), 23-24. Of the Connecticut Forest and Park Association, since 1929.

H334 Hick, R.M. "The Influence of Land History and Legislative Enactments on the Character and Condition of the State Forests in Massachusetts." Master's thesis, Harvard Univ., 1927.

H335 Hickel, Walter J. *Who Owns America?* Englewood Cliffs, New Jersey: Prentice-Hall, 1971. xii + 328 pp. Hickel was Secretary of the Interior in the Richard M. Nixon administration, 1970-1971. His book concerns resource problems, among others.

H336 Hickman, Clifford A. "Some Problems in Allocating Forest Service Stumpage to Heavily Dependent Users: The Lumber Industry of Eastern Oregon as a Case." Ph.D. diss., State Univ. of New York, College of Forestry, 1972. 399 pp.

H337 Hickman, Nollie W. "Mississippi Lumber Industry from 1840 to 1950." *Southern Lumberman* 193 (Dec. 15, 1956), 132-37.

H338 Hickman, Nollie W. "Logging and Rafting Timber in South Mississippi, 1840-1910." *Journal of Mississippi History* 19 (July 1957), 154-72.

H339 Hickman, Nollie W. "The Lumber Industry in South Mississippi, 1890-1915." *Journal of Mississippi History* 20 (Oct. 1958), 211-23.

H340 Hickman, Nollie W. "History of Forest Industries in the Longleaf Pine Belt of East Louisiana and Mississippi, 1840-1915." Ph.D. diss., Univ. of Texas, 1958. 492 pp.

H341 Hickman, Nollie W. *Mississippi Harvest: Lumbering in the Longleaf Pine Belt, 1840-1915.* University: University of Mississippi, 1962. x + 306 pp. Illus., maps, bib., notes, app., index. Restricted to the Mississippi Gulf Coast region, this standard work treats the lumber and forest products industries from the French colonial period through the 1920s.

H342 Hickman, Nollie W. "The Yellow Pine Industries in St. Tammany, Tangipahoa and Washington Parishes, 1840-1915." *Louisiana Studies* 5 (Summer 1966), 75-88. Chronicles the ever-increasing efficiency of the lumber machinery that by 1915 had exhausted the virgin timber stands of this area of Louisiana.

H343 Hickman, Nollie W. "Mississippi Forests." In *A History of Mississippi,* Volume 2, ed. by Richard Aubrey McLemore. Hattiesburg: University and College Press of Mississippi, 1973. Pp. 212-32. On the lumber industry, pulp and paper industry, forestry, and forest conservation since the 19th century.

H344 Hicks, Harry W. "Inferno in the Adirondacks." *North Country Life* 2 (Fall 1948), 7-8, 51-52. On a forest fire near Lake Placid, New York, 1903.

H345 Hicks, P.R. "Fifty Years in Wood Preservation." *Southern Lumberman* 144 (Dec. 15, 1931), 127-30. Includes some history of the American Wood Preservers' Association.

H346 Hicks, Philip Marshall. *The Development of the Natural History Essay in American Literature.* Philadelphia, 1924. 167 pp. Bib. Discusses John Burroughs, John Muir, and other nature writers.

H347 Hicks, William T. "Recent Expansion in the Southern Pulp and Paper Industry." *Southern Economic Journal* 6 (Apr. 1940), 440-48. Developments of the 1930s.

H348 Hicock, Henry W. "An Early History of Forests and Forestry in Connecticut." *Connecticut Woodlands* 35 (Spring 1970), 26-30, 40.

H349 Hicock, Henry W. "The Making of Charcoal: Man's Oldest Chemical Industry." *Connecticut Woodlands* 39 (Summer 1974), 11-15. Includes some history of the charcoal industry in Connecticut.

H350 Hidy, Ralph. "Lumbermen in Idaho: A Study in Adaptation to Change in Environment." *Idaho Yesterdays* 6 (Winter 1962), 2-17. On the history of Weyerhaeuser-related firms in Idaho since 1900.

H351 Hidy, Ralph; Hill, Frank Ernest; and Nevins, Allan. *Timber and Men: The Weyerhaeuser Story.* New York: Macmillan, 1963. xiv + 704 pp. Illus., notes, apps., list of sources cited, index. The landmark business history of the nation's largest forest products corporation, spanning a century of lumbering in the Great Lakes region, the South, and the Pacific Northwest.

H352 Hidy, Ralph W. "The Battle of the Bulge: A Memoir." *Proceedings of the Business History Conference* 14 (1967), 60-71. Includes an account of the problems faced in writing *Timber and Men: The Weyerhaeuser Story* (1963).

H353 Hidy, Ralph W. "Industry Response to Pressures and Opportunities for Change, 1840-1900." In *First National Colloquium on the History of the Forest Products Industries, Proceedings,* ed. by Elwood R. Maunder and Margaret G. Davidson. New Haven: Forest History Society, 1967. Pp. 10-24.

Hieronymus, G.H. See Kylie, Harry R., #K285

H354 Higbee, Edward Counselman. *The American Oasis: The Land and Its Uses.* New York: Alfred A. Knopf, 1957. xviii + 262 + vii pp. Illus., map, notes, bib. On American waste and destruction of land since 1607, and more recent efforts to conserve it and maintain its quality.

H355 Higgins, F. Hal. "Logging with Tractors in the '80s." *Timberman* 48 (May 1947), 68, 116. In the Pacific states.

H356 Higgins, F. Hal. "Steam Wagon Days." *Timberman* 48 (Aug. 1947), 48, 56-58. Logging equipment in the Pacific Northwest.

H357 Higgins, F. Hal. "P. Bunyan and His Blue Ox Babe Were Here in 1877!" *Engineers and Engines Magazine* 8 (Dec. 1962), 1-9. On the Sierra Flume & Lumber Company and its operations in northern California, 1870s.

H358 High, James. "Southern California Opinions Concerning Conservation of Forests, 1890-1905." *Historical Society of Southern California Quarterly* 33 (Dec. 1951), 291-312. A survey of newspaper opinion demonstrating public awareness of the relation between forest conservation and sustained water supply.

H359 Highsaw, Robert. *Mississippi's Wealth: A Study of the Public Administration of Natural Resources.* University of Mississippi, Bureau of Public Administration, State Administration Series, No. 2. University, 1947. 190 pp. Includes some history of federal, state, and local administration of natural resources.

H360 Highsmith, Richard Morgan, Jr., and Beh, John L. "Tillamook Burn: The Regeneration of a Forest." *Scientific Monthly* 75 (Sept. 1952), 139-48. On Oregon's Tillamook forest fires of 1933, 1939, and 1945, and subsequent reforestation efforts.

H361 Highsmith, Richard Morgan, Jr. "Resources and the Regional Economy." *Pacific Northwest Quarterly* 46 (Jan. 1955), 25-29. Incidental history of forest resources and their role in the economy of the Pacific Northwest.

H362 Highsmith, Richard Morgan, Jr.; Jensen, J. Granville; and Rudd, Robert D. *Conservation in the United States.* 1962. Second edition. Chicago: Rand McNally, 1969. 322 pp. Illus. A text which includes some history of conservation since the colonial period; there are several chapters on forests.

H363 Higley, Warren. "New York." *Proceedings of the American Forestry Association* 11 (1896), 99-103. History of New York's Adirondack Forest Preserve.

H364 Hilbert, G.E. "Twenty Years of Research by the Naval Stores Station." *American Turpentine Farmers Association Journal* 15 (Jan. 1953), 5-9.

H365 Hilfinger, George N. "Trade Promotion, Its Origin and Development in the Lumber Industry." Master's thesis, Yale Univ., 1947. 134 pp.

H366 Hill, Arthur. "The Pine Industry in Michigan." *Publications of the Michigan Political Science Association* 3 (Dec. 1898), 1-12.

H367 Hill, C.L. *Forests of Yosemite, Sequoia, and General Grant National Parks.* Washington: USDI, 1916. 39 pp. Illus., index. Incidental history.

H368 Hill, C.L. "The Development of Forest Research in California." *Journal of Forestry* 29 (Apr. 1931), 484-96. On forestry research by the USFS and the University of California since 1908.

Hill, C.L. See Weeks, David, #W133, W134

H369 Hill, Charles. "The Story of Standardization of Lumber." *New York Lumber Trade Journal* 74 (No. 881, 1923), 27-29.

Hill, Frank Ernest. See Hidy, Ralph W., #H351

Hill, Frank Ernest. See Holland, Kenneth, #H451

Hill, George W. See Kolehmainen, John I., #K233

H370 Hill, Irvin Bartle. "Timber Taxation." Master's thesis, Univ. of Oregon, 1934.

H371 Hill, Leslie G. "History of the Missouri Lumber and Mining Company, 1880-1909." Ph.D. diss., Univ. of Missouri, 1949. 307 pp. On an enterprise centered in Carter County, Missouri.

Hill, Ralph N. See Marvin, James W., #M258

H372 Hill, Robert C. "Wooden Ship Building in the Northwest." *Timberman* 43 (Apr. 1942), 54-56; 43 (Aug. 1942), 26-28. Since 1916.

H373 Hill, Robert C. "Old Lumber Vessels and New." *Timberman* 50 (Oct. 1949), 120-30. Pacific lumber trade and its vessels since 1895.

H374 Hill, Robert T. *The Public Domain and Democracy: A Study of Social, Economic and Political Problems in the United States in Relation to Western Development.* Columbia University Studies in History, Economics and Public Law, Volume 38, No. 1. New York: Columbia University, 1910. 253 pp. Tables, notes, bib. Includes some history of the public domain and federal land disposal.

H375 Hill, William Bancroft, and Weyerhaeuser, Louise L. *Frederick Weyerhaeuser, Pioneer Lumberman.* Minneapolis: McGill Lithograph Company, 1940. 62 pp. Illus. Biographical sketch of Frederick Weyerhaeuser and his family, prominent in the lumber business and related enterprises of the Great Lakes states and Pacific Northwest since the mid-19th century.

H376 Hilton, Cecil Max. *Rough Pulpwood Operating in Northwestern Maine, 1935-1940.* University of Maine Studies, Second Series, No. 57. Orono, Maine: University Press, 1942. 197 pp. Illus., maps, diags., tables, index. Incidental history.

H377 Hilton, Thomas. *High Water on the Bar.* Savannah: By the author, 1951. 22 pp. Brief history of the Hilton-Lachlison family and its lumber business in the Georgia coastal area since the 1850s.

H378 Hinckley, Ted C. "Alaska and the Emergence of America's Conservation Consciousness." In *Prairie Scout,* Volume 2. Abilene: Kansas Corral of The Westerners, 1974. Pp. 79-111. Emphasizes conservation of marine resources.

H379 Hines, Edward, Lumber Company. *50 Years: Edward Hines Lumber Co., Commemorating a Pioneer in the Nation's Oldest Industry, the Company Which Bears His Name, and Their First Half-Century of Accomplishment.* Chicago, 1942. 47 pp. Illus., map. Hines began as a Chicago wholesaler, turned to lumber production, and developed widespread timber holdings and woods operations in the Great Lakes states, the South, and Oregon.

H380 Hingston, William R. "Gifford Pinchot, 1922-1927." Ph.D. diss., Univ. of Pennsylvania, 1962. 421 pp. On his first term as governor of Pennsylvania.

H381 Hinsdale, Mary L. *A History of the President's Cabinet.* Ann Arbor: George Wohr, 1911. 355 pp. Index. See for mention of the agriculture and interior secretaries.

H382 Hipel, N.O. "The History and Status of Forestry in Ontario." *Canadian Geographic Journal* 25 (Sept. 1942), 110-45.

H383 Hirsch, S. Carl. *Guardians of Tomorrow: Pioneers in Ecology.* New York: Viking Press, 1971. 192 pp. Illus., bib. Juvenile or popular history of the conservation and ecology movements, including biographical sketches of Henry David Thoreau, George Perkins Marsh, Frederick Law Olmsted, John Muir, Gifford Pinchot, George Norris, Aldo Leopold, and Rachel Carson.

H384 Hirt, Ray R. "Fifty Years of White Pine Blister Rust in the Northwest." *Journal of Forestry* 54 (July 1956), 435-38. Since ca. 1900.

H385 *History News.* "Wisconsin Company Re-creates History of Logging Industry with Camp Five." *History News* 28 (Apr. 1973), 84-85. On the restored logging camp of Connor Industries near Laona, Wisconsin.

H386 *History Reference Bulletin.* "Labor Saving: Ford's River Rouge Plant, 1929; Mersey Paper Company, 1933." *History Reference Bulletin* 7 (Feb. 1934), 89-96.

H387 Hitchcock, Ethan Allen. *Fifty Years in Camp and Field.* Ed. by W.A. Croffut. New York: G.P. Putnam's Sons, 1909. 514 pp. Hitchcock served William McKinley and Theodore Roosevelt as Secretary of the Interior, 1899-1907.

H388 Hitching, R.G. "The Pulp and Paper Industry in North Carolina: An Economic Analysis." Master's thesis, Duke Univ., 1958.

H389 Hittell, John S. *The Commerce and Industries of the Pacific Coast of North America.* San Francisco: A.R. Bancroft and Company, 1882. 819 pp. Includes material on the lumber and paper industries of the Pacific Coast states.

H390 Hoaglund, H.E. "Early Transportation on the Mississippi." *Journal of Political Economics* 19 (Feb. 1911), 111-23. Includes reference to log rafting.

H391 Hoar, Walter G. *History is Our Heritage: A Chronology of Upper Wisconsin Lumbering History as It Related to the St. Croix and Chippewa Rivers, Including the Shell Lake Area, from 1879 to 1902, Recording the Coming of the*

Railroad, Development and Operation of the Shell Lake Lumber Company and the Crescent Springs Railroad, the Emerging of a Settlement with the Joys and Sorrows of a Pioneer People. Shell Lake, Wisconsin: By the author, 1968. 144 pp. Illus., maps, tables, index.

H392 Hobart, Seth G. "State Forestry in Virginia." *Virginia Forests* 17 (Fall 1962), 8-11, 18. A brief history of the Virginia Division of Forestry; subsequent issues carry related articles.

H393 Hobbs, John E. "The Beginnings of Lumbering as an Industry in the New World, and First Efforts at Forest Protection: A Historical Study." *Forestry Quarterly* 4 (Mar. 1906), 14-23. On commercial and private sawmills in 17th-century New England and legislation to protect forests.

H394 Hoch, Daniel K., and Shenton, Donald R. "Conservation in Berks County." *Historical Review of Berks County* 19 (Jan.-Mar. 1954), 34-36. On Pennsylvania's Berks County Conservation Association, organized in 1913, and the tree-planting activities of Solon L. Parkes, 1913-1921.

H395 Hochschild, Harold K. *Township 34: A History, with Digressions, of an Adirondack Township in Hamilton County in the State of New York.* New York: N.p., 1952. xxvi + 614 pp. Illus., maps, bib. On the lumber industry, resorts, and the settlement of Eagle Nest, New York, since 1771.

H396 Hochschild, Harold K. *Lumberjacks and Rivermen in the Central Adirondacks, 1850-1950.* Blue Mountain Lake, New York: Adirondack Museum, 1962. 88 pp. Illus., maps, app. A revised excerpt from his *Township 34* (1952).

H397 Hocker, Harold W., Jr. "Certain Aspects of Climate as Related to the Distribution of Loblolly Pine." *Ecology* 37 (Oct. 1956), 824-34. In the Southeastern states, 1921-1950.

H398 Hodge, Clarence L. *The Tennessee Valley Authority: A National Experiment in Regionalism.* Washington: American University Press, 1938. xii + 272 pp. Illus., bib. Incidental history.

H399 Hodge, F.W. "W J McGee." *American Anthropologist* 14 (Oct.-Dec. 1912), 683-87. An obituary and evaluation of McGee (1853-1912), a prominent conservation theorist and advisor to Theodore Roosevelt.

H400 Hodge, Jo Dent. "Lumbering in Laurel at the Turn of the Century." Master's thesis, Univ. of Mississippi, 1970.

H401 Hodge, Jo Dent. "The Lumber Industry in Laurel, Mississippi, at the Turn of the Nineteenth Century." *Journal of Mississippi History* 35 (Nov. 1973), 361-79. From 1882 to ca. 1915, with emphasis on the Eastman-Gardiner Company.

H402 Hodges, James C. "15 Million Dollars for Conservation." *Michigan Conservation* 25 (Mar.-Apr. 1956), 7-9. On the use of state and federal funds for conservation in Michigan since 1921.

Hodges, Ralph, Jr. See Seaton, Fred A., #S157

H403 Hodgetts, John Edwin. *Pioneer Public Service: An Administrative History of the United Canadas, 1841-1867.* Toronto: University of Toronto Press, 1956. xii + 292 pp. Diags., bib. Includes some history of the Woods and Forests Branch of the Crown Lands Department in Canada.

H404 Hodgson, Allen H. "An Oregon Pioneer in Forestry." *American Forests and Forest Life* 33 (Nov. 1927), 661-64. On reforestation and industrial forestry practiced by Crown Willamette Paper Company of Oregon since 1903.

H405 Hodson, Elmer R. *Rules and Specifications for the Grading of Lumber Adopted by the Various Lumber Manufacturing Associations of the United States.* USFS, Bulletin No. 71. Washington: GPO, 1906. 127 pp. Includes a brief history of lumber grading.

H406 Hoffman, Abraham. "Angeles Crest: The Creation of a Forest Highway System in the San Gabriel Mountains." *Southern California Quarterly* 50 (Sept. 1968), 309-45. On road building in the Angeles National Forest, 1919-1961, primarily for purposes of sightseeing and outdoor recreation.

H407 Hoffman, Daniel G. *Paul Bunyan: Last of the Frontier Demigods.* Philadelphia: University of Pennsylvania Press, for Temple University Publications, 1952. xiv + 215 pp. Bib. On the evolution of Paul Bunyan lore, 1910-1947.

H408 Hoffman, George C., Jr. "The Early Political Career of Charles McNary, 1917-1924." Ph.D. diss., Univ. of Southern California, 1952. 247 pp. Senator McNary of Oregon sponsored forestry legislation in the 1920s.

H409 Hoffnagle, Warren. "The Southern Homestead Act: Its Origins and Operations." *Historian* 32 (Aug. 1970), 612-29.

Hogan, C.S. See Best, Gerald M., #B251

H410 Hogg, J. Bernard. *The Allegheny Section of the Society of American Foresters: A Fifty Year History, 1922-1972.* N.p., Society of American Foresters, Allegheny Section, 1972. 27 pp. Notes. On the activities of professional foresters in Pennsylvania, New Jersey, Maryland, Delaware, and West Virginia.

H411 Hoglund, A. William. "Forest Conservation and Stove Inventors—1789-1850." *Forest History* 5 (Winter 1962), 2-8. On the technology of wood-burning stoves in relation to 19th-century fears of fuelwood shortage and forest devastation.

H412 Hogner, Dorothy (Childs). *Conservation in America.* Philadelphia: J.B. Lippincott, 1958. viii + 240 pp. Illus. Includes some history since the colonial period; there are chapters on wilderness, wildlife, and forests.

H413 Hoing, Willard Lee. "James Wilson as Secretary of Agriculture, 1897-1913." Ph.D. diss., Univ. of Wisconsin, 1964. 299 pp.

H414 Hokanson, Nels. "Swedes and the I.W.W." *Swedish Pioneer Historical Quarterly* 23 (Jan. 1972), 25-36. History and reminiscences of Swedish members of the Industrial Workers of the World, including lumberjacks in Oregon and Washington, 1910s-1920s.

H415 Holbrook, Charles V. "The Long-Bell Plantation: Wood as the Renewable Resource." *Forests & People* 24 (First quarter, 1974), 10-13. On the experimental plantation of the Long-Bell Lumber Company near Longville, Louisiana, since 1929.

H416 Holbrook, Stewart H. "Modern Loggers." *American Forests* 37 (Jan. 1931), 3-6, 62. On changes in loggers' working conditions and methods since the 1880s.

H417 Holbrook, Stewart H. "Port Ludlow, Oldest Sawmill town in Pacific Northwest Has Charm in Lumber Industry." *Four L Lumber News* 13 (July 1, 1931). Washington.

H418 Holbrook, Stewart H. "Gerrymandering the Tall Uncut." *American Forests* 42 (Aug. 1936), 360-62. On the gerrymandering of school districts in Tillamook County, Oregon, to include within them taxable timberlands.

H419 Holbrook, Stewart H. "Log Pirates of Puget Sound." *American Forests* 43 (Jan. 1937), 22-25. On the problem of log thefts from booms in Puget Sound, Washington, ca. 1917, and efforts to curtail the thievery through legislation and use of log marks or brands.

H420 Holbrook, Stewart H. "Ghost Towns Still Walk." *American Forests* 43 (May 1937), 216-17, 242, 258. On abandoned or declining lumber towns, including Port Ludlow, Cosmopolis, Three Lakes, Littell, Dryad, Doty, McCormick, and Walville, in Washington, and Cochran, Wheeler, Brighton, and Garibaldi, in Oregon, and the new sawmill town of Longview, Washington.

H421 Holbrook, Stewart H. "Timber Ships." *American Forests* 43 (Nov. 1937), 529-31, 562. On the Benson log rafts used to move logs in coastal waters between the Columbia River and California.

H422 Holbrook, Stewart H. *Holy Old Mackinaw: A Natural History of the American Lumberjack.* New York: Macmillan, 1938. viii + 278 pp. Bib., index. The classic popular history of the lumber industry, lumberjacks, and working conditions from Maine to the Great Lakes to the Pacific Northwest since ca. 1800. There are many later editions and reprints, including *The American Lumberjack* (1962).

H423 Holbrook, Stewart H. "When Peshtigo Burned." *American Forests* 45 (Apr. 1939), 158-60. On the forest fire in and near Peshtigo, Wisconsin, 1871.

H424 Holbrook, Stewart H. *Tall Timber*. New York: Macmillan, 1941. ix + 179 pp. Illus., maps. Popular or juvenile history of lumbering.

H425 Holbrook, Stewart H. "The Great Hinckley Fire." *American Mercury* 57 (Sept. 1943), 348-55. Minnesota, 1894.

H426 Holbrook, Stewart H. *Burning an Empire: The Story of American Forest Fires.* New York: Macmillan, 1943. 229 pp. Illus., notes, bib., index. A history of American and Canadian forest fires since the 1820s.

H427 Holbrook, Stewart H. *A Narrative of Schafer Bros. Logging Company's Half Century in Timber.* Seattle: Dogwood Press, for Schafer Bros. Logging Company, 1945. 110 pp. Illus., map. A logging firm of Aberdeen, Washington, since the 1890s.

H428 Holbrook, Stewart H. *Green Commonwealth: A Narrative of the Past and a Look at the Future of One Forest Products Community, 1895-1945.* Seattle: Simpson Logging Company, 1945. 163 pp. Illus., maps. On the history of the Simpson Logging Company, Shelton, Washington, and the forested area of the lower Olympic Peninsula, since 1895.

H429 Holbrook, Stewart H. "The Epic of Timber." In *Northwest Harvest: A Regional Stocktaking*, ed. by V.L.O. Chittick. New York: Macmillan, 1948. Pp. 83-100. On the traditions of the American lumberjack since the 17th century, especially in the Pacific Northwest.

H430 Holbrook, Stewart H. "A Century of Pope and Talbot." *American Forests* 55 (June 1949), 22-23, 43-45. On Pope & Talbot, Inc., a San Francisco-based lumber firm with operations on Puget Sound, Washington, since 1849.

H431 Holbrook, Stewart H. *The Yankee Exodus: An Account of Migration from New England.* New York: Macmillan, 1950. xii + 398 pp. Illus., maps, bib. Includes some history of lumbermen and lumberjacks.

H432 Holbrook, Stewart H. *Saga of a Saw Filer.* Portland: Armstrong Manufacturing Co., 1952. 42 pp. On the head filers of Pacific Northwest sawmills, especially Edward P. Armstrong.

H433 Holbrook, Stewart H. *Far Corner: A Personal View of the Pacific Northwest.* New York: Macmillan, 1952. viii + 270 pp. Index. Part 7, "The Changing Forest," contains some history and reminiscences of the lumber industry in the Pacific Northwest.

H434 Holbrook, Stewart H. "Sawdust on the Wind." *American Heritage* 4 (Summer 1953), 48-53. On logging, the lumber industry, and milltowns in Washington, since 1826.

H435 Holbrook, Stewart H. "A Lament for Things Gone." *American Forests* 60 (Sept. 1954), 28-29, 86-89. Memories of logging and the lumber industry in Oregon, 1920s.

H436 Holbrook, Stewart H., and McCready, Al. "Engine Smoke in the Big Woods." *Railroad Magazine* 66 (Aug. 1955), 12-23, 50. On locomotives designed by Ephraim Shay (1839-1916) to pull trainloads of logs up steep grades, ca. 1870.

H437 Holbrook, Stewart H. *The Columbia.* Rivers of America Series. New York: Rinehart and Company, 1956. 393 pp. Notes, bib., index. Includes incidental historical references to forests and forest industries of the Pacific Northwest.

H438 Holbrook, Stewart H. "Fire Makes Wind: Wind Makes Fire." *American Heritage* 7 (Aug. 1956), 52-57. On the forest fire which ravaged Peshtigo, Wisconsin, and surrounding areas, 1871.

H439 Holbrook, Stewart H. "Greeley Went West." *American Forests* 64 (Mar. 1958), 16-23, 52-62. Biographical sketch of William Buckhout Greeley (1879-1955), with emphasis on his influence on the development of industrial forestry in the Pacific Northwest as secretary-manager of the West Coast Lumbermen's Association in Seattle, 1928-1946.

H440 Holbrook, Stewart H. "Daylight in the Swamp." *American Heritage* 9 (Oct. 1958), 10-19, 77-80. On "old-time logging" in the Pacific Northwest and the working conditions there that contributed to the success of the Industrial Workers of the World, 1880s-1920s.

H441 Holbrook, Stewart H. "Log Drive." *Vermont Life* 13 (Spring 1959), 50-55. Reminiscences of log drives on the Connecticut River, ca. 1903.

H442 Holbrook, Stewart H., and Whisnant, Archie. "The First Fifty Years: A Brief Account of the Pacific Logging Congress During Half a Century." *Loggers Handbook* 19 (1959), 7-26. A lumbermen's organization of the Far West.

H443 Holbrook, Stewart H. *Yankee Loggers: A Recollection of Woodsmen, Cooks, and River Drivers.* New York: International Paper Company, 1961. 123 pp. Illus. Popular and anecdotal history of the forest products industries and lumberjacks in New England and New York, with some history of the International Paper Company.

H444 Holbrook, Stewart H. *The American Lumberjack.* New York: Collier Books, 1962. 254 pp. Bib., index. A revised edition of *Holy Old Mackinaw* (1938).

H445 Holcomb, Carl James. "A History of the Monongahela National Forest." *Davis and Elkins Historical Magazine* 5 (Apr. 1954), 29-34. West Virginia, since 1911.

H446 Holdsworth, R.P. "An Outline of Some Points in the History of American Forests and Forestry." Master's thesis, Yale Univ., 1928.

H447 Hole, Elmer C. " 'American Lumberman' on Parade." *American Lumberman* (July 10, 1943), 104-15. On W.B. Judson and James E. Defebaugh, former owner-editors of the trade journal, *American Lumberman.*

H448 Holland, I. Irving. "Some Factors Affecting the Consumption of Lumber in the U.S. with Emphasis on Demand." Ph.D. diss., Univ. of California, Berkeley, 1955. 105 pp.

H449 Holland, I. Irving. "An Explanation of Changing Lumber Consumption and Price." *Forest Science* 6 (June 1960), 171-92. Since 1922.

H450 Holland, I. Irving. "A Suggested Technique for Estimating the Future Price of Eastern White Pine Stumpage." *Forest Science* 6 (Dec. 1960), 369-96. Based on an analysis of statistics for the production and use of white pine lumber since 1905.

H451 Holland, Kenneth, and Hill, Frank Ernest. *Youth and the CCC.* Prepared for the American Youth Commission. Washington: American Council on Education, 1942. Reprint. New York: Arno Press, 1974. xv + 263 pp. Illus., tables. On the Civilian Conservation Corps as a youth-saving agency, 1930s.

H452 Holland, Reid A. "Life in Oklahoma's Civilian Conservation Corps." *Chronicles of Oklahoma* 48 (Summer 1970), 224-34. From 1933 to 1942.

H453 Holley, John Milton. "Waterways and Lumber Interests of Western Wisconsin." *Proceedings of the State Historical Society of Wisconsin,* 1906 (1907), 208-15. On pre-Civil War logging and log driving on the Wisconsin, Black, Chippewa, and St. Croix rivers.

H454 Holliday, J.S. "The Politics of John Muir." *Sierra Club Bulletin* 57 (Oct.-Nov. 1972), 10-13. On his political activities in behalf of wilderness preservation in California and other states, 1890s-1910s.

H455 Hollon, Gene. "The Kerrville Cedar Axe." *Southwestern Historical Quarterly* 50 (Oct. 1946), 241-50. On the invention in 1927 of a special axe used for chopping cedar trees and bushes in the hill country near Kerrville in south central Texas; includes some history of cedar "choppers" or cutters, the uses of cedar, and area ranchers' cedar eradication program in the 20th century.

Holloway, Garrett B. See Reynolds, George W., #R133

H456 Holm, Lovelock. "History of Paper Board." *Paper* 31 (No. 9, 1922), 9-10.

Holman, Barbara D. See Peterson, Virgil G., #P157

Holman, Barbara D. See Plant, Charles, #P244

H457 Holmes, Darrell O. "Railroad Logging." *The Big Smoke 1974* (Pend Oreille County [Washington] Historical Society) (1974), 18-22. Reminiscences of railroad logging near Elk and Ruby in northeastern Washington, 1920-1932.

H458 Holmes, David C. "An Examination of the Forest Management License System in British Columbia Together with an Analysis of Comparable Legislation in the U.S." Master's thesis, Yale Univ., 1949. Incidental history.

H459 Holmes, Frank, comp. *Minnesota in Three Centuries.* Volume 4. New York: Publishing Society of Minnesota, 1908. 458 pp. Illus., index. See chapter 29 for history of the lumber industry in Minnesota.

H460 Holmes, Frank R. "Palisades Interstate Park, New Jersey." *Americana* 17 (Apr. 1923), 186-91. A forested state park along the banks and cliffs of the Hudson River, New Jersey and New York.

H461 Holmes, Jack D.L. "Louisiana Trees and Their Uses: Colonial Period." *Louisiana Studies* 8 (No. 1, 1969), 36-67. Travelers' observations and commercial uses of forests in Spanish Louisiana.

H462 Holmes, John S., and Foster, J.H. *A Study of Forest Conditions of Southwestern Mississippi.* Mississippi State Geological Survey, Bulletin No. 5. Nashville: Brandon Printing Company, 1909. 56 pp. Map. Incidental history.

Holmes, John S. See Simmons, Roger E., #S285

H463 Holmes, John S. *Forest Conditions in Western North Carolina.* North Carolina Economic and Geological Survey, Bulletin No. 23. Raleigh: Edwards and Broughton, 1911. 116 pp. Incidental history.

H464 Holmes, Kenneth L. *Ewing Young, Master Trapper.* Portland: Binfords and Mort, 1967. viii + 180 pp. Illus., bib., index. Young established a sawmill in Oregon in 1838.

H465 Holmes, Michael S. "The New Deal and Georgia's Black Youth." *Journal of Southern History* 38 (Aug. 1972), 443-60. A case study of the Civilian Conservation Corps in Georgia.

H466 Holsoe, Torkel. "The Cooperative Association Approach to the Private Forestry Problem." *Journal of Forestry* 46 (July 1948), 511-13. On the work of the West Virginia Forest Products Association in bringing good forest practices to privately owned forest lands since 1937.

H467 Holst, Monterey Leman. "Zachariah Allen, Pioneer in Applied Silviculture." *Journal of Forestry* 44 (July 1946), 507-08. On Allen's silvicultural experiments in Rhode Island, 1820-1882.

H468 Holt, Clarence Eugene. *One Life in Maine.* Portland: Forest City Printing Company, 1940. 190 pp. The author (b. 1874) was a lumberman and dentist in Maine.

H469 Holt, William Arthur. *A Wisconsin Lumberman Looks Backward: An Intimate Glance into 100 Years of North Woods Lumbering by the Holt Family.* Oconto, Wisconsin: Privately printed, 1948. 81 pp. Illus. In northern Wisconsin and Ontonagon County, Michigan, 1847-1938.

H470 Holter, Anson M. "Pioneer Lumbering in Montana: Story of Early Days Graphically Told." *Timberman* 12 (Jan. 1911), 20-24.

H471 Holter, Anson M. "Pioneer Lumbering in Montana." *Contributions of the Montana Historical Society* 8 (1917), 251-81. The author's account of his timberlands and sawmill, 1863-1898.

H472 Holter, Anson M. *Pioneer Lumbering in Montana.* Portland: Timberman, n.d. 23 pp.

H473 Holter, Anson M. "Pioneer Lumbering in Montana." Ed. by Margaret E. Parsons. *Frontier* 8 (1928), 196-209.

H474 Holtz, William, "Homage to Joseph Wood Krutch: Tragedy and the Ecological Imperative." *American Scholar* 43 (Spring 1974), 267-79.

H475 Holtzclaw, Henry F. "The Lumber Industry and Trade." Ph.D. diss., Johns Hopkins Univ., 1917.

H476 Holtzclaw, Henry F. "Historical Survey of the Lumber Industry." *Southern Lumberman* 89 (Aug. 3, 1918), 32 ff; 89 (Aug. 10, 1918), 30 ff.

H477 Holtzclaw, Henry F. "History of Finance in Lumber Industry." *Southern Lumberman* 99 (June 11, 1921), 54, 56.

H478 Holz, Robert Kenneth. "The Area Organization of National Forests: A Case Study of the Manistee National Forest, Michigan." Ph.D. diss., Michigan State Univ., 1963. 260 pp. Includes some incidental history.

Honey, John C. See Wengert, Norman I., #W165

H479 Hood, A.B. "Logging Operations in the Black Hills." *Timberman* 29 (May 1928), 37, 162-67, 170. By the Warren-Lamb Lumber Company on South Dakota's Black Hills National Forest.

H480 Hood, Mary V. Jessup, and Haas, Robert B. "Eadweard Muybridge's Yosemite Valley Photographs, 1867-1872." *California Historical Society Quarterly* 42 (Mar. 1963), 5-26. Muybridge's photographs helped publicize the scenic beauty of Yosemite Valley.

H481 Hooker, Bill. "Fond du Lac, Its Sawmills and Freedmen—A Sketch." *Wisconsin Magazine of History* 16 (June 1933), 423-27. Reminiscences of the lumber industry in Fond du Lac, Wisconsin, 1850s-1860s.

H482 Hooper, John H. "Pine and Pasture Hills and the Part They Have Contributed to the Development of Medford." *Medford Historical Register* 18 (Apr. 1915), 25-32.

H483 Hoopes, Chad L. *Lure of Humboldt Bay Region.* 1966. Revised edition. Dubuque, Iowa: Kendall/Hunt Publishing Company, 1971. x + 299 pp. Illus., maps. Includes some history of the lumber industry in northwestern California.

H484 Hoover, Calvin B., and Ratchford, B.U. *Economic Resources and Policies of the South.* New York: Macmillan, 1951. xxvii + 464 pp. Tables, bib., index. Includes a chapter on forest resources and policies; incidental history.

H485 Hoover, Herbert C. *The Memoirs of Herbert Hoover.* Volumes 2 and 3. New York: Macmillan, 1952. Illus. On Hoover's services as Secretary of Commerce, as President, and in other public capacities, 1921-1941, including mention of conservation issues during this period.

H486 Hoover, Roy Otto. "The Public Land Policy of Washington State: The Initial Period, 1889-1921." Ph.D. diss., Washington State Univ., 1967. 268 pp.

H487 Hope, Jack. *Parks in Peril.* San Francisco: Sierra Club Books, 1972. 176 pp. Illus. Includes some incidental history of the national parks and National Park Service.

H488 Hopkins, Arthur S. "Land Acquisition for Forest Preserve Purposes in New York State, 1916-1944." *New York Forester* 1 (Dec. 1944), 2-6.

H489 Hopkins, Arthur S. *New York State's Reforestation Program.* New York Conservation Department, Bulletin No. 20. Albany, 1950. 36 pp. Review of New York public and private reforestation since 1899.

H490 Hopkins, Francis Washburn. *An Historical Sketch of Local Finance in Connecticut until 1930 with Special Reference to Forest Resources, Forest Industries and the Taxation of Forests.* Ann Arbor, Michigan: Edwards Brothers, 1936. vi + 209 pp. Map. History of forest taxation in Union, Lebanon, Killingworth, and North Haven, Connecticut.

H491 Hopkins, Howard G. "Accomplishments of the Timber Production War Project." *Journal of Forestry* 44 (May 1946), 330-34. On the work of the Timber Production War Project, administered by the USFS for the War Production Board, 1943-1945.

Hopkins, Howard G. See Resler, Rexford A., #R125

H492 Hopkins, William Clifford. "Stability of Forest Land Ownership in the United States: A Study of the Shifting Ownership of Forest Lands, of the Causes Back of It, and the Causes Thereof." Master's thesis, Yale Univ., 1941. 111 pp.

H493 Hopping, G.R., and McCardell, W.H. *A History of the Rocky Mountain Section, Canadian Institute of Forestry, 1948-1967.* Calgary: Rocky Mountain Section, Canadian Institute of Forestry, 1968. 40 pp.

H494 Horine, Irving. "History of Plywood in the South." *Southern Lumberman* (Dec. 15, 1943), 151-52.

H495 Horn, Allen F., Jr. "Resource Policy and Forest Industry Development in Pennsylvania." Ph.D. diss., State Univ. of New York, College of Environmental Science and Forestry, 1957.

Horn, Arthur G. See Cunningham, Russell N., #C749

H496 Horn, C. Lester. "Oregon's Columbia River Highway." *Oregon Historical Quarterly* 66 (Sept. 1965), 249-71. Reminiscence of the promoters, builders, and construction of the scenic highway through Oregon's Columbia Gorge, 1913-1915.

H497 Horn, Stanley F. *This Fascinating Lumber Business.* 1943. Second edition. Indianapolis: Bobbs-Merrill, 1951. 328 pp. Illus., app., index. General history and description of the lumber industry in the United States by the long-time editor of *Southern Lumberman.*

H498 Hornaday, William T. *Our Vanishing Wildlife; Its Extermination and Preservation.* New York: Charles Scribner's Sons, 1913. xv + 411 pp. Illus., maps.

H499 Hornaday, William T. *Thirty Years for Wildlife: Gains and Losses in the Thankless Cause.* New York: Charles Scribner's Sons, 1931. Wildlife conservation and preservation.

H500 Hornaday, William T. *Wild Life Conservation in Theory and Practice: Lectures Delivered Before the Forest School of Yale University, 1914.* New Haven, Connecticut: Yale University, Forest School, 1914. Reprint. New York: Arno Press, 1972.

H501 Hornbeck, L.W. "Looking Backward and Forward in the Lumber Industry." *Southern Lumber Journal* 49 (July 1, 1923), 48-49.

H502 Horne, G.F. "Watershed Management in the Department of Interior: Three Case Studies in Cooperation." *Journal of Forestry* 58 (Apr. 1960), 302-04. Watersheds of 15-Mile Creek, near Worland, Wyoming; Hill Creek, northeastern Utah; and Oregon City, Oregon, since 1916.

H503 Horne, Gilbert Richard. "The Receivership and Reorganization of the Abitibi Power and Paper Company, Limited." Ph.D. diss., Univ. of Michigan, 1954. 412 pp.

H504 Horner, Harlan H. "History of Forest Education in the State of New York." In *Dedication of the Louis Marshall Memorial.* Syracuse: New York State University, College of Forestry, 1933. Pp. 30-42. On the New York State College of Forestry at Cornell University, 1898-1903, and at Syracuse University since 1911.

H505 Horning, Walter H. "The O. and C. Lands: Their Role in Forest Conservation." *Journal of Forestry* 38 (May 1940), 379-83. Includes some history of the revested Oregon and California Railroad grant lands in western Oregon since 1916.

H506 Horning, Walter H. "The O. and C. Lands—An Adjustment and an Experiment in Forest Conservation." *Commonwealth Review* 22 (Jan. 1941), 237-52.

H507 Horning, Walter H. "The Oregon Checkerboard—Twenty Years Later." *American Forests* 63 (Dec. 1957), 10-13, 36-38. On the administration of the forested Oregon and California Railroad grant lands (O&C lands) in western Oregon since 1937.

H508 Horowitz, Morris. *Structure and Government of Carpenters' Union.* New York: John Wiley and Sons, 1962. 168 pp.

H509 Horsman, Reginald. *The Frontier in the Formative Years, 1783-1815.* New York: Holt, Rinehart and Winston, 1970. xii + 237 pp. Illus., maps, notes, bib., index. From the Appalachian Mountains to the Mississippi River; includes references to forests and forest uses.

H510 Horwitz, Eleanor C.J. *Clearcutting: A View from the Top.* Washington: Acropolis Books, 1974. 179 pp. Illus., bib., index. Includes five chapters by forestry educators. There are incidental historical references to logging and forestry.

H511 Horwitz, Robert H., and Mellor, Norman. *Land and Politics in Hawaii.* East Lansing: Bureau of Social and Political Research, Michigan State University, 1963. 59 pp. A case history of the struggle to influence the course of public land laws in Hawaii's first state legislature, 1959-1960.

H512 Hosie, R.C. *Forest Regeneration in Ontario.* Toronto: University of Toronto Press, 1953. 134 pp. Results of 57 regeneration surveys conducted between 1919 and 1951.

Hoskins, R.N. See Mobely, M.D., #M511

H513 Hosler, Wilbert. "History of Paper Making in Michigan." *Michigan History Magazine* 22 (Autumn 1938), 361-402. Since 1834.

H514 Hosley, N.W. "Black Rock Forest." *American Forests* 55 (Jan. 1949), 12-13, 48. On a 3,600-acre forest established in 1927 near Cornwall, New York, as a "research center for silviculture and forest utilization, and as a demonstration of sustained yield management."

H515 Hosmer, Paul. *Now We're Loggin'.* Portland: Metropolitan Press, 1930. 210 pp. Anecdotal essays on logging in the Pacific Northwest with some incidental historical references.

H516 Hosmer, Paul. "Driving the Deschutes." *Timberman* 40 (May 1939). Log driving on Oregon's Deschutes River.

H517 Hosmer, Ralph S. "The Progress of Education in Forestry in the United States." *Empire Forestry Journal* 2 (Apr. 1923), 1-24. History of the forestry movement and forestry education.

H518 Hosmer, Ralph S. "Dr. Fernow's Life Work as Seen by a Member of the Profession of Forestry." *Journal of Forestry* 21 (Apr. 1923), 320-23. Fernow (1851-1923) was chief of the Division of Forestry, 1886-1898, and subsequently a prominent forestry educator at the New York State College of Forestry at Cornell University and at the University of Toronto. He also edited the *Journal of Forestry* and its predecessor, *Forestry Quarterly,* from 1903 to 1923. This issue also contains many short tributes and a bibliography of Fernow's writings.

H519 Hosmer, Ralph S. "Fifty Years of Conservation." *Forest Leaves* 25 (Apr. 1935), 65-66. In New York.

H520 Hosmer, Ralph S. "The Society of American Foresters: An Historical Summary." *Journal of Forestry* 38 (Nov. 1940), 837-54. A charter member's account of the professional organization since its founding in 1900.

H521 Hosmer, Ralph S. "Some Recollections of Gifford Pinchot, 1898-1904." *Journal of Forestry* 43 (Aug. 1945), 558-62. As chief of the Division of Forestry and Bureau of Forestry; an abridged version appears in *Journal of Forestry* 63 (Aug. 1965), 587-89.

H522 Hosmer, Ralph S. "The Society of American Foresters—Notes on Its Progress, 1941-1945." *Journal of Forestry* 44 (June 1946), 426-31.

H523 Hosmer, Ralph S. "The National Forestry Program Committee, 1919-1928." *Journal of Forestry* 45 (Sept. 1947), 627-45. On the work of the committee which helped formulate national forest policy and legislation in the 1920s. The committee represented the major forestry organizations and forest product trade associations.

H524 Hosmer, Ralph S. *The Cornell Plantations; a History.* Ithaca, New York: Cornell University, 1947. xiv + 209 pp. Illus., maps, app., notes, index. On Cornell University's botanical garden and arboretum, conceived by horticulturalist Liberty Hyde Bailey, since 1868.

H525 Hosmer, Ralph S. "The Society of American Foresters: An Historical Summary." *Journal of Forestry* 48 (Nov. 1950), 756-77. From the origins of the organization in 1900. See also a list of "past and present officers," 1900-1950, pp. 778-80.

H526 Hosmer, Ralph S. *Forestry at Cornell: A Retrospect of Proposals, Developments, and Accomplishments in the Teaching of Professional Forestry at Cornell University.* Ithaca, New York: Cornell University, 1950. 64 pp. Illus., index. On forestry education at Cornell University since 1898.

H527 Hosmer, Ralph S. "Henry Solon Graves, 1871-1951." *Journal of Forestry* 49 (May 1951), 325. Brief sketch of his career as chief of the USFS (1910-1920) and as a forestry educator.

H528 Hosmer, Ralph S. "Franklin B. Hough: Father of American Forestry." *North Country Life* 6 (Summer 1952), 16-20. Hough (1822-1885) was the first federal forestry official, an advocate of forestry in New York, and a prolific writer on the subject.

H529 Hosmer, Ralph S. "Cornell University—Early Education in Professional Forestry." *Northeastern Logger* 4 (May 1956), 14, 62-65. From 1898 to 1903.

H530 Hosmer, Ralph S. "The Beginning Five Decades of Forestry in Hawaii." *Journal of Forestry* 57 (Feb. 1959), 83-89. Hosmer was the first territorial forester, 1904-1914.

H531 Hosmer, Ralph S., and Clepper, Henry. "The Society of American Foresters: An Historical Summary." *Journal of Forestry* 58 (Oct. 1960), 765-79. Since 1900.

H532 Hosmer, Ralph S. "Early Days in Forest School and Forest Service." OHI by Bruce C. Harding. *Forest History* 16 (Oct. 1972), 6-11. Reminiscences of work for the Division of Forestry, 1890s, and graduate study at Yale Forestry School, 1901-02.

H533 Hotchkiss, George W. *Industrial Chicago; the Lumber Interests.* Volume 5 of *Industrial Chicago.* Chocago, 1894. 580 pp. Illus. Includes history of the lumber industry and trade.

H534 Hotchkiss, George W. *History of the Lumber and Forest Industry of the Northwest.* Chicago: George W. Hotchkiss & Company, 1898. xv + 754 pp. Illus. A general history of the lumber and related industries in the Great Lakes states since pioneer settlement. It focuses on Michigan, Wisconsin, Minnesota, Illinois, and Iowa, but includes references to adjacent areas. There are many biographical sketches of prominent 19th-century lumbermen. The author was a Michigan lumberman and, in 1872, became a pioneer lumber journalist.

H535 Hough, Ashbel F. "The Early Americans and the Forest." *Empire Forester* 8 (1922), 68-72.

H536 Hough, Ashbel F. "Pioneer Tree Planter." *Pennsylvania Forests* 37 (Fall 1957), 88-90, 105-06. On Franklin R. Miller and his plantings in Warren County, Pennsylvania, 19th century.

H537 Hough, Donald. *The Cocktail Hour in Jackson Hole.* New York: W.W. Norton, 1956. 253 pp. Maps. On Jackson Hole in Grand Teton National Park, Wyoming, as observed by the author in visits from 1925 to 1950.

H538 Hough, Franklin Benjamin. "On the Duty of Governments in the Preservation of Forests." *Proceedings of the American Association for the Advancement of Science,* 1873 (1874), 1-22. The address which prompted the creation in 1876 of the first federal forestry agency.

H539 Hough, Franklin Benjamin. *Report Upon Forestry.* Volume 1. Washington: GPO, 1878. 650 pp. Tables, charts, index. Hough (1822-1885) prepared this and following reports under the direction of the Commissioner of Agriculture in pursuance of an Act of Congress approved August 15, 1876. His encyclopedic reports represented the federal government's first modern venture into the field of forestry. This first volume is a heterogeneous collection of information on the status of forested lands and forest industries, including many incidental historical references.

H540 Hough, Franklin Benjamin. *Report Upon Forestry.* Volume 2. Washington: GPO, 1880. 618 pp. Tables, charts, index. This volume of Hough's report contains much historical and contemporary material on the following subjects: North American exports and imports of timber and forest products; the Timber Culture Act of 1873; timber resources on the public lands; state and territorial forestry legislation; and the timber resources and trade of Canada.

H541 Hough, Franklin Benjamin. *Report Upon Forestry.* Volume 3. Washington: GPO, 1882. 318 pp. Tables, charts, index. Contains information on forested lands of the public domain, the influence of forests on climate, the charcoal and tanning industries, the need for forestry research, and the effect of forest fires. While most of this is contemporary description, there are numerous historical references to the subject in the United States and Canada. Volume 4 (1884) was written largely by Hough; see under Nathaniel H. Egleston.

H542 Hough, Franklin Benjamin. *The Elements of Forestry.* Cincinnati: Robert Clarke & Company, 1882. ix + 381 pp.

H543 Hough, Romeyn B. "The Incipiency of the Forestry Movement in America." *American Forestry* 19 (Aug. 1913), 547-50. On the efforts of Franklin B. Hough to establish a federal forestry agency in the 1870s.

H544 Houghton, Gilbert M. "Return to the Kaniksu: A Story of Blister Rust and Reforestation." *The Big Smoke 1972* (Pend Oreille County [Washington] Historical Society) (1972), 23-31. Reminiscences of USFS reforestation and blister rust control work on the Kaniksu National Forest of northern Idaho, including the role of German and Italian war internees.

H545 Houpt, William P. "Maine Long Logging and Its Reflection in the Works of Holman Francis Day." Ph.D. diss., Univ. of Pennsylvania, 1964. 406 pp. Day was a newspaperman (1865-1935) who described Maine logging and lumbermen in verse and novels.

H546 House, Frank H., comp. *Timber at War: An Account of the Organization and Activities of the Timber Control, 1939-1945.* London: Ernest Benn, 1965. 331 pp. Illus., tables, charts. Includes an account of British procurement of American timber through Lend Lease and the U.S. War Production Board.

H547 House, William P. "Forty Years of Forestry: A Sketch of the Society for the Protection of New Hampshire Forests." *Appalachia* (No. 90, 1940), 203-11.

H548 Houston, David Franklin. *Eight Years with Wilson's Cabinet, 1913 to 1920.* 2 Volumes. Garden City, New York: Doubleday, Page & Company, 1926. Houston was Woodrow Wilson's Secretary of Agriculture, 1913 to 1920.

Hove, John. See Dietrich, Irvine T., #D175

H549 Hovey, N.H. "Distribution Problems of the Red Cedar Shingle Industry." *Journal of Marketing* 4 (Oct. 1939), 157-67. Incidental history of the industry in the Pacific Northwest.

H550 Howard, Clinton Newton. *British Development of West Florida, 1763-1769.* Berkeley: University of California Press, 1947. viii + 166 pp. Includes mention of logging and the lumber trade.

H551 Howard, Frances L. "The Lumbermen of the Ottawa Valley." *Transactions of the Women's Canadian Historical Society of Ottawa* 3 (1910), 22-26.

H552 Howard, Irene. "Vancouver Swedes and the Loggers." *Swedish Pioneer Historical Quarterly* 21 (July 1970), 163-82. Includes some history of Swedish loggers in and near Vancouver, British Columbia, and especially of the efforts of organized labor to achieve better wages and working conditions in the lumber industry, since the 1920s.

H553 Howard, James O. *The Timber Resources of Central Washington.* USFS, Pacific Northwest Forest and Range Experiment Station, Resource Bulletin, No. 45. Portland, 1973. 68 pp. Illus., maps, tables, references. Includes some history of the region's lumber industry and of increasing demand for recreational use of forests.

H554 Howard, John A.C. "Trends in American Sawmilling." Master's thesis, Univ. of Minnesota, 1956.

H555 Howard, John A.C. "Trends in American Sawmilling." *Wood* 21 (Dec. 1956), 466-68; 22 (Jan. 1957), 15-17; 22 (Feb. 1957), 60-61. Recent history of sawmill design and power sources.

H556 Howard, John C. *The Negro in the Lumber Industry.* Racial Policies of American Industry, No. 19. Philadelphia: University of Pennsylvania Press, 1970. 97 pp. Tables, apps., index. Includes some history and statistics of the industry, especially in the South.

H557 Howard, Leland O. *A History of Applied Entomology.* Washington: Smithsonian Institution, 1930. viii + 564 pp. See for reference to forest entomology.

Howard, M.C. See Ferguson, Roland H., #F59

H558 Howard, Prescott L. "The Era of the Lombard Log Hauler." *Forest History* 6 (Spring/Summer 1962), 2-8. On the origins of mechanical log hauling in Maine, 1900-1920s.

H559 Howard, R.C. "Some Developments in the Protection of Forests from Fire." *Forestry Chronicle* 39 (Mar. 1963), 85-88. In Canada.

H560 Howard, Rosser Taylor. "The Gentry of Antebellum South Carolina." *North Carolina Historical Review* 17 (Apr. 1940). Includes reference to the lumber industry.

H561 Howard, William G. "Recreational Use of State Forests." *Forest Worker* 8 (Jan. 1932), 1-3. A brief history of the New York Forest Preserve and the development of recreational use.

H562 Howard, William G. "Forests and Parks of the Empire State." *American Forests* 38 (Mar. 1932), 165-68. New York, since 1885.

H563 Howard, William G. "New York State's Forest Preserve Policy." *Journal of Forestry* 35 (Aug. 1937), 762-68. Since 1885.

H564 Howard Publishing Company. *A History of the Wisconsin Paper Industry, 1848-1948.* Chicago, 1948. 76 pp. Illus.

H565 Howay, F.W. "A Short Historical Sketch of Jasper Park Region." *Sierra Club Bulletin* 14 (Feb. 1929), 28-33. Alberta.

H566 Howay, F.W. "Early Shipping on Burrard Inlet, 1863-1870." *British Columbia Historical Quarterly* 1 (Jan. 1937), 1-20. Pioneer sawmills and lumber exporting from the vicinity of Vancouver, British Columbia.

Howay, F.W. See Angus, Henry Forbes, #A459

H567 Howd, Cloice Ray. *Industrial Relations in the West Coast Lumber Industry.* U.S. Bureau of Labor Statistics, Miscellaneous Publication No. 349. Washington: GPO, 1924. 120 pp. On the period through World War I, including an account of the Loyal Legion of Loggers and Lumbermen, the government-inspired labor union intended to counteract radical organization of lumberjacks and millworkers.

H568 Howd, Cloice R. "Development of Lumber Industry of West Coast." *Timberman* 25 (Aug. 1924), 194-98.

H569 Howe, Clifton D., and White, J.H. *The Trent Watershed Survey: A Reconnaissance.* Ottawa: Canada Commission of Conservation, 1913. 156 pp. Illus., maps, tables, index. An analysis of forest and economic conditions in the Trent Valley of Ontario, including some history of the lumber industry since the 1840s.

H570 Howe, Clifton D. "Forests and Forest Industries of Canada." *Annals of the American Academy of Political and Social Science* 107 (1923), 95-101. Incidental history.

H571 Howe, Clifton D. "Debt Canada Owes to Pioneer Lumbermen—Influence of the Forests." *Canada Lumberman* 43 (July 15, 1923), 50-52.

H572 Howe, Clifton D. "Bernhard Eduard Fernow: An Appreciation." *Canadian Forestry Magazine* 19 (No. 3, 1923), 168-69.

H573 Howe, Clifton D. "The Work of Twenty-Five Years in Retrospect and in Prospect." *Canadian Forest and Outdoors* 21 (Feb. 1925), 77-79; 21 (Mar. 1925), 151-52; 21 (Apr. 1925), 211-14. On the Canadian Forestry Association since 1900.

H574 Howe, Clifton D. "Some Aspects of Forest Investigative Work in Canada." *Forestry Chronicle* 2 (Sept. 1926), 3-25. Some history of forest classification, survey, and regeneration studies in Canada since 1873.

H575 Howe, Clifton D. "Those Sixty Years." *Canadian Forest and Outdoors* 23 (1927), 369-71. On Canadian forests since 1867.

H576 Howell, H.A. "A Review of the Southern Pine Industry." Master's thesis, Yale Univ., 1931.

H577 Howell, Wilbur F. *A History of the Corrugated Shipping Container Industry in the United States.* Camden, New Jersey: Samuel M. Langston Company, 1940. 59 pp. Illus., diags., charts.

Howenstine, William L. See Frederick, Duke, #F223

H578 Howland, A.G. "The Story of Ground Wood Manufacture." *Paper Trade Journal* 74 (Apr. 13, 1922), 45-47.

Hoyle, Raymond J. See Reynolds, Robert V., #R138

H579 Hoyle, Raymond J. *Wood-Using Industries of New York.* New York State College of Forestry, Technical Publication No. 27. Syracuse, 1928. 160 pp. Illus. Includes comparisons with surveys made as early as 1912. A revised edition, coauthored by John R. Stillinger, was published in 1949. Another revised edition, coauthored by Russell C. Deckert, appeared in 1957.

H580 Hoyle, Raymond J., and Cox, Laurie D., eds. *The New York State College of Forestry at Syracuse University: A History of Its First Twenty-Five Years, 1911-1936.* Syracuse: New York State College of Forestry, 1936. 176 pp.

H581 Hoyle, Raymond J. "Changes in the Wood-Using Industries of New York State, 1912 to 1954." *Southern Lumberman* 192 (June 15, 1956), 30-31.

H582 Hoyle, Raymond J., and Deckert, Russell C. "Trends in the Wood-Using Industries of New York State since 1912." *Northeastern Logger* 5 (Dec. 1956), 36-37.

H583 Hoyle, Raymond J. "Developments, Trends, and Problems in Education and Research in Forest Utilization." *Journal of Forestry* 56 (Aug. 1958), 578-83. Since 1900.

H584 Hoyt, Ray. *'We Can Take It': A Short Story of the C.C.C.* New York: American Book Company, 1935. 128 pp. Illus. Includes an account of the origins of the Civilian Conservation Corps.

H585 Hoyt, William G., and Langenbein, Walter B. *Floods.* Princeton, New Jersey: Princeton University Press, 1955. 469 pp. Bib. On floods and flood control policies, including watershed management, since the colonial period.

H586 Hubbard, Alice H. *This Land of Ours: Community and Conservation Projects for Citizens.* New York: Macmillan, 1960. 272 pp. Includes some case histories of successful conservation projects relating to community forests, watersheds, roadside parks, soil conservation, and areas of natural beauty.

H587 Hubbard, Howard G. "The Lumberman." *Chronicle of the Early American Industries Association* 1 (Jan. 1936), 1-2; 1 (Mar. 1936), 7.

H588 Hubbard, Preston J. *Origins of the TVA: The Muscle Shoals Controversy, 1920-1932.* Nashville: Vanderbilt University Press, 1961. x + 340 pp. Notes, bib., index. Controversy over federal water power policy for the Tennessee River involved Gifford Pinchot and George Norris and had implications for the conservation movement generally.

H589 Hubert, Ernest E. *An Outline of Forest Pathology.* New York: John Wiley and Sons, 1931. 543 pp. Illus. Incidental history.

H590 Huberty, Martin R., and Flock, Warren L., eds. *Natural Resources.* New York: McGraw-Hill, 1959. 556 pp. Includes some history of natural resource problems.

H591 Huckleberry, E.R. "In Those Days . . . Tillamook County." *Oregon Historical Quarterly* 71 (June 1970), 116-40. Includes some references to logging and woods life in Tillamook County, Oregon, 1920s.

H592 Huckleberry, E.R. *The Adventures of Dr. Huckleberry: Tillamook County, Oregon.* Portland: Oregon Historical Society, 1970. xii + 272 pp. Illus., map. Includes accounts of logging, lumberjacks, and sawmills, by a country doctor, 1920s.

H593 Hudgins, Bert. *Michigan: Geographic Backgrounds in the Development of the Commonwealth.* Detroit, 1948. vii + 104 pp. Illus., maps. Includes a chapter on "The Lumbering Era."

H594 Hudgins, M.D. "The Farmer—Extraordinary." *Southern Lumberman* 192 (Jan. 1, 1956), 42-44. Sketch of William Logan Hall (1874-1960), his USFS career, 1899-1919, and as consulting forester and tree farmer in Arkansas since ca. 1900.

H595 Hudson, Fay Albert. "The Unexpected Guest." *New Mexico Magazine* 32 (Oct. 1954), 20, 50-51. Reminiscence of a visit of a forest ranger to the author's father's ranch near Magdalena, New Mexico, 1916.

H596 Hudson, G.M. "A Study of a Permanent Alabama Lumber Town." *Journal of Geography* 36 (1937).

H597 Hudson, J. Paul. *A Pictorial Booklet on Early Jamestown Commodities and Industries.* Jamestown 350th Anniversary Historical Booklet, No. 23. Williamsburg, Virginia: Virginia 350th Anniversary Celebration Corporation, 1957. 78 pp. Illus., bib. Brief historical sketches of 17th-century forest products industries in Jamestown, Virginia.

H598 Hudson, James J. "The McCloud River Affair of 1909: A Study in the Use of State Troops." *California Historical Society Quarterly* 35 (Mar. 1956), 29-35. On the intervention of the California National Guard in a strike of Italian laborers against the McCloud River Lumber Company in Siskiyou County, California.

H599 Huey, Ben M. "Elers Koch: 40-Year Federal Forester." *Journal of Forestry* 47 (Feb. 1949), 113-14. Sketch of Koch's career in the USFS, 1903-1944, especially in the national forests of his native Montana.

H600 Huey, Ben M. "Evan W. Kelly." *Journal of Forestry* 48 (July 1950), 499-500. Sketch of Kelly's career with the USFS, 1906-1944, as regional forester in Regions 1 and 7, and as director of the Guayule Emergency Rubber Project in California during World War II.

H601 Huey, Ben M. "Problems of Timber Products Procurement during World War II, 1941-1945." Master's thesis, Univ. of Montana, 1951.

H602 Huff, Boyd Francis. "The Maritime History of San Francisco Bay." Ph.D. diss., Univ. of California, Berkeley, 1956. Includes some history of the lumber trade.

H603 Huffington, Paul, and Clifford, J. Nelson. "Evolution of Shipbuilding in the Southeastern Massachusetts." *Economic Geography* 15 (1939), 362-78. Since the colonial period.

H604 Huffman, L.Q. "Huffman's Mills." *Indiana History Bulletin* 31 (June 1954), 113-14. On George Huffman (1784-1854) who built a grist mill and a sawmill on the Anderson River, Spencer County, Indiana, in 1816.

H605 Huffman, Robert O. *Drexel Enterprises, Inc.: A Brief History.* New York: Newcomen Society in North America, 1963. 24 pp. A North Carolina furniture manufacturer.

H606 Hughes, J. Donald. *The Story of Man in the Grand Canyon.* Grand Canyon, Arizona: Grand Canyon Natural History Association, 1967.

H607 Hughes, Jay Melvin. "Price and Prejudice in the Marketing of Colorado Lumber." Master's thesis, Colorado State Univ., 1958.

H608 Hughes, Jay Melvin. "Wilderness Land Allocation in a Multiple Use Forest Management Framework in the Pacific Northwest." Ph.D. diss., Michigan State Univ., 1964. 624 pp. Includes some account of USFS wilderness policy since the 1920s, especially in Oregon and Washington.

H609 Hughes, Thomas. *History of Minneopa State Park.* St. Paul: Minnesota Department of Conservation, Division of Forestry, 1932. 29 pp. Minneopa State Park, Blue Earth County, Minnesota.

H610 Hughson, John W., and Bond, Courtney, C.J. *Hurling Down the Pine: The Story of the Wright, Gilmour and Hughson Families, Timber and Lumber Manufacturers in the Hull and Ottawa Region and on the Gatineau River, 1800-1920.* Old Chelsea, Quebec: Historical Society of the Gatineau, 1964. vi + 130 pp. Illus., maps, index.

H611 Hughson, Oliver G. "Old Timer Tales." *Timberman* 28-29 (Dec. 1926-Sept. 1928). A series of reminiscences and anecdotes on the lumber industry in the Pacific Northwest, 1900s-1910s.

H612 Hughson, Oliver G. "When We Logged the Columbia." *Oregon Historical Quarterly* 60 (June 1959), 172-209. Reminiscences of logging camps on the lower Columbia River of Oregon and Washington, 1900-1910.

Hulbert, Richard C. See Hulbert, William D., #H613

H613 Hulbert, William D. *White Pine Days on the Taquamenon.* Ed. by Richard C. Hulbert. Lansing: Historical Society of Michigan, 1949. xx + 152 pp. Illus. Nine articles (eight reprinted from periodicals, 1900-1906) on natural history and lumbering along the Tahquamenon River of Michigan's Upper Peninsula, with a biographical sketch of the author (1868-1913).

H614 Hull, Clifton E. *Shortline Railroads of Arkansas.* Norman: University of Oklahoma Press, 1969. xvi + 416 pp. Illus., maps. Includes some history of logging railroads.

H615 Hull, William J., and Hull, Robert W. *The Origin and Development of the Waterways Policy of the United States.* Washington: National Waterways Conference, 1967. vii + 79 pp. Map, notes, index.

H616 Hult, Ruby El. *Steamboats in the Timber.* Caldwell, Idaho: Caxton Printers, 1952. Reprint. Portland: Binfords & Mort, 1969. 209 pp. Illus., map, app., index. On steamboats navigating Coeur d'Alene Lake and the St. Joe River, Idaho, and their use in logging, mining, and excursions, 1879-1945.

H617 Hult, Ruby El. *The Untamed Olympics: The Story of a Peninsula.* Portland: Binfords & Mort, 1954. 267 pp. Illus., map., bib., index. History of the heavily forested Olympic Peninsula of northwestern Washington, including the lumber industry and Olympic National Park.

H618 Hult, Ruby El. *Northwest Disaster: Avalanche and Fire.* Portland: Binfords & Mort, 1960. 238 pp. Illus., maps, bib. Includes an account of the forest fire in the Bitterroot Mountains of Idaho and Montana, 1910, as recorded in newspapers and public records and recalled by survivors.

H619 Humberger, Charles E. "Island Wilderness." *Bulletin of the Garden Club of America* 38 (Mar. 1950), 29-34. Isle Royale National Park, Michigan, since 1931.

H620 Hume, Robert D. *A Pygmy Monopolist: The Life and Doings of R.D. Hume, Written by Himself and Dedicated to His Neighbors.* Ed. by Gordon B. Dodds. Madison: State Historical Society of Wisconsin, for the Department of History, University of Wisconsin, 1961. viii + 87 pp. Map, notes. Hume, best known for his development of the salmon-canning industry in southern Oregon, owned a sawmill at Wedderburn in the 19th century.

H621 Humes, James C. "The Susquehanna Boom: A History of Logging and Rafting on the West Branch of the Susquehanna River." *Now and Then* 14 (No. 1, 1962), 4-14. From the 1830s to the 1880s, with emphasis on Williamsport, Pennsylvania, as a sawmill center and the site of the boom.

H622 Humphrey, Edward Frank. *An Economic History of the United States.* New York: Century, 1931. ix + 639 pp. Illus., maps, tables, charts, bib. See for mention of forest industries.

H623 Humphrey, Harry Baker. *Makers of North American Botany.* Chronica Botanica: An International Biological and Agricultural Series, No. 21. New York: Ronald Press, 1961. xi + 265 pp. Biographical sketches of 122 North American botanists, many having been involved in forestry or forest conservation.

H624 Humphreys, Hubert. "In a Sense Experimental: The Civilian Conservation Corps in Louisiana." *Louisiana History* 5 (Fall 1964), 345-67; 6 (Winter 1965), 27-52. Outlines effective CCC programs in soil conservation, flood control, erosion control, reforestation, and education, 1933-1942.

H625 Humphreys, Hubert. "Photographic Views of Red River Raft, 1873." *Louisiana History* 12 (Spring 1971), 101-08. On massive log jams which blocked navigation on Louisiana's Red River until removed by the U.S. Army Corps of Engineers in 1873.

H626 Hungerford, Norman D. "An Analysis of Forest Land Ownership Policies in the United States Pulp and Paper Industry." Ph.D. diss., State Univ. of New York, College of Environmental Science and Forestry, 1968. 261 pp. Incidental history.

H627 Hunken, Wilhelm A. "Landscape Changes and the Decision Making Process in the Whistler Mountains Area." Master's thesis, Simon Fraser Univ., 1969. British Columbia.

Hunn, Clark F. See Harding, T. Swann, #H99

H628 Hunsberger, Warren S. *Japan and the United States in World Trade.* New York: Harper and Row, for the Council on Foreign Relations, 1964. xvii + 492 pp. Includes some history of the trade in logs and lumber between the Pacific Northwest and Japan.

H629 Hunt, George M. *An Oral History Interview with George M. Hunt.* OHI by Donald G. Coleman. Madison: USFS, Forest Products Laboratory, 1961. 11 pp. Hunt was a former director of the USFS's Forest Products Laboratory, Madison, Wisconsin.

Hunt, George M. See Garratt, George A. #G31

H630 Hunt, Jack. "Land Tenure and Economic Development on the Warm Springs Indian Reservation." *Journal of the West* 9 (Jan. 1970), 93-109. Includes reference to the lumber industry on the Oregon reservation.

H631 Hunt, John Clark. "The Smith River Saga." *American Forests* 60 (July 1954), 20-23, 58. History and reminiscences of logging along Smith River in Oregon's Douglas County since the 1870s.

H632 Hunt, John Clark. "Burning Alaska." *American Forests* 64 (Aug. 1958), 12-15, 40-42. Includes some history of forest fires and fire protection and control in Alaska in the 20th century.

H633 Hunt, John Clark. "The Forest That Men Made." *American Forests* 71 (Nov. 1965), 18-21, 46-48; 71 (Dec. 1965), 32-35, 48-50. Charles E. Bessey, tree planting, and the Nebraska National Forest, Nebraska, since the 1890s.

Hunt, John Clark. See Heath, Virgil, #H261

H634 Hunt, Reed O. *Pulp, Paper and Pioneers: The Story of Crown Zellerbach Corporation.* New York: Newcomen Society in North America, 1961. 32 pp. Illus. A San Francisco-based forest products firm with operations in the Pacific Coast states and elsewhere.

H635 Hunt, William R. " 'I Chopped Wood': George M. Pilcher on the Yukon." *Pacific Northwest Quarterly* 63 (Apr. 1972), 63-68. Pilcher provided cordwood for steamboats on Alaska's Yukon River, 1898 to 1913.

Hunter, Beatrice J. See Hunter, Louis C., #H643

H636 Hunter, Dard. *Paper Making Through Eighteen Centuries.* New York: W.E. Rudge, 1930. 358 pp.

H637 Hunter, Dard. *Before Life Began, 1883-1923.* Cleveland: Rowfant Club, 1941. 115 pp. Autobiographical account by a paper manufacturer and leading authority on the history of papermaking.

H638 Hunter, Dard. *Papermaking: The History and Technique of an Ancient Craft.* New York: Alfred A. Knopf, 1943. 398 pp. Illus.

H639 Hunter, Dard. *Chronology of American Papermaking.* Holyoke, Massachusetts: B.F. Perkins & Son, 1948. 32 pp.

H640 Hunter, Dard. *My Life with Paper: An Autobiography.* New York: Alfred A. Knopf, 1958. 236 pp. Hunter was perhaps the leading authority on the history of papermaking and the paper industry.

H641 Hunter, Helen M. "The United States International Trade in Wood Pulp: A Case Study in International Trade." Ph.D. diss., Radcliffe College, 1952.

H642 Hunter, Helen M. "Innovation, Competition, and Locational Changes in the Pulp and Paper Industry, 1880-1950." *Land Economics* 31 (Nov. 1955), 314-27.

H643 Hunter, Louis C., and Hunter, Beatrice J. *Steamboats on the Western Rivers: An Economic and Technological History.* Cambridge: Harvard University Press, 1949, xiii + 684 pp. Illus., tables, maps, notes. Includes some account of traffic in lumber and the use of wood for fuel in 19th-century steamboats.

H644 Huntley, George W., Jr. *A Story of the Sinnamahone.* Williamsport, Pennsylvania: Williamsport Printing and Binding Company, 1936. 500 pp. Illus. White pine logging and the lumber industry on Sinnamahoning Creek, a tributary of the Susquehanna River in north central Pennsylvania, 1865-1885.

H645 Huntley, George W., Jr. *Sinnemahone, A Story of Great Trees and Powerful Men.* Boston: Christopher Publishing House, 1945. 411 pp. Lumbering in Pennsylvania along Sinnamahoning Creek.

H646 Hurd, Peter. "A Change in the Weather of Opinion." *Land* 9 (Spring 1950), 57-63. On the author's experience with conservation in southern New Mexico, 1933-1949.

H647 Hurst, Emmett B. "25 Years of Forestry Development: Consolidated Water Power and Paper Company." *Northeastern Logger* 5 (June 1957), 34-35, 46, 48.

H648 Hurst, James Willard. *Law and Economic Growth: The Legal History of the Lumber Industry in Wisconsin, 1836-1915.* Cambridge: Belknap Press of Harvard University Press, 1964. xx + 946 pp. Illus., maps, tables, notes, bib., index. A landmark study of the relation of law to economic growth and to social and institutional organization within the lumber industry.

H649 Hurst, Randle M. *The Smokejumpers.* Caldwell, Idaho: Caxton Printers, 1966. 284 pp. Illus., maps. Essentially a record of the author's parachute jumps to forest fires on the Gila National Forest, New Mexico, 1955.

H650 Hurst, Shirley, and Hussey, John A. *That the Past Shall Live: The History Program of the National Park Service.* Washington: U.S. National Park Service, 1959. 39 pp. Illus. The history program since 1872.

H651 Hurt, Bert. *A Sawmill History of the Sierra National Forest, 1852-1940.* San Francisco: USFS, California region, 1941. 51 pp. Illus. Processed.

H652 Hurt, Bert. "Sawmill History of the Sierra National Forest, California." *Timberman* 44 (Mar. 1943), 10-13, 30-32. Since 1852.

H653 Huser, Verne. "Yellowstone National Park: Use, Overuse & Misuse." *National Parks & Conservation Magazine* 46 (Mar. 1972), 8-17. Brief history of tourist use and other problems since 1872.

Hussey, John A. See Hurst, Shirley, #H650

H654 Hustich, Ilmari. "On the Forest Geography of the Labrador Peninsula: A Preliminary Synthesis." *Acta Geographica* 10 (No. 2, 1949), 1-63. Incidental history.

H655 Hustich, Ilmari. "Notes on the Forests of the East Coast of Hudson Bay and James Bay." *Acta Geographica* 11 (No. 1, 1950), 1-83. Incidental history.

H656 Huston, Harvey. *'93/'41: Thunder Lake Narrow Gauge.* Winnetka, Illinois: By the author, 1961. xii + 146 pp. Illus., maps, bib. On the Robbins Railroad Company, owned by the Robbins Lumber Company and Thunder Lake Lumber Company in northern Wisconsin, 1893-1941.

H657 Huston, Harvey. *The Roddis Line: The Roddis Lumber & Veneer Co. Railroad and the Dells & Northeastern Railway.* Winnetka, Illinois: By the author, 1972. 150 pp. Illus., notes. In northern Wisconsin since the 1890s.

H658 Hutchins, John G.B. "The Rise and Fall of the Building of Wooden Ships in America, 1607-1914." Ph.D. diss., Harvard Univ., 1937.

H659 Hutchins, John G.B. *The American Maritime Industries and Public Policy, 1789-1914: An Economic History.* Harvard Economic Studies, Volume 71. Cambridge: Harvard UniversityPress, 1941. xxi + 627 pp. Tables, bib. On the relationship of shipbuilding conditions to maritime policy, especially in connection with American abundance of timber and naval stores.

H660 Hutchinson, Bruce. *The Fraser.* Rivers of America Series. New York: Rinehart and Company, 1950. 368 pp. Illus., bib. Includes some account of British Columbia forests and forest industries.

H661 Hutchinson, Helen. "John Thomson, Canadian Pulp Pioneer." *Pulp and Paper Magazine of Canada* 61 (Mar. 1960), 81-83. Thomsom was the first Canadian manufacturer of chemical wood pulp, Windsor, Quebec, 1864.

H662 Hutchinson, I. "Some Aspects of Logging in the Coast Forest of British Columbia." *Empire Forestry Review* 37 (1958), 66-84; 37)1958), 165-87. Some historical references.

H663 Hutchinson, Thomas. *History of the Province of Massachusetts Bay from 1691-1750.* Third edition. 2 Volumes. Boston, 1795. Earlier and later editions appeared. Contains a royal governor's account of British forest and mast policy and of disputes arising from the Broad Arrow policy.

H664 Hutchinson, William H. *The California Investment: A History of the Diamond Match Company of California.* Chico, California: Diamond Match Company Lumber Division, 1957. 399 pp. Processed. Its lumber operations in Butte, Tehama, and Shasta counties since ca. 1900.

H665 Hutchinson, William H. *California Heritage: A History of Northern California Lumbering.* 1958. Reprint. Santa Cruz, California: Forest History Society, 1974. 32 pp. Illus., maps. History of the logging operations of Diamond International Company and its predecessors in the Red Bluff-Chico area since the 1860s.

H666 Hutchinson, William H. "The Caesarean Delivery of Paul Bunyan." *Western Folklore* 22 (Jan. 1963), 1-15. On the artistic renditions of Paul Bunyan by William B. Laughead, and their use in advertising by the Red River Lumber Company, Westwood, California, 1910s.

H667 Hutchinson, William H. *Oil, Land and Politics: The California Career of Thomas Robert Bard.* 2 Volumes. Norman: University of Oklahoma Press, 1965. Illus., maps, notes, apps., bib., index. Includes some account of his business interests in lumber.

H668 Hutchinson, William H."California's Economic Imperialism: An Historical Iceberg." In *Reflections of Western Historians,* ed. by John A. Carroll. Tucson: University of Arizona Press, 1969. Pp. 67-83. Includes mention of the lumber industry and trade.

H669 Hutchinson, William H. "The Sierra Flume & Lumber Company of California, 1875-1878." *Forest History* 17 (Oct. 1973), 14-20. The short-lived but large and complex operations of this company were located in Butte, Tehama, and Shasta counties of northern California.

H670 Hutchinson, William H. "Logging a Legend." *Westways* 66 (Nov. 1974), 23-25. On William B. Laughead's use of Paul Bunyan stories to advertise the Red River Lumber Company, Westwood, California, 1910s-1920s.

H671 Hutchison, O. Keith, and Winters, Robert K. *Kentucky's Forest Resources and Industries.* USFS, Forest Resource Report No. 7. Washington: GPO, 1953. iii + 56 pp. Illus., maps, tables, diags., notes, bib. Since 1899.

H672 Hutchison, O. Keith. *Indiana's Forest Resources and Industries.* USFS, Forest Resource Report No. 10. Washington: GPO, 1956. 44 pp. Incidental history.

H673 Hutchison, O. Keith, and Morgan, J.T. *Ohio's Forests and Wood-Using Industries.* USFS, Central States Forest Experiment Station, Forest Survey Release No. 19. Columbus, 1956. 40 pp. Incidental history.

H674 Hutchison, O. Keith. *Alaska's Forest Resource.* Resource Bulletin PNW-19. Portland: USFS, Pacific Northwest Forest and Range Experiment Station, 1967. 74 pp. Illus. Incidental history.

H675 Hutchison, S. Blair. "A Century of Lumbering in Northern Idaho." *Timberman* 39 (Aug. 1938), 20-21, 26; 39 (Sept. 1938), 14-15, 28; 39 (Oct. 1938), 34-39.

H676 Hutchison, S. Blair, and Winters, Robert K. *Northern Idaho Forest Resources and Industries.* USDA, Miscellaneous Publications, No. 508. Washington: GPO, 1942. 75 pp. Incidental history.

Hutchison, S. Blair. See Mathews, Donald N., #M288

Hutchison, S. Blair. See Dickerman, Murlyn Bennett, #D163

H677 Hutchison, S. Blair, and Kemp, Paul D. *Forest Resources of Montana.* USFS, Forest Resource Report No. 5. Washington: GPO, 1952. iii + 76 pp. Illus., maps, tables, notes. Since 1869.

H678 Hutchison, S. Blair. "Bringing Resource Conservation into the Main Stream of American Thought." *Natural Resources Journal* 9 (Oct. 1969), 518-36. Incidental historical references to the conservation movement.

H679 Huth, Hans. "Yosemite: The Story of An Idea." *Sierra Club Bulletin* 33 (Mar. 1948), 47-78. On the idea of national parks and its origin in early American attitudes toward nature, 1759-1908.

H680 Huth, Hans. "The American and Nature." *Journal of the Warburg and Courtauld Institutes* 13 (July 1950), 101-49.

H681 Huth, Hans. *Nature and the American: Three Centuries of Changing Attitudes.* Berkeley: University of California Press, 1957. Reprint. Lincoln: University of Nebraska Press, 1972. xvii + 250 pp. Illus., notes, bib., index. An important and pioneering study which examines the city park movement, romantic painters, summer vacations and tourism, the creation of national parks, and other topics related to American interest in forests, scenery, outdoor recreation, and conservation. The focus is on the 19th century.

H682 Hutslar, Donald A. "The Log Architecture of Ohio." *Ohio History* 80 (Summer-Autumn 1971), 172-271.

H683 Hutton, Gordon A. "Timber Mortality—A Loss to Montana's Economy." *Proceedings of the Montana Academy of Science* 13 (1953), 79-82. Effects of fire, insects, and disease on timber resources since 1910.

H684 Huyck, Dorothy Boyle. "Washington: City of Trees." *American Forests* 80 (Mar. 1974), 17-31. Many historical references to forested areas and individual trees of Washington, D.C.

H685 Huyck, Dorothy Boyle. "Voyageurs National Park—Birth Pangs." *American Forests* 81 (Mar. 1975), 22-25, 62-63. Includes some history of the movement to create Voyageurs National Park, Minnesota, established in 1971.

H686 Huyck, F.C., and Sons. *Paper: Pacemaker of Progress.* Albany, New York: F.C. Huyck and Sons, Kenwood Mills, 1946. 47 pp. Illus. A brief history of papermaking and the pulp and paper industry.

Hyde, Arthur M. See Wilbur, Ray Lyman, #W283

H687 Hyde, Phillip, and Leydet, Francois. *The Last Redwoods.* San Francisco: Sierra Club, 1963. 127 pp. Illus. Includes several chapters on the history of exploitation and preservation of California redwoods.

H688 Hyler, John E. "Log Handling: Historical and Present Skidding Practices." *Southern Lumberman* 193 (Dec. 15, 1956), 298-322; 194 (Jan. 1, 1957), 60, 64-67, 70. Since ca. 1880.

H689 Hyman, Harold M. *Soldiers and Spruce: Origins of the Loyal Legion of Loggers and Lumbermen.* Industrial Relations Monographs of the Institute of Industrial Relations, No. 10. Los Angeles: Institute of Industrial Relations, University of California, Los Angeles, 1963. viii + 341 pp. Notes. On the labor union sponsored by the U.S. Army during World War I, its purpose being to counteract the radical organizing of the Industrial Workers of the World and to accelerate the production in the Pacific Northwest of spruce lumber for aircraft.

Hyman, Harold M. See Parker, Carleton H., #P36

H690 Hynding, Alan A. "The Public Life of Eugene Semple: A Study of the Promoter-Politician on the Pacific Northwest Frontier." Ph.D. diss., Univ. of Washington, 1966. 342 pp.

H691 Hynding, Alan A. *The Public Life of Eugene Semple: Promoter and Politician of the Pacific Northwest.* Seattle: University of Washington Press, 1973. xiv + 195 pp. Illus., notes, bib., index. Semple, a territorial governor of Washington, was owner of the Lucia Mill Company in Vancouver in the 1880s.

H692 Hynning, Clifford J. *State Conservation of Resources.* Prepared for the National Resources Committee. Washington: GPO, 1939. x + 116 pp. Maps, tables, diags. Incidental history.

H693 Hyster Company. *One American Business: Hyster Company.* Portland, 1944. On a firm that produces lift trucks and other equipment widely used in the lumber industry.

I1 Ibberson, Joseph E.; Mickalitis, Albert B.; and Kurtz, Samuel. "The Cedar of Lebanon." *Pennsylvania Forests and Waters* 3 (Sept.-Oct. 1951), 96-104. Includes a partial census of exotic trees in Pennsylvania since 1807.

I2 Ickes, Harold L. "Not Guilty! Richard A. Ballinger—An American Dreyfus." *Saturday Evening Post* 212 (May 25, 1940), 9-11, 123-25, 128. Exonerates Secretary of the Interior Ballinger on charges made against him during the Taft administration.

I3 Ickes, Harold L. *Not Guilty: An Official Inquiry into the Charges Made by Glavis and Pinchot against Richard A. Ballinger, Secretary of the Interior, 1909-1911.* Washington: GPO, 1940. 58 pp. App. A defense of Richard A. Ballinger's role in the Ballinger-Pinchot controversy.

I4 Ickes, Harold L. *The Autobiography of a Curmudgeon.* New York: Reynal and Hitchcock, 1943. Reprint. Chicago: Quadrangle Books, 1969. xxv + 350 pp. Table, notes, index. Ickes served Presidents Franklin D. Roosevelt and Harry S Truman as Secretary of the Interior, 1933-1946, strongly favored conservation and wished to effect an executive department reorganization which would have transferred the USFS to the USDI.

I5 Ickes, Harold L. *The Secret Diary of Harold L. Ickes.* 3 Volumes. New York: Simon and Schuster, 1953, 1954. Reprint. New York: DaCapo Press, 1974. An intimate history of the New Deal administration of Franklin D. Roosevelt by Secretary of the Interior Ickes. Period covered is 1933 to 1941.

I6 Illick, John Rowland. "The Primary Wood-Using Industries of Northern New England: A Locational Study." Ph.D. diss., Harvard Univ., 1954.

I7 Illick, Joseph S. *Pennsylvania Trees.* Pennsylvania Department of Forestry, Bulletin No. 11. Harrisburg, 1914. 232 pp. Illus., index. Includes a brief historical sketch of forestry in Pennsylvania.

I8 Illick, Joseph S. "Forest Experiments in Pennsylvania." *Journal of Forestry* 17 (Mar. 1919), 297-311. Report on various experiments since 1897.

I9 Illick, Joseph S. "A Decade of Private Forest Planting in Pennsylvania." *American Forestry* 25 (Dec. 1919), 1538-41. On forest planting by private woodlands owners since 1909, with brief mention of 18th-century planting.

I10 Illick, Joseph S. "Twenty Years of Forest Tree Planting." *Canadian Forestry Journal* 16 (Aug.-Sept. 1920), 397-404.

I11 Illick, Joseph S. "Fifty Years Ago." *Forest Leaves* 21 (No. 1, 1927), 9-10.

I12 Illick, Joseph S. "A Few Trends in Pennsylvania Forestry." *Journal of Forestry* 25 (Mar. 1927), 338-48. A historical review and appraisal of accomplishments since the 1870s.

I13 Illick, Joseph S. "Planting Trees by Millions." *American Forests and Forest Life* 33 (May 1927), 275-78, 292. Includes some history of forest planting in Pennsylvania since 1899.

I14 Illick, Joseph S. "Joseph Trimble Rothrock, Father of Pennsylvania Forestry." *Proceedings and Addresses of the Pennsylvania German Society* 34 (1929), 83-94. Rothrock (1839-1922) was the first commissioner of forestry in Pennsylvania.

I15 Illick, Joseph S. *The State Forests of Pennsylvania.* Pennsylvania Department of Forests and Waters. Bulletin No. 37. Harrisburg, 1930. 83 pp. Incidental history.

I16 Illick, Joseph S. *Forest Research in Pennsylvania.* Harrisburg: Pennsylvania Department of Forests and Waters, 1930. 17 pp. Incidental history.

I17 Illick, Joseph S. "Some Trends in State Forest Administration." *Journal of Forestry* 33 (Mar. 1935), 310-19. In selected states since 1885.

I18 Illick, Joseph S. "Significant Forestry Trends in New York State." *Journal of Forestry* 35 (May 1937), 452-59. History of the administrative organization of forestry and resource conservation since 1885, including some comparisons with other states.

I19 Illick, Joseph S. "Administrative Setups for State Forestry." *Journal of Forestry* 35 (June 1937), 539-44. Incidental historical references to state forestry administration since the 1890s.

I20 Illick, Joseph S. "Forestry in Our Constitutions." *Journal of Forestry* 36 (Mar. 1938), 288-99. On provisions for forestry embodied in state constitutions, beginning with Colorado in 1876 and with special attention to New York.

I21 Illick, Joseph S. *After Forty Years in Forestry.* Syracuse: New York State College of Forestry, 1947. 23 pp. Illick (1884-1967) was state forester of Pennsylvania (1927-1931) and subsequently an educator at the New York State College of Forestry.

I22 Illick, Joseph S. "They Received Forestry Degrees." *Journal of Forestry* 51 (Nov. 1953), 809-14. Review of degrees in forestry granted in the United States or to Americans since 1897.

I23 Illick, Joseph S.; and Hopkins, Arthur S. *Chronological Recording of Important Legislative and Administrative Developments Relating to the State Forest Preserve... 1665-1952.* Albany, 1953. Also issued as Appendix A of the *Report of the [New York] Joint Legislative Committee on Natural Resources* (1953).

I24 Inches, H.C. "Wooden Ship Building." *Inland Seas* 7 (Spring 1951), 3-12. On the Great Lakes, 1814-1887.

I25 Inches, H.C. *The Great Lakes Wooden Shipbuilding Era.* Vermilion, Ohio: Great Lakes Historical Society, 1962.

I26 Industrial Workers of the World. *The Lumber Industry and Its Workers.* Chicago: I.W.W. Publishing Company, 1922.

I27 Industrial Workers of the World. *Twenty Five Years of Industrial Unionism.* Chicago, 1930.

Ineson, Frank. See Rettie, James C., # R127

Ingersoll, Fern. See Ringland, Arthur C., # R204

I28 Ingersoll, William T. "Lands of Change: Four Parks in Alaska." *Journal of the West* 7 (Apr. 1968), 178-92. A brief history of Glacier Bay, Katmai, and Sitka national monuments, and Mount McKinley National Park.

I29 Inglis, George. "An Honest Pile." *Beaver* 297 (Winter 1966), 34-35. On Edward Martin (d. 1928) who provided wood for Hudson's Bay Company steamboats on the Mackenzie River of the Northwest Territories.

I30 Ingram, Orrin H. *Autobiography.* Eau Claire, Wisconsin: By the author, 1912. 83 pp. Of a Wisconsin lumberman with experience in New York and Ontario.

I31 Ingram, Orrin H. *Letters of a Pioneer.* Eau Claire, Wisconsin: By the author, 1916.

I32 Innes, James. "The Cooperage Industry of Canada." *Canada Lumberman* 34 (Aug. 15, 1914), 106-07. Incidental history.

I33 Innes, James. "The Cooperage Industry in Ontario." *Canada Lumberman* 36 (May 15, 1916), 109-10.

I34 Innis, H.A. *Problems of Staple Production in Canada.* Toronto: Ryerson Press, 1933. xi + 124 pp. Illus. Includes reference to the timber trade in Canadian history.

I35 Innis, H.A. "The Economics of Conservation." *Geographical Review* 28 (Jan. 1938), 137-39. Canada.

I36 Insinger, F.N. "The History of Wood Pipe." *West Coast Lumberman* 44 (May 1, 1923), 195, 209, 220.

I37 Institute for Government Research. *The U.S. Geological Survey: Its History, Activities and Organization.* Service Monographs of the United States Government, No. 1. New York, 1919. Reprint. New York: AMS Press, 1973. 174 pp.

I38 Institute for Government Research. *The U.S. Reclamation Service: Its History, Activities and Organization.* Service Monographs of the United States Government, No. 2. New York, 1919. Reprint. New York: AMS Press, 1973. 190 pp.

I39 Inverarity, Robert B. "The Adirondack Museum." *New York History* 39 (July 1958), 261-67. Located at Elizabethtown, New York, the museum is largely concerned with the forest history of the Adirondack Mountain region.

I40 Iowa. State Board of Conservation. *Report on the Iowa Twenty-Five Year Conservation Plan.* Des Moines: Wallace-Homestead Company, 1933. 176 pp. Bib. Some references to forest conservation.

I41 Iowa Park and Forestry Association. *Major John F. Lacey: Memorial Volume.* Cedar Rapids: Iowa Park and Forestry Association, 1915. 454 pp. Lacey (1841-1913), as chairman of the House Committee on Public Lands, was author of much legislation affecting national parks, wildlife, and forests during his tenure as a congressman, 1889-1907.

I42 Iowa State College, Department of Forestry. *Fiftieth Anniversary, 1904-1954.* Ames, 1954. 24 pp.

I43 Irland, Lloyd C. "Louisiana's Forest Industries: 1946-1971." *Forests & People* 23 (First Quarter, 1973), 20-22, 37.

I44 Irland, Lloyd C. *Is Timber Scarce? The Economics of a Renewable Resource.* Bulletin No. 83. New Haven: Yale University, School of Forestry and Environmental Studies, 1974. xi + 97 pp. Illus. Based on statistics since the 1860s.

I45 Ironside, R.G. "The Territorial Status of the National Park in Canada." *Rocky Mountain Social Science Journal* 7 (Apr. 1970), 69-75. Includes some history and political geography of Canadian national parks, with emphasis on Banff National Park, Alberta.

I46 Isaac, Leo A. "The Seed-Flight Experiment: Policy Heeds Research." OHI by Amelia R. Fry. *Forest History* 16 (Oct. 1972), 54-60. On silvicultural research conducted by the Pacific Northwest Forest and Range Experiment Station, Portland, Oregon, and its influence on USFS timber sale policy, 1920s-1930s.

I47 Isaac, Sallie Carr (Drake). *The Story of Drakesboro and Its Founder.* N.p., 1952. v + 77 pp. Illus. Biography of the author's father, John Rice Drake (1847-1917), founder of a lumbering and mining community in Muhlenberg County, Kentucky.

I48 Isbell, Victor K. *Historical Development of the Spanish Fork Ranger District.* Provo, Utah: USFS, Uinta National Forest, 1972. x + 165 pp. Illus., maps, bib. Processed. Uinta National Forest, Utah.

I49 Ise, John. "The History of the Forestry Policy of the United States." Ph.D. diss., Harvard Univ., 1914.

I50 Ise, John. "A Chapter in the Early History of the United States Forest Policy." *Ames Forester* 3 (1915), 33-66.

I51 Ise, John. *The United States Forest Policy.* New Haven: Yale University Press, 1920. Reprint. New York: Arno Press, 1972. 395 pp. Maps, notes, bib., index. A pioneering scholarly study of the beginnings of American forestry and early national forest policy, especially from the 1870s to 1910s.

I52 Ise, John. *The United States Oil Policy.* New Haven: Yale University Press, 1926. xi + 547 pp. Illus., tables, charts, notes. Includes some history of federal land policy as applied to oil lands, with general relevance to the conservation movement.

I53 Ise, John. *Our National Park Policy: A Critical History.* Baltimore: Johns Hopkins Press, for Resources for the Future, 1961. xiii + 701 pp. Maps, notes, index. Contains history of each park and of each National Park Service administration since 1916. Includes analyses of special park problems: wildlife, concessions, finances, wilderness areas, and national parks in other countries.

I54 Isherwood, H.R., ed. *A Report on the Profitable Management of a Retail Lumber Business. . .Based on an Investigation of 491 Retail Lumber Yards in 38 States and Canada.* 5 Volumes. New York: A.W. Shaw Company, 1918. Illus., diags.

Iulo, William. See Guthrie, John A., #G386

I55 Ives, Edward D. "Larry Gorman and the Cante Fable." *New England Quarterly* 32 (June 1959), 226-37. On Lawrence Gorman (1846-1917), Maine lumberjack, humorist, and songwriter.

Ives, Edward D. See Gorman, Lawrence, #G194

I56 Ives, Edward D. "The Life and Work of Larry Gorman: A Preliminary Report." *Western Folklore* 19 (Jan. 1960), 17-23. Gorman, a native of Prince Edward Island, moved to Maine ca. 1885, worked as a lumberjack and yard hand in a paper mill, and composed satirical songs.

I57 Ives, Edward D. *Larry Gorman: The Man Who Made the Songs.* Bloomington: Indiana University Press, 1964. 225 pp. Maps, music, app., index. Also contains some general history of the folksongs and customs of lumberjacks in New England and the Maritime provinces.

Ives, Edward D. See Pride, Fleetwood, #P325

I58 Ives, Gideon Sprague. "William Gates Le Duc." *Minnesota History Bulletin* 3 (1919), 57-65. Le Duc was Rutherford B. Hayes's Commissioner of Agriculture, 1877-1881.

I59 Ivy, Thomas P. *Forestry Problems of the U.S.* Hendersonville, North Carolina: N.p., 1906. 47 pp. Treats three subjects: the USFS and the Civil Service; the effect of the Forest Reserve Act of 1891 upon eastern forests; and the Mississippi River and forestry.

Jackman, E.R. See Simpson, Charles D., #S294

J1 Jackson, Annette. *My Life in the Maine Woods: A Game Warden's Wife in the Allagash Country.* New York: W.W. Norton, 1954. vii + 236 pp. Reminiscences of life in northern Maine, 1932-1938.

J2 Jackson, Clarence S. *Picture Maker of the Old West: William H. Jackson.* 1947. Reprint. New York: Charles Scribner's Sons, 1971. x + 308 pp. Illus., index. Jackson (1843-1942) was a photographer with the Hayden expedition. His work assisted in the establishment of Yellowstone National Park and the promotion of other scenic areas of the Rocky Mountain region.

J3 Jackson, Dan D. "The International Woodworkers of America." Master's thesis, Univ. of California, Berkeley, 1953.

J4 Jackson, Harry F. "Lumbering on the Coal River." Master's thesis, West Virginia Univ., 1937.

J5 Jackson, Harry F. "Boom and Driving Days on Coal River and in the Adirondacks." *West Virginia History* 21 (Oct. 1959), 13-21. The Coal River of West Virginia, 1873-1906, and the Adirondack Mountains of New York, 1813-1901.

J6 Jackson, Harry F. "Branding and Driving in the Adirondacks." *North Country Life* 14 (Spring 1960), 18-22. Late 19th-century lumbering in New York.

J7 Jackson, J.C. "First Fifty Years, the Story of Reforestation in Ontario." *Forest and Outdoors* 52 (Apr. 1956), 15, 28-29.

J8 Jackson, Orlo M. "History of Government Timber Sale Number One." *Journal of Forestry* 49 (Apr. 1951), 281-83. On the Black Hills National Forest, South Dakota, 1899-1908.

J9 Jackson, William H., and Dassow, Ethel. *Handloggers.* Anchorage: Alaska Northwest Publishing Company, 1974. 251 pp. Illus. Reminiscences of William H. Jackson, mostly of handlogging in Alaska, since 1907.

J10 Jackson, William Henry. *Time Exposure.* New York: G.P. Putnam's Sons, 1940. Jackson (1843-1942) was a photographer with the Hayden expedition to the Yellowstone region and other Rocky Mountain areas, 1870-1878.

J11 Jackson, William Turrentine. "The Early Exploration and Founding of Yellowstone National Park." Ph.D. diss., Univ. of Texas, 1940. 114 pp.

J12 Jackson, William Turrentine. "The Washburn-Doane Expedition into the Upper Yellowstone, 1870." *Pacific Historical Review* 10 (June 1941), 189-208. Generated interest in a Yellowstone National Park.

J13 Jackson, William Turrentine. "The Cook-Folsom Exploration of the Upper Yellowstone, 1869." *Pacific Northwest Quarterly* 32 (July 1941), 307-22. The first purposeful exploration of what later became Yellowstone National Park, Wyoming. David E. Folsom and Charles W. Cook suggested that the natural and scientific wonders of the area be preserved.

J14 Jackson, William Turrentine. "Governmental Exploration of the Upper Yellowstone, 1871." *Pacific Historical Review* 11 (June 1942), 187-99. On the Barlow and Hayden expeditions to the region just prior to its establishment as Yellowstone National Park.

J15 Jackson, William Turrentine. "The Creation of Yellowstone National Park." *Mississippi Valley Historical Review* 29 (Sept. 1942), 187-206. Examines the citizen movement and political struggle in Montana as a factor in the park's creation, 1860s to 1872.

J16 Jackson, William Turrentine. "The Washburn-Doane Expedition of 1870." *Montana, Magazine of Western History* 7 (July 1957), 36-51.

J17 Jackson, William Turrentine. "The Creation of Yellowstone National Park." *Montana, Magazine of Western History* 7 (July 1957), 52-65.

J18 Jackson, William Turrentine, and Pisani, Donald J. *Lake Tahoe Water: A Chronicle of Conflict Affecting the Environment.* Davis: University of California, Davis, Institute of Governmental Affairs, 1972.

J19 Jackson, William Turrentine, and Pisani, Donald J. *From Resort Area to Urban Recreation Center: Themes in the Development of Lake Tahoe, 1946-1956.* Davis: University of California, Davis, Institute of Governmental Affairs, 1973. 85 pp.

J20 Jacobs, Mark L. "Harold Ickes: Progressive Administrator." Ph.D. diss., Univ. of Maine, 1973. 512 pp. On Ickes as Secretary of the Interior in the Franklin D. Roosevelt administration, 1933-1941.

J21 Jacbos, Wilbur R. "Frontiersmen, Fur Traders, and Other Varmints, An Ecological Appraisal of the Frontier in American History." *AHA Newsletter* 8 (Nov. 1970), 5-11. An appeal to study the conservation movement and to treat more critically the despoliation of the wilderness by frontiersmen and other groups whose extraction of natural resources has been glamorized by historians.

J22 Jacobsen, Edna L. "Franklin B. Hough, a Pioneer in Scientific Forestry in America." *New York History* 15 (July 1934), 311-325. Hough (1822-1885) was head of the Division of Forestry from 1876 to 1883, under Presidents Grant, Hayes, Garfield, and Arthur. The article also concerns his work in behalf of forestry in New York.

J23 Jain, Hem Chand. "Industrial Relations in the Pulp and Paper Industry in the Atlantic Region." Ph.D. diss., Univ. of Illinois, 1968. 284 pp.

J24 James, Arthur E. "The Paper Mills of Chester County, Pennsylvania, 1779-1967." *Paper Maker* 39 (Sept. 1970), 2-16.

J25 James, David. *Big Skookum: A Story of 100 Logging Years in Mason County, 1853-1953.* Seattle: Simpson Logging Company, 1953. 16 pp. Illus. Primarily the history of Simpson Logging Company, Shelton, Washington.

J26 James, George Wharton. *Reclaiming the Arid West: The Story of the United States Reclamation Service.* New York: Dodd, Mead and Company, 1917. 411 pp.

J27 James, George Wharton. *The Lake of the Sky, Lake Tahoe in High Sierras of California and Nevada: Its History . . . .* 1915. New edition. Chicago: C.T. Powner Company, 1956. xlvi + 414 pp. Illus., maps, tables, notes. Includes incidental historical references to logging and other uses of the surrounding national forest lands.

J28 James, Harlean. *Romance of the National Parks.* New York: Macmillan, 1939. Reprint. New York: Arno Press, 1972. xiv + 240 pp. Illus., map. Includes some history of the movement to create national parks and the subsequent development of the system.

J29 James, Harlean, ed. *Twenty-Fifth Anniversary Yearbook.* Washington: National Conference on State Parks, 1946. Includes some history of the state parks movement.

J30 James, Harry C. "The San Jacinto Winter Park Summer Resort Scheme." *Living Wilderness* 14 (Winter 1949-1950), 4-16. On a proposed resort complex in the primitive area on Mt. San Jacinto, California, opposed by conservationists since 1939.

J31 James, James Alton. "The Beginning of a State Park System for Illinois." *Transactions of the Illinois State Historical Society* (1936), 53-62. The system originated with the preservation of historic sites.

J32 James, Lee M. "Lumber Trade Policies." *Journal of Forestry* 43 (Feb. 1945), 81-87. Includes some account of the tariff on lumber since 1854.

J33 James, Lee M. "Lumber Consumption in the U.S." Ph.D. diss., Univ. of Michigan, 1945. 288 pp. Includes some history of the subject.

J34 James, Lee M. "Restrictive Agreements and Practices in the Lumber Industry, 1880-1939." *Southern Economic Journal* 13 (Oct. 1946), 115-25.

J35 James, Lee M. "The Trend of Lumber Prices." *Journal of Forestry* 45 (Sept. 1947), 646-49. General trends since 1860.

James, Lee M. See Korstian, Clarence F., #K240

J36 James, Lee M. "Timber Supplies for Industry in Mississippi." *Southern Economic Journal* 18 (July 1951), 61-71. From 1932 to 1948.

J37 James, Lee M. *Mississippi's Forest Resources and Industries.* USFS, Forest Resources Paper No. 4. Washington: GPO, 1951. iii + 92 pp. Illus., maps, tables, diags., notes. From 1880 to 1948.

J38 James, Lee M., and Yoho, James G. "Forest Taxation in the Northern Half of the Lower Peninsula of Michigan." *Land Economics* 33 (May 1957), 139-48. Since 1860.

J39 James, Lee M. "Property Taxes and Alternatives for Michigan." *Journal of Forestry* 58 (Feb. 1960), 86-92. On the taxation of forest lands since 1900.

J40 Jameson, John Robert, Sr. "Big Bend National Park of Texas: A Brief History of the Formative Years, 1930-1952." Ph.D. diss., Univ. of Toledo, 1974. 214 pp.

J41 Jameson, J.S. *Planting of Conifers in the Spruce Woods Forest Reserve, Manitoba, 1904-1929.* Forest Research Division, Technical Note No. 28. Ottawa: Forestry Branch, 1956. 19 pp.

J42 Jameson, S. "West Coast Strikes." *Pulp and Paper Magazine of Canada* 59 (July 1958), 105-08, 217-21. Concerns labor-management relations in the forest products industries of British Columbia.

J43 Jamieson, Allen. "Big Basin and Castle Rock." *Sierra Club Bulletin* (Sept. 1968). Includes some history of California's oldest and newest state parks, both located in the Santa Cruz Mountains: Big Basin Redwoods State Park and Castle Rock State Park.

J44 Jamison, James Knox. *This Ontonagon Country: The Story of an American Frontier.* Third Edition. Ontonagon, Michigan: Ontonagon Herald Company, 1948. viii + 240 pp. Illus., maps, bib. A major logging region on Michigan's Upper Peninsula since 1831.

J45 Jamison, James Knox. "Mountain Logging." *Michigan Conservation* 19 (Mar.-Apr. 1950), 31-32. In the Porcupine Mountains of Michigan's Upper Peninsula since 1845.

J46 Jamison, James Knox. "The Kaug." *Michigan Conservation* 22 (July-Aug. 1953), 27-29. On the Porcupine Mountains of Michigan's Upper Peninsula, and the mining, logging, and tourist industries, since 1844.

J47 Jamison, James Knox. "The Survey of the Public Lands in Michigan." *Michigan History* 42 (June 1958), 197-214. From 1815 to 1851.

J48 Janes, G.M. "Cooperative Production Among Shingle Weavers." *Quarterly Journal of Economics* 38 (May 1924), 530-36. Incidental history.

J49 Jarchow, Merrill E. *The Earth Brought Forth: A History of Minnesota Agriculture to 1885.* St. Paul: Minnesota Historical Society, 1949. Reprint. New York: Johnson Reprint Corporation, 1970. 314 pp. Illus. Includes reference to its relation to forested lands.

J50 Jardine, J.T. "Efficient Regulation of Grazing in Relation to Timber Production." *Journal of Forestry* 18 (Apr. 1920), 367-82. Includes some history of investigations conducted in the West since 1897.

J51 Jarrett, Henry, ed. *The Nation Looks at Its Resources: Report of the Mid-Century Conference on Resources for the Future.* Washington: Resources for the Future, 1954. 418 pp. App. Includes some incidental historical references to problems in the conservation of natural resources.

J52 Jarrett, Henry, ed. *Perspectives on Conservation: Essays on America's Natural Resources.* Baltimore: Johns Hopkins University Press, for Resources for the Future, 1958. 272 pp. Index. Essays on the changes in the conservation movement since the Governor's Conference on Conservation of 1908.

J53 Jarrett, Henry, ed. *Comparisons in Resource Management: Six Notable Programs in Other Countries and Their Possible U.S. Applications.* Baltimore: Johns Hopkins Press, for Resources for the Future, 1961. Reprint. Lincoln: University of Nebraska Press, 1965. 292 pp. A collection of essays on natural resource management, including many historical references.

J54 Jarrett, Henry, ed. *Environmental Quality in a Growing Economy.* Baltimore: Johns Hopkins University Press, for Resources for the Future, 1966. 190 pp. Twelve papers examine the dilemma of preserving environmental quality in a growing economy. There are incidental references to the history of conservation.

J55 Jarvis, J.M. *Forty-Five Years' Growth on the Goulais River Watershed.* Forestry Research Division, Technical Note No. 84. Ottawa: Forestry Branch, 1960. 31 pp. Forest regeneration on the cutover and burned-over Algoma District of Ontario, since the 1910s.

J56 Jeffers, Dwight S. "Free Land and Forest Policy in the United States." Ph.D. diss., Yale Univ., 1935.

J57 Jeffords, A.I., Jr. "Trends in Pine Pulpwood Marketing in the South." *Journal of Forestry* 54 (July 1956), 436-66. Since 1935.

J58 Jenkins, Charles F. "The Historical Background of Franklin's Tree." *Pennsylvania Magazine of History and Biography* 57 (July 1933), 193-208. On a variety of tree called "Franklinia Altamaha," introduced to Pennsylvania when seedlings were carried from Georgia in 1777.

J59 Jenkins, John Henry. "Developments in Kiln Drying Practice on the Pacific Coast." *Forestry Chronicle* 8 (Dec. 1932), 185-88. British Columbia.

J60 Jenkins, John Henry. *Kiln Drying B.C. Lumber.* Forest Service, Bulletin No. 86. Ottawa, 1934. 78 pp. Incidental history.

J61 Jenkins, John Henry. "Recent Developments in the Utilization of Sawmill Waste in the Southern Coast Region of British Columbia." *Forestry Chronicle* 12 (Feb. 1936), 72-78.

J62 Jenkins, John Henry. "Wood-Waste Utilization in British Columbia." *Forestry Chronicle* 15 (Dec. 1939), 192-99. Incidental history.

J63 Jenkins, John Henry. "Role of a Forest Products Research Laboratory in the Development of the National Economy." *Forestry Chronicle* 39 (Sept. 1963), 322-27.

J64 Jenkins, Sarah Agnes. "The Timber Interests of the Oregon and California Railroad." Master's thesis, Columbia Univ., 1937. See also under Sarah Jenkins Salo.

J65 Jenkins, Sidney C. "Permanent Production on Potlatch Forests." *American Forests* 44 (Aug. 1938), 360-62, 381. Includes some history of logging operations and industrial forestry practices on the Idaho lands of Potlatch Forests, Inc., and its predecessor, the Clearwater Timber Company, since 1920.

J66 Jenkins, Sidney C. "One Hundred Years of Lumbering." *Timberman* 41 (Mar. 1940), 14ff. In Idaho.

J67 Jenkins, Starr. "We Fly the Fire Patrol." *Pacific Northwest Quarterly* 46 (Jan. 1955), 12-18. Reminiscences of a forest fire spotter and smokejumper over the Coeur d'Alene National Forest of northern Idaho, 1952, with incidental history of fire spotting.

J68 Jenkins, Stella. "British Columbia Forest Service." *British Columbia Lumberman* 57-58 (May 1973-Oct. 1974). A series of fourteen articles on the history of the agency in the 1910s and 1920s. Topics include: early administrators (H.R. MacMillan and M. Allerdale Grainger), use of automobiles, forest fire suppression, forest research, public relations and education, marine vessels, and various incidents involving rangers in the field.

J69 Jenner, R.W. "Cliffs Dow Chemical Company: The World's Largest Charcoal Producer." *Northeastern Logger* 6 (Feb. 1958), 28-29, 48. Incidental history of the Michigan firm.

J70 Jensen, C.R. *Outdoor Recreation in America: Trends, Problems, and Opportunities.* Minneapolis: Burgess Publishing Company, 1973. ix + 284 pp. Illus.

J71 Jensen, Everett. "Principles of Land Management Planning for California National Forests." *Journal of Forestry* 48 (Sept. 1950), 435-38. Incidental history.

Jensen, J. Granville. See Highsmith, Richard Morgan, Jr., #H362

J72 Jensen, Merrill. "The Creation of the National Domain, 1781-1784." *Mississippi Valley Historical Review* 26 (Dec. 1939), 232-42.

J73 Jensen, Vernon H. "Labor Relations in the Northwest Lumber Industry." Ph.D. diss., Univ. of California, Berkeley, 1939. Includes a historical account of the Loyal Legion of Loggers and Lumbermen, a government-organized union which counteracted the radical organizers of lumberjacks and millworkers in the Pacific Northwest during the World War I era.

J74 Jensen, Vernon H. *Lumber and Labor.* Labor in Twentieth Century America Series. New York: Farrar &Rinehart, 1945. Reprint. New York: Arno Press, 1971. x + 314 pp. Maps, notes, bib., index. A standard history of labor in the lumber industry with chapters on the Northeast, Great Lakes region, South, and particular emphasis on the West and Pacific Northwest.

J75 Jensen, Vernon H. "Industrial Relations in the Lumber Industry." In *Labor in Postwar America,* ed. by Colston E. Warne. New York: Remsen Press, 1949. Recent history.

J76 Jepsen, Stanley M. *Trees and Forests.* South Brunswick, New York: A. S Barnes, 1969. 155 pp. Illus. Incidental history of forests, forestry, and forest industries.

J77 Jeremiah, D.B. "Financial Recovery of the Kraft Industry." *Annals of the American Academy of Political and Social Science* 193 (Sept. 1937), 14-21. Incidental history.

J78 Jesness, Oscar B.; Nowell, Reynolds I.; and associates. *A Program for Land Use in Northern Minnesota: A Type Study in Land Utilization.* Minneapolis: University of Minnesota Press, 1935. xvi + 338 pp. Maps, tables, notes, index. Includes some incidental history of the lumber industry and other land utilization in the forested areas of northern Minnesota.

J79 Jettmar, Karen, and Summers, Clarence. "Glacier Bay: Wilderness or Mining Boom?" *National Parks & Conservation Magazine* 49 (May 1975), 6-11. Brief history of the Alaska region established as a national monument in 1925.

J80 Joffe, Joseph. "John W. Meldrum: The Grand Old Man of Yellowstone National Park." *Annals of Wyoming* 13 (Jan. 1941), 5-47; 13 (Apr. 1941), 105-40. Meldrum was U.S. Commissioner for the park, 1894-1935.

J81 Johanneck, Donald P. *A History of Lumbering in the San Bernardino Mountains.* Redlands, California: San Bernardino County Museum, 1975. 130 pp. Illus.

J82 Johannessen, Carl L.; Davenport, William A.; and Millet, Artemus. "The Vegetation of the Willamette Valley." *Annals of the Association of American Geographers* 61 (No. 2, 1971), 286-302. Includes some history of forest expansion and lumbering in the Oregon valley.

J83 Johansen, Dorothy O., and Gates, Charles M. *Empire of the Columbia: A History of the Pacific Northwest.* 1957. Second edition, by Dorothy O. Johansen. New York: Harper & Row, 1967. xiii + 654 pp. Illus., maps, tables, graphs, notes, supplementary readings. The standard history of the Pacific Northwest; chapters 25 and 32 treat the lumber industry and its role in the regional economy.

J84 Johns, John O. "Tragedy of 'the Last Raft'." *Commonwealth* (Pennsylvania) 4 (Feb. 1950), 2-4. On the memorial voyage of a raft from Williamsport to Muncy, on the Susquehanna River of Pennsylvania. The raft, manned by retired lumberjacks and rivermen, overturned with a loss of seven lives.

J85 Johnsen, Bjarne. "Pulp and Paper Industry." *Industrial and Engineering Chemistry* 27 (May 1935), 514-18. A review of chemical and technical developments since the colonial period.

J86 Johnson, Amandus. "The Swedes Brought the Log Cabin to America." *American Swedish Monthly* 48 (Nov. 1954), 14-15, 29-30. On evidence that Swedish settlers constructed log structures along the Delaware River as early as 1638, while English and Dutch settlers did not build log structures before 1650.

J87 Johnson, Amandus. "Sweden Gave America the Rail Fence." *American Swedish Monthly* 49 (June 1955), 6-7, 29. On the first rail fences in America, built by Swedes in the Delaware Valley in 1640.

J88 Johnson, Bolling A. " 'The War of the Substitutes'—Past and Present." *Lumber World Review* (Oct. 25, 1914), 17-27. Owner-editor of *Lumber World Review* reviews his campaign against wood substitute building materials.

J89 Johnson, Bolling A. "A Short History of the James D. Lacey Co." *Lumber World Review* (Aug. 25, 1915), 17-20. Lacey was a broker in timberlands throughout the country.

J90 Johnson, Bolling A. "Story of the Big Salkehatchie Cypress Co. and the Black River Cypress Co. of South Carolina." *Lumber World Review* (Dec. 25, 1917), 23-27.

J91 Johnson, Bolling A. "History of the American Hardwood Manufacturers' Association." *Lumber World Review* (Aug. 25, 1918), 26-29.

J92 Johnson, Bolling A. "History and Achievements of the North Carolina Pine Association." *Lumber World Review* (Dec. 10, 1918), 23-26.

J93 Johnson, Bolling A.; Green, Thornton A.; and Fuller, Lucius E. "In the Realm of the Lumber Manufacturer: A History of European Saw Mills and the Story of the First Band Saw Mill in America." *Lumber World Review* 38 (No. 1, 1920), 27-30.

J94 Johnson, Bolling A. "Story of the Learned Mill—Longest Continuous Operation in the United States." *Lumber World Review* 38 (Jan. 10, 1920), 28-30. Founded in 1828 in Natchez, Mississippi, and known in early years as the Andrew Brown and Company, this firm specialized in cypress lumber operations.

J95 Johnson, Bolling A. "The First Things in Hoo Hoo—A Narrative." *Lumber World Review* (Sept. 10, 1924), 25-30. Historical sketch of the lumbermen's fraternal order by one of its founders.

J96 Johnson, C.N. "Wisconsin Lumber Industry in Days of Yore: Rafting Lumber to Market." *Disston Crucible* 9 (No. 1, 1920), 12-13, 15; 9 (No. 2, 1920), 30-31.

J97 Johnson, Cecil. *British West Florida, 1763-1783.* New Haven: Yale University Press, 1943. Reprint. Hamden, Connecticut: Shoe String Press, 1971. ix + 258 pp. Map, bib. Includes reference to logging and the lumber trade.

J98 Johnson, Charles W. "The Civilian Conservation Corps: The Role of the Army." Ph.D. diss., Univ. of Michigan, 1968. 246 pp. From 1933 to 1942.

J99 Johnson, Charles W. "The Army and the Civilian Conservation Corps, 1933-42." *Prologue* 4 (Fall 1972), 139-56.

J100 Johnson, Charles W. "The Army, the Negro and the Civilian Conservation Corps: 1933-1942." *Military Affairs* 36 (Oct. 1972), 82-88.

J101 Johnson, Clarence A. "Forestation in Nebraska." Master's thesis, Univ. of Nebraska, 1939.

J102 Johnson, Claudia Alta "Lady Bird." *White House Diary.* New York: Holt, Rinehart & Winston, 1970. ix + 806 pp. Illus. President Lyndon Johnson's wife was a leader in highway beautification and other environmental causes of the 1960s.

J103 Johnson, Corwin W. "Federal and State Control of Natural Resources." *Vanderbilt Law Review* 4 (June 1951), 739-65. Since 1930.

J104 Johnson, Erastus. *Autobiography of Erastus Johnson.* Los Angeles: Frank Wiggins Trade School, 1937. 82 pp. Johnson (1826-1912) was, among other things, a logger in Maine.

Johnson, Evert W. See Dana, Samuel Trask, #D31

J105 Johnson, George, and Macoun, J.M. *Pulpwood of Canada: The Forest Wealth of Canada.* Ottawa, 1904. 88 pp. Incidental history.

Johnson, Herman M. See Gibbons, William H., #G98

J106 Johnson, Herman M. "Log Prices by Regions, West Side Oregon, Washington and British Columbia, 1919-1935." *Timberman* 37 (July 1936), 60.

J107 Johnson, Herman M. "Trends in Lumber Production in Oregon and Washington." *Journal of Forestry* 37 (Aug. 1939), 620-22. Since 1904.

J108 Johnson, Herman M. *Production of Lumber, Shingles, and Lath in Washington and Oregon, 1869-1939.* USFS, Pacific Northwest Forest and Range Experiment Station, Products Paper No. 1. Portland, 1941. 11 pp. Graphs, tables.

J109 Johnson, Herman M. *Production of Logs in Washington & Oregon, 1925-1940.* USFS, Pacific Northwest Forest and Range Experiment Station, Products Paper No. 2. Portland, 1941. 11 pp.

J110 Johnson, Hildegard B. "Carl Schurz and Conservation." *American-German Review* 23 (Oct.-Nov. 1956), 4-8. Schurz was Secretary of the Interior in the Hayes administration 1877-1881.

J111 Johnson, Hugh A., and Jorgenson, Harold T. *The Land Resources of Alaska: A Conservation Foundation Study.* New York: University Publishers, for the University of Alaska, 1963. xiv + 551 pp. Maps, notes, tables, index. Includes some history of forested lands and forest industries, as well as other resource uses.

J112 Johnson, J.A. "History of Sawmilling in Arizona." *Timberman* 24 (Jan. 1923), 96.

J113 Johnson, Judith M. "Source Materials for Pacific Northwest History: Washington Mill Company Papers." *Pacific Northwest Quarterly* 57 (July 1960), 136-38. On the business records of the lumber company at Seabeck, Washington, 1857-1886.

J114 Johnson, Kenneth. "The Lost Eden: The New World in American Nature Writing.' Ph.D. diss., Univ. of New Mexico, 1973. 244 pp. Examines the nature writing of William Bartram, Henry D. Thoreau, John Muir, Joseph Wood Krutch, and Rachel Carson for insights into the changing course of the Edenic myth and the ecological history of North America.

Johnson, Maxine C. See Peters, William Stanley, #P136

J115 Johnson, N.B. "The American Indian as Conservationist." *Chronicles of Oklahoma* 30 (Autumn 1952), 333-40.

J116 Johnson, Olga Weydemeyer. *Early Libby and Troy, Montana.* Flathead and Kootenai Series. Rexford, Montana, 1958. 110 pp. Illus., map. Two lumber towns in the Kootenai Valley of northwestern Montana since the 1850s. A summary appeared in *Montana, Magazine of Western History* 16 (No. 3, 1966), 44-55.

J117 Johnson, Paul C. "Turn of the Wheel: The Motor Car Versus Yosemite." *California Historical Quarterly* 51 (Fall 1972), 205-12. On USDI regulations governing admission and use of automobiles in Yosemite National Park, California, 1914.

J118 Johnson, Ralph W. "Washington Timber Deeds and Contracts." *Washington Law Review and State Bar Journal* 32 (Spring 1957), 30-46. Since 1891.

J119 Johnson, Robert C. "Logs for Saginaw: The Development of Raft-Towing on Lake Huron." *Inland Seas* 5 (Spring 1949), 37-41; 5 (Summer 1949), 83-90. On the transportation of logs from the Canadian shores of Lake Huron and Lake Superior to the sawmills in Michigan's Saginaw Valley, 1883-1898.

J120 Johnson, Robert C. "Logs for Saginaw: An Episode in Canadian-American Tariff Relations." *Michigan History* 34 (Sept. 1950), 213-23. From 1862 to 1897.

J121 Johnson, Robert E. "Schooners Out of Coos Bay." Master's thesis, Univ. of Oregon, 1953. On shipbuilding and the lumber trade at Coos Bay, Oregon.

J122 Johnson, Robert Underwood. *Remembered Yesterdays.* Boston: Little, Brown and Company, 1923. xxi + 624 pp. Illus., index. Johnson's autobiography outlines his role in the conservation and preservation controversies of his era. As editor of *Century Magazine,* 1873-1913, he was associated with John Muir and others in the effort to preserve Yosemite National Park from various encroachments.

J123 Johnson, Roger T. "Charles L. McNary and the Republican Party during Prosperity and Depression." Ph.D. diss., Univ. of Wisconsin, 1967. 418 pp. Senator McNary of Oregon sponsored much forestry legislation in the 1920s.

J124 Johnson, Ronald C.A. "The Effect of Contemporary Thought upon Park Policy and Landscape Change in Canada's National Parks, 1885-1911." Ph.D. diss., Univ. of Minnesota, 1972. 288 pp. Includes some explanation and history of economic exploitation and urban development within Canada's national parks.

J125 Johnson, Shephard Sterling. "Ethics and Ecology in the Conservation Movement in the United States of America." Ph.D. diss., Boston Univ., 1970. 242 pp. Interprets the conservation movement as an expression of Christian social ethics.

J126 Johnson, Vernon W., and Barlowe, Raleigh. *Land Problems and Policies.* New York: McGraw-Hill, 1954. 422 pp. Notes, tables, charts, references. Includes some history of the public domain and its disposal.

J127 Johnson, Walter Samuel. *Twentieth Century Businessman.* OHI by Elwood R. Maunder. Santa Cruz, California: Forest History Society, 1974. vii + 127 pp. Illus., index. Processed. On Johnson's business career, including his role in the box lumber business of California and related trade associations.

J128 Johnston, Gary W. "Marvin Stone Gives an Inside Look at His Company's History and Growth." *Southern Pulp and Paper Manufacturer* 38 (Apr. 1975), 32-36. Stone Container Corporation, producers of corrugated containers at several plants in the South, since 1926.

J129 Johnston, Hank, and Law, James. *Railroads of the Yosemite Valley.* Long Beach, California: Johnston-Howe Publication, 1963. 208 pp. Illus. Includes some history of the lumber industry and logging railroads near Yosemite National Park and the Merced Valley of California.

J130 Johnston, Hank. *They Felled the Redwoods: Saga of Flumes and Rails in the High Sierra.* Los Angeles: Trans-Anglo Books, 1966. 166 pp. Illus., maps, bib., index. On the Kings River Lumber Company, Sanger Lumber Company, and Hume-Bennett Lumber Company, in the Kings River Canyon of California, 1888-1917.

J131 Johnston, Hank. *Thunder in the Mountains: The Life and Times of Madera Sugar Pine.* Los Angeles: Trans-Anglo Books, 1968. 128 pp. Illus., maps, tables, bib., index. On the Sierra Nevada logging operations, railroads, and flumes of the Madera Sugar Pine Company, Madera, California, 1873-1931.

J132 Johnston, R.N. "Aviation in the Ontario Forestry Branch." *Forestry Chronicle* 1 (Dec. 1925), 20-26. On the use of airplanes during the early 1920s.

J133 Johnston, Verna R. "The Ecology of Fire." *Audubon* 72 (Sept. 1970), 76-119. A historical view.

J134 Johnston, William F. "The Big Falls Forest." *Minnesota Volunteer* 36 (Sept.-Oct. 1973), 20-25. On cooperative research and management of the Big Falls Experimental Forest by the USFS and Minnesota Department of Natural Resources, since 1948.

J135 Jolley, Harley E. "The Blue Ridge Parkway: Origins and Early Development." Ph.D. diss., Florida State Univ., 1964. 253 pp.

J136 Jolley, Harley E. *The Blue Ridge Parkway.* Knoxville: University of Tennessee Press, 1969. xii + 172 pp. On the movement to construct a scenic highway connecting the Great Smoky Mountains and Shenandoah national parks, 1933-1937.

J137 Jolley, Harley E. "Biltmore Forest Fair, 1908." *Forest History* 14 (Apr. 1970), 6-17. Carl A. Schenck's conclave of foresters at George W. Vanderbilt's estate near Asheville, North Carolina, site of the nation's first forestry school.

J138 Jolley, Harley E. "The Cradle of Forestry: Where Tree Power Started." *American Forests* 76 (Oct. 1970), 16-21; 76 (Nov. 1970), 36-39. The Biltmore Forest School near Asheville, North Carolina, 1890s-1910s.

J139 Jolley, Harley E. "The Evolution of a Prescription: The Origin of the Cradle of Forestry." *Forest Farmer* 30 (Apr. 1971), 18-19, 41-42. Brief history of forestry education at the Biltmore Forest School.

J140 Jones, A. Durand. "Big Cypress Swamp and the Everglades: No Solutions Yet." *Living Wilderness* 37 (Winter 1973-1974), 28-36. Includes some history of the effort to preserve Florida's Big Cypress Swamp.

J141 Jones, Alden H. *From Jamestown to Coffin Rock: A History of Weyerhaeuser Operations in Southwest Washington.* Tacoma: Weyerhaeuser Company, 1974. x + 346 pp. Illus., maps, apps., index. A history of company operations centered about Longview, Washington, and elsewhere in southwestern Washington and northwestern Oregon in the 20th century.

J142 Jones, Arthur Francis. *Lumber Manufacturing Accounts.* New York: Ronald Press, 1914. 112 pp. Incidental history.

J143 Jones, Bassett. "Was Nantucket Ever Forested?" *Proceedings of the Nantucket Historical Association* 41 (1935), 19-27. On evidence that Nantucket Island, Massachusetts, once might have been forested and logged.

J144 Jones, Charles H. "The Lumber Industry in New Brunswick and Nova Scotia." Master's thesis, Univ. of Toronto, 1930.

J145 Jones, Dallas L. "The Survey and Sale of the Public Land in Michigan, 1815-1862." Master's thesis, Cornell Univ., 1952.

Jones, H.M. See Bennett, W.T., #B213

J146 Jones, Holway R. "Mysterious Origin of the Yosemite Park Bill." *Sierra Club Bulletin* 48 (Dec. 1963), 69-79. On the congressional act of 1890 which established Yosemite National Park, California.

J147 Jones, Holway R. *John Muir and the Sierra Club: The Battle for Yosemite*. San Francisco: Sierra Club, 1965. xvii + 207 pp. Illus., maps, apps., bib., notes, index. On the period from 1892 to 1913, focusing on the efforts of Muir and the Sierra Club to protect Yosemite National Park, California, especially from the damming of Hetch Hetchy Valley. It deals with some of the broad issues associated with the rise of preservationist doctrine and politics.

J148 Jones, Howard Mumford. *O Strange New World: American Culture, The Formative Years*. New York: Viking Press, 1964. xiv + 464 pp. Illus., notes, index. Includes some history of the influence of nature and landscape on American thought through the 19th century.

J149 Jones, Idwal. "Tin Can Harry." *Westways* 46 (Sept. 1954), 4-5. On the "tree garden" of Henry Ashland Greene at Pacific Grove, California, and his efforts to preserve the Monterey cypress from extinction, 1890s-1934.

J150 Jones, Lyle E. "Cadastral Surveys—The Rectangular System Surveys and Protractions." *Surveying and Mapping* 20 (Dec. 1960), 459-68. On the cadastral surveys of the Bureau of Land Management and predecessor agencies since 1784. Emphasis is on Alaska.

J151 Jones, Richard W. "Forest Policy in Georgia's Changing Socio-Political Environment." Ph. D. diss., Univ. of Georgia, 1968. 178 pp.

J152 Jones, Stanley Hugh. "The Development of State Forest Land Administration through Legislative Action." Master's thesis, Univ. of Washington, 1954.

J153 Jordan, Terry G. "Between the Forest and the Prairie." *Agricultural History* 38 (Oct. 1964), 205-16. On settlers' perceptions of forested lands.

J154 Jordan, Terry G. "Pioneer Evaluation of Vegetation in Frontier Texas." *Southwestern Historical Quarterly* 76 (Jan. 1973), 233-54. On 19th-century attitudes toward and uses of forested and prairie lands, especially the preference among pioneer settlers for the intermingling of forest and prairie.

J155 Jorgensen, Frederick E. *25 Years a Game Warden*. Brattleboro, Vermont: Stephen Daye Press, 1937. 168 pp. Autobiographical account of a Maine game warden.

Jorgenson, Harold T. See Johnson, Hugh A., #J111

J156 Josephson, Horace Richard. "Progress on the Forest Survey." *American Forests* 61 (June 1955), 8-12, 49-51. On a nationwide survey of forest resources, begun by the USFS in 1930.

Josephson, H.R. See Weeks, David, #W134

J157 *Journal of Forestry*. "Bernhard Eduard Fernow." *Journal of Forestry* 21 (Apr. 1923), 306-48. A collection of tributes, appreciations, and obituaries of Fernow (1851-1923), including a bibliography of his writings. Fernow was chief of the Division of Forestry, 1886-1898, and an eminent forestry educator.

J158 *Journal of Forestry*. "Man, Teacher, and Leader—Filibert Roth." *Journal of Forestry* 24 (Jan. 1926), 12-19. Biographical sketch and tribute to Roth (1858-1925), a prominent educator at the New York State College of Forestry and at the University of Michigan.

J159 *Journal of Forestry*. "Clifford R. Pettis." *Journal of Forestry* 25 (Mar. 1927), 257-59. Pettis (1877-1927) was a leader in New York state forestry.

J160 *Journal of Forestry*. "Report of the History of Forestry Committee." *Journal of Forestry* 37 (Feb. 1939), 140-43. A committee of the Society of American Foresters.

J161 *Journal of Forestry*. "Ferdinand Augustus Silcox." *Journal of Forestry* 38 (Jan. 1940), 4-5. Silcox (1882-1939) was USFS chief from 1933 to 1939.

J162 *Journal of Forestry*. "Forty Years of Forestry." *Journal of Forestry* 39 (Feb. 1941), 80-116. Nine papers presented at the fortieth annual meeting of the Society of American Foresters; historical aspects of federal, state, and private forestry policies are included.

J163 *Journal of Forestry*. "Gifford Pinchot—Eighty Years Young." *Journal of Forestry* 43 (Aug. 1945), 547-49. An editorial reflective of Pinchot's varied career in forestry and conservation; other pertinent articles follow.

J164 *Journal of Forestry*. "Carl A. Rishell: Timber Engineer." *Journal of Forestry* 48 (Feb. 1950) 133-34. Sketch of Rishell's career since 1923, especially as director of products research for the National Lumber Manufacturers Association and the Timber Engineering Company.

J165 *Journal of Forestry*. "Charter Members of the Society of American Foresters." *Journal of Forestry* 48 (Nov. 1950), 753-55. Biographical sketches of Gifford Pinchot, Edward Tyson Allen, Henry Solon Graves, William Logan Hall, Ralph Sheldon Hosmer, Overton Westfeldt Price, and Thomas Herrick Sherrard, organizers of the Society of American Foresters in 1900.

J166 *Journal of Forestry*. Special Golden Anniversary Issue. *Journal of Forestry* 53 (Feb. 1955), complete. On the fiftieth anniversary of the USFS; includes several historical articles.

J167 *Journal of Forestry*. "William Buckhout Greeley (1879-1955)." *Journal of Forestry* 54 (Jan. 1956), 34-44. On Greeley's career with the USFS (chief, 1920-1928), the West Coast Lumbermen's Association, and the American Forest Products Industries, since 1904.

J168 *Journal of Forestry*. "Committee on History of Forestry." *Journal of Forestry* 54 (Jan 1956), 58-59. A report of the Society of American Foresters' Committee on History of Forestry, which outlines some achievements and urges further research. See also *ibid.*, 55 (Jan. 1957), 57-58.

J169 *Journal of Forestry*. "Raphael Zon (1874-1956)." *Journal of Forestry* 54 (Dec. 1956), 850. On Zon's career with the USFS after 1901 and as editor of *Journal of Forestry,* 1917-1928.

J170 Joy, A.C. "History of Lumbering in California." *Timberman* 30 (Nov. 1929), 178.

J171 Judd, J.W. "Timber Land Taxation." *Southern Lumberman* 55 (Sept. 26, 1908), 25-31. Incidental history.

J172 Judd, Jacob. "America's Wooden Age: Sleepy Hollow Restorations, April 1973." *Technology and Culture* 15 (Jan. 1974), 64-69. A report on the Conference on America's Wooden Age, Tarrytown, New York, April 27-28, 1973, emphasizing research on wood supply and technology in the colonial period and early 19th century.

Juris, Frances. See Due, John F., #D279

K1 Kaatz, Martin Richard. "The Black Swamp: A Study in Historical Geography." *Annals of the Association of American Geographers* 45 (Mar. 1955), 1-35. On changes in a forested area south of the Maumee River in northwestern Ohio since 1791.

K2 Kahn, Herman. "The National Archives, Storehouse of National Park History." *Regional Review* 4 (Feb. 1940), 13-17.

K3 Kaiser, William K. "The Tackapausha Preserve." *Nassau County Historical Journal* 13 (Summer 1952), 96-103. History of a 65-acre park along Seaford Creek, Long Island, New York, since 1643.

K4 Kalbfus, Joseph H. *Dr. Kalbfus's Book*. Altoona, Pennsylvania: Times Tribune Company, 1926. 342 pp. Kalbfus (1852-1918) was secretary of the Pennsylvania State Game Commission for twenty-four years.

K5 Kalish, Richard J. "National Resource Planning: 1933-1939." Ph.D. diss., Univ. of Colorado, 1963. 353 pp. History of the National Resources Committee and its predecessor agency, the National Resources Planning Commission.

K6 Kalish, Richard J. "Environmental Protection." *Proceedings of the Academy of Political Science* 31 (May 1974), 250-62. Incidental history.

K7 Kalmbach, Edwin R. "In Memoriam: W.L. McAtee." *Auk* 80 (Oct. 1963), 474-85. Waldo Lee McAtee (1883-1962) was employed by the Biological Survey of the USDA and its successor agency, the Fish and Wildlife Service of the USDI, 1904-1947. An ornithologist, entomologist, and prolific writer, he was a leading conservationist.

K8 Kane, Lucile M. "Federal Protection of Public Timber in the Upper Great Lakes States." *Agricultural History* 23 (Apr. 1949), 135-39. Explains the failure of USDI agents to protect timber in the region, 1838-1898.

K9 Kane, Lucile M. "Touring with a Timber Agent." *Minnesota History* 31 (Sept. 1950), 158-62. Based on a report of J.S. Wallace, special timber agent of the U.S. government, April 21, 1890, regarding his recent inspection of Minnesota's Rainy River district.

K10 Kane, Lucile M. "Hersey, Staples, and Company, 1854-1860: Eastern Managers and Capital in Frontier Business." *Bulletin of the Business Historical Society* 26 (Dec. 1952), 199-213. A pioneer lumber company of Stillwater, Minnesota, 1854-1860.

K11 Kane, Lucile M. "Selling Cut-Over Lands in Wisconsin." *Business History Review* 28 (Sept. 1954), 236-47. On the activities of the American Immigration Company, jointly owned by nine lumber companies, which sold cutover lands in northwestern Wisconsin from 1906 to 1940.

K12 Kane, Lucile M. "Settling the Wisconsin Cutovers." *Wisconsin Magazine of History* 40 (Winter 1957), 91-98. On efforts to promote agricultural settlement on cutover lands of northern Wisconsin, 1889-1922.

K13 Kane, Lucile M. *The Waterfall That Built a City: The Falls of St. Anthony in Minneapolis*. St. Paul: Minnesota Historical Society, 1966. x + 224 pp. Illus., maps, notes, index. Includes some history of the lumber industry in Minneapolis, which began as the sawmilling village of St. Anthony.

K14 Kanneberg, Adolf. *Log Driving and the Rafting of Lumber in Wisconsin. Statutory Provisions and the Common Law of Wisconsin Pertaining to the Use of Navigable Water for Log Driving and the Rafting of Lumber.* Madison: Public Service Commission of Wisconsin, 1944. 84 pp. Processed.

K15 Kantner, Arthur H. *The Cypress Lumber Industry.* Federal Reserve Bank of Atlanta, Research Department, Economic Study No. 3. Atlanta, 1955. 34 pp. Diag., map, tables. On the industry in a band from Florida and the coast of southern Georgia to western Louisiana, and along the Mississippi River in Louisiana, Mississippi, and Arkansas, since 1940.

K16 Karges, Steven B. "David Clark Everest and Marathon Paper Mills Company: A Study of a Wisconsin Entrepreneur, 1909-1931." Ph. D. diss., Univ. of Wisconsin, 1968. 346 pp.

Karstetter, Albert D. See Walker, James Herbert, #W28

K17 Kasparek, Bob. "Dolly Sods: Picturesque and Primitive." *American Forests* 80 (Feb. 1974), 26-29. On an area of the Monongahela National Forest of West Virginia, logged and burned during the 19th century but under consideration in the 1970s for designation as a wilderness area.

K18 Kast, Fremont E. "Major Manufacturing Industries in Washington State: Changes in Their Relative Importance and Causes of Changes." D.B.A. diss., Univ. of Washington, 1956. 431 pp. Includes some account of the lumber and forest industries since 1939.

Kates, Robert W. See Burton, Ian, #B593

K19 Kaufert, Frank H. "Developments and Trends in Forest Products Research." *Journal of Forestry* 48 (Jan. 1950), 18-20. In the 20th century.

K20 Kaufert, Frank H., and Cummings, William H. *Forestry and Related Research in North America.* Washington: Society of American Foresters, 1955. viii + 280 pp. Tables, index. Includes some history of forestry research.

K21 Kauffmann, Carl. *Logging Days in Blind River*. Blind River, Ontario: By the author, 1970. xx + 124 pp. Illus., maps, apps., index. History of the lumber industry near Blind River, Ontario, since the 1850s.

K22 Kauffman, Erle. "Wood for Coal." *American Forests and Forest Life* 34 (July 1928), 401-03. On the use of wood by the Clearfield Bituminous Coal Corporation of Indiana, Pennsylvania, since 1883, and its practice of forestry on company timberlands since 1920.

K23 Kauffman, Erle. "A Tree Planter Reclaims a Town." *American Forests* 35 (Apr. 1929), 224-26. On efforts to restore the wood-using industries of Richmond, New Hampshire, through reforestation.

K24 Kauffman, Erle. "The Trees of Potowomut: An Early Private Tree Planting Experiment Has Given America a Unique Man-Made Forest." *American Forests and Forest Life* 35 (June 1929), 329-32. On the history of tree planting experiments (white pine and Douglas fir) on Potowomut Neck, Rhode Island, since the 1870s.

K25 Kauffman, Erle. "The Great Yellowstone Adventure." *American Forests and Forest Life* 35 (Aug. 1929), 457-61. On explorations leading to the establishment of Wyoming's Yellowstone National Park in 1872.

K26 Kauffman, Erle. *Trees of Washington: The Man—The City.* Washington: Outdoor Press, 1932. 90 pp. Illus. On the trees and forests of Mount Vernon, Virginia, and Washington, D.C.

K27 Kauffman, Erle. "They Had Faith in the Land." *American Forests* 56 (Mar. 1950), 6-11. Successful tree planting experiments by Great Southern Lumber Company and its successor, Gaylord Container Corporation, Bogalusa, Louisiana, since 1922.

K28 Kauffman, Erle. "The Southland Revisited." *American Forests* 61 (Aug. 1955), 33-40; 61 (Sept. 1955), 11-16; 61 (Oct. 1955), 14-19, 84-94, 112. On the Southern Forestry Educational Project of the American Forestry Association since 1928, and the general progress of forestry in the South.

K29 Kaufman, Clemens M. "Progress in Professional Education." *Journal of Forestry* 54 (Oct. 1956), 661-64, 668, 671-72. On forestry education in the South in the 20th century, with antecedents to 1795.

K30 Kaufman, Harold F. "Social Factors in the Reforestation of the Missouri Ozarks." Master's thesis, Univ. of Missouri, 1939.

K31 Kaufman, Harold F., and Kaufman, Lois C. *Toward the Stabilization and Enrichment of a Forest Community.* Missoula, Montana, 1946. 95 pp. Processed.

K32 Kaufman, Herbert. "Field Man in Administration: How the Administrative Behavior of a District Ranger is Influenced Within and by the United States Forest Service." Ph.D. diss., Columbia Univ., 1950. 281 pp.

K33 Kaufman, Herbert. *Forest Ranger: A Study in Administrative Behavior.* Baltimore: Johns Hopkins Press, for Resources for the Future, 1960. xviii + 259 pp. Diags., map, table, notes, index. On public administration in the USFS, with special attention to the field personnel. There are incidental historical references.

Kaufman, Lois C. See Kaufman, Harold F., #K31

K34 Kay, James. "Lumbering in British Columbia." *Empire Forestry Journal* 8 (1929), 238ff.

K35 Kayson, James P., comp. *The Railroads of Wisconsin, 1827-1937.* Boston: Railway & Locomotive Historical Society, 1937. Includes some history of logging railroads.

K36 Kazlauskas, Joseph Bernard. "School Forests of New York State, Their Number, Location, Size, Origin, Development and Use." Ph.D. diss., Cornell Univ., 1955. 188 pp. Since 1920.

K37 Kearns, Frank W. "An Economic Appraisal of the State Forests of Michigan." Ph.D. diss., Michigan State Univ., 1961. 204 pp. Including incidental history of timber, recreation, and wildlife.

K38 Kearns, Kevin C. "The History of the Acquisition, Development and Restoration of Forest Park, 1870-1910." Ph.D. diss., St. Louis Univ., 1966. 216 pp. Forest Park, St. Louis, Missouri.

K39 Kearns, Kevin C. "The Acquisition of St. Louis' Forest Park." *Missouri Historical Review* 62 (Jan. 1968), 95-106. On its purchase in the 1870s.

K40 Keck, Wendell M. *Great Basin Station—Sixty Years of Progress in Range and Watershed Research.* Ogden, Utah: USFS, Intermountain Forest and Range Experiment Station, 1972. 48 pp. Illus. Known as Utah Experiment Station, Great Basin Experiment Station, and, since 1930, Intermountain Forest and Range Experiment Station.

K41 Keeler, Vernon David. "An Economic History of the Jackson County Iron Industry." *Ohio Archaeological and His-*

torical Quarterly 42 (Apr. 1933), 133-238. Includes reference to the use of wood and charcoal.

Keen, Frederick Paul. See Snyder, Thomas E., #S417

Keen, Frederick Paul. See Craighead, Frank Cooper, #C693

Keen, Frederick Paul. See Miller, J.M., #M456

K42 Keenan, Hudson. "America's First Successful Logging Railroad." *Michigan History* 44 (Sept. 1960), 292-302. On the Lake George and Muskegon River Railroad, Clare County, Michigan, 1876-1887

K43 Keenleyside, Hugh L. "Forests of Canada." *Canadian Geographical Journal* 41 (July 1950), 2-15. Incidental history.

K44 Keenleyside, Hugh L. *The Place of the Forest Industry in the Canadian Economy.* Lecture Series No. 6. Vancouver: University of British Columbia, 1950. 16 pp. Incidental history.

K45 Kehl, James A., and Astorino, Samuel J. "A Bull Moose Responds to the New Deal: Pennsylvania's Gifford Pinchot." *Pennsylvania Magazine of History and Biography* 88 (Jan. 1964), 37-51. On Pinchot's acceptance and later rejection of the New Deal.

K46 Keil, Bill. "Wilderness in a City—Portland's Forest Park." *Pacific Search* 9 (Oct. 1974), 10. Brief history of Forest Park, Portland, Oregon, since the 1940s.

K47 Keir, Robert M. *Manufacturing Industries in America: Fundamental Economic Factors.* New York: Ronald Press, 1920. vii + 324 pp. Tables, notes. Includes a chapter on the history of the pulp and paper industry.

K48 Keir, Robert M. *The Epic of Industry.* The Pageant of America: A Pictorial History of the United States, Volume 5. New Haven: Yale University Press, 1926. 329 pp. Illus. Includes chapters entitled "Logging and Lumbermen" and "The Making of Paper."

K49 Keith, Herbert C., and Harte, Charles R. *Early Iron Industry of Connecticut.* New Haven: Mack and Noel Printers, 1935. 69 pp. Illus. On the charcoal iron industry.

K50 Keith, Herbert F. *Man of the Woods.* Blue Mountain Lake, New York: Adirondack Museum, 1972. xvi + 164 pp. Illus., maps. Autobiographical account of life in New York's Adirondack Mountains, including some history of the lumber town of Wanakena.

K51 Keith Paper Company. *Fifty Years of Quality: 1871-1921.* Turners Falls, Massachusetts, 1921. 15 pp. Illus.

K52 Keithahn, Edward L. *Monuments in Cedar.* Ketchikan, Alaska: Roy Anderson, 1945. 160 pp. On the history of the totem culture and the use of cedar along the coast of British Columbia and southeastern Alaska.

Keleher, William A. See Meyers, John M. #M668

K53 Keller, William F. "Henry Marie Brackenridge: First United States Forester." *Forest History* 15 (Jan. 1972), 12-23. On his role in the live oak culture experiment at Florida's Santa Rosa Island naval timber reserve, 1828-1832. Excerpted from Keller's *The Nation's Advocate: Henry Marie Brackenridge and Young America* (Pittsburgh: University of Pittsburgh Press, 1956).

K54 Kelley, Don Greame. "Trees of the Totem Culture." *American West* 8 (May 1971), 18-21, 63. On the use of cedar in Pacific Northwest coast Indian cultures—before and after arrival of Europeans.

K55 Kellogg, Royal S. *Forest Belts of Western Kansas and Nebraska.* USFS, Bulletin No. 66. Washington: GPO, 1905. 44 pp. Contains some history of tree planting under the Timber Culture Act of 1873.

K56 Kellogg, Royal S. *Forest Planting in Western Kansas.* Washington: GPO, 1909. 52 pp. Illus., map. On the Kansas National Forest.

K57 Kellogg, Royal S. *The Forests of Alaska.* USFS, Bulletin No. 81. Washington: GPO, 1910. 24 pp. Illus., map. Incidental history.

K58 Kellogg, Royal S. *The Lumber Industry.* New York: Ronald Press, 1914. 104 pp. Incidental history.

K59 Kellogg, Royal S. *Pulpwood and Wood Pulp in North America.* New York: McGraw-Hill, 1923. xii + 273 pp. Illus., maps, tables, apps., index. Includes some history and statistical data on the pulp and paper industry in the United States and Canada, since the 1870s.

K60 Kellogg, Royal S. *Newsprint Paper in North America.* New York: Newsprint Service Bureau, 1948. 94 pp. Diags., tables, map. Mainly statistical data relating to papermaking and the pulp and paper industry in Canada and the United States since 1788.

K61 Kellogg, Walter W. *The Kellogg Story: Fifty Years in Southern Hardwoods.* Monroe, Louisiana: Walter Kellogg Lumber Company, 1969. 179 pp. Illus. History of a Louisiana firm originally established as the C.M. Kellogg Lumber Company.

K62 Kellum, Ford. "Return to Deward." *Michigan Conservation* 28 (May-June 1959), 25-29. On a lumber town on the Manistee River near present Frederic, Michigan. Founded in 1854 and virtually abandoned in 1912, it is now included in Au Sable State Forest.

K63 Kelly, Daniel T. "Frontier Merchants: A Brief Sketch of Gross, Kelly, and Company." *Palacio* 65 (Feb. 1958), 7-15. History and reminiscences of a lumber and livestock firm of Las Vegas, New Mexico, since 1879.

K64 Kelly, Kenneth. "Damaged and Efficient Landscapes in Rural and Southern Ontario, 1880-1900." *Ontario History* 66 (Mar. 1974), 1-14. The effects of deforestation upon the agricultural areas of southern Ontario and the changing perceptions of the problems during the early years of the forestry movement.

K65 Kelly, Thomas Smith. "The Marketing of Redwood Lumber Through Wholesale Middlemen." Master's thesis, Univ. of California, Berkeley, 1939. Incidental history.

K66 Kelsey, Arthur E. "Golden Anniversary of Canada's First Fine Paper Mill." *Pulp and Paper Magazine of Canada* 33 (Sept. 1932), 323-26. Rolland Paper Company, St. Jerome, Quebec.

K67 Kelsey, Arthur E. "Price Brothers Company, Limited. for More than a Century One of Canada's Industrial Bulwarks." *Pulp and Paper Magazine of Canada* 34 (Oct. 1933), 585-88. A Quebec firm.

K68 Kelsey, F.W. *The First County Park System: A Complete History of the Inception and Development of the Essex County Parks of New Jersey.* New York: J.S. Ogilvie Publishing Company, 1905.

K69 Kelsey, Lucy N. *The September Holocaust, a Record of the Great Forest Fire of 1894, by One of the Survivors.* Minneapolis: A. Roper, 1894. 125 pp. On the Hinckley fire in Minnesota, 1894.

K70 Kelso, Elmer G. "Forestry on the Larger Holdings in the Northeast." *Journal of Forestry* 48 (Dec. 1950), 866-70. On privately owned forested lands in New Hampshire, Vermont, and Maine, with some history of forestry in the area since the 1890s.

K71 Kelso, Maurice M. "Current Issues in Federal Land Management in the Western United States." *Journal of Farm Economics* 29 (Nov. 1947), 1295-1313. Incidental history.

K72 Kelso, William M. "Shipbuilding in Virginia, 1763-1774." Master's thesis, College of William and Mary, 1964.

K73 Kelso, William M. "Shipbuilding in Virginia, 1763-1774." *Columbia Historical Society Records* 48 (1973), 1-13.

K74 Kemp, J. Larry. *Epitaph for the Giants: The Story of the Tillamook Burn.* Portland: Touchstone Press, 1967. 110 pp. Illus., maps. Popular history of Oregon's Tillamook Burn of 1933, with mention of subsequent forest fires in the region in 1939, 1945, and 1951.

Kemp, Paul D. See Hutchison, S. Blair, #S677

K75 Kemper, Jackson. *American Charcoal Making: In the Era of the Cold-Blast Furnace.* N.p.: Eastern National Park and Monument Association, 1936. 25 pp. Illus. History and description of the charcoal iron industry in Pennsylvania's Schuylkill Valley and at the Hopewell Village National Historic Site. References to collection of wood for making charcoal.

K76 Kendall, I.N. "Saw Milling Methods of Fifty Years Ago." *Canada Lumberman* 36 (May 15, 1916), 101-03. Canada, 1860s.

K77 Kendeigh, S.C. "History and Evaluation of Various Concepts of Plant and Animal Communities in North America." *Ecology* 35 (1954), 152-71.

K78 Kenderline, Thaddeus S. "Lumbering Days on the Delaware River." *Papers of the Bucks County Historical Society* 4 (1917), 239-52. Logging and rafting on the Delaware River of Pennsylvania.

K79 Kendig, John D. "Ghost City of Jamison." *American Forests* 53 (Oct. 1947), 453, 475-77. The lumber town of Jamison, Pennsylvania, 1880s-1920s.

K80 Kennedy, Allan. "Reminiscences of a Lumberjack." *Saskatchewan History* 19 (Winter 1966), 24-34. Logging in northern Saskatchewan with the Prince Albert Lumber Company and The Pas Lumber Company, 1910-1928.

K81 Kennedy, John R. *No Room for Discouragement: The Story of Federal Paper Board Company.* New York: Newcomen Society in North America, 1966. Federal Paper Board Company, New Jersey, with branch operations throughout the East and Midwest.

K82 Kennedy, John S. "The Forest Parks of New York." *American Forestry* 16 (Dec. 1910), 695-98. Historical sketch of Adirondack, Catskill, and Palisades parks.

K83 Kennedy, Robert L. "The Oak: Tree Beloved." *American Forests* 66 (July 1960), 30-31, 92-94. Lore and history of American oak trees since 1758.

K84 Kenney, Nathaniel T. "Our Green Treasury, the National Forests." *National Geographic Magazine* 110 (Sept. 1956), 287-324. Incidental history.

K85 Kenny, Norris G. "Federal Land Grants in Aid of Railroads." Ph.D. diss., American Univ., 1931. Including rich timberlands.

K86 Kenoyer, Leslie A. "Forest Distribution in Southwestern Michigan as Interpreted from Original Survey (1826-1832)." *Papers of the Michigan Academy of Science, Arts, and Letters* 19 (1933), 107-12.

K87 Kenoyer, Leslie A. "Forest Associations of Ottawa County, Michigan at the Time of the Original Survey." *Papers of the Michigan Academy of Science, Arts, and Letters* 28 (1942), 47-49. Early 19th century.

K88 Kensel, William H. "The Economic History of Spokane, Washington, 1881-1910." Ph.D. diss., Washington State Univ., 1962. 220 pp. The lumber and forest products industries were important factors in Spokane's growth after 1900.

K89 Kensel, William H. "The Early Spokane Lumber Industry, 1871-1910." *Idaho Yesterdays* 12 (Spring 1968), 25-31. Sawmills and the lumber industry in Spokane, Washington.

K90 Kent, Elizabeth Thacher. *William Kent, Independent: A Biography.* N.p., 1950. iii + 421 pp. Illus. On his career in Chicago, California, and national politics, 1890-1928, with special reference to his work for conservation. Kent donated the redwood forest which became California's Muir Woods National Monument.

K91 Kentucky. Department of Conservation. *Kentucky's Resources: Their Development and Use.* Frankfort, 1958. 342 pp. Includes some history of forest resources and industries.

K92 Kentucky. Department of Geology and Forestry. *The Mineral and Forest Resources of Kentucky.* Frankfort, 1919. 396 pp. Incidental history.

K93 Kephart, George S. "The Pulpwood Camps." *Forest History* 14 (July 1970), 27-34. Reminiscences of logging camps in Maine, 1920s.

K94 Kephart, George S. "Live Oak: The Tree with a Past." *American Forests* 78 (June 1972), 36-39, 58-61. On federal efforts to preserve Southern live oak stands for naval purposes, 1790s-1860s.

K95 Ker, J.W. "The Measurement of Forest Products in Canada: Past, Present and Future Historical and Legislative Background." *Forestry Chronicle* 42 (Mar. 1966), 29-38. Includes some history of scaling methods and legislation.

K96 Kerlee, Thomas M. "Some Chapters on the Forest Homestead Act with Emphasis on Western Montana." Master's thesis, Univ. of Montana, 1962.

K97 Kern, Jack C. "First Extensive Use of the Helicopter in Forest Fire Control." *Journal of Forestry* 46 (July 1948), 487-92. Reports on use of helicopter to suppress a forest fire on the Angeles National Forest, California, 1947.

K98 Kernan, Henry S. "Idaho Lumber-Jack Nicknames." *California Folklore Quarterly* 4 (July 1945), 239-44.

K99 Kernan, Henry S. "The Forest Frontier in Connecticut." *American Forests* 54 (Mar. 1948), 120-21, 142, 144. On forestry in Connecticut since 1896.

K100 Kernan, Henry S. *The World Is My Woodlot.* New York: Pageant Press, 1962. 210 pp. A collection of addresses, lectures, and essays, including some historical references to forests and forestry.

K101 Kerr, Clark, and Randall, Roger. *Crown Zellerbach and the Pacific Coast Pulp and Paper Industry.* Case Study No. 1 for the NPA Committee on the Causes of Industrial Peace under Collective Bargaining. Washington: National Planning Association, 1948. 78 pp. Tables, notes. Includes some history of industrial and labor relations.

K102 Kerr, Clark, and Randall, Roger. *Collective Bargaining in the Pacific Coast Pulp and Paper Industry.* Philadelphia: University of Pennsylvania Press, for the Labor Relations Council of the Wharton School of Finance and Commerce, 1948. 32 pp. Notes. Contains some incidental historical references to unionism and union activities.

K103 Kerr, Edward F. "Louisiana State Story." *American Forests* 59 (Apr. 1953), 24-26, 55; 59 (May 1953), 22-24, 47-48. On state forestry in Louisiana, especially since 1913.

K104 Kerr, Edward F., and Morgan, E. "Crossett—Swinging on a Star." *Forests & People* 6 (Second quarter, 1956), 8-15, 43. Some history of the Crossett Company, Crossett, Arkansas.

K105 Kerr, Edward F. "From Timber to Famine—And Back Again! The Story of Louisiana's Forest Industries." *Southern Lumberman* 193 (Dec. 15, 1956), 139-43. Since 1869.

K106 Kerr, Edward F. "They Spawned a Timber Empire." *Southern Lumberman* 195 (Dec. 15, 1957), 117-20. On lumbermen and the lumber industry of Alexandria, Louisiana, 1891-1925.

K107 Kerr, Edward F. "Southerners Who Set Woods on Fire." *Harper's Magazine* 217 (July 1958), 28-33. Incidental history of incendiary fires.

K108 Kerr, Edward F. *History of Forestry in Louisiana.* Baton Rouge: Louisiana Forestry Commission, 1958. 55 pp. The lumber industry and forestry since the 1870s.

K109 Kerr, Edward F. "Tree Farming Sweeps the Country." *American Forests* 66 (May 1960), 28-30, 52-55. Since 1941.

K110 Kerr, Edward R. "Tribute to a Senator." *American Forests* 68 (Feb. 1962), 23, 53. On the work of Senator Allen J. Ellender and others since 1947 in supporting the Alexandria Forestry Center in Louisiana, a project location of the USFS's Southern Forest Experiment Station.

K111 Kerr, Edward F. "1963: Fiftieth Anniversary of Forestry in Louisiana." *Southern Lumberman* 205 (Dec. 15, 1962), 135-38.

K112 Kerr, Edward F. "The History of Forestry in Louisiana." *Forests & People* 13 (First quarter, 1963), 11-39. Since 1875, including biographical sketches of state foresters, industrial leaders, and many references to the forest products industries.

K113 Kerr, Edward F. "Days of the Civilian Conservation Corps." *Forests & People* 22 (Second quarter, 1972), 22-23, 43. Brief account of CCC work in Louisiana, 1930s.

K114 Kerr, Kenneth C. "Evolution in Pacific Coast Shipbuilding: A Remarkable Story of Achievement from the Time of the Earliest Voyagers." *Railway and Marine News* 14 (Dec. 1916), 17-20.

K115 Kerr, R.E. "Lumbering in the Adirondack Foothills." *St. Lawrence County Historical Society Quarterly* 9 (Apr. 1964), 12-15. New York.

K116 Kerr, Robert S. *Land, Wood and Water.* Ed. by Malvina Stephenson and Tris Coffin. New York: Fleet Publishing Corporation, 1960. 380 pp. Illus., bib., index. A general history of conservation with emphasis on water policy.

K117 Kerr, W.F. "A Century of Timbering in Saskatchewan: Whip-Sawn Spruce Cut More Than 100 Years Ago—Now 400 to 500 Sawmills in the Province." *Canada Lumberman* 60 (Aug. 1, 1940), 65, 118.

K118 Kershaw, Gordon E. "Kennebeck Purchase: The Fortunes of a Land Company Extraordinary, 1749-1775." Ph.D. diss., Univ. of Pennsylvania, 1971. 475 pp. The Boston proprietors of the Kennebeck Purchase Company developed central Maine in this era and ran afoul of British authorities over possession of mast trees.

K119 Kershaw, Gordon E. "John Wentworth vs. Kennebeck Proprietors: The Formation of Royal Mast Policy, 1769-1778." *American Neptune* 33 (Apr. 1973), 95-119. Wentworth, royal governor of New Hampshire and surveyor-general of the King's Woods, was in conflict with Maine land speculators over British forest policy on the eve of the American Revolution.

K120 Kersten, Earl W., Jr. "Changing Economy and Landscape in a Missouri Ozarks Area." *Annals of the Association of American Geographers* 48 (Dec. 1958), 398-418. On Dent County, Missouri, and part of the Clark National Forest, since 1828.

Ketchledge, Edwin H. See Barrett, John W., #B97

K121 Ketchum, W.Q. "The Oldest Mill in Ottawa Valley: J.R. Booth, Limited, Claim the Distinction of Operating the Oldest Continuously Active Sawmill in the Ottawa Valley." *Canada Lumberman* 50 (Aug. 1, 1930), 147-49.

K122 Keuchel, Edward F. "A Purely Business Motive: German-American Lumber Company, 1901-1918." *Florida Historical Quarterly* 52 (Apr. 1974), 381-95. The German-American Lumber Company of Pensacola and St. Andrew Bay, Florida, was established in 1901 by Frederick Julius Schreyer, a German businessman. Its timberlands and exporting trade were extensive until 1918, when the firm was confiscated under the Trading-with-the-Enemy Act. American owners renamed the business the St. Andrew Bay Lumber Company in 1919.

K123 Key, Jack Brien. "John H. Bankhead, Jr., of Alabama: Creative Conservative." Ph.D. diss., Johns Hopkins Univ., 1964. Senator Bankhead sponsored forestry legislation in the 1930s and 1940s.

K124 Keyes, Nelson Beecher. *America's National Parks: A Photographic Encyclopedia of Our Magnificent Natural Wonderlands.* Garden City, New York: Doubleday, 1957. Incidental history.

K125 Keyes, Willard. "A Journal of Life in Wisconsin One Hundred Years Ago." *Wisconsin Magazine of History* 3 (Mar. 1920), 339-63; 3 (June 1920), 443-65. Keyes cut timber along the Black River of Wisconsin and rafted it down the Mississippi River, 1819.

K126 Keyser, C. Frank. *The Preservation of Wilderness Areas—An Analysis of Opinion on the Problem.* Washington: Legislative Reference Service, Library of Congress, 1949. 114 pp. Includes some history of wilderness preservation.

Keyser, C. Paul. See Munger, Thornton T., #M634

K127 Kidd, George P. "An Analysis of Forest Taxation in British Columbia." Master's thesis, Univ. of British Columbia, 1940.

Kiefer, Francis. See Harris, John Tyre, #H148

K128 Kiekenapp, Marian R. *Conservation in Minnesota.* University of Minnesota, Division of Library Instruction, Bibliographical Projects, No. 3. Minneapolis, 1936. 18 pp.

Kielbaso, J. James. See Cool, Robert A., #C583

K129 Kieley, James F., in cooperation with the Civilian Conservation Corps. *A Brief History of the National Park Service.* Washington: U.S. National Park Service, 1940. 56 pp. Processed.

K130 Kieley, James F. "William Henry Jackson: Yellowstone's Pioneer Photographer." *National Parks & Conservation Magazine* 46 (July 1972), 11-17. Jackson accompanied the Hayden expedition to Yellowstone in 1871 and he continued to promote Yellowstone and other national parks through his photographic work until the 1930s.

K131 Kienholz, Aaron Raymond. "William C. Shepard." *Connecticut Woodlands* 24 (Nov.-Dec. 1959), 87-88. William Chambers Shepard (1883-1958) was a forester with the Pennsylvania Railroad, the State Park and Forest Commission of Connecticut, and the Connecticut Forest and Park Association.

K132 Kienholz, Aaron Raymond. "Early Logging." *Connecticut Woodlands* 33 (Fall 1970), 6-9. By James B. Hall in northwestern Connecticut and adjacent Massachusetts, 1903-1925.

K133 Kieser, Paul W. "Dard Hunter: Artisan, Papermaker, Book Publisher." *Inland Printer* 112 (Nov. 1943), 29-31. An authority on the history of papermaking.

K134 Kilbourne, Frederick W. *Chronicles of the White Mountains.* Boston: Houghton Mifflin, 1916. xxii + 433 pp. Illus., map. Includes some history of the movement for a White Mountain National Forest in New Hampshire.

K135 Kilbourne, Richard. "A Quarter-Century of Forestry Progress in the Tennessee Valley." *Southern Lumberman* 195 (Dec. 15, 1957), 100-05. Since 1933.

K136 Kilbourne, Richard. "Watershed Improvement in the Tennessee Valley." *Journal of Forestry* 58 (Apr. 1960), 294-96. Watershed management by the Tennessee Valley Authority since 1934.

K137 Kilburn, Paul Dayton. "The Forest Prairie Ecotone in Northeastern Illinois." *American Midland Naturalist* 62 (July 1959), 206-17. On the original vegetation of Kane County, Illinois, from land survey records.

K138 Kilburn, Paul Dayton. "Effects of Logging and Fire on Xerophytic Forests in Northern Michigan." *Bulletin of the Torrey Botanical Club* 87 (Nov.-Dec. 1960), 402-05. In Cheboygan County since 1840.

K139 Kilgore, Bruce M. and L.S. "Forty Years Defending Parks: A History of the National Parks Association." *National Parks Magazine* 33 (May 1959), 13-15. Since 1919.

K140 Kimball, Lawrence M. "Lumber History of Vineland." *Vineland Historical Magazine* 42 (Jan.-Oct. 1957), 353-55. On lumber yards and associated businesses in Vineland, New Jersey, since 1862.

Kimball, Theodora. See Olmsted, Frederick Law, Jr., #O34

K141 Kimberly, J.A. "Seventy-Five Years' Operation of a Wisconsin Paper Mill." *Lake States Timber Digest* 2 (Mar. 11, 1948), 9-10. Kimberly-Clark Corporation, Neenah, Wisconsin.

K142 Kimberly, J.C., and Mahler, E. "Yesterday, Today and Tomorrow; Seventy-Five Years' Operation of a Wisconsin Paper Mill." *Lake States Timber Digest* 2 (Mar. 25, 1948), 10, 12; 2 (Apr. 8, 1948), 8, 19. Kimberly-Clark Corporation, Neenah, Wisconsin.

K143 Kimberly, John R. *Four Young Men Go in Search of a Profit: The Story of Kimberly-Clark Corporation (1872-1957).* New York: Newcomen Society in North America, 1957. 28 pp. On a paper manufacturing company in Neenah, Wisconsin, established in 1872 by John A. Kimberly, Charles B. Clark, Frank C. Shattuck, and Havilah Babcock.

K144 Kimberly-Clark Corporation. *Four Men and a Machine, Commemorating the Seventy-Fifth Anniversary of Kimberly-Clark Corporation.* Neenah, Wisconsin, 1947. 42 pp. Illus., charts. Of Neenah, Wisconsin, with operations throughout the United States and Canada since 1872.

K145 Kimes, William R. "John Muir, Champion of Trees." *Pacific Review* 4 (Summer 1970), 5-7. Muir (1838-1914) was a California naturalist and preservationist.

K146 Kincaid, Robert Lee. "Cumberland Gap National Park." *Commonwealth: The Magazine of Virginia* 26 (June 1959), 16-18, 72. On the history of the area since 1750 and the events since 1922 that led to its establishment as a national historical park in 1955 and its dedication in 1959.

K147 Kindle, E.M. "Notes on the Forests of Southeastern Labrador." *Geographical Review* 12 (1922), 57-71.

K148 King, Clarence. *Mountaineering in the Sierra Nevada.* London: S. Low, Marston, Low & Searle, 1872. 292

pp. King was an explorer and surveyor in the West for the federal government. Later editions of this work appeared in 1902 and 1935, the latter edited by Francis P. Farquhar.

K149 King, D.B. *Forest Resources and Industries of Missouri.* University of Missouri Agricultural Experiment Station, Research Bulletin No. 452. Columbia, 1949. Incidental history.

K150 King, D.B., and Winters, Robert K. *Forest Resources and Industries of Illinois.* Illinois Agricultural Experiment Station, Bulletin No. 562. Urbana: University of Illinois, 1952. 95 pp. Illus., diags., maps, tables, notes. Incidental history.

K151 King, Dale S., ed. *Arizona's National Monuments.* Santa Fe, New Mexico: Southwestern Monuments Association, 1945. 118 pp. Illus. Incidental history.

K152 King, Franklin A. "Logging Railroads of Northern Minnesota." *Bulletin of the Railway and Locomotive Historical Society* 93 (Oct. 1955), 94-115. Since 1888.

K153 King, Helene. "The Economic History of the Long-Bell Lumber Company." Master's thesis, Louisiana State Univ., 1936.

K154 King, John O. *The Early History of the Houston Oil Company of Texas, 1901-1908.* Texas Gulf Coast Historical Association Publications, Volume 3, No. 1. Houston, 1959. 100 pp. Illus., tables, notes, bib., index. Contains some history of John Henry Kirby, prominent lumberman of east Texas, and of the Kirby Lumber Company.

K155 King, Judson. *The Conservation Fight: Theodore Roosevelt to the Tennessee Valley Authority.* Washington: Public Affairs Press, 1959. xx + 316 pp. Illus., notes, index. A largely personal account of King's conservation activities from 1897 to the 1930s, especially in behalf of public power and the Tennessee Valley Authority.

K156 King, Ralph Terence; Dence, W.A.; and Webb, W.L. "History, Policy and Program of the Huntington Wildlife Forest Station." *Roosevelt Wildlife Bulletin* 7 (Sept. 1941), 393-460.

K157 King, Samuel A. "A Log Drive to Williamsport in 1868." *Pennsylvania History* 29 (Apr. 1962), 151-74. On a log drive by John DuBois and Hiram Woodward down Sinnemahoning Creek and the West Branch of the Susquehanna River to Williamsport, Pennsylvania.

King, Stella (Brooks). See Aber, Ted., #A6

K158 Kingery, Robert. "The State Parks and Illinois History." *Transactions of the Illinois State Historical Society* (1936), 63-67. On the creation of state parks at historic sites.

K159 Kingsbury, Arthur Murray. *Necedah, Juneau County, Wisconsin: The Story of the First 50 Years of the Village, 1853-1903.* St. Paul: Northland Press, n.d. 88 pp. Illus., tables, A 19th-century lumber town.

K160 Kingsford, W. *History, Structure, and Statistics of Plank Roads, in the United States and Canada.* Philadelphia: A. Hart, 1851. 40 pp. Table, diags.

Kingsley, Neal P. See Ferguson, Roland H., #F64

K161 Kingston, James T.B. *Statistical Record of the Pulp and Paper Industry in British Columbia.* Victoria: British Columbia Department of Trade and Industry, 1955. 60 pp. Processed.

K162 Kinnebrew, Randolph G. *Tung Oil in Mississippi.* University: University of Mississippi, Bureau of Business Research, 1952. Incidental history.

K163 Kinney, Abbot. *Forest and Water.* Los Angeles: Post Publishing Company, 1900. 250 + v pp. Illus., index. Includes some incidental history of forests and forestry, especially in California.

K164 Kinney, Jay P. *Forest Legislation in America Prior to March 4, 1789.* Cornell University, Agricultural Experiment Station of the New York State College of Agriculture, Department of Forestry, Bulletin No. 370. Ithaca, New York, 1916. p. 359-405. A colony-by-colony and state-by-state account of legislation regarding forest fires, timber trespass, forest conservation, and the regulation of the lumber industry.

K165 Kinney, Jay P. *The Development of Forest Law in America: A Historical Presentation of the Successive Enactments, by the Legislatures of the Forty-Eight States of the American Union and by the Federal Congress, Directed to the Conservation and Administration of Forest Resources.* New York: John Wiley & Sons, 1917. xviii + 254 + xxi pp. Notes, index. A reprint edition (New York: Arno Press, 1972, 405 pp.) includes Kinney's *Forest Legislation in America Prior to March 4, 1789* (1916).

K166 Kinney, Jay P. *The Essentials of American Timber Law.* New York; John Wiley & Sons, 1917. xix + 279 + x pp. Notes, index. Incidental history.

K167 Kinney, Jay P. "The Administration of Indian Forests." *Journal of Forestry* 28 (Dec. 1930), 1041-52. A historical account.

K168 Kinney, Jay P. *A Continent Lost–A Civilization Won: The Indian Land Tenure in America.* Baltimore: Johns Hopkins Press, 1937. xv + 366 pp. Illus., map, tables, app., bib., index. Contains several chapters on the development of natural resources, including forest resources, on Indian lands. The author includes many reminiscences of his own role in the Forestry Branch of the Indian Service, USDI.

K169 Kinney, Jay P. *Indian Forest and Range: A History of the Administration and Conservation of the Redman's Heritage.* Washington: Forestry Enterprises, 1950. ix + 357 pp. Illus., map, notes, index. A history of the federal government's management of timber sales and logging operations on Indian reservations, especially in the Great Lakes states and the Pacific Northwest.

K170 Kinney, Jay P. *The Office of Indian Affairs: A Career in Forestry.* OHI by Elwood R. Maunder, William Heritage, and George Thomas Morgan, Jr. New Haven: Forest History Society, 1969. 120 pp. Illus., index. Processed. On Kinney's career in forestry with the Indian Service since 1902.

K171 Kinney, Jay P. "Beginning Indian Lands Forestry." OHI by Elwood R. Maunder and George T. Morgan, Jr. *Forest History* 15 (July 1971), 6-15. Reminiscences of forestry in the Indian Service, 1910s-1920s.

K172 Kinney, Jay P. *My First Ninety-Five Years.* Hartwick, New York: By the author, 1972. 130 pp. Illus., index. Includes many reminiscences of the author's career in forestry, especially with the Forestry Branch of the Indian Service, USDI.

K173 Kinney, Jay P. *Facing Indian Facts.* Hartwick, New York: By the author, 1973. 206 pp. Illus., notes, index. A defense of the Indian Service, including many reminiscences of the author's role in the Forestry Branch, 1920s-1940s.

K174 Kirby Lumber Company. *The Kirby story.* Houston, 1951. 16 pp. Illus. Since 1901.

K175 Kirk, Ruth. *Exploring the Olympic Peninsula.* 1964. Revised edition. Seattle: University of Washington Press, 1967. 128 pp. Illus., maps. On the heavily forested Olympic Peninsula and Olympic National Park of Washington, including a brief historical chapter.

K176 Kirk, Ruth. *Exploring Mount Rainier.* Seattle: University of Washington Press, 1968. v + 104 pp. Illus., maps. Incidental history of Mount Rainier National Park.

K177 Kirk, Ruth. *Exploring Yellowstone.* Seattle: University of Washington Press, in cooperation with Yellowstone Library and Museum Association, 1972. vi + 120 pp. Illus., maps, index. Includes a section on the history and natural history of Yellowstone National Park.

K178 Kirk, Ruth. *Yellowstone: The First National Park.* New York: Atheneum Publishers, 1974. 98 pp. Illus.

K179 Kirk, Ruth. *Washington State: National Parks, Historic Sites, Recreation Areas, and Natural Landmarks.* Seattle: University of Washington Press, 1974. 64 pp. Illus., maps, app., index. Includes historical references to the North Cascades, Mount Rainier, and Olympic national parks, as well as other forested areas.

K180 Kirk, Ruth. *Exploring Crater Lake Country.* Seattle: University of Washington Press, with the Crater Lake Natural History Association, 1975. 96 pp. Illus., maps, index. Includes some history of the Oregon park and surrounding areas.

K181 Kirkland, Burt P. "Continuous Forest Production in the Pacific Northwest." *Commonwealth Review* 3 (1918), 63-78. Incidental history.

K182 Kirkland, Burt P. "Southern Forest Resources and Industries." *Southern Economic Journal* 6 (July 1939), 20-32.

K183 Kirkland, Burt P. *Forest Resources of the Douglas Fir Region.* Portland, 1946. 74 pp. Incidental history.

Kirkland, Burt P. See Foster, Clifford H., #F180

K184 Kirkland, Edward C. *A History of American Economic Life.* New York: Crofts, 1932. xv + 767 pp. Maps, tables, bib. See for references to forest industries.

K185 Kirkland, Edward C. *Industry Comes of Age: Business, Labor, and Public Policy, 1860-1897.* Economic History of the United States, Volume 6. New York: Holt, Rinehart and Winston, 1961. xiv + 445 pp. Illus., notes, bib., index. Contains some historical references to the forest reserves and the lumber industry, placed in the larger context of the business and industrial history of the era.

K186 Kirkland, Herbert Donald. "The American Forests, 1864-1898: A Trend Toward Conservation." Ph.D. diss., Florida State Univ., 1971. 301 pp. Treats individuals, organizations, and legislation in support of forest conservation.

K187 Kirkpatrick, John C., and Miles, R. Vance, Jr. "Alabama's Changing Forest Industry." *Journal of the Alabama Academy of Science* 29 (Oct. 1957), 28-32. Since 1946.

K188 Kirkwood, J.E. *Forest Distribution in the Northern Rocky Mountains.* Montana State University Studies, Series No. 2, Bulletin No. 247. Missoula: Montana State University, 1922. 180 pp. Illus., maps, tables, bib. Includes incidental historical references to forested lands.

K189 Kistler, Stan. "Loggers and Lokeys." *Trains* 20 (Apr. 1960), 44-52. On Rayonier, Inc., producer of chemical cellulose and fine papers, and its logging railroads in the areas of Grays Harbor and Clallam counties, Washington, built between 1903 and 1906.

K190 Kittredge, Joseph. *Forest Planting in the Lake States.* USDA, Bulletin No. 1497. Washington: GPO, 1929. 88 pp. Includes some history of forest planting.

K191 Kittredge, Joseph. *Forest Influences.* New York: McGraw-Hill, 1948. 394 pp. Bib. Incidental history.

K192 Klein, E.L. "An Analysis of the Christmas Tree Industry in Pennsylvania." Master's thesis, Pennsylvania State Univ., 1961.

K193 Kleinmaier, Judith. "Fire Ravaged Forests for Years Before Yielding to Control." *Wisconsin Then and Now* 19 (Feb. 1973), 2-5, 8. On the history of forest fire prevention and control in Wisconsin.

K194 Kleinmaier, Judith. "1915 Patrol Flight Was First Anywhere." *Wisconsin Then and Now* 19 (Feb. 1973).

K195 Kleinmaier, Judith. "Even Sawdust Finds a Use at the Forest Products Laboratory in Madison." *Wisconsin Then and Now* 19 (Feb. 1973).

K196 Kleinsorge, Paul Lincoln. "The Lumber Industry." *Monthly Labor Review* 82 (May 1959), 558-63. On labor-management relations in the lumber industry of California, Oregon, and Washington since 1933.

K197 Kleven, Bernhardt J. "Wisconsin Lumber Industry." Ph.D. diss., Univ. of Minnesota, 1941. Since the 1820s.

K198 Kleven, Bernhardt J. "The Mississippi River Logging Company." *Minnesota History* 27 (Sept. 1946), 190-202. On the organization by Frederick Weyerhaeuser and other lumbermen of log transportation on Wisconsin's Chippewa River, 1867-1904.

K199 Kleven, Bernhardt J. "Rafting Days on the Mississippi." *Proceedings of the Minnesota Academy of Science* 16 (1948), 53-56. Log rafting down the upper Mississippi River, 1831-1915.

K200 Kline, Benjamin F.G. *'Wild Catting' on the Mountain: The History of the Whitmer and Steele Lumber Companies.* Strasburg, Pennsylvania: By the author, 1971. 90 pp. Pennsylvania.

K201 Kline, Marcia B. *Beyond the Land Itself: Views of Nature in Canada and the United States.* Cambridge: Harvard University Press, 1970. 75 pp.

K202 Kling, Edwin M.; Nelson, S.W.; and Reed, Frank A. "Growth of the Lumber Industry in Northern New York." *Northern Logger* 12 (Apr. 1964), 16-17, 46-47, 52-55, 60-61.

K203 Kling, John B. *Cooperative Forest Fire Control: Policy Determination and Administration in the Clarke-McNary Grant-in-Aid Program.* New York State College of Forestry, Bulletin No. 25. Syracuse, 1951. 106 pp. On the operation of the program in the East.

K204 Klose, Nelson. *America's Crop Heritage: The History of Foreign Plant Introduction by the Federal Government.* Ames: Iowa State College Press, 1950. 156 pp. Bib., index.

K205 Knapp, David C. "Congressional Control of Agricultural Conservation Policy: A Case Study of the Appropriations Process." *Political Science Quarterly* 71 (June 1956), 257-81. From 1936 to 1951.

K206 Knapp, F.C. "Development of Lumbering in the Pacific Northwest." *American Lumberman* (Feb. 2, 1924), 47.

Knapp, J.G. See Lapham, Increase A., #L49

K207 Knappen, Theodore M. "An Old Industry Meets a New Age." *Southern Lumberman* (Mar. 1, 1932), 25-26, 44-45. Brief historical sketch of the lumber industry.

K208 Knechtel, Abraham. *The Dominion Forest Reserves.* Forestry Branch, Bulletin No. 3. Ottawa, 1908, 19 pp.

K209 Kneipp, Leon F. "Uncle Sam Buys Some Forests: How the Weeks Law of Twenty-Five Years Ago Is Building up a Great System of National Forests in the East." *American Forests* 42 (Oct. 1936), 443-46, 483-84.

K210 Kneiss, Gilbert H. *Redwood Railways: A Story of Redwoods, Picnics, and Commuters.* Berkeley: Howell-North, 1956. xviii + 165 pp. Illus., bib. On the Northwestern Pacific Railroad Company and its predecessors, operating from San Francisco to Eureka, California, through the redwood country since the 19th century.

Knetsch, Jack L. See Clawson, Marion, #C393

K211 Kniffen, Fred, and Glassie, Henry. "Building in Wood in the Eastern United States: A Time-Place Perspective." *Geographical Review* 56 (Jan. 1966), 40-66.

K212 Knight, E. Vernow, and Wulpi, Meinrad. *Veneers and Plywood.* New York: Ronald Press, 1927. Includes incidental history of these forest products.

K213 Knight, George Wells. *History and Management of Land Grants for Education in the Northwest Territory.* New York: G.P. Putnam's Sons, 1885. 175 pp. Including forested lands.

K214 Knight, Vernon James. "Forestry Agencies in Georgia, 1900-1951." Master's thesis, Univ. of Georgia, 1951.

K215 Knipe, William A. "The Mast Trade in New Hampshire." *American Neptune 22 (Jan. 1962).*

K216 Knittle, Walter A. *Early Eighteenth Century Palatine Emigration: A British Government Redemptioner Project to Manufacture Naval Stores.* Philadelphia: Dorrance and Company, 1937. Reprint. Baltimore: Genealogical Publishing Company, 1965. xxi + 320 pp. Illus., notes, maps, bib., app., index. In New York.

K217 Knoles, George H., ed. *Essays and Assays: California History Reappraised.* San Francisco: California Historical Society, 1973. viii + 132 pp. Illus., notes. Includes two interpretive essays on environmental themes: John W. Caughey's "The Californian and His Environment," and Andrew Rolle's "Brutalizing the California Scene."

K218 Knollenberg, Bernhard. *Origin of the American Revolution, 1759-1766.* New York: Macmillan, 1960. x + 486 pp. Notes, apps., bib., index. Includes discussion of British forest and mast policy as a grievance contributing to the American revolution.

K219 Knouf, Clyde E. "Early History and Development of Log Scaling Practice." *West Coast Lumberman* 36 (No. 427, 1919), 41, 54; 36 (No. 428, 1919), 40, 64-65.

K220 Knowlton, George W. "History of the Voelter Grinder." *Paper Mill and Wood Pulp News* 49 (Sept. 11, 1926).

K221 Knowlton, George W. "Early Manufacture of Newsprint from Wood Pulp, A Reminiscence." *Paper Industry* 8 (Sept. 1926), 967-68.

K222 Knowlton Brothers. *Knowlton Brothers: 150 Years of Craftmanship in Paper.* Watertown, New York, 1958. 26 pp. A Watertown firm since 1808.

K223 Koch, Elers. *When the Mountains Roared: Stories of the 1910 Fire.* Coeur d'Alene, Idaho: USFS, Coeur d'Alene National Forest, 1942. 39 pp. Reminiscences of the extensive forest fire of 1910 which ravaged much of northern Idaho and parts of western Montana.

K224 Koch, Elers. "Launching the U.S.F.S. in the Northern Region." *Forest History* 9 (Oct. 1965), 9-13. Memoir of the creation of a national forest system in Montana and Wyoming, 1903-08; adapted from USFS, Northern Region, *Early Days in the Forest Service,* 3 Volumes (1944-1962), q.v.

K225 Koch, Margaret. *Santa Cruz County—Parade of the Past.* Fresno: Valley Publishers, 1973. 264 pp. Illus., index. Includes some history of logging and the lumber industry in Santa Cruz County, California, since the 1850s.

K226 Koch, Michael. *The Shay Locomotive, Titan of the Timber.* Denver: World Press, 1971. ix + 488 pp. Illus., charts, bib., index. Widely used in logging operations and manufactured by the Lima Locomotive and Machine Company, Lima, Ohio, since 1880.

K227 Koehler, Arthur. *The Properties and Uses of Wood.* New York: McGraw-Hill, 1924. xiv + 354 pp. Illus. Incidental history of wood technology.

Koelsch, William A. See Rosenkrantz, Barbara Gutmann, #R300.

K228 Koen, Henry R. "Forestry Problems of the Ozarks." *Journal of Forestry* 37 (Feb. 1939), 168-73. Includes some history of the Ozark National Forest, Arkansas, since 1908.

K229 Koen, Henry R. "What Intensive Management of the Forests Will Mean to the Ozark Region of Oklahoma, Missouri, and Arkansas." *Journal of Forestry* 46 (Mar. 1948), 165-67. Includes some history of the region since 1908.

K230 Koenig, Karl F. "Sherburne's Community Forest." *American-German Review* 19 (Feb. 1953), 15, 37. Sherburne, New York, since 1909.

Kohara, Tom. See Richie, Jim, #R190

K231 Kohlmeyer, Fred W. "Northern Pine Lumbermen: A Study in Origins and Migrations." *Journal of Economic History* 16 (Dec. 1956), 529-38. Based on biographies of 131 lumbermen, nearly all born between 1810 and 1850, and operating in Minnesota, Michigan, and Wisconsin.

K232 Kohlmeyer, Fred W. *Timber Roots: The Laird, Norton Story, 1855-1905.* Winona, Minnesota: Winona County Historical Society, 1972. xvii + 382 pp. Illus., map, tables, notes, index. A business history of the Laird, Norton Company of Winona, Minnesota, and of its timberlands and lumber operations in Minnesota, Wisconsin, and Idaho. Also treated are members of the Laird and Norton families and their relations with the Weyerhaeuser interests.

K233 Kolehmainen, John I., and Hill, George W. *Haven in the Woods: The Story of the Finns in Wisconsin.* Madison: State Historical Society of Wisconsin, 1951. ix + 177 pp. Notes, app., bib., index. Includes some history of the lumber industry and settlements on cutover lands of northern Wisconsin.

Kollmorgen, Walter M. See Harrison, Robert W., #H165

K234 Komarek, E.V., Sr. "The Use of Fire: A Historical Background." *Tall Timbers Fire Ecology Conference Proceedings* 1 (1962), 7-10.

K234a Koppes, Clayton R. "Oscar L. Chapman: A Liberal at the Interior Department, 1933-1953." Ph.D. diss., Univ. of Kansas, 1974. 521 pp.

K235 Kornbluh, Joyce L., ed. *Rebel Voices: An I.W.W. Anthology.* Ann Arbor: University of Michigan Press, 1964. xii + 419 pp. Illus., notes, bib. Many lumberjacks in the Pacific Northwest belonged to the Industrial Workers of the World, a militant labor union of the early 20th century.

K236 Korson, George G., ed. *Pennsylvania Songs and Legends.* Philadelphia: University of Pennsylvania Press, 1949. 474 pp. Illus., music. Includes a chapter on lumberjacks and raftsmen by James Herbert Walker.

K237 Korstian, Clarence F. *The Economic Development of the Furniture Industry in the South and Its Future Dependence upon Forestry.* North Carolina Department of Conservation and Development, Economic Paper No. 57. Raleigh, 1926. 26 pp. Illus., tables, bib. Includes some history of the industry and its relation to wood supply.

K238 Korstian, Clarence F., and Maughan, William. *The Duke Forest: A Demonstration and Research Laboratory.* Duke University, Forestry Bulletin No. 1. Durham, North Carolina, 1935. 74 pp. Illus., maps, bib. Includes some history of the Duke Forest near Durham, North Carolina.

K239 Korstian, Clarence F. *Forestry on Private Lands in the United States.* Duke University, Forestry Bulletin No. 8. Durham, North Carolina, 1944. xiii + 234 pp. Illus., app., bib. Progress in private and industrial forestry in the important timber regions of the nation, with occasional historical references.

K240 Korstian, Clarence F., and James, Lee M. *Forestry in the South.* Southern Association of Science and Industry, Studies of Southern Resources, Monograph No. 1. Richmond, 1948. 57 pp. Incidental history.

K241 Korstian, Clarence F. *Clarence F. Korstian: Forty Years of Forestry.* OHI by Elwood R. Maunder. New Haven: Forest History Society, 1969. 74 pp. Illus., index. Processed. On Korstian's career in the USFS and as a forestry educator at Duke University, since ca. 1910.

K242 Korte, Karl H. "The History of Forestry in Hawaii: From World War II to the Present." *Aloha Aina* 1 (June 1970), 16-18.

K243 Kortum, Karl, and Olmsted, Roger. ". . . It Is a Dangerous Looking Place. Sailing Days on the Redwood Coast." *California Historical Quarterly* 50 (Mar. 1971), 43-58. Text and photographs describe the technology of loading redwood logs and lumber into ships along the northern California coast during the late 19th century.

Koser, Mary Ellen. See Koser, Wayne S., #K244

K244 Koser, Wayne S., and Koser, Mary Ellen. "Living With the Gypsy Moth." *Explorer* 16 (Summer 1974), 4-10. Includes some history of the gypsy moth and its infestations of Northeastern forests since its introduction to Massachusetts in 1869.

Kotok, E.I. See Show, Stuart B., #S250

K245 Kotok, E.I., and Hammath, R.F. "Ferdinand Augustus Silcox." *Public Administration Review* 2 (Summer 1942), 240-53. Biographical and career sketch of Silcox, who was chief of the USFS from 1933 to 1939.

K246 Kouba, Theodore F. *Wisconsin's Amazing Woods—Then and Now.* Madison: Wisconsin House, 1973. vii + 279 pp. Illus., maps, tables. Includes much history of logging, the forest products industries, the forestry movement, forest conservation, and other aspects of forest history in Wisconsin since the beginning of white settlement.

K247 Kousser, J. Morgan. "Ecological Regression and the Analysis of Past Politics." *Journal of Interdisciplinary History* 4 (Autumn 1973), 237-62.

K248 Kovach, Joseph. "The Lumber Industry of Georgia." *Southern Lumberman* 193 (Dec. 15, 1956), 156-58. Since the colonial period.

K249 Kraebel, Charles J. "Conquering Kennett's Gullies." *American Forests* 61 (Dec. 1955), 36-39, 42-44. On severe erosion in the vicinity of Kennett, California, produced by fumes of copper smelters from 1905-1919, and efforts after 1933 to remedy the erosion by reforestation.

K250 Kraenzel, Carl F. "Trees and People of the Plains." *Great Plains Journal* 6 (Fall 1966), 8-18. Incidental historical references to the influences of trees on the Great Plains.

K251 Kramer, Herman J. "A Brief History of the Pine Industry in Union, Wallowa, Baker, and Grant Counties in Oregon." Master's thesis, Univ. of Oregon, 1938.

K252 Kramer, Paul R. "The Texas Story." *Southern Lumberman* 193 (Dec. 15, 1956), 178C-178G. On the lumber industry since 1819.

K253 Kramer, William P. "Forestry Work in the Island of Puerto Rico." *Journal of Forestry* 24 (Apr. 1926), 419-25. On the work of the USFS and the Puerto Rico Forest Service since 1917.

K254 Kranz, Marvin W. "Pioneering in Conservation: A History of the Conservation Movement in New York State, 1865-1903." Ph.D. diss., Syracuse Univ., 1961. 634 pp.

Kranz, Marvin W. See Armstrong, George R., #A481

K255 Krauch, Hermann. "Managing Ponderosa Pine for Hewn Cross Ties." *Journal of Forestry* 47 (May 1949), 371-74. On the Rio Pueblo Ranger District of New Mexico's Carson National Forest, 1914-1944.

K256 Krauch, Hermann. "The Coronado National Forest as Don P. Johnston Knew It." *American Forests* 64 (Oct. 1958), 30-31, 63-69. On Arizona's Coronado National Forest since 1905, including some history of Don P. Johnston as forest supervisor after 1916.

K257 Kraus, Marcus. "Science Education in the National Parks of the United States: A Descriptive Study of the Development of Science Education Programs and Facilities by the National Park Service and the Relationship of These to the Advent of Nature Study and Conservation Education in America." Ph.D. diss., New York Univ., 1973.

K258 Krause, John, and Reid, H. *Rails Through Dixie.* San Marino, California: Golden West Books, 1965. 176 pp. Illus. Includes some history of West Virginia logging railroads.

K259 Krebs, Oliver M. "'I Remember—' Early Experiences in the Hardwood Lumber Business." *Southern Lumberman* 197 (Dec. 15, 1958), 153-55. Recollections of the industry in Buffalo, New York; Memphis, Tennessee; and in Indiana, Kentucky, and Arkansas, since 1893.

K260 Kreienbaum, C.H. "Forest Management and Community Stability: The Simpson Experience." OHI by Elwood R. Maunder. *Forest History* 12 (July 1968), 6-19. On Simpson Timber Company and sustained-yield forestry near Shelton, Washington, 1920s-1940s.

K261 Kreienbaum, C.H. *The Development of a Sustained-Yield Industry: The Simpson-Reed Lumber Interests in the Pacific Northwest, 1920s-1960s.* OHI by Elwood R. Maunder. Santa Cruz, California: Forest History Society, 1972. viii + 160 pp. Illus., app., index. Processed. On Kreienbaum's career in the lumber industry since 1912, especially with the Dempsey Lumber Company, Tacoma, Washington, and the Simpson Timber Company, Shelton, Washington, of which he was president during the 1940s.

K262 Kreps, Theodore J. "Vicissitudes of the American Potash Industry." *Journal of Economic and Business History* 3 (1931), 630-66. Including during the colonial period.

K263 Kreig, Allan. *Last of the 3 Foot Loggers.* San Marino, California: Golden West Books, 1962. 95 pp. Illus., maps. On the narrow-gauge logging railroad of West Side Lumber Company, Tuolumne County, California, 1900-1960.

K264 Kriesberg, Martin. "The Emergency Rubber Project." In *Inter-University Case Program Series, No. 3.* Washington: Inter-University Case Program, 1952. Pp. 635-48. On a project under the supervision of the USFS, 1942-1945.

K265 Krog, Carl E. "Marinette: Biography of a Nineteenth Century Lumbering Town, 1850-1910." Ph.D. diss., Univ. of Wisconsin, 1971. 330 pp. Lumbermen and their industry in Marinette, Wisconsin.

K266 Krog, Carl E. "Lumber Ports of Marinette-Menominee in the Nineteenth Century." *Inland Seas* 28 (Winter 1972), 272-80. Marinette, Wisconsin, and Meniminee, Michigan, 1850s-1910.

K267 Krog, Carl E. "Rails Across the Water." *Inland Seas* 29 (1973), 170-76. On the Wisconsin and Michigan Railroad, a narrow-gauge logging line organized by lumberman Isaac Stephenson, and the Lake Michigan Car Ferry Transportation Company, which barged railroad cars loaded with lumber from Peshtigo, Wisconsin, to Chicago, 1893-1907.

K268 Kromm, David E. "A Functional Geographic Approach to the Utilization of the Northern Michigan Forest Resource." Ph.D. diss., Michigan State Univ., 1967. 246 pp. Since the 1850s.

K269 Krueger, Myron E. "The Society of American Foresters and Forestry Education." *Journal of Forestry* 50 (Jan. 1952), 6-7. Since 1900.

Krueger, Myron E. See Dana, Samuel Trask, #D26

K270 Krutch, Joseph Wood. *Thoreau.* New York: William Sloane Associates, 1948. xiii + 298 pp. Illus. Thoreau's life and writings have inspired nature lovers and conservationists.

K271 Krutch, Joseph Wood, ed. *Great American Nature Writing.* New York: William Sloane Associates, 1950. 369 pp. A selection of nature writing since 1839, with an extensive prologue.

Krutch, Joseph Wood. See Thoreau, Henry David, #T89

K272 Kuenzel, John G. "Wood Requirements for Shipbuilding." *Journal of Forestry* 48 (Apr. 1950), 245-54. Especially during World War II.

K273 Kuenzel, John G., and Worth, Harold E. "The Use of Wood in Ships and Boats." *Journal of Forestry* 56 (Aug. 1958), 549-55. Historical sketch of shipbuilding in North America since the Viking era.

K274 Kuhns, Mrs. John F. "Loleta in 1910: Told by the Girl in the Office." *Pennsylvania History* 19 (Oct. 1952), 452-60. An abandoned lumber town in Elk County, Pennsylvania.

K275 Kuhns, Mrs. John F. "Arbor Day." *Bulletin of the Garden Club of America* 45 (Mar. 1957), 35-37. Its celebration since 1872.

K276 Kummerly, Walter, ed. *The Forest.* Translated by Ewald Osers. Washington: Robert B. Luce Company, 1973. 299 pp. Illus., notes. Many contributors write of the historic and economic importance of the forests of the world.

K277 Kuppens, Francis X. "On the Origin of the Yellowstone National Park." *Jesuit Bulletin* 41 (Oct. 1962), 6-7, 14.

K278 Kurjack, D.C. *Hopewell Village National Historic Site.* Washington: U.S. National Park Service, 1961. 44 pp. Illus. Includes some history of the charcoal iron industry in colonial Pennsylvania.

Kurtz, Samuel. See Ibberson, Joseph E., #I1

K279 Kury, Theodore William. "Historical Geography of the Iron Industry in the New York-New Jersey Highlands: 1700-1900." Ph.D. diss., Louisiana State Univ., 1968. 196 pp. A center of the charcoal iron industry.

K280 Kusnerz, Peg. "Winter Camp: Lumbering in Michigan, 1860-1900." *Chronicle* 8 (Third quarter, 1972), 10-19. A photographic essay.

Kutz, Donald B. See Mickalitis, Albert A., #M439

K281 Kutzleb, Charles R. "American Myth: Desert to Eden: Can Forests Bring Rain to the Plains?" *Forest History* 15 (Oct. 1971), 14-21. Tree planting experiments on the Great Plains, 1860s-1890s.

K282 Kuykendall, Ralph S. *Early History of Yosemite Valley, California.* Washington: GPO, 1919. 12 pp. A National Park Service publication reprinted from the *Grizzly Bear,* July 1919.

K283 Kvasnicka, Robert M. "The Timber is Mine." *Prologue* 3 (Spring 1971), 20-26. Government loggers vs. Chippewa Indians, Minnesota Territory, 1850.

K284 Kyle, John H. *The Building of TVA: An Illustrated History.* Baton Rouge: Louisiana State University Press, 1958. x + 162 pp. Illus., maps, diags., plans. Tennessee Valley Authority.

K285 Kylie, Harry R.; Hieronymus, G.H.; and Hall, A.G. *CCC Forestry.* Washington: GPO, 1937. 335 pp. Illus. Incidental history of forestry in the Civilian Conservation Corps.

L1 Labbe, John T., and Goe, Vernon. *Railroads in the Woods.* Berkeley: Howell-North, 1961. 269 pp. Illus., glossary, index. Illustrated history of railroad logging, especially in Oregon, Washington, and California since the 1880s.

Labbe, John T. See Carranco, Lynwood F., #C175

L2 Lacey, John F. "Forestry Legislation in the United States." *Gunton's Magazine* 24 (Feb. 1903), 125-37. Incidental history of federal and state legislation.

L3 Lachance, Paul-Emile. "Canadian-American Wood Trade." Master's thesis, Yale Univ., 1942.

L4 Lachance, Paul-Emile. "A Study of the Pulp and Paper Industry of the Province of Quebec in Relation to Its Present and Future Wood Supplies." Ph.D. diss., Univ. of Michigan, 1954. 203 pp. Includes some account of the 1940s.

L5 Lachance, Paul-Emile. "A Study of the Pulp and Paper Industry of the Province of Quebec in Relation to Its Present and Future Wood Supplies." *Pulp & Paper Magazine of Canada* 55 (Mar. 1954), 276-332.

L6 Lackey, Daniel Boone. "Cutting and Floating Red Cedar Logs in North Arkansas." *Arkansas Historical Quarterly* 19 (Winter 1960), 361-70. Reminiscence of logging and driving on the Big and Little Buffalo rivers, Newton County, Arkansas, 1906.

LaFaver, L.H. See Byrne, J.J., #B625

L7 Lafferty, J.B. *My Eventful Years.* Weiser, Idaho: Signal American Printers, 1963. 53 pp. Illus. Reminiscences of J.B. "Jake" Lafferty (1875-1965), who was supervisor of Idaho's Weiser (Payette) National Forest, 1906-1920.

L8 La Follette, Robert M. *La Follette's Autobiography: A Personal Narrative of Political Experiences.* 1911. Madison: La Follette Publishing Company, 1913. 807 pp. As governor and U.S. senator from Wisconsin, La Follette was involved in the conservation movement. There is a reprint by the University of Wisconsin Press, 1963.

Lagai, Rudolph L. See Glover, John G., #G158

L9 Lagerloef, E.G. "A Half Century of Progress in the Paper Industry." *Paper Mill News* 77 (Apr. 17, 1954), 48, 50, 119.

L10 *Lake States Timber Digest.* "Sixty Years of Progress." *Lake States Timber Digest* 2 (Dec. 18, 1947), 3-4. Nekoosa-Edwards Paper Company, Nekoosa and Port Edwards, Wisconsin.

L11 Lamar, Howard R. "Land Policy in the Spanish Southwest, 1846-1891: A Study in Contrasts." *Journal of Economic History* 22 (Dec. 1962), 498-515. Northern New Mexico and southern Colorado.

L12 Lamb, Charles R. "Sawdust Campaign." *Wisconsin Magazine of History* 22 (Sept. 1937), 6-14. Labor violence in the lumber industry of Eau Claire, Wisconsin, 1881.

L13 Lamb, Frank H. *Sagas of the Evergreens: The Story and the Economic, Social and Cultural Contribution of the Evergreen Trees and Forests of the World.* New York: W.W. Norton, 1938. 364 pp. Illus., map, bib., index. Includes some incidental history of the uses of particular species of evergreen trees in the United States, Canada and abroad. There is also a chapter on the conservation movement.

L14 Lamb, Frank H. *Book of the Broadleaf Trees: The Story of the Economic, Social and Cultural Contribution of the Temperate Broad-Leaved Trees and Forests of the World.* New York: W.W. Norton, 1939. 367 pp.

L15 Lamb, Frank H. "Paul Bunyan's Camp: A Proposed Logging Museum." *Timberman* 42 (Nov. 1940), 12-13, 26. On the need for such a museum in the Pacific Northwest, with references to logging history.

L16 Lamb, William Kaye. "Early Lumbering on Vancouver Island, 1844-1866." *British Columbia Historical Quarterly* 2 (Jan. 1938), 31-53; 2 (Apr. 1938), 95-121.

L17 Lambert, Darwin. "Patterns in National Parks Association History." *National Parks Magazine* 43 (May 1969), 4-8. Since its founding in 1919.

L18 Lambert, Darwin. *Timberline Ancients.* Photos by David Muench. Portland: Charles H. Belding, 1972. 128 pp. Illus., map. Contains some incidental historical information on stands of bristlecone pine in the Southwest.

L19 Lambert, Darwin. "Eastern Wilderness Wanted." *National Parks Magazine* 47 (Nov. 1973), 12-15. On efforts to create wilderness areas in the East since 1964.

L20 Lambert, John H., Jr. "Massachusetts Forests—Trends, Uses and Management." *Northern Logger and Timber Processor* 15 (Apr. 1967), 24-25, 58-59, 67. Includes some history of forestry and the forest industries since the colonial period.

L21 Lambert, Richard S., and Pross, A. Paul. *Renewing Nature's Wealth: A Centennial History of the Public Management of Lands, Forests & Wildlife in Ontario, 1763-1967.* Toronto: Ontario Department of Lands and Forests, 1967. xvi + 630 pp. Illus., maps, notes, bib., index. A thorough forest history of Ontario with emphasis on the work of the Ontario Department of Lands and Forests and its predecessor agencies.

L22 Lambert, Robert S. "Logging on Little River, 1890-1940." *East Tennessee Historical Society's Publications* 33 (1961), 32-42. Logging, log driving, and logging railroads of the Little River Lumber Company, Townsend, Tennessee. These operations in Blount and Sevier counties of eastern Tennessee were on land now part of the Great Smoky Mountains National Park.

L23 Lambert, Robert S. "Logging the Great Smokies, 1880-1930." *Tennessee Historical Quarterly* 20 (Dec. 1961), 350-63. Logging and logging railroads in the Great Smoky Mountains of western North Carolina and eastern Tennessee.

L24 Lambert, Robert S. "Income-Tax Records as Sources for Economic History." *American Archivist* 24 (July 1961), 341-44. On the use of them to document the activities of lumber companies between 1900 and 1930 on what is now the Great Smoky Mountains National Park in eastern Tennessee.

L25 Lamm, L.M. *Tariff History of the Paper Industry of the United States, 1789 to 1922.* American Paper and Pulp Association, Special Report No. 8. New York, 1927. 32 pp.

L26 Lamm, W.E. *Lumbering in Klamath.* Klamath Falls, Oregon: Lamm Lumber Company, n.d. 40 pp. Illus. History of the lumber industry in Klamath County, Oregon, since the 1860s.

L27 Lammi, Joe O. "Article X of the Lumber Code in the Douglas Fir Region." Master's thesis, Oregon State College, 1937.

L28 La Mothe, G.E. "Forests of Quebec." *Culture* 10 (Sept. 1949), 230-49. Incidental history.

Lampard, Eric E. See Perloff, Harvey S., #P119

L29 Lampen, Dorothy. *Economic and Social Aspects of Federal Reclamation.* Johns Hopkins University Studies in Historical and Political Science, Series 48, No. 1. Baltimore: Johns Hopkins Press, 1930.

L30 Lampman, Ben Hur. *The Centralia Tragedy and Trial.* Tacoma: American Legion, 1920. 80 pp. A pro-American Legion interpretation of the violent incident of 1919 between Legionnaires and members of the Industrial Workers of the World.

Landegger, Carl C. See Landegger, Karl F., #L31

L31 Landegger, Karl F., and Landegger, Carl C. *Growing with the Paper Industry Since 1853: The Parsons Whittemore Organization and The Black Clawson Company.* New York: Newcomen Society in North America, 1968. 24 pp. New York and Ohio producers of machinery and equipment for the pulp and paper industry.

L32 Lander, Richard N. "Sands' Mills." *Westchester County Historical Bulletin* 28 (Oct. 1952), 90-108. On a sawmill near Armonk, New York, 1737-1905.

L33 Landon, Harry F. *The North Country, a History, Embracing Jefferson, St. Lawrence, Oswego, Lewis and Franklin Counties, New York.* 3 Volumes. Indianapolis: Historical Publishing Company, 1932. Illus., map. Includes some history of the forest industries in northern New York. Volumes 2 and 3 are biographical.

Lane, Anne Wintermute. See Lane, Franklin K., #L36

L34 Lane, Edward Hudson. *Lane: '. . .Furniture with a Tradition and a Future.'* New York: Newcomen Society in North America, 1962. 28 pp. Illus. Lane Company of Virginia.

L35 Lane, Ferdinand C. *The Story of Trees.* Garden City, New York: Doubleday, 1952. 384 pp. Illus. Incidental history.

L36 Lane, Franklin K. *The Letters of Franklin K. Lane, Personal and Political.* Ed. by Anne Wintermute Lane and Louise Herrick Wall. Boston: Houghton Mifflin, 1922. xxiv + 473 pp. Illus. Lane was Woodrow Wilson's Secretary of the Interior, 1913-1920.

L37 Lane, Leighton Ernest. "Historical Study of the Formation and Development of Small Woodlots in the Northeast." Master's thesis, State Univ. of New York, College of Forestry, 1959.

L38 omitted.

L39 Lang, Aldon S. *Financial History of the Public Lands in Texas.* Baylor Bulletin, Volume 35, No. 3. Waco and Dallas, Texas: Baylor University, 1932. 262 pp. On the disposition of the Texas public domain since 1836. A Ph.D dissertation at the University of Texas, 1931.

L40 Lang, Fred H. "Two Decades of State Forestry in Arkansas." *Arkansas Historical Quarterly* 24 (Autumn 1965), 208-19. Concern over forest fires led to the formation of the Arkansas Forest Protective Association in 1928 and the Arkansas State Forestry Commission in 1931. Fire fighting has been the chief state activity.

L41 Langdale, Harley, Jr. "Naval Stores." *Journal of Forestry* 54 (Oct. 1956), 643-45. Incidental references to the industry in the South since 1665.

L42 Lange, Erwin F. "Pioneer Botanists of the Pacific Northwest." *Oregon Historical Quarterly* 57 (June 1956), 109-24. On Thomas Howell (1842-1912), Wilhelm N. Suksdorf (1850-1932), William C. Cusick (1842-1922), and Martin W. Gorman (1852-1926).

L43 Lange, Erwin F. "John Jeffrey and the Oregon Botanical Expedition." *Oregon Historical Quarterly* 68 (June 1967), 111-24. An English botanical explorer in the Pacific Northwest, 1850s.

L44 Langelier, J.C. *Les Arbres de commerce de la province de Québec.* Quebec: Dussault & Proulx, 1908. 106 pp. Incidental forest history.

Langenbein, Walter B. See Hoyt, William G., #H585

L45 Langford, Nathaniel Pitt. *The Discovery of Yellowstone Park: Journal of the Washburn Expedition to the Yellowstone and Firehole Rivers in the Year 1870.* 1905. Reprint. Foreword by Aubrey L. Haines. Lincoln: University of Nebraska Press, 1972. lxi + 125 pp. Illus., maps, table, app., notes, index. Langford's introduction contains some history of the movement to establish Yellowstone National Park.

L46 Langille, Harold D. "Mostly Division 'R' Days: Reminiscences of the Stormy, Pioneering Days of the Forest Reserves." *Oregon Historical Quarterly* 57 (Dec. 1956), 301-13. Reminiscences of the author's employment with the General Land Office and other federal agencies on the Cascade Range Forest Reserve of Oregon, 1890s-1905, including discussion of grazing controversies.

L47 Langlie, Arthur B. "The Northwest Way of American Forestry." *Journal of Forestry* 48 (Jan. 1950), 14-17. A historical overview of logging and forest industries and attendant problems in Oregon and Washington.

L48 Langton, John. *Early Days in Upper Canada; Letters of John Langton from the Backwoods of Upper Canada and the Audit Office of the Province of Canada.* Ed. by W.A. Langton. Toronto: Macmillan, 1926. xl + 310 pp. Illus., maps. Includes reminiscences and letters concerning the lumber industry in Ontario in the mid-19th century.

L49 Lapham, Increase A.; Knapp, J.G.; and Crocker, H. *Report on the Disastrous Effects of the Destruction of Forest Trees, Now Going on so Rapidly in the State of Wisconsin.* 1867. Reprint. Madison: State Historical Society of Wisconsin, 1967. 104 pp. Index. Includes some incidental history of forest devastation by members of the Wisconsin Forest Commission. This report marked the beginning of forest conservation in Wisconsin.

Lardieri, Nichols J. See Gehm, Harry W., #G77

L50 Larsen, Christian L. *South Carolina's Natural Resources: A Study in Public Administration.* Columbia: University of South Carolina Press, 1947. 211 pp. History and description of natural resources administration, including forests and state parks.

L51 Larsen, J.A., and Delavan, C.C. "Climate and Forest Fires in Montana and Northern Idaho, 1909-1919." *U.S. Monthly Weather Review* 50 (Feb. 1922), 55-68.

L52 Larsen, J. A. "Early Researches in the Relations of Forest Fires and Unusual Weather Conditions, Humidity, Duff Moisture, and Inflammability." *Iowa State College Journal of Science* 22 (July 1948), 405-13.

Larsen, J.A. See Hartman, George G., #H181

L53 Larsen, J.A. "Some Pioneers and Leaders in American Forestry and Conservation." *Iowa State Journal of Science* 34 (May 15, 1960), 521-44. Brief biographies of Samuel B. Green, John A. Warder, John F. Lacey, Joseph T. Rothrock, Filibert Roth, Franklin B. Hough, and Bernhard E. Fernow.

L54 Larsen, James Arthur. *Wisconsin's Renewable Resources.* Madison: University of Wisconsin, 1957. xvi + 160 pp. Illus., maps, tables, diags., bibs. A report on research at the University of Wisconsin into the renewable resources of field, forest, lake, and stream. Includes some history since 1852.

Larsen, Lawrence H. See Glaab, Charles N., #G138, G139

L55 Larson, A. Karl. "Zion National Park With Some Reminiscences Fifty Years Later." *Utah Historical Quarterly* 37 (Fall 1969).

L56 Larson, Agnes M. "On the Trail of the Woodsman in Minnesota." *Minnesota History* 13 (Dec. 1932), 349-66. History of the lumber industry and its contribution to Minnesota since 1836.

L57 Larson, Agnes M. "When Logs and Lumber Ruled Stillwater." *Minnesota History* 18 (June 1937), 165-79. Stillwater, Minnesota, 1844-1914.

L58 Larson, Agnes M. "Some Aspects of the History of the White Pine Industry in the Upper Mississippi Region, with Special Reference to Minnesota." Ph.D. diss., Radcliffe College, 1938.

L59 Larson, Agnes M. *History of the White Pine Industry in Minnesota.* Minneapolis: University of Minnesota Press, 1949. Reprint. New York: Arno Press, 1972. xv + 432 pp. Illus., map, tables, notes, bib., index. A thorough history of federal and state land policy, logging, the lumber industry, and forest conservation in Minnesota from the 1830s to the 1930s.

L60 Larson, Agnes M. "Early Logging in the St. Croix Delta." *Northern Logger* 12 (Mar. 1964), 12-13, 42-43, 54-55. History of logging and the lumber industry along the St. Croix River of Minnesota, 1839-1860s; derived from the author's *History of the White Pine Industry in Minnesota.*

L61 Larson, Charles C. "Government and the Small Forest Holding: A Study of the Administration of the Governmental Program for Farm and Small Nonfarm Woodland Owners." Ph.D. diss., State Univ. of New York, College of Forestry, 1952.

L62 Larson, Charles C. *Timber Resources and the Economy of the Saranac Lake—Lake Placid Area.* State University of New York, College of Forestry, Bulletin No. 32. Syracuse, 1954. 42 pp. Illus., maps, diags., tables. Incidental history.

L63 Larson, Charles C. *Forest Economy of the Adirondack Region.* State University of New York, College of Forestry, Bulletin No. 39. Syracuse, 1956. vi + 48 pp. Illus., maps, tables, diags., notes. Incidental history.

L64 Larson, Edwin vH. *The Forest Resources of New Hampshire.* USFS, Forest Resource Report No. 8. Washington: GPO, 1954. iv + 39 pp. Illus., maps, diags., tables, notes, bib. Since 1830.

Larson, Edwin vH. See Schrepfer, Susan R., #S119

L65 Larson, Esther E. *Tales from the Minnesota Forest Fires: A Personal Experience of a Rural School Teacher.* St. Paul: Webb Publishing Company, 1912. 94 pp. Illus.

L66 Larson, George H. "Winnfield: City of Many Faces." *Forests & People* 23 (Fourth quarter, 1973), 28-38. Includes some history of the lumber, plywood, and pulp and paper industries in Winnfield, Louisiana.

L67 Larson, Joseph S. "Wildlife Forage Clearings on Forest Lands—A Critical Appraisal and Research Needs." Ph.D. diss., Virginia Polytechnic Institute and State Univ., 1966. 146 pp. Includes analysis of forest-wildlife habitat improvement practices in the East since 1935.

L68 Larson, Robert W. *The Timber Supply Outlook in South Carolina.* USFS, Forest Resource Report No. 3. Washington: GPO, 1951. iii + 66 pp. Illus., maps, diags., tables, notes, bib. Some history since 1936.

L69 Larson, Robert W. *The Timber Supply in Florida.* USFS, Forest Resource Report No. 6. Washington: GPO, 1952. iii + 60 pp. Illus., maps, tables, diags., notes, bib. Since 1909.

L70 Larson, Robert W. *Timber Supply Situation in Georgia.* USFS, Forest Resource Report No. 12. Washington: GPO, 1956. 44 pp.

L71 Larson, Robert W. *North Carolina's Timber Supply, 1955.* USFS, Southeastern Forest Experiment Station, Forest Survey Release No. 49. Asheville, North Carolina, 1957. 71 pp. Illus. Incidental history.

L72 Larson, Robert W. *Virginia's Timber.* USFS, Southeastern Forest Experiment Station, Forest Survey Release No. 54. Asheville, North Carolina, 1959. 72 pp. Incidental history.

L73 Larson, Robert W. *South Carolina's Timber.* USFS, Southeastern Forest Experiment Station, Forest Survey Release No. 55. Asheville, North Carolina, 1960. 38 pp. Illus.

L74 Larson, Robert W., and Goforth, M.H. *Florida's Timber.* USFS, Southeastern Forest Experiment Station, Forest Survey Release No. 57. Asheville, North Carolina, 1961. 32 pp. Since 1936.

L75 Larson, Robert W., and Spada, B. *Georgia's Timber.* USFS, Southeastern Forest Experiment Station, Resource Bulletin No. SE-1. Asheville, North Carolina, 1963. 39 pp.

L76 Larson, Robert Walter. "Ballinger vs. Rough Rider George Curry: The Other Feud." *New Mexico Historical Review* 43 (Oct. 1968), 271-90. On Territorial Governor George Curry's attitudes toward conservation in New Mexico and his relations with Gifford Pinchot, the USFS, Secretary of the Interior Richard A. Ballinger, and former president Theodore Roosevelt, 1907-1912.

L77 Laslett, John H.M. *Labor and the Left: A Study of Socialist and Radical Influences in the American Labor Movement, 1881-1924.* New York: Basic Cooks, 1970. vi + 326 pp. See for references to the Industrial Workers of the World.

L78 Lasswell, Mary. *John Henry Kirby, Prince of the Pines.* Austin: Encino Press, 1967. xv + 203 pp. Illus. Kirby was an east Texas lumberman.

L79 Latham, Bryan. "The Growth of the Timber Industry in Canada." *Wood* (Oct. 1955-Feb. 1956). Since the 18th century.

L80 Latham, Bryan. "The Development of the American Timber Trade." *Wood* (Sept. 1956-Jan. 1957). A general survey of the lumber industry in the United States since the colonial period.

L81 Latham, Bryan. "Hands Across the Sea: Seventy-Five Years of Anglo-American Lumber Trade—How It All Began." *Southern Lumberman* 193 (Dec. 15, 1956), 259-64. Emphasis on the period since 1881, but with brief references to the 17th century.

L82 Latham, Bryan. *Timber, Its Development and Distribution. A Historical Survey.* London: George G. Harrap & Company, 1957. xxvii + 303 + xxvii pp. Illus., index. Includes chapters on the history of the lumber industry in the United States and Canada.

L83 Latham, Bryan. *Wood: From Forest to Man.* London: George G. Harrap and Company, 1964. 192 pp. Illus., glossary, index. Essentially descriptive of wood technology and utilization, but includes many historical references to forestry, the lumber industry, and sawmill technology in North America.

L84 Latham, Bryan. *History of the Timber Trade Federation of the United Kingdom: The First Seventy Years.* London: Ernest Benn, 1965. 176 pp. Illus., index. Contains incidental references to the lumber trade with the United States and Canada.

L85 Latta, Robert Ray. *Reminiscences of Pioneer Life.* Kansas City, Missouri: F. Hudson, 1912. 186 pp. Latta was a logger and sawmill operator in the Midwest in the 1850s and 1860s.

L86 Lauber, Patricia. *Everglades Country: A Question of Life or Death.* New York: Viking Press, 1973. 125 pp. Incidental history of Everglades National Park, Florida.

L87 Laughead, William B. "Old Timers Will Remember the High Wheels." *Timberman* 48 (Jan. 1947), 213. Used in the transportation of logs.

L88 Laughead, William B. "The Birth of Paul Bunyan." OHI by W.H. Hutchinson. *Forest History* 16 (Oct. 1972), 44-49. Author-illustrator recalls the origins of Paul Bunyan lore, and his use of it for advertising the Red River Lumber Company, Westwood, California, 1910s.

L89 Lauver, Mary E. "A History of the Use and Management of the Forested Lands of Arizona, 1862-1936." Master's thesis, Univ. of Arizona, 1938.

L90 Lavender, David S. *The Big Divide.* Garden City, New York: Doubleday, 1948. 321 pp. Bib. A general history of Colorado, Wyoming, eastern Utah, and northern New Mexico, including chapters on conservation, reclamation, and tourism.

L91 Lavender, David S. *Land of Giants: The Drive to the Pacific Northwest, 1750-1950.* Garden City, New York: Doubleday, 1958. x + 468 pp. Maps, notes, bib. A general history of the region with some treatment of logging, the lumber industry, and forestry.

L92 Lavender, David S. "The Accessible Wilderness." *American West* 11 (Jan. 1974), 19-26. Incidental history.

Law, James. See Johnston, Hank, #J129

L93 Lawler, James. "Aperçu historique sur l'exploitation des forêts au Canada." *Bulletin of the Société de Géographique de Québec* 10 (1916), 271-81. Historical outline or summary of forest exploitation in Canada.

L94 Lawler, James. *Historical Sketch of Canada's Timber Industry.* Forestry Branch, Circular No. 15, Ottawa, 1922. 12 pp.

L95 Lawrence, Scott. "Railways in the Woods." *Raincoast Chronicles* 1 (Winter 1973), 16-23, 26. Some history and reminiscence of logging railroads in British Columbia since 1900.

L96 Lawrence, Joseph Collins. "Markets and Capital: A History of the Lumber Industry of British Columbia." Master's thesis, Univ. of British Columbia, 1957. Since 1778.

L97 Lawrence, Richard W., Jr. "The Adirondack Center." *New York History* 39 (July 1958), 256-60. On a museum established at Elizabethtown, New York, in 1955. It is partly concerned with the forest history of the Adirondack Mountain region.

L98 Lawson, George W. *History of Labor in Minnesota.* St. Paul: Minnesota State Federation of Labor, 1955. 623 pp. Includes references to labor in the forest industries.

L99 Lawson, Publius V. "Paper-Making in Wisconsin." *Proceedings of the Wisconsin Historical Society* 57 (1910), 273-80. Reprinted in *Lake States Timber Digest* 1 (Aug. 28, 1947), 9-10; 1 (Sept. 11, 1947), 9-10.

L100 Lawson, William P. *The Log of a Timber Cruiser.* New York: Duffield, 1915. 214 pp. Illus. Account of a six-month's timber cruise on the Gila National Forest, New Mexico.

L101 Laxton, Josephine. "Pioneers in Forestry at Biltmore." *American Forests* 37 (May 1931), 269-72, 319. On the men who pioneered American forestry at George W. Vanderbilt's Biltmore Estate near Asheville, North Carolina: Vanderbilt, Gifford Pinchot, and Carl Alwin Schenck.

L102 Laxton, Josephine. "Pisgah—A Forest Treasureland." *American Forests* 37 (June 1931), 339-42. On the movement to preserve part of the southern Appalachians since 1885, resulting in the creation of the Pisgah National Forest in 1916.

L103 Leach, Charles W. "Public Forestry Education under Federal Authority." Master's thesis, Yale Univ., 1947.

L104 Leach, Douglas E. *The Northern Colonial Frontier, 1607-1763.* New York: Holt, Rinehart and Winston, 1966. xviii + 266 pp. Illus., notes, bib., index. Includes reference to the forests and forest industries.

L105 Leach, M.C. *History of the 800th Engineer Forestry Company, World War II.* Century, Florida, 1954.

L106 Leach, Morgan L. *A History of the Grand Traverse Region.* Traverse City, Michigan: Grand Traverse Herald, 1883. Reprint. Mount Pleasant, Michigan: Central Michigan University, n.d. 162 pp. Includes some history of the lumber industry in and near Traverse City, Michigan, in the mid-19th century.

L107 Leach, Walter. "Old Time Lumber Days Along the West Branch." Pennsylvania Department of Forests and Waters, *Service Letter* 9 (July 1938), 99-105. West Branch of the Susquehanna River.

L108 Leadabrand, Russ. *A Guidebood to the Mojave Desert of California, Including Death Valley, Joshua Tree National Monument, and the Antelope Valley.* Los Angeles: Ward Ritchie Press, 1966. xii + 180 pp. Illus., maps, bib., index. Includes some history of the Joshua Tree National Monument.

L109 Leane, John J. *The Oxford Story, 1847-1958.* Rumford, Maine: Oxford Paper Company, 1958. 40 pp. Illus.

L110 Learned, Henry B. *The President's Cabinet: Studies in the Origin, Formation, and Structure of an American Institution.* New Haven: Yale University Press, 1912. 471 pp. Index. Includes chapters on the USDA, USDI, and their secretaryships.

L111 Leavitt, Clyde. "The Progress of Forestry in Canada." *Canadian Forestry Journal* 16 (No. 3, 1920), 130-38; 16 (No. 5, 1920), 259-62.

L112 Leavitt, Clyde. "Railway Fire Protection in Canada." *Journal of Forestry* 26 (Nov. 1928), 871-77. On the work of the chief fire inspector, Canadian Board of Railway Commissioners, since 1903.

L113 Leavitt, John F. *Wake of the Coasters.* Middletown, Connecticut: Wesleyan University Press, for the Marine Historical Association, 1970. xvii + 201 pp. Illus. Reminiscences of Maine coastal schooners and their lumber cargoes.

L114 LeBel, E.A. "Once Twenty Cargoes of Lumber a Day—Now None." *Canadian Lumberman* 60 (May 15, 1940), 17. On the lumber trade in southwestern Ontario since 1888.

L115 LeBarron, Russell K. "The History of Forestry in Hawaii: From the beginning through World War II." *Aloha Aina* I (Apr. 1970), 12-14.

L116 LeConte, Joseph N. "The Sierra Club." *Sierra Club Bulletin* 10 (Jan. 1917), 135-45. Club history by a charter member.

L117 LeDuc, Thomas. "The Maine Frontier and the Northeastern Boundary Controversy." *American Historical Review* 53 (Oct. 1947), 30-41. Includes some account of forest resources and timber trespass as factors in the boundary dispute between Maine and New Brunswick in the 1830s and 1840s.

L118 LaDuc, Thomas. "The Disposal of the Public Domain on the Trans-Mississippi Plains: Some Opportunities for Investigation." *Agricultural History* 24 (Oct. 1950), 199-204. On methods of investigation required for proper study of this subject.

L119 LeDuc, Thomas. "The Historiography of Conservation." *Forest History* 9 (Oct. 1965), 23-28. Argues that scholars should define the field of conservation history in broader terms, paying more attention to voluntary actions and state activities and less attention to the political facts of conservation.

L120 LeDuc, William Gates. *Recollections of a Civil War Quartermaster: The Autobiography of William Le Duc.* St. Paul: North Central Publishing Company, 1963. 167 pp. LeDuc was Commissioner of Agriculture, 1877-1881, treated here in two brief chapters.

L121 Lee, Francis B. "Forests of Colonial Jersey as the Settlers Found Them." *Forester* 1 (1895), 30.

L122 Lee, Guy A. "The General Records of the United States Department of Agriculture in the National Archives." *Agricultural History* 19 (Oct. 1945), 242-49. A summary of the principal correspondence files of the Office of the Secretary of Agriculture since 1893.

L123 Lee, Lawrence B. "William Ellsworth Smythe and the Irrigation Movement: A Reconsideration." *Pacific Historical Review* 41 (Aug. 1972), 289-311. Smythe's role in the irrigation and reclamation movements, 1880s-1900s, is revealing of the importance of popular agitation and support for conservation generally.

L124 Lee, Lawrence B. "Environmental Implications of Governmental Reclamation in California." *Agricultural History* 49 (Jan. 1975), 223-29. Includes some history of the irrigation and reclamation movements in California.

L125 Lee, Ronald F. *Public Use of the National Park System, 1872-2000.* Washington: U.S. National Park Service, 1968. 93 pp. Includes some history of national park policy.

L126 Lee, Ronald F. *Family Tree of the National Park System.* Philadelphia: Eastern National Park and Monument Association, 1972. 99 pp. Charts, index.

L127 Leech, Carl A. "Sharon Hollow: Story of an Early Mulay Sawmill of Michigan." *Michigan History Magazine* 17 (Summer-Autumn 1933), 377-92. Built in 1835-1836.

L128 Leech, Carl A. "Lumbering Days." *Michigan History Magazine* 18 (Spring 1934), 135-42. Reminiscences of Michigan lumbering, 1860s-1870s, based on an interview with John J. Higgins.

L129 Leech, Carl A. "Paul Bunyan's Land and the First Sawmills of Michigan." *Michigan History Magazine* 20 (Winter 1936), 69-89. Early sawmills and the lumber industry in eastern Michigan, 19th century.

L130 Leech, Carl A. "Pictures of Michigan Lumbering." *Michigan History Magazine* 23 (Autumn 1939), 337-49. On the need to preserve such history in photographs, including some examples.

L131 Leech, Carl A. "Deward, a Lumberman's Ghost Town." *Michigan History Magazine* 28 (Jan.-Mar. 1944), 5-19. Deward, Michigan, prospered from 1902 to 1912 during the logging of the David Ward estate, ca. 90,000 acres of pine and hardwood.

L132 Leedy, Daniel L. "The Wildlife Society." *Journal of Forestry* 54 (Dec. 1956), 821-23. On its work in conservation since established in 1937.

L133 Leeper, David Rohrer. *The Argonauts of 'Forty-Nine: Some Recollections of the Plains and the Diggings.* South Bend, Indiana: J.B. Stall & Company, 1894. 146 pp. Also includes references to logging in California.

L134 Lee-Whiting, Brenda B. "Saga of a Nineteenth Century Sawmill." *Canadian Geographical Journal* 74 (No. 2, 1967), 46-51. On a water-powered sawmill at Balaclava, Ontario, operated continuously since the 1850s.

L135 Leffelman, L. J. "Early Ohio Forest Plantings." *Ohio Forest News* 2 (1928), 6-8.

L136 Leger, Mary Celeste. "A Study of the Public Career of Ethan Allen Hitchcock." Ph.D. diss., City Univ. of New York, 1971. 373 pp. Secretary of the Interior, 1899-1907, in the administrations of William McKinley and Theodore Roosevelt.

L137 Leggett, Robert F. "A Prophet of Conservation." *Dalhousie Review* 45 (No. 1, 1965), 34-42. On George Perkins Marsh (1801-1882), his seminal study, *Man and Nature,* and his influence on the conservation movement.

Le Grande, W.P. See Bethune, J.E., #B255

L138 Lehde, Norman B. "Origins of the Pinchot Family." *Journal of Forestry* 63 (Aug. 1965), 582-84. Since 1816.

L139 Lehman, John W., and Vogenberger, Ralph A. "The Role of a Regional Agency in Forest Fire Control." *American Forests* 53 (June 1955), 430-35. The Tennessee Valley Authority and cooperating local forestry agencies since 1933.

L140 Lehman, John W. *The Changing Sawmill Industry: A Status Report on 58 Circular Sawmills in the Tennessee Valley, 1950-1960.* Norris, Tennessee: Tennessee Valley Authority, Division of Forestry Relations, 1961. 23 pp. Tables, diags.

L141 Leighly, John B. "John Muir's Image of the West." *Annals of the Association of American Geographers* 48 (Dec. 1958), 309-18. On Muir's work as a geographer and naturalist in California, Nevada, Oregon, Washington, and Alaska, his exaltation of the wilderness, and his dislike of urban life and industrialization, 1868-1914.

L142 Leland, Waldo Gifford. "Newton Bishop Drury." *National Parks Magazine* 25 (Apr.-June 1951), 42-44, 62-66. On his term as director of the National Park Service, 1940-1951.

L143 Le Master, Dennis Clyde. "Recent Merger Activity of the Largest Firms in the Forest Products Industries." Ph.D. diss., Washington State Univ., 1974. 268 pp. Economic study of 420 mergers in the industry, 1950-1970, with special attention to Oregon and Washington.

Lemay, Neil. See Mitchell, J. Alfred, #M499

L144 Lemly, James H. *The Gulf, Mobile, and Ohio: A Railroad That Had to Expand or Expire.* Indiana University, School of Business, Study No. 36. Homewood, Illinois: R.D. Irwin, 1953. viii + 347 pp. Illus., maps, diags., notes, bib. Includes the relation of the Mississippi lumber industry to railroad and transportation development in that state since 1847.

L145 Lemmer, George F. "Norman J. Colman, and *Colman's Rural World:* A Study in Agricultural Leadership." Ph.D. diss., Univ. of Missouri, 1947. 407 pp.

L146 Lemmer, George F. *Norman J. Colman and 'Colman's Rural World': A Study in Agricultural Leadership.* University of Missouri Studies, Volume 25, No. 3. Columbia: Curators of the University of Missouri, 1953. 168 pp. Illus., notes, bib. Colman was Grover Cleveland's Commissioner of Agriculture, 1885-1889, and in the latter year became the first Secretary of Agriculture.

L147 Lent, Henry Bolles. *From Trees to Paper: The Story of Newsprint.* New York: Macmillan, 1952. 149 pp.

L148 Lentz, Artie F. "Potlatch Forests, Incorporated." *Pacific Historian* 16 (Summer 1972), 37-46. Log drives on the main branch and North Fork of the Clearwater River to Lewiston, Idaho, 1928-1971.

Lentz, A.N. See Moore, E.B., #M535

Leonard, E.C. See Allard, Harry A., #A69

L149 Leopold, Aldo. "Wilderness as a Form of Land Use." *Journal of Land and Public Utility Economics* 1 (1925), 398-404.

L150 Leopold, Aldo. "The Conservation Ethic." *Journal of Forestry* 31 (Oct. 1933), 634-43. An influential argument for a philosophical or ethical approach to land use.

L151 Leopold, Aldo. *Game Management.* New York: Charles Scribner's Sons, 1933. xxi + 481 pp. Illus. A standard text in wildlife management and conservation; includes a chapter on the history of ideas in game management.

L152 Leopold, Aldo. "Origin and Ideals of Wilderness Areas." *Living Wilderness* 5 (July 1940), 7-9. On the wilderness movement in the Southwestern states since 1920, by one of its leaders.

L153 Leopold, Aldo. "Flambeau: The Story of a Wild River." *American Forests* 49 (Jan. 1943), 12-14, 47. Some history of the Flambeau River region of northern Wisconsin and a plea for its protection.

L154 Leopold, Aldo. "P.S. Lovejoy." *Journal of Wildlife Management* 7 (Jan. 1943), 125-28. Parrish Storrs Lovejoy (1884-1942) worked for the USFS, 1905-1912, taught forestry at the University of Michigan, 1912-1920, and subsequently made many contributions to forest and wildlife conservation in Michigan.

L155 Leopold, Aldo. *A Sand County Almanac, and Sketches Here and There.* New York: Oxford University Press, 1949. xiii + 226 pp. Illus. Includes some historical references to the conservation movement, although it is better known as a classic statement of Leopold's philosophy of ecology and a land ethic. There are later editions and reprints.

L156 Leopold, Aldo. *Round River, from the Journal of Aldo Leopold.* Ed. by Luna B. Leopold. New York: Oxford University Press, 1953. Reprint. 1972. xi + 173 pp. Excerpts relating to the author's travels to study nature in Canada, Mexico, and the western United States, and illustrating his devotion to the conservation of wildlife, 1922-1937.

L157 Leopold, Aldo Starker. "Too Many Deer." *Sierra Club Bulletin* 38 (Oct. 1953), 51-57. On policies for regulating the size of deer herds, 1904-1953.

L158 Leopold, Aldo Starker. *Wildlife in Alaska: An Ecological Reconnaissance.* New York: Ronald Press, 1953. 129 pp. Incidental history.

L159 Leopold, Luna B. "Vegetation of Southwestern Watersheds in the Nineteenth Century." *Geographical Review* 41 (Apr. 1951), 295-316. A study of travelers' accounts and photographs, 1826-1946, indicating watershed cover and vegetation density in New Mexico and Arizona.

Leopold, Luna B. See Leopold, Aldo, #L156

L160 Leopold, Luna B., and Maddock, Thomas. *The Flood Control Controversy: Big Dams, Little Dams and Land Management.* New York: Ronald Press, 1954. 278 pp. On flood control programs and controversies involving the U.S. Army Corps of Engineers, USDA, and USFS.

L161 Lepine, Paul. *The Life Story of a Lumberjack: The Hardships, Fights, Loves and Adventures of a Wanderer from Coast to Coast: The True Story of Paul Lepine Written by Himself.* N.p., 1924. 164 pp. Includes some account of logging in the Rocky Mountain states.

L162 Leslie, A.P. "Some Historical Aspects of Forestry in Ontario." *Forestry Chronicle* 26 (Sept. 1950), 243-50.

L163 Leslie, A.P. "Forest Research in Ontario." *Canadian Geographical Journal* 44 (Feb. 1952), 71-91. Incidental history.

L164 Leslie, Donald S. *The Story of Hammermill Paper Company.* New York: Newcomen Society in North America, 1964. Hammermill Paper Company of Erie, Pennsylvania, with operations in New York and Pennsylvania.

L165 Leslie, James W. "The Arkansas Lumber Industry." Master's thesis, Univ. of Arkansas, 1938.

L166 Leue, Adolph. *Carriage Timber, or the Carriage Manufacturing Interest and Forestry.* Columbus, Ohio, 1889. 28 pp.

L167 Leuthold, Walter M. "NLMA—Past and Present." *Southern Lumberman* 193 (Dec. 15, 1956), 122-24. On the National Lumber Manufacturers Association since 1902.

Levin, Oscar R. See Cheyney, Edward G., #C288

L168 Levin, Oscar R. "The South Olympic Tree Farm." *Journal of Forestry* 52 (Apr. 1954), 243-49. Washington.

L169 LeWarne, Charles P. "The Aberdeen, Washington, Free Speech Fight of 1911-1912." *Pacific Northwest Quarterly* 66 (Jan. 1975), 1-12. Part of an effort of the Industrial Workers of the World to organize lumberjacks and millworkers in Washington's Grays Harbor lumber region.

L170 Lewis, E.A., comp. *Laws Relating to Forestry, Game Conservation, Flood Control and Related Subjects.* Washington: GPO, 1944. 228 pp.

L171 Lewis, Ferris E. "Frederic: A Typical Logging Village in the Twilight of the Lumbering Era, 1912-1918." *Michigan History* 32 (Dec. 1948), 321-39; 33 (June 1949), 131-40; 34 (Mar. 1950), 34-49.

L172 Lewis, Henry T. *Patterns of Indian Burning in California: Ecology and Ethnohistory.* Ballena Anthropological Papers, No. 1. Ramona, California: Ballena Press, 1973. 148 pp. Includes reference to forest burning.

L173 Lewis, Martin D. "Lumberman from Flint: The Michigan Career of Henry H. Crapo, 1855-1869." Ph.D. diss., Univ. of Chicago, 1957. 287 pp.

L174 Lewis, Martin D. *Lumberman from Flint: The Michigan Career of Henry H. Crapo, 1855-1869.* Detroit: Wayne State University Press, 1958. x + 289 pp. Illus., maps, tables, notes, apps., index. Crapo (1804-1869) was a governor of Michigan and had extensive interests in land speculation and the lumber industry, especially involving his sawmill in Flint.

L175 Lewis, Oscar. *High Sierra Country.* American Folkways Series. New York: Duell, Sloan & Pearce, 1955. ix + 291 pp. Maps. Includes references to the forests and the forest industries of the Sierra Nevada and adjacent areas of California and Nevada, 1772-1915.

L176 Lewis, R.G. *Wood-Using Industries of Ontario.* Forestry Branch, Bulletin No. 36. Ottawa, 1913. 127 pp. Incidental history.

L177 Lewis, R.G., and Boyce, W.G.H. *Wood-Using Industries of the Maritime Provinces.* Forestry Branch, Bulletin No. 44. Ottawa, 1914. 100 pp.

L178 Lewis, R.G., and Boyce, W.G.H. *Wood-Using Industries of the Prairie Provinces.* Forestry Branch, Bulletin No. 50. Ottawa, 1915. 75 pp.

L179 Lewis, R.G., and Doucet, J.A. *Wood-Using Industries of Quebec.* Forestry Branch, Bulletin No. 63. Ottawa, 1918. 89 pp.

L180 Lewis, R.G. *Wood-Using Industries of Ontario — II.* Forestry Branch, Bulletin No. 75. Ottawa, 1924. 106 pp.

L181 Lewis, R.G. "The Development of the Pulp and Paper Industry from 1900-1925." *Pulp and Paper Magazine,* International Number. (Feb. 1927), 185-92.

L182 Lewis, Richard W.B. *The American Adam: Innocence, Tragedy, and Tradition in the Nineteenth Century.* Chicago: University of Chicago Press, 1955. ix + 204 pp. Notes. On the relationship of wilderness and civilization, as revealed in American literature, 1820-1860.

Lewis, Verne B. See Gaus, John M., #G72

L183 Leydet, Francois. *The Last Redwoods and the Parkland of Redwood Creek.* San Francisco: Sierra Club, 1969. 160 pp. Illus. Includes some history of the region now preserved as Redwoods National Park, California.

L184 L'Heureux, Eugene. *Le Probleme des chantiers.* Chicoutimi, Quebec, 1927. 31 pp. On the life, manners, customs, and social conditions of Quebec lumberjacks.

L185 Lichtenberg, Caroline. "Beginnings of the United States Military Land Bounty Policy, 1637-1812." Master's thesis, Univ. of Wisconsin, 1945.

L186 Lieber, Emma. *Richard Lieber, by His Wife, Emma.* Indianapolis: Privately printed, 1947. 170 pp. Illus. Lieber (1869-1944) was director of the Indiana Department of Conservation, 1919-1933, a long-time advocate of state parks, and consultant to the National Park Service.

L187 Lieber, Richard. *America's Natural Wealth: A Story of the Use and Abuse of Our Resources.* New York: Harper, 1942. 245 pp. Illus. Includes a chapter on forests.

L188 Liimatta, Into Maine. "Problems and Development of Logging to Supply the First Pulp Mill in Alaska." Master's thesis, Univ. of Washington, 1953.

L189 Lilienthal, David E. *TVA — Democracy on the March.* New York: Harper & Brothers, 1944. xiv + 248 pp. Illus., app., bib., index. Contains some account of conservation under the Tennessee Valley Authority by the chairman of that agency. A revised edition appeared in 1953.

L190 Lilienthal, David E. *The Journals of David E. Lilienthal: The TVA, 1939-1945.* 5 Volumes. New York: Harper Row, 1964.

L191 Lillard, Richard G. "Timber King." *Pacific Spectator* 1 (Winter 1947), 14-26. Biographical sketch of Frederick Weyerhaeuser (1834-1914), prominent lumberman of Illinois, Minnesota, and Washington, and president of Weyerhaeuser Timber Company of Tacoma.

L192 Lillard, Richard G. *The Great Forest.* New York: Alfred A. Knopf, 1948x + 399 + xivpp. Illus., notes, apps, bib., index. A broad survey of American forest history since the colonial period, including history of the forests and frontier settlement, the forest industries, forest conservation, and forestry.

L193 Lillard, Richard G. *Eden in Jeopardy: Man's Prodigal Meddling with His Environment: The Southern California Experience.* New York: Alfred A. Knopf, 1966. xii + 329 + vi pp. Illus., map, bib., index. Broadly concerned with man's impact on the natural environment of southern California, including a section on forest fires and their consequences.

L194 Lillard, Richard G. "The Siege and Conquest of a National Park." *American West* 5 (Jan. 1968), 28-32, 67-72. Controversies over the introduction of automobiles to California's Yosemite National Park, especially in the 1900s-1910s.

L195 Lincoln, Ceylon Childs. "Personal Experiences of a Wisconsin River Raftsman." *Proceedings of the State Historical Society of Wisconsin*]8 (1911), 181-89. In 1868.

L196 Lincoln, Charles Z. *The Constitutional History of New York from the Beginning of the Colonial Period to the Year 1905.* 5 Volumes. Rochester: Lawyers Co-Operative Publishing, 1906. Constitutional history of the New York Forest Preserve is treated in Volume 3, pp. 391-454.

L197 Lincoln, Proctor. "New England's Lumber Business." *Southern Lumberman* 193 (Dec. 15, 1956), 169-72. Since 1719.

L198 Lincoln, Robert P. "Hot Springs National Park." *Fur-Fish-Game* 89 (July 1949), 3-6, 23. Arkansas, since 1804.

K199 Lincoln, Robert P. "Charles Hallock." *Fur-Fish-Game* 46 (May 1951), 7-9, 22-24. Hallock (1834-1917) was the founder of *Forest and Stream.*

L200 Lind, Anna M. "Women in Early Logging Camps: A Personal Reminiscence." *Journal of Forest History* 19 (July 1975), 128-35. In California, Oregon, and Washington, 1920s-1930s.

L201 Linder, Arthur. "Forest Fire Fighting." *The Big Smoke 1974* (Pend Oreille County Historical Society) (1974), 33-37. Pend Oreille County, Washington, since 1910.

L202 Lindh, C. Otto. "The Aztec Case: A Story of Two Centuries." *American Forests* 65 (Dec. 1959), 24-27, 48-49. On litigation between the Santa Fe Pacific Railroad Company and the Aztec Land and Cattle Company concerning ownership of railroad grant lands in Arizona, and the eventual acquisition of part of the lands by the USFS, since 1886.

L203 Linn, Ed R. "Florida's Forests Are Different." *American Forests* 54 (Feb. 1948), 76-78, 94. On forestry in Florida since 1935.

L204 Linn, Ed R. *A Forest Industries Survey of Oklahoma.* Oklahoma Agricultural Experiment Station, Bulletin B-325. Stillwater, 1948. 35 pp. Map, diags., tables, notes. Since 1889.

Linn, Thomas G. See Wood, Andrew D., #W425

L205 Lipscomb, E.M. "Henry Ford — Timber Farmer." *Southern Lumberman* (Dec. 15, 1942), 202-04. Richmond Hill Plantation, Bryan County, Georgia.

L206 Littlejohn, Bruce M. *Quetico-Superior Country: Wilderness Highway to Wilderness Recreation.* Toronto: Quetico Foundation of Ontario, 1965. 31 pp. Illus. History and significance of the Quetico-Superior area of northern Minnesota and western Ontario; reprinted from the *Canadian Geographical Journal* (Aug.-Sept. 1965).

L207 Little, Charles E. *The New Oregon Trail.* Washington: Conservation Foundation, 1974. 37 pp. Map. An account of the development and passage of state land-use legislation in Oregon, 1960s-1970s.

L208 Little, James. *The Lumber Trade of the Ottawa Valley with a Description of Some of the Principal Manufacturing Establishments.* Ottawa: Times Printing and Publishing Company, 1872. Includes some history of the lumber industry.

L209 Little, James. *The Timber Supply Question of Canada and the United States of America.* Montreal, 1876. 23 pp.

L210 Little, John J. "The 1910 Forest Fires in Montana and Idaho: Their Impact on Federal and State Legislation." Master's thesis, Univ. of Montana, 1968.

L211 Little, Silas. "Ellwood B. Moore." *Journal of Forestry* 46 (Oct. 1948), 769-70. Sketch of Moore's career with the New Jersey Department of Conservation, especially since 1926.

L212 Littlefield, Daniel F., Jr., and Underhill, Lonnie E. "Hildebrand's Mill Near Flint, Cherokee Nation." *Chronicles of Oklahoma* 48 (Spring 1970), 83-94. On a sawmill established in the 1870s in Indian Territory and operated until recent years by Aaron Headin Beck and his descendants.

L213 Littlefield, Daniel F., Jr., and Underhill, Lonnie E. "Timber Depredations and Cherokee Legislation, 1869-1881." *Journal of Forest History* 18 (Apr. 1974), 4-13. The lumber industry, timber depredations, and regulatory legislation by the Cherokee nation of the Indian Territory (Oklahoma).

Littlefield, Edward W. See York, Harlan H., #Y21

L214 Littlefield, Edward W. "Bill Howard—An Appreciation." *Journal of Forestry* 47 (Apr. 1949), 303-04. A long-time state forestry official in New York; also appeared in *Adirondac* 22 (Mar.-Apr. 1958), 33-34.

L215 Littlefield, Edward W. "Small Woodlands and the Public Services: A Retrospect." *Journal of Forestry* 53 (Feb. 1955), 125-28. On the services of federal and New York agencies to small woodlands in New York since 1899.

L216 Lively, Charles E., and Preiss, Jack J. *Conservation Education in American Colleges.* New York: Ronald Press, for the Conservation Foundation, 1957. ix + 267 pp. Diags., tables, notes, bibs. Since 1867.

L217 *Living Wilderness.* "1889. . .Olaus J. Murie. . . 1963." *Living Wilderness* No. 84 (Summer-Fall 1963), 3-14. Murie was a naturalist, preservationist, and officer of the Wilderness Society. Other tributes follow in the same issue.

L218 *Living Wilderness.* "1906. . .Howard Clinton Zahniser. . .1964." *Living Wilderness* No. 85 (Winter-Spring 1964), 3-6. Zahniser was a preservationist and officer of the Wilderness Society.

L219 *Living Wilderness.* "Harvey Broome, July 15, 1902-March 8, 1968." *Living Wilderness* 31 (Winter 1967-1968), 4-6. Broome was a preservationist and a founder and officer of the Wilderness Society.

L220 Livingston-Little, Dallas E. *An Economic History of North Idaho, 1890-1900.* Los Angeles: Journal of the West, 1972. xx + 134 pp. Illus., maps, tables, bib., index. Includes some history of the lumber industry and the use of wood in other economic enterprises; first published serially in the *Journal of the West* (1963-1964).

Lockard, C.R. See Behre, C. Edward, #B177

L221 Locke, P.R. "Washington Log Brands." *Timberman* 27 (June 1926), 48-50. Incidental history.

L222 Lockhart, J.W. "The I.W.W. Raid at Centralia." *Current History* 17 (Oct. 1922), 55-57. On violence between Industrial Workers of the World and American Legionnaires, Centralia, Washington, 1919.

L223 Lockley, Fred. *History of the Columbia River Valley from the Dalles to the Sea.* 3 Volumes. Chicago: S.J. Clarke Publishing Company, 1928. Includes some history of lumbering along the lower Columbia River.

L224 Lockmann, Ronald F. "Changing Evaluations of Resources and the Establishment of National Forests in California's Transverse Ranges, 1875 to 1911." Ph.D. diss., Univ. of California, Los Angeles, 1972. 219 pp. Los Padres, Angeles, and San Bernardino national forests.

L225 Lockwood, Frank C. *The Life of Edward E. Ayer.* Chicago: A.C.McClurg, 1929. 300 pp. Lumberman.

Lockwood, Frank C. See Barnes, Will Croft, #B85

L226 Lockwood, James H. "Early Times and Events in Wisconsin." *Collections of the Wisconsin State Historical Society* 2 (1856), 98-196. Includes references to his early logging on the Red Cedar branch of the Chippewa River, 1822.

L227 Lockwood Trade Journal Company. *1690-1940, 250 Years of Papermaking in America.* New York, 1940. 180 pp. Illus.

L228 Lockwood Trade Journal Company. *Technical Association of the Pulp and Paper Industry. . .1916-1941: Tappi, Twenty-fifth Anniversary Supplement, Paper Trade Journal, September 25, 1941.* New York, 1941. 63 pp.

L229 Lockwood Trade Journal Company. *The Progress of Paper; With Particular Emphasis on the Remarkable Industrial Development in the Past 75 Years and the Part That 'Paper Trade Journal' Has Been Privileged to Share in That Development.* New York, 1947. 392 pp. Illus. A history of papermaking and the pulp and paper industry; also appeared as Volume 124, No. 27 of *Paper Trade Journal* (1947).

L230 Lodewick, J.E. *Wood-Using Industries of Virginia.* Virginia Polytechnic Institute, Engineering Extension Division Series, Bulletin No. 21. Blacksburg, 1929. 169 pp. Illus. There is a revised version by John B. Grantham (q.v.).

L231 Loehr, Rodney C. "Caleb D. Dorr and the Early Minnesota Lumber Industry." *Minnesota History* 24 (June 1943), 125-41. Dorr started sawing lumber at St. Anthony Falls (Minneapolis) in 1848, and was boom master of the Mississippi & Rum River Boom Company, organized in 1856.

L232 Loehr, Rodney C. "Franklin Steele, Frontier Businessman." *Minnesota History* 27 (Dec. 1946), 309-18: A lumberman and speculator in Minnesota timberlands in the mid-19th century.

L233 Loehr, Rodney C. "Preserving the History of the Forest Products Industries." *Southern Lumberman* 17 (Dec. 15, 1948), 271-72, 274-76. On the program of the Forest Products History Foundation since 1946.

L234 Loehr, Rodney C. "Saving the Kerf: The Introduction of the Band Saw Mill." *Agricultural History* 23 (July 1949), 168-72. On the manufacture of band saws by Jacob R. Hoffman and his firm at Fort Wayne, Indiana, 1868-1886. Also appeared in *Southern Lumberman* 178 (June 15, 1949), 43-46.

Loehr, Rodney C. See Girard, James W., #G133

L235 Loehr, Rodney C. "Softwood Lumber Grading." *Southern Lumberman* 180 (May 1, 1950), 50, 52; (May 15, 1950), 48, 50; (June 1, 1950), 50, 52. On the origins of lumber grading in the 1830s, the adoption of lumber standards in the 1920s, and federal antitrust suits against trade association grade marking.

L236 Loehr, Rodney C. "Some More Light on Paul Bunyan." *Journal of American Folklore* 64 (Oct.-Dec. 1951), 405-07. Lumberjack folklore, ca. 1880s.

Loehr, Rodney C. See Mason, David T., #M266

L237 Loft, Genivera Edmunds. "The Evolution of the Wood-Working Industries of Wisconsin." Master's thesis, Univ. of Wisconsin, 1916.

L238 Logan, Cyrus Field. "Reminiscences of an Old-Timer." *Paper Industry and Paper World* 24 (Aug. 1942), 503-05.

L239 Logan, Harold A. *Trade Unions in Canada: Their Development and Functioning.* Toronto: Macmillan, 1948. 656 pp. Bib. See for reference to organized labor in forest industries.

L240 Logan, Robert R. "Notes on the First Land Surveys in Arkansas." *Arkansas Historical Quarterly* 19 (Autumn 1960), 260-70. Made in 1815-1816.

L241 Lo Jacono, S., comp. *Establishment and Modification of National Forest Boundaries: A Chronologic Record, 1891-1959.* Washington: USFS, 1959. 92 + vii pp. Tables. Essentially a compilation of proclamations, executive orders, public laws, administrative orders, and other documents concerning national forests and their boundaries.

L242 Lokke, Carl Ludwig. "A French Appreciation of New England Timber." *New England Quarterly* 8 (1935), 409-11. A translated letter (1787) of the French intendancy of marine at Toulon, remarking "in favor of procuring ships' knees from New England."

L243 Lokken, Roscoe L. "The Administration and Disposal of the Public Lands in Iowa." Ph.D. diss., Univ. of Iowa, 1940. 166 pp.

L244 Lokken, Roscoe L. *Iowa—Public Land Disposal.* Iowa City: State Historical Society of Iowa, 1942. Reprint. New York: Arno Press, 1972. 318 pp.

L245 Lomax, Alfred L. "Hawaii-Columbia River Trade in Early Days." *Oregon Historical Quarterly* 43 (Dec. 1942), 328-38. Includes reference to lumber exports from Oregon to the Hawaiian Islands, 1830s-1850s.

L246 Lomax, Alfred L. "Early Shipping and Industry in the Lower Siuslaw Valley." *Lane County Historian* 16 (Summer 1971), 32-39. Includes some history of the lumber industry in coastal Lane County, Oregon, 1870s-1890s.

L247 Long, E. John. "The National Arboretum." *American Forests* 60 (May 1954), 10-14, 50. In Washington, D.C., since 1927.

L248 Long-Bell Lumber Company. *From Tree to Trade.* Kansas City, Missouri, 1920. 46 pp. Illus. Incidental history.

L249 Longwood, Franklin R. "A Land Use History of Benton County, Oregon." Master's thesis, Oregon State College, 1940.

Longwood, Franklin R. See Ferguson, Roland H., #F63

L250 Loomis, R.D. "Alberta's Forest Resources and Forest Management." *Pulp and Paper Magazine of Canada* 58 (June 1957), 108-13. Includes some history of Alberta Department of Lands and Forests.

L251 Loomis, W.E., and McComb, A.L. "Recent Advances of the Forest in Iowa." *Proceedings of the Iowa Academy of Science* 51 (1944), 217-24.

L252 Lord, Eleanor L. *Industrial Experiments in the British Colonies of North America.* Johns Hopkins University Studies in Historical and Political Science, Extra Volume No. 17. Baltimore: Johns Hopkins Press, 1898. x + 154 pp. Tables, bib. Includes some history of the lumber trade, naval stores, and the crown forests, 1696-1729. Originally a Ph.D. dissertation at Bryn Mawr College (1896). There is a modern reprint (1969).

L253 Lord, Russell. *Behold Our Land.* Boston: Houghton Mifflin, 1938. Reprint. New York: Da Capo Press, 1974. ix + 309 pp. Includes some history of natural resource use in North America since the beginning of settlement; emphasis is on soil depletion and conservation.

L254 Lord, Russell. *The Wallaces of Iowa.* Boston: Houghton Mifflin, 1947. Reprint. New York: Da Capo Press, 1972. xiii + 615 pp. Illus., notes, bib., index. Henry C. Wallace was Secretary of Agriculture in the Warren G. Harding and Calvin Coolidge administrations, 1921-1924, and Henry A. Wallace was Secretary of Agriculture in the Franklin D. Roosevelt administration, 1933-1940.

L255 Lord, Russell R. "Chief of the Soil Revival: Some Notes for Historians of Agriculture." *Land* 10 (Spring 1951), 83-89. On Hugh H. Bennett, chief of the Soil Conservation Service, 1935-1951.

L256 Lord, Russell R. "Forester Apprentices." *American Forests* 66 (Jan. 1960), 26-27, 59-63; (Feb. 1960), 30-32, 58-60. On the author's life in Baltimore County, Maryland, and the program of woods and forestry camping instruction he organized there.

L257 Lothian, W.F. *A Brief History of National Park Administration in Canada.* Ottawa: Department of Northern

Affairs and National Resources, National Parks Branch, 1955. Processed.

L258 Louisiana. Department of Conservation. *Brief History of Conservation in Louisiana, and Facts Regarding Contract with Urania Lumber Company.* New Orleans, 1928. 32 pp.

L259 Louisiana. Forestry Commission. *Forest Resources of Louisiana.* Baton Rouge: Louisiana Forestry Commission and the American Forestry Association, 1947. 86 pp. Incidental history.

L260 Louisiana State University and Agricultural and Mechanical College, School of Forestry. *Modern Forest Fire Management in the South.* 4th Forestry Symposium. Baton Rouge: Louisiana State University, 1955. Includes several articles of a historical nature on forest fires in the South.

L261 Love, Hamilton. "Experiences of a Lumber Salesman." *Southern Lumberman* 61 (Dec. 25, 1909), 49-50. In the South.

L262 Love, Lawrence D., and Goddell, B.C. "Watershed Research on the Fraser Experiment Station." *Journal of Forestry* 58 (Apr. 1960), 272-75. On USFS studies of snow accumulation and streamflow in relation to patterns of timber harvesting in the Arapaho National Forest, Colorado, since 1929.

L263 Lovejoy, Parrish Storrs. "Michigan's Fight for Forests." *American Forestry* 28 (Dec. 1922), 749-53. History of forest conservation and the forestry movement in Michigan since ca. 1905.

L264 Lovejoy, Parrish Storrs. "In the Name of Development." *American Forestry* 29 (July 1923), 387-93, 447. Some history since 1905 of timber depredations of "homesteaders" on national forests; also concerned with cutover lands policy in the Great Lakes states.

L265 Lovejoy, Parrish Storrs. "Legend Unending, Beginnings Unknown: P.S. Lovejoy on 'Paul Bunyon'." *Forest History* 11 (Jan. 1968), 37-38. A letter (1921) to Esther Shepard concerning the origins of Paul Bunyan tales.

L266 Loveridge, Earl W. "The Fire Suppression Policy of the U.S. Forest Service." *Journal of Forestry* 42 (Aug. 1944), 549-54. On its development since the early 20th century.

L267 Lowdermilk, Walter C. *Conquest of the Land Through 7,000 Years.* Washington: Soil Conservation Service, 1953. 30 pp. Illus.

L268 Lowe, Vyron D. "Gumming and Guiding in the White Mountains." OHI by Richard G. Wood. *Appalachia* 29 (June 1953), 370-73. On the author's work of collecting gum from spruce trees in New Hampshire and Maine and refining and marketing it as chewing gum, 1890s.

L269 Lowell, John W. "The First Ranger Station." *American Forests and Forest Life* 36 (July 1930), 396-97. Established by the USDI on the Bitterroot Forest Reserve of Montana in 1899 and built by H.C Tuttle and Than Wilkerson, rangers in the employ of Division R.

L270 Lowenthal, David. "The Maine Press and the Aroostook War." *Canadian Historical Review* 32 (Dec. 1951), 315-36. A border incident (Maine-New Brunswick) involving Canadian lumberjacks, 1838-1839.

L271 Lowenthal, David. "George Perkins Marsh and the American Geographical Tradition." *Geographical Review* 43 (April 1953), 207-13. On Marsh's interest in erosion and "forest economy," 1847-1864.

L272 Lowenthal, David. "George Perkins Marsh." Ph. D. diss., Univ. of Wisconsin, 1954.

L273 Lowenthal, David. *George Perkins Marsh, Versatile Vermonter.* New York: Columbia University Press, 1958. xii + 442 pp. Illus., notes, bib. Marsh (1801-82) had a broad career in law, business, national politics, foreign service, and academics, but he is best remembered for his work in natural history and especially for his *Man and Nature* (1864), which had great implications for the forestry and conservation movements.

L274 Lowenthal, David. "Nature and the American Creed of Virtue." *Landscape* 9 (Winter 1959-1960), 24-25. Examines some ideas or assumptions of the "conservationist creed," including the "balance of nature" concept.

Lowenthal, David. See Marsh, George Perkins, #M218

L275 Lower, Arthur R.M. "The Assault on the Laurentian Barrier, 1850-1870." *Canadian Historical Review* 10 (Dec. 1929), 294-307. Including efforts of lumbermen to exploit and open for settlement the Laurentian Shield area of Quebec and Ontario.

L276 Lower, Arthur R.M. "The Forest in New France: A Sketch of Lumbering in Canada before the English Conquest." *Canadian Historical Association, Annual Report, 1928* (1929), 78-90.

L277 Lower, Arthur R.M. "Lumbering in Canada: A Study in Economics and Social History." Ph.D. diss., Harvard Univ., 1929. 699 pp.

L278 Lower, Arthur R.M. "The Trade in Square Timber." *University of Toronto Studies in History and Economics* 6 (1933), 40-61. From 1784 to 1908, and its relation to Canadian economic development. This essay also appears in *Approaches to Canadian Economic History,* ed. by W.T. Easterbrook and M.H. Watkins (Toronto, 1967).

L279 Lower, Arthur R.M. *Settlement and the Forest Frontier in Eastern Canada.* Canadian Frontiers of Settlement, Volume 9, Part 1. Toronto: Macmillan, 1936 xiv + 166 pp. Illus., maps, notes, tables, apps., index. On the relation of settlement to provincial land policies and the lumber and pulp and paper industries in the Maritime provinces, Quebec, and Ontario, since the early 19th century.

L280 Lower, Arthur R.M. "A History of the Lumber Trade between Canada and the United States." In Lower, Arthur R.M.; Carrothers, W.A.; and Saunders, A.A. *The North American Assault on the Canadian Forest,* ed. by Harold A. Innis. Relations of Canada and the United States Series. Toronto: Ryerson Press; New Haven: Yale University Press, 1938. Reprint. New York: Greenwood Press, 1968. pp. 1-223. Maps, notes, tables, graphs, bib., index. Also concerned with the Canadian forests, logging, lumberjacks, and sawmills. The geographical focus on the Great Lakes area, St. Lawrence Valley, and the Maritime provinces. All but one of fifteen chapters treat the lumber industry of the 19th century.

L281 Lower, Arthur R.M. *My First Seventy-Five Years.* Toronto: Macmillan, 1967. 384 pp. Illus., index. Autobiography of a leading Canadian forest historian.

L282 Lower, Arthur R.M. *Great Britain's Woodyard: British America and the Timber Trade, 1763-1867.* Montreal: McGill-Queens University Press, 1973. xiv + 271 pp. Illus., notes, tables, charts, index. On the Canadian lumber industry and the timber trade with Great Britain.

Lowes, John Livingston. See Bartram, William, #B117

L283 Lowitt, Richard. *A Merchant Prince of the Nineteenth Century: William E. Dodge.* New York: Columbia University Press, 1954. xii + 384 pp. Dodge held timberlands in addition to his many other manufacturing and mercantile interests.

L284 Lowry, Alexander, and Verardo, Denzil. *Big Basin.* Los Altos, California: Sempervirens Fund, 1973. 50 pp. Illus. Verardo's text outlines the movement led by Andrew P. Hill to preserve redwoods by establishing Big Basin State Park, California, in 1902. Subsequent park history is also included.

L285 Lowry, Alexander, and Earnshaw, Deanne. *Castle Rock—West of Skyline.* Los Altos, California: Sempervirens Fund, 1973. 32 pp. Illus. Earnshaw's text outlines the movement led by Russell Varian to establish Castle Rock State Park in California's Santa Cruz Mountains.

Lowry, Robert E. See Durisch, Lawrence L., #D315

L286 Lowry, Stanley Todd. "Henry Hardtner, Pioneer in Southern Forestry: An Analysis of the Economic Bases of His Reforestation Program." Master's thesis, Louisiana State Univ., 1956.

L287 Lucas, Alec. "Thoreau, Field Naturalist." *University of Toronto Quarterly* 23 (Apr. 1954), 227-32. Henry David Thoreau's work as a naturalist, 1840-1860.

L288 Lucas, Edward R. "Swedish Gang Saws in the Pacific Northwest." *American Swedish Monthly* 46 (Mar. 1952), 8-9, 29; 46 (Apr. 1952), 19, 28. On the Mill Engineering & Supply Company, founded by Carl S. Sundbom, Swedish engineer, in 1929.

L289 Lucas, Rex A. *Minetown, Milltown, Railtown: Life in Canadian Communities of Single Industry.* Toronto: University of Toronto Press, 1972. xiii + 433 pp. Includes some historical references to life in Canadian lumber towns.

L290 Lucas, Robert C. "The Quetico-Superior Area: Recreational Use in Relation to Capacity." Ph.D. diss., Univ. of Minnesota, 1962. 390 pp. Incidental history.

L291 Lucas, Robert C. "The Contribution of Environmental Research to Wilderness Policy Decisions." *Journal of Social Issues* 22 (Oct. 1966), 116-26. Examines several aspects of public dispute over wilderness use and the need for further research on the environment.

L292 Lucas, Robert C. "Forest Service Wilderness Research in the Rockies: What We've Learned So Far." *Western Wildlands* 1 (Spring 1974), 4-12. A summary of USFS research on wilderness areas in the northern Rocky Mountains, especially Montana.

L293 Lucia, Ellis. "Sequoia Johnny and His Nuggets." *American Forests* 66 (June 1960), 34-35, 55-56. John Porter planted cones of the California *Sequoia gigantea* near Forest Grove, Oregon, in the late 1840s. The trees are now 200 feet tall.

L294 Lucia, Ellis. *Tough Men, Tough Country.* Englewood Cliffs, New Jersey: Prentice-Hall, 1963. 336 pp. Illus. Includes reference to lumberjacks and forest industries of the Pacific Northwest.

L295 Lucia, Ellis. *The Big Blow: Story of the Pacific Northwest's Columbus Day Storm.* 1963. Enlarged edition. Portland: Overland West Press, 1967. 80 pp. Illus., map. Heavily illustrated account of the freak storm on October 12, 1962, which devastated much standing timber in northern California, Oregon, and Washington.

L296 Lucia, Ellis. *Head Rig, Story of the West Coast Lumber Industry.* Portland: Overland West Press, in cooperation with the West Coast Lumbermen's Association, 1965. vii + 238 pp. Illus., bib., index. A popular history of the lumber industry of the Pacific Northwest and of the role of the West Coast Lumbermen's Association since 1911.

L297 Lucia, Ellis. *The Big Woods: Logging and Lumbering—from Bull Teams to Helicopters—in the Pacific Northwest.* Garden City, New York: Doubleday & Company, 1975. xii + 222 pp. Illus., bib., index. A popular history of Pacific Northwest forests and forest industries. There are chapters on special topics such as Portland's Forestry Building, lumberjacks and the Industrial Workers of the World, the Yacolt and Tillamook burns, tree farming, the Keep Green movement, power saws, company towns, and modern logging technology.

L298 Luck, H.F. "Ontario Wholesale Lumber Dealers Have an Effective Association: Although the Present Organization Originated Only Twenty-Three Years Ago, There is Record of Wholesalers Cooperating as Early as 1886." *Canada Lumberman* 60 (Aug. 1, 1940), 92-93.

L299 Lufkin, Dan W. *'The Spoiler's Hand–The Rage of Gain': Social, Political and Environmental Considerations of Land Use.* New York: Newcomen Society in North America, 1974. 18 pp. Illus. Includes some history of land-use legislation and practice, especially in Grant County, Oregon, and in Connecticut.

L300 Lull, Howard W. "Forest Influences: Growth of a Concept." *Journal of Forestry* 47 (Sept. 1949), 700-05. On the mythical and scientific bases (since 1785) for the belief that forests have physical effects on soil, climate, stream flow, and man. Scientifically documented cases of forest influences have had impact on conservation programs and legislation.

L301 Lull, Howard W., and Storey, Herbert C. "Factors Influencing Streamflow from Two Watersheds in Northeastern Pennsylvania." *Journal of Forestry* 55 (Mar. 1957), 198-200. Since 1928.

L302 Lull, Howard W. "Forested Municipal Watersheds in the Northeast." *Journal o Forestry* 58 (Feb. 1960), 83-86. Includes reference to forest-bounded reservoirs since 1882.

L303 Lull, Howard W. "Watershed Management Research in the Northeast." *Journal o Forestry* 58 (Apr. 1960), 285-87. On studies to determine how forest cover affects streamflow, since 1912.

L304 Lull, Howard W., and Reinhart, Kenneth G. *Forests and Flood in the Eastern United States.* USFS, Research Paper NE-226. Upper Darby, Pennsylvania: Northeast Forest Experiment Station, 1972. 94 pp. Includes a historical treatment of study on the influence of forests on floods, erosion, and sedimentation.

L305 *Lumber.* "History of the Timber Supply in the New England States." *Lumber* (Jan. 24, 1918), 12ff.

L306 *Lumber.* "Motor Trucks in the Lumber Industry." *Lumber* (Oct. 28, 1918), 17-42. Many illustrations and short articles on the introduction and increasing use of trucks in the lumber industry.

L307 *Lumberman.* "Who Was the Coast's First Woodworker?" *Lumberman* 49 (Mar. 1948), 44-45, 78, 80. On the E.C. Miller Cedar Lumber Company, Aberdeen, Washington, founded in 1915.

L308 *Lumberman.* "Surveyors of the Wilderness." *Lumberman* 49 (July 1948), 52-53, 92-94, 96, 98. On timber surveying in Oregon since 1851.

L309 *Lumber Trade Journal.* "Interesting Data on Canadian Lumber Industry, 1917-1929." *Lumber Trade Journal* (Apr. 15, 1931), 23ff.

L310 *Lumber World Review.* "Rise and Progress of the Kaul Lumber Co." *Lumber World Review* (Apr. 10, 1914), 35-50. An Alabama firm.

L311 *Lumber World Review.* "Interesting Dry Kiln History." *Lumber World Review* 34 (June 25, 1918), 29-31. On the inventions of the Moore Dry Kiln Company, Jacksonville, Florida.

L312 *Lumber World Review.* "History of the Shingle Branch, West Coast Lumbermen's Association." *Lumber World Review* (Dec. 25, 1918), 79-80.

L313 *Lumber World Review.* "An Evolution in Iron and Steel: Being a Story of the Rise and Progress of the W.K. Henderson Iron Works and Supply Company of Shreveport, La." *Lumber World Review* (Nov. 10, 1920), 59-118. Heavily illustrated article on a Southern manufacturer of sawmill machinery.

L314 *Lumber World Review.* "Pacific Spruce Corporation and Subsidiaries." *Lumber World Review* 46 (Feb. 10, 1924), 35-124. History of the firms which purchased the properties of the U.S. Spruce Production Corporation in Oregon's Lincoln County, 1920s. Biographical sketches of company officials of the Pacific Spruce Corporation, C.D. Johnson Lumber Company, Manary Logging Company, and Pacific Spruce Railway Company are included.

L315 Lundie, J.A. "British Columbia's First Paper Mill." *Pulp and Paper Magazine of Canada* 33 (Dec. 1932), 473.

L316 Lundsted, James E. "Log Marks, Forgotten Lore of the Logging Era." *Wisconsin Magazine of History* 39 (Autumn 1955), 44-46. On the use of these marks and their analogy to cattle brands.

L317 Luntey, Robert S. "Katmai National Monument." *National Parks Magazine* 30 (Jan.-Mar. 1956), 7-15, 36-37. On the volcanic eruption of Mount Katmai in 1912, the creation of a national monument in 1918, and its subsequent enlargement.

L318 Lussier, O., and Maheux, Georges. *Forests of the Southern United States and Their Influence on the Quebec Forest Industry.* Quebec: Laval University Forest Research Foundation, 1959. 23 pp. Map. Incidental history.

L319 Luther, Thomas F. "This Business of Private Forestry." *Journal of Forestry* 45 (July 1947), 492-93. On private forestry as practiced at the Luther Forest, Saratoga County, New York, since 1898.

Lutz, E.A. See Zimmerman, W.E., #Z16

L320 Lutz, George, and Vetleson, Jack. "Working and Quitting on the B.C. Coast." OHI by Derek Reimer. *Sound Heritage* 3 (No. 2, 1974), 25-33. Reminiscences of logging and working conditions in coastal British Columbia, 1940s.

L321 Lutz, Harold J. *Trends and Silvicultural Significance of Upland Forest Successions in Southern New England.* Yale University, School of Forestry, Bulletin No. 22. New Haven, 1928. 68 pp.

L322 Lutz, Harold J. "The First-Born of Alaskan Forests: How the Russian Ship 'Phoenix' Was Hewn from the Sylvan Monarchs of Resurrection Bay." *American Forests and Forest Life* 35 (July 1929), 403-05, 408. Completed in 1794, it was the first ship built on the Pacific Coast of North America.

L323 Lutz, Harold J. "Original Forest Composition in Northwestern Pennsylvania as Indicated by Early Land Survey Notes." *Journal of Forestry* 28 (Dec. 1930), 1098-1103. Based on land surveys conducted in 1814-1815.

L324 Lutz, Harold J., and Chandler, Robert F., Jr. *Forest Soils.* New York: John Wiley and Sons, 1946. xi + 514 pp. Incidental history.

Lutz, Harold J. See Hawley, Ralph C., #H229

L325 Lutz, Harold J., and Cline, A.C. *Results of the First Thirty Years of Experimentation in Silviculture in the Harvard Forest, 1908-1938.* Harvard Forest, Bulletin No. 23. Petersham, Massachusetts, 1947. 182 pp.

L326 Lutz, Harold J. *Ecological Effects of Forest Fires in the Interior of Alaska.* USDA, Technical Bulletin No. 1133. Washington: GPO, 1956. 121 pp. Illus., map, diags., tables, notes, bib. Incidental history.

L327 Lutz, Harold J. "Applications of Ecology in Forest Management." *Ecology* 38 (Jan. 1957), 46-49. Since 1914.

L328 Lutz, Harold J. *Aboriginal Man and White Man as Historical Causes of Fires in the Boreal Forest, with Particular Reference to Alaska.* Yale University, School of Forestry, Bulletin No. 65. New Haven, 1959. 55 pp. Bib. On the period from 1868 to 1915.

L329 Lutz, Harold J. *Early Forest Conditions in the Alaska Interior: An Historical Account with Original Sources.* Juneau: USFS, Northern Forest Experiment Station, 1963. 74 pp. Bib. Based mostly on observations made in the 19th century.

L330 Lutz, Harold J. *Sitka Spruce Planted in 1805 at Unalaska Island by the Russians.* Juneau: USFS, Northern Forest Experiment Station, 1963. 25 pp.

L331 Lyman, Henry H. *Memories of the Old Homestead.* Oswego, New York: R.J. Oliphant, 1900, 181 pp. Autobiography of a member of the New York Forest Commission.

L332 Lyon, Alanson Forman. "A Trip up the Menominee River in 1854." Ed. by Lewis Beeson. *Michigan History* 47 (Dec. 1963), 301-11. Diary of a trip to investigate logging possibilities along the Menominee River of Michigan and Wisconsin.

L333 Lyon, Charles J. "Hemlocks and History in New Hampshire." *New England Quarterly* 8 (1935), 567-72. A study of hemlock stumps, cut in 1790, helps recreate history of Wolfeboro, New Hampshire.

L334 Lyon, Grace E. "New Jersey Forests and Forestry." *Forestry Quarterly* 8 (Dec. 1910), 450-61. Includes some incidental historical references.

L335 Lyon, R.W. "Reforestation and Pulp Companies." *Canadian Forestry Journal* 16 (July 1920), 340-42. Incidental history.

M1 Maass, Arthur. *Muddy Waters: The Army Engineers and the Nation's Rivers.* Cambridge: Harvard University Press, 1951. Reprint. New York: Da Capo Press, 1974. xiv + 306 pp. On the history of the Corps of Engineers and its role in the development of water resources, especially California's Kings River project and rivalries with the Bureau of Reclamation.

M2 McAllister, Walter A. "A Study of Railroad Land Grant Disposal in California with Reference to the Western Pacific, the Central Pacific, and the Southern Pacific Railroad Companies." Ph.D. diss., Univ. of Southern California, 1940. 190 pp.

M3 McArdle, Richard C. "History of Forest Fire Prevention in the U.S." Master's thesis, Univ. of Michigan, 1952.

M4 McArdle, Richard E. "The Forest Service's First Fifty Years." *Journal of Forestry* 53 (Feb. 1955), 99-106. USFS chief writes of forestry work conducted by his agency and its predecessors since 1876.

M5 McArdle, Richard E. "Fifty Years of Firest Progress." *Soil Conservation* 29 (Feb. 1955), 147-52. With the USFS.

M6 McArdle, Richard E. "Looking at Forestry over 50 Years." *Pulp & Paper* 29 (July 1, 1955), 88-90.

M7 McArdle, Richard E. "Seventy-Five Years in Southern Forestry." *Southern Lumberman* 193 (Dec. 15, 1956), 119-21. On the work of the USFS and other advances in forestry in the South since the 1880s.

M8 McArdle, Richard E. "Earle Hart Clapp, 1877-1970." *Journal of Forestry* 68 (Aug. 1970), 498. Clapp was USFS chief, 1939-1943.

M9 McArdle, Richard E. *Dr. Richard E. McArdle: An Interview with the Former Chief, U.S. Forest Service, 1952-1962.* OHI by Elwood R. Maunder, Santa Cruz, California: Forest History Society, 1975. 252 pp. Illus., apps., index. Processed. On McArdle's career wtih the USFS at the Pacific Northwest Forest and Range Experiment Station and as chief.

M10 McAtee, Waldo Lee. "In Memoriam: Theodore Sherman Palmer." *Auk* 73 (July 1956), 367-77. On his service with the Division of Ornithology and Mammalogy, USDA, Biological Survey, U.S Fish and Wildlife Service, and his lifelong interest in wildlife conservation, 1889-1955.

McBain, B.T. See Snell, Ralph M., #S411

M11 McBee, May (Wilson). *The Life and Times of David Smith: Patriot, Pioneer, and Indian Fighter.* Kansas City, Missouri: N.p., 1959. 84 pp. Smith (1753-1835) also owned a sawmill in the Natchez District (Mississippi) during the 1780s.

M12 McCaffrey, Joseph E. "Go South, Young Man." OHI by Elwood R. Maunder. *Forest History* 8 (Winter 1965), 2-18. On his career in industrial forestry and public service, especially with International Paper Company in the South.

M13 McCaffrey, Joseph E. "State Forestry in the South." OHI by Elwood R. Maunder. *Forest History* 16 (Oct. 1972), 50-53. Reminiscences of state forestry in Florida, Georgia, South Carolina, Mississippi, and Arkansas, 1920s-1940s.

M14 McCaffrey, Joseph E. *Go South, Young Man.* OHI by Elwood R. Maunder. Santa Cruz, California: Forest History Society, 1973. 270 pp. Illus., index. Processed. On McCaffrey's career in forestry, primarily with International Paper Company in the South, since the 1930s.

M15 McCarthy, G. Michael. "Colorado Confronts the Conservation Impulse, 1891-1907." Ph.D. diss., Univ. of Denver, 1969. 538 pp.

M16 McCarthy, G. Michael. "White River Forest Reserve: The Conservation Conflict." *Colorado Magazine* 49 (Winter 1972), 55-67. On the establishment of the White River Forest Reserve, Colorado, 1891, and subsequent political opposition to it through the 1890s, led by H.H. Eddy.

M17 McCarthy, G. Michael. "Retreat from Responsibility: The Colorado Legislature in the Conservation Era, 1876-1908." *Rocky Mountain Social Science Journal* 10 (Apr. 1973), 27-36. On its original ambivalent attitude and its opposition after 1891 to federal and state conservation efforts.

M18 McCarthy, G. Michael. "The Pharisee Spirit: Gifford Pinchot in Colorado." *Pennsylvania Magazine of History and Biography* 97 (July 1973), 362-78. On Pinchot's public appearances in Colorado in behalf of federal forest policy and reserves, and the anticonservationist reaction of Colorado congressmen and stockmen, 1896-1907.

M19 McCarthy, Keith R. "Forever Wild?" *American Forests* 58 (Aug. 1952), 6-9, 43. On a provision of the New York Constitution of 1885 regarding the Forest Preserve in the Adirondack and Catskill mountains, and proposals for changing forest management practices.

M20 McCarthy, Keith R. "The Steward of 'Seward's Folly'." *American Forests* 59 (Apr. 1953), 8-9, 41-45. On Frank Heintzleman as a USFS forester in Alaska since 1918, and his appointment as territorial governor.

M21 McClay, T.A. "Seasonal Fluctuations in Ponderosa Pine Lumber Prices." *Journal o Forestry* 57 (Sept. 1959), 644-47. Since 1936.

M22 McClellan, J.C. "The Growing American Tree Farm System." *Connecticut Woodlands* 20 (May 1955), 48-50. Since 1941.

M23 McClellan, Sylvia. *Timber: The Story of McPhee.* Dolores, Colorado: Dolores Star, 1971. On the New Mexico Lumber Company and its operations at McPhee, Colorado, since the 1920s.

M24 McClelland, B. Riley. "The Courts and the Conservation of Natural Beauty." *Western Wildlands* 1 (Spring 1974), 20-26. A summary of 20th-century legislation and litigation.

M25 McClelland, John M., Jr. *Longview: The Remarkable Beginnings of a Modern Western City.* Portland: Binfords & Mort, 1949. x + 158 pp. Illus., maps, index. On the planned city of Longview, Washington, built between 1919 and 1923 by Robert Alexander Long (1850-1934) to accommodate the Long-Bell Lumber Company and later the site of many forest products industries.

M26 McClelland, John M., Jr. "Terror on Tower Avenue." *Pacific Northwest Quarterly* 57 (Apr. 1966), 65-72. On the violent incident at Centralia, Washington, 1919, involving the Industrial Workers of the World and the American Legion.

M27 McClement, Fred. *The Flaming Forests.* Toronto: McClelland & Stewart, 1969. 220 pp. A history of forest fires in North America (emphasis on Canada) since the fire of 1871 at Peshtigo, Wisconsin.

M28 McCloskey, Michael. "Note and Comment: Natural Resources—National Forests—The Multiple Use-Sustained Yield Act of 1960." *Oregon Law Review* 41 (Dec. 1961), 49-78. Includes some legislative history of the Multiple Use-Sustained Yield Act and of USFS law and administration since the Forest Reserve Act of 1897.

M29 McCloskey, Michael. "The Wilderness Act of 1964: Its Background and Meaning." *Oregon Law Review* 45 (June 1966), 288-321. Legislative analysis and history of the Wilderness Act.

M30 McCloskey, Michael. "The Last Battle of the Redwoods." *American West* 6 (Sept. 1969), 55-64. On the movement to preserve redwoods and to create Redwood National Park, California, since 1918.

M31 McCloskey, Michael. "Wilderness Movement at the Crossroads, 1945-1970." *Pacific Historical Review* 41 (Aug. 1972), 346-61. A Sierra Club leader examines the wilderness preservation movement.

M32 McCluney, William R., ed. *The Environmental Destruction of South Florida.* Coral Gables, Florida: University of Miami Press, 1971. viii + 134 pp. Maps.

M33 McClung, L. "A Brief History of Forestry in West Virginia." *West Virginia Conservation* 23 (Sept. 1959), 1-3.

M34 McClung, Robert M. *Lost Wild America: The Story of Our Extinct & Vanishing Wildlife.* New York: William Morrow, 1969. 240 pp. Illus.

M35 McClure, Albert J. "A Rare Forester." *Journal of Forestry* 71 (Nov. 1973), 716-17. On the career of Thomas H. Gill (1891-1972), a versatile forester and writer of international experience and reputation.

M36 McClure, J.W. "The Development of Hardwood Inspection Rules." *Southern Lumberman* (Dec. 17, 1921), 130-31. Includes some history of the National Hardwood Lumber Association.

McComb, A.L. See Loomis, W.E., #L251

M37 McConkey, O.M. *Conservation in Canada.* Toronto: J.M. Dent, 1952. 215 pp. Includes some history.

M38 McConnell, Grant. "The Conservation Movement: Past and Present." *Western Political Quarterly* 7 (Sept. 1954), 463-78. On the development of the conservation movement, its underlying principles, and challenges to them since 1907.

M39 McConnell, Grant. "The Multiple-Use Concept in Forest Service Policy." *Sierra Club Bulletin* 44 (Oct. 1959), 14-28. Incidental history.

M40 McCord, Charles R. "A Brief History of the Brotherhood of Timber Workers." Master's thesis, Univ. of Texas, 1958.

M41 McCormick, Dell. *Tall Timber Tales—More Paul Bunyan Stories.* Caldwell, Idaho: Caxton Printers, 1939. 155 pp.

McCormick, Jack. See Potzger, John E., #P290

M42 McCormick, Jack. *The Living Forest.* New York: Harper and Row, in cooperation with the American Museum of Natural History, 1959. 127 pp. Includes incidental history of American forests.

M43 McCormick, Jack. *The Life of the Forest.* New York: McGraw-Hill, in cooperation with World Book Encyclopedia and USDI, 1966. 232 pp. A text in forest ecology with incidental historical references.

M44 McCown, William H. "Early Day Sawmills on Russian River: An Account of the 'Inexhaustible' Redwood Forests That Once Lay Within Fifty Miles of San Francisco." *Timberman* 32 (Feb. 1931), 28-29, 60.

M45 McCoy, George W. *A Brief History of the Great Smoky Mountains National Park Movement in North Carolina.* Asheville, North Carolina, 1940.

M46 McCoy, Robert B. "Rendezvous with Red Death." *American Forests* 63 (July 1957), 34-35, 44-45. On a forest fire in Blackwater Canyon, Wyoming, 1937.

M47 McCracken, Harold. *George Catlin and the Old Frontier.* New York: Dial Press, 1958. 216 pp. Illus., notes, bib. Catlin (1796-1892) was a landscape artist whose work influenced American attitudes toward nature and wilderness.

M48 McCraw, W.E. "The History and Development of Rubber-Tired Skidders." *Northern Logger* 14 (May 1966), 12-13, 34-35. In Canada, since 1951.

M49 McCrea, R.C. "Tax Discrimination in the Paper and Pulp Industry." *Quarterly Journal of Economics* 21 (Aug. 1907), 632-44. Incidental history.

McCready, Al. See Holbrook, Stewart H., #H436

M50 McCreight, Israel. *Cook Forest Park: Story of the Sixteen Year Battle to Save the Last Stand of Historic Penn's Woods.* Du Bois, Pennsylvania: Gray Printing, 1936. 109 pp.

M51 McCulloch, Walter F. "Leo A. Isaac." *Journal of Forestry* 47 (May 1949), 401-02. On Isaac's career with the USFS, especially in silviculture research in Oregon and Washington, since 1920.

M52 McCulloch, Walter F. *Forest Management Education in Oregon.* Corvallis: Oregon State College, 1949. 135 pp. Incidental history.

M53 McCulloch, Walter F. "The 'Firstest' Logger." *American Forests* 58 (Nov. 1952), 10-11, 32-36. On William K. Dyche, a logger in various parts of the West, 1899-ca. 1910, and subsequently a superintendent for the Algoma Lumber Company, Klamath Falls, Oregon.

M54 McCulloch, Walter F., comp. *Woods Words: A Comprehensive Dictionary of Loggers Terms.* Portland: Oregon Historical Society and Champoeg Press. vi + 219 pp. Illus. Terminology collected since 1919 in logging camps in the Pacific Northwest.

M55 McCulloch, Walter F. "Big Men, Big Camps, Big Timber." *Timberman* 60 (Nov. 1959), 48-51, 82. Pacific Northwest.

M56 McCulloch, Walter F. "The Making of 'Woods Words'." OHI by Amelia Fry. *Forest History* 16 (Oct. 1972), 34-37. McCulloch compiled a dictionary of West Coast logging terms, *Woods Words* (1958).

M57 McCulloch, Winifred. *The Glavis-Ballinger Dispute.* Inter-University Case Program, ICP Case No. 4. Washington, 1952. 20 pp. On criticism by Louis Glavis in 1908-1909 against Richard A. Ballinger, Commissioner of the General Land Office and later Secretary of the Interior, regarding his supervision of Alaskan coal lands and other natural resource matters.

M58 McCune, Jason C. "The Port of Oakland, California, with Reference to Lumber Handling Facilities." *Timberman* 27 (Mar. 1925), 40-41, 166, 170. Since 1919.

M59 McCune, Wesley. *Ezra Taft Benson: A Man with a Mission.* Washington: Public Affairs Press, 1958. 123 pp. Index. Benson was Secretary of Agriculture, 1953-1961.

M60 Macdaniels, E.H. "Twenty-Five National Forests of North Pacific Region." *Oregon Historical Quarterly* 42 (Sept. 1941), 247-55. A chronology of forest reserve withdrawals and the establishment of national forests in Oregon since 1892.

Macdaniels, E.H. See DeVries, Wade, #D157

M61 Macdaniels, E.H. *A Decade of Progress in Douglas-Fir Forestry.* Seattle: Joint Committee on Forest Conservation of the West Coast Lumbermen's Association and the Pacific Northwest Loggers Association, 1943. 64 pp. Illus., tables, graphs. A general account of forestry progress in western Oregon and western Washington during the 1930s.

McDermott, Clare (Anderson). See Anderson, Antone A., #A423

M62 McDiarmid, J.S. "Railroads Gave Impetus to Manitoba Lumbering." *Canada Lumberman* 60 (Aug. 1, 1940), 74-75.

M63 MacDonald, Alexander Norbert. "Seattle's Economic Development, 1880-1910." Ph.D. diss., Univ. of Washington, 1959. 356 pp. Logging, forest industries, and lumber exports were a factor in Seattle's growth.

M64 MacDonald, Alexander Norbert. "The Business Leaders of Seattle, 1880-1910." *Pacific Northwest Quarterly* 50 (Jan. 1959), 1-13. Statistical profiles of 87 prominent business leaders, 13 of whom were lumbermen.

M65 Macdonald, Austin F. *Federal Aid: A Study of the American Subsidy System.* New York: Thomas Crowell, 1928. 285 pp. Bib. Includes a chapter on forest fire control.

M66 MacDonald, D.A. "Canada's Forest Sky Fleet." *American Forests* 37 (Apr. 1931), 203-05, 254. On Canadian use of airplanes in forestry, 1920s.

M67 Macdonald, Dwight. *Henry Wallace: The Man and the Myth.* New York: Vanguard Press, 1948. 187 pp. Henry A. Wallace was Secretary of Agriculture, 1933-1940.

M68 MacDonald, F.A. "A Historical Review of Forest Protection in British Columbia." *Forestry Chronicle* 5 (Dec. 1929), 31-35.

M69 MacDonald, Gilmore B. "Wood-Using Industries of Iowa." *Ames Forester* 11 (1923), 103-12. Incidental history.

M70 MacDonald, Gilmore B. "The Beginning of a National and State Forestry Program in Iowa." *Ames Forester* 25 (1935), 15-20.

M71 MacDonald, J.E. *Shantymen and Sodbusters: An Account of Logging and Settlement of Kirkwood Township, 1869-1928.* N.p.: By the author, 1966. vii + 134 pp. Illus., index. In northern Ontario near the North Channel of Lake Huron.

M72 McDonald, John. "Georgia-Pacific: It Grows Big on Trees." *Fortune* 65 (May 1962), 111-17. Includes some history of the Portland-based Georgia-Pacific Corporation.

M73 McDonald, John K. "Dr. C. Audrey Richards." *Journal of Forestry* 49 (Dec. 1951), 918-19. On her career, especially as a wood products pathologist at the Forest Products Laboratory, Madison, Wisconsin, since 1917.

M74 MacDonald, Onilee. "A History of Au Sable and Oscoda." Master's thesis, Wayne State Univ., 1942. 65 pp. A lumber region in Michigan.

M75 McDonald, P.A. "The Origin of Forest Protection in Wisconsin." *Wisconsin Conservation Bulletin* 8 (Apr. 1943), 8-11.

M76 MacDonald, R. Douglas. "Multiple Use Forestry in Megalopolis: A Case Study of the Evolution of Forest Policies and Programs in Connecticut." Ph.D. diss., Yale Univ., 1969. 355 pp. Since 1901.

M77 MacDougall, Alex. " 'Crooked Brook': A Song of the Maine Woods." *Northeast Folklore* 2 (Summer 1959), 24-26. Concerns a lumberjack ballad composed ca. 1858.

M78 MacDougall, Walter M. "Lombard's Iron Horse." *Northeastern Logger* 11 (Mar. 1963), 10-11, 34-35, 38-39. On the log hauler or steam tractor invented by Alvin O. Lombard of Maine.

M79 McEntee, James J. *Now They Are Men: The Story of the CCC.* Washington: National Home Library Foundation, 1940. 69 pp. Civilian Conservation Corps since 1933.

M80 McFeely, Otto. "Oak Park's Bit of the Forest Primeval." *Journal of the Illinois State Historical Society* 53 (Winter

1960), 412-14. On four acres of forested land in "Austin Garden," formerly owned by Henry Warren Austin and given by his son to the park district of Oak Park, Illinois, 1947.

M81 Macfie, Matthew. *Vancouver Island and British Columbia: Their History, Resources, and Prospects.* 1865. Reprint. New York: Arno Press, 1973. xxi + 574 pp. Illus., maps, tables, app., index. Incidental history of forested lands.

M82 McGarr, Philip M. "Emmet Hoyt Scott, 1842-1924: Middle West Businessman." Ph.D. diss., Indiana Univ., 1972. 337 pp. Scott was associated with timber, mining, and land enterprises in Michigan's Upper Peninsula.

M83 McGaugh, Maurice E. *The Settlement of the Saginaw Basin.* University of Chicago, Department of Geography, Research Paper No. 16. Chicago, 1950. xviii + 407 pp. Diags., tables, maps, notes, bib. Includes some history of the lumber and forest industries in Michigan's Saginaw Basin since 1819.

M84 McGeary, Martin Nelson. "Gifford Pinchot's Years of Frustration, 1917-1920." *Pennsylvania Magazine of History and Biography* 83 (July 1959), 327-42. Primarily concerned with Pinchot's activities in Pennsylvania and national politics.

M85 McGeary, Martin Nelson. *Gifford Pinchot, Forester-Politician.* Princeton, New Jersey: Princeton University Press, 1960. xii + 481 pp. Illus., notes, bib., index. A biography of Pinchot (1865-1946), emphasizing his career in forestry and his involvement in state and national politics and government service.

M86 McGeary, Martin Nelson. "Pinchot's Contributions to American Forestry." *Forest History* 5 (Summer 1961), 2-5.

M87 McGee, Emma R. *Life of W J McGee.* Farley, Iowa: Privately printed, 1915. 240 pp. William John McGee (1853-1912) was a prominent theoretician of the conservation movement, 1880s-1912, especially during the Theodore Roosevelt administration.

M88 McGee, W J, ed. *Proceedings of a Conference of Governors in the White House, Washington, D.C., May 13-15, 1908.* Washington: GPO, 1909. xxxvii + 451 pp. Illus., tables, graphs, index. A collection of printed speeches, some including historical references to state and national conservation efforts.

M89 McGehee, Judson Dodds. "The Nature Essay as a Literary Genre: An Intrinsic Study of the Works of Six English and American Nature Writers." Ph.D. diss., Univ. of Michigan, 1958. 246 pp. Includes studies of Henry David Thoreau, John Burroughs, John Muir, William Beebe, and Donald Culross Peattie.

M90 McGiffert, J.R. "Development of Logging Machinery." *Southern Lumberman* 103 (Oct. 22, 1921), 41-42. Review of skidders and skidding machinery since 1883.

M91 MacGillivray, James. "When Pine Was King." *American Forests* 39 (Apr. 1933), 151-54, 184-85. On logging and lumberjacks in Michigan in the late 19th century.

M92 McGowin, Earl M. "History of Conservation in Alabama." *Alabama Review* 7 (Jan. 1954), 42-52.

M93 McGrath, P.T. "Newfoundland's Advantages in the Production of Pulp and Paper." *Paper Trade Journal* 56 (Nov. 8, 1913), 103-21. Incidental history.

M94 McGrath, Thomas S. *Timber Bonds.* Chicago: Craig-Wayne Company, 1911. 504 pp. Incidental history.

M95 McGuire, Harvey P. "The Civilian Conservation Corps in Maine, 1933-1942." Master's thesis, Univ. of Maine, 1966.

McGuire, John R. See Ferguson, Roland H., #F60

M96 McHale, W.L. "The Paper Industry and the Southland Mill." *Journal of Forestry* 50 (July 1952), 536-38. Southland Paper Mills, Lufkin, Texas, the South's first newsprint mill, was established in 1940.

M97 McHenry, Robert, and VanDoren, Charles, eds. *A Documentary History of Conservation in America.* New York: Praeger Publishers, 1972. 422 pp. Bib., index. A collection of edited documents reflecting the variety of the movement.

M98 McIntire, G.S. "100 Years of Michigan Forests." As told to Russell McKee. *Michigan Conservation* 26 (Mar.-Apr. 1957), 2-8. Since 1850.

M99 McIntire, G.S. "A History of State Forest Management." *Michigan Conservation* 27 (Mar.-Apr. 1958), 22-26. In Michigan, since 1903.

M100 McIntosh, C. Barron. "Forest Lieu Selections in the Sand Hills of Nebraska." *Annals of the Association of American Geographers* 64 (Mar. 1974), 87-99. On the operation of the Forest Lieu Act of 1897 and the Kinkaid Act of 1904 in the Sand Hills of Nebraska.

M101 McIntosh, James H. "Thoreau's Shifting Stance Toward Nature: A Study in Romanticism." Ph.D. diss., Yale Univ., 1967. 239 pp.

M102 McIntosh, R.P. "The Forest Cover of the Catskill Mountain Region, New York, as Indicated by Land Survey Records." *American Midland Naturalist* 68 (Oct. 1962), 409-23.

M103 McIntyre, H.L. "White Pine Blister Rust Control Policies in New York State." *Journal of Forestry* 40 (Oct. 1942), 782-85. Since 1906.

M104 McIntyre, Robert Norman. "A Brief Administrative History of Mount Rainier National Park, 1899-1952." Master's thesis, Univ. of Washington, 1952.

M105 MacKaye, Benton. *The New Exploration: A Philosophy of Regional Planning.* New York: Harcourt, Brace and Company, 1928. Reprint. Intro. by Lewis Mumford. Urbana: University of Illinois Press, 1962. xxx + 243 pp. Maps, app., index. Includes some history of the conservation movement. Mumford's introduction contains a biographical sketch and evaluation of MacKaye's career in conservation.

M106 MacKaye, Benton. *From Geography to Geotechnics.* Ed. and with an intro. by Paul T. Bryant. Urbana: University of Illinois Press, 1968. 194 pp. Notes, index. Selected essays, some historical, by a distinguished conservationist-forester-planner. Bryant's introduction includes some biographical information on MacKaye.

M107 McKean, Fleetwood K. "Early Parry Sound and the Beatty Family." *Ontario History* 56 (No. 3, 1964), 167-84. On the family's timber and sawmill interests at Parry Sound, Ontario, on the north shore on Georgian Bay, 1863-1898.

M108 McKean, Gertrude L. "Tacoma, Lumber Metropolis." *Economic Geography* 17 (July 1941), 311-20.

M109 McKean, Herbert B. "Have the Forest Products Industries Lagged Behind Other Industries in Research and Development." In *First National Colloquium on the History of the Forest Products Industries, Proceedings*, ed. by Elwood R. Maunder and Margaret G. Davidson. New Haven: Forest History Society, 1967. Pp. 176-98. In the 1950s and 1960s.

M110 McKee, R.G. "Canada's Pacific Forests." *Unasylva* 13 (No. 4, 1959), 174-83. British Columbia forests and forestry since 1949.

M111 McKee, Russell. "25 Busy Years in Conservation." *Michigan Conservation* 26 (Jan.-Feb. 1957), 2-7. The Michigan Department of Conservation since 1931.

McKee, Russell. See McIntire, G.S., #M98

McKee, Russell. See Daw, T.E., #D88

M112 McKellar, A.D. "Twenty Years of Plywood." *American Forests* 54 (Jan. 1948), 29, 39, 41. On the growth of the plywood industry since 1925.

M113 McKelvey, Susan D. *Botanical Exploration of the Trans-Mississippi West, 1790-1850.* Jamaica Plain, Massachusetts: Arnold Arboretum of Harvard University, 1955. 1144 pp.

M114 MacKenzie, Donald E. "Logging Equipment Development in the West." OHI by Elwood R. Maunder. *Forest History* 16 (Oct. 1972), 30-33. Logging equipment developments in Idaho and Montana, 1911-1940s.

M115 McKibbin, Clifford W. "William James Beal, Michigan's Pioneer Forester." *American Forests and Forest Life* 30 (Apr. 1924), 216-17. An advocate of forestry and forest conservation in Michigan since the 1870s.

M116 McKim, C.R. *50 Year History of Monongahela National Forest.* Elkins, West Virginia: USFS, Monongahela National Forest, 1970. 66 pp. Illus.

M117 McKinley, Charles. *Uncle Sam in the Pacific Northwest: Federal Management of Natural Resources in the Columbia River Valley.* Publications of the Bureau of Business and Economics Research, University of California. Berkeley: University of California Press, 1952. xx + 673 pp. Diags., maps, tables, notes, index. On the activities and functions of the Corps of Engineers, Bureau of Reclamation, Bureau of Land Management, Bonneville Power Administration, USFS, Soil Conservation Service, National Park Service, Fish and Wildlife Service, Geological Survey, Bureau of Mines, and the proposed Columbia Valley Authority, especially since 1932.

M118 McKinley, Charles. *The Management of Land Related Water Resources in Oregon: A Case Study in Administrative Federalism.* Washington: Resources for the Future, 1965. 552 pp. Illus., index. Processed. Includes some history of forest resources and conservation policy in Oregon.

M119 McKinley, Donald C. "Wilderness Politics of the North Cascades." Master's thesis, Reed College, 1967.

M120 McKinsey and Company. *Evaluation of Forest Service Timber Sales Activities, Department of Agriculture, Forest Service.* Washington, 1955. v + 101 pp. Diags., tables, notes. Since 1930.

M121 Mackintosh, W.A. "The Laurentian Plateau in Canadian Economic Development." *Economic Geography* 2 (Oct. 1926). Includes some history of forest industries.

M122 McKittrick, Reuben. *The Public Land System of Texas, 1823-1910.* Madison: University of Wisconsin, 1918. 172 pp. Map. Also published as the *Bulletin of the University of Wisconsin,* No. 905, Economics and Political Science Series, Volume 9, No. 1.

M123 McKnight, Tom L. "Recreational Use of the National Forests of Colorado." *Southwestern Social Science Quarterly* 32 (1952), 264-70. Incidental history.

M124 McLaughlin, Doris B. *Michigan Labor: A Brief History from 1818 to the Present.* Ann Arbor: Institute of Labor and Industrial Relations, University of Michigan-Wayne State University, 1970. xii + 179 pp. Includes a chapter on the Saginaw Valley lumber strike of 1885.

M125 McLaughlin, Willard Thomas. "Northwest Botanical Expedition." *Pacific Northwesterner* 3 (Summer 1959), 33-38. On the botanical travels of David Douglas to the Pacific Northwest, 1820s, with mention of other botanical travelers in the region between 1778 and 1855.

M126 MacLaurin, W. Rupert. "Wages and Profits in the Paper Industry, 1929-1939." *Quarterly Journal of Economics* 58 (Feb. 1944), 196-228. Includes some history of the industry in the United States and Canada.

M127 MacLea Lumber Company. *Hewing to the Line, the MacLea Lumber Company.* Baltimore, 1943. 47 pp. Illus. A Baltimore firm since 1893.

M128 McLendon, Samuel G. *History of the Public Domain of Georgia.* Atlanta: Foote and Davies, 1924. 200 pp.

M129 McLeod, D.H. "The Southern Pine Paper Pioneers." *Southern Lumberman* (Dec. 20, 1924), 147-48. On the Yellow Pine Paper Mill, Orange, Texas, built in 1904.

M130 McLeod, G.M. "River Driving in New Brunswick." Master's thesis, Univ. of New Brunswick, 1900.

M131 MacLeod, William Christie. "Fuel and Early Civilization." *American Anthropologist,* New Series, 27 (1925), 344-46. On Indian communities and the availability of firewood.

M132 McLeskey, Howard M. *Forestry Law and Organization in Mississippi.* Mississippi State University, Social Science Center, Report No. 23. State College, 1968. Since 1926.

M133 McMacken, David. "Jim Carr, Desperado." *Michigan History* 55 (Summer 1971), 121-40. A notorious brawler and bordello operator in the lumber towns of Meredith and Harrison, Michigan, 1880s-1890s.

M134 McManus, Irene. "A Living Christmas Tree for the Nation—Once Again." *American Forests* 79 (Dec. 1973), 16-19. The custom of National Christmas Trees at the White House began in 1923 and was promoted in following years by the American Forestry Association.

M135 Macmillan, David S. "Shipbuilding in New Brunswick, from the Clyde to Saint John River, 1798." *Canadian Banker* 77 (1970), 34-36.

M136 MacMillan, H.R. *The Profession and Practice of Forestry in Canada, 1907-1957.* Toronto: University of Toronto, Faculty of Forestry, 1958. 15 pp.

M137 McMillan, Horace Greeley. "The Appointment of James Wilson as Secretary of Agriculture." *Iowa Journal of History* 56 (Jan. 1958), 77-88. The Republican state chairman of Iowa recalls political maneuvers leading to Wilson's appointment, 1896-1897. Reprinted from the Traer, Iowa, *Clipper,* April 14, 1933.

M138 MacMillan, Bloedel and Powell River Limited. Chemainus Division. *A Century of Sawmilling.* Chemainus, British Columbia, 1963. 4 pp. A brief history of the Chemainus mill since its establishment by Thomas Askew in 1862.

M139 McMillen, Jack R. "The Trees of Arlington National Cemetery." *Arlington Historical Magazine* 1 (1960), 3-13. Since the 17th century.

MacMullen, Jerry. See McNairn, Jack, +M142

M140 McMurry, J.H. "Some Forest Fire Problems in Florida." *Papers of the Michigan Academy of Science, Arts, and Letters* 40 (1954), 189-97. On the organization of forest fire control in Florida.

M141 McMurtry, Grady Shannon. "The Redwood National Park: A Case Study of Legislative Compromise." Master's thesis, Syracuse Univ., 1972.

M142 McNairn Jack, and MacMullen, Jerry. *Ships of the Redwood Coast.* Stanford, California: Stanford University Press, 1945. 156 pp. Illus., bib., index. Description and history of sailing vessels and "steam schooners" involved in the lumber trade along the California coast.

M143 McNall, Neil A. *An Agricultural History of the Genesee Valley, 1790-1860.* Philadelphia: University of Pennsylvania Press, 1952. xii + 276 pp. Maps, tables, bib. Ch. 12, "Lumbering and Agriculture in the Southern Tier," pertains to the New York-Pennsylvania border region.

M144 MacNamara, Charles. "The Camboose Shanty." *Ontario History* (Spring 1959), 73-78. Reminiscences by an Ontario logger.

M145 McNary, James G. *This Is My Life.* Albuquerque: University of New Mexico Press, 1956. ixx + 271 pp. Illus. McNary was a partner with William M. Cady in the W.M. Cady Lumber Company of Louisiana. In 1924 the firm was moved to McNary, Arizona, where it continued operations under several name changes (McNary Lumber Company, Southwest Lumber Mills, and Southwest Forest Industries). McNary was president of the National Lumber Manufacturers Association, 1937-1939.

M146 McNary, James G. *Briefly: The Story of a Life.* New York: Newcomen Society in North America, 1957. 24 pp. Includes some account on his lumbering interests in Louisiana and Arizona.

M147 MacNaughton, Victor B. "Something of Value." *Southern Lumberman* 193 (Dec. 15, 1956), 187-88. On reforestation and conservation activities by the USFS in the Yazoo-Little Tallahatchie watershed of northern Mississippi since 1947.

M148 MacNaughton, Victor B. "For Land's Sake." *American Forests* 66 (Jan. 1960), 34-35, 46-48. On flood and erosion control through reforestation in Lafayette County, Mississippi, since 1946.

M149 McNeil, Eddy. "Le Commerce d'exportation du bois canadian." In *Publications de l'école des hautes études commerciales de Montreal.* Études economiques, Volume 4. Montreal: L'école des hautes études commerciales de Montreal, 1934. Pp. 133-85. Includes some history of Canadian timber exports and trade.

M150 McNulty, W.J. "The Kings and Snowballs: Historic Families of New Brunswick." *Canada Lumberman* 50 (Aug. 1930), 175-76. Lumbermen.

M151 MacNutt, W.S. "Politics of the Timber Trade in Colonial New Brunswick, 1825-1840." *Canadian Historical Review* 30 (Mar. 1949), 47-65. Examines struggles between the executive and legislature over timber rights.

M152 MacNutt, W.S. *New Brunswick, A History: 1784-1867.* Toronto: Macmillan, 1963. Includes some general forest history of the province.

M153 MacNutt, W.S. *The Atlantic Provinces; the Emergence of Colonial Society, 1712-1857.* Canadian Centenary Series. Toronto: McClelland & Stewart, 1965. Includes some history of the lumber industry, the timber trade, and their relationship with New Brunswick politics.

Macon, Hershal L. See Durisch, Lawrence L., #D314

M154 MacPhee, Donald A. "The Centralia Incident and the Pamphleteers." *Pacific Northwest Quarterly* 62 (July 1971), 110-16. On the literary and journalistic consequences of the violent encounter between members of the Industrial Workers of the World and the American Legion, Centralia, Washington, 1919.

M155 McPhee, John A. *The Pine Barrens.* New York: Farrar, Straus & Giroux, 1967. 167 pp. Illus. Includes some historical references to forest industries in this wooded area of New Jersey.

M156 McRae, J. Finley. *Paper Making in Alabama.* New York: Newcomen Society in North America, 1956. 32 pp. On the pulp and paper industry, including some history of International Paper Company's operations in Alabama.

M157 McRae, James A. *Call Me Tomorrow.* Toronto: Ryerson Press, 1960. 248 pp. Mainly reminiscences of the author's life as a lumberjack and miner in Canada before 1925; includes some account of his work in lumbering and railroad construction in Montana and Washington before returning home to Sarnia, Ontario, in 1908.

M158 McTiver, Bernard. "Con Culhane, Real-Life Paul Bunyan." *Michigan Conservation* 23 (Jan.-Feb. 1954), 11-14. Lore collected from men who worked under Culhane, a lumberman in Luce and Chippewa counties, Michigan, c 1890-1906.

M159 McWhiney, Grady. "Louisiana Socalists in the Early Twentieth Century: A Study of Rustic Radicalism." *Journal of Southern History* 20 (Aug. 1954), 315-36. Deals in part with the Brotherhood of Timber Workers and the Industrial

Workers of the World. These organizations supported Eugene V. Debs and other Socialist candidates in 1912.

M160 Madden, James L. *A History of Hollingsworth & Whitney Company, 1862-1954.* N.p., 1954. 16 pp. A pulp and paper firm in Maine, now merged with Scott Paper Company.

M161 Madden, Richard B. *'Tree Farmers and Wood Converters': The Story of Potlatch Corporation.* New York: Newcomen Society in North America, 1975. 24 pp. Illus. On the Potlatch Lumber Company, Clearwater Timber Company, and Edward Rutledge Timber Company, lumber firms founded in Idaho ca. 1900 and merged in 1930 as Potlatch Forests, Inc., now Potlatch Corporation. The San Francisco-based lumber and paper company has major mills in Lewiston, Idaho, and other operations in Minnesota and Arkansas.

Maddock, Thomas. See Leopold, Luna B., #L160

M162 Maddox, Harry A. *Paper! Its History, Sources and Manufacture.* Sixth edition. London: Sir I. Pitman and Sons, 1939. x + 180 pp.

M163 Magee, David F. "Rafting on the Susquehanna." *Papers of the Lancaster County Historical Society* 24 (Nov. 1920), 193-202. Pennsylvania.

M164 Mahaffey, R.B. "A Pioneer in Southern Kraft." *Paper Mill* 58 (Feb. 9, 1935), 23-26.

M165 Mahaffay, Robert E. "Timber Cinderella." *American Forests* 54 (Jan. 1948), 24-25, 43. On the lodgepole pine industry in South Dakota's Black Hills, Colorado, California, and Alaska since 1889.

M166 Mahar, Franklyn D. "Douglas McKay and the Issues of Power Development in Oregon, 1953-1956." Ph.D. diss., Univ. of Oregon, 1968. 351 pp. McKay, a former governor of Oregon, was Secretary of the Interior, 1953-1956.

M167 Mahar, Franklyn D. "The Politics of Power: The Oregon Test for Partnership." *Pacific Northwest Quarterly* 65 (Jan. 1974), 29-37. On political battles concerning hydroelectric power policies in Oregon during the 1950s, and their implications for the conservation movement.

Maheux, Georges. See Lussier, O., #L318

M168 Maheux, Georges. "Half a Century in Retrospect: 1910-1960," *Forêt-Conservation* 26 (Oct. 1960), 17-18. On the forest conservation movement in Canada.

Mahler, E. See Kimberly, J.C., #K142

M169 Mahoney, J. Rolla. *Natural Resources Activity of the Federal Government: Historical, Descriptive, Analytical.* Public Affairs Bulletin No. 76. Washington: U.S Library of Congress, 1950. 261 pp. Diags., tables. On the public domain and the control of soils, forested lands, mineral resources, water resources, and fish and wildlife. Prepared for the Task Force on Natural Resources of the Hoover Commission.

M170 Mailhot, Charles-Édouard. *Les Bois-Francs.* Arthabaska, Quebec: La Cie d'Imprimerie D'Arthabaskaville, 1914. 474 pp. Includes some forest history of southern Quebec.

M171 Maine. Forest Commission. *Maine Forest Service: Brief History and Organization.* Augusta, 1929. 23 pp. Illus.

M172 Maine, University of, Forestry Department. *Forestry Department Golden Anniversary, 1903-1953; Theme: The Role of Forestry in the Development and Future of Maine.* Orono, 1954. 92 pp.

M173 Maisch, Theodore. "Episode in Lumber History." *Southern Lumberman* (June 15, 1929), 64. Financial arrangements of colonial lumber manufacturers.

M174 Maki, Lillian. "Lincoln County Homesteader." *Oregon Historical Quarterly* 71 (Sept. 1970), 265-73. Includes some reminiscences of logging in Lincoln County, Oregon, 1900s-1930s.

M175 Malin, James C. *The Grassland of North America: Prologomena to Its History.* 1947. Reprint. Gloucester, Massachusetts: Peter Smith, 1967. viii + 490 pp. Bib. See for reference to wood and forests (and their absence) on the Great Plains.

M176 Malin, James C. *Grassland Historical Studies: Natural Resources Utilization in a Background of Science and Technology. Volume 1. Geology and Geography.* Lawrence, Kansas: By the author, 1950. xii + 377 pp. Illus., notes. Includes some history of the use of wood and minerals for fuel and building material, especially in Kansas City.

M177 Malin, James C. "Ecology and History." *Scientific Monthly* 70 (May 1950), 295-98. On an ecological approach to American history.

Malkin, B. See Bennett, W.T., #B213

M178 Mallory, Enid S. "Ottawa Lumber Era." *Canadian Geographical Journal* 68 (Feb. 1964), 60-73. The Ottawa Valley of Ontario and Quebec in the 19th century.

M179 Malone, Joseph J. *Pine Trees and Politics: The Naval Stores and Forest Policy in Colonial New England, 1691-1775.* Seattle: University of Washington Press, 1964. xi + 219 pp. Apps., notes, bib., essay, index. On British policy and colonial response in the years leading to the American Revolution.

M180 Malone, Michael P. *C. Ben Ross and the New Deal in Idaho.* Seattle: University of Washington Press, 1970. 191 pp. Illus., notes, bib., index. Incidental references to conservation programs.

M181 Malone, Michael P. "The Gallatin Canyon and the Tides of History." *Montana, Magazine of Western History* 23 (Summer 1973), 2-17. The Gallatin Canyon, a gateway to Yellowstone National Park, has been the site of logging operations, USFS land exchanges, and Big Sky of Montana, a controversial resort complex opposed by many conservationists.

Maloney, Keith. See Allen, Richard S., #A88

M182 Malouf, Carling. "The Coniferous Forests and Their Use in the Northern Rockies Through 6000 Years of Prehistory." In *Coniferous Forest of the North Rocky Mountains.* Missoula, Montana, 1969. Pp. 270-90.

M183 Malsberger, H.J. "Fifteen Years of Forestry Progress by the Southern Pulp and Paper Industry." *Southern Pulp and Paper Manufacturer* 17 (Jan. 11, 1954), 64-68. On the work of the Southern Pulpwood Conservation Association.

M184 Malsberger, H.J. "The Pulp and Paper Industry in the South." *Journal of Forestry* 54 (Oct. 1956), 639-42. Includes some history of paper manufacturing from Southern wood pulp since 1878.

M185 Malsberger, H.J. "Seventy-Five Year History of the Wood-Pulp and Paper Industry in the South." *Southern Lumberman* 193 (Dec. 14, 1956), 182-84.

M186 Manchester, Herbert. *The Diamond Match Company: A Century of Service, of Progress, and of Growth, 1835-1935.* New York: Diamond Match Company, 1935. 108 pp. On its operations throughout the United States.

M187 Mancil, Ervin. "Some Historical and Geographical Notes on the Cypress Lumbering Industry in Louisiana." *Louisiana Studies* 8 (No. 1, 1969), 14-25. Especially from the 1880s to 1940s.

M188 Mancil, Ervin. "An Historical Geography of Industrial Cypress Lumbering in Louisiana." Ph.D. diss., Louisiana State Univ., 1972. 299 pp. From 1890 to 1925.

M189 Manley, L. "Industry Had Early Start in British Columbia." *Pulp & Paper Magazine of Canada* 54 (May 1953), 155-56.

M190 Manley, R.S. "History of Wood Block Paving in the South." *Wood Preservers' Bulletin* 1 (No. 3, 1914), 20.

M191 Mann, Dean E. *The Politics of Water in Arizona*. Tucson: University of Arizona Press, 1963. 317 pp. Illus., maps, notes, bib. Includes some history of irrigation, land management, and conservation.

M192 Mann, Robert. *The Autobiography of Robert Mann*. Philadelphia: J.B. Lippincott Company, 1897. 83 pp. The author (b. 1824) was a manufacturer of axes in Pennsylvania.

M193 Mann, Roberts. "Aldo Leopold: Priest and Prophet." *American Forests* 60 (August. 1954), 23, 42-43. On his work as a forester, game manager, and essayist on conservation and a "land ethic," 1909-1948.

M194 Manners, William. *TR and Will: A Friendship That Split the Republican Party*. New York: Harcourt, Brace & World, 1969. xiv + 335 pp. Illus. A popular account, including references to the Ballinger-Pinchot controversy and other conservation issues.

M195 Manning, Harvey. *The Wild Cascades: Forgotten Parkland.* San Francisco: Sierra Club, 1965. 160 pp. Illus. Includes some history of the region now preserved as the North Cascades National Park, Washington.

M196 Manning, Thomas G. "The United States Geological Survey, 1867-1894." Ph.D. diss., Yale Univ., 1941. 432 pp.

M197 Manning, Thomas G. *Government in Science: The U.S. Geological Survey, 1867-1894.* Lexington: University of Kentucky Press, 1967. xiv + 257 pp. Maps, notes, bib., index. Much of the Geological Survey's work was carried out in the forested areas of the West and with regard to conservation of natural resources. John Wesley Powell, Clarence King, and Arnold Hague are treated.

M198 Manny, Louise. *Ships of the Miramichi: A History of Shipbuilding on the Miramichi River, New Brunswick, Canada, 1773-1919.* Saint John, New Brunswick: New Brunswick Museum, 1960. 84 pp.

M199 Mansfield, J.H. "Woodworking Machinery, History of Development from 1852-1952." *Mechanical Engineering* 74 (Dec. 1952), 983-95.

M200 Mansueti, Romeo. "Battle Creek Cypress Swamp in Southern Maryland." *Atlantic Naturalist* 10 (May-Aug. 1955), 248-57. On the natural history of this Calvert County area since 1907.

M201 Marchi, John R. "Conservation in Montana." *Montana Law Review* 17 (Fall 1955), 100-07. Since 1917.

M202 Marckworth, Gordon D. "Western Logging Engineering Schools: University of Washington." *Loggers Handbook* 11 (1951), 85-86. Education in logging engineering since 1909.

M203 Marckworth, Gordon D. "Pulp Co. Expands from Washington to Alaska: History of Puget Sound Pulp and Timber Shows Progress." *Western Conservation Journal* (Nov.-Dec. 1959), 1-6.

M204 Marckworth, Gordon D. "Chelan Box Still Runs with Steam Power." *Western Conservation Journal* (Mar.-Apr. 1961), 2-4. Brief history of a box manufacturing company in Chelan, Washington, since 1931.

M205 Marckworth, Gordon D. "Taylor Logging Co. Started by Driving on Rivers." *Western Conservation Journal* 21 (Feb.-Mar. 1964), 40-41, 45. A central Washington firm since 1908.

Marckworth, Gordon D. See Schmitz, Henry, #S101

M206 Margolin, Louis. "The 'Hand-Loggers' of British Columbia." *Forestry Quarterly* 9 (Dec. 1911), 563-67. Incidental history.

M207 Mark, Irving. "The Homestead Ideal and Conservation of the Public Domain." *American Journal of Economics and Sociology* 22 (Apr. 1963). A general outline of federal land policy and the influence of conservation in the 20th century.

M208 Markham, Alton, and Schrack, Robert A. "Was This the First Farm Forestry in America?" *Journal of Forestry* 45 (Dec. 1947), 900-02. On the plantation of Francis Willett and descendants, Saunderstown, Rhode Island, since the 18th century.

M209 Markham, Wilbur A. "Private Land Forestry in Mason County, Wash." Master's thesis, Univ. of Washington, 1933.

M210 Markwell, Katherine. "Recent State Laws on Forestry." *Journal of Forestry* 36 (Mar. 1938), 300-05. A survey of state forestry laws adopted during the 1930s.

M211 Marquess, E. Lawrence. "The West Virginia Venture: Empire Out of Wilderness." *West Virginia History* 14 (Oct. 1952), 5-27. On the West Virginia Central and Pittsburgh Railway Company, organized by U.S. senators Henry G. Davis and Stephen B. Elkins to develop the coal and timber resources of northeastern West Virginia, 1881-1902.

M212 Marquis, Ralph W. *Economics of Private Forestry*. American Forestry Series. New York: McGraw-Hill, 1939. viii + 219 pp. Tables, bib., index. Includes a chapter on the "Evolution of the Forest Problem," as well as many incidental historical references to forest economics.

M213 Marquis, Ralph W. "Log Production and Sustained Yield." *Timberman* 48 (Apr. 1947), 36ff. In the Pacific Northwest since 1925.

M214 Marquis, Ralph W. "Forest Research in the Northeast." *Southern Lumberman* 193 (Dec. 15, 1956), 193-94. Since 1876.

M215 Marriner, Ernest C. *Kennebec Yesterdays.* Waterville, Maine: Colby College Press, 1954. 320 pp. Illus. Includes some forest history of Kennebec County, Maine.

M216 Marriner, Ernest C. *History of Colby College.* Waterville, Maine: Colby College Press, 1962. 659 pp. Illus. Includes the college's timber sales.

M217 Marsh, Alton. "Last Stand at Red River Gorge." *National Parks and Conservation Magazine* 48 (Aug. 1974), 18-22. On a long-time controversy over a proposed dam on the Red River on Daniel Boone National Forest, Kentucky, since 1937.

M218 Marsh, George Perkins. *Man and Nature: Or Physical Geography as Modified by Human Action.* Ed. with an intro. by David Lowenthal. Cambridge: Belknap Press of Harvard University Press, 1965. xxix + 472 pp. Notes, index. Marsh's classic work, first published in 1864 and subsequently revised in 1874 and 1885, is a powerful indictment of man's abuse of his physical environment, especially through deforestation and overgrazing. The work had a large impact on the conservation movement.

M219 Marsh, John S. "Glacier National Park, British Columbia, 1880 to the Present." *British Columbia Historical News* (Feb. 1970), 7-10.

M220 Marsh, John S. "Man, Landscape and Recreation in Glacier National Park, 1880 to Present." Ph.D. diss., Univ. of Alberta, 1971.

M221 Marsh, Raymond E. "Timber Cruising in the Early Days." *American Forests* 74 (Feb. 1968), 28-31, 43-44. A USFS party on the Missoula and Lolo national forests of Montana, 1909.

M222 Marsh, Raymond E. "Timber Cruising on the National Forests of the Southwest." *Forest History* 13 (Oct. 1969), 22-32. Memoir of a USFS forester in Arizona, New Mexico, and Arkansas, 1910-1912.

M223 Marsh, Seward H. "Arthur A. Wood: Forest Officer and Public Servant." *Journal of Forestry* 47 (Sept. 1949), 750-51. On Wood's career with the USFS since 1913, especially on the Shenandoah (George Washington), Nantahala, and Monongahela national forests.

M224 Marsh, William Barton. *Philadelphia Hardwood, 1798-1948: The Story of the McIlvains of Philadelphia and the Business They Founded*. New York: W.E. Rudge's Sons, 1948. 99 pp. Illus., maps, table. On the J. Gibson McIlvain Company and its lumber business.

M225 Marshall, George. "Adirondacks to Alaska: A Biographical Sketch of Robert Marshall." *Ad-i-ron-dac* 15 (May-June 1951), 44-45, 59. On the career of Marshall (1901-1939), a USFS forester and wilderness enthusiast.

M226 Marshall, George. "Robert Marshall as a Writer." *Living Wilderness* 16 (Autumn 1951), 14-23; 19 (Summer 1954), 31-35. Includes a bibliography of Robert Marshall 1901-1939), with reviews of his published works and a list of his writings, mainly on forestry, the wilderness, and conservation, 1922-1939.

Marshall, George. See Marshall, Robert, #M234

M227 Marshall, George. "Bob Marshall and the Alaska Arctic Wilderness." *Living Wilderness* 34 (Autumn 1970), 29-32. On Marshall's studies of Alaska's Brooks Range, 1929-1939, and subsequent efforts to preserve that wilderness.

M228 Marshall, George A. "Watershed and Reservoir Control in the Pacific Northwest . . . Historical Development." *Journal, American Water Works Association* 46 (Aug. 1954), 723-28. Since 1890.

M229 Marshall, Hubert, and Young, Robert J. *Public Administration of Florida's Natural Resources*. Gainesville: University of Florida, Public Administration Clearing Service, 1953. 257 pp. Since 1845.

M230 Marshall, Robert. "Influence of Precipitation Cycles on Forestry." *Journal of Forestry* 25 (Apr. 1927), 415-29. Includes some history of climatic cycles in northern Idaho, based on tree-ring data since 1706 and travelers accounts since 1805.

M231 Marshall, Robert. "A Contribution to the Life History of the Lumberjack." *Pulp and Paper Magazine of Canada* 31 (May 21, 1931), 641-42.

M232 Marshall, Robert. *The People's Forests*. New York: Harrison Smith and Robert Haas, 1933. ix + 233 pp. Tables, references, index. Includes some history of forest devastation as part of a larger argument for public ownership of forests.

M233 Marshall, Robert. "Mills Blake: Adirondack Explorer." *Ad-i-ron-dac* 15 (May-June 1951), 46-48. On his surveys in New York's Adirondack Mountains, 1872-1900.

M234 Marshall, Robert. *Arctic Wilderness.* Ed. with an intro. by George Marshall. Berkeley: University of California Press, 1956. xxvi + 171 pp. Illus., maps. Marshall (1901-1939) was a wilderness enthusiast. His diary account of observations and explorations in northern Alaska, 1929-1939, contains much of his wilderness preservation philosophy. A second edition under the title *Alaska Wilderness: Exploring the Central Brooks Range* was published in 1970.

M235 Marshall, Roujet DeLisle. *Autobiography of Roujet D. Marshall.* 2 Volumes. Ed. by Gilson G. Glasier. Madison: Privately printed, 1923, 1931. Illus., index. Marshall, a prominent Wisconsin lawyer, was employed by Frederick Weyerhaeuser and other lumbermen in the upper Mississippi Valley during the 1870s and 1880s.

Marshall, William H. See Burcalow, Donald W., #B563

M236 Martin, C.W. "History of Acquisition of Connecticut State Forests." *Wooden Nutmeg* 19 (Dec. 1943), 9.

M237 Martin, Calvin. "Fire and Forest Structure in Aboriginal Eastern Forest." *Indian Historian* 6 (Summer 1973), 23-26; 6 (Fall 1973), 38-42, 54. On the use of fire by Indians in the original forests of eastern North America.

M238 Martin, Chester. *'The Natural Resources Question': The Historical Basis of Provincial Claims.* Winnipeg: King's Printer, for the Province of Manitoba, 1920. 148 pp. On the legal ownership of natural resources in Manitoba.

M239 Martin, Chester. *Dominion Land Policy.* Toronto: Macmillan, 1938. A study of Canadian land policy and settlement.

M240 Martin, Clyde S. *History and Influence of the Western Forestry and Conservation Association on Cooperative Forestry in the West.* Portland: Western Forestry and Conservation Association, 1944. 12 pp.

M241 Martin, Clyde S. "History and Influence of the Western Forestry and Conservation Association on Cooperative Forestry in the West." *Journal of Forestry* 43 (Mar. 1945), 165-70. Since 1909.

M242 Martin, Deborah B. *History of Brown County, Wisconsin, Past and Present.* 2 Volumes. Chicago, 1913. Includes some history of forest industries in the Fox Valley.

M243 Martin, Dorothy M. "Foresters, Fair." *American Forests* 56 (Dec. 1950), 14-15, 36-37, 42. On twenty-eight women foresters, 1915-1950.

Martin, Edward C. See Myers, Clifford A., #M663

M244 Martin, Helen M. "Michigan Story." *Michigan Conservation* 19 (Mar.-Apr. 1950), 5-8. On the history of natural resource use and conservation.

Martin, James W. See Briscoe, Vera, #B434

M245 Martin, James William. "History of Forest Conservation in Texas, 1900 to 1935." Master's thesis, Stephen F. Austin State College, 1966.

Martin, James William. See Maxwell, Robert S., #M356

M246 Martin, John B. *Call It North Country: The Story of Upper Michigan.* New York: Alfred A. Knopf, 1944. 281 pp. Bib. Includes some history of forest industries.

M247 Martin, Roscoe G. *From Forest to Front Page: How a Paper Corporation Came to East Tennessee.* Inter-University Case Program, ICP Case No. 34. University: University of Alabama Press for ICP, 1956. 66 pp. Illus. On the Bowaters Southern Paper Company, which the federal government permitted to build and operate a pulp paper mill in Tennessee using publicly owned resources in the early 1950s.

M248 Martin, Roscoe C., ed. *TVA—The First Twenty Years: A Staff Report.* University: University of Alabama Press; Knoxville: University of Tennessee Press, 1956. xiii + 282 pp. Map, diags., tables. A semiofficial history of the Tennessee Valley Authority by staff members and longtime observers; includes some account of forestry and conservation.

M249 Martin, Roscoe C. "The Tennessee Valley: A Study of Federal Control." *Law and Contemporary Problems* 22 (Summer 1957), 351-77. Reviews development of the Tenessee Valley Authority.

M250 Martin, Sydney Walter. "The Public Domain in Territorial Florida." *Journal of Southern History* 10 (May 1944), 174-87.

M251 Martin, W.L. "The Evolution of Grades for Hardwoods." *Canada Lumberman* 38 (Jan. 1, 1918), 31-32.

Martin, Wallace E. See Genzoli, Andrew M., #G85, G86

M252 Martin, William R. "Transportation Planning by the U.S. Forest Service in the Pacific Northwest: A Case Study." Master's thesis, Univ. of Washington, 1970.

M253 Martinson, Arthur D. "The Influence of the Longmire Family upon the Early History of Mount Rainier National Park." Master's thesis, Washington State Univ., 1961.

M254 Martinson, Arthur D. "Mountain in the Sky: A History of Mount Rainier National Park." Ph.D. diss., Washington State Univ., 1966. 182 pp.

M255 Martinson, Arthur D. "Mount Rainier National Park: First Years." *Forest History* 10 (Oct. 1966), 26-33. On problems of administration and development in the Washington park, 1899-1915.

M256 Martinson, Arthur D. "The Story of a Mountain: A Pictorial History of Mount Rainier National Park." *American West* 8 (Mar. 1971), 34-41. On its discovery, exploration, and the movement leading to its establishment as a national park in 1899.

M257 Marty, Robert. "Timber Prices and Timber Policies." *Michigan Academician* 3 (Summer 1970). Incidental history.

Martzolff, Clement L. See Ewing, Thomas, #E136

M258 Marvin, James W., and Hill, Ralph N. "Taming the Wild Maple." *Vermont Life* 10 (Spring 1956), 2-7. On University of Vermont studies of the sugar maple and the making of maple sugar since 1896.

M259 Marx, Leo. *The Machine in the Garden: Technology and the Pastoral Ideal in America.* New York: Oxford University Press, 1964. 392 pp. Illus., notes, index. A study of the conflict between technology and the pastoral ideal in American history, especially as seen through literature. There are many implications here for American attitudes toward wilderness.

M260 Maryland. Board of Natural Resources. *Conservation Progress in Maryland, 1941-1947.* Bulletin 2. Second Edition. Annapolis, 1951. 47 pp. Illus. Includes forest conservation.

M261 Mason, Alpheus Thomas. *Bureaucracy Convicts Itself: The Ballinger-Pinchot Controversy of 1910.* New York: Viking Press, 1941. 224 pp. Illus., bib., index.

M262 Mason, David T. *Timber Ownership and Lumber Production in the Inland Empire.* Portland: Western Pine Manufacturers Association, 1920. 110 pp. Illus. Includes some incidental historical references to the lumber industry and forest ownership in the interior Pacific Northwest; originally prepared in 1914 as Part 5 of a series of USFS studies of the lumber industry.

M263 Mason, David T., and Stevens, Carl M. "Progress of Private Reforestation in the West." *Lumber World Review* 45 (Dec. 10, 1923), 27-30.

M264 Mason, David T. *The Lumber Code.* New Haven: Yale University Press, 1935. 30 pp. Incidental history.

M265 Mason, David T. "Present Trends in Private Forest Ownership." *Journal of Forestry* 35 (Feb. 1937), 105-07. Trends in the private acquisition of timberland in the West since 1907.

M266 Mason, David T. *Forests for the Future: The Story of Sustained Yield as Told in the Diaries and Papers of David T. Mason, 1907-1950.* Ed. with intro. and notes by Rodney C. Loehr. Forest Products History Foundation Series, No. 5. St. Paul: Forest Products History Foundation and Minnesota Historical Society, 1952. xi + 283 pp. Illus., notes, app., index. Mason (1883-1973), an eminent forestry consultant, long advocated sustained-yield forest management and influenced or actually wrote much of the legislation which implemented it. After work for the USFS and the University of California, he was associated with the Portland firm of Mason, Bruce & Girard for fifty years.

M267 Mason, David T. "Changing Economic Conditions and Forest Practices on Privately Owned Lands." *Journal of Forestry* 51 (Nov. 1953), 803-08. Since ca. 1900.

M268 Mason, David T. *The Development of Industrial Forestry in the United States.* Winton Lecture in Industrial Forestry. Berkeley: University of California, School of Forestry, 1957. 14 pp. Notes.

M269 Mason, David T. "Time to a Forester." *Oregon Historical Quarterly* 58 (Dec. 1957), 358-69. Reminiscences of his forestry career, mostly in the Pacific Northwest, since 1905.

M270 Mason, David T., and Henze, Karl D. "The Shelton Cooperative Sustained Yield Unit." *Journal of Forestry* 57 (Mar. 1959), 163-68. On the background of the Sustained Yield Unit Law (Public Law 273) of 1944 and an evaluation of the sustained yield contract between Simpson Timber Company and the USFS on the Shelton Unit in Washington since 1947.

M271 Mason, David T. "Pinchot, Cary, Greeley: Architects of American Forestry." *Forest History* 5 (Summer 1961), 6-8. An evaluation of the forestry contributions of Gifford Pinchot, Austin F. Cary, and William B. Greeley, by a forester who worked with all three.

M272 Mason, David T. "The Effect of O & C Management on the Economy of Oregon." *Oregon Historical Quarterly* 64 (Mar. 1963), 55-67. On the economic impact of federal management of the revested Oregon and California Railroad grant lands since 1916.

M273 Mason, David T. "Memoirs of a Forester." OHI by Elwood R. Maunder. *Forest History* 10 (Jan. 1967), 6-12, 29-35. On his early career with the USFS in the West and as forestry educator at the University of California, 1905-1920s.

M274 Mason, David T. "Six Decades of Change in the Forest Products Industries: Personal Recollections." In *First National Colloquium on the History of the Forest Products Industries, Proceedings,* ed. by Elwood R. Maunder and Margaret G. Davidson. New Haven: Forest History Society, 1967. Pp. 81-91.

M275 Mason, David T. "Memoirs of a Forester, Part II." OHI by Elwood R. Maunder. *Forest History* 13 (Apr./July 1969), 28-39. On his career as a Portland consulting forester and advocate of sustained-yield forestry, 1921-1940s.

M276 Mason, David T.; Henze, Karl D.; and Briegleb, Philip A. "The Shelton Cooperative Sustained Yield Unit—The First 25 Years." *Journal of Forestry* 70 (Aug. 1972), 462-67. Since 1946.

M277 Mason, Glenn. "River Driving in Lane County." *Lane County Historian* 18 (Summer 1973), 23-38. On the Willamette and McKenzie rivers, Lane County, Oregon, 1870s-1910s.

M278 Mason, Herbert Louis. "A Pleistocene Flora from the Tomales Bay Region and Its Bearing on the History of the Coastal Pine Forests of California." Ph.D. diss., Univ. of California, Berkeley, 1933.

M279 Mason, Phillip P. *Lumbering Era in Michigan History (1860-1900).* Lansing: Michigan Historical Commission, 1956.

M280 Mason, Robert L. *The Lure of the Great Smokies.* Boston: Houghton Mifflin, 1927. xix + 320 pp. Incidental references to its forest history.

M281 Massachusetts. State Board of Agriculture. *The Gypsy Moth: A Report of the Work of Destroying the Insect in the Commonwealth of massachusetts, Together with an Account of Its History and Habits both in Massachusetts and Europe.* Boston: Wright and Potter, 1896. 495 pp.

M282 Massey, Richard W. "A History of the Lumber Industry in Alabama and West Florida, 1880-1914." Ph.D. diss., Vanderbilt Univ., 1960. 233 pp.

M283 Massey, Richard W. "Logging Railroads in Alabama, 1880-1914." *Alabama Review* 14 (Jan. 1961), 41-50.

M284 Massey, Richard W. "Labor Conditions in the Lumber Industry in Alabama, 1880-1914." *Journal of the Alabama Academy of Science* 37 (No. 2, 1966), 172-81. On working conditions, wage rates, attempts at unionization, and the paternalism of owners.

M285 Massie, Michael, and Haring, Robert C. *The Forest Economy of Haines, Alaska: A Study of Current Forest Utilization, Forest Management and Utilization Alternatives and Resultant Economic Impact.* Seattle: University of Washington Press, 1969. 203 pp. Incidental history.

M286 Mastelotto, Virgil. "Lumberjack Language in Northeastern California." *Western Folklore* 9 (July 1950), 380-82. Loggers' terms gathered from Yuba River and Feather River areas.

M287 Masten, Arthur H. *The Story of Adirondac.* New York: Privately printed, 1923. Reprint. Intro. and notes by William K. Verner. Blue Mountain Lake, New York: Adirondack Museum, 1968. 240 pp. Illus., notes, index. History of 19th-century exploration and development in New York's Adirondack Mountains.

M288 Mathews, Donald N., and Hutchison, S. Blair. *Development of a Blister Rust Control Policy for the National Forests in the Inland Empire.* USFS, Northern Rocky Mountain

Forest and Range Experiment Station, Paper No. 16. Missoula, Montana, 1948. 116 pp. Interior Pacific Northwest.

M289 Mattes, Merrill J. "Jackson Hole, Crossroads of the Western Fur Trade, 1807-1829." *Pacific Northwest Quarterly* 37 (Apr. 1946), 87-108. On the discovery and early history of the area which is now part of the Grand Teton National Park, Wyoming.

M290 Mattes, Merrill J. "Jackson Hole, Crossroads of the Western Fur Trade, 1830-1840." *Pacific Northwest Quarterly* 39 (Jan. 1948), 3-32.

M291 Mattes, Merrill J. "Behind the Legend of Colter's Hell: The Early Exploration of Yellowstone National Park." *Mississippi Valley Historical Review* 36 (Sept. 1949), 251-82.

M292 Matthew, L.S. "Christmas Trees, Profitable Woodland Crops." *American Forests* 62 (Dec. 1956), 14-16, 63. On the Douglas fir Christmas tree industry in Montana since 1927.

M293 Matthews, Albert. "The Word 'Park' in the United States." *Publications of the Colonial Society of Massachusetts* 8 (Apr. 1904), 373-99. Includes some history of the national parks idea and movement, including reference to Yellowstone and Yosemite national parks.

M294 Matthews, Jeremiah H. *Reminiscences of a Life-Time Experience.* Nashville: By the author, 1924. 227 pp. The author (b. 1847) operated sawmills and transported lumber on the Ohio and Tennessee rivers.

M295 Matthews, William H., III. *A Guide to the National Parks—Their Landscape and Geology.* Garden City, New York: Doubleday & Company, 1973. 529 pp. Illus., glossary, index. Incidental history.

M296 Matthiessen, Peter. *Wildlife in America.* New York: Viking Press, 1959. 304 pp. Illus. A history of man's impact on wildlife in North America.

M297 Mattila, Walter, ed. *The Finnish Paul Bunyans.* Portland: Finnish American Historical Society of the West, 1973. 110 pp. Illus. Includes some history of Finnish-American loggers in Oregon and elsewhere in the West.

M298 Mattingly, Arthur H. "Plank Roads and Corduroy Trails: The Farmer's Railroads." *Kansas Quarterly* (Spring 1973), 62-69. Especially in Missouri in the 1850s.

M299 Mattison, C.W. "Conservation's Ever Widening Stream." *American Forests* 61 (Mar. 1955), 28-29, 92-95. On the growth of the conservation movement since 1875, as reflected by the proliferation of public agencies and private organizations espousing resource conservation.

M300 Mattison, Ray H. "Devils Tower National Monument—A History." *Annals of Wyoming* 28 (Apr. 1956), 2-20. Since 1859.

M301 Mattoon, M.A. "Robie Mason Evans." *Journal of Forestry* 49 (Feb. 1951), 116-17. On his career in the USFS since 1910, especially as regional forester of Region 7 (Northeastern states) since 1934.

M302 Mattoon, Wilbur R. "Twenty Years of Slash Pine." *Journal of Forestry* 34 (June 1936), 562-70. On its increasing use as a commercial species and on slash pine plantations in the South since 1916.

Maughan, William. See Korstian, Clarence F., #K238

M303 Maughan, William, ed. *Guide to Forestry Activities in North Carolina, South Carolina and Tennessee.* Asheville, North Carolina: Appalachian Section, Society of American Foresters, 1939. 287 pp. Bib. Of the USFS, National Park Service, Tennessee Valley Authority, Civilian Conservation Corps, and other agencies of the USDA and USDI. There are also sections on state, private, and industrial forestry organizations, including some historical references.

Mauk, Charlotte E. See Muir, John, #M617

M304 Maunder, Elwood R. "History—Is It Bunk or Good Business?" *Southern Lumberman* 185 (Dec. 15, 1952), 140-43. On the activities of the Forest Products History Foundation (Forest History Society).

M305 Maunder, Elwood R. "Dr. Carl Alwin Schenck: German Pioneer in the Field of American Forestry." *Paper Maker* 23 (Sept. 1954), 17-30. Schenck (1868-1955) directed the Biltmore Forest School near Asheville, North Carolina, 1898-1909.

M306 Maunder, Elwood R. "For the History of Logging." *Pacific Northwest Quarterly* 46 (Oct. 1955), 113-14. Brief history of the American Forest History Foundation (Forest History Society) since 1947.

M307 Maunder, Elwood R. "The *Southern Lumberman* and American History." *Southern Lumberman* 193 (Dec. 15, 1956), 124-26. On the work of this trade journal since 1881, and its importance as a source of forest history.

M308 Maunder, Elwood R. "Writing the History of Forest Industries." *Pacific Northwest Quarterly* 48 (Oct. 1957), 127-33. On the work of the Forest History Foundation in preserving the records and promoting the history of the forest industries and forest history in general since 1947.

Maunder, Elwood R. See Cowan, Charles S., #C640

M309 Maunder, Elwood R. "Building on Sawdust." *Pacific Northwest Quarterly* 51 (Apr. 1960), 57-62. On forest industries as an element in the growth of Seattle and Washington since 1851.

Maunder, Elwood R. See Eldredge, Inman F., #E54, E55

Maunder, Elwood R. See Robertson, Reuben B., #R238

M310 Maunder, Elwood R. "Milestones of Louisiana Forestry: Henry Hardtner Signs the First Reforestation Contract." *Forests & People* 13 (First quarter, 1963), 56-57, 124-25. On the career of Henry Hardtner (1870-1935) of Urania Lumber Company, and especially his promotion of industrial forestry, beginning with a reforestation contract with the state of Louisiana in 1913.

Maunder, Elwood R. See Stevens, James F., #S625

Maunder, Elwood R. See McCaffrey, Joseph E., #M12, M13, M14

M311 Maunder, Elwood R. "Our Forest Industries' History: It Must Be Preserved." *Southern Lumberman* 211 (Dec. 15, 1965), 185-87.

Maunder, Elwood R. See Dana, Samuel Trask, #D35, D36

Maunder, Elwood R. See Mason, David T., #M273, M275

M312 Maunder, Elwood R., and Davidson, Margaret G., eds. *First National Colloquium on the History of the Forest Products Industries, Proceedings, Boston, Massachusetts, May 17-18, 1966.* New Haven: Forest History Society, 1967. vii + 221 pp. Illus., map, tables, graphs, index.

M313 Maunder, Elwood R. "The History of Land Use in the Housatonic Valley." *Connecticut Woodlands* 33 (Spring 1968), 10-14. Including some account of forest industries and forestry in the Connecticut valley since the 17th century.

Maunder, Elwood R. See Kreienbaum, C.H., #K260

Maunder, Elwood R. See Kinney, Jay P, #K170, K171

Maunder, Elwood R. See Korstian, Clarence F., #K241

Maunder, Elwood R. See Fritz, Emanuel, #F254

Maunder, Elwood R. See Grover, Frederick W., #G366

Maunder, Elwood R. See Harper, Verne L., #H140, H141

Maunder, Elwood R. See Stone, J. Herbert, #S664

M314 Maunder, Elwood R. "David Townsend Mason: A Personal Appraisal." *Journal of Forestry* 71 (Oct. 1973), 670. Career sketch of Mason (1883-1973), a leading forestry consultant and advocate of sustained-yield forestry.

Maunder, Elwood R. See Garratt, George A., #G37

Maunder, Elwood R. See Schrepfer, Susan R., #S119

Maunder, Elwood R. See Gutermuth, Clinton Raymond, #G380, G381

Maunder, Elwood R. See Redfield, Alfred C., #R84

Maunder, Elwood R. See Reckord, Milton A., #R75

Maunder, Elwood R. See Johnson, Walter Samuel, #J127

Maunder, Elwood R. See McArdle, Richard E., #M9

Maunder, Elwood R. See Smith, Paul R., #S396

Maunder, Elwood R. See Peterson, Virgil G., #P157

Maunder, Elwood R. See Plant, Charles, #P244

M315 Maus, Calvin D. "The Development of Recreational Use on the National Forests of the U.S." Master's thesis, Yale Univ., 1946.

M316 Maxey, Carl C. *The Forest Economy of the South in Transition*. Special Study Series, No. 6. State College: Business Research Station, School of Business and Industry, Mississippi State College, 1950. 148 pp. Diags., maps, tables, notes, bib. Since 1935.

M317 Maxwell, C.W. "Steps Taken to Preserve Old Up-and-Down Sawmill." *West Virginia Conservation* 14 (Feb. 1951), 18-19.

M318 Maxwell, C. W. "Old Timer Recalls First Sawmill West of the Mountains." *West Virginia Conservation* 15 (Feb. 1952), 8-9.

M319 Maxwell, E.G. *Twenty-Five Years of Clarke-McNary Tree Distribution*. Extension Circular No. 1728. Lincoln: University of Nebraska, College of Agriculture, Extension Service, 1951.

M320 Maxwell, Hu. *History of Tucker County*. Kingwood, West Virginia: Preston Publishing Company, 1884. 574 pp. Includes a section on the "lumber interests" of Tucker County, West Virginia, pp. 139-66. See also Maxwell's histories of Hampshire County (1897), Randolph County (1898), and Barbour County (1899) for similar references to West Virginia lumber interests.

M321 Maxwell, Hu. *A Study of Massachusetts Wood-Using Industries.* Boston: Wright & Potter, 1910. 38 pp. This and the following studies of wood-using industries in various states contain incidental history.

M322 Maxwell, Hu, and Besley, Fred W. *The Wood-Using Industries of Maryland, with a Chapter on Maryland's Lumber and Timber Cut and the Timber Supply*. Baltimore, 1910. 58 pp.

M323 Maxwell, Hu. "The Use and Abuse of Forests by the Virginia Indians." *William and Mary College Quarterly* 19 (Oct. 1910), 73-103.

M324 Maxwell, Hu. "The Wood-Using Industries of Louisiana." *Lumber Trade Journal* (Jan. 1, 1912), 19-33.

Maxwell, Hu. See Gould, Clark W., #G201, G202

Maxwell, Hu. See Hatch, C.F., #H196

Maxwell, Hu. See Harris, John Tyre, #H149

M325 Maxwell, Hu, and Hatch, C.F. "The Wood-Using Industries of Texas." *Lumber Trade Journal* 61 (May 15, 1912), 27-44.

M326 Maxwell, Hu. *Wood-Using Industries of Florida.* Tallahassee, Florida: T.J. Appleyard, 1912. 85 pp.

M327 Maxwell, Hu. *Wood-Using Industries of Michigan.* Lansing: Wynkoop Hallenbeck Crawford Company, 1912. 101 pp.

M328 Maxwell, Hu, and Harris, John Tyre. *The Wood-Using Industries of Iowa.* Iowa Agricultural Experiment Station, Bulletin 142. Ames, 1913. 68 pp. Illus.

M329 Maxwell, Hu. *The Wood-Using Industries of Vermont.* Vermont Forestry Publication No. 11. Rutland: Vermont Forestry Service, 1913. 119 pp. Illus.

M330 Maxwell, Hu, and Harris, John Tyre. *Wood-Using Industries of Minnesota*. St. Paul: Minnesota State Board of Forestry, 1913. 87 pp.

M331 Maxwell, Hu. "The Story of White Pine." *American Forestry* 21 (Jan. 1915), 34-46. On the use of white pine since the colonial period.

M332 Maxwell, Hu. "The Development of Logging Operations." *American Forestry* 24 (May 1918), 272-79. Logging methods since the colonial period.

M333 Maxwell, Hu. "The Uses of Wood: The Sawing and Transportation of Lumber." *American Forestry* 24 (June 1918), 333-42. This and the following articles in the series contain incidental historical information.

M334 Maxwell, Hu. "The Uses of Wood: Wood Used in Rough Construction." *American Forestry* 24 (July 1918), 419-27.

M335 Maxwell, Hu. "The Uses of Wood: The Place of the Wooden Roof in Civilization." *American Forestry* 24 (Aug. 1918), 473-82.

M336 Maxwell, Hu. "The Uses of Wood: Wood in the Manufacture of Boxes and Crates." *American Forestry* 24 (Sept. 1918), 533-40.

M337 Maxwell, Hu. "The Uses of Wood: The Employment of Wood as House Finish." *American Forestry* 24 (Oct. 1918), 593-602.

M338 Maxwell, Hu. "The Uses of Wood: Woods Used in the Manufacture of Handles." *American Forestry* 24 (Nov. 1918), 679-87.

M339 Maxwell, Hu. "The Uses of Wood: Wooden Furniture and the Place It Fills." *American Forestry* 24 (Dec. 1918), 731-41.

M340 Maxwell, Hu. "The Uses of Wood: Wooden Artificial Limbs." *American Forestry* 25 (Jan. 1919), 807-16.

M341 Maxwell, Hu. "The Uses of Wood: Wood Used in Vehicle Manufacture." *American Forestry* 25 (Feb. 1919), 845-52.

M342 Maxwell, Hu. "The Uses of Wood: Fencing Materials from Forests." *American Forestry* 25 (Mar. 1919), 923-30.

M343 Maxwell, Hu. "The Uses of Wood: Wooden Boats and Their Manufacture." *American Forestry* 25 (Apr. 1919), 973-83.

M344 Maxwell, Hu. "The Uses of Wood: Wood Used in the Cooperage Industry." *American Forestry* 25 (July 1919), 1208-16.

M345 Maxwell, Hu. "The Uses of Wood: Floors Made of Wood." *American Forestry* 25 (Sept. 1919), 1343-49.

M346 Maxwell, Hu. "The Uses of Wood: Wood in Agricultural Implements." *American Forestry* 26 (Mar. 1920), 148-55.

M347 Maxwell, Hu. "The Uses of Wood: Wood for Musical Instruments." *American Forestry* 26 (Sept. 1920), 532-39.

M348 Maxwell, Hu. "The Uses of Wood: Wood in Games and Sports." *American Forestry* 27 (July 1921), 431-38.

M349 Maxwell, Hu. "The Uses of Wood: Wood for Professional Scientific Instruments." *American Forestry* 28 (Mar. 1922), 151-58.

M350 Maxwell, Hu. "Pioneer Sawmill of Applachian Lumber Region." *American Lumberman* (Mar. 7, 1925), 56-57.

M351 Maxwell, J.A. "Alienation of the Federal Domain in Canada." *Land Economics* 12 (Nov. 1936), 398-409. On the Canadian government's disposal of 116 million acres of land between Ontario and Alberta, formerly land of the Hudson's Bay Company, 1869-1880s.

M352 Maxwell, Robert S. *La Follette and the Rise of the Progressives in Wisconsin.* Madison: State Historical Society of Wisconsin, 1956. viii + 271 pp. Illus., notes, bib. Includes references to conservation and natural resource issues in Wisconsin politics, ca. 1900-1915.

M353 Maxwell, Robert S. *Whistle in the Piney Woods: Paul Bremond and the Houston, East and West Texas Railway.* Texas Gulf Coast Historical Association, Publication Series, Volume 7, No. 2. Houston, 1963. 77 pp. Illus., maps, notes, bib., index. History of a railroad in the logging region of eastern Texas, 1875-1899.

M354 Maxwell, Robert S. "Lumbermen of the East Texas Frontier." *Forest History* 9 (Apr. 1965), 12-16. Lumbermen and the lumber industry, 1870s-1930s.

M355 Maxwell, Robert S. "Manuscript Collections at Stephen F. Austin State College." *American Archivist* 28 (July 1965), 421-26. Includes descriptions of the collections documenting lumbering in eastern Texas, 1880-1930.

M356 Maxwell, Robert S., and Martin, James Williams. *A Short History of Forest Conservation in Texas.* Nacogdoches, Texas: Stephen F. Austin State University, 1970.

M357 Maxwell, Robert S. "Researching Forest History in the Gulf Southwest: The Unity of the Sabine Valley." *Louisiana Studies* 10 (Summer 1971), 109-22. A general account of the forests, lumbermen, and forest products industries in western Louisiana and eastern Texas since the late 19th century.

M358 Maxwell, Robert S. "The Impact of Forestry on the Gulf South." *Forest History* 17 (Apr. 1973), 30-35. Especially its influence on the forest products industries of Texas, Louisiana, Arkansas, and Mississippi in the 20th century.

M359 Maxwell, Robert S. "One Man's Legacy: W. Goodrich Jones and Texas Conservation." *Southwestern Historical Quarterly* 77 (Jan. 1974), 355-80. William Goodrich Jones (1860-1950) helped to organize the Texas Forestry Association and the Texas Department of Forestry in the 1910s and was an effective promoter of forestry and forest conservation.

M360 Maxwell, Robert S. "Mystery at East Mayfield: The Knox Family Saga." *Texas Forestry* 14 (Apr. 1974), 28-29. On the William Hiram Knox family and the lumber industry of eastern Texas, 1900-1920s.

M361 May, Richard H. "Recent Trends of Lumber Production in California." *Lumberman* 49 (Mar. 1948), 56, 58, 60.

M362 May, Richard H. *A Century of Lumber Production in California and Nevada.* Forest Survey Release No. 2. Berkeley: USFS, California Forest and Range Experiment Station, 1953. 33 pp. Illus., bib. Processed. Includes some history of the lumber industry, with statistics by species and year.

M363 May, Richard H. "Early History of the Redwood Lumber Industry." *Redwood Region Logging Conference Bulletin* 15 (1953), 8-9. In California since 1842.

M364 May, Richard H. "History of California Mills." *Timberman* 54 (Apr. 1953), 128.

M365 May, Richard H. *Notes on the History of Charcoal Production and Use in California.* Berkeley: USFS, California Forest and Range Experiment Station, 1956. 8 pp.

M366 May, Richard H. *Development of the Veneer and Plywood Industry in California.* Forest Survey Release No. 34. Berkeley: USFS, California Forest and Range Experiment Station, 1958. 26 pp. Map, graphs, tables.

M367 Maybee, Rolland H. "David Ward: Pioneer Timber King." *Michigan History* 32 (Mar. 1948), 1-14. In Michigan, 1836-1834.

M368 Maybee, Rolland H. "Michigan's White Pine Era, 1840-1900." *Michigan History* 43 (Dec. 1959), 385-432. Reprinted as John M. Munson History Fund Pamphlet No. 1 (Lansing: Michigan Historical Commission, 1960), 55 pp. Illus., map, bib.

Mayer, Carl E. See Ferguson, Roland H., #F65, F66

M369 Mayer, Harold M. "Politics and Land Use: The Indiana Shoreline of Lake Michigan." *Annals of the Association of American Geographers* 54 (Dec. 1964), 508-23. On the struggle between industrialists and conservationists over the Indiana Dunes project.

M370 Mayes, Edward. *Lucius Q.C. Lamar: His Life, Times, and Speeches.* Nashville: Publishing House of the Methodist Episcopal Church, South, 1896. 820 pp. Lamar was Grover Cleveland's Secretary of the Interior, 1885-1888.

M371 Mayhew, Dean R. "The Wooden Sailing Barges of Maine, 1886-to-1945." Master's thesis, Univ. of Maine, 1959.

M372 Mayo, Lawrence S. "'The King's Woods." *Proceedings of the Massachusetts Historical Society* 54 (Oct. 1920-June 1921), 50-61. On the British forest policy in northern New England and the efforts of New Hampshire Governor John Wentworth to protect the forests from timber trespass, 1769-1770; includes some account of colonial forest industries.

M373 Mayo, Lawrence S. *John Wentworth, Governor of New Hampshire, 1767-1775.* Cambridge: Harvard University Press, 1921. Includes a chapter on "The King's Woods."

M374 Mbogho, Archie W. "Sawmilling in Lane County, Oregon: A Geographical Examination of Its Development." Master's thesis, Univ. of Oregon, 1965.

M375 Mead, Walter J. "The Forest Products Economy of the Pacific Northwest." *Land Economics* 32 (May 1956), 127-33. Incidental history.

M376 Mead, Walter J. "Changing Pattern of Cycles in Lumber Production." *Journal of Forestry* 59 (Nov. 1961), 808-13. Douglas fir lumber production, 1919-1938 and 1945-1960.

M377 Mead, Walter J. *Mergers and Economic Concentration in the Douglas-Fir Lumber Industry.* Research Paper PNW-9. Portland: USFS, Pacific Northwest Forest and Range Experiment Station, 1964. vi + 81 pp. Tables, graphs, bib. Especially since 1950.

M378 Mead, Walter J. "Seasonal Variation in Lumber Prices." *Journal of Forestry* 62 (Feb. 1964), 89-95. Ponderosa pine and Douglas fir price behavior since 1921.

M379 Mead, Walter J. *Competition and Oligopsony in the Douglas Fir Lumber Industry.* Berkeley: University of California Press, 1966. xiv + 276 pp. Tables, notes, index. Incidental history of the industry in the Pacific states.

M380 Mead Corporation. *In Quiet Ways: George H. Mead, The Man and the Company.* Dayton: Privately printed, 1970. 301 pp. Illus., notes, apps., index. On George Houck Mead (1877-1963) and the Mead Corporation, Dayton, Ohio, a giant of the pulp and paper industry, with mills at Chillicothe, Ohio, and elsewhere in the East.

M381 Meany, Edmond S., ed. *Mount Rainier, a Record of Exploration.* New York: Macmillan, 1916. xi + 325 pp. Illus. Includes several articles on the history of the mountain and the establishment of Mount Rainier National Park, Washington.

M382 Meany, Edmond S. "Western Spruce and the War." *Washington Historical Quarterly* 9 (Oct. 1918), 255-58. Brief history of spruce in the Pacific Northwest and its use in the construction of aircraft during World War I.

M383 Meany, Edmond S., Jr. "The History of the Lumber Industry in the Pacific Northwest to 1917." Ph.D. diss., Harvard Univ., 1936. 412 pp.

M384 Mears, Carrie E. "Charles Mears, Lumberman." *Michigan History Magazine* 30 (July-Sept. 1946), 535-45.

M385 Mears, Eliot Grinnell. *Maritime Trade of the Western United States.* Stanford, California: Stanford University Press, 1935. xvii + 538 pp. Maps, tables, charts. Includes some history of the Pacific lumber trade.

M386 Mecklin, A.H. "Starting on the Second 100 Years: Bagdad Operation Has Been Making Lumber Since 1830." *Southern Lumberman* 144 (Dec. 15, 1931), 86-88. Bagdad Land and Lumber Company of Florida.

M387 Meelig, Martha. "Theodore Roosevelt, Forester." *Journal of Forestry* 56 (June 1958), 387-92. On his interest in forest conservation and his encouragement of the young profession of forestry, 1881-1919.

M388 Meinecke, Emilie P.M. "Forest Protection—Diseases." *Journal of Forestry* 23 (Mar. 1925), 260-69. Includes some history of forest pathology in the USDA and USFS since 1887.

M389 Meinig, Donald W. *The Great Columbian Plain: A Historical Geography, 1805-1910.* Seattle: University of Washington Press, 1968. xxi + 576 pp. Illus., maps, notes, tables, app., bib., index. Includes brief mention of the lumber industry and sawmills in this prairie region of eastern Washington, northeastern Oregon, and northern Idaho.

M390 Melbo, Irving Robert. *Our Country's National Parks.* 2 Volumes. Indianapolis: Bobbs-Merrill, 1941. Illus., maps. Contains descriptions and some history of each park.

M391 Melendy, Howard Brett. "One Hundred Years of the Redwood Lumber Industry, 1850-1950." Ph.D. diss., Stanford Univ., 1953. 377 pp.

M392 Melendy, Howard Brett. "Two Men and a Mill: John Dolbeer, William Carson, and the Redwood Lumber Industry in California." *California Historical Society Quarterly* 38 (Mar. 1959), 59-71. On the Dolbeer & Carson Lumber Company, Eureka, 1864-1950, and the activities of its founders in the Humboldt County lumber industry since 1853.

M393 Mell, Clayton D. "Historical Aspects of the Tropical American Forests." *Hardwood Record* 61 (Dec. 1926), 26, 28, 48.

Mellor, Norman. See Horwitz, Robert H., #H511

M394 Melton, William Ray. *The Lumber Industry of Washington, Including Logging, Sawmills, Shingle Mills, Plywood, Pulp & Paper, Specialties, Distribution.* National Youth Administration of Washington, Industrial Study No. 1. Tacoma: National Youth Administration of Washington, 1938. 160 pp. Illus., tables. Processed. Incidental history.

M395 Mendenhall, Herbert D. "The History of Land Surveying in Florida." *Surveying and Mapping* 10 (Jan.-Mar. 1950), 278-83. From 1513 to 1931.

M396 Mendocino County Historical Society. *Logging with Ox Teams: An Epoch in Ingenuity*. Ukiah, California, n.d. 15 pp. Processed. Some history of logging, especially with ox teams, in the redwood region of Mendocino County, California, 19th century.

M397 Mercer, Henry C. *The Origin of Log Houses in the United States*. Doylestown, Pennsylvania: Bucks County Historical Society, 1926. 40 pp. Illus.

M398 Merchant, E.O. "The Government and the Newsprint Paper Manufacturers." *Quarterly Journal of Economics* 32 (Feb. 1918), 236-56; 34 (Feb. 1920), 313-28. Since 1915.

M399 Mercier, H. "Forest Problems and Newsprint." *Pulp and Paper Magazine of Canada* 37 (Feb. 1936), 85-86.

M400 Merk, Frederick. *Economic History of Wisconsin During the Civil War Decade*. 1916. Reprint. Madison: Wisconsin State Historical Society, 1971. 414 pp. Illus., chart, table, notes, index. Includes a chapter on the lumber industry in the 1860s.

M401 Merrens, Harry R. "The Changing Geography of the Colony of North Carolina During the Eighteenth Century." Ph.D. diss., Univ. of Wisconsin, 1962. 495 pp.

M402 Merrens, Harry R. *Colonial North Carolina in the Eighteenth Century: A Study in Historical Geography*. Chapel Hill: University of North Carolina Press, 1964. 293 pp. Contains a chapter on forest utilization with emphasis on the naval stores, lumber, and wood-using industries.

M403 Merriam, Charles E. "National Resources Planning Board: A Chapter in American Planning Experiences." *American Political Science Review* 38 (Dec. 1944), 1075-88. And its predecessor agency, the National Planning Board, since 1933.

M404 Merriam, John Campbell. *Parks, National and State*. Washington: W.F. Roberts Company, 1933. By a leading conservationist.

M405 Merriam, John Campbell. *The Highest Uses of the Redwoods: Messages to the Council of the Save-the-Redwoods League, 1922-1941, by John C. Merriam*. Berkeley, California: Save-the-Redwoods League, 1941. 47 pp. Illus.

M406 Merriam, Lawrence C., Jr. "Forest Management Procedures for the Timbered Lands of the Oregon State Highway Commission." Master's thesis, Oregon State College, 1958.

Merriam, Lawrence C., Jr. See Brockman, Christian Frank, #B451

M407 Merriam, Lawrence C., Jr. "An Application of Recreational Forestry." *Journal of Forestry* 58 (Oct. 1960), 810, 813-15. On the Henry B. Van Duzer Forest Corridor along state highway 18 in Tillamook and Polk counties, Oregon, since 1946.

M408 Merriam, Lawrence C., Jr. "The Bob Marshall Wilderness Area of Montana, A Study in Land Use." Ph.D. diss., Oregon State Univ., 1963. 235 pp. Includes some history of USFS wilderness policy.

M409 Merriam, Lawrence C., Jr. "The National Park System: Growth and Outlook." *National Parks & Conservation Magazine* 46 (Dec. 1972), 4-12. Some general history of the system since 1872.

M410 Merriam, Paul G. "Portland, Oregon, 1840-1890: A Social and Economic History." Ph.D. diss., Univ. of Oregon, 1971. 358 pp. Includes reference to forest industries.

M411 Merrick, George Byron. *Old Times on the Mississippi*. Cleveland: Arthur H. Clark Company, 1909. 323 pp. Includes some history of log and lumber rafting on the Upper Mississippi River.

M412 Merrick, Gordon D. *Trends in Lumber Distribution, 1922-1943*. Washington: USFS, Division of Forest Economics Research, 1948. 13 pp.

M413 Merrick, Gordon D. "Trends in Distribution of Lumber, 1922-1943." *Southern Lumberman* 177 (Sept. 1, 1948), 56, 58, 60, 62.

M414 Merrill, Anthony F. *Our Eastern Playgrounds: A Guide to the National and State Parks and Forests of Our Eastern Seaboard*. New York: Whittlesey House, 1950. 353 pp. Incidental history.

M415 Merrill, Horace S. *William Freeman Vilas: Doctrinaire Democrat.* Madison: State Historical Society of Wisconsin, 1954. vii + 310 pp. Illus., notes, bib. Vilas was Grover Cleveland's Secretary of the Interior, 1888-1889, and U.S. senator from Wisconsin, 1891-1897.

M416 Merrill, Margaret (Becker). *Bears in My Kitchen.* New York: McGraw-Hill, 1956. x + 249 pp. On the author's life with her husband, Billy Merrill, a ranger in the Yosemite and General Grant national parks of California, 1930-1949.

M417 Merrill, Perry Henry. *History of Forestry in Vermont, 1909-1959.* Montpelier: Vermont State Board of Forests and Parks, 1959. 66 pp. Illus., map, diags., tables. A chronological survey of forest and forestry legislation and especially the Vermont Forest Service.

M418 Merrill, Sereno Taylor. *Narrative.* Milwaukee: Press of the Evening Wisconsin Company, 1900. 91 pp. Merrill (b. 1816) was a manufacturer of paper products in Wisconsin.

M419 Mershon, E.C. "Band Sawing." *Hardwood Record* (Feb. 10, 1911), 69-71. On manufacturing and refining of band resaws by William B. Mershon & Company since 1888.

M420 Messing, John. "Public Lands, Politics, and Progressives: The Oregon Land Fraud Trials, 1903-1910." *Pacific Historical Review* 35 (Feb. 1966), 35-66. Timberlands and public office holders were involved.

M421 Metcalf, Woodbridge. "History of Forestry Progress in California." *Timberman* 28 (Jan. 1927), 98-101. On forestry legislation and forest policy since 1905.

Metz, Louis J. See Hewlett, John D., #H324

M422 Metzger, Esta E. "Gifford Pinchot." Master's thesis, Lehigh Univ., 1933.

M423 Metzger, H.S. "The Susquehanna Boom." *Pennsylvania Forests and Waters* 15 (May-June 1944), 35-37.

M424 Metzler, Ken. "From the Forests to the Sea." *Ships and the Sea* 6 (Spring 1957), 20-23. Coos Bay, Oregon, as a lumber port since 1859.

Meyer, Arthur B. See Clepper, Henry, #C434, C440

M425 Meyer, Arthur B. *Forestry as a Profession.* Washington: Society of American Foresters, 1962. 16 pp. Illus. Incidental history.

Meyer, Carl E. See Barnard, Joseph E., #B76

M426 Meyer, Hans Arthur; Recknagel, Arthur B.; and Stevenson, Donald D. *Forest Management.* New York: Ronald Press, 1952. xii + 290 pp. Illus., tables, diags., references, index. Includes a chapter on the "History and Present Status of Forest Management." Ronald A. Bartoo joined the above three as coauthor of the second edition, 1961.

M427 Meyer, Marie E. "Rafting on the Mississippi." *Palimpsest* 8 (Apr. 1927), 121-31. On the rafting of logs to Iowa sawmills in the late 19th century.

M428 Meyer, Roy W. "The Story of Forest Mills, A Midwest Milling Community." *Minnesota History* 35 (Mar. 1956), 11-21. Forest Mills, a lumber town in Goodhue County, Minnesota, 1867-1945.

M429 Meyer, Roy W. "Forestville: The Making of a State Park." *Minnesota History* 44 (Fall 1974), 82-95. On the state park movement in Minnesota and the establishment in 1963 of Forestville State Park, a 2500-acre tract of hardwood forest in Fillmore County.

M430 Meyer, Walter H., and Plusnin, Basil A. *The Yale Forest in Tolland and Windham Counties, Connecticut.* Yale University, School of Forestry, Bulletin No. 55. New Haven, 1945. 54 pp. A forest land-use history of the region and an account of the demonstration forest.

Meyer, Walter H. See Chapman, Herman Haupt, #C254

M431 Meyer, Walter H. "Impressions of Industrial Forestry in Southeastern U.S." *Journal of Forestry* 58 (Mar. 1960), 179-87. On changes noted during the 1940s and 1950s.

M432 Michaud, Howard H. "A Brief History of the Conservation of Natural Resources in Indiana." *Proceedings of the Indiana Academy of Science* 58 (1948), 257-62. Since 1869.

M433 Michaud, Howard H. "Conservation of Recreational and Scenic Resources." *Proceedings of the Indiana Academy of Science* 66 (1957), 268-74. On state parks, beaches, and recreation areas in Indiana since 1915.

M434 Michaud, Howard H. "History of Early Development of Game Regulations in Indiana." *Proceedings of the Indiana Academy of Science* 67 (1958), 256-59.

M435 Michigan-California Lumber Company. *The El Dorado; Historic Land of Gold Where Giant Pines Stand.* Camino, California, 1940. 16 pp. On the history of El Dorado County and of the lumber company's role in it. An excerpt appears in *Timberman* 42 (Oct. 1941), 50-53.

M436 *Michigan Conservation.* "Pine Days: Being an Album of Pictures and Some Account of the Early Days of Logging in Northern Michigan." *Michigan Conservation* (Mar.-Apr. 1962), 17-32. Supplement.

M437 *Michigan History Magazine.* "Dedication of Memorial to Michigan's Pioneer Lumbermen." *Michigan History Magazine* 16 (Autumn 1932), 497-504. On the Au Sable River near East Tawas, Michigan; includes a list of pioneer lumbermen from the region.

M438 Michigan State College, Department of Forestry. *Fifty Years of Forestry at Michigan State College.* East Lansing, 1953.

M439 Mickalitis, Albert B., and Kutz, Donald B. "Experiments and Observations on Planting Areas 'Stripped' for Coal in Pennsylvania." *Pennsylvania Forests and Waters* 1 (Jan.-Feb. 1949), 62-66.

Mickalitis, Albert B. See Ibberson, Joseph E., #I1

Mickalitis, Albert B. See Rowland, H.B., #R342

M440 Mickey, Carrol Milton. "A Sociological Analysis of the Conservation Movement." Ph.D. diss., Univ. of Iowa, 1949. 279 pp. Since 1908.

Middleton, William. See Snyder, Thomas E., #S417

M441 Middour, J.C. "Reclamation of Strip Mined Areas." *Pennsylvania Forests and Waters* 2 (Sept.-Oct. 1950), 98-99, 110-11. Reforestation in Pennsylvania since 1945.

M442 Mignery, A.L. "Factors Affecting Small-Woodland Management in Nacogdoches County, Texas." *Journal of Forestry* 54 (Feb. 1956), 102-05. On forestry and forestry education since the early 1930s.

Milde, Gorden T. See Fabos, Julius G., #F1

M443 Miles, A.W. "The End of the Drive." *Michigan History Magazine* 20 (Spring-Summer 1936), 221-29. A reminiscence of the last log drive on the Muskegon River; includes lists of pioneer lumbermen and lumber companies of Michigan's Muskegon Valley.

M444 Miles, John G. "The Redwood Park Question." *Forest History* 11 (Apr. 1967), 6-11, 31. Includes some history of the movement to establish Redwood National Park, California.

Miles, R. Vance, Jr. See Kirkpatrick, John C., #K187

M445 *Milford (New Jersey) Leader.* "Lumbering on the Delaware." *Proceedings of the New Jersey Historical Society* 71 (July 1953), 212-14. The lumber industry and rafting in the Delaware Valley, New Jersey-Pennsylvania-New York, since 1763. Reprinted from the *Milford Leader,* May 10, 1883.

M446 Millar, Charles E., and Galloway, Harry M. "Cooperative Effort of the State, Counties, and Municipalities in the Disposal of Tax-Reverted Lands in Northern Michigan." *Papers of the Michigan Academy of Science, Arts, and Letters* 33 (1949), 195-213. Since 1939.

M447 Millar, W.N. "Status of Forestry in Western Canada." *Journal of Forestry* 20 (Jan. 1922), 10-17. Surveys important developments since 1915.

M448 Miller, Alton. "White Pine, a Self Replenishing Resource." *Connecticut Woodlands* 23 (Nov.-Dec. 1958), 96-97. Since 1645.

M449 Miller, Charles I. "History of Chain Saws." *Southern Lumberman* 178 (Apr. 15, 1949), 52, 54, 56. Since 1858.

M450 Miller, E. Willard. "The Industrial Structure of the Bradford Oil Region." *Western Pennsylvania Historical Magazine* 26 (Mar.-June 1943), 59-78. Includes some history of the lumber industry in McKean County, Pennsylvania.

Miller, Ernest C. See Putnam, Theodore L., #P356

M451 Miller, Eunice. "The Timber Resources of Nevada." *Nevada Historical Society Papers* 5 (1926), 375-457.

M452 Miller, F.G. "Twenty Years of Forestry at the University of Idaho." *Idaho Forester* 11 (1929), 10-12, 46-48.

M453 Miller, G.L. "Forest Fire Protection in New Brunswick." *Pulp and Paper Magazine of Canada* 37 (Apr. 1936), 268-70. Incidental history.

M454 Miller, Gordon K. *A Biographical Sketch of Major Edward E. Hartwick.* Detroit: N.p., 1921. 135 pp. Illus. Hartwick (1871-1918) was a Michigan lumberman and served with the 20th Engineers (forestry) in France during World War I.

M455 Miller, Howard A. "Game Populations on Southern National Forests." *Southern Lumberman* 195 (Dec. 15, 1957), 106-08. Since 1915.

M456 Miller, J.M., and Keen, Frederick Paul. *Biology and Control of the Western Pine Beetle, a Summary of the First Fifty Years of Research.* USFS, Miscellaneous Publication No. 800. Washington: GPO, 1960. 381 pp. Illus., tables.

Miller, J.M. See Craighead, Frank Cooper, #C693

Miller, Joseph A. See Flynn, Ted P., #F128

M457 Miller, Joseph A. *Pulp and Paper History: A Selected List of Publications on the History of Industry in North America.* Bibliographic Series No. 1. St. Paul: Forest History Society, 1963. 41 pp. Processed. Includes approximately 350 annotated entries for books and articles on the history of the industry.

M458 Miller, Joseph A. "The Changing Forest: Recent Research in the Historical Geography of American Forests." *Forest History* 9 (Apr. 1965), 18-24. A review essay of doctoral dissertations revealing the historical geography of forest vegetation in several areas of the East and Midwest.

M459 Miller, Joseph A. "Forests and the Regional Landscape." *Forest History* 9 (July 1965), 24-29. A review of doctoral dissertations revealing the historical geography of forests and forest industries in various regions of the United States.

M460 Miller, Joseph A. "A New Generation of Forest Scientists." *Yale Scientific Magazine* (Dec. 1965). Includes some history of forestry research.

Miller, Joseph A. See Collingwood, George Harris, #C524

M461 Miller, Joseph A. "Congress and the Origins of Conservation: Natural Resource Policies, 1865-1900." Ph.D. diss., Univ. of Minnesota, 1973. 504 pp. A study of the legislative origins of federal resource and conservation policy, with special attention to forest reserves.

M462 Miller, Lena Fastabend. "Letter to the Editor." *Oregon Historical Quarterly* 66 (June 1965), 178-82. Brief account of John Antone Fastabend, builder of ocean-going log rafts, 1890s-1930s, especially for the Benson Timber Company on the lower Columbia River.

M463 Miller, Leslie A., and Albright, Horace M. *Natural Resources: Organization and Policy in the Field of Natural Resources.* U.S. Commission on Organization of the Executive Branch of the Government, Task Force Report, Appendix L. Washington: GPO, 1949. 244 pp.

M464 Miller, Mike, and Wayburn, Peggy. *Alaska: The Great Land.* San Francisco: Sierra Club, 1974. 152 pp. Illus. Includes some history of Alaska, especially of its scenic and wilderness areas.

M465 Miller, Paul I. "Thomas Ewing, Last of the Whigs." Ph.D. diss., Ohio State Univ., 1934. Ewing was the first Secretary of the Interior, 1849-1850.

M466 Miller, R.B. "Review of Illinois Lumbering and Forestry." *Lumber World Review* 41 (Nov. 10, 1921), 60-62. Incidental history.

M467 Miller, R.L., and Rothrock, C.W. "A History of Chip Shredding." *Tappi* 46 (July 1963), 174A-78A.

M468 Miller, Robert L., and Choate, Grover A. *The Forest Resource of Colorado.* Resource Bulletin INT-3. Fort Collins, Colorado: USFS, Rocky Mountain Forest and Range Experiment Station, 1964. 55 pp. Incidental history.

M469 Miller, Thomas Lloyd. *The Public Lands of Texas, 1519-1970.* Norman: University of Oklahoma Press, 1972. xxii + 341 pp. Maps, illus., tables, apps., notes, bib., index.

M470 Miller, Thomas W. "The Genesis and Programs of the Nevada State Park System." *Planning and Civic Comment* 24 (June 1958), 47-49. Since 1923.

M471 Miller, William D., comp. *The Hofmann Forest: A History of the North Carolina Forestry Foundation.* Raleigh: North Carolina Forestry Foundation, 1970. xviii + 177 pp. Illus., maps, bib., app. History of the Hofmann Forest, an industrial and demonstration forest in Jones and Onslow counties, North Carolina, since 1936.

Millet, Artemus. See Johannessen, Carl L., #J82

M472 Millet, Donald J. "The Economic Development of Southwest Louisiana, 1865-1900." Ph.D. diss., Louisiana State Univ., 1964. 467 pp.

M473 Millet, Donald J. "The Lumber Industry of 'Imperial' Calcasieu: 1865-1900." *Louisiana History* 7 (Winter 1966), 51-69. On the logging of longleaf pine and cypress in the vicinity of Lake Charles in southwestern Louisiana. Attention is given to log driving, logging railroads, timber trespass, sawmills, labor relations, lumber exports, and individual lumbermen and companies.

M474 Milligan, G. "Lumber and Rumors of War Along Lake Champlain." *North Country Life* 16 (Summer 1962), 42-44. A letter from Keeseville, New York, March 16, 1839, during a period of American-Canadian hostility.

M475 Mills, Borden H., Sr. "David McClure and the Forest Preserve." *Ad-i-ron-dac* 13 (Nov.-Dec. 1949), 116-19. On McClure's life (1848-1912) and his work in the New York constitutional convention of 1894 for the preservation of the state forests.

M476 Mills, C.R. "Ontario's Forest Fires: An Analysis of the Cause and Effect of Conflagrations from 1881 up to the Present Time." *Forests and Outdoors* 22 (Feb. 1926), 85-86.

M477 Mills, Enos A. *The Story of a Thousand-Year Pine.* Boston: Houghton Mifflin, 1909. 38 pp. Illus. On the life history of an ancient pine tree near Mesa Verde National Park, Colorado, as determined from its growth rings.

M478 Mills, Enos A. *The Rocky Mountain Wonderland.* Boston: Houghton Mifflin, 1915. xiii + 362 pp. Includes some history of Colorado mountains and forests.

M479 Mills, Enos A. *Your National Parks.* Boston: Houghton Mifflin, 1917. xxi + 431 pp. Illus., apps., bib., index. History and description of American and Canadian national parks and monuments.

M480 Mills, Enos A. *The Rocky Mountain National Park.* New York: Doubleday, Page & Company, 1924. Boston: Houghton Mifflin, 1932. xxiii + 239 pp. By the Colorado conservationist who was a leader in establishing the park.

M481 Mills, Enos A. *Early Estes Park.* Foreword and biographical sketch by Esther B. Mills. Estes Park, Colorado: Mrs. E.A. Mills, 1959. xx + 52 pp. Illus., map, diags. Includes a sketch of the author (1870-1922), lecturer and writer on conservation and active in the establishment of the Rocky Mountain National Park, by his widow. First printed in 1911.

Mills, Esther B. See Mills, Enos A., #M481

Mills, Esther B. See Hawthorne, Hildegarde, #H233

M482 Mills, Randall V. "Prineville's Municipal Railroad." *Oregon Historical Quarterly* 42 (Sept. 1941), 256-62. On the railroad serving the central Oregon lumber town of Prineville.

M483 Mills, Randall V. *Railroads Down the Valleys: Some Short Lines of the Oregon Country.* Palo Alto, California: Pacific Books, 1950. ix + 151 pp. Illus., maps, tables, bib. Includes some history of logging railroads and railroad transportation of lumber in Oregon.

M484 Miner, J.H. "Early Lumbering in the South." *Southern Lumberman* (Jan. 15, 1940-Feb. 15, 1941). Series of articles in the form of reminiscences, by the president of the J.H. Miner Saw Manufacturing Company, Meridian, Mississippi, concerning his travels throughout the South as an itinerant saw filer, 1870-1900.

M485 Minnesota and Ontario Paper Company. *The Great Mando Log Drive.* International Falls, Minnesota, 1940.

M486 Minnesota. Committee on Land Utilization. *Land Utilization in Minnesota: A State Program for the Cut-Over Lands.* Final Report of the Committee on Land Utilization. Minneapolis: University of Minnesota Press, 1934. xiv + 289 pp. Maps, tables, bib., index. Includes some history of forested lands, cutover lands, land policies, and the lumber industry.

M487 Minnesota. Division of Forestry. *Forestry in Minnesota.* St. Paul, 1961. 76 pp. Incidental history of state forestry.

M488 Minnesota. Forestry Board. *Forest Fires in Minnesota.* St. Paul, 1927. 74 pp. Incidental history.

M489 Minnesota Historical Society. "What is Forest History? And Why a Forest History Center?" *Timber Producers Association Bulletin* 29 (Apr.-May 1974), 16-17. Includes a brief sketch of Minnesota's forest history, especially since the 1880s.

M490 Minnesota, University of. School of Forestry. *Forestry in Minnesota—Past, Present, and Future.* St. Paul, 1953. 100 pp. A historical sketch of the School of Forestry at the University of Minnesota and of the development of federal, state, county, and private activities in forestry.

M491 Minville, Esdras, ed. *La Forêt: étude preparée avec la collaboration de l'école de génie forestier de Québec.* Montreal: École des Hautes Études Commerciales, 1944. 414 pp. Tables. A collection of articles, many concerning the history of the forests, forest industries, and forestry in Quebec.

M492 Misfeldt, O.H. "Timberland Terminology." *American Speech* 26 (Oct. 1941), 232-34.

M493 *Missouri Historical Review.* "Views from the Past Lumber Industry." *Missouri Historical Review* 63 (Jan. 1969), 248-50. Photographs of logging and sawmills in Missouri, 1890s-1900s.

Mitchell, Annie R. See Griggs, Monroe Christopher, #G347

M494 Mitchell, George W. "History of the Kraft Paper Industry in Louisiana." *Paper Industry* 19 (July 1937), 426-30. Especially the Bogalusa Paper Company, Bogalusa, Louisiana.

M495 Mitchell, Howard T. "A Half Century of Marketing Forest Products." *British Columbia Lumberman* 50 (Aug. 1966), 28-30. Of British Columbia.

M496 Mitchell, J. Alfred. "A Review of Forest Fires in the South." *Lumber World Review* 42 (No. 4, 1922), 44-45.

M497 Mitchell, J. Alfred. "Accomplishments in Fire Protection in the Lake States." *Journal of Forestry* 37 (Sept.

1939), 748-50. Some history of forest fires and forest protection since the 1870s.

M498 Mitchell, J. Alfred. "Forest Fires in Indiana." *Journal of Forestry* 45 (Aug. 1947), 570-74. Since 1930.

M499 Mitchell, J. Alfred, and Lemay, Neil. *Forest Fires and Forest-Fire Control in Wisconsin.* Madison: Wisconsin Conservation Commission, in cooperation with the USFS, 1952. 75 pp. Illus., maps, tables, graphs. Includes some history of forest fire legislation and protection.

M500 Mitchell, J. Roy. "The Story of the Smoke Jumpers." *Okanogan County Heritage* (Fall 1972), 3-12. On USFS experiments in smokejumping on Washington's Chelan (Okanogan) National Forest, 1939.

M501 Mitchell, James. *Lumbering and Rafting in Clearfield County, Pennsylvania, on the West Branch of the Susquehanna River.* Clearfield, Pennsylvania: N.p., n.d. 76 pp.

M502 Mitchell, John G. "The Bitter Struggle for a National Park." *American Heritage* 21 (Apr. 1970), 97-108. On the movement to establish and protect Everglades National Park, Florida.

M503 Mitchell, John G. "Best of the S.O.B.s." *Audubon* 76 (Sept. 1974), 48-62. On the Weyerhaeuser Company, Tacoma, Washington, including some history of its leadership in the field of industrial forestry.

M504 Mitchell, Wesley C. *History of Prices during the War: International Price Comparisons.* Washington: GPO, 1919. 395 pp. A War Industries Board investigation of commodity prices, 1913-1918, including those for 21 forest products. These are analyzed in a review in *Journal of Forestry* 17 (Nov. 1919), 843-50.

M505 Mitler, Charles I. "History of Chain Saws." *Southern Lumberman* (Apr. 15, 1949), 52-56. Since the first American patent for an "endless" saw in 1858.

Mitson, Betty E. See Peterson, Virgil G., #P157

Mitson, Betty E. See Plant, Charles, #P244

M506 Mittelman, Edward B. "The Loyal Legion of Loggers and Lumbermen—An Experiment in Industrial Relations." *Journal of Political Economy* 31 (June 1923), 313-41. On a government-inspired labor union to counteract radical organizers among lumberjacks and millworkers of the Pacific Northwest during the World War I era.

M507 Mittelman, Edward B. "The Gyppo System." *Journal of Political Economy* 31 (Dec. 1923), 840-51. Includes some history of independent logging and sawmilling in the Pacific Northwest.

M508 Mixon, James E. "Progress of Protection from Forest Fires in the South." *Journal of Forestry* 54 (Oct. 1956), 649-52. Since 1905.

M509 Moak, James E. "The Pine Lumber Industry in Mississippi: Its Changing Aspects." Ph.D. diss., State Univ. of New York, College of Environmental Science and Forestry, 1965. 173 pp.

M510 Moak, James E. *Forestry: Its Economic Importance to Mississippi.* Bulletin No. 785. State College: Mississippi Agricultural and Forestry Experiment Station, 1971. Incidental history.

M511 Mobely, M.D., and Hoskins, R.N. *Forestry in the South.* Atlanta: Turner E. Smith and Company, 1956. vii + 440 pp. Illus., maps, tables, glossary, apps., index. Incidental history of forestry and forest industries.

M512 Mohler, Levi L.; Wampole, John H.; and Fichter, Edson. "Mule Deer in Nebraska National Forest." *Journal of Wildlife Management* 15 (Apr. 1951), 129-57. Since 1804.

M513 Moir, William H. "Natural Areas." *Science* 177 (Aug. 4, 1972), 396-400. Definitions and history of the "natural area" concept in the 20th century.

M514 Moltke, Alfred W. *Memoirs of a Logger.* College Place, Washington: College Press, 1965. 415 pp. Illus. Moltke began logging near Wenatchee, Washington, ca. 1920, and later was the owner-operator of the Pilot Rock Lumber Company, Pilot Rock, Oregon.

M515 Monahan, Robert S. "Sherman Adams." *Journal of Forestry* 47 (Apr. 1949), 307-08. On his career in industrial forestry culminating in his election as governor of New Hampshire in 1948.

Monahan, Robert S. See Appalachian Mountain Club, #A465

M516 Montague, Margaret. *Up Eel River.* New York: Macmillan, 1928. 225 pp. Folklore and legends of West Virginia lumberjacks.

M517 Montague, Richard W., and the editors of Alaska Travel Publications, 1973. 294 pp. Illus., maps, bib. Includes *Car/Camper, Bus, Bicycle or on Foot.* Anchorage: Alaska Travel Publications 1973. 294 pp. Illus., maps, bib. Includes a section on the history of Mount McKinley National Park.

M518 *Montana, Magazine of Western History.* Yellowstone Park Centennial. *Montana, Magazine of Western History* 22 (Summer 1972), complete. Articles on the history of the national park.

M519 Montell, William Lynwood. *The Saga of Coe Ridge: A Study in Oral History.* Knoxville: University of Tennessee Press, 1970. xxi + 231 pp. Illus., charts, notes, apps., index. Contains much folk history of Negro and mulatto woodcutters and raftsmen in the Cumberland Hills of southern Kentucky, early 20th century.

Montgomery, David. See Chapman, Herman Haupt, #C263

M520 Montgomery, W.F. "Pioneer Lumber Dealers in Los Angeles." *Historical Society of Southern California Quarterly* 24 (June 1942), 66-68. Brief sketch of wholesale and retail firms established between the 1860s and 1880s.

M521 Moody, Linwood W. *The Maine Two-Footers: The Story of the Two-Foot Gauge Railroads of Maine.* Berkeley: Howell-North Books, 1959. xv + 203 pp. Illus., maps, tables, notes. Reminiscences and descriptions of locomotives and cars, including some used in logging.

M522 Moomaw, Jack C. *Recollections of a Rocky Mountain Ranger.* Longmont, Colorado: Times-Call Publishing, 1963. 216 pp.

M523 Moon. D.G. "The Southern Pulp and Paper Industry—Past, Present, and Future." *Southern Pulp & Paper Manufacturer* 17 (Jan. 11, 1954), 52-62, 128.

M524 Moon, Frederick F., and Brown, Nelson Courtlandt. *Elements of Forestry.* New York: John Wiley & Sons, 1914. xvii + 392 pp. Illus., tables, diags. Includes incidental references to the history of forestry and the forest product industries. Subsequent editions appeared in 1924 and 1937.

M525 Moore, A. Milton. "Conservation and Taxation of the Natural Resource." *Canadian Tax Journal* 5 (Sept.-Oct. 1957), 334-43.

M526 Moore, A. Milton. *Forestry Tenures and Taxes in Canada: The Economic Effects of Taxation and the Regulating of the Crown Forests by the Provinces.* Toronto: Canadian Tax Foundation, 1957. 315 pp. Includes some history of forest taxation and an account for each province.

M527 Moore, A. Milton. "Forestry Taxation Problems." *Canadian Tax Journal* 8 (Sept.-Oct. 1960), 353-57.

M528 Moore, Andrew G.T. *Transportation as a Factor in Forest Conservation and Lumber Distribution.* Lumber Industry Series No. 12. New Haven: Yale University, School of Forestry, 1937. 30 pp. Map. Incidental history.

M529 Moore, Andrew G.T. *Grow Green Gold: The Development of Forest Conservation Thought and Practices by Southern Lumbermen.* New Orleans: Southern Pine Association, 1938. 20 pp. Illus., tables, graph. Includes some account of the Southern Pine Association's work in promoting forest conservation.

M530 Moore, Andrew G.T. "Traffic and Transportation in Lumber Merchandising." *Southern Lumberman* 177 (Dec. 15, 1948), 187-90.

M531 Moore, Arthur K. *The Frontier Mind: A Cultural Analysis of the Kentucky Frontiersman.* Lexington: University of Kentucky Press, 1957. x + 264 pp. Notes. Explores the impact of wilderness on European man in America during the 18th and 19th centuries.

M532 Moore, Barrington. "Working Plans: Past History, Present Situation, and Future Development." *Proceedings of the Society of American Foresters* 10 (July 1915), 217-58. By the USFS since 1905.

M533 Moore, E.B. *Forest Management in New Jersey.* Trenton: New Jersey Department of Conservation and Development, Division of Forests and Parks, 1939. 54 pp. Incidental history.

M534 Moore, E.B. "Timber Rafting on the Delaware." *American Forests* 47 (Nov. 1941), 515-17, 542.

M535 Moore, E.B., and Lentz, A.N. "The Cooperative Forest Management Program in New Jersey." *Journal of Forestry* 49 (Jan. 1951), 31-34. On the New Jersey Bureau of Forest Management's program of forestry assistance to woodland owners since 1938.

M536 Moore, Frank L. "Development of Interest in Forestry by Manufacturers of Paper." *Paper Trade Journal* 71 (Nov. 18, 1921), 36-37.

M537 Moore, James Henry, Jr. "The Transfer of Title to Timber in West Virginia," *West Virginia Law Review* 52 (June 1950), 228-36. Since 1892.

M538 Moore, John Hebron. "Simon Gray, Riverman: A Slave Who Was Almost Free." *Mississippi Valley Historical Review* 49 (Dec. 1962), 472-84. Gray supervised rafting operations and made some lumber purchases for Andrew Brown and Company, a cypress lumber business with operations in the Yazoo River basin of Mississippi, Natchez, and New Orleans, 1835-1862.

M539 Moore, John Hebron. "William H. Mason, Southern Industrialist." *Journal of Southern History* 27 (May 1961), 169-83. On his career in Mississippi as an inventor, lumberman, wood technologist, and founder of the Masonite Corporation, manufacturer of wallboard from sawmill wastes, 1920s-1930s.

M540 Moore, John Hebron. *Andrew Brown and Cypress Lumbering in the Old Southwest.* Baton Rouge: Louisiana State University Press, 1967. xv + 180 pp. Illus., notes, bib., index. A history of Brown's logging operations in the Yazoo-Mississippi Delta, his steam sawmill at Natchez, Mississippi, and his slave labor force, 1829-1865.

M541 Moore, John R., ed. *The Economic Impact of TVA.* Knoxville: University of Tennessee Press, 1967. Includes some history of the Tennessee Valley Authority.

M542 Moore, R.J. *A Directory of Botanists in Canada.* Ottawa: Department of Agriculture, Plant Research Institute, Research Branch, 1959. 99 pp. Includes some history of forestry and botanical research in Canada.

M543 Moore, Robert M., and Buchwalter, Nichelsen E. "Collective Bargaining in the Southern Lumber Industry." *Southern Lumberman* 185 (July 15, 1952), 42-50. Since 1910.

M544 Moore, S.T. "Personal Observations and Experiences of an Old Forest Surveyor." *Journal of Forestry* 22 (Apr. 1924), 413-21. Includes some reminiscences of the late 19th-century forestry movement and forest surveying in Pennsylvania.

M545 Moore, William C. *Eastern Oregon Lumber Survey.* Moscow, Idaho: Queen City Printing Company, 1941. 143 pp. Illus., tables.

M546 Moran, Joe A. "When the Chippewa Forks Were Driving Streams." *Wisconsin Magazine of History* 26 (June 1943), 391-407. In Wisconsin's Chippewa Valley, late 19th century.

Moravets, F.L. See Cowlin, Robert W., #C645

M547 Moravets, F.L. *Lumber Production in Oregon and Washington, 1869-1948.* Forest Survey Report No. 100, Portland: USFS, Pacific Northwest Forest and Range Experiment Station, 1949.

M548 Moravets, F.L. *Production of Logs in Oregon and Washington, 1925-1948.* Forest Survey Report No. 101. Portland: USFS, Pacific Northwest Forest and Range Experiment Station, 1950. 15 pp. Graphs, tables.

M549 Morden, C.M. "History of Pacific Coast Paper Industry: Industry's Beginning Traced Back Seventy-five Years to Samuel P. Taylor Mill in California."

Timberman 32 (Mar. 1931), 62-63. The Pioneer Paper Mill, Paperville, California, 1856.

M550 Moreell, Ben. *Our Nation's Water Resources-Policies and Politics.* Chicago: University of Chicago Law School, 1956. v + 266 pp. Index. Includes some incidental historical references to the influence of forests and forestry on water resources, hydroelectric power, reclamation, navigation, flood control, and other aspects of resource conservation.

M551 Morgan, A.W. *Fifty Years in Siletz Timber.* Portland: By the author, 1959. xiii + 82 pp. Illus. Reminiscences of the lumber industry and timber transactions in the Siletz Valley of Lincoln County, Oregon, since the 1900s.

M552 Morgan, Arthur E. *Dams and Other Disasters: A Century of the Army Corps of Engineers in Civil Works.* Boston: Porter Sargent, 1971. xxiii + 422 pp. Illus., bib., index. A highly critical account of the agency and its abuse of the environment.

M553 Morgan, Charles M. "Arizona's Busiest Ambassador." *Arizona Highways* 26 (Sept. 1950), 10-13. On Leslie N. Gooding's career as an Arizona forester, botanist, and promoter of the Arizona cypress, since 1907.

Morgan, E. See Kerr, Edward F., #K104

M554 Morgan, George Thomas, Jr. "William B. Greeley, Practical Forester, 1904-1928." Master's thesis, Univ. of Oregon, 1960.

Morgan, George Thomas, Jr. See Greeley, William B., #G308

M555 Morgan, George Thomas, Jr. *William B. Greeley, A Practical Forester, 1879-1955.* St. Paul: Forest History Society, 1961. 82 pp. Illus., notes, bib., index. Emphasizes Greeley's early career in forestry, 1904-1928, especially his tenure as chief of the USFS, 1920-1928.

M556 Morgan, George Thomas, Jr. "The Fight Against Fire, the Development of Cooperative Forestry in the Pacific Northwest." *Idaho Yesterdays* 6 (Winter 1962), 20-30. On the origins and history of cooperative forest protection in Oregon, Washington, and Idaho, since ca. 1900.

M557 Morgan, George Thomas, Jr. "The Fight Against Fire, the Development of Cooperative Forestry in the Pacific Northwest, 1900-1950." Ph.D. diss., Univ. of Oregon, 1964. 283 pp.

M558 Morgan, George Thomas, Jr. "No Compromise—No Recognition: John Henry Kirby, the Southern Lumber Operators' Association, and Unionism in the Piney Woods, 1906-1916." *Labor History* 10 (Spring 1969), 193-204. Eastern Texas and Louisiana.

Morgan, George Thomas, Jr. See Kinney, Jay P., #K170, K171

M559 Morgan, George Thomas, Jr. "The Gospel of Wealth Goes South: John Henry Kirby and Labor's Struggle for Self-Determination, 1901-1916." *Southwestern Historical Quarterly* 75 (Oct. 1971), 186-97. Kirby was a Texas lumberman of national prominence. Despite his claims to be the "Peon's Pal," Kirby was strongly opposed to organized labor.

Morgan, J.T. See Hutchison, O. Keith, #H673

Morgan, J.T. See Thornton, P.L., #T92

M560 Morgan, James T., and Dickerman, Murlyn B. "Return to the Forest." *Michigan Conservation* 29 (Jan.-Feb. 1960), 18-21. In Michigan since 1839.

M561 Morgan, Murray. *Skid Road: An Informal Portrait of Seattle.* New York: Viking Press, 1951. 280 pp. Includes some history of the lumber industry and lumberjacks in Seattle, Washington, since the mid-19th century.

M562 Morgan, Murray. *The Last Wilderness.* New York: Viking Press, 1955. xi + 275 pp. Illus., maps. On the Olympic Peninsula of northwestern Washington, its lumber industry, forests, and national park.

M563 Morgan, Murray. *The Northwest Corner: The Pacific Northwest, Its Past and Present.* New York: Viking Press, 1962. 168 pp. Illus., map, index. Incidental history of the forests and forest industries.

M564 Morgan, Robert J. *Governing Soil Conservation: Thirty Years of the New Decentralization.* Baltimore: Johns Hopkins Press, for Resources for the Future, 1966. xiv + 399 pp. Maps, tables, graphs, notes, apps., index. An administrative history of the U.S. Soil Conservation Service and the operations of soil conservation districts.

Morison, Elting E. See Roosevelt, Theodore, #R289

M565 Morison, M.B. *The Forests of New Brunswick.* Forest Service, Bulletin No. 91. Ottawa, 1938. 112 pp. Incidental history.

M566 Morison, Samuel Eliot. *The Maritime History of Massachusetts, 1783-1860.* Boston: Houghton Mifflin, 1921. xiv + 401 pp. Illus., apps., bib. Includes some history of shipbuilding and the trade in lumber and forest products.

M567 Morison, Samuel Eliot. *The Story of Mount Desert Island, Maine.* Boston: Little, Brown, 1960. viii + 81 pp. Illus., map, notes, bib. On which is located Acadia National Park.

M568 Morley, Alan. *Vancouver: From Milltown to Metropolis.* Vancouver: Mitchell Press, 1961. xiii + 234 pp. Illus., index. Includes some history of the lumber industry in Vancouver, British Columbia, since 1862.

M569 Morley, Jim. *Muir Woods: The History, Sights, and Seasons of the Famous Redwood Forest Near San Francisco, a Pictorial Guide.* Berkeley: Howell-North Books, 1968. 68 pp. Muir Woods National Monument, California.

M570 Morrel, E. "Papermaking in the South—A Brief History." *Manufacturers' Record* 108 (Apr. 1939), 34-35, 44, 54, 56.

M571 Morrill, John B. "Forest Preserve District of Cook County, Illinois: An Outer Park and Reservation System for Chicago." *Landscape Architecture* 38 (July 1948), 139-44. Incidental history.

M572 Morrill, Walter J. "History of the Forestry Department of the Colorado Agricultural College." *Colorado Forester* (1925), 7-10.

Morrill, Walter J. See Baker, James H., #B30

M573 Morrill, Walter J. "Birth of the Roosevelt National Forest." *Colorado Magazine* 20 (Sept. 1943), 178-81. Colorado.

M574 Morris, Caleb L. "The Dutch Elm Disease." *Pennsylvania Forests and Waters* 3 (May-June 1951), 52-55. Since 1930.

M575 Morris, John Milton. "History of Las Posadas Forest." Comp. by Edith Gregory. *California Historical Society Quarterly* 35 (Mar. 1956), 1-9; 35 (June 1956), 155-60. On a tract of virgin redwood in Napa County, California, where the author lived from 1878 to 1885.

M576 Morris, Penrose C. "The Land System of Hawaii." *American Bar Association Journal* 21 (1935), 649-52. A history of land tenure since the "primitive period."

M577 Morris, William G. "Forest Fires in Western Oregon and Western Washington." *Oregon Historical Quarterly* 35 (Dec. 1934), 313-39. Since 1826.

M578 Morris, William W. *Forest Plantations of Wisconsin.* Wisconsin Department of Agriculture, Bulletin No. 232. Madison, 1942. 62 pp. Includes some history of forest plantations and reforestation.

M579 Morris, William W. "An Early Forest Plantation in Wisconsin." *Wisconsin Magazine of History* 27 (June 1944), 436-38. A reforestation experiment conducted by Walter Ware in 1869.

M580 Morrison, Hugh M. "The Crown Land Policies of the Canadian Government, 1838-1872." Ph.D. diss., Clark Univ., 1933.

M581 Morrison, J.W. "Lumberjack Rhetoric: The Terse and Picturesque Language of the Lumber Woods." *American Forests and Forest Life* 30 (Dec. 1924), 722-24, 754. Includes some history of logger terms.

Morrison, John G. See Bourgeois, Euclid J., #B372

M582 Morrison, John H. *History of New York Ship Yards.* New York: William F. Sametz & Company, 1909. 165 pp. Illus. From 1784 to the decline of wooden shipbuilding.

M583 Morrison, Paul W. "The Establishment of the Forest Service within the Roosevelt National Forest." Master's thesis, Univ. of Colorado, 1945.

Morse, Chandler. See Barnett, Harold J., #B88

M584 Morton, J. Sterling. "Effects of Arbor Day upon Economic Forest Planting." *Forester* 4 (1898), 72-73. By the originator of Arbor Day.

M585 Morton, W.L. "Arthur Lower and the Timber Trade." *Queen's Quarterly* 80 (Winter 1973), 616-19. A tribute and review article on Lower's research and writing on forest history, especially his *Great Britain's Woodyard: British America and the Timber Trade, 1763-1867* (1973).

M586 Moser, A.J. "History of Hallock Manufacture." *Timberman* 15 (Dec. 1913), 51-52. Manufacturer of berry boxes in Portland, Oregon, since 1885.

M587 Moser, Robert J. ". . .And Then There Was One." *American Forests* 65 (Dec. 1959), 21, 43-44. On a "fast water" flume, allegedly the longest ever built, owned by Broughton Lumber Company, Underwood, Washington, since 1923.

M588 Moses, Leslie. "The Louisiana Department of Conservation." *Southwestern Law Journal* 5 (Spring 1951), 170-79. Since 1921, with emphasis on the conservation of oil and gas.

M589 Mosher, M.M. "A Study of Boards and Commissions in State Forest Administration." Master's thesis, State Univ. of New York, College of Forestry, 1937.

M590 Mosk, Sanford A. "Land Policy and Stock Raising in the Western United States." *Agricultural History* 17 (Jan. 1943), 14-30. Especially Arizona and New Mexico.

M591 Moslemi, Ali A. *Particleboard.* 2 Volumes. Carbondale: Southern Illinois University Press, 1974. Incidental history of particleboard since the 1940s.

M592 Mosley, T.J. "A Forest That 'Just Growed'." *American Forests* 37 (Oct. 1931), 588-91. On the growth of aspen on the cutover lands of the Great Lakes states.

M593 Moss, Albert E. "Reminiscences of Early Days in Connecticut." *Connecticut Woodlands* 35 (Spring 1970), 24-25. By a state forester and forestry educator.

M594 Moss, Albert E. "Chestnut and Its Demise in Connecticut." *Connecticut Woodlands* 38 (Spring 1973), 7-13. On its varied commercial uses since the colonial period.

Moungovan, Julia L. See Moungovan, Thomas O., #M595

M595 Moungovan, Thomas O.; Moungovan, Julia L.; and Escola, Nannie. *Where There's a Will There's a Way: Unusual Logging and Lumbering Methods on the Mendocino Coast.* Logging in Mendocino County Series, No. 2. Fort Bragg, California: Mendocino County Historical Society, 1968. 33 pp. Illus., diags. Since the 1850s.

M596 Mount, Peter R. "The Appalachian Pallet Industry." *Northern Logger* 20 (Feb. 1971), 22-23, 60-61. Maryland, Pennsylvania, and West Virginia since the 1960s.

M597 *Mountaineer.* "Fifty Golden Years of Mountaineering." *Mountaineer* 50 (Dec. 28, 1956), 6-75. A general history of the Seattle-based Mountaineers, a climbing group active in conservation.

M598 Mouzon, Olin Terrill. "The Social and Economic Implications of Recent Developments within the Wood Pulp and Paper Industry in the South." Ph.D. diss., Univ. of North Carolina, 1941. 332 pp.

M599 Mowat, Charles L. "The Land Policy in British East Florida." *Agricultural History* 14 (Apr. 1940), 75-77.

M600 Mowry, George E. *Theodore Roosevelt and the Progressive Movement.* Madison: University of Wisconsin Press, 1946. 405 pp. Includes an account of the Ballinger-Pinchot controversy of 1910 and other conservation issues.

Moyle, John B. See Nielsen, Etlar L., #N116

M601 Mucklow, Walter. *Lumber Accounts.* New York: American Institute Publishing Company, 1936. 458 pp. Incidental history.

Muench, David. See Lambert, Darwin, #L18

M602 Muffly, Edgar Lee. "Timber Trespasses on the St. Joe National Forest, 1906-1920." Master's thesis, Washington State Univ., 1959.

Muir, Jean. See Emmerson, Irma L., #E92

M603 Muir, John, ed. *Picturesque California and the Region West of the Rocky Mountains, from Alaska to Mexico.* 2 Volumes. New York: J. Dewing Publishing Company, 1888. This and the following books by Muir contain incidental historical references. More importantly, they reflect the growth of the conservation, preservation, and wilderness movements in which Muir played both active and symbolic roles.

M604 Muir, John. *The Mountains of California.* New York: Century Company, 1894. 381 pp. Illus., maps. There are later editions and reprints.

M605 Muir, John. "The American Forests." *Atlantic Monthly* 80 (Aug. 1897), 145-57. An argument for forest protection containing some history of American forest policy.

M606 Muir, John. "The Wild Parks and Forest Reservations of the West." *Atlantic Monthly* 81 (Jan. 1898), 15-28. Incidental history and reminiscence of forest reserves and national parks.

M607 Muir, John. *Our National Parks.* 1901. Enlarged edition. Boston: Houghton Mifflin, 1909. x + 382 pp. Illus.

M608 Muir, John. *My First Summer in the Sierra.* Boston: Houghton Mifflin, 1911. Reprint. Dunwoody, Georgia: N.S. Berg, 1972. 353 pp. Illus.

M609 Muir, John. *The Yosemite.* New York: Century Company, 1912. x + 284 pp. Illus. Later editions and reprints. Includes some history of Yosemite Valley and Yosemite National Park.

M610 Muir, John. *The Story of My Boyhood and Youth.* Boston: Houghton Mifflin, 1913. v + 293 pp. Illus. In Wisconsin.

M611 Muir, John. *Travels in Alaska.* Boston: Houghton Mifflin, 1915. ix + 326 pp.

M612 Muir, John. *Writings.* Ed. by William Frederic Badè. 10 Volumes. Boston: Houghton Mifflin, 1915-1924. Illus. Badè's *Life and Letters of John Muir* (q.v.) comprise volumes 9 and 10.

M613 Muir, John. *A Thousand Mile Walk to the Gulf.* Ed. by William F. Badè. Boston: Houghton Mifflin, 1916. xxvi + 219 pp. Illus.

M614 Muir, John. *Steep Trails: California, Utah, Nevada, Washington, Oregon, the Grand Cañon.* Boston: Houghton Mifflin, 1918. ix + 390 pp.

Muir, John. See Badè, William Frederic, #B11

M615 Muir, John. *John of the Mountains: The Unpublished Journals of John Muir.* Ed. by Linnie Marsh Wolfe. Boston: Houghton Mifflin, 1938. xxii + 458 pp. Illus. From 1867 to 1911.

M616 Muir, John. "The Creation of Yosemite National Park." *Sierra Club Bulletin* 29 (Oct. 1944), 49-60.

M617 Muir, John. *Yosemite and the Sierra Nevada. . . . Selections from the Works of John Muir.* Ed. by Charlotte E. Mauk. Boston: Houghton Mifflin Co., 1948. xix + 132 pp. Illus. Muir's descriptions and observations, 1860s-1870s.

M618 Muir, John. *The Wilderness World of John Muir.* With an intro. and interpretive comments by Edwin Way Teale. Boston: Houghton Mifflin, 1954. xx + 332 pp. Illus. Undated selections from his writings, grouped topically, dealing mainly with California mountains.

M619 Muir, John. *Gentle Wilderness: The Sierra Nevada.* Ed. by David Brower. Exhibit Format Series. San Francisco: Sierra Club, 1964. 167 pp. Text is from Muir's *My First Summer in the Sierra* (1911).

M620 Mulford, Walter. "The Opening Years of Forestry at Cornell as Viewed by One of the Boys." *Proceedings of the Society of American Foresters,* 1948 (1949), 335-37.

M621 Mulholland, F.D. "The Progress of Forestry in British Columbia." *Empire Forestry Journal* 15 (1936), 160-73.

M622 Mulholland, F.D. *The Forest Resources of British Columbia.* Victoria: British Columbia Forest Service, 1937. 153 pp. Illus., maps. Incidental history of forests and forest industries.

M623 Mulier, Joseph L. "The Standardization of Hardwood Lumber Grades." *Southern Lumberman* 185 (Dec. 15, 1952), 112-13. Since 1898.

M624 Muller, H.N. "Floating a Lumber Raft to Quebec City, 1805: The Journal of Guy Catlin of Burlington." *Vermont History* 39 (Spring 1971), 116-24. Includes some history of lumber rafting and the timber trade from Burlington, Vermont, and Lake Champlain, down the Richelieu and St. Lawrence rivers to Quebec, 1800s-1820s.

M625 Muller, Herman J. "The Civilian Conservation Corps, 1933-1942." *Historical Bulletin* 28 (Mar. 1950), 55-60.

M626 Mulligan, Brian O. "The Washington Arboretum." *Garden Journal of the New York Botanical Garden* 2 (Sept.-Oct. 1952), 147-50. Seattle, Washington, since 1935.

M627 Mullins, Doreen K. "Changes in Location and Structure in the Forest Industry of North Central British Columbia: 1909-1966." Master's thesis, Univ. of British Columbia, 1967.

M628 Mumey, Nolie. *The Teton Mountains.* Denver: Artcraft Press, 1947. xxiii + 462 pp. Includes some history of the Wyoming range and Grand Teton National Park.

M629 Munger, David A. "A Survey of the Western Red Cedar Shake Industry of the Pacific Northwest." Master's thesis, Univ. of Washington, 1970.

M630 Munger, Hiram. *The Life and Religious Experiences of Hiram Munger.* Chiopee (Chicopee) Falls, Massachusetts: N.p., 1861. 215 pp. Munger (b. 1806) was a Massachusetts sawmill owner.

M631 Munger, Thornton T. "Out of the Ashes of Nestucca." *American Forests* 50 (July 1944), 342-45, 366-68. History of a region along the central Oregon coast which was swept by fire in the 1840s, with emphasis on natural forest growth and subsequent fires.

M632 Munger, Thornton T. "A Look at Selective Cutting in Douglas-Fir." *Journal of Forestry* 48 (Feb. 1950), 97-99. On experiments in selective cutting and attendant controversies in Oregon and Washington, 1930s-1940s.

M633 Munger, Thornton T. "Fifty Years of Forest Research in the Pacific Northwest." *Oregon Historical Quarterly* 56 (Sept. 1955), 226-47. Particularly by the USFS's Pacific Northwest Forest and Range Experiment Station in Portland.

M634 Munger, Thornton T., in collaboration with C. Paul Keyser. *History of Portland's Forest-Park.* Portland: Forest-Park Committee of Fifty, 1960. 37 pp. Illus., map, apps. Processed. On the largest forested city park in the United States since the 1850s.

M635 Munger, Thornton T. "Recollections of My Thirty-Eight Years in the Forest Service, 1908-1946." *Timber Lines* 16 (Dec. 1962), Supplement, 1-30. Munger was director of the Pacific Northwest Forest Experiment Station, Portland, 1924-1938, and chief of forest management research in Portland, 1938-1946.

M636 Munger, Thornton T. "Trees in Hazard: Oregon's Myrtle Groves." *Oregon Historical Quarterly* 67 (Mar. 1966), 40-53. On efforts to preserve groves of Oregon myrtle in Coos and Curry counties of southwestern Oregon since the 1920s.

Munns, Edward N. See Frank, Bernard, #F208

M637 Munns, Edward N., and Stoeckeler, Joseph H. "How Are the Great Plains Shelterbelts?" *Journal of Forestry* 44 (Apr. 1946), 237-57. On the progress of shelterbelts planted from 1935 to 1943 by the USFS and the Soil Conservation Service.

M638 Munns, Edward N. "The Next Fifty Years in Watershed Management." *Journal of Forestry* 49 (June 1951), 419-24. Includes some history of watershed management.

M639 Munns, Edward N. "Homage to John Muir (1838-1914)." *Journal of Forestry* 62 (May 1964), 307-08. Biographical sketch and evaluation of the California preservationist.

M640 Munsell, Joel. *A Chronology of Paper and Paper-Making.* 1856. Fifth edition. Albany, New York: Munsell, 1876. 263 pp.

M641 Muntz, Alfred Philip. "The Changing Geography of the New Jersey Woodlands, 1600-1900." Ph.D. diss., Univ. of Wisconsin, 1959. 313 pp. Originally regarded as an obstacle, New Jersey forests became a valuable resource until the late 19th century.

M642 Muntz, Alfred Philip. "Forests and Iron: The Charcoal Iron Industry of the New Jersey Highlands." *Geografiska Annaler* 42 (1960), 315-23.

M643 Murbarger, Nell. "Charcoal, the West's Forgotten Industry." *Desert Magazine* 19 (June 1956), 4-9. On the industry in California, Nevada, Utah, and Colorado, 1860-1880.

M644 Murdoch, Richard K., ed. "Report of the Forest Resources of Spanish East Florida in 1792." *Agricultural History* 27 (July 1953), 147-51. A report by Governor Juan Nepomuceno de Quesada to the governor general of Cuba.

M645 Murdock, Harold R. "A Record of Progress: James L. Coker, Jr., Founder of Carolina Fiber Company." *Southern Pulp and Paper Journal* 3 (June 1940), 7-9.

M646 Murie, Olaus J. "Return to Denali." *Sierra Club Bulletin* 38 (Oct. 1953), 29-34. On changes in Alaska's Mount McKinley National Park since 1923.

M647 Murie, Olaus J. "Our Farthest North National Park." *National Parks Magazine* 33 (Dec. 1959), 8-10. Mount McKinley National Park, Alaska, since 1921.

M648 Murphy, James B. "L.Q.C. Lamar: Pragmatic Patriot." Ph.D. diss., Louisiana State Univ., 1968, 432 pp.

M649 Murphy, James B. *L.Q.C. Lamar: Pragmatic Patriot.* Baton Rouge: Louisiana State University Press, 1973. vii + 294 pp. Lamar was Grover Cleveland's Secretary of the Interior, 1885-1888.

M650 Murphy, Louis S. *Forests of Porto Rico, Past, Present and Future, and Their Physical and Economic Development.* USDA, Bulletin No. 354. Washington: GPO, 1916. 99 pp. Illus.

M651 Murphy, Louis S. "Forest Taxation Experience of the States and Conclusions Based on Them." *Journal of Forestry* 22 (May 1924), 453-63. Incidental historical references to forest taxation since the 1860s.

M652 Murphy, P.A. *Alabama Forests: Trends and Prospects.* Resource Bulletin, SO-42. New Orleans: USFS, Southern Forest Experiment Station, 1973. 36 pp. Statistics on forest acreage and production since 1963.

M653 Murphy, Richard Ernest. "Land Ownership in the Wilderness Areas of the United States National Forests." Ph.D. diss., Clark Univ., 1957. 587 pp. Incidental history.

M654 Murphy, Robert. "John James Audubon (1785-1851): An Evaluation of the Man and His Work." *New-York Historical Society Quarterly* 40 (Oct. 1956), 315-60. On the noted ornithologist and precursor of the conservation movement.

M655 Murphy, Robert. *Wild Sanctuaries: Our National Wildlife Refuges—A Heritage Restored.* New York: E.P. Dutton & Company, 1968. 288 pp. Illus., maps, index. Includes many incidental references to the history of wildlife conservation and the national wildlife refuges, with descriptions of the latter.

M656 Murray, James Edward. *Theodore Roosevelt and the Conservation Movement: Addresses . . . in the United States Senate, May 13 and March 4, 1958.* 85th Congress, 2nd Session, Senate Document 121. Washington: GPO, 1958. On Roosevelt's establishment of conservation as the responsibility of the federal government.

M657 Murray, Robert K. *The Harding Era: Warren G. Harding and His Administration.* Minneapolis: University of Minnesota Press, 1969. xii + 626 pp. Illus., notes, essay on sources, index. Includes many incidental references to the USFS, USDI, conservation, and natural resource policy during the Harding years, 1921-1923.

Murray, William G. See Timmons, John F., #T213

M658 Musselman, Lloyd D. "Rocky Mountain National Park, 1915-1965: An Administrative History." Ph.D. diss., Univ. of Denver, 1969. 363 pp.

M659 [Musser family]. *Peter Miller Musser.* Muscatine, Iowa: N.p., [1919]. 38 pp. Illus. Musser (1841-1919) was active in the lumber industry from 1863 to 1905, mostly as president of the Musser Lumber Company, Muscatine, Iowa.

M660 Muth, Phil D. "New Orleans— Important Lumber Center." *Southern Lumberman* 193 (Dec. 15, 1956), 241-42. Since 1850.

Muth, Richard F. See Perloff, Harvey S., #P119

M661 Myatt, Wilfred G. "The Willamette Valley." Ph.D. diss., Clark Univ., 1958. 253 pp. Logging and the forest industries of Oregon are prominent subjects of this geographical study.

M662 Myers, Charles A., and Shultz, George P. *Nashua Gummed and Coated Paper Company and Seven AFL Unions.* Washington: National Planning Association, 1950.

M663 Myers, Clifford A., and Martin, Edward C. "Fifty Years Progress in Converting Virgin Southwestern Ponderosa Pine to Managed Stands." *Journal of Forestry* 61 (Aug. 1963), 583-86. At the USFS's Fort Valley Experiment Forest near Flagstaff, Arizona.

M664 Myers, Delia J. "Grandfather Was Known as a 'Born Kiln Burner'." *Reveille* 2 (Sept. 1957), 2-3. Reminiscences of the charcoal industry in northern New York.

M665 Myers, Gustavus. *History of the Great American Fortunes.* 1907-1910. Reprint. New York: Modern Library, 1937. 732 pp. Notes, index. Includes unflattering sketches of James J. Hill and Frederick Weyerhaeuser—their timberlands and their rise to wealth and influence.

M666 Myers, J. Walter, Jr. *Opportunities Unlimited: The Story of Our Southern Forests.* Chicago: Illinois Central Railroad, 1950. Incidental history.

M667 Myers, J. Walter, Jr. "Thirty Years of Continuous Service." *Forest Farmer* 30 (Apr. 1971), 24-25, 40-41. On the Forest Farmers Association.

M668 Myers, John M., comp. and ed. *The Westerners: A Roundup of Pioneer Reminiscences.* Englewood Cliffs, New Jersey: Prentice-Hall, 1969. xiv + 258 pp. Index. Includes reminiscences by Elliott S. Barker, William A. Keleher, and Edwin Bennett, former USFS rangers in Colorado and New Mexico early in the 20th century.

Myers, Wayne L. See Cool, Robert A., #C583

M669 Myrick, David F. *Railroads of Nevada and Eastern California.* 2 Volumes. Berkeley: Howell-North Books, 1962-1963. Illus. Includes some history of logging railroads and lumber companies on the eastern slope of the Sierra Nevada in California and Nevada since the 19th century.

N1 Namejunas, Alfred. "A Plantation with a Past." *Land* 10 (Winter 1951-1952), 397-99. A forest plantation near Hancock, Wisconsin, since 1876.

N2 Nash, Gerald D. "Problems and Projects in the History of Nineteenth-Century California Land Policy." *Arizona and the West* 2 (Winter 1960), 327-40. On the various conditions which promoted concentration of land ownership in California.

N3 Nash, Gerald D. *State Government and Economic Development: A History of Administrative Policies in California, 1849-1933.* Berkeley: University of California, Institute of Government Studies, 1964. xii + 379 pp. Notes, bib., index. Includes some history of efforts to develop and conserve such natural resources as forests.

N4 Nash, Gerald D. "The California State Land Office, 1858-1898." *Huntington Library Quarterly* 27 (Aug. 1964), 347-56. On its distribution of state lands.

N5 Nash, Gerald D. "The California State Board of Forestry, 1883-1960." *Historical Society of Southern California Quarterly* 47 (Sept. 1965), 291-301.

Nash, Gerald D. See Pinchot, Gifford, #P209

N6 Nash, Gerald D. *United States Oil Policy, 1890-1964: Business and Government in Twentieth Century America.* Pittsburgh: University of Pittsburgh Press, 1968. xiii + 286 pp. Notes, app., bib., index. Includes some general history of federal conservation policy, with reference to natural resources other than oil.

N7 Nash, Gerald D. *The American West in the Twentieth Century: A Short History of an Urban Oasis.* Englewood Cliffs, New Jersey: Prentice-Hall, 1973. viii + 312 pp. Illus., notes, bib., index. Includes some history of lumbering, forestry, and conservation.

N8 Nash, Jay B., ed. *Recreation: Pertinent Readings.* Dubuque, Iowa: W.C. Brown Company, 1965. xiv + 265 pp. Includes an article by Roderick Nash on "Recreation and the Wilderness," pp. 234-40.

N9 Nash, Roderick W. "The American Wilderness: A History of Its Preservation." Master's thesis, Univ. of Wisconsin, 1960.

N10 Nash, Roderick W. "The Wisdom of Aldo Leopold." *Wisconsin Academy Review* 8 (Fall 1961), 161-67. A brief sketch of Leopold (1886-1948) as a forester, conservationist, ecologist, and philosopher of a land ethic.

N11 Nash, Roderick W. "The American Wilderness in Historical Perspective." *Forest History* 6 (Winter 1963), 2-13. Surveys 300 years of changing attitudes toward wilderness.

N12 Nash, Roderick W. "Wilderness and the American Mind." Ph.D. diss., Univ. of Wisconsin, 1965. 316 pp.

Nash, Roderick W. See Nash, Jay B., #N8

N13 Nash, Roderick. "The Strenuous Life of Bob Marshall." *Forest History* 10 (Oct. 1966), 18-25. Robert Marshall, an ardent conservationist and a founder of the Wilderness Society, headed the USFS's Division of Recreation and Lands, 1937-1939, when much wilderness was preserved.

N14 Nash, Roderick. "The American Cult of the Primitive." *American Quarterly* 18 (Fall 1966), 517-37. On early interest in wilderness preservation, as reflected in such issues as the Hetch Hetchy Valley controversy in California, the wilderness venture of Joseph Knowles in Maine in 1913, and the general fascination with things primitive in the late 19th and early 20th centuries.

N15 Nash, Roderick. "John Muir, William Kent, and the Conservation Schism." *Pacific Historical Review* 36 (Nov. 1967), 423-33. On California congressman William Kent, whose philanthropy made possible Muir Woods National Monument, and his opposition to John Muir and others who sought to prevent building of Hetch Hetchy Dam in Yosemite National Park, 1900s-1910s.

N16 Nash, Roderick. *Wilderness and the American Mind.* New Haven: Yale University Press, 1967. viii + 256 pp. Notes, a note on the sources, index. An important study which treats the changing American attitudes toward wilderness and their political implications. The concept of wilderness as a barrier to material progress gave way gradually to increasing enthusiasm for wilderness bordering on a damaging overpopularity. Those who influenced the change in attitudes, such as Henry David Thoreau, John Muir, Aldo Leopold, and Robert Marshall, receive special attention. A revised and enlarged edition was issued in 1973.

N17 Nash, Roderick, ed. *The American Environment: Readings in the History of Conservation.* Reading, Massachusetts: Addison-Wesley, 1968. xix + 236 pp. Illus., table, notes, bib. A collection of essays ranging chronologically from George Catlin (1832) to Stewart Udall and other conservationists of the 1960s; includes a "Chronology of Important Events."

Nash, Roderick. See Nelson, J.G., #N61

N18 Nash, Roderick. "The Potential of Conservation History." *Environmental Education* 1 (Spring 1970), 83-88.

N19 Nash, Roderick. "The American Invention of National Parks." *American Quarterly* 22 (Fall 1970), 726-35. On the 19th-century background and origins of national parks in the United States and on the influence of this experience upon other countries.

N20 Nash, Roderick. "The State of Environmental History." In *The State of American History,* ed. by Herbert J. Bass. Chicago: Quadrangle Books, 1970. Pp. 249-60. Examines achievements in the broadly defined field of environmental history and suggests areas for scholarly inquiry.

N21 Nash, Roderick. "A Home for the Spirit: A Brief History of the Wilderness Preservation Movement—The Story of an Idea Given the Strength of Law." *American West* 8 (Jan. 1971), 40-47. From John Muir to the Wilderness Act of 1964 and the Wild and Scenic Rivers Act of 1968.

N22 Nash, Roderick. "American Environmental History: A New Teaching Frontier." *Pacific Historical Review* 41 (Aug. 1972), 362-72.

N23 Nash, Roderick, ed. *Environment and Americans: The Problem of Priorities.* New York: Holt, Rinehart, and Winston, 1972. 119 pp. An anthology.

N24 Nash, Roderick. *The American Conservation Movement.* Forums in History Series. St. Charles, Missouri: Forum Press, 1974. 16 pp. An interpretive essay on changing attitudes toward the environment since the 17th century.

N25 Naske, Claus-M. "103,350,000 Acres." *Alaska Journal* 2 (Autumn 1972), 2-13. On the land grant provisions of Alaska statehood bills since 1916.

N26 National Hardwood Lumber Association. *A Half Century of Progress, 1898-1947.* Chicago, 1947. 88 pp. Illus.

N27 *National Hardwood Magazine.* "Who Says Money Doesn't Grow on Trees?" *National Hardwood Magazine* 49 (Feb. 1975), 40-41, 44, 66. On the use of brass coins minted and distributed by the Goodman Lumber Company, Goodman, Wisconsin, 1920s-1959.

N28 National Lumber Manufacturers Association. *Highlights of a Decade of Achievement.* Washington, 1929. 67 pp. A summary of association activities since its reorganization in 1918.

N29 National Lumber Manufacturers Association. *Proceedings of the Golden Anniversary Meeting of the National Lumber Manufacturers Association at St. Louis, Missouri, May 8-10, 1952.* Washington, 1952. 211 pp. Includes historical sketches of the industry, private forestry by region, lumber production and distribution, and of the NLMA.

N30 *National Parks Magazine.* "National Parks of Canada." *National Parks Magazine* 22 (July-Sept. 1948), 6-30.

N31 National Research Council and the Society of American Foresters, Joint Committee on Forestry. *Problems and Progress of Forestry in the United States: Report of the Joint Committee.* Washington: Society of American Foresters, 1947. 112 pp. Includes many historical references.

N32 Neal, Dorothy Jensen. *The Cloud Climbing Railroad: A Story of Timber, Trestles and Trains.* Alamogordo, New Mexico: Alamogordo Printing Company, 1966. 81 pp. Illus., map, notes, app., index. On the Alamogordo and Sacramento Mountain Railway, which was used to supply logs from camps in the Sacramento Mountains of south central New Mexico, ca. 1890s-1940s; includes some history of the Alamogordo, Sacramento Mountain, and Southwest lumber companies.

N33 Nearing, Helen, and Nearing, Scott. *The Maple Sugar Book: Together with Remarks on Pioneering as a Way of Living in the Twentieth Century.* 1950. Reprint. New York: Schocken Books, 1970. 273 pp. Notes, bib., index. Includes some history and lore of the maple sugar industry of the Northeast.

Nearing, Scott. See Nearing, Helen, #N33

N34 Nearn, William T. "Lumber Export Trade of the U.S." Master's thesis, Yale Univ., 1947.

N35 Neasham, Aubrey. "Sutter's Sawmill." *California Historical Society Quarterly* 26 (June 1947), 109-33. History of John Sutter's sawmill at Coloma, California, site of first gold discovery in 1848.

N36 Neiderheiser, Clodaugh M., comp. *Forest History Sources of the United States and Canada: A Compilation of the Manuscript Sources of Forestry, Forest Industry, and Conservation History.* St. Paul: Forest History Foundation, 1956. xiii + 140 pp. Index.

N37 Neiland, Bonita J. "Forest and Adjacent Burn in the Tillamook Burn Area of Northwestern Oregon." *Ecology* 39 (Oct. 1958), 660-71. A survey of forests in and near the Tillamook Burn, which suffered from fire in 1933, 1939, and 1945.

N38 Neill, Wilfred T. "Surveyor's Field Notes as a Source of Historical Information." *Florida Historical Quarterly* 34

(Apr. 1956), 329-33. Includes some incidental history and references to the "Big Scrub" country, now the Ocala National Forest, and other forested areas in Florida in the 1830s and 1840s.

N39 Neils, Selma M. *So This is Klickitat*. Portland: Metropolitan Press, for Klickitat Woman's Club, 1967. 176 pp. Illus., apps., index. Includes some history of the J. Neils Lumber Company (1922-1957) of Klickitat, Washington, now operated by St. Regis Paper Company.

N40 Neilson, Barry J. "The Drowned Forest of the Columbia Gorge." *Washington Historical Quarterly* 26 (1935), 119-22. A list of references, 1805-1923.

N41 Nelles, Henry Vivian. "The Politics of Development: Forests, Mines and Hydro-Electric Power in Ontario, 1890-1939." Ph.D. diss., Univ. of Toronto, 1970.

N42 Nelles, Henry Vivian. "Empire Ontario: The Problems of Resource Development." In *Oliver Mowat's Ontario*, ed. by Donald Swainson. Toronto, 1972. Pp. 189-210.

N43 Nelles, Henry Vivian. *The Politics of Development: Forests, Mines & Hydro-Electric Power in Ontario, 1849-1941*. Hamden, Connecticut: Archon Books of Shoe String Press, 1974. xii + 514 pp. Illus., notes, note on sources, index. A study of the involvement of the state in the development and regulation of three natural resource industries of Ontario—pulp and paper, mining, and hydroelectric power.

N44 Nelligan, John Emmett. *The Life of a Lumberman*. Ed. by Charles M. Sheridan. Madison: By the author, 1929. 202 pp. Illus. Reminiscences of a logger in New Brunswick, Maine, Pennsylvania, and especially the Great Lakes states of Minnesota, Michigan, and Wisconsin, 1860s-1890s. Also appeared in *Wisconsin Magazine of History* 13 (Sept. 1929), 3-65; 13 (Dec. 1929), 131-85; 13 (Mar. 1930), 241-304.

N45 Nellis, Jesse Charles. "Wood-Using Industries in Maine." *Report of the Maine Forest Commissioner* 9 (1912), 83-188. This and following entries by Nellis contain incidental history.

N46 Nellis, Jesse Charles, and Harris, John Tyre. *Wood-Using Industries of West Virginia*. West Virginia Department of Agriculture, Bulletin No. 10. Charleston, 1915. 144 pp.

N47 Nellis, Jesse Charles. *The Wood-Using Industries of Indiana*. Washington: GPO, 1916. 37 pp.

N48 Nelson, Alfred L. "Department of Conservation: The History Behind Minnesota's Resource Management." *Conservation Volunteer* 12 (Mar.-Apr. 1949), 1-14.

N49 Nelson, Alfred L. "Minnesota School of Forestry—Fifty Years of Conservation." *Conservation Volunteer* 16 (July-Aug. 1953), 1-7.

N50 Nelson, Arthur H. "A Study of Conservation Education in the Junior Colleges of the U.S." Ph.D. diss., Cornell Univ., 1949. 80 pp.

N51 Nelson, Arthur L. "Beetles Kill Four Billion Board Feet of Engelmann Spruce in Colorado." *Journal of Forestry* 48 (Mar. 1950), 182-83. Since 1939, with mention of other bark beetle infestations in Colorado since 1855.

N52 Nelson, Beatrice M. *State Recreation: Parks, Forests and Game Preserves*. Washington: National Conference on State Parks, 1928. 436 pp. Includes some history of the state parks movement.

N53 Nelson, C.N. "History of Forestry in North Dakota." *Cross Tie Bulletin* 50 (Nov. 1969), 23, 26, 28-29.

N54 Nelson, Charles A. "A History of the Forest Products Laboratory." Ph.D. diss., Univ. of Wisconsin, 1964. 465 pp.

N55 Nelson, Charles A. "Born and Raised in Madison: The Forest Products Laboratory." *Forest History* 11 (July 1967), 6-14. Some history of early forest products research and of the movement leading to the establishment of the USFS's Forest Products Laboratory at Madison, Wisconsin, 1909.

N56 Nelson, Charles A. *History of the U.S. Forest Products Laboratory (1910-1963)*. Madison: USFS, Forest Products Laboratory, 1971. 177 pp. Illus., tables, notes, index.

N57 Nelson, Charles C. *Wisconsin & Northern Railroad*. Railway and Locomotive Historical Society, Bulletin No. 116. Cambridge: Baker Library, Harvard Business School, 1967. 48 pp. Illus. On a northern Wisconsin railroad, primarily a carrier of lumber, 1907-1921.

N58 Nelson, George W. "Plain and Fancy." *San Diego Historical Society Quarterly* 1 (July 1955). A brief history of redwood shingles.

N59 Nelson, Gerhart Helmer. "Economic Aspects of Cutover Land Use in Western Montana." Master's thesis, Univ. of Montana, 1955.

N60 Nelson, J.G., and Byrne, A.R. "Man as an Instrument of Landscape Change—Fires, Floods and National Parks in the Bow Valley, Alberta." *Geographical Review* 56 (No. 2, 1966), 226-38. On forest fires and floods in this part of Banff National Park since the arrival of railroads in 1880s.

N61 Nelson, J.G., and Scace, Robert C., eds. *The Canadian National Parks: Today and Tomorrow*. Proceedings of Conference held at Calgary, Alberta. 2 Volumes. Calgary, 1968. Collection of papers concerned with the history, extent, and use of Canadian and other national parks. See, for example, Roderick Nash, "Wilderness and Man in North America," Volume I, pp. 66-93.

N62 Nelson, L.A. "Lumber History and Grading Rules." *Timberman* 25 (Jan. 1924), 57, 58, 60.

N63 Nelson, Milton O. "The Lumber Industry in America." *Review of Reviews* 35 (Nov. 1907), 561-75. Incidental history.

N64 Nelson, Neal M. "A Historical Inquiry into the Civilian Conservation Corps, with Special Reference to the Ninth Corps Area." Master's thesis, Univ. of Idaho, 1938.

N65 Nelson, Ralph K. "Development of Forest Management on the Crossett Lumber Company Properties." Master's thesis, Yale Univ., 1942. Arkansas.

N66 Nelson, Sharlene P. "Camp Six: The Tacoma Logging Museum." *Forest History* 9 (Jan. 1966), 24-27. Operated by the Western Washington Forest Industries Museum, Tacoma, Washington.

Nelson, S.W. See Kling, Edwin M., #K202

N67 Nelson, Thomas C. "The Original Forests of the Georgia Piedmont." *Ecology* 38 (July 1957), 390-97. Based on early accounts and travelers' descriptions since 1539.

N68 Nesbit, Robert C. *Wisconsin: A History.* Madison: University of Wisconsin Press, 1973. xiv + 607 pp. Illus., maps, tables, apps., bib., index. Includes a chapter on the lumber industry and other sections on forestry and conservation.

N69 Nesbitt, William A., and Netboy, Anthony. "The History of Settlement and Land Use in Bent Creek Forest." *Agricultural History* 20 (Apr. 1946), 121-27. Near Asheville, North Carolina.

Netboy, Anthony. See Nesbitt, William A., #N69

Netboy, Anthony. See Frank, Bernard, #F210

N70 Netboy, Anthony. "The Indian and the Forest." *American Forests* 60 (Oct. 1954), 24-25, 63. A brief historical survey of Indian use of forests, including the use of fire for clearing land and hunting game.

N71 Netboy, Anthony. "Uproar on Klamath Reservation." *American Forests* 63 (Jan. 1957), 20-21, 61-62. Includes some history of the lumber industry and forest policy on Oregon's Klamath Indian Reservation.

N72 Nettels, Curtis P. *The Emergence of a National Economy, 1775-1815.* Economic History of the United States, Volume 2. New York: Holt, Rinehart and Winston, 1962. xviii + 424 pp. Illus., tables, notes, bib., app., index. Includes some history of the lumber and other forest industries.

N73 Neubauer, Gerhardt William "The Significance of Selected Aspects of Wood Technology for Western Culture." Ph.D. diss., Univ. of Minnesota, 1956. 956 pp. Includes a section on North America in the 20th century.

N74 Neuberger, Richard L. "How Oregon Rescued a Forest." *Harper's* 218 (Apr. 1959), 48-52. On the termination of the Klamath Indian Reservation and its establishment as a national forest, 1950s.

N75 Neutson, K. *Memoirs of a Pioneer.* New Orleans: Peerless Printing Company, 1938. 38 pp. Includes an account of lumber rafting on the Red River of Louisiana, 1870-1873.

Nevins, Allan. See Hidy, Ralph W., #H351

N76 Nevue, Wilfred. "A Winter in the Woods." *Minnesota History* 34 (Winter 1954), 149-53. Reminiscences of logging for the T.B. Walker Lumber Company near Akeley, Minnesota, 1905-1906.

N77 Newberry, Farrar. "The Concatenated Order of Hoo-Hoo." *Arkansas Historical Quarterly* 22 (Winter 1963), 301-10. On the origins of the secret society of lumbermen, founded by Bolling A. Johnson and five others at Gurdon, Arkansas, 1892, and its subsequent history.

N78 New Brunswick. Forest Service. *Forests and Forestry in New Brunswick.* Fredericton, 1928. 70 pp. Incidental history.

N79 New Brunswick. Provincial Archives. *The Wood Industries of New Brunswick.* Fredericton, 1969. iv + 25 pp. Illus., map, tables. A reprint of the New Brunswick section of *The Wood Industries of Canada,* published first in 1897 by the *Timber Trades Journal.* The introduction to the reprint constitutes a concise forest history of the province since 1779.

N80 Newburg, Alvin A., comp. *Statistical History of the Paperboard Industry.* Chicago: National Paperboard Association, 1952. 32 pp. Diags., tables. Since 1925.

N81 Newcomb, R.M. "Timber Industry of Vancouver Island." *Timberman* 26 (July 1925), 194-98. Includes some history of forest resources, land policy, and timber sales on Vancouver Island, British Columbia.

N82 Newell, Gordon R. *Ships of the Inland Sea: The Story of the Puget Sound Steamboats.* Portland: Binfords & Mort, 1951. ix + 241 pp. Illus., map, bib. Since 1836, with many references to the lumber industry and trade.

N83 Newell, Gordon R. *Pacific Tugboats.* Seattle: Superior Publishing Company, 1957. 191 pp. Illus., bib. An illustrated history of tugboats used along the Pacific coast, including in the lumber trade.

N84 Newell, Gordon R., and Williamson, Joe. *Pacific Lumber Ships.* Seattle: Superior Publishing Company, 1960. 192 pp. Illus. An illustrated history of ships used in the Pacific lumber trade since 1843.

N85 Newell, Gordon R., ed. *The H.W. McCurdy Marine History of the Pacific Northwest: An Illustrated Review of the Growth and Development of the Maritime Industry from 1895.* Seattle: Superior Publishing Company, 1966. Includes many references to the lumber trade.

N86 Newfoundland. Royal Commission of Forestry. *Report of the Newfoundland Royal Commission on Forestry.* St. Johns, 1955. 240 pp. Illus., tables, graphs, app. Includes some history of forest industries, forest resources, and forestry in Newfoundland and Labrador.

N87 Newhall, Beaumont, and Newhall, Nancy, eds. *Masters of Photography.* New York: George Braziller, 1958. 192 pp. Illus., notes, bib. Includes a section on Ansel Adams, photographer, conservationist, and friend of the national parks.

N88 Newhall, Nancy; Osborn, Fairfield; and Albright, Horace M. *A Contribution to the Heritage of Every American: The Conservation Activities of John D. Rockefeller, Jr.* New York: Alfred A. Knopf, 1957. 179 pp. The national park system benefited immensely from the philanthropy of Rockefeller (1874-1960).

Newhall, Nancy. See Newhall, Beaumont, #N87

N89 New Hampshire. Forestry Department. *Forest Resources of New Hampshire.* Concord, 1923. 59 pp. Incidental history.

N90 *New Hampshire Profiles.* "Fight for Franconia." *New Hampshire Profiles* 1 (May 1952), 16-18, 52-53. On the efforts of the Society for the Protection of New Hampshire Forests to save Franconia Notch from "commercial devastation," 1923-1928.

N91 Newkirk, Arthur Edward. "The Birth and Growth of the Forest Preserve." *Ad-i-ron-dac* 24 (Nov.-Dec. 1960), 108-10. New York, since the 19th century.

N92 Newland, Harrod B. "Wood Using Industries of Kentucky." *Northern Logger* 13 (Mar. 1965), 18-19, 44-45, 53. Since the early 19th century.

N93 Newman, F.S. "The Development of a Forest Nursery and Its Problems." *Forestry Chronicle* 3 (Mar. 1927), 19-30. Canada.

N94 Newman, James Gilbert. "The Menominee Forest of Wisconsin: A Case History in American Forest Management." Ph.D. diss., Michigan State Univ., 1967. 159 pp. Since 1854.

N95 Newman, Oliver P. "A Forest Fire That Cost Uncle Sam Fifteen Million Dollars." *American Forests and Forest Life* 31 (June 1925), 323-26. On claims against the federal government stemming from the forest fire which swept the area near Cloquet, Minnesota, in 1918.

Newman, Thomas Stell. See De Camp, David, #D106

N96 Newport, Carl A. "A Summary of Forest Taxation Laws by States." *Journal of Forestry* 49 (Mar. 1951), 196-200. Incidental history.

N97 Newport, Carl A. *Forest Service Policies in Timber Management and Silviculture as They Affect the Lumber Industry: A Case Study of the Black Hills.* Pierre: South Dakota Department of Game, Fish and Parks, 1956. v + 112 pp. Illus., maps, tables. The cover title of this doctoral dissertation (State Univ. of New York, College of Forestry, 1954) is "A History of Black Hills Forestry." The emphasis is on relationships between the USFS and the lumber industry in the 20th century.

N98 Newport, Carl A. "Forest Service Policies as They Affect the Lumber Industry: A Case Study of the Black Hills." *Journal of Forestry* 54 (Jan. 1956), 17-20. Since 1897 on South Dakota's Black Hills National Forest.

N99 Newton, Craig A., and Sperry, James R. *A Quiet Boomtown: Jamison City, Pa., 1889-1912.* Bloomsburg, Pennsylvania: Columbia County Historical Society, 1972. xviii + 148 pp. Illus., maps, notes, index. Jamison City was a center of the lumber and tanning industries during the period.

N100 Newton, Norman T. *Design on the Land: The Development of Landscape Architecture.* Cambridge: Harvard University Press, 1971. xxiv + 714 pp. Illus., maps, notes, bib., index. See for reference to the conservation movement.

N101 Newton, W.M. "Wiarton's Busy Mills Were Centre of Industry in Bruce Peninsula." *Western Ontario Historical Notes* 4 (Dec. 1946), 84-91. Concerns the rise and demise of the lumber industry near Wiarton on Ontario's Bruce Peninsula.

N102 Newton-White, E. *Canadian Reforestation.* Toronto: Ryerson Press, 1944. 227 pp. Incidental history.

N103 New York and New Jersey. Palisades Interstate Park Commission. *Palisades Interstate Park, 1900-1960.* Bear Mountain, New York: Palisades Interstate Park Commission, 1960. 106 pp. Illus., maps, diags., tables. A history of the park commission, comprising reprints of two previous pamphlets with same title, covering the years 1900-1929 and 1929-1947, with an added account of the period 1947-1960.

N104 New York. Forest Commission. *Compendium of Laws Relative to the Adirondack Wilderness from 1774 to 1894.* New York Forest Commission, Annual Report, 1893, Volume 2. Albany: J.B. Lyon Company, 1894. 468 pp.

N105 New York. State Council of Parks. *New York State Parks: Twenty-Fifth Anniversary Report, 1924-1949.* Albany, 1949. 40 pp. Maps, illus.

N106 *New York Public Library Bulletin.* "When Did Newspapers Begin to Use Wood Pulp?" *New York Public Library Bulletin* 33 (Oct. 1929), 743-49.

N107 New York State College of Forestry. "Dedication, Conference and History: The New York State Ranger School, Wanakena, N.Y., August, 1928." *Bulletin of the New York State College of Forestry at Syracuse University* 2 (Jan. 1929), 1-84. Includes some history of the ranger school since 1912.

N108 Nibley, Charles W. *Reminiscences.* Salt Lake City: N.p., 1934. 193 pp. The author (b. 1849) was in the lumber business in Idaho and Oregon.

N109 Nicholas, Anna. "Beginning of Conservation in America." *Journal of American History* 4 (No. 4, 1910), 571-80. Concerning a manuscript written by William Tatham in Madrid, Spain, in 1796.

N110 Nicholls, William Hord. "Some Foundations of Economic Development in the Upper East Tennessee Valley, 1850-1900." *Journal of Political Economy* 64 (Aug. 1956), 277-302; 64 (Oct. 1956), 400-15. Includes references to the lumber industry in Tennessee and Virginia.

N111 Nichols, Claude W., Jr. "Brotherhood in the Woods: The Loyal Legion of Loggers and Lumbermen, a Twenty Year Attempt at 'Industrial Cooperation'." Ph.D. diss., Univ. of Oregon, 1959. 193 pp. On an organization established to counteract a strike of the Industrial Workers of the World against the lumber operators in the Pacific Northwest during World War I, 1917-1918, continued under Brice P. Disque for the U.S. War Department, and disbanded in 1935.

N112 Nichols, Robert. "The Victoria Lumber Mill." *Louisiana Studies* 5 (Summer 1966), 156-58. Manufacturer of cypress and pine lumber in Natchitoches Parish, Louisiana, ca. 1880-1900.

N113 Nicholson, Katherine Stanley. *Historic American Trees.* New York: Frye Publishing Company, 1922. 104 pp. Illus.

N114 Nickerson, Edward F. "The Story of the Vermont Forest and Farm Land Foundation." *Vermont Life* 8 (Autumn 1954), 2-7.

Nicolaiff, J.E. See Babcock, H.M., #B2

N115 Neider, Paul. "The Timber Culture Laws in Western Kansas, 1873-1891." Master's thesis, Univ. of Kansas, 1966.

N116 Nielsen, Etlar L., and Moyle, John B. "Forest Invasion and Succession on the Basins of Two Catastrophically Drained Lakes in Northern Minnesota." *American Midland Naturalist* 25 (May 1941), 564-79. Bass Lake in St. Louis County and Sunken Lake in Itasca County.

N117 Nisbet, Fred J. "Biltmore House and Its Gardens." *Garden Journal of the New York Botanical Garden* 9 (Mar.-Apr. 1959), 56-59. On the George Vanderbilt estate near Asheville, North Carolina, built between 1889 and 1895, and especially its landscape gardening, experimental forest, and other attractions.

N118 Nix, L.A. "Forestry Consulting in Canada." *Forestry Chronicle* 36 (Sept. 1960), 252-59.

Nixon, Edgar B. See Roosevelt, Franklin Delano, #R285

N119 Nixon, John Harmon. "Government Control of Lumber Prices in World War II." Ph.D. diss., Harvard Univ., 1953.

N120 Nixon, Stuart. *Redwood Empire.* New York: E.P. Dutton, 1966. 256 pp. Illus., index. A pictorial history of the California region, including references to the lumber industry and movements to preserve redwoods in state and national parks.

N121 Nobles, Robert W. "Forestry in the U.S. Virgin Islands." *Journal of Forestry* 58 (July 1960), 524-27. Since 1930.

N122 Nobles, William Scott. "Harold L. Ickes: New Deal Hatchet Man." *Western Speech* 22 (Summer 1958), 158-64. On the partisan and antagonistic rhetoric of Ickes, Secretary of the Interior in the Roosevelt and Truman administrations, 1933-1946.

Noblet, U.J. See Hesterberg, Gene A., #H319

N123 Noggle, Burl L. "Conservation in Politics: A Study of Teapot Dome." Ph.D. diss., Duke Univ., 1956.

N124 Noggle, Burl L. "The Origins of the Teapot Dome Investigation." *Mississippi Valley Historical Review* 44 (Sept. 1957), 237-66. Concerns investigations of scandal in the Department of the Interior, 1920-1923, and implications for the conservation movement.

N125 Noggle, Burl L. *Teapot Dome: Oil and Politics in the 1920s.* Baton Rouge: Louisiana State University Press, 1962. 234 pp. Includes many references to the conservation movement in this decade.

N126 Norcross, Charles P. "Weyerhaeuser—Richer Than John D. Rockefeller." *Cosmopolitan* 42 (Jan. 1967), 252-59. On Frederick Weyerhaeuser, lumberman with vast timberlands and sawmill interests in the Great Lakes and the Pacific Northwest.

N127 Norcross, F.G. "Evolution of the Log Band Mill." *Southern Lumberman* 118 (Jan. 10, 1925), 42-44. Archer, Mancourt Company, Burnside, Kentucky, 1880.

N128 Norcross, Nicholas W. "Lumbering on the Merrimack River." *Lowell Historical Society, Contributions* 2 (No. 3, 1926), 441-50. Massachusetts.

N129 Nord, Sverre. *A Logger's Odyssey.* Caldwell, Idaho: Caxton Printers, 1943. 255 pp. Includes reminiscences of work in the lumber industry of the Pacific Northwest, 1900s-1920s.

N130 Norden, Roger Lawrence. "An Inventory and Study of the Historical Development of the Major Resources of Marquette County, Michigan." Ph.D. diss., Michigan State Univ., 1960. 298 pp. Includes forest resources.

N131 Norgress, Rachel Edna. "The History of the Cypress Lumber Industry in Louisiana." *Louisiana Historical Quarterly* 30 (July 1947), 979-1059. Since 1849, but with mention of earlier uses of cypress.

N132 Norman, Charles. *John Muir, Father of Our National Parks.* New York: Julian Messner, 1957. 191 pp. Bib., index. Muir (1838-1914) was a California preservationist and wilderness philosopher.

N133 Norris, George W. *Fighting Liberal: The Autobiography of George W. Norris.* New York: Macmillan, 1945. 419 pp. Includes reminiscences of the Nebraska senator's involvement in the politics of conservation.

N134 Norris, James D. "The Maramec Iron Works, 1826-1876: The History of a Pioneer Iron Works in Missouri." Ph.D. diss., Univ. of Missouri, 1961. 325 pp.

N135 Norris, James D. *Frontier Iron: The Maramec Iron Works, 1826-1876.* Madison: State Historical Society of Wisconsin, 1964. vii + 206 pp. Illus., maps, bib., index. A case study of the charcoal iron industry in Missouri and of the company's use of slaves as woodcutters.

N136 North Carolina. Department of Conservation and Development. *Conservation and Development in North Carolina.* Comp. with an intro. by Charles Sylvester Green. 2 Volumes. Raleigh, 1953. 372 pp. Conference papers on state parks, forestry, wildlife, etc.; some are historical.

N137 *North Country Life.* "The Enchanted Forest of the Adirondacks." *North Country Life* 11 (Summer 1957), 15-22. Old Forge, New York.

N138 *Northeastern Logger.* "Old Days at Lincoln, New Hampshire." *Northeastern Logger* 3 (Apr. 1955), 16-17, 52.

N139 *Northeastern Logger.* "Old Days in Potter County." *Northeastern Logger* 4 (Dec. 1955), 23, 64. Logging in north central Pennsylvania.

N140 *Northeastern Logger.* New York issue. *Northeastern Logger* 4 (May 1956), complete. Includes historical articles on forest resources, forestry education, industrial forestry, wood-using industries, and the pulp and paper industry of New York.

N141 *Northeastern Logger.* "Paper Making in Northern New York." *Northeastern Logger* 4 (May 1956), 10-11, 58-61. Since 1768.

N142 *Northeastern Logger.* "The Oswego River Paper Industry Development." *Northeastern Logger* 4 (May 1956), 12-13, 53, 61. Especially the Oswego Falls-Sealright Company.

N143 *Northeastern Logger.* "International Paper Company Woodlands State of Maine, Past and Present." *Northeastern Logger* 5 (July 1956), 16-17, 42.

N144 *Northeastern Logger.* "Fisher Forestry and Realty Corporation." *Northeastern Logger* 5 (July 1956), 21, 29. On their practice of industrial forestry in New York since the 1880s.

N145 *Northeastern Logger.* Vermont issue. *Northeastern Logger* 5 (Sept. 1956), complete. Includes articles on the history of forestry, state forests, and the Eaton Lumber Company.

N146 *Northeastern Logger.* New Hampshire issue. *Northeastern Logger* 5 (Oct. 1956), complete. Includes historical articles on state forests, wood-using industries, and other topics.

N147 *Northeastern Logger.* Wisconsin issue. *Northeastern Logger* 5 (Feb. 1957), complete. Includes historical articles on state forestry, national forests, the Forest Products Laboratory, the lumber industry, and other aspects of forest use in Wisconsin.

N148 *Northeastern Logger.* "The Connor Lumber and Land Company." *Northeastern Logger* 5 (Apr. 1957), 16-17, 34-35. Operating in Wisconsin since 1872 and considered a pioneer in industrial forestry.

N149 *Northeastern Logger.* Michigan issue. *Northeastern Logger* 5 (June 1957), complete. Includes historical articles on forest resources, education, and research, as well as the forest products industries and their operations.

N150 *Northeastern Logger.* Adirondack issue. *Northeastern Logger* 6 (July 1957), complete. Includes historical articles on the pulp and paper industry, outdoor recreation, International Paper Company, and other aspects of forest history in the Adirondacks.

N151 *Northeastern Logger.* Pennsylvania issue. *Northeastern Logger* 6 (Sept. 1957), complete. Includes historical articles on state forestry, forestry education, reforestation, and the lumber industry.

N152 *Northeastern Logger.* "The Story of 'The Louisville Slugger' and Pennsylvania Ash." *Northeastern Logger* 6 (Sept. 1957), 28-29, 76-77. On the Pennsylvania firm of Hillerick and Bradsby, makers of baseball bats.

N153 *Northeastern Logger.* Ohio issue. *Northeastern Logger* 6 (Oct. 1957), complete. Includes articles on the Ohio Agricultural Experiment Station, Ohio Forestry Association, Muskingum watershed control, D.B. Frampton Company, Mead Corporation, and forest resources and research.

N154 *Northeastern Logger.* "The Mower Lumber Company Story." *Northeastern Logger* 6 (Dec. 1957), 38ff. West Virginia.

N155 *Northeastern Logger.* "Cherry River Boom and Lumber Company." *Northeastern Logger* 6 (Dec. 1957), 42-43, 85. Some history of a West Virginia firm.

N156 *Northeastern Logger.* Illinois issue. *Northeastern Logger* 6 (Mar. 1958), complete. Includes articles on the Shawnee National Forest, Sinissippi State Forest, Crab Orchard Wildlife Refuge, forest resources, forestry research, forestry education, and forest products industries.

N157 *Northeastern Logger.* "Cotton-Hanlon: A Growth Story of Over 35 Years." *Northeastern Logger* 6 (May 1958), 22-23, 45. A lumber company of Schuyler County, New York, since the 1920s.

N158 *Northeastern Logger.* Maryland issue. *Northeastern Logger* 7 (Nov. 1958), complete. Includes historical articles on the Gladfelter Pulp Wood Company, wood-using industries, industrial forestry, and other aspects of Maryland forest history.

N159 *Northeastern Logger.* "The Adirondack Museum, Blue Mountain Lake, N.Y." *Northeastern Logger* 9 (Oct. 1960), 18-19. And its interest in forest history.

N160 *Northeastern Logger.* Indiana issue. *Northeastern Logger* 9 (Jan. 1961), complete. Includes articles on state and national forests, wood-using industries, tree farming, and the Indiana Hardwood Lumbermen's Association.

N161 *Northeastern Logger.* "The Paper Industry in Northern New England." *Northeastern Logger* 9 (May 1961), 8-11, 26-27, 47, 53, 60, 65. Brief historical sketches of companies in the region.

N162 *Northeastern Logger.* Pennsylvania issue. *Northeastern Logger* 10 (May 1962), complete. Includes articles on the paper, lumber, and charcoal industries.

N163 *Northeastern Logger.* "Jasper, Indiana: A Forest Products Center." *Northeastern Logger* 11 (Jan. 1963), 10-13, 32. Includes some history of the furniture industry of Jasper.

N164 *Northern Logger.* Minnesota issue. *Northern Logger* 12 (Mar. 1964), complete. Contains historical articles on logging, log marks, forest products industries, forestry research, forestry education, and other aspects of Minnesota's forest history.

N165 *Northern Logger.* "Minnesota School of Forestry." *Northern Logger* 12 (Mar. 1964), 14-15, 48-51. History of forestry education at the University of Minnesota since the 1880s.

N166 *Northern Logger.* Northern New York issue. *Northern Logger* 12 (Apr. 1964). Includes articles on the history of the lumber and pulp and paper industries.

N167 Northrup, B.G. "Arbor Day: Its History and Aims and How to Secure Them." *Report of the Connecticut Board of Agriculture* (1886), 39-51.

N168 Northrup, H.R. "The Retail Dealers' National Association." *Southern Lumberman* 193 (Dec. 15, 1956), 269-70. On the National Retail Lumber Dealers' Association since 1917.

N169 Northrup, Herbert R. *The Negro in the Paper Industry.* Racial Policies of American Industry, No. 8. Philadelphia: University of Pennsylvania Press, 1969. 233 pp. Tables, apps., index. Includes some history and statistics of the industry, especially in the South.

N170 *Northwestern Lumberman.* "Logs Driven in the Saginaw District, 1864-1892." *Northwestern Lumberman* (Feb. 18, 1893), 16.

N171 *Northwestern Lumberman.* "The Tree." *Northwestern Lumberman* 1 (July 1974), 10-12. Includes some history of the redwood and sequoia trees of California and their uses.

N172 *Northwestern Lumberman.* "Sequoias of Sonoma." *Northwestern Lumberman* 1 (July 1974), 26-28. On individual giant redwoods cut in Sonoma County, California, late 19th century; reprinted from *Wood and Iron* (1895).

N173 *Northwestern Lumberman.* "Sierra Nevada Big Trees." *Northwestern Lumberman* 1 (July 1974), 40-42. Reminiscences of the lumber industry in the sequoia groves of Tulare County, California, 1850s-1890s; adapted from articles in the bulletin of the Tulare County Historical Society.

N174 *Northwestern Lumberman.* "Bullwhacking in the Del Norte." *Northwestern Lumberman* 1 (July 1974), 44-45. Reminiscences of redwood logging in Del Norte County, California.

N175 *Northwestern Lumberman.* "Redwood . . . and the City." *Northwestern Lumberman* 1 (July 1974), 46-49. On San Francisco as a market for redwood and as a playground for lumberjacks in the late 19th century.

N176 Norton, Boyd. "The Oldest Established Perennially Debated Tree Fight in the West." *Audubon* 74 (July 1972), 60-69. On controversies among developers, preservationists, and the USFS over Idaho's Magruder Corridor in the Nez Perce and Bitterroot national forests since the 1930s.

N177 Norton, Boyd. *Snake Wilderness.* San Francisco: Sierra Club, 1972. 159 pp. Illus., map. Includes some incidental history of forested and wilderness areas in the Snake River drainage of Idaho, western Wyoming, and eastern Oregon.

N178 Norton, Matthew G. *The Mississippi River Logging Company: An Historical Sketch.* N.p., 1912. 97 pp. Account of the organization of log transportation on Wisconsin's Chippewa River, 1870s to 1909, by the attorney for the company.

Nowell, Reynolds I. See Jesness, Oscar B., #J78

N179 Nowlin, Rankin S. "Economic Development of the Kirby Lumber Co. of Houston, Texas." Master's thesis, George Peabody College for Teachers, 1930.

N180 Nunis, Doyce B., Jr. *The Trials of Isaac Graham.* Los Angeles: Dawson's Book Shop, 1967. xv + 129 pp. Illus., maps, notes, apps., bib. Graham was one of the earliest lumbermen in California's Santa Cruz Mountains, 1830s-1840s.

N181 Nuschke, Marie K. "Hicks, Fighters and Clog Dancers: Early Lumber Camps in Freeman Run Valley." *Pennsylvania History* 19 (Oct. 1952), 436-51. Reminiscences of the Brisbois brothers and their logging camp in Potter County, Pennsylvania, 1880s.

N182 Nute, Grace L. *The Voyageur's Highway: Minnesota's Border Lake Land.* St. Paul: Minnesota Historical Society, 1941. xiii + 113 pp. Illus., maps, bib. Includes some history of logging between Rainy Lake and Lake Superior in northern Minnesota.

N183 Nute, Grace L. *Lake Superior.* American Lakes Series. New York: Bobbs-Merrill, 1944. 376 pp. Illus., map. Includes references to the lumber trade.

N184 Nute, Grace L. *Rainy River Country: A Brief History of the Region Bordering Minnesota and Ontario.* St. Paul: Minnesota Historical Society, 1950. xiii + 143 pp. Illus., maps, bib., index. Includes some history of logging, the lumber industry, and the pulp and paper industry.

Nutting, A.D. See Freedman, L.J., #F226

N185 Nye, Russell B. *This Almost Chosen People: Essays in the History of American Ideas.* East Lansing: Michigan State University Press, 1966. x + 374 pp. Bib., index. Includes a chapter on "The American View of Nature."

O1 Oakleaf, Howard B. *Wood-Using Industries of Oregon.* Portland: Oregon Conservation Association, 1911. 46 pp.

O2 Oakleaf, Howard B. "Washington's Secondary Wood-Using Industries." *Pacific Lumber Trade Journal* (Nov. 1911), 22-30.

O3 Oakleaf, Howard B. *Lumber Manufacture in the Douglas Fir Region.* Chicago: Commercial Journal, 1920. 182 pp. Illus., diags. Incidental history.

O4 Ober, Michael J. "Glacier's Skyland Camps: A Culver Colonel's Abortive Dream." *Montana, Magazine of Western History* 23 (Summer 1973), 30-39. On L.R. Gignilliat and his camps for cadets of an Indiana military academy in Glacier National Park, Montana, 1921-1940.

O5 Oberholtzer, Ernest C. "The Chronicle of the Olmsteds." *Living Wilderness* 64 (Spring 1958), 1-4. On the contributions of Frederick Law Olmsted, Sr. (1822-1903), John Charles Olmsted (his nephew and stepson, d. 1920), and Frederick Law Olmsted, Jr. (1870-1957), as promoters of conservation and landscape architecture.

O6 O'Brien, Bob Randolph. "The Yellowstone National Park Road System: Past, Present and Future." Ph.D. diss., Univ. of Washington, 1965. 276 pp.

O7 O'Brien, Bob Randolph. "The Roads of Yellowstone, 1870-1915." *Montana, Magazine of Western History* 17 (July 1967), 30-39.

O8 O'Callaghan, Jerry A. "The Disposition of the Public Domain in Oregon." Ph.D. diss., Stanford Univ., 1952. 247 pp.

O9 O'Callaghan, Jerry A. "Klamath Indians and the Oregon Wagon Road Grant, 1864-1938." *Oregon Historical Quarterly* 53 (Mar. 1952), 23-38. Concerns in part the misappropriation of timberlands on Oregon's Klamath Indian Reservation.

O10 O'Callaghan, Jerry A. "Senator Mitchell and the Oregon Land Frauds, 1905." *Pacific Historical Review* 21 (Aug. 1952), 255-61. John H. Mitchell (1835-1905) was implicated in a number of timberland frauds and died while appealing a conviction for bribery.

O11 O'Callaghan, Jerry A. "The War Veteran and the Public Lands." *Agricultural History* 28 (Oct. 1954), 163-68. On the granting of lands as a reward for military service, 1776-1873.

O12 O'Callaghan, Jerry A. *The Disposition of the Public Domain in Oregon.* Washington: GPO, 1960. xii + 113 pp. Tables, notes, bib. A history of federal lands in Oregon since the 1840s, including chapters on timberland disposals and frauds.

O13 O'Callaghan, Jerry A. "Significance of the United States Public Land Survey." *Western Pennsylvania Historical Magazine* 50 (Jan. 1967), 51-59. On the origins of the public land survey, 1785 and later.

O14 O'Callaghan, Jerry A. "BLM: 1946-1971." *Our Public Lands* 21 (Fall 1971), 4-15. Series of articles commemorating the 25th anniversary of the U.S. Bureau of Land Management.

O15 O'Connor, Harvey. *Revolution in Seattle: A Memoir.* New York: Monthly Review Press, 1964. xv + 300 pp. Illus., notes, bib., index. O'Connor, a logger, socialist, and member of the Industrial Workers of the World, recalls lumber strikes and related violence in Washington, 1914-1924.

O16 O'Dell, W.V. "The Gypsy Moth Control Program." *Journal of Forestry* 57 (Apr. 1959), 271-72. Since 1869.

O17 Oden, Jack Porter. "Development of the Southern Pulp and Paper Industry, 1900-1970." Ph.D. diss., Mississippi State Univ., 1973. 664 pp.

O18 Oehler, C.M. *Time in the Timber.* Forest Products History Foundation Series, No. 2. St. Paul: Forest Products History Foundation, Minnesota Historical Society, 1948. 56 pp. Illus., map. On the author's experience as clerk for a logging crew of the Virginia and Rainy Lake Lumber Company at Cusson, Minnesota, 1928.

O19 Oehser, Paul H. "Pioneers in Conservation: Footnote to the History of an Idea." *Nature Magazine* 38 (Apr. 1945), 188-90.

O20 Oehser, Paul H. "W J (no period) McGee." *Land* 7 (Summer 1948), 216-19. On William John McGee (1853-1912), geologist, ethnologist, conservationist, and advisor to Theodore Roosevelt.

O21 Officer, James E. "The New Conservation." *New Jersey History* 87 (Spring 1969), 35-44. A general discussion of the conservation movement in the 1960s.

O22 Ogden, Gerald R., comp. *The United States Forest Service: A Historical Bibliography, 1876-1972.* Santa Cruz, California: Forest History Society, 1973. Index, apps. Contains over 7,000 entries for published items.

O23 Ogden, J. Gordon, III. "Forest History of Martha's Vineyard, Massachusetts. I. Modern and Pre-Colonial Forests." *American Midland Naturalist* 66 (Oct. 1961), 417-30.

O24 O'Gorman, James F. "The Hoo Hoo House, Alaska-Yukon-Pacific Exposition, Seattle, 1909." *Journal of the Society of Architectural Historians* 19 (Sept. 1960), 123-25. On a clubhouse of the International Order of Hoo Hoo, the lumbermen's fraternity, built in Seattle in 1909.

O25 *Olde Ulster.* "The Building of Plank Roads." *Olde Ulster* 8 (Oct. 1912), 289-97. Ulster County, New York, in the mid-19th century.

O26 Oldham Saw Works. *The Oldham History of Saws.* New York, 1918.

O27 Olds, Fred C. "Michigan's Lumber Pikes." *Railroad Magazine* 61 (July 1953), 50-69. On logging railways in Michigan, 1876-1907.

O28 Olds, Nicholas V. "Public Rights in Conservation—Their Protection and Preservation." *Michigan Conservation* 21 (May-June 1952), 3-4, 23-25. Since 1920.

O29 Oliver, A.C., and Dudley, Harold M. *This Is America, the Story of the CCC.* New York: Longmans, 1937. 188 pp. Incidental history of the Civilian Conservation Corps.

O30 Oliver, George D. "History of the Sierra Nevada Lumbering Industry." *Intercollegiate Forest Club Annual* 1 (1921), 30-34.

O31 Oliverio, Jean E. *Footprints in the Soil and Reflections on the Water: Thirty Years of Soil and Water Conservation in West Virginia.* Compiled for the West Virginia Soil and Water Conservation District Supervisors Association. Parsons, West Virginia: McClain Printing, 1972. 241 pp.

O32 Olmsted, Frederick Law. "The Yosemite Valley and the Mariposa Big Trees: A Preliminary Report (1865)." Intro. by Laura Wood Roper. *Landscape Architecture* 43 (Oct. 1952), 12-25. A reprint of Yosemite State Park Commissioner Olmsted's report to the California Legislature, 1865.

O33 Olmsted, Frederick Law. *Landscape into Cityscape: Frederick Law Olmsted's Plans for a Greater New York City.* Ed. with an introductory essay and notes by Albert Fein. Ithaca, New York: Cornell University Press, 1968. x + 490 pp. Illus., maps, notes, index. The introductory essay on Olmsted offers biographical material and an analysis of his influence on environmental attitudes.

O34 Olmsted, Frederick Law, Jr., and Kimball, Theodora, eds. *Frederick Law Olmsted, Landscape Architect, 1822-1903.* 2 Volumes. New York: G.P. Putnam's Sons, 1922, 1928. Illus. Includes a brief biography of Olmsted and some of his professional papers.

Olmsted, Roger. See Kortum, Karl, #K243

O35 Olsen, Charles O. "Logging in the Good Old Days." Oregon Agricultural College, Forest Club, *Annual Cruise* 8 (1927), 29-34, 100-04.

O36 Olsen, Charles O. "Hunt's Mill." *Timberman* 43 (Aug. 1942), 62-63. A pioneer sash sawmill in Oregon, 1844-1850s.

O37 Olsen, Charles O. "Old Hobsonville: Ghost Town Once Center of Lumber Activity in Tillamook County, Oregon." *Timberman* 44 (Nov. 1942), 18-22.

O38 Olsen, W.H. *Water Over the Wheel.* Chemainus, British Columbia: Chemainus Valley Historical Society, 1963. 169 pp. Illus., maps, apps. Includes some history of the lumber industry in and near Chemainus, Vancouver Island, British Columbia.

O39 Olson, D.S., and Fahnestock, G.R. *Logging Slash: A Study of the Problem in Inland Empire Forests.* Moscow: University of Idaho, College of Forestry, 1955. 51 pp. Incidental history.

O40 Olson, David S. "The Savenac Nursery." *Forest History* 11 (July 1967), 15-21. Memoir of a USFS nurseryman on Montana's Lolo National Forest, 1915.

O41 Olson, Harold. "Forest Management in Action: The Forty-Year Program of the Collins Pennsylvania Forest." *American Forests* 51 (Aug. 1945), 390-91, 405-08. A private forest in Pennsylvania.

O42 Olson, Harold. "West is Birthplace of Tree Farm Plan." *Western Conservation Journal* 11 (July-August 1954), 16, 40. Since the 1940s.

O43 Olson, Harold. "Penta Comes of Age." *American Forests* 62 (Nov. 1956), 34-35, 46. On the use of pentachlorophenol as a preservative of wood since 1936.

O44 Olson, James C. *J. Sterling Morton.* Lincoln: University of Nebraska Press, 1942. Reprint. Lincoln: Nebraska State Historical Society Foundation, 1972. xvi + 451 pp. Illus., notes, bib., index. Morton was Grover Cleveland's Secretary of Agriculture, 1893-1897, a power in Nebraska and national Democratic politics, the "father" of Arbor Day, and an early enthusiast of forestry. Also issued as a Ph.D. dissertation at the University of Nebraska, 1942.

O45 Olson, James C. "Arbor Day—A Pioneer Expression of Concern for the Environment." *Nebraska History* 53 (Spring 1972), 1-13. Arbor Day was first celebrated in Nebraska in 1872. Its founder was J. Sterling Morton, later Secretary of Agriculture. Other Nebraskans active in tree planting were George L. Miller and Robert W. Furnas.

O46 Olson, Keith W. "Franklin K. Lane: A Biography." Ph.D. diss., Univ. of Wisconsin, 1964. 367 pp. Lane was Secretary of the Interior, 1913-1920.

O47 Olson, Keith W. "Woodrow Wilson, Franklin K. Lane, and the Wilson Cabinet Meetings." *Historian* 32 (Feb. 1970), 270-75.

O48 Olson, Sherry Hessler. "Commerce and Conservation: A History of Railway Timber." Ph.D. diss., Johns Hopkins Univ., 1965. 284 pp.

O49 Olson, Sherry Hessler. "Commerce and Conservation: The Railroad Experience." *Forest History* 9 (Jan. 1966), 2-15. On the railroads' use of wood and its relationship to forest conservation in the 19th and early 20th centuries. Wood preservation, tree planting, forestry, and the work of Ernest A. Sterling and Hermann von Schrenk are topics considered.

O50 Olson, Sherry Hessler. *The Depletion Myth: A History of Railroad Use of Timber.* Cambridge: Harvard University Press, 1971. xvi + 228 pp. Illus., maps, graphs, notes, index. A fuller exposition of the above entry; emphasis is on the period from the 1870s to the 1930s.

O51 Olson, Sigurd F. "Swift as the Wild Goose Flies." *National Parks Magazine* 23 (Oct.-Dec. 1949), 3-9. On the Superior National Forest, Minnesota, and Quetico Provincial Park, Ontario, since 1909. The following works by Olson pertain largely to the Quetico-Superior area and various efforts to preserve it.

O52 Olson, Sigurd F. "Late Frontier: Quetico-Superior." *American Heritage* 1 (Spring 1950), 48-50.

O53 Olson, Sigurd F. *The Singing Wilderness.* New York: Alfred A. Knopf, 1955. 244 pp. Illus.

O54 Olson, Sigurd F. *Listening Point.* New York: Alfred A. Knopf, 1958. 243 pp. Illus.

O55 Olson, Sigurd F. "Six Decades of Progress." *American Forests* 68 (Oct. 1962), 16-19.

O56 Olson, Sigurd F. "Battle for a Wilderness." *Living Wilderness* 32 (Winter 1968-1969), 4-13.

O57 Olson, Sigurd F. *Open Horizons.* New York: Alfred A. Knopf, 1969. 277 pp. Illus. Autobiographical.

O58 O'Meara, Walter. *We Made It Through the Winter: A Memoir of Northern Minnesota Boyhood.* St. Paul: Minnesota Historical Society, 1974. xi + 128 pp. Illus., notes, index. Reminiscences of the northern Minnesota logging town of Cloquet, ca. 1906.

O59 Oneal, Marion Sherrard. "A School Ma'am in Louisiana's Piney Woods, 1902-1903." *Louisiana History* 5 (Spring 1964), 135-42. Reminiscences of Hill Switch, a Louisiana lumber town.

O60 O'Neil, Clinton De Witt. *Timber n'Injuns.* Kalispell, Montana: Kalispell News—Farm Journal Print, 1955. 216 pp. Illus. On the author's life in Kalispell and the Flathead Valley of Montana since 1894, with some account of his sawmill and lumber yard.

O61 O'Neill, Eugene James. "Parks and Forest Conservation in New York, 1850-1920." Ed. D. diss., Columbia Univ., 1963. 231 pp.

O62 O'Neil, W.J. "Logging in the Lake States in 1910." *Northern Logger* 12 (June 1964), 32, 51-52. Reminiscence of logging for the Pine Tree Manufacturing Company near Hill City, Minnesota.

O63 O'Neill, H.B. "Early Days at Grand'mere." *Pulp and Paper Magazine of Canada* 43 (Dec. 1942), 941-43. Laurentide Pulp Company, Grand'mere, Quebec.

O64 Ontario. Department of Lands and Forests. "History and Status of Forestry in Ontario." *Canadian Geographical Journal* 25 (Sept. 1942), 110-45.

O65 Ontario. Department of Lands and Forests. *A History of Geraldton Forest District.* District History Series, No. 2. Toronto, 1963. 50 pp. Map, tables.

O66 Ontario. Department of Lands and Forests. *A History of White River Forest District.* District History Series, No. 5. Toronto, 1963. 19 pp. Map.

O67 Ontario. Department of Lands and Forests. *A History of Fort Frances Forest District.* District History Series, No. 8. Toronto, 1963. 16 pp. Map.

O68 Ontario. Department of Lands and Forests. *A History of Kenora Forest District.* District History Series, No. 10. Toronto, 1963. 22 pp. Map.

O69 Ontario. Department of Lands and Forests. *A History of Gogama Forest District.* District Histor Series, No. 11. Toronto, 1964. 37 pp. Illus., map.

O70 Ontario. Department of Lands and Forests. *A History of Cochrane Forest District.* District History Series, No. 14. Toronto, 1964. 51 pp. Maps.

O71 Ontario. Department of Lands and Forests. *A History of Swastika Forest District.* District History Series, No. 15. Toronto, 1964. 82 pp. Map.

O72 Ontario. Department of Lands and Forests. *A History of Chapleau Forest District.* District History Series, No. 16. Toronto, 1964. 54 pp. Illus., maps, tables.

O73 Ontario. Department of Lands and Forests. *A History of Tweed Forest District.* District History Series, No. 19. Toronto, 1965. 54 pp. Illus., maps, tables.

O74 Opie, John, ed. *Americans and Environment: The Controversy over Ecology.* Lexington, Massachusetts: D.C. Heath and Company, 1971. xiv + 203 pp. Tables, notes, bib. An anthology with some historical articles.

O75 Orcutt, Wright T. "The Minnesota Lumberjacks." *Minnesota History* 6 (Mar. 1925), 3-19.

O76 Ordway, Samuel H., Jr. "The Law and Progress in Conservation." *Journal of Forestry* 57 (June 1959), 403-08. Conservation law since 1805.

O77 Oregon. Department of Forestry. *Forest Resources of Oregon: Their Management and Use.* Salem, 1961. Incidental history.

O78 Oregon. State Board of Forestry. *Forest Resources of Oregon.* Salem, 1943. 62 pp. Illus., maps, tables. Contains some history of forest industries, state and federal forestry, and forest fires.

O79 Oregon, University of. Bureau of Municipal Research and Service. *O & C Counties: Population, Economic Development, and Finance.* Prepared for the Association of O & C Counties. Eugene, 1957. 138 pp. Map, tables, bib. On eighteen Oregon counties that include lands granted to the Oregon and California Railroad in 1866 but revested to the U.S. government in 1916. The counties receive income from the sale of timber on these lands, which are administered by the USFS and the USDI.

O80 Oregon, University of. Bureau of Governmental Research and Service. *The Significance of the O & C Forest Resource in Western Oregon.* Eugene, 1968. xii + 169 pp. Maps, tables, graphs, apps. Includes some history of the Oregon and California Railroad grant lands, heavily forested and important to the economies of eighteen western Oregon counties.

O81 Oregon State College. *50 Years of Forestry at Oregon State College.* Corvallis, 1956. 24 pp. Illus., tables.

O82 Orfield, Matthias Nordberg. *Federal Land Grants to the States, with Special Reference to Minnesota.* Studies in the Social Sciences, No. 2. Minneapolis: University of Minnesota, 1915. v + 275 pp. Bib. A historical treatment of American policy, with part 3 being devoted to the administration of public lands in Minnesota. Also issued as the *Bulletin of the University of Minnesota* (Mar. 1915).

O83 Organisation for European Economic Co-operation. *The Pulp and Paper Industry in the USA: A Report by a Mission of European Experts.* Paris, 1951. 378 pp. Map, tables, diags., index. Incidental history.

O84 O'Riordan, Timothy. "The Third American Conservation Movement: New Implications for Public Policy." *Journal of American Studies* 5 (Aug. 1971), 155-71. A comparison of the underlying philosophies and public policies of the contemporary conservation movement with movements of the Progressive and New Deal periods.

O85 O'Riordan, Timothy. "An Analysis of the Use and Management of Campgrounds in British Columbia Provincial Parks." *Economic Geography* 49 (Oct. 1973), 298-308. Incidental history.

O86 Ormsby, Warren. "Peeling the Tanoak." *Forest History* 15 (Jan. 1972), 6-10. Reminiscences of peeling bark from tanoak trees along the California coast, 1890s. Tanoak bark was used in leather tanning.

O87 Osborn, Fairfield. *Our Plundered Planet.* Boston: Little, Brown and Company, 1948. xiv + 217 pp. Bib. A broad survey of the history of land abuse, including a chapter on North America.

Osborn, Fairfield. See Newhall, Nancy, #N88

O88 Osborn, William C. *The Paper Plantation: Ralph Nader's Study Group Report on the Pulp and Paper Industry in Maine.* Ne York: Grossman Publishers, 1974. xx + 300 pp. Illus., map, tables, apps., notes, index. An argumentative study of the pulp and paper industry and its effect on the Maine environment and people. It contains some incidental historical information.

O89 Osborne, Maurice M. "The Last of the Big Trees." *Appalachia* 28 (June 1951), 398-407. Reminiscences of Pemigewasset and Twin Mountain Range, forested regions of New Hampshire, 1906-1908.

O90 Osborne, William B. "Review of Progress in Fire Fighting and Protection." *Timberman* 20 (Oct. 1919), 94-95. Pacific Northwest.

O91 O'Shaughnessy, M.M. *Hetch Hetchy: Its Origin and History.* San Francisco: Recorder Printing and Publishing Company, 1934. 134 pp. Map. On the valley and reservoir in California's Yosemite National Park which became the focus of a conservation controversy early in the 20th century.

O92 Ostrander, George N. "A Discussion of Article VII, Section 7 of the State Constitution." *New York Forestry* 1 (Oct. 1914), 11-14. Includes some history of changing attitudes toward the practice of forestry on New York state lands.

O93 Ostrander, George N. "History of the New York Forest Preserve." *Journal of Forestry* 40 (Apr. 1942), 301-04. Administrative history of the preserve in the Adirondack and Catskill Mountains of New York since 1885.

O94 Ostrom, Vincent A. "State Administration of Natural Resources in the West." *American Political Science Review* 47 (June 1953), 478-93. Since 1850.

O95 Ostrom, Vincent A. "The Social Scientist and the Control and Development of Natural Resources." *Land Economics* 29 (May 1953), 105-16. On the social consequences of the various ways in which resources are exploited.

O96 Ostrum, Carl E. "History of the Gum Naval Stores Industry." *Chemurgic Digest* 4 (July 15, 1945), 217-23. South.

O97 Otjen, C.J. "History of the Forestry Conservation Movement in the State of Wisconsin." Master's thesis, Univ. of Wisconsin, 1914.

O98 Otter, Floyd L. *The Men of Mammoth Forest: A Hundred-Year History of a Sequoia Forest and Its People in Tulare County, California.* Ann Arbor, Michigan: Edwards Brothers, 1963. viii + 169 pp. Illus., notes, apps., index. History of forested lands and the lumber industry near the North Fork of the Tule River, now part of Sequoia National Forest and Mountain Home State Forest, California, since the 1850s.

O99 Ottoson, Howard W., ed. *Land Use Policy and Problems in the United States.* Lincoln: University of Nebraska Press, 1963. x + 470 pp. Charts, graphs, notes. Contains three essays on the history of land policy.

O100 Ouderkirk, Nancy Myster. "The Civilian Conservation Corps: Roosevelt's Robin Hoods." *Northern Logger and*

Timber Processor 23 (May 1975), 22-23, 56-58. General history and description of woods operations.

O101 Ouellet, Fernand. *Histoire économique et sociale du Québec, 1760-1850: structures et conjoncture.* Montreal: Fides, 1966. 671 pp. Maps, charts, bib. Includes some history and economic analysis of the lumber industry in Quebec.

O102 *Out West.* "A Record of Achievement: The Epoch-Making Work of the Water and Forest Association." *Out West* 16 (Feb. 1902), 209-20. A California association which promoted irrigation and forestry.

O103 Overpack, Roy M. "The Michigan Logging Wheels." *Michigan History* 35 (June 1951), 222-25. On logging equipment manufactured by Sylas C. Overpack of Manistee, Michigan, 1870-1935.

O104 Overstreet, Daphne. "The Man Who Told Time by the Trees: Dr. Andrew Ellicott Douglass, Father of the Science of Dendrochronology." *American West* 11 (Sept. 1974), 26-29, 60-61. Douglass (1867-1962) originated the science of dendrochronology in 1901 and pursued tree-ring research in Arizona until his death.

O105 Owen, Marguerite. *The Tennessee Valley Authority.* New York: Praeger Publishers, 1973. xi + 275 pp. Illus., map, tables, apps., notes, bib., index. Includes some history of natural resource development and conservation.

O106 Owen, O.S. *Natural Resource Conservation: An Ecological Approach.* New York: Macmillan, 1971. 593 pp. Incidental history.

O107 Owsley, Clifford D. "Famous U.S. Forest Service Firsts." *Southern Lumberman* 227 (Dec. 15, 1973), 119-21.

Owsley, Clifford D. See Graham, Suzan, #G240

O108 Owsley, Clifford D. "The Forest Service History Program at Age Four." *Southern Lumberman* 229 (Sept. 1, 1974), 8-12. Review of historical writing sponsored or encouraged by the USFS History Program since 1970.

O109 Owsley, Frank L. *Plain People of the Old South.* Baton Rouge: Louisiana State University Press, 1949. 235 pp. Maps, tables, bib., notes. Contains some history of forests as a factor in pioneer settlement in the South.

O110 Oxford Paper Company. *Oxford Paper Company, 1949, Annual Report: Fiftieth Anniversary, 1899-1949.* New York, 1950.

O111 Oxford Paper Company. *The Oxford Story: A History of the Oxford Paper Company, 1847-1958.* New York, 1958.

O112 Oxholm, Axel H. "How French Forests Kept the American Army Warm." *Journal of Forestry* 44 (May 1946), 326-29. On the work of the Army Quartermaster Corps to provide fuelwood for American troops in France during World War II.

P1 Pacific Lumber Company. *Scotia: Home of the Pacific Lumber Company.* San Francisco, 1957. 31 pp. Illus.

P2 Pacific Mills, Limited. *A Graphic History of Pacific Mills, Limited, Manufacturers of Pulp and Paper Products.* Vancouver, British Columbia, 1945. 104 pp. Illus.

P3 Pacific Northwest Regional Planning Commission. *Forest Resources of the Pacific Northwest: A Report by the Pacific Northwest Regional Planning Commission.* Washington: GPO and National Resources Committee, 1938. vii + 86 pp. Maps, tables, graphs. Incidental history.

P4 *Pacific Pulp and Paper Industry.* "A New Decade: Principal Achievements of the Last Decade." *Pacific Pulp and Paper Industry* 14 (May 1940), 36-37.

P5 *Pacific Pulp and Paper Industry.* "Story of Tileston and Hollingsworth Co.: 'Oldest U.S. Mill' Is One of Most Progressive." *Pacific Pulp and Paper Industry* 18 (July 1944), 7-9. Hyde Park, Massachusetts, since 1728.

P6 Pack, Arthur N. *Our Vanishing Forests.* New York: Macmillan, 1923. 189 pp. Illus., graphs, tables. Incidental history of forest depletion.

P7 Pack, Charles Lathrop, and Gill, Thomas H. *Forests and Mankind.* New York: Macmillan, 1929. x + 250 pp. Illus., maps, bib. A broad survey of forests and forestry, including much history of both.

P8 Packard, Fred M. "The Past Decade in Our National Parks." *Garden Journal of the New York Botanical Garden* 2 (July-Aug. 1952), 103-04, 116.

P9 Packer, B.G. *Wisconsin's Wood-Using Industries.* Wisconsin Department of Agriculture, Bulletin No. 67. Madison, 1925. 58 pp.

P10 Padover, Saul K. "Ickes: Memoir of a Man without Fear." *Reporter* 6 (Mar. 4, 1952), 36-38. The author, a former assistant to Harold L. Ickes, Secretary of the Interior, 1933-1946, judges him to have been a superb administrator.

P11 Page, Rosewell, Jr. "Forest Conservation and Development (as Seen through the Eyes of a Virginia Landholder and Former Fire-Warden)." *Virginia Record* 79 (Mar. 1957), 8-9, 15-21. Since 1900.

P12 Paisley, Clifton. "Wade Leonard, Florida Naval Stores Operator." *Florida Historical Quarterly* 51 (Apr. 1973), 381-400. On Wade Hampton Leonard, his brothers, and their logging and naval stores business in Calhoun County, Florida, 1900-1930s.

P13 Palais, Hyman, and Roberts, Earl. "The History of the Lumber Industry in Humboldt County." *Pacific Historical Review* 19 (Feb. 1950), 1-16. California, since 1850.

P14 Palais, Hyman. "Pioneer Redwood Logging in Humboldt County." *Forest History* 17 (Jan. 1974), 18-27. California, 1850s-1910s.

P15 Palley, Marshall N. "Ideals and Purposes in American Forestry." Master's thesis, Univ. of Michigan, 1941.

P16 Palm, Harry W. *Lumberjack Days in the St. Croix Valley: Hitherto Unpublished Tales of the Early Pioneers along the St. Croix River in Minnesota.* Bayport, Minnesota: Bayport Printing House, 1969. 51 pp.

P17 Palmer, Charlotte. "Conservation and the Camera." *Prologue* 3 (Winter 1971), 183-96. On a USFS photographic collection being transferred from the USDA to the National Archives. The collection was started by Gifford Pinchot and some photos date back to 1895.

P18 Palmer, E.J. "The Red River Forest at Fulton, Arkansas." *Journal of the Arnold Arboretum* 4 (No. 1, 1923), 8-33.

P19 Palmer, Howard J. "History of Paper Making in Northern New York." *Paper* 27-28 (Jan.-July 1921). A series.

P20 Palmer, Richard F. "Williamstown and Redfield Railroad." *Northern Logger and Timber Processor* 18 (Oct. 1969), 14-15, 53. A logging railroad in northern New York, 1860s-1870s.

P21 Palmer, Richard F. "Logging Railroads in the Adirondacks." *Northern Logger and Timber Processor* 18 (Feb. 1970), 28-29, 47; 18 (Mar. 1970), 14-15; 18 (Apr. 1970), 26-27, 47; 19 (Mar. 1971), 16-17, 26; 19 (Apr. 1971), 14-15, 37. The series concerns logging railroads of the late 19th century, including those of the Brooklyn Cooperage Company, the Fulton Chain Railroad, and others.

Palmer, Richard F. See Allen, Richard S., #A88

P22 Pammel, Louis H. "What the College Has Done for Park and Forestry during the Last Thirty Years." *Proceedings of the Iowa Park and Forest Association* 4 (1904), 51-70.

P23 Pammel, Louis H. "The Beginnings of Forestry Instruction at Ames." *Ames Forester* 15 (1927), 39-47. Iowa State College.

P24 Pammel, Louis H. "Theodore Roosevelt, the Conservationist." *Ames Forester* 17 (1929), 62-68.

P25 Pammel, Louis H. *The Arbor Day, Park and Conservation Movements in Iowa.* Des Moines, 1929. 104 pp.

Pamplin, Robert B. See Cheatham, Owen R., #C276

P26 Panshin, Alexis John; Harrar, E.S.; Baker, W.J.; and Proctor, P.B. *Forest Products: Their Source, Production, and Utilization.* American Forestry Series. 1950. Second edition. New York: McGraw-Hill, 1962. ix + 549 pp. Illus., notes, tables, diags., index. Includes some incidental historical references to the manufacture of a wide variety of wood products and to the economic importance of such industries.

P27 Pape, Fred E. "Half a Century of Fire Prevention Progress." *West Coast Lumberman* 50 (1926), 201, 204.

P28 *Paper and Printing Digest.* "America's First Ground-Wood Mill and America's First Paper Mill." *Paper and Printing Digest* 4 (Jan. 1938), 11-14.

P29 *Paper Mill and Wood Pulp News.* "Paper in the Middle West: History of the Industry in the Miami Valley." *Paper Mill and Wood Pulp News* 24 (Feb. 16, 1901), 40-46. Ohio.

P30 *Paper Trade Journal.* "250 Years of Paper Making in America: A History of the Industry from the Times of William Rittenhouse." *Paper Trade Journal* 111 (Nov. 28, 1940), supplement, pp. 5-135. Includes sketches of individuals and firms important to the industry's history.

P31 *Paper Trade Journal.* "Achievements of Technical Association of the Pulp and Paper Industry." *Paper Trade Journal* 111 (Nov. 28, 1940), 72-75.

P32 *Paper Trade Journal.* "Three Score Years and Ten: The 'Paper Trade Journal,' 1872-1942." *Paper Trade Journal* 114 (May 21, 1942), 19-50.

P33 Parchomchuk, William. "Truck, Rail and Water Transport of Raw Wood in the British Columbia Forest Industry." Master's thesis, Univ. of British Columbia, 1968.

P34 Parent, Annette Richards. "Guadalupe: Barrier Reef in the Desert." *National Parks & Conservation Magazine* 48 (Oct. 1974), 4-9. Includes some incidental history of Guadalupe Mountains National Park in west Texas, established in 1966.

P35 Pares, Richard. *Yankees and Creoles: The Trade between North America and the West Indies before the American Revolution.* Cambridge: Harvard University Press, 1956. vii + 168 pp. Maps, notes, bib., index. Includes some history of the trade in lumber and forest products.

P36 Parker, Carleton H. *The Casual Laborer and Other Essays.* New York: Harcourt, Brace and Howe, 1920. Reprint. Intro. by Harold M. Hyman. Seattle: University of Washington Press, 1972. xxiv + 199 pp. Includes some references to unionizing activities of the Industrial Workers of the World among lumber and millworkers of the Pacific states.

P37 Parker, Joseph Allen. "An Investigation of the Ten Most Active Acquisitors in the Paper and Allied Products Industry, 1950-1965." Ph.D. diss., Univ. of Oklahoma, 1968. 245 pp.

P38 Parker, Keith Alfred. "The Staple Industries and Economic Development, Canada, 1841-1867." Ph.D. diss., Univ. of Maryland, 1966. 322 pp. Especially the forest industries.

P39 Parker, Robert J. "Larkin's Monterey Business: Articles of Trade, 1833-1839." *Historical Society of Southern California Quarterly* 24 (June 1942), 54-60. Includes reference to Thomas O. Larkin's trade in lumber in Monterey, California.

P40 Parker, Robert J. "Larkin, Anglo-American Businessman in Mexican California." In *Greater America: Essays in Honor of Herbert Eugene Bolton,* ed. by Adele Ogden and Engel Sluiter. Berkeley: University of California Press, 1945. Pp. 415-29.

P41 Parkins, Almon E., and Whitaker, J. Russell, eds. *Our Natural Resources and Their Conservation.* New York: John Wiley and Sons, 1936. 650 pp. Illus. Includes some history; there are later editions.

P42 Parkins, Almon E. *The South: Its Economic-Geographic Development.* London: Chapman and Hall, 1938. 528 pp. Bib. Includes a chapter on the history of forests and the lumber industry.

Parks, W. Robert. See Walker, Herman, Jr., #W27

P43 Parks, W. Robert. *Soil Conservation Districts in Action.* Ames: Iowa State University Press, 1952. 253 pp. On the administrative powers and practices of soil conservation districts and their relations with the parent federal agency, the Soil Conservation Service.

P44 Parman, Donald L. "The Indian Civilian Conservation Corps." Ph.D. diss., Univ. of Oklahoma, 1967. 267 pp.

P45 Parman, Donald L. "The Indian and the Civilian Conservation Corps." *Pacific Historical Review* 40 (Feb. 1971), 39-56. During the 1930s.

P46 Parson, Dudley L., and Sudarsky, Peter. *The Moore Story, 1882-1957.* Toronto: Moore Corporation, 1957. 40 pp. On The Moore Corporation, a Toronto manufacturer of paper boxes and business forms.

P47 Parson, Ruben L. *Conserving American Resources.* 1955. Third edition. Englewood Cliffs, New Jersey: Prentice-Hall, 1972. 608 pp. Illus., maps, graphs, index. Contains some history of natural resource conservation, including several chapters on forests and forest conservation.

Parsons, James J. See Ciriacy-Wantrup, Siegfried von, #C326

P48 Partain, Gerald L. "Timber Taxation in Humboldt County, California: A Case Study." Ph.D. diss., State Univ. of New York, College of Environmental Science and Forestry, 1972. 138 pp. Incidental history.

P49 Patten, Phyllis. *Oh, That Reminds Me.* Felton, California: Big Trees Press, 1969. xii + 114 pp. Illus., maps. Includes reminiscences of logging camp life near Soquel, California, early 20th century.

P50 Patterson, Lester W. "History of Lumbering in the Douglas Fir Region (Particularly Western Washington)." Master's thesis, College of Puget Sound, 1951.

P51 Pattison, William D. *The Beginnings of the American Rectangular Land Survey System, 1784-1800.* University of Chicago, Department of Geography, Research Paper No. 50. Chicago: University of Chicago Press, 1957. vii + 248 pp. Issued also as a Ph.D. dissertation.

P52 Pattison, William D. "The Survey of the Seven Ranges." *Ohio Historical Quarterly* 68 (Apr. 1959), 115-40. In southeastern Ohio by Thomas Hutchins and government surveyors, 1785-1788.

P53 Patton, C.R. "Biltmore Estate—Cradle of Conservation in America." *Journal of Soil and Water Conservation* 13 (May 1958), 101-05. A brief historical sketch of the Biltmore Estate near Asheville, North Carolina, and of the Biltmore Forest School and its impact on forestry and conservation.

P54 Paulsen, David F. *Natural Resources in the Governmental Process: A Bibliography, Selected and Annotated.* American Government Studies No. 3. Tucson: University of Arizona Press, 1970. 99 pp. Author index. The bibliography is organized by subject and contains 304 annotated entries. Intended primarily for use by political scientists and students of public administration, it also contains entries of interest to the forest and conservation historian.

P55 Payne, John W. "David Franklin Houston: A Biography." Ph.D. diss., Univ. of Texas, 1953. 318 pp. Houston was Secretary of Agriculture, 1913-1920.

P56 Payne, Monty. "J. Brooks Toler (1906-1949)." *Journal of Forestry* 47 (Nov. 1949), 920-21. Sketch of Toler's career in state forestry (Mississippi and Alabama), with the Southern Pine Association in New Orleans, and with the Masonite Corporation at Laurel, Mississippi, since 1928.

P57 Peabody, M.B. "125 Years of Sawmilling." *Southern Lumberman* 193 (Dec. 15, 1956). 151-52. On the R.F. Learned Company, a sawmill successively operated near Natchez, Mississippi, by Peter Little, Andrew Brown, Rufus F. Learned, and Andrew Brown Learned, since 1825.

P58 Peacock, Blanche G. "Reelfoot Lake State Park." *Tennessee Historical Quarterly* 32 (Fall 1973), 205-32. On the history of the forested state park in western Tennessee, including some account of logging operations in the 1870s and 1880s.

P59 Pearce, William. "Establishment of National Parks in the Rockies." *Alberta Historical Review* 10 (Summer 1962), 8-17. On Banff, Jasper, Waterton Lakes, and other national parks and forest reserves in Alberta and British Columbia since the 1880s.

P60 Pearl, Milton A. "Historical View of Public Land Disposal and the American Land Use Pattern, California." *Western Law Review* 4 (Spring 1968), 65-75.

P61 Pearlman, Albert L. "History of Box Tariff No. 1." *Timberman* 33 (Oct. 1932), 16-18. Adopted by the National Association of Wooden Box Manufacturers in 1928 as a sales and manufacturing guide.

P62 Pearse, John B. *A Concise History of the Iron Manufacture of the American Colonies up to the Revolution and of Pennsylvania until the Present Time.* Philadelphia: Allen, Lane & Scott, 1876. 282 pp. Illus., maps, table, bib. Includes some history of the charcoal iron industry by colony and state.

P63 Pearson, C.W. *England's Timber Trade in the Last of the 17th and First of the 18th Century, More Especially with the Baltic Sea. . . .* Goettingen, Germany: W.F. Kaestner, 1869. Includes a chapter on the timber trade with North America.

P64 Pearson, Grant H. *A History of Mount McKinley National Park, Alaska.* Washington: U.S. National Park Service, 1953. iii + 91 pp. Illus., notes, bib. Written by the park superintendent, it includes some history of the area since 1896.

P65 Pearson, Grant H. *My Life of High Adventure.* Englewood Cliffs, New Jersey: Prentice-Hall, 1962. 234 pp. Illus. An account of Mount McKinley National Park, Alaska, and experiences of the author while a park ranger and superintendent, 1925-1960.

P66 Pearson, Gustaf A. "Autobiographical Sketch." *Plateau* 28 (Apr. 1956), 86-90. Pearson (1880-1949) writes of his own career in forestry. A chronological list of his writings, compiled by Horace S. Haskell and Chester F. Deaver, is appended.

P67 Pease, Lute. "The Way of the Land Transgressor." *Pacific Monthly* 18 (Aug. 1907), 145-63; 18 (Sept. 1907), 345-63; 18 (Oct. 1907), 485-91; 18 (Nov. 1907), 547-51. A series concerned with federal land policies during the Theodore Roosevelt era. Individual articles are subtitled: "The West and the President's Land Policies," "Gifford Pinchot and the National Forests," "Ethan Allen Hitchcock and the Lieu Land Operators," and "Some Queer Operations in the Rocky Mountain States."

Pease, Theodore Calvin. See Browning, Orville Hickman, #B514

P68 Peattie, Donald Culross. *Green Laurels: The Lives and Achievements of the Great Naturalists.* New York: Simon and Schuster, 1936. xxiii + 368 pp. Illus. Includes sketches of John and William Bartram, André and François André Michaux, and John James Audubon.

P69 Peattie, Donald Culross. *The Road of a Naturalist.* Boston: Houghton Mifflin, 1941. 315 pp. Incidental history.

P70 Peattie, Donald Culross. "White Pine." *Scientific American* 178 (June 1948), 48-53. Since the 17th century.

P71 Peattie, Donald Culross. "The Elms Go Down." *Atlantic Monthly* 182 (July 1948), 21-24. On the Dutch elm disease, introduced in 1930.

P72 Peattie, Donald Culross. "Shagbark Hickory." *Scientific American* 179 (Sept. 1948), 40-43. Since the 17th century.

P73 Peattie, Donald Culross. "Spruce, Balsam and Birch." *Scientific American* 179 (Nov. 1948), 20-23. On their use since the 16th century.

P74 Peattie, Donald Culross. *American Heartwood.* Boston: Houghton Mifflin Co., 1949. x + 307 pp. Reprinted articles, some of which deal with American forests and woodsmen prior to the 20th century.

P75 Peattie, Donald Culross. *A Natural History of Trees of Eastern and Central North America.* Boston: Houghton Mifflin, 1950. xv + 606 pp. Illus. Essentially descriptive but contains information on the history and use of each species.

P76 Peattie, Donald Culross. "The Sugar Pine." *Natural History* 61 (Apr. 1952), 182-88. On the Pacific Coast, since 1825.

P77 Peattie, Donald Culross, *A Natural History of Western Trees.* Boston: Houghton Mifflin, 1953. xiv + 751 pp. Illus. Includes some history of the use of many species.

P78 Peattie, Roderick. *The Incurable Romantic.* New York: Macmillan, 1941. 220 pp. Autobiography of a geologist-geographer-naturalist.

P79 Peattie, Roderick, ed. *The Friendly Mountains, Green, White, and Adirondacks.* New York: Vanguard Press, 1942. 341 pp. Illus., tables, diags. New York, New Hampshire, and Vermont.

P80 Peattie, Roderick, ed. *Great Smokies and the Blue Ridge.* New York: Vanguard Press, 1943. x + 372 pp. North Carolina, Tennessee, and Virginia.

P81 Peattie, Roderick, ed. *Pacific Coast Ranges.* New York: Vanguard Press, 1946. xviii + 402 pp. Illus., maps.

P82 Peattie, Roderick, ed. *The Sierra Nevada—The Range of Light.* New York: Vanguard Press, 1947. 398 pp. Illus., map. California and Nevada.

P83 Peattie, Roderick, ed. *The Cascades: Mountains of the Pacific Northwest.* New York: Vanguard Press, 1949. 417 pp. Illus., map. Oregon and Washington.

P84 Pecenka, Joseph Otto. "A Financial Analysis of Selected Major Pulp, Paper and Paperboard Producers, 1947-1964." Ph.D. diss., Univ. of Illinois, 1967. 354 pp.

P85 Pechanec, J.F. "The History and Accomplishments of Our Range Society." *Journal of Range Management* 10 (July 1957), 189-93. The Society of Range Management since 1946.

P86 Peck, Ellen B. "The Founder of Arbor Day." *New England Magazine* 22 (1900), 269-75. Birdsey Grant Northrop.

P87 Peet, Creighton. "12 Billion Dollar Jackpot?" *American Forests* 66 (Oct. 1960), 12-15, 42. On industrial uses of lignin since ca. 1869.

P88 Peffer, Al. "Evolution of the Fire Danger Meter." *American Forests* 66 (Oct. 1960), 12-15, 42. On industrial ment and other fire research by Harry T. Gisborne at the USFS's Northern Rocky Mountain Forest and Range Experiment Station, Missoula, Montana, since 1919.

P89 Peffer, E. Louise. "The Closure of the Public Domain, 1902-1936." Ph.D. diss., Univ. of California, Berkeley, 1942. 154 pp.

P90 Peffer, E. Louise. "Which Public Domain Do You Mean?" *Agricultural History* 23 (Apr. 1949), 140-46. On definitions of "public domain" since 1789.

P91 Peffer, E. Louise. *The Closing of the Public Domain: Disposal and Reservation Policies, 1900-1950.* Stanford University Food Research Institute, Miscellaneous Publications, No. 10. Stanford, California: Stanford University Press, 1951. Reprint. New York: Arno Press, 1972. xi + 372 pp. Tables, notes. The focus of this work is on the grazing lands of the West, although considerable attention is given to forested lands.

P92 Peirce, Earl S., and Stahl, William J., comps. *Cooperative Forest Fire Control: A History of Its Origin and Development under the Weeks and Clarke-McNary Acts.* Washington: USFS, 1964. 94 pp. Tables, graphs.

P93 Peirce, Earl S. *Multiple Use and the U.S. Forest Service, 1910 to 1950.* OHI by Susan R. Schrepfer. Santa Cruz, California: Forest History Society, 1972. x + 118 pp. Illus., apps., bib., index. Processed. On Peirce's career in the USFS, especially in the Rocky Mountain region and in a variety of administrative positions leading to chief of the Division of State Cooperation.

Peirce, G.R. See Bates, Carlos G., #B120

P94 Pell, Stuyvesant Morris. *Scribblings of an Outdoor Boy.* Princeton, New Jersey: Princeton University Press, 1945. 75 pp. Pell (b. 1905) was a ranger and naturalist in Maine.

P95 Pelzer, Louis. "The Public Domain as a Field for Historical Study." *Iowa Journal of History and Politics* 12 (Oct. 1914), 568-78.

P96 Pendleton, Maurice B. "A Short History of the National Hardwood Lumber Association." *Southern Lumberman* 193 (Dec. 15, 1956), 127-30. Since 1898.

P97 Pendleton, Maurice B. *The People, Policies and Progress of the National Hardwood Lumber Association, Celebrating the Diamond Anniversary of Its Founding, 1898.* Chicago: National Hardwood Lumber Association, 1972. 24 pp. Includes a brief history of the trade association and biographical sketches of its presidents since 1898.

P98 Pendleton, Nathaniel. "Nathaniel Pendleton's 'Short Account of the Sea Coast of Georgia with Respect to Agriculture, Navigation, and the Timber Trade'." Ed. by Theodore Thayer. *Georgia Historical Quarterly* 41 (Mar. 1957), 70-81. The manuscript, dated February 28, 1800, includes reference to pine lands along Georgia's St. Mary's River.

Penfound, William T. See Rice, Elroy Leon, #R153

P99 Penick, James L., Jr. "The Ballinger-Pinchot Controversy." Ph.D. diss., Univ. of California, Berkeley, 1962.

P100 Penick, James L., Jr. "The Age of the Bureaucrat: Another View of the Ballinger-Pinchot Controversy." *Forest History* 7 (Spring/Summer 1963). 15-21. On the controversy between Secretary of the Interior Richard A. Ballinger and Gifford Pinchot of the USFS in 1910, and its implications for the conservation movement.

P101 Penick, James L., Jr. "Louis Russell Glavis: A Postscript to the Ballinger-Pinchot Controversy." *Pacific Northwest Quarterly* 55 (Apr. 1964), 67-75. On his role in the controversy of 1910 and his subsequent career in California with the USDI in the 1930s.

P102 Penick, James L., Jr. *Progressive Politics and Conservation: The Ballinger-Pinchot Affair.* Chicago: University of Chicago Press, 1968. xv + 207 pp. Notes, bib., index. Explores the political controversy of 1910 which split the conservation movement, the Taft administration, and the Republican party.

Penick, James L., Jr. See Pinchot, Gifford, #P210

P103 Penick, James L., Jr. "The Progressives and the Environment: Three Themes from the First Conservation Movement." In *The Progressive Era,* ed. by Lewis L. Gould. Syracuse: Syracuse University Press, 1974. Pp. 115-31.

P104 Pennsylvania. Department of Forests and Waters. "The Charcoal Burning Industry—'A Threadbare Shadow of Its Former Self'." *Pennsylvania Forests and Waters* 4 (Jan.-Feb. 1952), 8-9, 23. From 1716 to 1870.

P105 Pennsylvania. Department of Forests and Waters. "Ships of Trees: Strange Vessels That Sailed from Pennsylvania Ports." *Pennsylvania Forests and Waters* 4 (Sept.-Oct. 1952), 114-15. On ocean-going raft ships built for the purpose of carrying logs and lumber, 1791-1906.

P106 Pennsylvania. State Planning Board. "Forest Resources." *Pennsylvania Planning* 10 (Aug. 1946), 1-44. Includes some history of the lumber and other forest industries.

P107 *Pennsylvania Forests.* "Mont Alto 50th, 1903-1953." *Pennsylvania Forests* 38 (Spring 1953), 28-57. Includes six articles on the history of forestry education at the Pennsylvania State Forest School at Mont Alto.

P108 *Pennsylvania Forests.* "50 Years of Forestry Education." *Pennsylvania Forests* 38 (Fall 1953), 92-118. Includes eleven articles on all aspects of forestry education in Pennsylvania.

P109 *Pennsylvania Forests.* "The P.H. Glatfelter Company—100 Years of Papermaking." *Pennsylvania Forests* 54 (Fall 1964), 44-46. Spring Grove, Pennsylvania.

P110 Penny, J. Russell, and Clawson, Marion. "Administration of Grazing Districts." *Land Economics* 29 (Feb. 1953), 23-24. On the history of grazing and the provisions and administration of the Taylor Grazing Act of 1934.

P111 Penrose, Edith T. "The Growth of the Firm—A Case Study: The Hercules Powder Company." *Business History Review* 34 (Spring 1960), 1-23. Concerns the Delaware manufacturer of paper-making chemicals, 1912-1954.

P112 Peplow, Edward H., Jr. "A Tribute to the National Park Service: 1919—Grand Canyon National Park—1969." *Arizona Highways* 45 (Mar. 1969).

P113 Percival, W.C., and Carvell, K.L. "The History of Forestry Education at West Virginia University." *Northeastern Logger* 9 (Apr. 1961), 12-13, 50.

P114 Perkins, Nelson S. *Plywood in Retrospect: Washington Veneer Company.* Monographs on the History of West Coast Plywood Plants, No. 11. Tacoma: Plywood Pioneers Association, 1971. 6 pp. Illus. A plywood mill in Olympia, Wahington, since 1924.

P115 Perkins, Nelson S. *Plywood in Retrospect: Vancouver Plywood Company.* Monographs on the History of West Coast Plywood Plants, No. 12. Tacoma: Plywood Pioneers Association, 1972. 6 pp. Illus. And its predecessors in Vancouver, Washington, since the 1920s.

P116 Perkins, Nelson S. *Plywood in Retrospect: Robinson Plywood & Timber Co.* Monographs on the History of West Coast Plywood Plants, No. 13. Tacoma: Plywood Pioneers Association, 1973. 6 pp. Illus. On a family-run enterprise in Everett, Washington, known as Robinson Manufacturing Company, 1889-1948, and Everett Plywood & Door Company since 1959.

P117 Perkins, Nelson S. *Plywood in Retrospect: Harbor Plywood Corporation.* Monographs on the History of West Coast Plywood Plants, No. 14. Tacoma: Plywood Pioneers Association, 1974. 10 pp. Illus. Of Aberdeen and Hoquiam, Washington, with operations elsewhere in the Pacific Northwest, 1920s-1950s.

P118 Perkins, Nelson S. *Plywood in Retrospect: Aircraft Plywood Company.* Monographs on the History of West Coast Plywood Plants, No. 15. Tacoma: Plywood Pioneers Association, 1975. 9 pp. Illus. Of Seattle, Washington, since 1929. The firm was later purchased by U.S. Plywood Company.

Perlman, Selig. See Commons, John R., #C543

P119 Perloff, Harvey S.; Dunn, Edgar S., Jr.; Lampard, Eric E.; and Muth, Richard F. *Regions, Resources, and Economic Growth.* Baltimore: Johns Hopkins Universtiy Press, for Resources for the Future, 1960. xxv + 742 pp. On regional economic growth since the 1870s, with reference to the availability of natural resources and the economic importance of the forest industries.

P120 Pernin, Peter. *The Finger of God is There! or Thrilling Episode of a Strange Event Related by an Eye-Witness, Rev. P. Pernin, United States Missionary, Published with the Approbation of His Lordship the Bishop of Montreal.* Montreal, 1874. On the forest fire which devastated Peshtigo, Wisconsin, in 1871.

P121 Pernin, Peter. "The Finger of God Is There." *Wisconsin Magazine of History* 2 (Dec. 1918), 158-80; 2 (Mar. 1919), 274-93.

P122 Pernin, Peter. "The Great Peshtigo Fire: An Eyewitness Account." *Wisconsin Magazine of History* 54 (Summer 1971), 246-72. A reprint (with an editorial introduction) of Father Pernin's account of the forest fire which in 1871 destroyed Peshtigo, Wisconsin.

P123 Perrin, Charles N. "The Development of Hardwood Grading." *Southern Lumberman* (Dec. 17, 1927), 155-57.

P124 Perrin, Richard W.E. "Wisconsin 'Stovewood' Walls: Ingenious Forms of Early Log Construction." *Wisconsin Magazine of History* 46 (Spring 1963), 215-19.

P125 Perrin, Robert Anthony, Jr. "Two Decades of Turbulence, a Study of the Great Lumber Strikes of North Idaho (1916-1936)." Master's thesis, Univ. of Idaho, 1961.

P126 Perry, Percival. "The Naval Stores Industry in the Ante-Bellum South, 1789-1861." Ph. D. diss., Duke Univ., 1947. 313 pp. Emphasizes North Carolina, the center of the industry.

P127 Perry, Percival. "The Naval-Stores Industry in the Old South, 1790-1860." *Journal of Southern History* 34 (Nov. 1968), 509-26.

P128 Perry, Thomas D. "Seventy-Five Years of Plywood." *Southern Lumberman* 193 (Dec. 15, 1956), 280-90.

P129 Person, Hubert Lawrence. *Commercial Planting on Redwood Cutover Lands.* USDA, Circular No. 434. Washington: GPO, 1937. 40 pp. Illus., diags. Incidental history.

P130 Pesch, A.W. "An Example in Mississippi of How Science Has Aided in the Development of a Great Industrial Asset." *Journal of the Mississippi Academy of Sciences* 4 (1948-1950), 98-103. On the manufacture of paper from wood pulp since 1908.

P131 Peters, Bernard C. "No Trees on the Prairie: Persistence of Error in Landscape Terminology." *Michigan History* 54 (Spring 1970), 19-28. On rival theories concerning vegetation in southern Michigan, 1820s-1830s.

P132 Peters, Bernard C. "Pioneer Evaluation of the Kalamazoo County Landscape." *Michigan Academician* 3 (Fall 1970), 15-25. 19th century.

P133 Peters, George Hugo. *The Trees of Long Island: A Short Account of Their History, Distribution, Utilization, and Significance in the Development of the Region. Also, the Results of the First Systematic and Comprehensive Census of the 'Big Trees' of Long Island, Including a List of the Largest Specimens of the Species Reported.* Publication No. 1. Farmingdale, New York: Long Island Horticultural Society, 1952. 63 pp. Illus. Long Island, New York, since 1609.

P134 Peters, James Girvin. "Development of Fire Protection in the States." *American Forestry* 19 (Nov. 1913), 721-36. Other articles in the issue contain incidental history of forest fires and fire protection.

P135 Peters, William E. *Ohio Lands and Their History.* Athens, Ohio: W.E. Peters, 1930. 401 pp. Maps.

P136 Peters, William Stanley, and Johnson, Maxine C. *Public Lands in Montana: Their History and Current Significance.* Montana State University, Bureau of Business and Economic Research, Regional Study No. 10. Missoula, 1959. x + 69 pp. Map, tables, notes, bib. Since the 1860s.

P137 Petersen, Eugene T. "The History of Wild Life Conservation in Michigan, 1859-1921." Ph.D. diss., Univ. of Michigan, 1953. 342 pp.

P138 Petersen, Eugene T. "The Michigan Sportsmen's Association: A Pioneer in Game Conservation." *Michigan History* 37 (Dec. 1953), 355-72. On its work from 1874 to 1921; includes information on lumber camps as markets for slaughtered deer.

P139 Petersen, Eugene T. "Wildlife Conservation in Michigan." *Michigan History* 44 (June 1960), 129-46. Since 1859.

P140 Petersen, Eugene T. *Conservation of Michigan's Natural Resources.* Lansing: Michigan Historical Commission, 1960. 29 pp.

P141 Petersen, Peter L. "A Publisher in Politics: Edwin T. Meredith, Progressive Reform, and the Democratic Party, 1912-1928." Ph.D. diss., Univ. of Iowa, 1971. 366 pp. Meredith was Woodrow Wilson's Secretary of Agriculture, 1920-1921.

P142 Petersen, William J. *Steamboating on the Upper Mississippi.* 1937. Reprint. Iowa City: State Historical Society of Iowa, 1968. 575 pp. Illus., maps, notes, index. Includes some history of rafting and the lumber trade after 1840.

P143 Petersen, William J. "Rafting on the Mississippi: Prologue to Prosperity." *Iowa Journal of History* 58 (Oct. 1960), 289-320. On the Iowa lumber trade and the Mississippi River lumber traffic as observed from Iowa ports, 1833-1870.

P144 Peterson, Arthur D. "Arboreal Law in Iowa." *Iowa Law Review* 44 (Summer 1959), 680-92. On legal questions involving trees since 1856.

P145 Peterson, Charles E. "Sawdust Trail: Annals of Sawmilling and the Lumber Trade from Virginia to Hawaii via Maine, Barbados, Sault Ste. Marie, Manchac and Seattle to the Year 1860." *Bulletin of the Association of Preservation Technology* 5 (No. 2, 1973), 84-152.

P146 Peterson, Charles S. "Albert F. Potter's Wasatch Survey, 1902: A Beginning for Public Management of Natural Resources in Utah." *Utah Historical Quarterly* 39 (Summer 1971), 238-53. Prompted by overgrazing and erosion problems, the USDI's Forestry Division sponsored a survey of Utah's natural resources.

P147 Peterson, Charles S. "Small Holding Land Patterns in Utah and the Problem of Forest Watershed Management." *Forest History* 17 (July 1973), 4-13. On the problems of livestock grazing on forests and watersheds, 1870s-1920s, and attempts by the USFS to regulate grazing on Utah's national forests in the 20th century.

P148 Peterson, Charles S. *Look to the Mountains: Southeast Utah and the La Sal National Forest.* Provo, Utah: Brigham Young University Press, 1975. 261 pp. Illus., maps, notes, bib., index. On the history of the region since 1760, with Part 2 focusing on the impact of the Manti-La Sal National Forest on the resources and people of southeastern Utah and extreme western Colorado. There are chapters on forest administration, boundary and claim adjustments, grazing and range management, wildlife, timber, watershed, and other federal programs.

P149 Peterson, Dale Arthur. "Lumbering on the Chippewa: The Eau Claire Area, 1845-1885." Ph.D. diss., Univ. of Minnesota, 1970. 739 pp.

P150 Peterson, Edwin L. *Penn's Woods West.* Pittsburgh: University of Pittsburgh Press, 1958. Illus. Incidental history of the forests of western Pennsylvania.

P151 Peterson, Ellen Z. "The Hanging Flume of Dolores Canyon." *Colorado Magazine* 40 (Apr. 1963), 128-31. Includes some account of lumber cut and used in its construction in western Colorado, 1889-1891.

P152 Peterson, George L. "Speed-Free Wilderness." *Land* 9 (Spring 1950), 72-76. On successful efforts to preserve the wilderness qualities of the Superior National Forest, Minnesota, since 1922.

P153 Peterson, H. Gardner. "The Ship That Walked." *Ships and the Sea* 3 (Nov. 1953), 8-9, 44-45. On the Pacific lumber carrier *North Bend,* and her misfortunes in the Columbia River and on the Oregon coast, 1919-1940.

P154 Peterson, J. Merriam. "History of the Naval Stores Industry in America." *Journal of Chemical Education* 16 (1939), 203, 212, 317-22.

P155 Peterson, Raymond Arnold. "An Analysis of Land Use Change in Selected North Florida Counties." Ph.D. diss., Univ. of Florida, 1967. 231 pp. Includes some history of tree planting by pulp and paper companies and other forest uses in Columbia, Madison, Hamilton, and Suwannee counties since 1938.

P156 Peterson, Roger Tory. "The Evolution of a Magazine." *Audubon* 75 (Jan. 1973), 46-51. History of *Audubon* and its predecessor, *Bird-Lore,* since 1899.

P157 Peterson, Virgil G.; Smith, Paul R.; and Plant, Charles. "Red Cedar Shingles & Shakes: Unique Wood Industry." OHI by Elwood R. Maunder. Ed. by Betty E. Mitson and Barbara D. Holman. *Journal of Forest History* 19 (Apr. 1975), 56-71. Reminiscences and history of the red cedar shingle and shake industry in Washington and British Columbia, since ca. 1900, and of its trade association, the Red Cedar Shingle and Handsplit Shake Bureau.

Peterson, William. See Cook, Charles W., #C575

P158 Petit, Thomas A. "The Economics of the Softwood Plywood Industry." Ph.D. diss., Univ. of California, Berkeley, 1957. Includes some history of the industry.

P159 Petit, Thomas A. "Some Economic Characteristics of the Pacific Coast Softwood Plywood Industry." *Journal of Forestry* 55 (Feb. 1957), 124-29. Since 1905.

P160 Petite, Irving. "Science and Mythology Manage a Forest." *American Forests* 62 (Sept. 1956), 30-31, 54-56. On the Yakima Indian Reservation, Washington, since 1941.

P161 Pettis, Clifford R. *Resources of the Forest Preserve.* New York Conservation Commission, Division of Land and Forests, Bulletin No. 12. Albany, 1915. 41 pp. Diags. Incidental history.

P162 Pettis, Clifford R. "Forest Provisions of New York State Constitution." *Forestry Quarterly* 14 (Mar. 1916), 50-60. Includes some history of forest conservation as exhibited by the constitution of 1894 and other laws affecting the Forest Preserve.

P163 Pettis, Clifford R. "Conservation in New York State." *Quarterly Journal of the New York State Historical Association* 3 (Apr. 1922), 77-82.

P164 Pettit, Theodore S. "A Half-Century of Service to Conservation." *Pennsylvania Game News* 21 (Feb. 1960), 2-4. On the conservation activities of the Boy Scouts of America since 1910.

P165 Peyton, Jeannie S. "Forestry Movement of the Seventies, in the Interior Department, under Schurz." *Journal of Forestry* 18 (Apr. 1920), 391-405. On the advocacy of a forestry policy by Carl Schurz, Secretary of the Interior, and James A. Williamson, Commissioner of the General Land Office, as reconstructed from annual reports, 1870s.

P166 Pfister, Richard. "History and Evaluation of the Shelterbelt Project." Master's thesis, Univ. of Kansas, 1950.

P167 Philbrick, Francis S. *Rise of the West, 1754-1830.* New York: Harper & Row, 1965. xvii + 398 pp. Illus., map. Includes some history of forested lands in relation to settlement.

P168 Philips, David. "When Loggers Sang for Their Supper." *Down East* 1 (Apr. 1955), 34-37. On singing in 19th-century Maine lumber camps, with reprinted texts of three ballads.

P169 Phillips, Fred. "Boss Gibson, Lumber Baron of the Eighties, Who Left His Mark on New Brunswick." *Canada Lumberman* 57 (Jan. 15, 1937), 20-22. Alexander Gibson.

P170 Phillips, Grace D. "Guardian of the Rockies." *National Parks Magazine* 29 (Jan.-Mar. 1955), 8-15. On Enos A. Mills, Colorado lodge-keeper, nature writer, and promoter of Rocky Mountain National Park, 1884-1922.

P171 Phillips, John C. "The Passing of the Maine Wilderness." *American Forests and Forest Life* 34 (Apr. 1928), 195-98, 232. On the problems of forest devastation, with some history of logging and forest uses in Maine since the early 19th century.

P172 Phillips, Max. "Benjamin Chew Tilghman, and the Origin of the Sulphite Process for the Delignification of Wood." *Journal of Chemical Education* 20 (Sept. 1943), 444-47. A new process for converting wood pulp to paper, 1880s.

P173 Phipps, R.W. "Across the Watershed of Eastern Ontario." *Forest History* 9 (Oct. 1965), 2-8. A descriptive account of lumbering operations in Ontario's Ottawa Valley; excerpted from his *Report on Forestry, 1884* (1885).

P174 Piché, Gustave C. "L'Histoire de l'administration des forêts dans la province de Québec." *Canadian Forestry Journal* 7 (1911), 20-21.

P175 Piché, Gustave C. "The Forests of Quebec." *Journal of Forestry* 20 (Jan. 1922), 25-43. Includes some history of the lumber industry and reforestation efforts since the late 19th century.

P176 Piché, Gustave C. "The Role of the Government Forester." *Forestry Chronicle* 4 (Mar. 1928), 7-15. Based on the author's experiences since 1905 in the Quebec Forest Service.

P177 Pickall, Adolph J. "Bark Utilization in the U.S." Master's thesis, Yale Univ., 1947.

P178 Pierce, Arthur Dudley. *Iron in the Pines: The Story of New Jersey's Ghost Towns and Bog Iron.* New Brunswick, New Jersey: Rutgers University Press, 1957. ix + 244 pp. Illus., maps, notes, bib. On the charcoal iron industry in and near the Wharton Estate in the New Jersey Pine Barrens, 1765-1868, and the later use of the properties for paper mills.

P179 Pierce, Arthur Dudley. *Family Empire in Jersey Iron: The Richards Enterprises in the Pine Barrens.* New Brunswick, New Jersey: Rutgers University Press, 1964. 303 pp. Illus., bib. Charcoal iron industry.

P180 Pierce, Bessie L. *A History of Chicago. Volume II, The Rise of a Modern City, 1871-1893.* New York: Alfred A. Knopf, 1957. xii + 575 + xxxvi pp. Includes an account of Chicago as a lumber center during this period.

P181 Pierce, Tom. "Some Like It Hot! The Story of Charcoal." *Forests & People* 24 (First quarter, 1974), 22-23, 32-33. On the manufacture of charcoal by Royal Oak Charcoal Company, Memphis, Tennessee, since 1931.

P182 Pierson, Albert H. *Wood Using Industries of Connecticut.* Connecticut Agricultural Experiment Station, Bulletin No. 174. New Haven, 1913. 96 pp. Tables, app.

P183 Pierson, Albert H. *Wood Using Industries of New Jersey.* Report of the New Jersey Forest Park Reservation Commission, Union Hill, New Jersey, 1914. 63 pp.

Pierson, Albert H. See Reynolds, Robert V., #R139, R140, R142, R143, R144, R145, R146, R147, R148

P184 Pierson, Lloyd M. "Wildlife—A Part of History." *Historical Preservation* 25 (July-Sept. 1973), 22-23.

P185 Pierson, William H. *The Geography of the Bellingham Lowland, Washington.* University of Chicago, Department of Geography, Research Paper No. 28. Chicago, 1953. ix + 159 pp. Illus., maps, tables, bib. Includes a section on forestry in the Nooksack Valley, Whatcom County, Washington, since 1950.

P186 Pietrak, Paul. *The History of the Buffalo & Susquehanna.* North Boston, New York: By the author, 1967. 130 pp. Illus., maps, tables. Includes several chapters on railroad logging in Pennsylvania on lines subsidiary to the Buffalo & Susquehanna Railway, 1880s-1920s. The emphasis is on the Goodyear Lumber Company and the Sinnemahoning Valley Railroad.

P187 Pike, Galen W. "Recreation Plans for the Superior National Forest." *Journal of Forestry* 51 (July 1953), 508-11. Minnesota, since 1930.

P188 Pike, Robert E. *Tall Trees, Tough Men.* New York: W.W. Norton and Company, 1967. 288 pp. Illus., map, index. Informal history and reminiscences of lumberjacks, log drivers, and the lumber industry in northern New England, ca. 1870-1920.

P189 Pike, Robert E. "Hell and High Water." *American Heritage* 18 (Feb. 1967), 64-70. A general account of log driving, especially on the rivers of New England.

P190 Pike, Robert E. *Spiked Boots: Sketches of the North Country.* Eatontown, New Jersey: By the author, 1961. 193 pp. Reminiscences of logging and other wood occupations in Coos County, New Hampshire, and elsewhere in northern New England.

P191 Pikl, Ignatz James, Jr. "Economic Problems of Pine Pulpwood Production in the South and in the Hiwassee Region." Ph.D. diss., Vanderbilt Univ., 1958. 505 pp. Includes some history of the pulp and paper industry.

P192 Pikl, Ignatz James, Jr. "Southern Forest-Products and Forestry: Development and Prospects." *Journal of Farm Economics* 42 (May 1960), 268-81. Since the 1890s.

P193 Pikl, Ignatz James, Jr. "Southern Woods-Labor 'Shortage' of 1955." *Southern Economic Journal* 27 (July 1960), 43-50. On the pulpwood shortage of 1955.

P194 Pikl, Ignatz James, Jr. *A History of Georgia Forestry.* Research Monograph No. 2. Athens: University of Georgia, Bureau of Business and Economic Research, 1966. iv + 91 pp. Illus., maps, tables, apps. Forest policy and forestry in Georgia, especially in the 20th century.

P195 Pikl, Ignatz James, Jr. "Pulp and Paper and Georgia: The Newsprint Paradox." *Forest History* 12 (Oct. 1968), 6-19. Emphasizes Charles H. Herty's efforts to attract the newsprint industry to Georgia, 1920s-1930s.

P196 Pinchot, Cornelia B. "Gifford Pinchot and the Conservation Ideal." *Journal of Forestry* 48 (Feb. 1950), 83-86. Address delivered at the dedication of the Gifford Pinchot National Forest, Washington, 1949.

P197 Pinchot, Gifford, and Ashe, William W. *Timber Trees and Forests of North Carolina.* North Carolina Geological Survey, Bulletin No. 6. Raleigh, 1897. 227 pp. Illus. Includes some history of forest industries.

P198 Pinchot, Gifford. *The Adirondack Spruce: A Study of the Forest in Ne-Ha-Sa-Ne Park with Tables of Volume and Yield and a Working Plan for Conservative Lumbering.* New York: Critic Company, 1898. Reprint. New York: Arno Press, 1970. vii + 157 pp. Illus., tables. New York; incidental history.

P199 Pinchot, Gifford. "The New Hope for the West: Progress in the Irrigation and Forest Reserve Movements." *Century Magazine* 68 (June 1904), 309-13.

P200 Pinchot, Gifford. "Sir Dietrich Brandis." *Proceedings of the Society of American Foresters* 3 (Oct. 1908), 54-66. An appreciation of the German-born forester, his work in India, and his influence on Pinchot and other American foresters.

P201 Pinchot, Gifford. "Southern Forest Products, and Forest Destruction and Conservation Since 1865." In *The South in the Building of the Nation,* Volume 6, ed. by J.C. Ballagh. Richmond, 1910. Pp. 151-58.

P202 Pinchot, Gifford. *The Training of a Forester.* 1914. Fourth edition, revised. Philadelphia: J.B. Lippincott, 1937. xii + 129 pp.

P203 Pinchot, Gifford. "Roosevelt's Part in Forestry." *Journal of Forestry* 17 (Feb. 1919), 122-24. On the contributions of Theodore Roosevelt to forestry and the national forests.

P204 Pinchot, Gifford. "The Economic Significance of Forestry." *North American Review* 213 (Feb. 1921), 157-67. Incidental history.

P205 Pinchot, Gifford. "The Blazed Trail of Forest Depletion." *American Forestry* 29 (June 1923), 323-28, 374. History of forest depletion in Pennsylvania.

Pinchot, Gifford. See Drinker, Henry S., #D262

P206 Pinchot, Gifford. "How the National Forests Were Won." *American Forests and Forest Life* 36 (Oct. 1930), 615-19, 674. An account of the struggle to establish, preserve, and protect national forests since 1878.

P207 Pinchot, Gifford. "How Conservation Began in the United States." *Agricultural History* 11 (Oct. 1937), 255-65. On the 19th-century origins of forest conservation and reminiscences of his own role in coining the term "conservation" in 1907.

P208 Pinchot, Gifford. "The Public Good Comes First." *Journal of Forestry* 39 (Feb. 1941), 208-12. Reminiscences of a former USFS chief on the achievements of forestry since 1900.

P209 Pinchot, Gifford. *The Fight for Conservation.* New York: Doubleday, Page and Company, 1910. Reprint. Intro. by Gerald D. Nash. Seattle: University of Washington Press, 1967. xxvii + 152 pp. Index. Essays written before Pinchot was removed as chief of the USFS in 1910. There are historical references throughout.

P210 Pinchot, Gifford. *Breaking New Ground.* New York: Harcourt, Brace and Company, 1947. Reprint. Intro. by James L. Penick, Jr. Seattle: University of Washington Press, 1972. xxvi + 522 pp. Illus., index. Pinchot's personal story of forestry and the conservation movement in the United States, 1885-1910s.

P211 Pine County Historical Society. *One Hundred Years in Pine County.* Askov, Minnesota: American Publishing Company, 1949. 144 pp. Illus., map. Includes some history of the lumber industry in Pine County, Minnesota, since 1856.

P212 Pine, Joshua. "A Rafting Story of the Delaware River." *Papers of the Bucks County Historical Society* 6 (1932). Pennsylvania.

P213 Pinkett, Harold T. "Records of Research Units of the United States Forest Service in the National Archives." *Journal of Forestry* 45 (Apr. 1947), 272-75. Describes historical source material in the USFS Research Compilation File, 1897-1941.

P214 Pinkett, Harold T. *Preliminary Inventory of the Records of the Civilian Conservation Corps.* U.S. National Archives, Preliminary Inventory No. 11. Washington, 1948. iii + 16 pp. Records of Emergency Conservation Work, 1933-1937, and of the CCC, 1937-1943.

P215 Pinkett, Harold T. "Records in the National Archives Relating to the Social Purposes and Results of the Operation of the Civilian Conservation Corps." *Social Science Review* 22 (Mar. 1948), 46-53. From 1933 to 1942.

P216 Pinkett, Harold T. *Preliminary Inventory of the Records of the Forest Service.* U.S. National Archives, Preliminary Inventory No. 18. Washington, 1949. iii + 17 pp. Describes records from 1882 to 1946.

P217 Pinkett, Harold T. "Gifford Pinchot and the Early Conservation Movement in the United States." Ph.D. diss., American Univ., 1953. 323 pp.

P218 Pinkett, Harold T. "Records of the First Century on the Interest of the United States Government in Plant Industries." *Agricultural History* 29 (Jan. 1955), 38-45. Includes some description of records concerning the history of forest diseases.

P219 Pinkett, Harold T. "Gifford Pinchot at Biltmore." *North Carolina Historical Review* 34 (July 1957), 346-57. On his management of an experimental forest on the estate of George W. Vanderbilt near Asheville, North Carolina, 1892-1898.

P220 Pinkett, Harold T. "Gifford Pinchot, Consulting Forester, 1893-1898." *New York History* 39 (Jan. 1958), 34-49. On his career as a consulting forester in New York.

P221 Pinkett, Harold T. "The Forest Service, Trail Blazer in Recordkeeping Methods." *American Archivist* 22 (Oct. 1959), 419-26. Since 1898.

P222 Pinkett, Harold T. "Gifford Pinchot, Forester-Politician." *Pennsylvania Forests* 51 (Jan.-June 1961), 124-25. His Pennsylvania career.

P223 Pinkett, Harold T. "The First Federal 'Expert in Forest History'." *Forest History* 6 (Fall 1962), 10. Treadwell Cleveland, Jr., a Pinchot protégé and publicist of forestry in the Division of Forestry, 1900s.

P224 Pinkett, Harold T. "Early Records of the U.S. Department of Agriculture." *American Archivist* 25 (Oct. 1962), 407-16. 19th century.

P225 Pinkett, Harold T. "The Archival Product of a Century of Federal Assistance to Agriculture." *American Historical Review* 69 (Apr. 1964), 689-706.

P226 Pinkett, Harold T. "The Keep Commission, 1905-1909: A Rooseveltian Effort for Administrative Reform." *Journal of American History* 52 (Sept. 1965), 297-312. Gifford Pinchot and other conservationists were involved.

P227 Pinkett, Harold T. "Federal Agricultural Records: Preserving the Valuable Core." *Agricultural History* 42 (Apr. 1968), 139-46.

P228 Pinkett, Harold T. "Forest Service Records as Research Material." *Forest History* 13 (Jan. 1970), 18-29. Describes USFS records at the National Archives and at the regional federal records centers.

P229 Pinkett, Harold T. *Gifford Pinchot: Private and Public Forester.* Urbana: University of Illinois Press, 1970. 167 pp. Illus., notes, bib., index. On Pinchot's career in forestry and the conservation movement, 1890s-1940s.

P230 Pintarich, Paul. "A Man Called Finley." *American Forests* 80 (May 1974), 12-15. On William Lovell Finley (1876-1953), Oregon naturalist, wildlife conservationist, and friend of Theodore Roosevelt.

P231 Pirie, F.W. "Changes Since the Early Days in New Brunswick Lumbering." *Canada Lumberman* 60 (Aug. 1, 1940), 103, 106.

P232 Pirsko, Arthur R. "The History, Development, and Current Use of Forest Fire Danger Meters in the U.S. and Canada." Master's thesis, Univ. of Michigan, 1950.

Pisani, Donald J. See Jackson, William Turrentine, #J18, J19

P233 Pittman, F.B. "Mississippi's Forests of Yesteryear." *Forest Farmer* 22 (Apr. 1963), 24-26.

P234 Pitzer, Paul C. "Hamlin Garland and Burton Babcock." *Pacific Northwest Quarterly* 56 (Apr. 1965), 86-88. On Garland's futile efforts to aid his friend Babcock in retrieving land in Washington lost through failure to observe ownership provisions of the Forest Homestead Act of 1906.

P235 Place, Howard, and Place, Marian T. *The Story of Crater Lake National Park*. Caldwell, Idaho: Caxton Printers, 1974. 84 pp. Illus., maps. A history of the region, especially from discovery in 1853 to establishment of Crater Lake National Park, Oregon, in 1902.

P236 Place, I.C.M., and Thomson, C.C. *History and Description of the Acadia Forest Experiment Station.* Canadian Pulp and Paper Association, Woodlands Section, Index No. 1333. Montreal, 1953. 17 pp.

P237 Place, Marian T. [Dale White], and Florek, Larry. *Tall Timber Pilots*. New York: Viking Press, 1953. 222 pp. Illus. On the Johnson Flying Service of Missoula, Montana, organized by Bob and Dick Johnson, and their contracts with the USFS and lumber companies in Montana and Idaho since 1923. Includes reminiscences by Florek, one of the pilots.

P238 Place, Marian T. [Dale White]. "Cavalcade to Hell." *American Heritage* 5 (Spring 1954), 40-43, 56-57. On the Washburn expedition of 1870, which contributed to the creation of Yellowstone National Park.

P239 Place, Marian T. [Dale White]. *Gifford Pinchot: The Man Who Saved the Forests*. New York: Julian Messner, 1957. 192 pp. Bib., index. A popular account of Pinchot's career in forestry and conservation.

Place, Marian T. See Place, Howard, #P235

P240 Plair, T.B. "Snohomish Forestry." *Journal of Forestry* 47 (Nov. 1949), 876-81. On cooperative farm forestry in Snohomish County, Washington, between farm woodland owners and tree farmers (Washington Forest Products Cooperative Association) and lumber mill owners (Cascade Forest Industries, Inc.) since 1938.

P241 Plair, T.B. "John Frederick Preston." *Journal of Forestry* 49 (July 1951), 528-29. On Preston's career with the USFS, 1907-1920s, and the Forestry Division of the Soil Conservation Service, 1936-1947.

P242 Plair, T.B. "How the CCC Has Paid Off." *American Forests* 60 (Feb. 1954), 28-30, 44-45. On the results of tree planting programs of the Civilian Conservation Corps, 1930s.

P243 *Planning and Civic Comment*. "State Parks Growing in Importance and Numbers—The National Conference on State Parks." *Planning and Civic Comment* 25 (June 1959), 5-6. Since 1921.

Plant, Charles. See Peterson, Virgil G., #P157

P244 Plant, Charles; Stilson, Harold L.; and Smith, Paul R. "Red Cedar Shingles & Shakes: The Labor Story." OHI by Elwood R. Maunder. Ed. by Betty E. Mitson and Barbara D. Holman. *Journal of Forest History* 19 (July 1975), 112-27. On working conditions and the role of labor in the industry in the Pacific Northwest and British Columbia.

P245 Platt, Rutherford. *Wilderness—The Discovery of a Continent of Wonder*. New York: Dodd, Mead & Company, 1961. 310 pp. On the wilderness regions successively encountered in the exploration and settlement of North America.

P246 Platt, Rutherford. *The Great American Forest.* Englewood Cliffs, New Jersey: Prentice-Hall, 1965. xii + 271 pp. Illus., map, index. Mostly natural history of North American forests; includes some reminiscences of forest exploration and some history of forest policy.

P247 Platt, Virginia B. "Tar, Staves, and New England Rum: The Trade of Aaron Lopez of Newport, Rhode Island, with Colonial North Carolina." *North Carolina Historical Review* 48 (Jan. 1971), 1-22. Trade in naval stores, 1760s-1770s.

P248 Pleasants, Henry, Jr. *Three Scientists of Chester County.* West Chester, Pennsylvania: H.F. Temple, 1936. 49 pp. Three leaders of early Pennsylvania forestry: Humphrey Marshall (1722-1801), William Darlington (1782-1863), and Joseph T. Rothrock (1839-1922).

P249 Plowman, Amon B. "The Work of the Spoilers; How the Finest Hardwood Forests on the Continent, in Western Ohio, Have Been Ravished." *Forestry and Irrigation* 14 (July 1908), 363-69.

P250 Plummer, Fred G. *Forest Fires: Their Causes, Extent and Effects, with a Summary of Recorded Destruction and Loss.* USFS Bulletin No. 117. Washington: GPO, 1912. 39 pp. Data on forest fire causes and damages since 1880s; includes tables broken down by state.

Plusnin, Basil A. See Meyer, Walter H., #M430

P251 Poatgieter, A. Hermina. *Pioneer Lumbering*. Gopher Historian Leaflet Series, No. 2. St. Paul: Minnesota Historical Society, 1969. 15 pp.

P252 Pohl, Thomas Walter. "Seattle, 1851-1861: A Frontier Community." Ph.D. diss., Univ. of Washington, 1970. 367 pp. Includes some history of the lumber industry.

P253 Polenberg, Richard. "Conservation and Reorganization: The Forest Service Lobby, 1937-38." *Agricultural History* 39 (Oct. 1965), 230-39. Analyzes the USFS opposition to the proposal that the USDI should become the Department of Conservation. Fearing that the USFS would be transferred to the new agency, professional foresters joined conservation groups, farm organizations, the lumber industry, and grazing interests to block the plan.

P254 Polenberg, Richard. *Reorganizing Roosevelt's Government: The Controversy over Executive Reorganization, 1936-1939*. Cambridge: Harvard University Press, 1966. viii + 275 pp. Bib., notes, index. Includes an account of efforts to create a Department of Conservation, and the successful opposition of the USFS, lumber industry, and forestry and conservation groups.

P255 Polenberg, Richard. "The Great Conservation Contest." *Forest History* 10 (Jan. 1967), 13-23. Details the struggle for the USFS waged between the USDA and the USDI from 1933 to 1941. Major elements in the clash were the ambitions of Secretary of the Interior Harold Ickes, Franklin D. Roosevelt's administrative procedures, and New Deal interest in expanding conservation activities.

P256 Poli, Adon, and Baker, H.L. *Ownership and Use of Forest Land in the Redwood-Douglas Fir Subregion of California*. Technical Paper No. 7. Berkeley: USFS, California Forest and Range Experiment Station, 1954. 76 pp.

P257 Poli, Adon. "Ownership and Use of Forest Land in Northwestern California." *Land Economics* 32 (May 1956), 144-51. Since 1848.

P258 Polkinghorn, R.S. *Pino Grande: Logging Railroads of the Michigan-California Lumber Company*. Berkeley: Howell-North Books, 1966. 144 pp. Illus., maps, diags., index. Illustrated history of the logging railroads of the Michigan-California Lumber Company and its predecessors in California's Sierra Nevada—American River Land and Lumber Company and El Dorado Lumber Company, 1888-1951.

P259 Polley, C.A. "Lumber and Timber." In *Lawrence County*, ed. by Mildred Fielder. Lead, South Dakota: N.p., 1960. Pp. 115-23. In Lawrence County and the Black Hills of South Dakota since 1876.

P260 Pollock, George Freeman. *Skyland: The Heart of the Shenandoah National Park.* Ed. by Stuart E. Brown, Jr. N.p., 1960. xv + 283 pp. Illus., notes. On the author's summer resort at Stony Man Mountain, 1894-1937, and his part in the establishment of the Shenandoah National Park, Virginia.

P261 Polos, Nicholas C. "John Muir, a Stranger in the Southland." *Pacific Historian* 18 (Summer 1974), 2-12. On Muir's travels and impressions of the San Gabriel Mountains of southern California in 1877.

P262 Pomerleau, Rene. "History of the Dutch Elm Disease in the Province of Quebec, Canada." *Forestry Chronicle* 37 (Dec. 1961), 356-67. Since 1940.

P263 Pomeroy, Earl S. *In Search of the Golden West: The Tourist in Western America.* New York: Alfred A. Knopf, 1957. xii + 233 + vi pp. Illus., notes. On the growth of tourism in the West, including the movement to preserve scenery and wilderness, the influence of the automobile, and the growing popularity of camping, outdoor recreation, and the national parks.

P264 Pomeroy, Earl S. *The Pacific Slope: A History of California, Oregon, Washington, Idaho, Utah, and Nevada.* New York: Alfred A. Knopf, 1965. Reprint. Seattle: University of Washington Press, 1973. xv + 412 + xvii pp. Illus., maps, notes, notes on further reading, index. An interpretive history with many references to lumbering, conservation, and the changing patterns of outdoor recreation and tourism.

P265 Pomeroy, Kenneth B. "Modern Trends in an Ancient Industry." *Journal of Forestry* 50 (Apr. 1952), 297-99. On research and woods operations in the gum naval stores industry since ca. 1900.

P266 Pomeroy, Kenneth B. "Pinpointing the Fire Problem." *American Forests* 62 (Oct. 1956), 50-52, 54. On Dr. John P. Shea's studies to determine the reasons for incendiary fires in the South, 1938-1939, and the subsequent progress in prevention and suppression of man-made fires.

P267 Pomeroy, Kenneth B. "Origins and Objectives of the American and Pennsylvania Forestry Associations." *Pennsylvania Forests* 39 (Summer 1959).

P268 Pomeroy, Kenneth B. "The American Forestry Association." *Southern Lumberman* 193 (Dec. 15, 1956), 208-10. Since 1875.

P269 Pomeroy, Kenneth B., "Time to 'Pull a New Streak'." *American Forests* 66 (Dec. 1960), 2, 45-46. On the production of gum naval stores from Southern pine since 1934.

P270 Pomeroy, Kenneth B., and Yoho, James G. *North Carolina Lands: Ownership, Use, and Management of Forest and Related Lands.* Washington: American Forestry Association, 1964. xx + 372 pp. Illus., maps, tables, apps., notes, bib., index. Includes much history of forest ownership, management, industries, and conservation since the colonial period.

Pomeroy, Kenneth B. See Dana, Samuel Trask, #D32

P271 Pommer, Patricia J. "Plank Roads: A Chapter in the Early History of Wisconsin Transportation (1846-1871)." Master's thesis, Univ. of Wisconsin, 1950.

P272 Pool, Raymond John. "A Brief Sketch of the Life and Work of Charles Edwin Bessey." *American Journal of Botany* 2 (Dec. 1915), 505-18. Bessey (1845-1915), a University of Nebraska botanist, led the movement to establish the Nebraska National Forest and was involved in forestry and conservation activities elsewhere in the nation.

P273 Pool, Raymond John. "Fifty Years on the Nebraska National Forest." *Nebraska History* 34 (Sept. 1953), 139-79. On the development of Nebraska's "man-made" national forest since 1902.

Pope, Jennie Barnes. See Albion, Robert G., #A47

P274 Popham, W.L. "Development of the Cooperative Effort in Forest Pest Control." *Journal of Forestry* 48 (May 1950), 321-23. Advances since passage of the Forest Pest Control Act of 1947.

P275 Popoff, Nicholas. "A Study of Management Policies and Practices of the Bureau of Indian Affairs on Lands Held in Federal Trust on the Quinault Indian Reservation." Master's thesis, Univ. of Washington, 1970. Washington.

P276 Porter, Earl. "James Francis Dubar: A Man and a School." *Journal of Forestry* 53 (Aug. 1955), 590-91. New York State Ranger School.

P277 Porter, Earl. "A History of Forest Fire Prevention on International Paper Company Southern Timberlands in the Past 35 Years." *Journal of Forestry* 66 (Aug. 1968), 619-21.

P278 Porter, Henry Holmes. *A Short Autobiography.* Chicago: R.R. Donnelley, 1915. 40 pp. Porter (1835-1910) was involved in the lumber business in Illinois.

Porter, Jane M. See Baker, Gladys L., #B27

P279 Porter, Kenneth Wiggins. "John Jacob Astor and the Sandalwood Trade of the Hawaiian Islands, 1816-1828." *Journal of Economic and Business History* 2 (May 1930), 495-519. On Astor's ships and agents in the Pacific lumber trade.

P280 Porter, Marjorie L. *Lem Merrill, Surveyor-Conservationist.* Plattsburgh, New York: Clinton Press, 1944. 45 pp. Illus. Elmer Marcellus Merrill of New York.

P281 Porter, Marjorie L. "Old Time Uses of Wood." *Northeastern Logger* 9 (June 1961), 20, 55. Sketch of various domestic and industrial uses of wood in the 19th century.

P282 Porter, William W., II. "The Public Domain—Heart of the Republic." *American Forests* 72 (Jan. 1966), 12-15, 46-48; 72 (Feb. 1966), 26-29, 46-51; 72 (Mar. 1966), 34ff.; 72 (Apr. 1966), 28-31, 53-59. A general history of the public domain and federal land policy.

Porterfield, Nolan. See Swank, Roy, #S752

P283 Porteus, Alexander. *Forest Folklore, Mythology, and Romance.* New York, 1928.

P284 Potter, Barrett George. "The Civilian Conservation Corps in New York State: Its Social and Political Impact (1933-1942)." Ph.D. diss., State Univ. of New York at Buffalo, 1973. 278 pp.

P285 Potter, G.R.L. "Our Forest Heritage." *Canadian Banker* 60 (Autumn 1953), 50-60; 61 (Winter 1954), 36-48.

P286 Potter, Loren D. "North Dakota's Heritage of Pine." *North Dakota History* 19 (July 1952), 157-66. On stands of ponderosa pine in the valley of the Little Missouri River, 1877-1917.

P287 Potter, Neal, and Christy, Francis T., Jr. *Trends in Natural Resource Commodities: Statistics of Prices, Output, Consumption, Foreign Trade, and Employment in the United States, 1870-1957.* Baltimore: Johns Hopkins Press, for Resources for the Future, 1962. ix + 568 pp. Includes a section on lumber and timber.

P288 Potts, Merlin K. "Rocky Mountain National Park." *Colorado Magazine* 42 (Summer 1965), 217-23. On the history of the park since its establishment in 1915, including the promotional work of Enos A. Mills.

P289 Pottsmith, Marie Holst. "Pioneering Years in Hamlet, Oregon: A Finnish Community." *Oregon Historical Quarterly* 61 (Mar. 1960), 5-45. Reminiscence and history of a Finnish logging community in Clatsop County, Oregon, early 20th century.

P290 Potzger, John E.; Potzger, Margaret E.; and McCormick, Jack. "The Forest Primeval of Indiana As Recorded in the Original U.S. Land Surveys, and an Evaluation of Previous Interpretations of Indiana Vegetation." *Butler University Botanical Studies* 13 (Dec. 1956), 95-111. Land surveys were made between 1799 and 1846.

Potzger, John E. See Blewett, Marilyn B., #B314

Potzger, John E. See Rohr, Fred W., #R276

Potzger, Margaret E. See Potzger, John E., #P290

P291 Pouliot, L.J. *Canada's Pulp and Paper Industry: A Review of Its Growth, Production and Exports.* Ottawa: Department of Trade and Commerce, 1945. 24 pp.

P292 Powell, Fred W. *The Bureau of Plant Industry—Its History, Activities and Organization.* Institute for Government Research, Service Monographs of the United States Government, No. 47. Baltimore: Johns Hopkins University Press, 1927. Reprint. New York: AMS Press, 1973. xii + 121 pp.

P293 Powell, John Wesley. *Report on the Lands of the Arid Region of the United States, With a More Detailed Account of the Lands of Utah.* Ed. with intro. by Wallace Stegner. Cambridge: Belknap Press of Harvard University Press, 1962. xxvii + 202 pp. Maps, tables, notes, index. Powell's report of 1878 on irrigation possibilities and proposed land reforms for the West.

P294 Powers, Alfred. *Redwood Country: The Lava Region and the Redwoods.* New York: Duell, Sloan & Pearce, 1949. xvii + 292 pp. Map. Incidental history of lumbering in California.

P295 Powers, Ralph A. *The Old Stone Mill on the Oxoboxo: Robertson Paper and Boxes since 1850.* New York: Newcomen Society in North America, 1957. 24 pp. Illus. On the Robertson Paper Company and the Robertson Paper Box Company, New London, Connecticut, since 1850. The founder was Carmichael Robertson (1828-1880).

P296 Prater, Leland J. "Historical Forest Service Photo Collection." *Journal of Forest History* 18 (Apr. 1974), 28-31. On the USFS photo collection (500,000 + negatives), covering the period from 1896 to 1966, and the author's effort to catalogue the collection prior to permanent deposit in the National Archives.

P297 Pratt, Edward E. *The Export Lumber Trade of the United States.* U.S. Bureau of Foreign and Domestic Commerce, Miscellaneous Series, No. 67. Washington: GPO, 1918. 117 pp. Illus., maps, tables. Includes incidental history of trade organization and methods in the export of lumber and logs.

P298 Pratt, George DuPont. "The Use of the New York State Forests for Public Recreation." *Proceedings of the Society of American Foresters* 11 (July 1916), 281-85.

P299 Pratt, Joseph Hyde. "Twelve Years of Preparation for the Passage of the Weeks Law." *Journal of Forestry* 34 (Dec. 1936), 1028-32. On the movement to create national forests in the Appalachian Mountains, 1899-1911.

P300 Pratt, M.B. "The California State Forest Service: Its Growth and Its Objectives." *Journal of Forestry* 29 (Apr. 1931), 497-504. On the history of state forestry in California since 1885.

P301 Prebble, M.L. "Development of Forest Insect Research and Control in Canada." *Journal of Forestry* 57 (Apr. 1959), 255-59. Forest entomology in Canada since 1909.

Preiss, Jack J. See Lively, Charles E., #L216

Prentice, Lee. See Blanchard, Louis, #B308

P302 Prentis, Henning W., Jr. *Thomas Morton Armstrong (1836-1908): Pioneer in Cork.* New York: Newcomen Society in North America, 1950. 32 pp. Illus. On the Armstrong Cork Company, Lancaster, Pennsylvania, founded in Pittsburgh in 1856.

P303 Prentiss, A.M. "Our Military Forests." *American Forests and Forest Life* 31 (Dec. 1925), 744-48. On the establishment of "national forests" on American military reservations in the 1920s.

P304 Prescott, DeWitt C. *The Evolution of Modern Band Saw Mills for Sawing Logs.* Menominee, Michigan: Prescott Company, 1910. 49 pp. Illus. Since 1880.

P305 Prescott, Philander. "Autobiography and Reminiscences of Philander Prescott." *Collections of the Minnesota Historical Society* 6 (1894), 475-91. Incidental history of lumbering in Minnesota.

P306 Preston, John F. "Control of Bark Beetles on the National Forests." *Journal of Forestry* 23 (Jan. 1925), 49-61. Many incidental historical references to bark beetle epidemics and a general account of USFS insect control work since 1906.

P307 Preston, John F. "Silvicultural Practice in the United States during the Past Quarter Century." *Journal of Forestry* 23 (Mar. 1925), 236-44.

P308 Preston, John F. "Lessons from the Farm Forestry Projects." *Journal of Forestry* 44 (Jan. 1946), 9-12. On farm forestry demonstration projects conducted by the Soil Conservation Service since 1935.

P309 Preston, John F. *Farm Wood Crops.* New York: McGraw-Hill, 1949. viii + 302 pp. Illus. Incidental history.

P310 Preston, John F. *Developing Farm Woodlands.* New York: McGraw-Hill, 1954. 386 pp. Incidental history.

P311 Preston, R.J. "History and Development of Instructional Programs in Wood Technology." *Society of American Foresters' Meeting, Proceedings, 1948* (1949), 2-11.

P312 Prestridge, J.A. "Cypress from Ancient to Modern Times." *Southern Lumberman* 193 (Dec. 15, 1956), 166-69. On its use and production since 1577, with an account of the Southern Cypress Manufacturers Association since 1905.

P313 Priaulx, Arthur W. "The Story of Keep Oregon Green." *Journal of Forestry* 48 (Feb. 1950), 87-91. On an industrially sponsored forest fire prevention program since 1939.

P314 Priaulx, Arthur W. "Oregon's Planted Cottonwoods." *American Forests* 58 (Oct. 1952), 10-11, 66-70. On the planting of northern black cottonwood along the Willamette River by the Willamette Pulp and Paper Company since 1893.

P315 Price, F.A. "At Great Lakes Paper: A Classic Example of Planned Expansion." *Pulp and Paper Magazine of Canada* 58 (Dec. 1957), 110-19. Development of Great Lakes Paper Company, Ltd., Fort William (Thunder Bay), Ontario, since 1936.

P316 Price, F.A. "Fifty Years Progress at Grand Falls, the Impact of Anglo-Newfoundland Development Co. Ltd., on the Economy of Newfoundland." *Pulp and Paper Magazine of Canada* 60 (Oct. 1959), 69-148. A pulp and paper plant since 1909.

P317 Price, J.H. "The Clarke-McNary Act and Federal Responsibility in California's State Forestry Program." *Journal of Forestry* 29 (May 1931), 731-36. Includes some history of federal-state cooperation in forestry in California since 1912, and on the impact of the Clarke-McNary Act of 1924.

P318 Price, Jenny Ellsworth. "Duane Bliss and the Development of the Lake Tahoe Region." *Daughters of the American Revolution Magazine* 92 (Mar. 1958), 245-52. Including his involvement in the lumber industry near Lake Tahoe, California-Nevada, late 19th century.

P319 Price, Overton W., and others. *The Forests of the United States: Their Use.* USFS, Circular No. 171. Washington: GPO, 1909. 25 pp. Incidental history.

P320 Price, Overton W. *The Land We Live In: The Book of Conservation.* Boston: Small, Maynard and Company, 1911. viii + 242 pp. Illus., index. Includes some popular history of the conservation movement.

P321 Price, Overton W. "Geo. W. Vanderbilt, Pioneer in Forestry." *American Forestry* 20 (June 1914), 420-25. On his Biltmore Estate near Asheville, North Carolina, where private forestry was first practiced in the 1890s, and on Vanderbilt's general support of the forestry movement.

P322 Price, Paul Holland. "The Evolution of the Geological Survey." *West Virginia History* 13 (Oct. 1951), 20-32. In West Virginia since 1897.

P323 Price, R.E. "Naval Stores Industry in Southwest Louisiana." *McNeese Review* 2 (1949), 9-12. On Crosby Chemicals, De Ridder, Louisiana, manufacturer of products from pine stumps, 1940s.

P324 Price, Robert. *Johnny Appleseed: Man and Myth.* Bloomington: Indiana University Press, 1954. xv + 320 pp. Illus., maps, notes, bib. On the life of John Chapman (1774-1845) and the myths and traditions associated with him.

Price, Victoria. See Haskell, Elizabeth H., #H191

P325 Pride, Fleetwood. *Fleetwood Pride, 1864-1960: The Autobiography of a Maine Woodsman.* Ed. by Edward D. Ives and David C. Smith. Northeast Folklore, Volume 9. Orono: Northeast Folklore Society, University of Maine, 1968. 60 pp. Illus., notes, bib. Pride was a Maine river driver and lumberman.

P326 Primack, Martin L. "Land Clearing Under Nineteenth Century Techniques: Some Preliminary Calculations." *Journal of Economic History* 22 (Dec. 1962), 484-97. Includes some history of girdling and controlled burning and cost estimates of clearing forested lands for agriculture. Comments follow on pp. 516-19.

P327 Prince, Gilbert H. "Forestry Administration in New Brunswick." *Journal of Forestry* 20 (Jan. 1922), 54-61. Includes some history of forestry and of the New Brunswick Forest Service, particularly since 1908.

P328 Pringle, Henry F. *Theodore Roosevelt, a Biography.* New York: Harcourt, Brace, 1931. 627 pp. Illus. See for an account of Roosevelt and the conservation movement.

P329 Pringle, Henry F. *The Life and Times of William Howard Taft: A Biography.* 2 Volumes. New York: Farrar and Rinehart, 1939. Illus. See for an account of the Ballinger-Pinchot controversy of 1910 and for reference to other conservation issues.

P330 Pritchett, Charles Herman. *The Tennessee Valley Authority: A Study in Public Administration.* Chapel Hill: University of North Carolina Press, 1943. xiii + 333 pp. Illus., diags. A general history of the TVA, including reference to its natural resource and conservation programs.

Proctor, Lewis A. See Aikens, Andrew J., #A37

Proctor, P.B. See Panshin, Alexis John, #P26

P331 Progulske, Donald R. *Yellow Ore, Yellow Hair, Yellow Pine: A Photographic Study of a Century of Forest Ecology.* South Dakota State University, Agricultural Experiment Station, Bulletin 616. Brookings, 1974. 169 pp. Illus., maps, bib., apps., tables. Photographs made in 1874 by the Custer expedition are compared with contemporary shots to demonstrate changes in the forest ecology of South Dakota's Black Hills.

P332 Prokop, John. "Do We Need Another CCC?" *American Forests* 68 (Mar. 1962), 20-23, 48-51. An argument for a Youth Conservation Corps—based in part on some history of the Civilian Conservation Corps of the 1930s.

P333 Pross, A. Paul. "The Development of a Forest Policy: A Study of the Ontario Department of Lands and Forests." Ph.D. diss., Univ. of Toronto, 1967.

Pross, A. Paul. See Lambert, Richard S., #L21

P334 Pross, A. Paul. "The Development of Professions in the Public Service: The Foresters in Ontario." *Canadian Public Administration* 10 (Sept. 1967), 376-404.

P335 Pross, A. Paul, ed. "Historical Memorandum on the Management of the Crown Lands . . . ." *Forest History* 15 (Apr. 1971), 22-29. Edited memorandum (1867) by Alexander J. Russell, chief officer of Crown Lands Department of United Canadas (Quebec and Ontario), on forests of the Ottawa Valley.

P336 Pross, A. Paul. "Input versus Withinput: Pressure Group Demands and Administrative Survival." In *Pressure Group Behaviour in Canadian Politics,* ed. by A. Paul Pross. Toronto: McGraw-Hill, 1975. Pp. 147-71. A case study of a policy dispute within the Ontario Department of Lands and Forests, 1941-1946.

P337 Proudfit, S.V. *Public Land System of the United States, Historical Outline.* Washington: GPO, 1924. 18 pp.

P338 Prout, Clarence. "Building Minnesota." *Conservation Volunteer* 18 (Sept. 1955), 26-35; 18 (Nov. 1955), 21-26; 19 (Jan. 1956), 37-42. Historical sketch of state forestry.

P339 Prucha, Francis P. *Broadax and Bayonet: The Role of the United States Army in the Development of the Northwest, 1815-1860.* Madison: State Historical Society of Wisconsin, 1953. Reprint. Lincoln: University of Nebraska Press, 1967. xii + 263 pp. Illus., notes, bib., index. Includes some references to lumbering in relation to army posts in Wisconsin, Minnesota, Iowa, and Illinois.

P340 Prunty, Merle C., Jr. "Recent Expansions in the Southern Pulp-Paper Industries." *Economic Geography* 32 (Jan. 1956), 51-57. Since 1946.

P341 Prunty, Merle C. "The Woodland Plantation as a Contemporary Occupance Type in the South." *Geographical Review* 53 (No. 1, 1963), 1-21. On the uses of private woodlands since the 1940s.

P342 Prunty, Merle C. "Some Geographic View of the Role of Fire in Settlement Processes in the South." *Tall Timbers Fire Ecology Conference, Proceedings* 4 (1965), 161-68.

P343 Pulliam, Grace. "Pioneer Days of Lumber Manufacturing in Territory of Washington." *West Coast Lumberman* 29 (1916), 81, 86.

P344 Pullman, Raymond. "Destroying Mt. Mitchell." *American Forestry* 21 (Feb. 1915), 83-93. On the devastation caused by logging and forest fires to this North Carolina mountain, and efforts to preserve it.

P345 *Pulp and Paper Magazine of Canada.* "Historical Notes on British Columbia's Pulp and Paper Industry." *Pulp and Paper Magazine of Canada* 49 (July 1948), 67-82. Brief histories of individual companies and a map indicating mill locations.

P346 *Pulp and Paper Magazine of Canada.* "Forty Years of Progress." *Pulp and Paper Magazine of Canada* 43 (July 1942), 574-622. Articles on technical progress in every phase of mill operations.

P347 *Pulp and Paper Magazine of Canada.* "Forestry Commissions in Canada." *Pulp and Paper Magazine of Canada* 59 (Nov. 1958), 160, 166, 168-76, 182-84. Incidental history.

P348 *Pulp and Paper Magazine of Canada.* "Dramatic Growth of Crown Zellerbach Canada Limited." *Pulp and Paper Magazine of Canada* 59 (Dec. 1958), 98-105.

P349 *Pulp and Paper Magazine of Canada.* "Logging at Iroquois Falls." *Pulp and Paper Magazine of Canada* 61 (Oct. 1960), 173-83. Abitibi Power and Paper Company, Iroquois Falls, Ontario.

P350 *Pulp and Paper Magazine of Canada.* "Technology's Time Capsule . . . From the Magazine, 1915-73." *Pulp and Paper Magazine of Canada* 75 (Jan. 1974), 28-34, 43-107 passim. A month-by-month chronology of events affecting the technology of the Canadian pulp and paper industry.

P351 Pursell, Carroll W., Jr. "Administration of Science in the Department of Agriculture, 1933-40." *Agricultural History* 42 (July 1968), 231-40. Includes incidental references to the USFS.

P352 Pursell, Carroll W., Jr., ed. *From Conservation to Ecology: The Development of Environmental Concern.* New York: Thomas Y. Crowell, 1973. 148 pp. Illus., bib. An anthology.

Pusateri, Samuel J. See White, John R., #W249

P353 Puter, Stephen Arnold Douglass, and Stevens, Horace. *Looters of the Public Domain.* Portland: Portland Printing House, 1908. Reprint. New York: Arno Press, 1972. 495 pp. Illus. Composed in jail by one of the more prominent "looters," this book discloses the techniques used by individuals and corporations to defraud the government of timberlands in the 19th and early 20th centuries, especially in the Pacific Northwest and California.

P354 Putman, John J. "Timber: How Much Is Enough?" *National Geographic Magazine* 145 (Apr. 1974), 484-511. Includes many incidental references to the history of forestry and the forest industries.

P355 Putnam, F. "What's the Matter with New England? I. Maine." *New England Magazine* 36 (1907), 515-40. On deforestation and the post-Civil War sale of public lands in Maine.

Putnam, J.A. See Winters, Robert K., #W376

Putnam, J.A. See Sternitzke, Herbert S., #S605

P356 Putnam, Theodore L. "Down the Rivers: A Rafting Journal of 1859—from Warren, Pa., to Louisville, Ky." Ed. by Ernest C. Miller. *Western Pennsylvania Historical Magazine* 40 (Fall 1957), 149-62. A record of the shipment of boards, shingles, and lath via the Allegheny and Ohio rivers.

P357 Pyke, Don O. "90 Years Old . . . and Going Strong." *American Paper Merchant* 43 (Apr. 1946), 26-27. Graham Paper Company, St. Louis, Missouri.

P358 Pyle, Joseph G. *The Life of James J. Hill.* 2 Volumes. Garden City, New York: Doubleday, Page & Co. 1917. Reprint. New York: Peter Smith, 1936. Illus. As owner of the Great Northern and Northern Pacific railroads, Hill had control of vast timberlands in Montana, northern Idaho, and Washington near the turn of the century.

P359 Pyles, Hamilton K. "Fire Prevention Strategy in the East." OHI by Susan R. Schrepfer. *Forest History* 16 (Oct. 1972), 22-23. On fire prevention programs promoted by the USFS's Division of Information and Education, 1950s.

P360 Pyles, Hamilton K. *Multiple Use of the National Forests.* OHI by Susan R. Schrepfer. Santa Cruz, California: Forest History Society, 1972. xi + 211 pp. Illus., apps., bib., index. Processed. On his career in the USFS from 1931 to 1966, involving forest engineering, fire control, and forest administration in California and the East. Pyles became deputy chief of the USFS.

Q1 Quaife, Milo M. "Increase Allen Lapham, Father of Forest Conservation." *Wisconsin Magazine of History* 5 (Sept. 1921), 104-08. Lapham, chairman of a special forestry commission created by the Wisconsin legislature, issued a report in 1867 showing the evil consequences of deforestation.

Quam, Louis Otto. See Foscue, Edwin J., #F173

Q2 Quick, Amy. "The History of Bogalusa: The 'Magic City' of Louisiana." Master's thesis, Louisiana State Univ., 1942. A center of the lumber industry.

Quick, Edward. See Quick, Herbert, #Q3

Q3 Quick, Herbert, and Quick, Edward. *Mississippi Steamboatin': A History of Steamboating on the Mississippi and Its Tributaries.* New York: H. Holt and Company, 1926. xiv + 342 pp. Illus., bib. See for reference to the use of fuel wood.

Quinney, Dean N. See Cunningham, Russell N., #C749

R1 Rabbit, J.C., and Rabbit, M.C. "The U.S. Geological Survey: 75 Years of Service to the Nation, 1879-1954." *Science* 119 (May 28, 1954), 741-58.

Rabbit, M.C. See Rabbit, J.C., #R1

R2 Raber, Oran. "The History of Shipmast Locust." *Journal of Forestry* 36 (Nov. 1938), 1116-19. On the introduction of this tree on Long Island, New York, probably ca. 1700.

R3 Rader, Benjamin G. "The Montana Lumber Strike of 1917." *Pacific Historical Review* 36 (May 1967), 189-207. On the labor troubles among lumberjacks and millworkers, including the role of the Industrial Workers of the World.

R4 Radtke, Leonard B. "Pioneer New Mexico Sawmill." *Timberman* 28 (June 1927), 192. On the Mescalero Indian Reservation, 1862-1896.

R5 Rae, John B. "Commissioner Sparks and the Railroad Land Grants." *Mississippi Valley Historical Review* 25 (Sept. 1938), 211-30. William A.J. Sparks, Commissioner of the General Land Office, 1885-1887.

R6 Rae, John B. "The Great Northern's Land Grant." *Journal of Economic History* 12 (Spring 1952), 140-45. On lands granted by Minnesota to the St. Paul and Pacific Railroad, 1857, later assumed by the Great Northern Railway.

R7 Rafferty, Milton D. "Missouri's Black Walnut Kernel Industry." *Missouri Historical Review* 63 (No. 2, 1969), 214-26. On a 20th-century industry based on the surviving black walnut forests of the Ozark Mountains.

R8 Rafferty, Milton D. "The Black Walnut Kernel Industry: The Modernization of a Pioneer Custom." *Pioneer America* 5 (Jan. 1973), 23-32.

R9 Raftery, John H. "Historical Sketch of Yellowstone National Park." *Annals of Wyoming* 15 (Apr. 1943), 101-32. Exploration and early history to the 1880s.

R10 Raines, William McLeod, and Barnes, Will Croft. *Cattle, Cowboys, and Rangers.* New York: Grosset and Dunlap, 1930. Incidental history of the USFS and grazing policy in the West.

R11 Rajala, Ben. *The Saga of Ivar Rajala: Logging in Busticagan Township, Bigfork River Valley.* Grand Rapids, Minnesota: Ensign Press, 1972. 40 pp. Illus. Biographical sketch of a Finnish immigrant lumberman in Minnesota since 1910.

R12 Rakestraw, Lawrence. "Uncle Sam's Forest Reserves." *Pacific Northwest Quarterly* 44 (Oct. 1953), 145-51. On Western attitudes toward the federal forest reserves, 1891-1910.

R13 Rakestraw, Lawrence. "A History of Forest Conservation in the Pacific Northwest, 1891-1913." Ph.D. diss., Univ. of Washington, 1955. 358 pp.

R14 Rakestraw, Lawrence. "Urban Influences on Forest Conservation." *Pacific Northwest Quarterly* 46 (Oct. 1955), 108-13. A study of urban groups—municipal officials, recreational organizations, and university faculties—and their support for federal forest reserves in the West, 1891-1905.

R15 Rakestraw, Lawrence. "The West, States' Rights, and Conservation: A Study of Six Public Lands Conferences." *Pacific Northwest Quarterly* 47 (July 1957), 89-99. Argues that the six conferences studies—Denver (1907), St. Paul (1910), Denver (1913), Washington (1913), Denver (1914), and Seattle (1915)—were not spontaneous in origin, that they revealed a variety of Western attitudes on conservation, and that states' rights were not synonymous with anticonservation.

R16 Rakestraw, Lawrence. "Sheep Grazing in the Cascade Range: John Minto vs. John Muir." *Pacific Historical Review* 27 (Nov. 1958), 371-82. Between 1897 and 1900 John Muir opposed John Minto in the latter's effort to open the Cascade Range Forest Reserve of Oregon to sheep grazing. Although a compromise was worked out, the struggle over grazing foreshadowed a split between preservationists and utilitarians in the conservation movement.

R17 Rakestraw, Lawrence. "Forest Missionary: George Patrick Ahern, 1894-1899." *Montana, Magazine of Western History* 9 (Oct. 1959), 36-44. On Ahern's contributions to forest protection, forest management, and forestry education in Montana.

R18 Rakestraw, Lawrence. "Before McNary: The Northwestern Conservationist, 1889-1913." *Pacific Northwest Quarterly* 51 (Apr. 1960), 49-56. On the local leaders of the forest conservation movement in the Pacific Northwest.

R19 Rakestraw, Lawrence. "George Patrick Ahern and the Philippine Bureau of Forestry, 1900-1914." *Pacific Northwest Quarterly* 58 (July 1967), 142-50. Ahern (1859-1942) was involved in forest conservation and forestry education in Montana in the 1890s, prior to his work in the Philippines.

R20 Rakestraw, Lawrence. "Conservation Historiography: An Assessment." *Pacific Historical Review* 41 (Aug. 1972), 271-88. An analysis of recent historical writing on the subject of conservation.

R21 Radestraw, Lawrence. "Forest History in Alaska: A Regional Approach." In *Writing Alaska's History: A Guide to Research, Volume 1.* Ed. by Robert A. Frederick and Patricia A. Jelle. Anchorage: Alaska Historical Commission and Anchorage Higher Education Consortium Library, 1974. Pp. 121-28. An exploratory article which considers four topics: the Indian as conservationist; early concern for environmental, historic, and scenery preservation; cooperation and conflict among federal and state resource agencies, especially the USFS and National Park Service; and the role of women in forestry and conservation.

R22 Ramsdell, W.F. *Township Government and the Exploitation of Timber and Wild Land Resources in Northern Michigan.* Detroit: Michigan Commission of Inquiry into County, Township, and School District Government, 1933. 49 pp. Includes some history of the lumber industry and forest depletion in northern Michigan.

R23 Ramsden, H.F. "Down a Lumber Sluice." *Northwestern Lumberman* 1 (July 1974), 30-34. On the use of flumes for transportation of logs in California's Sierra Nevada; reprinted from *Wood and Iron,* March 1888.

R24 Rand, Ernest. "The Twentieth Engineers: A Forestry Regiment in World War I." *Northeastern Logger* 11 (Feb. 1963), 12-13, 28-29, 32-33.

R25 Randall, Charles E. "The Centennial of John Muir, Man of the Mountains." *American Forests* 44 (Apr. 1938), 162-64, 190-92. Biographical sketch of Muir (1838-1914), ardent defender and interpreter of wilderness, especially in California.

R26 Randall, Charles E., and Heisley, Marie Foote. *Our Forests: What They Are and What They Mean to Us.* USDA, Miscellaneous Publication No. 162. Washington: GPO, 1933. 34 pp. Illus., map. Includes a brief historical sketch of federal and state forestry; there are later editions in 1940 and 1944.

R27 Randall, Charles E., and Edgerton, Daisy Priscilla. *Famous Trees.* USDA, Miscellaneous Publication No. 295. Washington: GPO, 1938. 115 pp. Illus., bib. On historic, memorial, and record-sized or shaped trees throughout the United States.

R28 Randall, Charles E. "Erle Kauffman." *Journal of Forestry* 48 (Apr. 1950), 284-85. On Kauffman's career with the USFS, as a writer, and on the editorial staff of *American Forests,* since 1927.

R29 Randall, Charles E. "Hough: Man of Approved Attainments." *American Forests* 67 (May 1961), 10-11, 41-44. On Franklin B. Hough (1822-1885), who became in 1876 the first federal forestry official and served as head of the Division of Forestry, 1881-1883.

R30 Randall, Charles E. "George Perkins Marsh: Conservation's Forgotten Man." *American Forests* 71 (Apr. 1965), 20-23. Marsh (1801-1882), author of *Man and Nature* (1864), significantly influenced the conservation movement in the 19th century.

R31 Randall, Charles E. "Them Were the Good Old Days." *American Forests* 73 (May 1967), 26-29, 61-63. Working conditions and practices in the USFS, ca. 1905.

R32 Randall, Charles E. "In Praise of the Mule." *American Forests* 74 (July 1968), 24-25, 37-41. On the use of mules as pack animals on national forests since 1920.

R33 Randall, Charles E., ed. *Enjoying Our Trees.* Washington: American Forestry Association, 1969. v + 122 pp. Illus. Brief articles about individual species. Some are historical; all were previously published in *American Forests.*

R34 Randall, Charles E. "Shasta-Trinity: Life-Giving Forests." *American Forests* 80 (May 1974), 8-11, 52-54. Some history of the country in and near the Shasta and Trinity national forests in northern California, since the mid-19th century.

Randall, James G. See Browning, Orville Hickman, #B514

R35 Randall, Peter. *Mount Washington: A Guide and Short History.* Hanover, New Hampshire: University Press of New England, 1974. Illus., maps, bib., index. Mount Washington, New Hampshire, on the White Mountain National Forest.

R36 Randall, Roger. *Labor Relations in the Pulp and Paper Industry of the Pacific Northwest.* Portland: Northwest Regional Council, 1942. 107 pp. Includes some history of the International Brotherhood of Pulp, Sulphite and Paper Mill Workers and the International Brotherhood of Paper Makers.

Randall, Roger. See Kerr, Clark, #K101, K102

R37 Randall, Robert Henry. "Federal Surveys: Coast and Geodetic, Geological, etc." *Surveying and Mapping* 18 (Apr.-June 1958), 207-12. Since 1879.

R38 Randall, Thomas E. *History of Chippewa Valley.* Eau Claire, Wisconsin: Free Press Print, 1875. 207 pp. Includes some history of lumbering in the Wisconsin Valley.

R39 Rane, F.W. "The Evolution of American Forestry." *Ohio Forester* 7 (No. 2, 1915), 3-6.

R40 Raney, William F. "The Timber Culture Acts." *Mississippi Valley Historical Association, Proceedings* 10 (1919-1920), 219-29. On the Timber Culture Act of 1873 and its amendments in 1874, 1876, and 1878.

R41 Raney, William F. "Pine Lumbering in Wisconsin." *Wisconsin Magazine of History* 19 (Sept. 1935), 71-90. From 1835 to 1911.

R42 Raney, William F. *Wisconsin: A Story of Progress.* New York: Prentice-Hall, 1940. xvii + 554 pp. Includes a chapter on the history of the lumber industry, 1830s to ca. 1910.

R43 Ranger, Dan, Jr. *Pacific Coast Shay: Strong Man of the Woods.* San Marino, California: Golden West Books, 1964. 103 pp. Illus. The Shay geared locomotive was used in logging operations from the 1880s until the 1940s.

R44 Ranger, Dan, Jr. "Shay: The Folly That Was Worth a Fortune." *Trains* 27 (Aug. 1967), 32-43, 46-49. On the use of Shay locomotives in logging since the 1880s.

R45 Rankin, John. *A History of Our Firm, Being Some Account of the Firm of Pollock, Gilmour and Company and Its Offshoots and Connections, 1804-1920.* Liverpool: Henry Young, 1921. 330 pp. During the period from 1812 to 1879, this English firm had several branches engaged in the timber and shipbuilding industry in eastern Canada, especially New Brunswick.

Rankin, M.C. See Simmons, Roger E., #S286

R46 Ransmeier, Joseph S. *The Tennessee Valley Authority: A Case Study in the Economics of Multiple Purpose Stream Planning.* Nashville: Vanderbilt University Press, 1942. xx + 487 pp.

R47 Ransom, Frank E. *The City Built on Wood: A History of the Furniture Industry in Grand Rapids, Michigan, 1850-1950.* Ann Arbor: Edwards Brothers, 1955. viii + 101 pp. Illus., diags., map, tables.

R48 Ransom, James M. *Vanishing Ironworks of the Ramapos: The Story of the Forges, Furnaces, and Mines of the New Jersey-New York Border Area.* New Brunswick, New Jersey: Rutgers University Press, 1966. 397 pp. Illus., maps, bib. Includes some history of the charcoal iron industry.

Ransom, Jay Ellis. See Ransom, Jay George, #R49

R49 Ransom, Jay George. "Timber! Log Drive to Boise: 1909." As told to Jay Ellis Ransom. *Montana, Magazine of Western History* 19 (Apr. 1969), 2-9. Reminiscence of a log drive on Fall Creek and the South Fork of the Boise River, Idaho.

R50 Rapraeger, E.F. "Movement of Douglas Fir Log Prices on Puget Sound, 1896-1933." *Timberman* 35 (Aug. 1934), 12-13. Washington.

Rapraeger, E.F. See Anderson, I.V., #A435

R51 Rapraeger, E.F. "Frontier Sawmill." *Timberman* 42 (May 1941), 14-15, 36. The Montana sawmill of Anson M. Holter.

Rapraeger, E.F. See Billings, C.L., #B267

R52 Rasmussen, Jewell J. *Utah's Public Lands: Their Status and Fiscal Significance.* Salt Lake City: University of Utah, Bureau of Economic and Business Research, 1962. 71 pp. Tables, notes. On state and federal land policy.

R53 Rasmussen, Wayne D. "Forty Years of Agricultural History." *Agricultural History* 33 (Oct. 1959), 177-84. On the progress of agricultural historiography since the establishment of the Agricultural History Society in 1919.

R54 Rasmussen, Wayne D., ed. *Growth Through Agricultural Progress; Lecture Series in Honor of the United States Department of Agriculture Centennial Year.* Washington: USDA, Graduate School, 1961. 74 pp. There are several essays of interest. See, for example, Vernon R. Carstenson, "Profile of the USDA—First Fifty Years," pp. 3-17.

Rasmussen, Wayne D. See Baker, Gladys L., #B27

R55 Rasmussen, Wayne D., and Baker, Gladys L. *The Department of Agriculture.* Praeger Library of U.S. Government Departments and Agencies, No. 32. New York: Praeger Publishers, 1972. 257 pp. Illus., apps., bib., index. Includes much history of the USDA, parent agency to the USFS, and of its natural resource policies.

Ratchford, B.U. See Hoover, Calvin B., #H484

R56 Rathbun, Lawrence W. "New Hampshire Taxes the Ax." *American Forests* 59 (Mar. 1953), 14-15, 48-49. On the taxation of growing timber in New Hampshire since 1941.

R57 Raup, Hugh M. "Old Field Forests of Southeastern New England." *Journal of the Arnold Arboretum* 21 (1940), 266-73. On the composition of the forests in the early 20th century.

R58 Raup, Hugh M., and Carlson, Reynold E. *The History of Land Use in the Harvard Forest.* Harvard Forest, Bulletin No. 20. Petersham, Massachusetts, 1941. 64 pp. Maps, tables, app., bib. On the Harvard Forest since 1907, and under its previous owners since 1733.

R59 Raup, Hugh M. "Forests and Gardens Along the Alaska Highway." *Geographical Review* 35 (Jan. 1945), 22-48. Incidental history of forests along the highway in the Yukon Territory and Alaska.

R60 Raup, Hugh M. "The View from John Sanderson's Farm: A Perspective for the Use of the Land." *Forest History* 10 (Apr. 1966), 2-11. History of forest land use near Petersham, Massachusetts, since the 18th century.

Raup, H.F. See Freeman, O.W., #F228

R61 Rawick, George P. "The New Deal and Youth: The Civilian Conservation Corps, the National Youth Administration, and the American Youth Congress." Ph.D. diss., Univ. of Wisconsin, 1957. 416 pp.

Rawlings, C.C. See Strait, Robert F., #S683

R62 Ray, Joseph M., and Worley, Lillian. *Alabama's Heritage: A Study of the Public Administration of Natural Resources.* University of Alabama, Bureau of Public Administration, Publication No. 27. University: University of Alabama Press, 1947. 186 pp. History and description of federal, state, and local resource management activities; includes a chapter on forests.

R63 Ray, Joseph M. "The Influence of the Tennessee Valley Authority on the Government of the South." *American Political Science Review* 43 (Oct. 1949), 922-32. Account of TVA relationships with state and local governments, especially in the administration of natural resources.

R64 Raymond, William O. *The River St. John.* Saint John, New Brunswick, 1910. Includes a chapter on the mast and naval stores industry in New Brunswick, especially in the 18th and 19th centuries.

R65 Read, Arthur D. *The Profession of Forestry.* New York: Macmillan, 1935. viii + 68 pp. Illus., app., index. Includes some history of forestry.

R66 Read, Arthur D. "Forestry's Progress in Louisiana." *Southern Lumberman* 193 (Dec. 15, 1956), 143-47.

R67 Read, Ralph A. *The Great Plains Shelterbelt in 1954. (A Re-evaluation of Field Windbreaks Planted between 1935 and 1942 and a Suggested Research Program).* Lincoln: Experiment Station, University of Nebraska, 1958. 125 pp. Illus., tables.

R68 Recknagel, Arthur B. *The Theory and Practice of Working Plans (Forest Organization).* New York: John Wiley and Sons, 1913. ix + 235 pp. Illus., index. Contains some incidental history of working plans. The second edition was published in 1917.

R69 Recknagel, Arthur B., and Bentley, John, Jr. *Forest Management.* New York: John Wiley and Sons, 1919. xiii + 269 pp. Illus., tables, charts. Incidental history; the second edition, wiht Cedric H. Guise, was published in 1926.

R70 Recknagel, Arthur B. *The Forests of New York State.* New York: Macmillan, 1923. xiii + 167 pp. Illus., tables, bib., index. Includes some history of forests and forest policy, as well as briefer sketches of the lumber, pulp and paper, and other forest industries.

R71 Recknagel, Arthur B., and Spring, Samuel N. *Forestry: A Study of its Origin, Application and Significance in the United States.* New York: Alfred A. Knopf, 1929. xiii + 255 + xxxvii pp. Illus., app., bib., index. Includes a chapter on the history of forestry.

R72 Recknagel, Arthur B. "Private Forestry in the Northeastern States with Special Reference to Northern New York." *Journal of Forestry* 37 (Jan. 1939), 22-24. Finch, Pruyn & Company, Glens Falls, New York, since 1911.

Recknagel, Arthur B. See Meyer, Hans Arthur, #M426

R73 Recknagel, Arthur B. "How Far Have We Come in Industrial Forestry?" *Northeastern Logger* 4 (May 1956), 18-19, 72-73.

Recknagel, Arthur B. See Brown, Nelson Courtlandt, #B505

R74 Recknagel, Arthur B. "The Pulp and Paper Industry in Northern and Central New York." *Northeastern Logger* 8 (May 1960), 16-17, 62. A historical sketch of the industry.

Recknagel, Arthur B. See Good, Thomas, #G173

R75 Reckord, Milton A. *Lieutenant General Milton A. Reckord: A Personal Memoir of Ninety-Five Years.* OHI by Elwood R. Maunder. Santa Cruz, California: Forest History Society, 1974. vii + 59 pp. Illus., index. Processed. On Reckord's career with the National Rifle Association and its involvement in the conservation movement.

R76 Rector, William G. "From Woods to Sawmill: Transportation Problems in Logging." *Southern Lumberman* 179 (July 15, 1949), 54, 56, 60, 62. On water transportation of logs in the Great Lakes states, 1840-1914, particularly on the St. Croix River, Wisconsin and Minnesota.

R77 Rector, William G. "From Woods to Sawmill: Transportation Problems in Logging." *Agricultural History* 23 (Oct. 1949), 239-44. See above.

R78 Rector, William G. "Lumber Barons in Revolt." *Minnesota History* 31 (Mar. 1950), 33-39. On the selection of a surveyor general of logs and timber for the St. Croix Valley of Minnesota, and the opposition of lumbermen, 1884-1885.

R79 Rector, William G. "Working with Lumber Industry Records." *Wisconsin Magazine of History* 33 (June 1950), 472-78. On the early work of the Forest History Foundation, St. Paul, Minnesota.

R80 Rector, William G. "The Development of Log Transportation in the Lake States Lumber Industry, 1840-1918." Ph.D. diss., Univ. of Minnesota, 1951.

R81 Rector, William G. "Railroad Logging in the Lake States." *Michigan History* 36 (Dec. 1952), 351-62. From 1863 to 1900.

R82 Rector, William G. *Log Transportation in the Lake States Lumber Industry, 1840-1918: The Movement of Logs and Its Relationship to Land Settlement, Waterway Development, Railroad Construction, Lumber Production, and Prices.* American Waterways Series, No. 4. Glendale, California: Arthur H. Clark Company, 1953. 352 pp. Illus., tables, notes, apps., bib., index. A thorough study focusing on Wisconsin, Minnesota, and Michigan.

R83 Rector, William G. "The Birth of the St. Croix Octopus." *Wisconsin Magazine of History* 40 (Spring 1957), 171-77. On the lumber industry in the St. Croix Valley of Minnesota and Wisconsin, 1837-1914, particularly the St. Croix Boom Company, incorporated in 1856.

R84 Redfield, Alfred C. *The Recollections of an Ecologist on the Origins of the Natural Resources Council of America.* OHI by Elwood R. Maunder. Santa Cruz, California: Forest History Society, 1974. 74 pp. Illus., apps., index. Processed. Redfield, an oceanographer, was chairman of the Natural Resources Council of America and president of the Ecological Society of America in the 1940s.

R85 Redington, Paul G. "Fifty Years of Forestry." *American Forests and Forest Life* 32 (Dec. 1926), 719-24, 750. Federal, state, and private forestry since ca. 1875.

R86 Reed, Frank. "Old Days in Steuben County." *Northeastern Logger* 3 (Feb. 1955), 12, 32. Steuben County, New York.

R87 Reed, Frank. "Old Days in the Chemung Watershed." *Northeastern Logger* 6 (May 1958), 18-19, 41, 61. Chemung County, New York.

R88 Reed, Frank. "Old Days in the Mohawk Valley." *Northeastern Logger* 8 (May 1960), 14, 64ff. Mohawk Valley, New York.

Reed, Frank A. See Stone, Flora Pierce, #S663

Reed, Frank A. See Braden, Leo, #B391

Reed, Frank A. See Good, Thomas, #G173

Reed, Frank A. See Kling, Edwin M., #K202

R89 Reed, Frank A. "Forty-Eight Years in the North Country." New York State Ranger School, *Alumni News* (1964), 9-11. Reminiscences of logging in New York's Adirondack Mountains.

R90 Reed, Frank A. *Lumberjack Sky Pilot.* Old Forge, New York: North Country Books, 1965. xi + 155 pp. Illus., maps. Reminiscences of logging camps in New York, northern New England, and Minnesota. The bulk of the book is autobiographical, although there are biographical sketches of five other "sky pilots" or ministers to lumberjacks, since 1895.

R91 Reed, Franklin W. "The United States Timber Conservation Board: Its Origin and Organization; Its Purpose and Progress."*Journal of Forestry* 29 (Dec. 1931), 1202-05. On the early work of a presidentially appointed board mandated in 1930 to investigate the problems facing the lumber industry.

R92 Reed, Howard S. *A Short History of the Plant Sciences.* Waltham, Massachusetts: Chronica Botanica, 1942. 320 pp. Illus.

R93 Reed, Merl E. "The IWW and Individual Freedom in Western Louisiana, 1913." *Louisiana History* 10 (Winter 1969), 61-69. Edited documents illustrating lumbermen's successful thwarting of labor activities of the Industrial Workers of the World, particularly by acts of violence against lumberjacks and millworkers and by denial of press freedoms.

R94 Reed, Merl E. "Lumberjacks and Longshoremen: The IWW in Louisiana." *Labor History* 13 (Winter 1972), 41-59. Industrial Workers of the World, 1907-1912.

R95 Reeder, Faye B. "The Evolution of the Virginia Land Grant System in the Eighteenth Century." Ph.D. diss., Ohio State Univ., 1937.

R96 Reetz, Byron Samuel. "A History of a Lumbered County." Master's thesis, Michigan State Univ., 1951. 139 pp.

Reeves, J.E. See Briscoe, Vera, #B434

R97 Reeves, John Henry, Jr. "The History and Development of Wildlife Conservation in Virginia: A Critical Review." Ph.D. diss., Virginia Polytechnic Institute and State Univ., 1960.

R98 Reeves, Thomas C. "President Arthur in Yellowstone National Park." *Montana, Magazine of Western History* 19 (Summer 1969), 18-29. On the visit of Chester A. Arthur and Senator George G. Vest to Yellowstone National Park in 1883. Publicity given to the expedition helped bring further support for the park's preservation.

R99 Rehder, Alfred. "Charles Sprague Sargent."*Journal of the Arnold Arboretum* 8 (Apr. 1927), 69-86. A horticulturalist and landscape gardener, Sargent (1841-1927) became the first director of the Arnold Arboretum at Harvard University in 1873. He influenced forest policy in New York and in the federal government and advanced knowledge of the science through the columns of *Garden and Forest,* a magazine which he founded and edited, 1887-1897.

R100 Reich, Charles A. "The Public and the Nation's Forests." *California Law Review* 50 (Aug. 1962), 381-97. Incidental history.

R101 Reich, Charles A. *Bureaucracy and the Forests.* Santa Barbara, California: Center for the Study of Democratic Institutions, 1962. 13 pp. Includes incidental history of laws and policies governing federal forest lands.

R102 Reich, Nathan. *The Pulp and Paper Industry in Canada: With Special Reference to the Export of Pulpwood.* McGill University, Economic Studies No. 7. Toronto, 1926. 77 pp. Illus.

Reid, H. See Krause, John #K258

R103 Reid, Russell. "The North Dakota State Park System." *North Dakota Historical Quarterly* 8 (Oct. 1940), 63-78. Emphasis on the preservation of historic and scenic sites, including a roster of state parks, by the superintendent of the system.

R104 Reiger, John F. "George Bird Grinnell and the Development of American Conservation, 1870-1901." Ph.D. diss., Northwestern Univ., 1970. 268 pp. Grinnell (1849-1938) was an ornithologist, explorer, writer, national parks advocate, editor and publisher of *Forest and Stream,* and advisor to Theodore Roosevelt and others on conservation matters.

Reiger, John F. See Grinnell, George Bird, #G355

R105 Reiger, John F. *American Sportsmen and the Origins of Conservation.* New York: Winchester Press, 1975. 352 pp. Illus. A general history of the conservation movement with emphasis on the role of hunting and sportsmen's groups since the 1870s.

R106 Reimann, Lewis Charles. *Between the Iron and the Pine: A Biography of a Pioneer Family and a Pioneer Town.* Ann Arbor: Northwoods Publishers, 1951. x + 225 pp. Illus. Reminiscences of Iron River, a logging town on Michigan's Upper Peninsula, late 19th century.

R107 Reimann, Lewis Charles. *When Pine Was King.* Ann Arbor, Michigan: Edwards Brothers, 1952. vii + 163 pp. Illus. Reminiscences and history of the lumber industry in Michigan's Upper Peninsula, 1875-1900.

R108 Reimann, Lewis Charles. *Incredible Seney: The First Complete Story of Michigan's Fabulous Lumber Town.* Ann Arbor: Northwoods Publishers, 1953. vii + 190 pp. Illus. On Michigan's Upper Peninsula, 1880s-1890s.

R109 Reimer, Derek. "Aural History and British Columbia's Forest Industry." *Sound Heritage* 3 (No. 2, 1974), 11-12.

Reimer, Derek. See Lutz, George, #L320

R110 Reinert, Guy F. "The Vanishing Craft of Birch Oil Distilling—An Old Pennsylvania-Dutch Industry." *American-German Review* 6 (No. 6, 1940), 30-31.

R111 Reinertsen, Philip J. "The Pulp and Paper Industries in Sweden and Canada." Ph.D. diss., Univ. of Chicago, 1959. Incidental history.

R112 Reinhardt, Richard. "The Case of the Hard-Nosed Conservationists." *American West* 4 (Feb. 1967), 52-54, 85-92. On the work of the Sierra Club, especially in California, since 1892.

R113 Reinhardt, Richard. "What Became of Arbor Day?" *Cry California* 5 (Spring 1970), 26-33.

Reinhart, Kenneth G. See Lull, Howard W., #L304

R114 Reitze, Arnold W., Jr. *Environmental Planning: Law of Land & Resources.* Washington: North American International, 1974. 926 pp. Chapters on public lands, national forests, national parks, grazing, scenic and wild rivers, and the Wilderness Act contain incidental historical references.

R115 Rempel, John I. "The History and Development of Early Forms of Building Construction in Ontario." *Ontario History* 52 (Autumn 1960), 235-44; 53 (Mar. 1961), 1-35.

R116 Rempel, John I. *Building with Wood and other Aspects of Nineteenth-Century Building in Ontario.* Toronto: University of Toronto Press, 1967. 287 pp. Illus., apps., bib., index. Includes some history of pioneer forest utilization and the construction of log cabins.

R117 Renne, Roland R. "State Conservation and the Development of Natural Resources." *State Government* 23 (June 1950), 134-37. Since 1900.

R118 Renner, George Thomas. *Conservation of National Resources, An Educational Approach to the Problem.* New York: Wiley, 1942. 228 pp. Illus. Incidental history.

R119 Rensch, Hero Eugene. *Mount Rainier: Its Human History Associations.* Berkeley: U.S. National Park Service, Field Division of Education, 1935. 50 pp. Processed. Mount Rainier National Park, Washington.

R120 Rensch, Hero Eugene. *Historical Background for the Rocky Mountain National Park, Colorado.* Berkeley: U.S. National Park Service, Field Division of Education, 1935. 49 pp. Processed.

R121 Renshaw, Patrick. *The Wobblies: The Story of Syndicalism in the United States.* Garden City, New York: Doubleday and Company, 1967. 312 pp. Illus., notes, index. Includes some history of the Industrial Workers of the World labor activity among the lumberjacks and millworkers of the Pacific Northwest, especially from 1905 to 1924.

R122 Renshaw, Patrick. "The IWW and the Red Scare, 1917-1924." *Journal of Contemporary History* 3 (Oct. 1968), 63-72.

R123 Replinger, Peter J. "Simpson Timber Co." *Pacific News* 14 (Aug. 1974), 4-14. On logging railroads operated by Simpson Timber Company, Shelton, Washington, since the 1880s.

R124 Reppeto, Paul. *Way of the Logger.* Chehalis, Washington: Loggers World, 1970. 42 pp. Illus. A collection of short articles describing many aspects of early logging in Washington.

R125 Resler, Rexford A., and Hopkins, Howard G. "Case History of a Watershed." *American Forests* 61 (Mar. 1955), 56-60, 62-64. On the watershed of Corvallis, Oregon, located near Marys Peak on the Siuslaw National Forest, since 1920; emphasis is on a timber salvage problem undertaken cooperatively by the city of Corvallis and the USFS.

R126 Rettie, Dwight Fay. "National Forest Timber Sale Policy: A Case Study of the Disposal of Federally Owned Natural Resources Severable from Land." Master's thesis, Univ. of California, Berkeley, 1955.

R127 Rettie, James C., and Ineson, Frank. *Otsego Forest Products Cooperative Association of Cooperstown, New York.* Washington: USDA, 1950. 42 pp. Incidental history.

Rettig, E.C. See Billings, C.L., #B267

R128 Reynolds, Arthur R. "Rafting Down the Chippewa and the Mississippi: Daniel Shaw Lumber Company, a Type Study." *Wisconsin Magazine of History* 32 (Dec. 1948), 143-52. A firm of Eau Claire, Wisconsin, 1856-1912.

R129 Reynolds, Arthur R. "The Kinkaid Act and Its Effects." *Agricultural History* 23 (Jan. 1949), 20-29. The Kinkaid Act of 1904 applied to western Nebraska.

R130 Reynolds, Arthur R. "Wisconsin Lumbering Frontier: The Daniel Shaw Lumber Company." Ph.D. diss., Univ. of Minnesota, 1950.

R131 Reynolds, Arthur R. "Sources of Credit for a Frontier Lumber Company: The Daniel Shaw Lumber Company as a Type Company." *Bulletin of the Business Historical Society* 24 (Dec. 1950), 184-95.

R132 Reynolds, Arthur R. *The Daniel Shaw Lumber Company: A Case Study of the Wisconsin Lumbering Frontier.* New York University, Graduate School of Business Administration, Business History Series, No. 5. New York: New York University Press, 1957. x + 177 pp. Illus., diags., tables, notes, apps., bib., index. On its operations in Wisconsin's Chippewa Valley, 1856-1912.

R133 Reynolds, George W.; Holloway, Garrett B.; and Halm, Joe B. "Death of a Forest." *Montana, Magazine of Western History* 10 (Autumn 1960), 45-58. Three separate reminiscences of the forest fire which burned much of northern Idaho and western Montana in 1910.

R134 Reynolds, Harris A. "New England Forestry Foundation." *Journal of Forestry* 45 (Feb. 1947), 89-91. On the work of this conservation organization since its founding in 1944.

R135 Reynolds, Jeanne. " 'Emporium' Was the Word for Lumbering." *St. Lawrence County Historical Society Quarterly* 3 (Oct. 1958), 6-8. On the Emporium Forestry Company in northern New York.

Reynolds, Morris. See Cochrell, Albert N., #C493

R136 Reynolds, Richard D. "Fire Protection Law Enforcement Trends in California." *Journal of Forestry* 48 (Oct. 1950), 696-99. Specifically in the redwood belt since the 1930s.

R137 Reynolds, Richard D. "Effects of Natural Fires and Aboriginal Burning upon the Forests of the Central Sierra Nevada." Master's thesis, Univ. of California, Berkeley, 1959.

R138 Reynolds, Robert V., and Hoyle, Raymond J. *Wood-Using Industries of New York.* Technical Publication No. 14. Syracuse: New York State College of Forestry, 1921.

R139 Reynolds, Robert V., and Pierson, Albert H. *Lumber Cut of the United States, 1870-1920: Declining Production and High Prices as Related to Forest Exhaustion.* USDA, Bulletin No. 1119. Washington: GPO, 1923. 63 pp. Illus., tables, charts. Largely statistical.

R140 Reynolds, Robert V., and Pierson, Albert H. "Tracking the Sawmill Westward: The Story of the Lumber Industry in the U.S. as Unfolded by Its Trail across the Continent." *American Forests* 31 (Nov. 1925), 643-48. On the historical relationship between centers of production and consumption.

R141 Reynolds, Robert V. "How Long Will Our Sawtimber Last?" *American Forests and Forest Life* 34 (May 1928), 259-62, 307-08. Includes some history of timber supply and forecasts of timber famine since the colonial period.

R142 Reynolds, Robert V., and Pierson, Albert H. "The Aggregate Cut of American Lumber, 1801-1935." *Journal of Forestry* 35 (Dec. 1937), 1099-1101. Statistical charts.

R143 Reynolds, Robert V., and Pierson, Albert H. *Statistics of Forest Products in the Rocky Mountain States.* USDA, Statistical Bulletin No. 64. Washington, 1938. 29 pp. Graphs, tables. Since 1899.

R144 Reynolds, Robert V., and Pierson, Albert H. *Forest Products Statistics of the Pacific Coast States.* USDA, Statistical Bulletin No. 65. Washington, 1938. 30 pp. Graphs, tables, Since 1899.

R145 Reynolds, Robert V., and Pierson, Albert H. *Forest Products Statistics of the Lake States.* USDA, Statistical Bulletin No. 68. Washington, 1939. 40 pp. Tables.

R146 Reynolds, Robert V., and Pierson, Albert H. *Forest Products Statistics of the Southern States.* USDA, Statistical Bulletin No. 69. Washington, 1939. 106 pp. Graphs, tables. Since 1889.

R147 Reynolds, Robert V., and Pierson, Albert H. *Forest Product Statistics of Central and Prairie States.* USDA, Statistical Bulletin No. 73. Washington, 1941. 94 pp. Tables. From 1770 to 1939.

R148 Reynolds, Robert V., and Pierson, Albert H. *Fuel Wood Used in the United States, 1630-1930.* USDA, Circular No. 641. Washington, 1942. 20 pp. Tables, graphs. Statistical tables showing consumption by regions and decades, preferred wood species, and lumber equivalents of trees used for fuel.

R149 Reynolds, Russell Roy. *Eighteen Years of Selection: Timber Management on the Crossett Experimental Forest.* USDA, Technical Bulletin No. 1206. Washington, 1959. iv + 68 pp. Diags., maps, tables, notes, bib. Arkansas, since 1938.

R150 Rhodenbaugh, Beth, and Goertzen, Dorine. *The History of State Forestry in Idaho.* Boise: Idaho Department of Forestry, 1961. 33 pp. Processed. A brief historical sketch of forestry, with emphasis on forest protection.

R151 Rhodes, Marion E. "Big Flats Saw Mills Flourished in 1850s." *Chemung County Historical Journal* 8 (June 1963), 1133-35. New York.

Ricards, Sherman L., Jr. See Blackburn, George M., #B292, B293

R152 Rice, Barton & Fales. *A Line of Men One Hundred Years Long: The Story of Rice, Barton & Fales, Incorporated.* Worcester: Rice, Barton & Fales, 1937. 69 pp. Illus. A papermaking firm of Worcester, Massachusetts.

R153 Rice, Elroy Leon, and Penfound, William T. "The Upland Forests of Oklahoma." *Ecology* 40 (Oct. 1959), 594-608. Study results of the 1950s compared with accounts since 1911.

R154 Rice, Otis K. *The Allegheny Frontier: West Virginia Beginnings, 1730-1830.* Lexington: University Press of Kentucky, 1970. xiii + 438 pp. Illus., maps, notes, bib., index. Includes some history of forest use in relation to settlement.

R155 Rice, Richard. "The Wrights of Saint John: A Study of Shipbuilding and Shipowning in the Maritimes, 1839-1855." In *Canadian Business History: Selected Studies, 1497-1971,* ed. by David S. Macmillan. Toronto: McClelland and Stewart, 1972. Pp. 317-37. Saint John, New Brunswick.

R156 Rich, Louise Dickinson. *We Took to the Woods.* Philadelphia: J.B. Lippincott, 1942. 322 pp. Illus. Account of life in the Maine woods.

R157 Rich, Louise Dickinson. *The Natural World of Louise Dickinson Rich.* New York: Dodd, Mead, 1962. 195 pp. More natural history and reminiscences of the Maine wilderness.

R158 Richards, Annette H. "Smokey, the Little Bear That Didn't Stay Home." *Frontiers* 20 (June 1956), 142-43, 146-47. On Smokey and the USFS's fire prevention campaign since 1942.

R159 Richards, E.C.M. "New York—Port of Fancy Hardwoods." *American Forests* 32 (Nov. 1926), 649-54. On New York as a sawmill center for imported hardwoods.

R160 Richards, E.C.M. "Raphael Zon—The Man." *Journal of Forestry* 24 (Dec. 1926), 850-57. Of the USFS and editor-in-chief of the *Journal of Forestry.*

R161 Richards, Edward S. "Forest of the Navajo." *Loggers Handbook* 32 (1972), 23-24, 154, 158, 161-63. Logging, sawmilling, and forest management on the Navajo Indian Reservation, Arizona-New Mexico, since the 1880s, including history of the Indian-owned and operated Navajo Forest Products Industries since 1959.

R162 Richards, Gilbert. *Crossroads: People and Events of the Redwoods of San Mateo County.* Woodside, California: Gilbert Richards Publications, 1973. x + 128 pp. Illus., maps, bib., index. Includes a chapter on the history of redwood logging in the county and many other historical references to forests and forest industries.

R163 Richards, Henry. *Ninety Years On, 1848-1940.* Augusta, Maine: By the author, 1940. 502 pp. Includes reminiscences of the author's pulp and paper mill in Gardiner, Maine, in the 19th century.

R164 Richards, J. Howard. "Lands and Policies: Attitudes and Controls in the Alienation of Lands in Ontario during the First Century of Settlement." *Ontario Historical Society Papers and Records* 50 (1958), 193-208. From 1780 to 1875, with incidental reference to the lumber industry and forested lands.

R165 Richards, John. *Treatise on the Construction and Operation of Wood-Working Machines: Including a History of the Origin and Progress of the Manufacture of Wood-Working Machinery.* London: F.N. Spon, 1872. 283 pp. Includes American inventions and methods in the lumber and forest products industries.

R166 Richards, John. "A Treatise on the Construction and Operation of Woodworking Machines." *Forest History* 9 (Jan. 1966), 16-23. Selections from his critique of American wood technology, 1872.

R167 Richards, Laura E. *Stepping Westward.* New York: Appleton, 1931. x + 405 pp. Autobiography of Henry Richards' wife, with references to the pulp and paper industry in 19th-century Maine.

R168 Richards, Laura Barbara. "George Frederick Jewett: Lumberman and Conservationist." Master's thesis, Univ. of Idaho, 1969.

R169 Richardson, Arthur Herbert. *Forestry in Ontario.* Toronto, 1928. 73 pp. Illus. Incidental history.

R170 Richardson, Arthur Herbert. "Twenty-Five Years' Reforestation in Ontario." *Canada Lumberman* 51 (Apr. 1, 1931), 23-24.

R171 Richardson, Arthur Herbert. "Early Lumbering Days in Simcoe County." *Canada Lumberman* 54 (Dec. 15, 1934), 15-16. Sawmills and the lumber industry near Angus, Ontario, 1830s to 1900.

R172 Richardson, Arthur Herbert. "A Comparative Historical Study of Timber Building in Canada." *Bulletin of the Association for Preservation Technology* 5 (No. 3, 1973), 77-102.

R173 Richardson, Arthur Herbert. "The Earliest Wood-Processing Industry in North America, 1607-23." *Bulletin of the Association for Preservation Technology* 5 (No. 4, 1973), 81-84.

R174 Richardson, Arthur Herbert. *Conservation by the People: The History of the Conservation Movement in Ontario to 1970.* Ed. by A.S.L. Barnes. Toronto: University of Toronto Press, for the Conservation Authorities of Ontario, 1974. xi + 154 pp. Illus., maps, tables, apps., index. The author and editor, directors of the Conservation Authorities Branch of the Ontario Department of Planning and Development, 1944 to 1970, relate the history of the conservation authorities movement since the 1930s. Principal work of the community and regional organizations has been in reforestation, wildlife management, flood control, historical preservation, and outdoor recreation.

R175 Richardson, Elmo R. "The Politics of the Conservation Issue in the Far West, 1896-1913." Ph.D. diss., Univ. of California, Los Angeles, 1958.

R176 Richardson, Elmo R. "Conservation as a Political Issue: The Western Progressives' Dilemma, 1909-1912." *Pacific Northwest Quarterly* 49 (Apr. 1958), 49-54. The conservation issue failed to provide expected support for Theodore Roosevelt in 1912, indicating the complexity of economic and political opinion among progressives in the West.

R177 Richardson, Elmo R. "George Curry and the Politics of Forest Conservation in New Mexico." *New Mexico Historical Review* 33 (Oct. 1958), 277-84. During Curry's term as territorial governor, 1907-1909.

R178 Richardson, Elmo R. "The Struggle for the Valley: California's Hetch Hetchy Controversy, 1905-1913." *California Historical Society Quarterly* 38 (Sept. 1959), 249-58. John Muir and Robert U. Johnson led the unsuccessful opposition to forces wishing to convert Hetch Hetchy Valley in Yosemite National Park into a reservoir for San Francisco's water supply. The controversy further split the conservation movement into preservationist and utilitarian camps.

R179 Richardson, Elmo R., and Farley, Alan W. *John Palmer Usher: Lincoln's Secretary of the Interior.* Lawrence: University of Kansas Press, 1960. xii + 152 pp. Illus., notes, bib., index. Usher served from 1863 to 1865.

R180 Richardson, Elmo R. *The Politics of Conservation: Crusades and Controversies, 1897-1913.* University of California Publications in History, Volume 70. Berkeley: University of California Press, 1962. ix + 207 pp. Notes, bib., index. On the conservation movement and its role in Western and national politics, culminating in the Ballinger-Pinchot controversy.

R181 Richardson, Elmo R. "Western Politics and New Deal Policies: A Study of T.A. Walters of Idaho." *Pacific Northwest Quarterly* 54 (Jan. 1963), 9-18. Walters was first assistant secretary in the Department of the Interior, 1933-1937. He was involved in making conservation policy but more interested in Idaho patronage problems.

R182 Richardson, Elmo R. "Federal Park Policy in Utah: The Escalante National Monument Controversy of 1935-1940." *Utah Historical Quarterly* 33 (Spring 1965), 109-33. Describes the conflict between state and federal officials over a USDI proposal to establish Escalante National Monument in southeastern Utah.

R183 Richardson, Elmo R. "The Civilian Conservation Corps and the Origins of the New Mexico State Park System." *Natural Resources Journal* 6 (Apr. 1966), 248-67.

Richardson, Elmo R. See Tanasoca, Donald, #T12

R184 Richardson, Elmo R. "Olympic National Park: Twenty Years of Controversy." *Forest History* 12 (Apr. 1968), 6-15. An analysis of the competing interests favoring and opposing the Washington park, the expansion of its boundaries, and the status of logging within it, 1933-1953.

R185 Richardson, Elmo R. "Was There Politics in the Civilian Conservation Corps?" *Forest History* 16 (July 1972), 12-21. Asserts that Republican allegations of CCC involvement in political affairs of the 1930s were unfounded.

R186 Richardson, Elmo R. "The Interior Secretary as Conservation Villain: The Notorious Case of Douglas 'Give-away' McKay." *Pacific Historical Review* 41 (Aug. 1972), 333-45. McKay was Secretary of the Interior, 1953-1956.

R187 Richardson, Elmo R. *Dams, Parks, & Politics: Resource Development and Preservation in the Truman-Eisenhower Era.* Lexington: University Press of Kentucky, 1973. 247 pp. Illus., notes, sources, index. On the politics of conservation and resource development during the period from 1945 to 1960, including much attention to the controversial role of Secretary of the Interior Douglas McKay.

R188 Richardson, F.M., and Wand, Ben. "The Story of the 'Southern Lumber Journal,' 40 Years." *Southern Lumber Journal* 40 (Dec. 1936), 12-13. A trade journal since 1898.

R189 Richardson, Herbert W. "The Northeastern Minnesota Forest Fires of October 12, 1918." *Geographical Review* 7 (Apr. 1919), 220-32. The Cloquet Fire.

R190 Richie, Jim, and Kohara, Tom. "Woodworth, The Story of Alexander State Forest." *Forests & People* 2 (No. 2, 1952), 18-21. Louisiana.

Richter, D.S. See Best, Gerald M., #B251

R191 Rickaby, Franz L., ed. *Ballads and Songs of the Shanty-Boy.* Cambridge: Harvard University Press, 1926. xli + 244 pp. Notes, glossary, index. The introduction contains some history of lumberjack songs of the United States and Canada.

R192 Rickard, T.E. "Gilbert Malcolm Sproat." *British Columbia Historical Quarterly* 1 (Jan. 1937), 21-32. Sproat was involved in lumbering in 19th-century British Columbia.

R193 Ricker, M.B. "The Influence of the Use of Substitutes for Wood upon the Practices of Forestry in the U.S." Master's thesis, Yale Univ., 1930.

R194 Riddell, R.G. "A Study in the Land Policy of the Colonial Office, 1763-1855." *Canadian Historical Review* 18 (Dec. 1937), 385-405.

R195 Ridgway, Robert. "Forests of the Lower Wabash Bottomlands During the Period 1870-1890." *Proceedings of the Indiana Academy of Science* 67 (1958), 244-48. A letter from Ridgway to Charles C. Deam, dated 1919, containing reminiscences of the forests of Knox and Gibson Counties, Indiana, and some photographs taken in the 1880s.

R196 Riebold, R.J. "Wildlife and Timber, Too, on the Francis Marion National Forest." *Southern Lumberman* 193 (Dec. 15, 1956), 254-58. South Carolina, since 1936.

Ries, H. See Gustafson, Alex Ferdinand, #G378

R197 Riesch, Anna Lou. "Conservation under Franklin D. Roosevelt." Ph.D. diss., Univ. of Wisconsin, 1952. On Roosevelt's lifelong interest in conservation and his efforts to implement a conservation program during his presidency.

R198 Riis, John. *Ranger Trails.* Richmond, Virginia: Dietz Press, 1937. 160 pp. Illus. Recollections of life as a USFS ranger on the national forests of the West, early 20th century.

R199 Riley, George A. "A History of Tanning in the State of Maine." Master's thesis, Univ. of Maine, 1935.

R200 Riley, R.D., and Daly, Dorothy. "A Pioneer Timberman Looks Back to the Days of Bulls and Horses." *Loggers Handbook* 24 (1964), 15, 97-98. Recollections of logging in Wisconsin and the Pacific Northwest.

R201 Riley, R.D. "Old Timer Reports: Early Days of Timber and Men." *Loggers Handbook* 26 (1966), 19-20, 125. Recollections of lumbermen in the Great Lake states and Pacific Northwest, 1890s-1920s.

R202 Rincliffe, Roy George. *'Conowingo!' The History of A Great Development on the Susquehanna.* New York: Newcomen Society in North America, 1953. 28 pp. Illus. Susquehanna Water Power and Paper Company, Pennsylvania, since 1884.

R203 Ring, Elizabeth. "Fanny Hardy Eckstorm: Maine Woods Historian." *New England Quarterly* 26 (Mar. 1953), 45-64.

R204 Ringland, Arthur C. "Pioneering in Southwest Forestry." OHI by Fern Ingersoll. *Forest History* 17 (Apr. 1973), 4-11. Reminiscences of the Fort Valley Ranger School, Arizona, 1909-1910, of USFS problems with New Mexico Senator Albert Fall, and of Arizona Senator Ralph Cameron concerning development of Grand Canyon National Monument, 1914-1915.

R205 Riordon, Carl. "Notes on the Founding of the Technical Section." *Pulp and Paper Magazine of Canada* 41 (Aug. 1940), 534. Technical section of the Canadian Pulp and Paper Association.

R206 Ripley, Thomas Emerson. "Shakespeare in the Logging Camp." *Forest History* 8 (Fall 1964), 7-10. An excerpt from the author's autobiographical account of life in Tacoma, Washington, 1890s; this sketch is of lumberjack George Moore.

R207 Ripley, Thomas Emerson. "Shakespeare in the Logging Camp." *American West* 4 (May 1967), 12-16. George Moore.

R208 Ripley, Thomas Emerson. *Green Timber: On the Flood Tide to Fortune in the Great Northwest.* Palo Alto, California: American West Publishing Company and Washington State Historical Society, 1968. 126 pp. Illus. Reminiscences of his association with Wheeler Osgood Company, a lumber and millwork concern of Tacoma, Washington, and of life in that burgeoning lumber town, 1890-1893.

R209 Ripton, Michael J. "Pennsylvania Lumber Museum." *Forest History* 17 (Oct. 1973), 29-32. The museum, located in the Susquehannock State Forest and administered by the Pennsylvania Historical and Museum Commission, was established in 1967 to preserve the heritage of Pennsylvania logging.

Rischen, H.W.L. See Van Groenou, H. Broese, #V22

R210 Rishell, Carl A. "Postwar Developments in Forest Products Research." *Journal of Forestry* 48 (Oct. 1950), 685-89. And trends in wood utilization, 1940s.

R211 Risi, Joseph. *L'Industrie de la carbonisation du bois dans la province de Québec.* Quebec Forest Service, Bulletin New Series, "e". Quebec, 1942. 144 pp. On the charcoal industry.

R212 Rising Paper Company. *Rising Papers: Fine Paper at Its Best and How It Gets That Way!* Housatonic, Massachusetts, 1951.

R213 Risser, James. "The U.S. Forest Service: Smokey's Strip Miners." *Washington Monthly* 3 (Oct. 1971), 16-26. Examines the link between the USFS and the National Forest Products Association, the industry lobby, especially in relation to the clearcutting issue.

R214 Rittenhouse & Embree Company. *We Rate Diamonds, 1883-1958.* Chicago, 1958. 44 pp. Illus. A retail lumber firm of Chicago.

R215 Ritter, F.M. "Oldest Lumber Association in Canada." *Canada Lumberman* 51 (Sept. 1, 1931), 59-60. Western Retail Lumberman's Association since 1890.

R216 Ritter, William M. *The Lumber Business: Organization, Production, Distribution. Observations and Comments on Efficiency and Service.* Nashville: Southern Lumberman, 1920. 108 pp. Incidental history.

R217 Ritter, William M. "Early Days in West Virginia." *Southern Lumberman* (Dec. 15, 1943), 167-68.

R218 Ritter, W.M., Lumber Company. *The Romance of Appalachian Hardwood Lumber, 1890—Fifty Years of Service—1940.* Richmond, Virginia, 1940. 37 pp.

R219 Ritzenthaler, Robert E. "The Menominee Indian Sawmill: A Successful Community Project." *Wisconsin Archeologist* 32 (June 1951), 39-44. Neopit, Wisconsin, since 1908.

R220 Rivkin, Allen. "Loggers and Logging in the 1860's." *American Forests* 61 (Feb. 1955), 36-38, 46-47. Logging technology and working conditions in the northern states during the Civil War era.

R221 Robbins, Michael Warren. "The Principio Company: Iron-Making in Colonial Maryland, 1720-1781." Ph.D. diss., George Washington Univ., 1972. 398 pp. Charcoal iron industry.

R222 Robbins, Peggy. "John James Audubon: 'The American Woodsman'." *American History Illustrated* 9 (Oct. 1974), 4-9, 38-44. Audubon (1785-1851), a naturalist, ornithologist, and artist, was of symbolic importance to later conservationists.

R223 Robbins, Roy M. "History of the Preemption of the Public Lands, 1780-1890." Ph.D. diss., Univ. of Wisconsin, 1929.

R224 Robbins, Roy M. "The Federal Land System in an Embryo State." *Pacific Historical Review* 4 (Dec. 1935), 356-75. Washington Territory, 1853-1889.

R225 Robbins, Roy M. "The Public Domain in the Era of Exploitation, 1862-1901." *Agricultural History* 13 (Apr. 1939), 97-108.

R226 Robbins, Roy M. *Our Landed Heritage: The Public Domain, 1776-1936.* Princeton, New Jersey: Princeton University Press, 1942. Reprint. Lincoln: University of Nebraska Press, 1962. x + 450 pp. Maps, illus., notes, bib., index. A standard history of federal land policy, including sections on forests and conservation.

R227 Roberge, Earl. *Timber Country: Logging in the Great Northwest.* Caldwell, Idaho: Caxton Printers, 1973. xvi + 182 pp. Illus., glossary, index. A lavishly illustrated work containing many incidental references to the history of logging and the forest industries in Oregon, Washington, Idaho, British Columbia, and Alaska.

R228 Roberge, Roger A. "The Timing, Type, and Location of Adaptive Inventive Activity in the Eastern Canadian Pulp and Paper Industry: 1806-1940." Ph.D. diss., Clark Univ., 1972. 270 pp.

Roberts, Earl. See Palais, Hyman, #P13

R229 Roberts, Elliot. *One River—Seven States: TVA-State Relations in the Development of the Tennessee River.* Knoxville: Bureau of Public Administration, University of Tennessee, 1955. vii + 100 pp.

R230 Roberts, N. Keith, and Gardner, B. Delworth. "Livestock and the Public Lands." *Utah Historical Quarterly* 32 (Summer 1964), 285-300.

R231 Roberts, Omer Lonnie, Jr. "Cypress Land and Flooding: Environmental Change and the Development of Land Utilization in the Atchafalaya Basin, Louisiana." Ph.D. diss., Univ. of Tennessee, 1974. 108 pp. Includes some history of cypress logging.

R232 Roberts, Paul B. "White Pine Industry of Maine." Master's thesis, Colorado Agricultural and Mechanical College, 1949.

R233 Roberts, Paul H. *Hoof Prints on Forest Ranges: The Early Years of National Forest Range Administration.* San Antonio: Naylor, 1963. xv + 151 pp. Illus., notes. History of range management and grazing policy on the national forests, 1905-1930.

R234 Roberts, Paul H. *Them Were the Days.* San Antonio: Naylor, 1965. 134 pp. Illus. Reminiscences and some history of the author's forestry career on the national forests of Arizona and New Mexico, 1915-1931.

R235 Roberts, Walter K. "The Political Career of Charles L. McNary, 1924-1944." Ph.D. diss., Univ. of North Carolina, 1954. The U.S. senator from Oregon sponsored forestry and conservation legislation.

R236 Robertson, Alexander F. *Alexander Hugh Holmes Stuart, 1807-1891: A Biography.* Richmond: William Byrd Press, 1925. xix + 484 pp. Illus. Stuart was Secretary of the Interior, 1850-1853.

R237 Robertson, Reuben B. "Recent Developments in Southern Forestry." *Georgia Review* 5 (Fall 1951), 362-68. Since 1898.

R238 Robertson, Reuben B. "Trailblazing in the Southern Paper Industry." OHI by Elwood R. Maunder and Elwood L. Demmon. *Forest History* 5 (Spring 1961), 6-12. A Champion Paper and Fibre Company executive recalls forest operations in Texas and the Carolinas and technical advances in the manufacture of paper.

R239 Robertson, William Beckwith, Jr. *Everglades—The Park Story.* Coral Gables, Florida: University of Miami Press, 1959. 88 pp. Illus., diags. Includes some history of Everglades National Park, Florida.

R240 Robertson, William B. "Canada's Forest Resources: Their Production and Growth." *Forestry Chronicle* 6 (No. 1, 1930), 34-39.

R241 Robinson, C.D. "The Northern Wisconsin Fires of 1871." *Wisconsin Academy Review* 8 (Fall 1961), 153-56.

R242 Robinson, Charles D. "The Wooden Age." *Scribner's Monthly* 15 (Dec. 1877), 145-56. Includes some history of wood utilization.

R243 Robinson, Donald H. "The Glacier Moves Tortuously." Ed. by Harry B. Robinson. *Montana, Magazine of*

Western History 7 (July 1957), 12-25. On early surveys and the movement to establish Montana's Glacier National Park, 1853-1914.

Robinson, Donald H. See Bowers, Maynard O., #B375

Robinson, Edgar Eugene. See Wilbur, Ray Lyman, #W284

R244 Robinson, Elizabeth A.F. "The Evaluation of Forest Land in British Columbia." Master's thesis, Univ. of British Columbia, 1968.

R245 Robinson, Elmo A. "Prolegomena to a Philosophy of Mountaineering." *Sierra Club Bulletin* 23 (1938), 50-64. Includes some history of man's attitude toward wilderness.

R246 Robinson, Glen O. *The Forest Service: A Study in Public Land Management.* Baltimore: Johns Hopkins University Press, for Resources for the Future, 1975. 368 pp. Illus., tables, index. Emphasizes contemporary USFS organization, administration, problems, and controversies, but also includes some history of the agency and its predecessors since 1876.

Robinson, Harry B. See Robinson, Donald H., #R243

R247 Robinson, J.M. "Agriculture and Forests of Yukon Territory." *Canadian Geographical Journal* 31 (Aug. 1945), 55-72.

R248 Robinson, J.M. "Forest Resources of the Mackenzie River Basin, Northwest Territories." *Polar Record* 10 (Sept. 1960), 231-36. Incidental history.

R249 Robinson, J.M. *The History of Wood Measurement in Canada.* Forest Management Research and Services Institute, Internal Report FMR-7. Ottawa: Forestry Branch, 1967. 185 pp. Tables. Processed.

R250 Robinson, R.R. "Forest and Range Fire Control in Alaska." *Journal of Forestry* 58 (June 1960), 448-53. Since the 1920s.

R251 Robinson, William W. *The Forest and the People: The Story of the Angeles National Forest.* Los Angeles: Title Insurance and Trust Company, 1946. ix + 45 pp. Illus., maps. History of the national forest and the region since the 19th century; includes accounts of lumbering, pioneer settlement, forest fires, outdoor recreation, etc.

R252 Robinson, William W. *Land in California: The Story of Mission Lands, Ranchos, Squatters, Mining Claims, Railroad Grants, Land Scrip, Homesteads.* Berkeley: University of California Press, 1948. xiii + 291 pp. Illus., maps, bib. Land policy since the Spanish period.

R253 Robison, William C. "Historical Geography of the Santa Cruz Mountain Redwoods." Master's thesis, Univ. of California, Berkeley, 1949. Especially the lumber industry.

R254 Robison, William C. "Cultural Plant Geography of the Middle Appalachians." Ph.D. diss., Boston Univ., 1960. 329 pp. On the original forest cover of the Virginia and Maryland Appalachian region, and the extent and nature of man's modification of it.

Robnett, Ronald H. See Fowler, William A., #F195

R255 Robson, Durward. "Conservation on the Wing." *Michigan Conservation* 25 (July-Aug. 1956), 8-11. On the use of airplanes by the Michigan Department of Conservation since 1933.

R256 Rock, David. "A Program of Sustained Yield Forestry in the Appalachian Hardwood Region." *Journal of Forestry* 59 (Feb. 1961), 114-18. The demonstration forest at Berea College, Kentucky, since 1907.

R257 Rockwell, Arthur E. "The Lumber Trade and the Panama Canal, 1921-1940." *Economic History Review,* Second series, 24 (Aug. 1971), 445-62. On the impact of the canal upon lumber shipments from the Pacific Northwest.

R258 Rockwell, Landon Gale. "National Resources Planning: The Role of the National Resources Planning Board in the Process of Government." Ph.D. diss., Princeton Univ., 1942. 253 pp. National Resources Planning Board since 1939.

R259 Rodgers, Andrew Denny. *John Torrey, A Story of North American Botany.* Princeton, New Jersey: Princeton University Press, 1944. 340 pp. Bib. Torrey (1796-1873) was a leading botanist and natural scientist.

R260 Rodgers, Andrew Denny. *American Botany, 1873-1892: Decades of Transition.* Princeton, New Jersey: Princeton University Press, 1944. 340 pp.

R261 Rodgers, Andrew Denny. *Liberty Hyde Bailey: A Story of American Plant Sciences.* Princeton, New Jersey: Princeton University Press, 1949. 506 pp. Bailey (1858-1954), a botanist and horticulturalist, also made contributions to forestry and conceived of the Cornell University arboretum.

R262 Rodgers, Andrew Denny. *Bernhard Eduard Fernow: A Story of North American Forestry.* Princeton, New Jersey: Princeton University Press, 1951. 623 pp. Illus., notes, index. Fernow (1851-1923) headed the Division of Forestry, 1886-1898, and was a forestry educator at the New York State College of Forestry, Pennsylvania State College, and the University of Toronto. A prolific writer and active promoter, he did much to shape the science and profession of forestry in the United States and Canada.

R263 Rodney, William. *Kootenai Brown: His Life & Times, 1839-1916.* Sidney, British Columbia: Gray's Publishers, 1969. 251 pp. Illus., maps. John George Brown was a promoter of Waterton Lakes National Park, Alberta.

R264 Roelofs, Vernon W. *100 Years of Leadership in the Millwork Industry.* Oshkosh, Wisconsin: Paine Lumber Company, 1953. 52 pp. The Paine Lumber Company gradually came to specialize in millwork and door manufacturing.

R265 Roelofs, Vernon W. *Fifty Colorful Years, 1907-1957.* Appleton, Wisconsin: Appleton Coated Paper Company, 1957. 48 pp.

R266 Rogers, Benjamin F., Jr. "The United States Department of Agriculture (1862-1889): A Study in Bureaucracy." Ph.D. diss., Univ. of Minnesota, 1951.

R267 Rogers, Benjamin F. "William Gates Le Duc, Commissioner of Agriculture." *Minnesota History* 34 (Autumn 1955), 287-95. In the Rutherford B. Hayes administration, 1877-1881.

R268 Rogers, Edmund B., comp. *History of Legislation Relating to the National Park System Through the 82nd Congress.* 3 Volumes. Washington: USDI, n.d.

R269 Rogers, Edmund B. "Notes on the Establishment of Mesa Verde National Park." *Colorado Magazine* 29 (Jan. 1952). 10-17. From 1886 to 1913.

R270 Rogers, George W. *Alaska in Transition: The Southeast Region.* Baltimore: Johns Hopkins Press, for Resources for the Future, 1960. xiii + 384 pp. Illus., maps, tables, notes, apps., index. Includes some history of forest resources and their management by the USFS.

R271 Rogers, George W. *The Future of Alaska: Economic Consequences of Statehood.* Baltimore: Johns Hopkins University Press, for Resources for the Future, 1962. 326 pp. Includes some history of natural resource development and conservation.

R272 Rogers, James P. "The Battle for Public Timber." *American Forests* 61 (July 1955), 8-12, 42-44. Includes some history of federal management of the revested Oregon and California Railroad grant lands in western Oregon since 1916.

R273 Rogers, Lore A., and Scribner, Caleb W. "The Log Haulers." *Northern Logger and Timber Processor* 16 (Aug. 1967), 8-9, 34-38. Especially the Lombard log hauler in Maine and elsewhere since ca. 1900.

R274 Rogers, Nelson S. "Who Shall Regulate Our Forests?" *Journal of Forestry* 40 (May 1942), 384-87. Includes some history of state forestry and state regulation of private forest management in Oregon since 1867.

R275 Rogers, Nelson S. "Objectives and Accomplishments of the Oregon Forest Laws." *Journal of Forestry* 42 (July 1944), 480-82. Includes some incidental history of forest legislation in Oregon since 1913.

Rogers, R.D. Shaw, Charlie, #S192

R276 Rohr, Fred W., and Potzger, John E. "Forest and Prairie in Three Northwestern Indiana Counties." *Butler University Botanical Studies* 10 (1950), 80-89.

R277 Rohrbough, Malcolm J. "The General Land Office, 1812-1826: An Administrative Study." Ph.D. diss., Univ. of Wisconsin, 1963. 270 pp.

R278 Rohrbough, Malcolm J. *The Land Office Business: The Settlement and Administration of American Public Lands, 1789-1837.* New York: Oxford University Press, 1968. xiii + 331 pp. Maps, tables, notes, bib., index. A general history of federal land policy and its administration by the General Land Office.

R279 Rolfe, Mary A. *Our National Parks.* 2 Volumes. 1927-1928. Third edition. New York: Benjamin H. Sanborn and Company, 1937.

R280 Rolland, Jean. "The Cradle of Canada's Paper Industry: Province of Quebec is Proud of Historic 'Firsts' in Paper Making." *Pulp and Paper Magazine of Canada* 40 (Oct. 1939), 619-20.

Rolle, Andrew. See Knoles, George H., #K217

R281 Roller, Harry M., Jr. "The Pulp and Paper Industry in the Southwest Alabama Forest Empire." *Journal of the Alabama Academy of Science* 30 (Jan. 1959), 67-72.

R282 Rollins, George W. "Land Policies of the United States as Applied to Utah to 1910." *Utah Historical Quarterly* 20 (July 1952), 239-51.

R283 Roloff, Clifford E. "The Mount Olympus National Monument." *Washington Historical Quarterly* 25 (1934), 214-28. On the discovery and exploration of Washington's Olympic Peninsula and the creation (1909) and administration of the national monument, later Olympic National Park.

R284 Romancier, Robert M. "Natural Area Programs." *Journal of Forestry* 72 (Jan. 1974), 37-42. Historical survey of public and private efforts to preserve undisturbed forested areas for scientific or educational uses at the state, regional, national, and international levels, since the 1910s and 1920s.

R285 Roosevelt, Franklin Delano. *Franklin D. Roosevelt & Conservation, 1911-1945.* Comp. and ed. by Edgar B. Nixon. 2 Volumes. Hyde Park, New York: U.S. National Archives and Records Service, Franklin D. Roosevelt Library, 1957. Reprint. New York: Arno Press, 1972. Illus., notes, index. Contains selected incoming and outgoing correspondence; the documents and letters selected represent about one-third of the total number that have been identified among the Roosevelt Papers bearing on the subject of resource conservation.

R286 Roosevelt, Nicholas. *Theodore Roosevelt, The Man as I Knew Him.* New York: Dodd, Mead & Company, 1967. 205 pp. Illus. Includes some history and reminiscence of Roosevelt's role in the conservation movement.

R287 Roosevelt, Nicholas. *Conservation: Now or Never.* New York: Dodd, Mead & Company, 1970. x + 238 pp. Index. Includes case histories of efforts to preserve scenery and wilderness in the United States since ca. 1900. Special topics include the USFS; U.S. National Park Service; Jackson Hole, Grand Teton National Park, Wyoming; Oregon's state park system; U.S. Bureau of Land Management; Society for the Protection of New Hampshire Forests; Storm King Mountain, New York; and others.

R288 Roosevelt, Theodore. *Theodore Roosevelt: An Autobiography.* New York: Macmillan, 1913. xii + 647 pp. Illus. There are other editions.

R289 Roosevelt, Theodore. *The Letters of Theodore Roosevelt.* Ed. by Elting E. Morison. 8 Volumes. Cambridge: Harvard University Press, 1951-1954. Many reveal Roosevelt's interest in natural resources, forestry, and conservation.

R290 Roosevelt, Theodore. *Theodore Roosevelt's America: Selections from the Writings of the Oyster Bay Naturalist.* Ed. by Farida A. Wiley. New York: Devin-Adair, 1955. xxiii + 418 pp. Illus., bib., index. Includes some writings pertaining to natural resources and conservation.

Roper, Laura Wood. See Olmsted, Frederick Law, #O32

R291 Roper, Laura Wood. *FLO: A Biography of Frederick Law Olmsted.* Baltimore: Johns Hopkins University Press, 1973. xx + 555 pp. Illus., apps., notes, index. Olmsted (1822-1903), a leading landscape architect, planner, and advocate of parks, was a seminal figure in the American conservation movement.

R292 Roppel, Patricia. "Gravina." *Alaska Journal* 2 (Summer 1972), 13-15. On a sawmill and lumber town near Ketchikan, Alaska, 1892-1904.

R293 Roppel, Patricia. "Alaskan Lumber for Australia." *Alaska Journal* 4 (Winter 1974), 20-24. Lumber exports from southeastern Alaska to Australia, 1919-1923.

R294 Rosbach, Ronald L. "The Lieu Land Controversy in Eastern Washington." Master's thesis, Washington State Univ., 1957.

R295 Rose, B.B., and Bramlett, G.A. *Timber and Wood Products in the Economic Development of the Coosa Valley Area of Georgia.* Georgia Agricultural Experiment Station, Bulletin No. 91. Athens, 1962. 33 pp.

R296 Rose, Gerald A. "The Westwood Lumber Strike." *Labor History* 13 (Spring 1972), 171-99. On relations between lumber workers and the Red River Lumber Company, Westwood, California, especially the strikes of 1938 and 1939 led by the International Woodworkers of America.

R297 Rose, Harold Milton. "An Analysis of Land Use in Central-North Florida." Ph.D. diss., Ohio State Univ., 1960. 275 pp. Includes some history of this mixed farm and forest region.

R298 Rosenberger, Jesse Leonard. *Law for Lumbermen: A Digest of Decisions of Courts of Last Resort on Matters of Interest to Lumbermen, Arranged by Subject.* Chicago: American Lumberman, 1902. 275 pp. Reprinted from the columns of the *American Lumberman.*

R299 Rosenbluth, Robert. "The Many Lives of Robert Rosenbluth: Excerpts from His Autobiography." *Forest History* 8 (Spring/Summer 1964), 17-21. From his student days at the Yale Forest School, with the USFS in the West, and in state forestry in New York, 1904-1914.

R300 Rosenkrantz, Barbara Gutmann, and Koelsch, William A., eds. *American Habitat: A Historical Perspective.* New York: Free Press, 1973. x + 372 pp. Illus. Includes some essays on conservation history.

R301 Ross, Alexander H.D. "The Forest Resources of the Labrador Peninsula." *Canadian Forestry Journal* 1 (1905), 28-37.

R302 Ross, Alexander H.D. *Ottawa Past and Present.* Ottawa: Thorburn and Abbott, 1927. 224 pp. Illus. Includes some history of the lumber industry and the timber trade in the Ottawa Valley.

R303 Ross, C.R. "From Teacher to Timberman." *American Forests* 60 (Oct. 1954), 14-15, 42-44. T.J. Starker, forestry educator at Oregon State University and practitioner of forestry on his own timberlands in Oregon.

R304 Ross, Charles R. "An Old Sash Saw in the Southern Appalachians." *Southern Lumberman* (Sept. 15, 1940), 47. West Virginia.

R305 Ross, Charles S. "Robert Walter Graeber." *Journal of Forestry* 46 (Dec. 1948), 932-33. On Graeber's career, especially as an extension forester in North Carolina.

R306 Ross, Donald. *A History of the Shipbuilding Industry in New Brunswick.* Fredericton: University of New Brunswick, 1933. 100 pp.

R307 Ross, Earle D. "Squandering Our Public Lands." *American Scholar* 2 (Jan. 1933), 77-86. Includes some general history of federal land disposal.

R308 Ross, Earle D. "The United States Department of Agriculture During the Commissionership: A Study in Politics, Administration, and Technology, 1862-1889." *Agricultural History* 20 (Apr. 1946), 129-43.

R309 Ross, Helen Beaumont. "A Study of Conservation Education in the Elementary Schools of the U.S." Ph.D. diss., Cornell Univ., 1950. 117 pp.

R310 Ross, John R. " 'Pork Barrels' and the General Welfare: Problems in Conservation, 1900-1920." Ph.D. diss., Duke Univ., 1969. 592 pp. On the development of multipurpose resource management policies, with special attention to North Carolina.

R311 Ross, John R. "Conservation and Economy: The North Carolina Geological Survey, 1891-1920." *Forest History* 16 (Jan. 1973), 20-27. On the survey of natural resources and the developing conservation movement.

R312 Ross, John R. "Conservation and the Congress, 1900-1920." *Virginia Social Science Journal* 8 (July 1973), 41-46. An analysis of conservationists and anticonservationists in Congress according to such factors as political affiliation, lodge membership, education, rural vs. urban origins, and vocation.

R313 Ross, John R. "Benton MacKaye: The Appalachian Trail." *Journal of the American Institute of Planners* 41 (Mar. 1975), 110-14. Biographical sketch of MacKaye (b. 1879), with emphasis on his proposal for an Appalachian Trail and related conservation activities, 1920s.

R314 Ross, John Simpson, II. *A Pioneer Lumberman's Story: Autobiography of John Simpson Ross II.* Ed. by Dorothy Ross Balaam. Fort Bragg, California: Mendocino County Historical Society, 1972. 42 pp.

R315 Ross, Mildred I. "*Pinus Virginiana* in the Forest Primeval of Five Southern Indiana Counties." *Butler University Botanical Studies* 10 (Aug. 1951), 80-89. Studies of the extent of "pitch pine," based on land-survey records made between 1799-1809, and other early sources.

R316 Ross, Norman M. *The Tree-Planting Division: Its History and Work.* Ottawa: Forestry Branch, 1923. 14 pp. Illus.

Rossman, Allen. See Rossman, George. #R317

R317 Rossman, George, and Rossman, Allen. *The Chippewa National Forest.* Grand Rapids, Minnesota: Herald-Review, 1956. 22 pp. Minnesota.

R318 Rossman, Laurence Alonzo. *The Lumberjack.* Grand Rapids, Minnesota: By the author, 1948. 12 pp. Informal description of Minnesota lumberjacks in the 1880s and 1890s.

R319 Rostlund, Erhard. "The Myth of a Natural Prairie Belt in Alabama: An Interpretation of Historical Records." *Annals of the Association of American Geographers* 47 (Dec.

1957), 392-411. On the original vegetation of the Black Belt of Mississippi and Alabama as indicated by travelers' accounts and other records from as early as 1544.

R320 Roth, Filibert. *On the Forestry Conditions of Northern Wisconsin.* Madison: Wisconsin Geological and Natural History Survey, 1898. 78 pp. Includes an account of the original forest.

R321 Roth, Filibert. *A First Book of Forestry.* Boston: Ginn & Company, 1902. ix + 291 pp. Illus., maps, tables, apps., index. Includes references to the history of forestry.

R322 Roth, Filibert. "Great Teacher of Forestry Retires." *American Forestry* 26 (Apr. 1920), 209-12. A biographical sketch and appreciation of Bernhard Eduard Fernow as an educator at the New York State College of Forestry at Cornell University and at the University of Toronto, from which he retired in 1919.

R323 Roth, Filibert. *Forest Regulation, or the Preparation and Development of Forest Working Plans.* 1914. Second edition. Ann Arbor: George Wahr, 1925. ix + 239 pp. Incidental history.

R324 Roth, Filibert. *Forest Valuation.* 1916. Second edition. Ann Arbor: George Wahr, 1926. v + 176 pp. Incidental history. This and the previous entry comprise the *Michigan Manual of Forestry.*

R325 Roth, Hal. *Pathway in the Sky: The Story of the John Muir Trail.* Berkeley: Howell-North Books, 1965. vii + 231 pp. Illus., maps, notes, index. In California's Sierra Nevada.

R326 Rothery, Julian E. "Forest Resources of Labrador Peninsula." *Paper Trade Journal* 56 (No. 8, 1913), 199-205.

R327 Rothery, Julian E. *A Study of Forest Taxation in the Pacific Northwest.* Portland: Industrial Forestry Association, 1952. 47 pp. Illus., tables.

Rothra, Elizabeth Ogren. See Gifford, John C. #G111

Rothrock, C.W. See Miller, R.L., #M467

R328 Rothrock, Joseph T. "On the Growth of the Forestry Idea in Pennsylvania." *Proceedings of the American Philosophical Society* 32 (1893), 332-42.

R329 Rothrock, Joseph T. "Forests of Pennsylvania." *Proceedings of the American Philosophical Society* 33 (1894), 114-33. Incidental history.

R330 Rothrock, Joseph T. "A Century of American Lumbering." Connecticut State Board of Agriculture, *Annual Report* 28 (1894), 197-216.

R331 Rothrock, Joseph T. *Areas of Desolation in Pennsylvania.* Philadelphia: H. Welsh, 1915. 30 pp. Illus.

R332 Round, Harold F. "The Wilderness." *Virginia Cavalcade* 14 (No. 3, 1964) 4-9. The site of Civil War battles was turned into tangled thickets as a consequence of timber cutting for the nearby iron industry in the 18th century.

R333 Roundy, Charles G. "Changing Attitudes Toward the Maine Wilderness." Master's thesis, Univ. of Maine, 1970. 217 pp. On the forestry, conservation, and preservation movements since the late 19th century.

R334 Rousseau, B.G. "Early Lumbering in California." *Overland Monthly* 82 (Oct. 1924), 451-52, 480. Brief account of a sawmill erected at Bodega Corners in 1843 and of logging in the Oakland Hills.

R335 Rousseau, Jacques. "De la forêt hudsonienne à Madagascar avec le citoyen Michaux." *Cahiers des Dix* 29 (1964), 223-45. Includes an account of André Michaux's travels and botanical work in North America, 1780s-1790s.

R336 Rousseau, L.Z. "Les Sciences et la technique forestiere." *Culture* 10 (Jan. 1949), 129-33; 11 (Mar. 1950), 29-38.

R337 Rowan, James. *The I.W.W. in the Lumber Industry.* Seattle: Lumber Workers Industrial Union No. 500, 1919. 64 pp.

R338 Rowe, Samuel M. *Handbook of Timber Preservation.* Chicago: Atchison, Topeka, and Santa Fe Railroad, 1904. Includes some history of the railroad's Wellhouse process of timber preservation.

R339 Rowe, William Hutchinson. *Shipbuilding Days in Casco Bay, 1727-1890: Being Footnotes to the Maritime History of Maine.* Yarmouth, Maine: By the author, 1929. xii + 222 pp.

R340 Rowe, William Hutchinson. *The Maritime History of Maine: Three Centuries of Shipbuilding & Seafaring.* New York: W.W. Norton, 1948. 333 pp. Illus., maps, bib. From 1614 to ca. 1900, with emphasis on wooden shipbuilding.

R341 Rowland, H.B., and Smith, W.H. "Reclaiming 'Pennsylvania's Desert'." *Pennsylvania Forests and Waters* 3 (Jan.-Feb. 1951), 4-7, 20-21. Reforestation on burned and cutover lands since 1897.

R342 Rowland, H.B., and Mickalitis, Albert B. *Historical Highlights in Pennsylvania Forestry.* Harrisburg: Pennsylvania Department of Forests and Waters, 1955. 7 pp. Processed. A chronological listing of events since 1662.

R343 Rowland, Leon. "Lewis and Clark, Timber Cruisers." *Timberman* 24 (Mar. 1923), 47-48. Data taken from the journals of their expedition to the Pacific Northwest, 1805-1806.

R344 Rowland, Leon. "Pioneers of the Redwood." *Timberman* 30 (Mar. 1929), 44. On lumbermen Isaac Graham, William Sawyer, R.O. Tripp, and Dennis Martin, pre-1850 sawmill owners in California's Santa Cruz Mountains.

R345 Rowley, Virginia M. "J. Russell Smith, Geographer, Teacher, and Conservationist." Ph.D. diss., Columbia Univ., 1961. 261 pp.

R346 Rowley, Virginia M. *J. Russell Smith, Geographer, Educator, and Conservationist.* Philadelphia: University of Pennsylvania Press, 1964. 247 pp. Biography of Joseph Russell Smith (1874-1966), who was involved in forestry and the promotion of tree nut crops in the Appalachian region and elsewhere.

R347 Rowson, Leonard. "Minnesota's County Forests." *Northern Logger* 12 (Mar. 1964), 26-27, 44-45, 47. Includes some history of county forests or tax-forfeited lands in Minnesota since ca. 1920.

R348 Roy, Henri. "Ranger School Education in Quebec." *Forestry Chronicle* 4 (Mar. 1928), 41-48.

R349 Roy, Henri. "Log Scaling in Quebec." *Journal of Forestry* 36 (Oct. 1938), 969-75. The evolution of log scaling rules since 1865.

R350 Royce, Ann L. "Monadnock: Past and Present." *Forest Notes* (Summer 1970), 2-6. History of a New Hampshire state park.

Rudd, Robert D. See Highsmith, Richard Morgan, Jr., #H362

R351 Rudd, V.E. "Botanical Contributions of the Lewis and Clark Expedition." *Journal of the Washington Academy of Science* 44 (Nov. 1954), 351-56. To the Pacific Northwest in 1805-1806.

Rudnicki, Judith C. See Collingwood, George Harris, #C524

R352 Rudolph, Paul O. "History of Forest Planting in the Lake States." *Minnesota Conservationist*, No. 29 (1935), 12-13, 23-24. A survey of reforestation work in Michigan, Minnesota, and Wisconsin since 1888.

R353 Rudolf, Paul O. *Forest Plantations in the Lake States*. USFS, Technical Bulletin No. 1010. Washington: GPO, 1950. 171 pp. Illus., maps, tables, bib. Includes some history of reforestation in Minnesota, Wisconsin, and Michigan since the 1870s.

R354 Rue, J.D. "The Development of Pulp and Paper-making in the South." *Southern Lumberman* 117 (Dec. 20, 1924), 141-44, 148. Some history and comparison with other regions.

R355 Rue, J.D. "Fifty Years' Progress in the Pulp Industry." *Industrial and Engineering Chemistry* 18 (1926), 917-19.

R356 Ruenheck, Wilbert H. "Business History of the Robert Gair Company, 1864 to 1927." Ph.D. diss., New York Univ., 1951. A paper firm.

R357 Ruhle, George C. *Guide to Glacier National Park*. Minneapolis: Campbell Withun, 1949. Includes some history of the Montana park.

R358 Rule, Edwin R. "Incident Near Dirty Shirt Peak." *American Forests* 58 (Aug. 1952), 10-12, 38-41. A forest fire in the Piute Range east of Bakersfield, California, 1947.

R359 Rummell, R.S. "An Abridged History of Southern Range Research." *Iowa State Journal of Science* 34 (No. 4, 1960), 749-59.

R360 Runeberg, L. *Trade in Forest Products between Finland and the United States of America*. Helsinki, 1946. 166 pp. Diags., bib. Incidental history.

R361 Runte, Alfred. "Yellowstone: It's Useless, So Why Not a Park?" *National Parks & Conservation Magazine* 46 (Mar. 1972), 5-7. Theorizing on the establishment of Yellowstone National Park in 1872.

R362 Runte, Alfred. "How Niagara Falls Was Saved: The Beginning of Esthetic Conservation in the United States." *Conservationist* 26 (Apr.-May 1972), 32-35, 43. On the impulse that also resulted in the preservation of wilderness.

R363 Runte, Alfred. "Beyond the Spectacular: The Niagara Falls Preservation Campaign." *New-York Historical Society Quarterly* 57 (No. 1, 1973), 30-50. On the efforts of Charles Eliot Norton, Frederick Law Olmsted, and others to preserve the natural beauty of Niagara Falls against commercial encroachment, 1869-1885.

R364 Runte, Alfred. " 'Worthless' Lands—Our National Parks." *American West* 10 (May 1973), 4-11. Includes some history of national parks and their promotion and preservation as lands being worthless or only marginally valuable for other than recreational purposes, since 1872.

R365 Runte, Alfred. "Pragmatic Alliance: Western Railroads and the National Parks." *National Parks & Conservation Magazine* 48 (Apr. 1974), 14-21. On the role of the railroad companies in publicizing national parks and capitalizing on the tourist trade since the 1870s.

R366 Runte, Alfred. "Yosemite Valley Railroad: Highway to History, Pathway to Promise." *National Parks & Conservation Magazine* 48 (Dec. 1974), 4-9. On the railroad between Merced and El Portal, California, and its impact upon tourist use of Yosemite National Park, 1907-1944.

R367 Runyon, K.L., and others. *Economic Impact of the Nova Scotia Forest Industry*. Fredericton, New Brunswick: Maritimes Forest Research Centre, 1973. 78 pp.

R368 Rupp, Alfred E. "History of Land Puchase in Pennsylvania." *Journal of Forestry* 22 (May 1924), 490-97. Emphasizes the origins of a state forest system after 1895.

R369 Rush, William Marshall. *Wild Animals of the Rockies: Adventures of a Forest Ranger*. New York: Harper, 1942. xxiii + 296 pp. Reminiscences of a USFS ranger in the West since ca. 1910.

R370 Russell, Carl P. "Early Years in Yosemite." *California Historical Society Quarterly* 5 (Dec. 1926), 328-41. Yosemite National Park, California.

R371 Russell, Carl P. *A Concise History of Scientists and Scientific Investigations in Yellowstone National Park With a Bibliography of the Results of Research and Travel in the Park Area*. Washington: USDI, Office of National Parks, Buildings and Reservations, 1934. 144 pp. Processed.

R372 Russell, Carl P. "Wilderness Preservation." *National Parks Magazine* 71 (Apr.-June 1944), 3-6, 26-28. Incidental history.

R373 Russell, Carl P. *One Hundred Years in Yosemite: The Story of a Great Park and Its Friends*. 1931. Revised edition. Berkeley: University of California Press, 1947. 226 pp. Illus., bib. The Yosemite Natural History Association published a reprint in 1957.

R374 Russell, Carl P. "Centennial in Yosemite." *Sierra Club Bulletin* 36 (May 1951), 72-74. Yosemite National Park, California.

R375 Russell, Carl P. "A History of the National Park Service." *National Parks Magazine* 33 (May 1959), 6-11. Includes some history of the movement leading to its establishment in 1916, as well as subsequent history.

R376 Russell, Carl P. "Birth of the National Park Idea." In *Yosemite, Saga of a Century, 1864-1914*. Oakhurst, California, 1964.

R377 Russell, Charles E. "The Mysterious Octopus." *World Today* 21 (Feb. 1912), 1735-50; 21 (Mar. 1912), 1960-72; 21 (Apr. 1912), 2074-85. Muckraking view of timber acquisition by the lumber industry.

R378 Russell, Charles Edward. *A-rafting on the Mississipp'*. New York: Century Company, 1928. xii + 357 pp. Illus., maps, apps., index. Rafting, steamboating, and the lumber industry on the Upper Mississippi River to about 1915.

R379 Russell, Curran N., and Baer, Donna Degen. *The Lumbermen's Legacy*. Manistee, Michigan: Manistee County History Society, 1954. 67 pp. Map, illus. On Manistee, Michigan, and its lumber industry since 1841.

R380 Russell, Franklin. "The Vermont Prophet: George Perkins Marsh." *Horizon* 10 (No. 3, 1968), 16-23. On Marsh (1801-1882) and his influence on the conservation movement.

R381 Russell, Jason Almus. "Gathering Sawdust." *Chronicle of Early American Industries* 3 (Mar. 1950), 201, 204. On uses for sawdust in farming in New England, ca. 1900.

R382 Russell, Jason Almus. "Wooden Trunks from American Forests." *American Forests* 65 (Sept. 1959), 32-33, 42-43. On the manufacture of wooden trunks and luggage, 18th-19th centuries.

R383 Russell, John Andrew. *Joseph Warren Fordney: An American Legislator*. Boston: Stratford Company, 1928. 247 pp. U.S. representative from Michigan, 1899-1923, Fordney earned his fortune in the lumber industry near Saginaw.

R384 Russell, Paul H. "We Still Have Chestnut." *Southern Lumberman* 179 (Dec. 15, 1949), 236-38. On the commercial uses of the American chestnut in the southern Appalachian Mountains and the effects of chestnut blight since 1915.

R385 Rutledge, Peter J. "Genesis of the Steam Logging Donkey." *Timberman* 34 (Mar. 1933), 9ff. In California.

R386 Rutledge, Peter J., and Tooker, Richard H. "Steam Power for Loggers: Two Views of the Dolbeer Donkey." *Forest History* 14 (Apr. 1970), 18-29. Reminiscences of donkey engines in California redwood logging since the 1880s.

R387 Ruttan, V.W., and Callahan, J.C. "Resource Inputs and Output Growth: Comparisons between Agriculture and Forestry." *Forest Science* 8 (Mar. 1962), 68-82. On price movements and productivity in the lumber industry from 1868 to 1959.

R388 Ryan, Charles B. *Molding Its Future with Wood and Plastics: The Story of MPI Industries, Inc.* New York: Newcomen Society in North America, 1968. 24 pp. Illus.

R389 Ryan, D.E. "Forestry Public Relations in New England." Master's thesis, New York State College of Forestry, 1941.

R390 Ryan, J.C. "Power in the Forest: Railroad Logging in Minnesota." *Conservation Volunteer* 6 (May 1943), 31-36.

R391 Ryan, J.C. "Minnesota Logging Railroads." *Minnesota History* 27 (Dec. 1946), 300-08.

R392 Ryan, J.C. "Loggers of the Past." *Timber Producers Association Bulletin* 24-29 (1969-1975). A series of short articles on the lumber industry and loggers of Minnesota since the late 19th century.

R393 Ryan, J.C. *Early Loggers in Minnesota*. Duluth: Minnesota Timber Producers Association, 1973. 47 pp. Illus., map. A reprint of the first twenty articles in his "Loggers of the Past" series from the *Timber Producers Association Bulletin*.

R394 Ryan, J.C. "Loggers of the Past." *Northwestern Lumberman* 2 (1974-1975). Reprints from the author's *Early Loggers in Minnesota* (1973).

R395 Ryan, Victor A. *Some Geographic and Economic Aspects of the Cork Oak*. Baltimore: Crown Cork and Seal Company, 1948. 116 pp.

R396 Ryder, David Warren. *Memories of the Mendocino Coast: Being a Brief Account of the Discovery, Settlement, and Development of the Mendocino Coast, Together with the Correlated History of the Union Lumber Company and How Coast and Company Grew up Together*. San Francisco: Union Lumber Company, 1948. xiv + 81 pp. Illus., map. On the lumber industry in the redwood belt of the northern California coast since 1851. The Union Lumber Company was established at Ft. Bragg in the 1880s by Charles R. Johnson.

R397 Rymon, Larry M. "A Critical Analysis of Wildlife Conservation in Oregon." Ph.D. diss., Oregon State Univ., 1969. 441 pp. Including some history since the 19th century, with emphasis on the Oregon Game Commission.

S1 Saalberg, John J. "Roosevelt, Fechner and the CCC—A Study in Executive Leadership." Ph.D. diss., Cornell Univ., 1962. 223 pp. Civilian Conservation Corps, 1933-1942.

S2 Sachse, Nancy D. "Madison's Public Wilderness: The University of Wisconsin Arboretum." *Wisconsin Magazine of History* 44 (Winter 1960).

Sackett, H.S. See Simmons, Roger E., #S285, S286

Sage, W.N. See Angus, Henry Forbes, #A459

S3 St. Barbe Baker, Richard. *Green Glory: The Forests of the World*. New York: A.A. Wyn, 1949. 253 pp. Illus., index. Includes some description and history of forested lands, the lumber industry, and other aspects of forest use in North America.

S4 Sakarias, Michael. "The Cleveland National Forest: San Diego's Watershed." *Journal of San Diego History* 21 (Fall 1975), 54-65.

S5 Sakolski, Aaron M. *The Great American Land Bubble: The Amazing Story of Land-Grabbing, Speculations, and Booms from Colonial Days to the Present Time*. New York: Harper and Brothers, 1932. Reprint. New York: Johnson Reprint Corporation, 1966. xii + 373 pp. Illus., maps, notes, diags. Includes some history of the public domain and its disposal, with emphasis on land companies, speculators, and frauds.

S6 Sakolski, Aaron M. *Land Tenure and Land Taxation in America*. New York: Robert Schalkenbach Foundation, 1957.

xii + 316 pp. Notes, bib., index. A general history of federal land policy to 1879, providing a historical background for the single-tax doctrine of Henry George. There is a chapter on forested lands.

S7 Saley, Met Lawson. *Realm of the Retailer: The Retail Lumber Trade, Its Difficulties and Successes, Its Humor and Philosophy, Its Theory and Practice, with Practical Yard Ideas.* Chicago: American Lumberman, 1902. 386 pp. Illus. Derived from the author's weekly column in *American Lumberman;* includes some history of the early retail lumber trade.

S8 Salmond, John A. "The Civilian Conservation Corps and the Negro." *Journal of American History* 52 (June 1965), 75-88. On the CCC's failure to provide for full participation by Negroes, 1933-1942. The author divides blame between Franklin D. Roosevelt and CCC director Robert Fechner.

S9 Salmond, John A. " 'Roosevelt's Tree Army': A History of the Civilian Conservation Corps, 1933-1942." Ph.D. diss., Duke Univ., 1964. 311 pp.

S10 Salmond, John A. *The Civilian Conservation Corps, 1933-1942: A New Deal Case Study.* Durham, North Carolina: Duke University Press, 1967. vi + 240 pp. Notes, bib., index. A general history which also evaluates the specific reforestation and recreational achievements of the CCC.

S11 Salo, Sarah Jenkins. *Timber Concentration in the Pacific Northwest; with Special Reference to the Timber Holdings of the Southern Pacific Railroad, the Northern Pacific Railroad and the Weyerhaeuser Timber Company.* Ann Arbor, Michigan: Edwards Brothers, 1945. 79 pp. A doctoral dissertation submitted at Columbia University.

S12 Saloutos, Theodore. "The New Deal and Farm Policy in the Great Plains." *Agricultural History* 43 (July 1969), 345-55. Includes reference to shelterbelt and soil conservation programs of the 1930s.

S13 Salter, Leonard A., Jr. *A Critical Review of Research in Land Economics.* Madison: University of Wisconsin Press, 1967. xix + 258 pp. Index. Includes some history of forest conservation and land utilization on the cutover areas of the Great Lakes states.

S14 Sampson, Arthur W. *Range and Pasture Management.* New York: John Wiley and Sons, 1923. xix + 421 pp. Illus., tables, index. Includes incidental references to the history of national forests, range lands, and grazing policy.

S15 Sampson, Clayburne B. "Jamestown's Store-Boats." *New York Folklore Quarterly* 13 (Autumn 1957), 163-69. On river transportation of lumber and other cargo on the Allegheny and Ohio Rivers, 1805-1885.

S16 Sanborn, John Bell. "Congressional Grants of Land in Aid of Railways." Ph.D. diss., Univ. of Wisconsin, 1899.

S17 Sanborn, Theodore A. *Credit and Collection Policies of Kansas Retail Lumber Dealers.* Kansas Studies in Business No. 4. Lawrence: University of Kansas, School of Business, 1926. 28 pp. Incidental history.

S18 Sand, George X. *The Everglades Today: Endangered Wilderness.* New York: Four Winds Press, 1971. 191 pp. Illus., maps. Incidental history of Florida's Everglades National Park.

S19 Sanderson, Mary H. "A Comparative Study of French Canadian Loggers and Their Families." Ph.D. diss., Boston Univ., 1967. 233 pp. Incidental history of lumberjacks in Quebec and New Hampshire.

Sanford, Albert H. See Sherman, Simon Augustus, #S216

S20 Sanford, Albert H. "The Beginnings of a Great Industry at La Crosse." *Wisconsin Magazine of History* 18 (June 1935), 375-88. Colman Lumber Company, La Crosse, Wisconsin, 1854-1858.

S21 Sanford, Albert H. "Log Marking and Scaling." *La Crosse County Historical Sketches,* series 3 (1937), 29-34.

S22 Sanford, Charles L. *The Quest for Paradise: Europe and the American Moral Imagination.* Urbana: University of Illinois Press, 1961. x + 282 pp. Notes, index. Includes some treatment of the relationships between Americans (transplanted Europeans) and the New World environment— both as hostile wilderness and as sublime scenery.

S23 Sanford, Everett R. "A Short History of California Lumbering Including a Descriptive Bibliography of Material on Lumbering and Forestry in California." Master's thesis, Univ. of California, Berkeley, 1931.

S24 Saposs, David J. *Communism in American Unions.* New York: McGraw-Hill, 1959. xii + 279 pp. Notes. Includes some references to workers in the lumber and pulp and paper industries.

S25 Sardo, William H., Jr. "Wooden Pallet Industry Marches Ahead." *Southern Lumberman* 193 (Dec. 15, 1956), 217-19. Since the 1930s.

S26 Sargent, Charles Sprague. "The Protection of the Forests." *North American Review* 135 (Oct. 1882), 386-401. Recommendations for a national forest policy, including some historical references.

S27 Sargent, Charles Sprague. *Report on the Forests of North America (Exclusive of Mexico).* Washington: GPO, 1884. 612 pp. Volume 9 of the *Tenth Census of the United States, 1880;* includes many incidental historical references to the trees, woods, and forests of North America.

S28 Sargent, Charles Sprague. *The Silva of North America: A Description of the Trees Which Grow Naturally in North America Exclusive of Mexico.* Boston: Houghton Mifflin, 1891-1902. 14 Volumes. Incidental history.

S29 Sargent, Charles Sprague. "The First Fifty Years of the Arnold Arboretum." *Journal of the Arnold Arboretum* 3 (1922), 127-71. Jamaica Plain, Massachusetts.

Sargent, Grace Tompkins. See Shinn, Julia Tyler, #S228

S30 Sargent, Shirley. *Galen Clark, Yosemite Guardian.* San Francisco: Sierra Club, 1964. 176 pp. Illus., notes, bib., index. Clark (1814-1910) was appointed guardian of the Yosemite Valley and Big Tree Grove by the Yosemite commissioners in 1866. He supervised recreational activities in the valley for many years in the late 19th century.

S31 Sargent, Shirley. "Pictures from Yosemite's Past: Galen Clark's Photograph Album." *California Historical Society Quarterly* 45 (Mar. 1966), 31-40.

S32 Sargent, Shirley. *Theodore Parker Lukens: Father of Forestry.* Los Angeles: Dawson's Book Shop, 1969. 91 pp. Illus. Lukens (1848-1918), a pioneer conservationist and associate of John Muir, was an agent of the Division of Forestry and participated in reforestation experiments in southern California.

S33 Sargent, Shirley. "John Muir in Yosemite." *Pacific Review* 4 (Summer 1970), 8-9.

S34 Sargent, Shirley. *John Muir in Yosemite.* Yosemite, California: Flying Spur Press, 1971. 48 pp. Illus., index. On John Muir's devotion to Yosemite Valley and Yosemite National Park, California, 1868-1914.

S35 Sargent, Shirley. "Wellllcome to Camp Curry." *California History Quarterly* 53 (Summer 1974), 131-38. An illustrated history of the Yosemite Park and Curry Company, innkeepers and tour guides in Yosemite National Park since 1899.

S36 Saskatchewan. Department of Natural Resources. *Saskatchewan's Forests: A Report.* Regina, 1955. 129 pp. Incidental history.

S37 Saskatchewan. Department of Natural Resources. *Saskatchewan's Forest Inventory, 1947 to 1956.* Regina, 1959. 127 pp.

S38 Sato, Shosuke. *History of the Land Question in the United States.* Johns Hopkins University, Studies in Historical and Political Science, Series 4, No. 7-9. Baltimore, 1886. A doctoral dissertation.

Satterlund, Donald R. See Barrett, John W., #B97

S39 Satterlund, Donald R. *Wildland Watershed Management.* New York: Ronald Press, 1972. vii + 370 pp. Illus. Incidental history.

S40 Sauerlander, Annemarie Margaret. "Henry L. Yesler in Early Seattle." *American-German Review* 26 (Feb.-Mar. 1960), 7-12. Yesler (1810-1892) owned the first sawmill in Seattle and eventually became mayor.

S41 Saunders, Audrey. *Algonquin Story.* 1947. Reprint. Toronto: Ontario Department of Lands and Forests, 1963. xii + 196 pp. Illus., maps. History of the region which became Ontario's (and Canada's) first provincial park in 1893. There is history of lumbering since the 1850s and of park administration subsequent to its preservation.

S42 Saunders, Ivan W. "Geared Iron Horses in Pennsylvania." *Pennsylvania History* 32 (Oct. 1965), 355-65. Geared locomotives, of which Shay was the most popular type, were used extensively in Pennsylvania logging operations in the late 19th and early 20th centuries.

S43 Saunders, S.A. "Forest Industries in the Maritime Provinces." In Lower, Arthur R.M.; Carrothers, W.A.; and Saunders, S.A. *The North American Assault on the Canadian Forest,* ed. by Harold A. Innis. Relations of Canada and the United States Series. Toronto: Ryerson Press; New Haven: Yale University Press, 1938. Pp. 347-71. Tables, bib., index. Resources, industries, trade, and conservation since the 1870s.

S44 Saunderson, Mont H. *Western Land and Water Use.* Norman: University of Oklahoma Press, 1950. 217 pp. Illus., bib., index. Includes some history of federal lands and natural resource policies of the West.

S45 Savage, Henry, Jr. *Lost Heritage.* New York: William Morrow and Company, 1970. 329 pp. Illus., bib., index. History and biographical sketches of seven naturalists who explored the wilderness of eastern North America in the 18th and early 19th centuries: John Lawson, Mark Catesby, John Bartram, William Bartram, André Michaux, François André Michaux, and Alexander Wilson.

S46 Savard, F.A. "Le Patrimoine forestier du Quebec." *Revue de l'Université Laval* 5 (Jan. 1951), 416-24.

S47 Sawaya, C.P. "The Employment Effect of Minimum Wage Regulation in the Southern Pine Lumber Industry." Ph.D. diss., Indiana Univ., 1959. 400 pp. Since 1938.

S48 Sawyer, Alvah L. *A History of the Northern Peninsula of Michigan and Its People, Its Mining, Lumber and Agricultural Industries.* 3 Volumes. Chicago: Lewis Publishing Company, 1911.

S49 Sawyer, Alvah L. "The Forests of the Upper Peninsula and Their Place in History." *Michigan History Magazine* 3 (July 1919), 367-83.

S50 Sawyer, Carl J. *History of Lumbering in Delta County.* Escanaba, Michigan, 1949. On Michigan's Upper Peninsula.

S51 Sawyer, Josephine. "Personal Reminiscences of the Big Fire of 1871." *Michigan History Magazine* 16 (Autumn 1932), 422-30. On the forest fire which ravaged part of Michigan's Upper Peninsula and the region near Peshtigo, Wisconsin.

S52 Sax, I.M. *Songs of the Lumberjacks.* New York: Macmillan, 1927.

S53 Say, Harold Bradley, and Baser, Nort. "They Plan by the Century." *American Forests* 58 (Feb. 1952), 27-34. On conservation of timber in western Oregon by Pope & Talbot, Inc.

S54 Sayers, Wilson B. "The Changing Land Ownership Patterns in the United States." *Forest History* 9 (July 1965), 2-9. A review of federal land policy with emphasis on forested lands administered by the federal government.

S55 Sayers, Wilson B. ". . . . . To Tell the Truth: 25 Years of American Forest Products Industries, Inc." *American Forests* 64 (Oct. 1966), 657-63. The public relations arm of the forest industries since 1941.

S56 Saylor, David J. *Jackson Hole, Wyoming: In the Shadow of the Tetons.* Norman: University of Oklahoma Press, 1970. 268 pp. Illus., notes, maps, bib., index. Includes much history of the region and of the movement to establish Grand Teton National Park.

S57 Sayn-Wittgenstein, L. "Forestry: From Branch to Department." *Canadian Public Administration* 6 (Dec. 1963), 434-52. On the organization of national forestry in Canada since 1930.

S58 Scace, Robert C. *Banff: A Cultural-Historical Study of Land Use and Management in a National Park Community to 1945.* Studies in Land Use History and Landscape Change, National Park Series, No. 2. Calgary: University of Calgary, 1968. vii + 154 pp. Illus., maps. Banff National Park, Alberta.

Scace, Robert C. See Nelson, J.G., #N61

S59 Scammon, C.M. "Lumbering in Washington Territory." *Overland Monthly* 5 (1870), 55-60. Incidental history.

S60 Schaefer, Paul A. "The Adirondacks: Pattern of a Battle of Woods and Water." *Living Wilderness* (Mar. 1946), 13-22. On the movement to preserve wilderness in northern New York.

S61 Schafer, Joseph. *History of Agriculture in Wisconsin.* Wisconsin Domesday Book, General Studies, Volume 1. Madison: State Historical Society of Wisconsin, 1922. 212 pp. Illus., bib. Includes a chapter on "Lumbering and Farming."

S62 Schafer, Joseph. "Great Fires of Seventy-One." *Wisconsin Magazine of History* 11 (Sept. 1927), 96-106. On the forest fire in northern Wisconsin which destroyed the town of Peshtigo, October, 1871.

S63 Schafer, Joseph. *Four Wisconsin Counties: Prairie and Forest.* Wisconsin Domesday Book, General Studies, Volume 2. Madison: State Historical Society of Wisconsin, 1927. vii + 429 pp. Maps, tables. Includes some history of forest utilization in Kenosha, Racine, Milwaukee, and Ozaukee counties.

S64 Schafer, Louise. "Report from Aberdeen." *Pacific Northwest Quarterly* 47 (Jan. 1956), 9-14. Includes some history of the forests and forest industries as factors in the growth of Aberdeen, Washington, since the 1870s.

S65 Schantz, Homer L. "Big Game Populations in the National Forests, 1921 to 1950." *Forest Science* 2 (Mar. 1956), 7-17.

Schantz-Hansen, Thorvald. See Cheyney, Edward G., #C289

S66 Schapsmeier, Edward L. "Henry A. Wallace: The Origins and Development of His Political Philosophy, The Agrarian Years, 1920-1940." Ph.D. diss., Univ. of Southern California, 1965. 509 pp.

S67 Schapsmeier, Edward L., and Schapsmeier, Frederick H. *Henry A. Wallace of Iowa: The Agrarian Years, 1910-1940.* Ames: Iowa State University Press, 1968. xii + 327 pp. Illus., notes, bib., index. Wallace was Franklin D. Roosevelt's Secretary of Agriculture, 1933-1940.

S68 Schapsmeier, Edward L., and Schapsmeier, Frederick H. *Prophet in Politics: Henry A. Wallace and the War Years, 1940-1965.* Ames: Iowa State University Press, 1970. xiv + 268 pp. Illus., bib., notes, index.

S69 Schapsmeier, Edward L., and Schapsmeier, Frederick H. *Ezra Taft Benson and the Politics of Agriculture: The Eisenhower Years, 1953-1961.* Danville, Illinois: Interstate Printers and Publishers, 1975. xviii + 374 pp. Bib., index. Benson was Secretary of Agriculture.

Schapsmeier, Frederick H. See Schapsmeier, Edward L., #S67, S68, S69

S70 Scharff, Robert. *The Yellowstone and Grand Teton National Parks.* New York: David McKay and Company, 1966. vi + 209 pp. Includes some history of the Wyoming national parks.

S71 Schauffler, R.E., ed. *Arbor Day: Its History, Observance, Spirit and Significance; with Practical Selections on Tree-Planting and Conservation, and a Nature Anthology.* New York, 1909. 360 pp.

S72 Scheffer, Theodore C., and Hedgcock, George G. *Injury to Northwestern Forest Trees by Sulfur Dioxide from Smelters.* USFS, Technical Bulletin No. 1117. Washington, 1955. 49 pp. Illus., maps, diags., tables, notes, bib. On "smelter injury" to forest trees in Washington and Montana, 1928-1936.

S73 Scheffey, Alice W. "The Origin of Recreation Policy in the National Forests—A Case Study of the Superior National Forest." Master's thesis, Univ. of Michigan, 1958. On the movement leading to the establishment of the "roadless area" on the Superior National Forest, Minnesota.

S74 Scheffey, Andrew J.W. *Conservation Commissions in Massachusetts.* Washington: Conservation Foundation, 1969. 216 pp. Incidental history; includes supplement on local conservation commissions in the Northeast.

S75 Schefft, Charles Ernest. "The Tanning Industry in Wisconsin: A History of Its Frontier Origins and Its Development." Master's thesis, Univ. of Wisconsin, 1938.

S76 Schell, Herbert S. "Early Manufacturing Activities in South Dakota, 1857-1875." *South Dakota Historical Review* 2 (Jan. 1937), 73-95. Includes some history of sawmilling.

S77 Schell, Herbert S. *South Dakota Manufacturing to 1900.* University of South Dakota, Business Research Bureau, Bulletin No. 40. Vermillion, 1955. 87 pp. Includes some history of the lumber industry.

S78 Schenck, Carl Alwin. *Biltmore Doings, 1898-1914.* Asheville, North Carolina: Biltmore Forest School, 1914. 47 pp. On forestry education at the Biltmore Forest School near Asheville, North Carolina.

S79 Schenck, Carl Alwin, ed. *The Biltmore Immortals.* 2 Volumes. Darmstadt, Germany: L.C. Wittich, 1953, 1957. Illus., indexes. Autobiographies of graduates, educators, and friends of the Biltmore Forest School near Asheville, North Carolina.

S80 Schenck, Carl Alwin. *The Biltmore Story: Recollections of the Beginning of Forestry in the United States.* Ed. by Ovid Butler. St. Paul: American Forest History Foundation and Minnesota Historical Society, 1955. xiv + 224 pp. Illus., index. This book was reprinted under the title *The Birth of Forestry in America: Biltmore Forest School, 1898-1913* (Santa Cruz, California: Forest History Society and Appalachian Consortium, 1974). Schenck (1868-1955), a German forester, was director of the Biltmore Forest School near Asheville, North Carolina.

S81 Schenck, Vernor. "Logging Railroads Chug Down Memory Lane." *Western Conservation Journal* 30 (Jan.-Feb. 1973), 32-33. On the demise of logging railroads.

S82 Schenck, Vernor. "Old Logging Railroads: Reaching the End of the Line." *Forests & People* 23 (First quarter, 1973), 18-19, 40-41, 44. Brief sketch on the decline of logging railroads, especially those operated by the Georgia-Pacific Corporation in the South and Pacific states.

S83 Schenck, Vernor. "Logging Railroads Reach End of Line." *Forest Farmer* 33 (Sept. 1974), 8-10. On the demise of logging railroads operated by the Georgia-Pacific Corporation, especially in Oregon and California.

S84 Schieder, Rupert. "Martin Allerdale Grainger: Woodsman of the West." *Forest History* 11 (Oct. 1967), 6-13. On his career as logger, novelist, and chief forester of the British Columbia Forest Service, 1874-1941.

S85 Schierbeck, Otto. "Lumbering is Nova Scotia's Second Industry." *Canada Lumberman* 50 (Aug. 1, 1930), 137-38.

S86 Schiff, Ashley Leo. "The United States Forest Service: Science and Administration." Ph.D. diss., Harvard Univ., 1959.

S87 Schiff, Ashley Leo. *Fire and Water: Scientific Heresy in the Forest Service.* Cambridge: Harvard University Press, 1962. xiv + 225 pp. Notes, index. Examines controversies between administrative and research personnel over controlled burning as a forest management practice and over the relation of forests to flood control.

S88 Schiff, Ashley Leo. "Innovation and Administrative Decision-Making: The Conservation of Land Resources." *Administrative Science Quarterly* 11 (June 1966), 1-30. On the influence of conservation ideology on the administration of natural resources.

S89 Schleef, Margaret L. "Rival Unionism in the Lumber Industry." Master's thesis, Univ. of California, Berkeley, 1950.

S90 Schmaltz, Norman J. "Cutover Land Crusade: The Michigan Forest Conservation Movement, 1899-1931." Ph.D. diss., Univ. of Michigan, 1972. 503 pp.

S91 Schmaltz, Norman J. "P.S. Lovejoy: Michigan's Cantankerous Conservationist." *Journal of Forest History* 19 (Apr. 1975), 72-81. Parrish Storrs Lovejoy (1884-1942) worked for the USFS in the West, taught forestry at the University of Michigan, and worked for the Michigan Department of Conservation in several capacities. He was an effective promoter of forestry, wildlife management, and land-use planning.

S92 Schmautz, Jack E. "A Study of Shelterbelts in Eastern Montana." Master's thesis, Univ. of Montana, 1948.

S93 Schmeckebier, Laurence F. *The Office of Indian Affairs: Its History, Activities and Organization.* Institute for Government Research, Service Monographs of the United States Government, No. 48. Baltimore: Johns Hopkins University Press, 1927. Reprint. New York: AMS Press, 1973. 605 pp. Includes some reference to forestry in the Indian Service.

S94 Schmidt, Herbert G. *Agriculture in New Jersey: A Three-Hundred-Year History.* New Brunswick, New Jersey: Rutgers University Press, 1973. 335 pp. Illus., notes, bib., index. Includes reference to forests.

S95 Schmiel, Eugene D. "The Career of Jacob Dolson Cox, 1828-1900: Soldier, Scholar, Statesman." Ph.D. diss., Ohio State Univ., 1969. 534 pp. Cox was Ulysses S. Grant's Secretary of the Interior, 1869-1870.

S96 Schmitt, Peter J. "Call of the Wild: The Arcadian Myth in Urban America, 1900-1930." Ph.D. diss., Univ. of Minnesota, 1966. 386 pp.

S97 Schmitt, Peter J. "The Arcadian Myth." *Forest History* 13 (Apr./July 1969), 18-26. On the "back-to-nature" movement of the early 20th century, with emphasis on the role of individual leaders such as Theodore Roosevelt.

S98 Schmitt, Peter J. *Back to Nature: The Arcadian Myth in Urban America.* New York: Oxford University Press, 1969. xxiii + 230 pp. A study of how the pressures of urban life turned middle-class Americans in the early 20th century to a new appreciation, sometimes romanticized, of nature and the outdoor life. This was manifested by increased attention to resource conservation, preservation of scenery and wildlife, support for parks and gardens, enthusiasm for outdoor literature, and the like.

S99 Schmitz, Henry. "Shelter Belt Planting Revealed in Early Minnesota Forestry." *Minnesota Conservationist* No. 21 (1935), 2-5, 17. The growth and development of the shelterbelt idea, 1866-1898, as seen from the records of the Minnesota State Horticultural Society.

S100 Schmitz, Henry. " 'To Whom the Nation Owes Most' " *Journal of Forestry* 43 (Aug. 1945), 563-65. On Gifford Pinchot's contributions to the forest conservation movement.

S101 Schmitz, Henry. *The Long Road Travelled: An Account of Forestry at the University of Washington.* Ed. by Gordon D. Marckworth. Seattle: Arboretum Foundation, University of Washington, 1973. xiv + 273 pp. Illus., maps, apps., index. Forestry education since 1894.

S102 Schmoe, F.W. *Our Greatest Mountain, A Handbook for Mount Rainier National Park.* New York: G.P. Putnam's Sons, 1925. xii + 366 pp. Incidental history.

S103 Schmon, Arthur A. *Papermakers and Pioneers: The Ontario Paper Company Limited and Quebec North Shore Paper Company.* New York: Newcomen Society in North America, 1962. 24 pp. The Ontario Paper Company, Ltd., Thorold, Ontario, and the Quebec North Shore Paper Company, Baie Comeau, Quebec.

S104 Schneider, Arthur E. "The Management of County Forest Lands in Minnesota." Ph.D. diss., Univ. of Washington, 1953. 209 pp. Includes some history of Minnesota's public lands, especially tax-defaulted county lands in northeastern Minnesota.

S105 Schnure, William M. "The Maine Saw Mill." *Snyder County Historical Society Bulletin* 3 (No. 6, 1952), 16-21. Selinsgrove, Pennsylvania, 1852-1907.

S106 Schoenfeld, Clay, ed. *Outlines of Environmental Education.* Madison: Dembar Educational Research Service, 1971. 246 pp. Articles, some of a historical nature, reprinted from the *Journal of Environmental Education,* 1969-1971.

S107 Schoenfeld, Clay, ed. *Interpreting Environmental Issues: Research and Development in Conservation Communications.* Madison: Dembar Educational Research Services, 1972. 300 pp. Includes some historical articles reprinted from the *Journal of Environmental Education,* 1969-1972.

S108 Schon, D.O.L. "The Handlogger: Unique British Columbia Pioneer." *Forest History* 14 (Jan. 1971), 18-20. Handlogging, 1860s-1966.

S109 Schonning, Egil. "Union-Management Relations in the Pulp and Paper Industry of Ontario and Quebec." Ph.D. diss., Univ. of Toronto, 1955. 105 pp. Incidental history.

S110 Schontz, J.L. "The Development of the Pulp and Paper Industry in the U.S." Master's thesis, Yale Univ., 1931.

Schores, D.D. See Stevenson, Donald D., #S630

S111 Schorger, Arlie W., and Betts, H.S. *The Naval Stores Industry.* USDA, Bulletin No. 229. Washington: GPO, 1915. 58 pp. Illus. Incidental history.

Schrack, Robert A. See Markham, Alton, #M208

S112 Schrader, O. Harry, Jr. "New Developments in the Forest Products Industries." *Journal of Forestry* 48 (June 1950), 425-28. Technological developments in the 1940s, especially in Washington.

S113 Schramm, Jacob Richard. "The Memorial to Francois André Michaux at the Morris Arboretum, University of Pennsylvania." *Proceedings of the American Philosophical Society* 100 (Apr. 1956), 145-49. On his bequest in 1855 to the American Philosophical Society for the promotion of silvicultural research, with some account of botanical collections made in the United States by him and by his father, André Michaux.

S114 Schramm, Jacob Richard. "Influence—Past and Present—of François André Michaux on Forestry and Forest Research in America." *Proceedings of the American Philosophical Society* 101 (Aug. 1957), 336-43. On the uses made of his bequest to the American Philosophical Society for the "extension and progress of agriculture, and more especially of sylviculture, in the United States."

S115 Schreiner, Ernst J. "Genetics in Relation to Forestry." *Journal of Forestry* 48 (Jan. 1950), 33-38. Includes some history of tree breeding and forest genetics research since 1845.

S116 Schrenk, Hermann von. "The History and Use of Zinc Chloride as a Wood PReservative." *Proceedings of the American Wood Preservers' Association* 20 (1924), 188-91.

S117 Schrenk, Hermann von. "An Historical Statement of the Use of Straight Coal Tar for Tie Treatment." *Bulletin of the American Railway Engineering Association* 51 (1949), 387-400. On the preservation of wooden crossties.

S118 Schrepfer, Susan R. "A Conservative Reform: Saving the Redwoods, 1917 to 1940." Ph.D. diss., Univ. of California, Riverside, 1971. 403 pp. On the work of the Save-the-Redwoods League in California.

Schrepfer, Susan R. See Crafts, Edward C., #C674

Schrepfer, Susan R. See Peirce, Earl S., #P93

Schrepfer, Susan R. See Pyles, Hamilton K., #P359, P360

Schrepfer, Susan R. See Steen, Harold K., #S567

S119 Schrepfer. Susan R.; Larson, Edwin vH; and Maunder, Elwood R. *A History of the Northeastern Forest Experiment Station, 1923 to 1973.* General Technical Report NE-7. Upper Darby, Pennsylvania: USFS, Northeastern Forest Experiment Station, 1973. 52 pp. Illus., maps, notes, bib., list of personnel. The original Northeastern Forest Experiment Station opened at Amherst, Massachusetts, in 1923. The Allegheny Forest Experiment Station was established in 1927 in Philadelphia. The two stations merged under the former name in 1945 and is now located at Upper Darby, Pennsylvania.

S120 Schroeder, George H. *The Art and Science of Protecting Forest Lands from Fire.* Corvallis, Oregon: Oregon State College Cooperative Association, 1938. 184 pp. Illus. Processed. Includes some history of forest fires and fire legislation in Oregon, Washington, and California.

S121 Schuessler, Raymond. "The Story of the Paper Bag." *American Forests* 61 (Nov. 1955), 18-19, 47. On the manufacture of bags from wood pulp paper since 1852.

S122 Schuessler, Raymond. "The Story of the Match." *American Forests* 62 (Jan. 1956), 24-26, 52-53. Includes some history of the manufacture of wooden matches, with emphasis on the Diamond Match Company's Spokane division and its woods operations and log drives in Idaho, since 1882.

S123 Schulman, Edmund. "Tree-Rings and History in the Western United States." *Economic Botany* 8 (July-Sept. 1954), 234-50. On the development of dendrochronology since 1904 and its applications in dendroclimatology and dendroarcheology. Also appeared in the *Annual Report of the Board of Regents of the Smithsonian Institution* (1955), 459-73.

S124 Schulman, Steven A. "The Lumber Industry of the Upper Cumberland River Valley." *Tennessee Historical Quarterly* 32 (Fall 1973), 255-64. Of Kentucky and Tennessee, with some account of log rafting downriver to Nashville, 1870s-1930s.

S125 Schulman, Steven A. "Rafting Logs on the Upper Cumberland River." *Pioneer America* 6 (Jan. 1974), 14-26. Rafting logs to the sawmills of Nashville, Tennessee, 1870s-1930s, with emphasis on the technology of log rafts and rafting.

Schultz, George P. See Myers, Charles A., #M662

S126 Schulz, William F., Jr. *Conservation Law and Administration: A Case Study of Law and Resource Use in Pennsylvania.* New York: Ronald Press, 1953. 607 pp. Includes several chapters on forestry with historical references to the movement in Pennsylvania.

Schumacher, Francis X. See Bruce, Donald, #B517

S127 Schurr, Sam H., ed. *Energy, Economic Growth, and the Environment.* Baltimore: Johns Hopkins University Press, for Resources for the Future, 1972. 240 pp. Incidental history.

S128 Schurz, Carl. *Reminiscences.* 3 Volumes. New York: McClure Company, 1907-1908. As Rutherford B. Hayes's Secretary of the Interior, 1877-1881, Schurz showed particular concern for forest protection and conservation.

S129 Schurz, Carl. *Speeches, Correspondence, and Political Papers of Carl Schurz.* Ed. by Frederick Bancroft. 6 Volumes. New York: G.P. Putnam's Sons, 1913. Reprint. Westport, Connecticut: Greenwood Press, 1974. The papers cover the period from 1852 to 1906, including his term as Secretary of the Interior, 1877-1881.

S130 Schutza, Judy. "History—The Great Fire in Idaho." *American Forests* 80 (Oct. 1974), 32-35, 45-47. On the devastating forest fire which burned much of northern Idaho in 1910.

S131 Schwarz, G.F. *Forest Trees and Forest Scenery.* New York: Grafton Press, 1901. 183 pp. Incidental history.

S132 *Scientific Monthly.* "John Muir and the National Monument in His Honor." *Scientific Monthly* 46 (May 1938), 490-94. On Muir's role in the conservation movement and the Muir Woods National Monument, a grove of California redwoods donated by William Kent.

S133 Scott, Anthony. "Conservation Policy and Capital Theory." *Canadian Journal of Economics and Political Science* 20 (Nov. 1954), 504-13. Incidental history.

S134 Scott, Anthony. *Natural Resources: The Economics of Conservation.* Canadian Studies in Economics, No. 3 Toronto: University of Toronto Press, 1955. 184 pp. Includes a historical survey of natural resource policies in Europe, Canada, and the United States.

S135 Scott, C.A. "Historic Sketch of the Nebraska Sand Hill Tree Planting Project." *Colorado Forester* 5 (1929), 31-33.

S136 Scott, David W. "American Landscape: A Changing Frontier." *Living Wilderness* 33 (Winter 1969), 3-13. A brief history of landscape and wilderness painting in the United States.

S137 Scott, Douglas W. "The Origins and Development of the Wilderness Bill, 1930-1956." Master's thesis, Univ. of Michigan, 1973.

S138 Scott, Edgar C., Jr. "The St. Louis Lumber Market: An Historical Sketch." *Southern Lumberman* 193 (Dec. 15, 1956), 178M-178O. From 1839 to 1915.

S139 Scott, Edward B. *The Saga of Lake Tahoe: A Complete Documentation of Lake Tahoe's Development over the Last One Hundred Years.* Crystal Bay, Nevada: Sierra-Tahoe Publishing Company, 1957. xii + 519 pp. Illus., maps, notes, bib. Concerned in part with the forest, scenic, and recreational resources of the area.

S140 Scott, Edward B. *Squaw Valley.* Lake Tahoe, Nevada: Sierra-Tahoe Publishing Company, 1960. 88 pp. Illus., maps, notes. Includes references to logging in the Squaw Valley region of the Sierra Nevada, California.

S141 Scott, Ferris Huntington. *The Yosemite Story.* Santa Ana, California: Western Resort Publications, 1954. 64 pp. Illus., maps, diags., tables. On the valley and Yosemite National Park, California, since 1833.

S142 Scott, Walter E. "Conservation History." *Wisconsin Conservation Bulletin* 2-3 (Mar. 1937-Apr. 1938). A series of articles emphasizing the history of wildlife conservation.

S143 Scott, Walter E. *Conservation's First Century in Wisconsin: Landmark Dates and People.* Madison: Wisconsin Conservation Department, 1967. 26 pp. Processed. A brief narrative history followed by a chronology and list of 200 Wisconsin conservationists.

Scott, Walter E. See Aberg, William J.P., #A7

S144 Scott Paper Company. *Paper and Scott Paper Company: The History of Paper, the Manufacture of Paper, the Story of Scott Paper Company.* Chester, Pennsylvania, 1956. 24 pp.

S145 Scoyen, E.T., and Taylor, Frank J. *Rainbow Canyons.* Stanford, California: Stanford University Press, 1931. ix + 105 pp. Includes some history of Zion and Bryce national parks, Utah.

S146 Scoyen, E.T. "Policies and Objectives of the National Park Service." *Journal of Forestry* 44 (Sept. 1946), 641-46. A historical sketch of national parks, the National Park Service, and the legislation authorizing them.

Scribner, Caleb W. See Rogers, Lore A., #R273

S147 Searle, R. Newell. "Minnesota National Forest: The Politics of Compromise, 1898-1908." *Minnesota History* 42 (Fall 1971), 242-57. Explores the movement for a national park which resulted instead in the Minnesota (now the Chippewa) National Forest.

S148 Searle, R. Newell. "Minnesota Forestry Comes of Age: Christopher C. Andrews, 1895-1911." *Forest History* 17 (July 1973), 14-25. On the origins of state forestry in Minnesota, especially the role of Andrews and other "amateur" forest conservationists.

S149 Searle, R. Newell. "Minnesota State Forestry Association: Seedbed of Forest Conservation." *Minnesota History* 44 (Spring 1974), 16-29. On the work of the Minnesota State Forestry Association, 1876-1903, placed in the context of the forestry and forest conservation movements statewide and nationally.

S150 Searle, R. Newell. "A Land Set Apart." Ph.D. diss. Univ. of Minnesota, 1975. 647 pp. A history of the movement to preserve the Quetico Provincial Park of Ontario and the Superior National Forest and Boundary Waters Canoe Area of northern Minnesota, since the early 20th century. This crusade is treated as a prototype for similar wilderness preservation efforts in the United States.

S151 Sears, Paul B. "The Natural Vegetation of Ohio: A Map of Virgin Forest." *Ohio Journal of Science* 25 (1925), 139-49.

S152 Sears, Paul B. *Deserts on the March.* 1935. Third edition, revised. Norman: University of Oklahoma Press, 1959. xiii + 178 pp. On waste of natural resources in America since the arrival of the white man, and the progress of measures for conservation.

S153 Sears, Paul B. "Forest Sequence and Climatic Change in Northeastern North America since Early Wisconsin Time." *Ecology* 29 (July 1946), 326-33.

S154 Sears, Paul B. "Man and Nature in Modern Ohio." *Ohio State Archaeological and Historical Quarterly* 56 (Apr. 1947), 144-53. Includes a brief history of forest devastation in Ohio.

S155 Sears, Paul B. "Science and Natural Resources." *American Scientist* 44 (Oct. 1956), 330-46. On the author's interest in resource conservation as an aspect of his work as an ecologist since the 1920s.

S156 Sears, Paul B. "Botanists and the Conservation of Natural Resources." *American Journal of Botany* 43 (Nov. 1956), 731-35. Since 1873.

S157 Seaton, Fred A.; Clawson, Marion; Hodges, Ralph, Jr.; Spurr, Stephen; and Zinn, Donald. *Report of the Presi-*

dent's Advisory Panel on Timber and the Environment. Washington: GPO, 1973. 541 pp. Illus. A thorough study of the nation's forest situation, linking timber supply and forest policy with environmental concerns. It includes some history of the USFS and other forest-related agencies.

S158 Seelav, Robert. *Is There a Lumber Trust?* New York: Editorial Review Company, 1912. 16 pp. A review of the industry and of its investigation by the U.S. Bureau of Corporations.

S159 Seeman, Albert L. "Economic Adjustments on the Olympic Peninsula." *Economic Geography* 8 (July 1932), 299-310. Includes reference to forest industries.

S160 Seeman, Albert L. "Regions and Resources of Alaska." *Economic Geography* 13 (Oct. 1937), 334-46. Incidental history of forest resources.

S161 Segrest, J.L. "Resume of Alabama State Park History." *Alabama Historical Quarterly* 10 (1948), 77-80. Since 1927.

S162 Seigworth, Kenneth J. "Reforestation in the Tennessee Valley." *Public Administration Journal* 8 (Autumn 1948), 280-85. Since 1933.

S163 Seigworth, Kenneth J., and Barton, James H. *Initial Forest Management in the Tennessee Valley.* Norris, Tennessee: Tennessee Valley Authority, Division of Forestry Relations, 1961. 150 pp. On forestry extension work of the Tennessee Valley Authority since the 1930s.

S164 Seigworth, Kenneth J. "Forestry Plus . . . In the Tennessee Valley." *Journal of Forestry* 66 (Apr. 1968), 324-28. History of forestry and forest industries in the seven-state region, including mention of the Tennessee Valley Authority and cooperating state agencies, since the 1930s.

S165 Seitz, May Albright. *The History of the Hoffman Paper Mills in Maryland.* Towson, Maryland: By the author, 1946. 63 pp. From late 18th century to 1908.

S166 Sellars, Richard W. "Early Promotion and Development of Missouri's Natural Resources." Ph.D. diss., Univ. of Missouri-Columbia, 1972. 235 pp. A survey of 19th-century exploitation of natural resources, with some account of the origins of scientific conservation of resources.

S167 Sellers, C.H. *Eucalyptus: Its History, Growth and Utilization.* Sacramento, 1910. 93 pp. Illus. California.

S168 Selmeier, Lewis L. "First Camera on the Yellowstone a Century Ago." *Montana, Magazine of Western History* 20 (Summer 1972), 42-53. Photographs taken by William Henry Jackson during the Hayden expedition's tour of Yellowstone in 1871 were circulated in Congress to enlist support for the park's establishment.

S169 Selwyn-Brown, Arthur. *American Paper and Pulp Association: Fiftieth Anniversary, 1878-1927.* East Stroudsburg, Pennsylvania: Press Printing Company, 1927. 102 pp. Also published as a special section of *Paper Trade Journal,* February 24, 1927.

S170 Selznick, Philip. *TVA and the Grass Roots: A Study in the Sociology of Formal Organization.* Berkeley: University of California Press, 1949. Reprint. New York: Harper and Row, 1966. xvi + 274 pp. Tables, charts, notes, bib., index. A general history of the Tennessee Valley Authority, 1933-1943.

Semingsen, Earl M. See Beard, Daniel B., #B147

S171 Sendak, Paul E. "Why the Tariff Failed." *National Maple Syrup Digest* 11 (Dec. 1972), 18-23. Reviews the history of the American tariff levied on Canadian maple syrup since the 1940s and its removal in 1972.

S172 Sendak, Paul E. *The Effect of the Tariff on the Maple Industry.* Research Note NE-148. Upper Darby, Pennsylvania: USFS, Northeastern Forest Experiment Station, 1972. 5 pp. Illus.

S173 Senninger, Earl J., Jr. "The Chicory Industry of Michigan." *Papers of the Michigan Academy of Science, Arts, and Letters* 45 (1960), 145-53. On an industry extinct in Michigan but formerly localized in the Saginaw Valley and "the Thumb," 1890-1954.

S174 Sequin, R.L. "Les Premières Scieries dans la presqu'il de Vaudreil et de Soulanges." *Bulletin des Recherches Historiques* 59 (Apr.-June 1953), 85-89. Early sawmills in Quebec, near the confluence of the St. Lawrence and Ottawa rivers.

S175 Setterington, Eatha G. "Pioneer Days in the Canadian Woods." *Canadian Banker* 63 (Winter 1956), 47-64. On the relation of forests to early settlement in Canada.

S176 Setzer, Curt F. *Plane 'em Thick, Rip 'em Wide.* N.p.: By the author, 1968. iv + 114 pp. Illus. Autobiography of a lumberman, box manufacturer, producer of Presto-Logs, and other wood products in southern Oregon and northern California, 1890s-1950s. He owned Setzer Box Company, Sacramento, California; Chelsea Box and Lumber Company, Klamath Falls, Oregon; and other firms.

S177 Sewell, Joseph Herbert. "A Comparative Study of the Development of Forest Policy in Maine and New Brunswick." Master's thesis, Univ. of Maine, 1958. 200 pp.

Sexsmith, E.R. See Bonner, E., #B339

S178 Shaler, N.S. "Forests of North America." *Scribner's Magazine* 1 (1887), 561-80. Incidental history.

S179 Shampeny, Worth. "Sawmills in Upper White River Valley, Vermont—1786 to 1863." *Northeastern Logger* 11 (Apr. 1963), 18-19, 40-41.

S180 Shand, Eden Arthur. "The Development of the Japanese Market for Pacific Northwest Lumber: A Historical Survey." Master's thesis, Univ. of British Columbia, 1968.

Shank, Henry M. See Byrne, J.J., #B625

S181 Shankland, Robert. *Steve Mather of the National Parks.* 1951. Third edition. New York: Alfred A. Knopf, 1970. xii + 370 + xxiii pp. Map, illus., tables, selected bib., index. Mather (1867-1930), the first director of the National Park Service, served from 1916 to 1929. The author devotes five chapters to the administration of the park system by Mather's successors.

S182 Shanklin, John F. *Highlights in the History of Forest and Related Natural Resource Conservation in the Department of the Interior.* Washington: USDI, 1956. 37 pp. Processed. A chronology since 1783.

S183 Shanklin, John F. *United States Department of the Interior, Forest Conservation.* USDI, Conservation Bulletin No. 42. Washington: GPO, 1960. 86 pp. Illus., maps. Includes some history of forestry and forest conservation in the Bureau of Indian Affairs, Bureau of Land Management, National Park Service, and Bureau of Sport Fisheries and Wildlife.

Shanklin, John F. See Fitch, Edwin M., #F110

S184 Shannon, Fred A. *The Farmer's Last Frontier: Agriculture, 1860-1897.* The Economic History of the United States, Volume 5. New York: Rinehart and Company, 1945. xiv + 434 pp. Illus., maps, tables, notes, app., bib., index. Contains some history of the disposal of public lands and federal land policy, including reference to forested lands.

S185 Sharp, Paul F. "The Tree Farm Movement: Its Origin and Development." *Agricultural History* 23 (Jan. 1949), 41-45. Begun by the Weyerhaeuser Timber Company in the Pacific Northwest in 1940, the tree farm movement expanded rapidly. Initially attractive as a means of forest protection and of forestalling government regulation, the movement later focused on growing more trees.

S186 Sharp, Paul F. "The War of the Substitutes: The Reaction of the Forest Industries to the Competition of Wood Substitutes." *Agricultural History* 23 (Oct. 1949), 274-79. As revealed by trade journal literature, ca. 1900-1920.

S187 Sharp, S.J., and Wade, B. "Redwood Industry—1922/1947." *California Lumber Merchant* 26 (July 1, 1947), 102-04.

Sharpe, G.W. See Allen, Shirley W., #A93

S188 Shattuck, C.H. "Early Days in Forestry at the University of Idaho." *Idaho Forester* 9 (1927), 3-4, 46-47.

S189 Shaw, A.C. "The Pulp and Paper Industry's Role in American Forestry." *Southern Pulp and Paper Manufacturer* 20 (Nov. 1957), 118-21.

S190 Shaw, C.D. "Some Facts Relating to the Early History of Greenville and Moosehead Lake." *Collections of the Piscataquis County Historical Society* 15 (1910), 52-56. Towing logs on Moosehead Lake, Maine.

S191 Shaw, C.D. "History of the Shaw Family with a Sketch of Milton G. Shaw of Greenville." *Collections of the Piscataquis County Historical Society* 15 (1910), 424-33. Shaw was a lumberman in the Moosehead Lake-Kennebec River area of Maine.

S192 Shaw, Charlie. *The Flathead Story.* Assisted by R.D. Rogers. Kalispell, Montana: USFS, Flathead National Forest, 1967. ii + 145 pp. Illus., maps, tables. A general history of Montana's Flathead National Forest and surrounding areas since 1809.

Shaw, Elmer W. See Gaines, Edward M., #G6, G7

S193 Shaw, Noah. "Early Reminiscences of Sawmill History." *Mississippi Valley Lumberman* 26 (Feb. 1, 1895).

S194 Shaw, Ralph H. "History of the Acquisition of the Butano." *National Shade Tree Conference Proceedings* 30 (1954), 293-96. Butano redwood forest, now Butano State Park, California.

S195 omitted

S196 Shea, John P. " 'Our Pappies Burned the Woods'." *American Forests* 46 (Apr. 1940), 159-62, 174. Report by a USFS psychologist on the practice of woods burning in the South, with incidental history since the 19th century.

S197 Shebl, James M. *King, of the Mountains.* Pacific Center Monograph Series, No. 5. Stockton, California: Pacific Center for Western Historical Studies, 1974. 76 pp. Illus., maps, app., notes, index. On Clarence King (1842-1901), director of the U.S. Geological Survey, writer, and western explorer.

S198 Sheffer, George P., ed. and comp. *True Tales of the Clarion River.* Clarion, Pennsylvania: Northwestern Pennsylvania Raftmen's Association, 1933. Log rafting in Pennsylvania.

S199 Sheldon, Addison E. "Silver Anniversary of the Nebraska National Forest." *Journal of Forestry* 25 (Dec. 1927), 1020-23. A brief history of the movement to establish the forest in Nebraska's Sand Hills.

S200 Sheldon, Addison E. *Land Systems and Land Policies in Nebraska: A History of Nebraska Land, Public Domain and Private Property, Its Title, Transfers, Ownership, Legislation, Administration, Prices, Values, Productions, Uses, Social Changes, Comparisons, from the Aboriginal Period to 1936.* Nebraska State Historical Society Publications, Volume 22. Lincoln, 1936. xvi + 383 pp. Illus., maps.

Sheldon, Charles. See Grinnell, George Bird, #G353

S201 Sheldon, G.W. "The Old Shipbuilders of New York." *Harper's* 65 (1882), 223-41.

S202 Sheldon, Roger. "Texas Big Thicket." *American Forests* 58 (Sept. 1952), 22-24, 46. Lore about the Big Thicket, a forest and swamp in southeastern Texas, since 1836.

S203 Shelford, Victor E. "Deciduous Forest Man and the Grassland Fauna." *Science* 100 (Aug. 18, 1944), 135-62.

S204 Shelford, Victor E. *The Ecology of North America.* Urbana: University of Illinois Press, 1963. xxii + 576 pp. Illus., maps, tables, notes, index. Description of the ecology and natural history of the continent before European settlement.

S205 Shelton, Ronald Lee. "The Environmental Era: A Chronological Guide to Policy and Concepts, 1962-1972." Ph.D. diss., Cornell Univ., 1973. 584 pp.

Shenton, Donald R. See Hoch, Daniel K., #H394

S206 Shepard, Harold B. *Forest Fire Insurance in the Pacific Coast States.* USDA, Technical Bulletin No. 511. Washington: GPO, 1937. 168 pp. Bib. Incidental history.

S207 Shepard, Paul. *Man in the Landscape: A Historic View of the Esthetics of Nature.* New York: Alfred A. Knopf, 1967. xx + 290 + v pp. Illus., bib., index. A historical interpretation of the ways in which man has perceived and acted upon his environment, including forested and wilderness lands.

S208 Shepherd, F.C. "History and Practice of Tie Preservation." *Cross Tie Bulletin* 5 (No. 12, 1924), 20-26.

S209 Sheppard, C.C. *Wages and Hours of Labor in the South.* New Orleans: Southern Pine Association, 1933. A statement in behalf of the Southern lumber industry before the National Industrial Recovery Administration; contains some historical information.

S210 Sherfesee, W.F. *Wood Preservation in the United States.* USFS, Bulletin No. 78. Washington: GPO, 1909. 31 pp. Incidental history.

Sheridan, Charles M. See Nelligan, John Emmett, #N44

S211 Sherman, Dorothy M. "A Brief History of the Lumber Industry in the Fir Belt of Oregon." Master's thesis, Univ. of Oregon, 1934.

S212 Sherman, Edward A. "Thirty Five Years of National Forest Growth." *Journal of Forestry* 24 (Feb. 1926), 129-35. On the expansion of national forests since 1891, and the legislation, reports, and other decisions affecting their growth in area.

S213 Sherman, Ivan C. "The Life and Work of Holman Francis Day." Master's thesis, Univ. of Maine, 1932. Poet of the Maine Woods.

S214 Sherman, John. *Twenty Years of Collective Bargaining and Twenty Years of Peace.* Glens Falls, New York, 1954. The author was active in union affairs of the pulp and paper industry in the Pacific Northwest.

S215 Sherman, Rexford B. "The Bangor and Aroostook Railroad and the Development of the Port of Searsport." Master's thesis, Univ. of Maine, 1966.

S216 Sherman, Simon Augustus. "Lumber Rafting on Wisconsin River (1849-1850)." Ed. by Albert H. Sanford. *Proceedings of the Wisconsin State Historical Society* 58 (1918), 171-80.

S217 Sherrard, William R. "Measuring Labor Productivity for the Firm: A Case Study." D.B.A. diss., Univ. of Washington, 1965. 341 pp. One chapter offers some history of the St. Paul and Tacoma Lumber Company, Tacoma, Washington.

S218 Sherrard, William R. "Labor Productivity for the Firm: A Case Study." *Quarterly Review of Economics and Business* (Spring 1967). St. Paul and Tacoma Lumber Company, Tacoma, Washington, 1889-1938.

S219 Shertzer, Leonard L. "Reminiscences of an Oldtime Hardwood Exporter." *Southern Lumberman* (Dec. 15, 1959), 134-35.

S220 Sherwood, Malcolm H. *From Forest to Furniture: The Romance of Wood.* New York: W.W. Norton, 1936. 284 pp. Illus., index. Incidental history.

S221 Sherwood, Morgan B. "American Scientific Exploration of Alaska, 1865-1900." Ph.D. diss., Univ. of California, Berkeley, 1962.

S222 Shideler, Frank J. "Custer Country: One Hundred Years of Change." *American West* 10 (July 1973), 25-31. Photographic evidence of the changing forests of South Dakota's Black Hills.

S223 Schideler, James H. "Opportunities and Hazards in Forest History Research." *Forest History* 7 (Spring/Summer 1963), 10-14. And the kinship of agricultural and forest history.

S224 Shih, Yang-ch'eng. "The Theory and Practice of the Conservation of Water Resources of the United States with Special Reference to Federal Administration." Ph.D. diss., Univ. of Washington, 1953. 606 pp. Includes some history of federal water conservation through several agencies since ca. 1900.

S225 Shimek, Bohumil. "The Pioneer and the Forest." *Proceedings of the Mississippi Valley Historical Association* 3 (1911), 96-105.

S226 Shinn, Charles H. "Shakes and Shake-Making in a California Forest." *Proceedings of the Society of American Foresters* 4 (1909), 151-71. Includes some history of shake-making on California's Sierra National Forest.

S227 Shinn, Charles H. "The Giver of Muir Woods." *American Forests and Forest Life* 30 (Mar. 1924), 147-48. On William Kent, whose gift of a redwood forest in Marin County, California, became Muir Woods National Monument in 1907.

S228 Shinn, Julia Tyler. "Forgotten Mother of the Sierra: Letters of Julia Tyler Shinn." Ed. by Grace Tompkins Sargent. *California Historical Quarterly* 38 (June 1959), 157-63; 38 (Sept. 1959), 219-28. Letters concerning her life on the Sierra National Forest, California, 1905-1911. Her husband, Charles Howard Shinn, was the first forest supervisor and a friend of Gifford Pinchot.

S229 Shipley, Donald D. "A Study of the Conservation Philosophies and Contributions of Some Important American Conservation Leaders." Ph.D. diss., Cornell Univ., 1954.

S230 Shipley, Grant B. "Trend of the Wood Preserving Industry in the United States." *Proceedings of the American Wood Preservers' Association* 25 (1929), 76-97.

S231 Shipley, Grant B. *A Review of the Lumber, Cross Tie, and Wood Preserving Industry in the United States, 51 Year Period (1890-1940).* N.p., 1942, 31 pp. Illus. Also appeared in the *Proceedings of the American Wood Preservers' Association* 38 (1942), 534-59.

S232 Shipley, Grant B. "Wood Preservers' Contribution to Conservation: A Study Based on the 41-Year Period, 1909-1949." *Southern Lumberman* 183 (July 1, 1951), 47-48.

S233 Shipley, John W. *Pulp and Paper in Canada.* Toronto: Longmans, Green, 1929. 139 pp. Incidental history.

S234 Shirley, Hardy L. *Forestry and Its Career Opportunities.* American Forestry Series. 1952. Third edition. New York: McGraw-Hill, 1973. xi + 464 pp. Includes a chapter on the history of forestry.

S235 Shirley, Hardy L. "Ralph Sheldon Hosmer." *Journal of Forestry* 55 (May 1957), 380-81. On his work as a forester since 1898.

S236 Shirley, Hardy L. "Forestry in an Era of Change." *Journal of Forestry* 55 (Oct. 1957), 707-10. Incidental historical references to forestry in New York.

S237 Shirley, James C. *The Redwoods of Coast and Sierra.* 1937. Fourth Edition. Berkeley: University of California Press,

1947. 84 pp. Includes some popular history of the uses of preservation of redwoods and sequoias in California.

S238 Shiverick, Nathan C. "Virginia and the Western Land Problems, 1776-1800." Ph.D. diss., Harvard Univ., 1965.

S239 Shoemaker, Carl D. *The Stories Behind the Organization of the National Wildlife Federation and Its Early Struggles for Survival*. Washington: National Wildlife Federation, 1960. 47 pp. Since its founding in 1936.

S240 Shoemaker, Florence J. "The Pioneers of Estes Park." *Colorado Magazine* 24 (Jan. 1947), 15-23. Early residents, exploration, and tourism in an area later to become part of Colorado's Rocky Mountain National Park, 1860s-1890s.

S241 Shoemaker, Henry W.; French, John C.; and Chatham, John H. *North Pennsylvania Minstrelsy, as Sung in the Backwood Settlements, Hunting Cabins, and Lumber Camps in the 'Black Forest' of Pennsylvania, 1840-1910*. Altoona, Pennsylvania: Times Tribune Company, 1923. 228 pp. Illus.

S242 Shoemaker, Leonard C. "National Forests." *Colorado Magazine* 21 (Sept. 1944), 182-84. Brief chronology of national forests in Colorado, including name and boundary changes since 1891.

S243 Shoemaker, Leonard C. "The First Forest Ranger." *Westerners Brand Book* (Denver) 7 (1952), 95-121. On William Richard Kreutzer as USFS ranger in Colorado, 1898-1939.

S244 Shoemaker, Leonard C. *Saga of a Forest Ranger: A Biography of William R. Kreutzer, Forest Ranger No. 1, and a Historical Account of the U.S. Forest Service in Colorado*. Boulder: University of Colorado Press, 1958. xii + 216 pp. Illus. On his career as a ranger in Colorado, and particularly during the years 1905 to 1921, when he was supervisor of Gunnison National Forest.

S245 Shofner, Jerrell H. "Militant Negro Laborers in Reconstruction Florida." *Journal of Southern History* 39 (Aug. 1973), 397-408. Largely concerned with black laborers in Florida's forest industries.

S246 Shorey, Archibald T. "Albert Tatum Davis, Woodsman and Surveyor." *Ad-i-ron-dac* 25 (Mar.-Apr. 1961), 36-38. New York.

S247 Shortridge, Wilson P. "Henry Hastings Sibley and the Minnesota Frontier." *Minnesota History Bulletin* 3 (Aug. 1919), 115-25. Sibley began cutting pine along Minnesota's St. Croix River in 1837.

S248 Shortt, Adam. "Down the St. Lawrence on a Timber Raft." *Queen's Quarterly* 10 (July 1902), 16-34. In the 19th century.

S249 Shoulders, Eugene. "Timber Stand Improvement in Ozark Forests—An Appraisal after 15 Years." *Journal of Forestry* 54 (Dec. 1956), 824-27. On the Ozark National Forest of Arkansas, commenced by the Civilian Conservation Corps in 1934.

S250 Show, Stuart B., and Kotok, E.I. *Forest Fires in California, 1911-1920: An Analytical Study*. USDA, Circular No. 243. Washington: GPO, 1923. 80 pp. Illus. An analysis of 10,500 forest fires on twelve national forests of California.

S251 Show, Stuart B. "Modifications in Forests of the Pacific Slope Due to Human Agencies." *Journal of Forestry* 24 (May 1926), 500-06. Includes some incidental history of the use of forest resources and the reduction of forest by fire since the 19th century.

Show, Stuart B. See Brown, William S., #B511

S252 Shuman, Stanley B. "The Forest Resource Situation of the Eastern Part of the Upper Peninsula of Michigan." Ph.D. diss., Univ. of Illinois, 1957. 227 pp. The historical geography of forest protection and utilization.

S253 Shurtleff, Flavel. *Digest of Laws Relating to State Parks*. Washington: National Conference on State Parks, 1955. 256 pp.

S254 Shurtleff, Harold R. *The Log-Cabin Myth: A Study of the Early Dwellings of the English Colonists in North America*. Cambridge: Harvard University Press, 1939. xxi + 243 pp. Illus. Argues that log cabin construction was introduced to North America by Swedish settlers in the early 17th century. There is a modern reprint.

S255 Shutts, Elmer E. "Industrial History of Southwest Louisiana." *McNeese Review* 6 (1954), 93-97. Includes reference to the lumber and wood-using industries since the 1890s.

S256 Siau, John F. "A History of Paul Smith's College." *Ad-i-ron-dac* 24 (Mar.-Apr. 1960), 30-31, 35. Forestry education in New York.

S257 Sieber, George W. "Sawmilling on the Mississippi: The W.J. Young Lumber Company, 1858-1900." Ph.D. diss., Univ. of Iowa, 1960. 636 pp. One of the largest sawmills in the country at Clinton, Iowa.

S258 Sieber, George W. "Sawlogs for a Clinton Sawmill." *Annals of Iowa* 37 (Summer 1964), 348-59. On the competitive acquisition of Wisconsin logs for the W.J. Young & Company, Clinton, Iowa, 1860s.

S259 Sieber, George W. "Railroads and Lumber Marketing, 1858-78: The Relationship Between an Iowa Sawmill Firm and the Chicago & Northwestern Railroad." *Annals of Iowa* 39 (Summer 1967), 33-46. W.J. Young and Company's relations with the Chicago & Northwestern Railroad involved disputes over rates and loading weights.

S260 Sieber, George W. "Wisconsin Pine Land and Logging Management." *Transactions of the Wisconsin Academy of Sciences, Arts & Letters* 56 (Summer 1968), 65-72. Absentee ownership, business techniques, and contract logging on Wisconsin timberlands owned by W.J. Young & Company of Clinton, Iowa, 1860s-1880s.

S261 Sieber, George W. "Lumbermen at Clinton: Nineteenth Century Sawmill Center." *Annals of Iowa* 41 (Fall 1971), 779-802. Sawmills at Clinton, Iowa, and the socioeconomic role of lumbermen and millworkers, with special attention to W.J. Young of W.J. Young & Company, 1860s-1890s.

S262 Siecke, E.O. "Development of Oregon's Forest Policy." *Commonwealth Review* 1 (1916), 189-99.

S263 Siefkin, Gordon, et al. *The Place of the Pulp and Paper Industry in the Georgia Economy*. Emory University,

School of Business Administration, Studies in Business and Economics, No. 8. Atlanta, 1958. 140 pp.

S264 Siegel, William C. "Environmental Law—Some Implications for Forest Resource Management." *Environmental Law* 4 (Fall 1973), 115-34. On the impact of environmental law and litigation since the early 1960s, with references to specific cases and decisions.

S265 Siegel, William C. "State Forest Practice Laws Today." *Journal of Forestry* 72 (Apr. 1974), 208-11. Incidental historical references to state forest practice laws, especially in Oregon and California.

S266 Sieker, John. "The Future of Forest Recreation." *Journal of Forestry* 49 (July 1951), 503-06. Incidental history.

S267 Sievers, F.R. "History and Romance of Paper." *Pacific Pulp & Paper Industry* 18 (Mar. 1944), 35-37; 18 (Apr. 1944), 42,45.

S268 Sievers, Ruth. " '. . . to Check the Action of Destructive Causes. . . .' " *1975 NRA Conservation Yearbook* (1974), 20-35. A brief general history of federal legislation to conserve natural resources and wildlife and to foster their wise management.

S269 Sights, Mrs. Warren. "The Land Between the Rivers." *Bulletin of the Garden Club of America* 48 (Jan. 1960), 30-32. On the Kentucky Woodland National Wildlife Refuge, on the ridge separating the lower courses of the Tennessee and Cumberland rivers, since 1920.

S270 Silcox, Ferdinand A. "Our Adventure in Conservation: The CCC." *Atlantic Monthly* 160 (Dec. 1937), 714-22. Includes some history of forest conservation and the Civilian Conservation Corps since 1933.

S271 Silen, R.R., and Woike, L.R. *The Wind River Arboretum from 1912 to 1956*. Research Paper No. 33. Portland: USFS, Pacific Northwest Forest and Range Experiment Station, 1959. 50 pp. Located on the Gifford Pinchot National Forest, Washington.

S272 Silver, David M., ed. "Richard Lieber and Indiana's Forest Heritage." *Indiana Magazine of History* 67 (Mar. 1971), 45-55. Lieber (1869-1944) was a national leader in the state parks movement and served Indiana as secretary of the Board of Forestry, 1917-1919, and director of the Department of Conservation, 1919-1933. A biographical sketch of Lieber is followed by his report as chairman of the Committee on Indiana State Centennial Memorial (1916), describing the acquisition of Turkey Run, later established as a state park.

S273 Silver, James W. "Paul Bunyan Comes to Mississippi." *Journal of Mississippi History* 19 (Apr. 1957), 93-119. On the operations of the Carrier Lumber and Manufacturing Company in the swamps of Panola and Quitman counties, Mississippi, 1898-1929, including some lore about local lumberjacks.

S274 Silver, James W. "The Hardwood Producers Come of Age." *Journal of Southern History* 55 (Nov. 1957), 427-53. On the activities of the Hardwood Manufacturers Association of the United States, 1902-1918, the American Hardwood Manufacturers Association, 1919-1922, and the origins of the Hardwood Manufacturers Institute in 1922, with particular reference to the issues of competition and monopoly.

Silver, James W. See Brockway, Chauncey, #B452

S275 Sim, Robert J., and Weiss, Harry B. *Charcoal Burning in New Jersey from Early Times to the Present*. Trenton: New Jersey Agricultural Society, 1955. 62 pp.

S276 Simmons, Fred C. "Forest Policies of New York State." Master's thesis, Yale Univ., 1931.

S277 Simmons, Fred C. "956 Years of Lumbering in the Northeast." *Southern Lumberman* 193 (Dec. 15, 1956), 227-29. Since the arrival of Vikings, ca. 1000.

S278 Simmons, Fred C. "The Impact of New Methods and Machinery on Forest Management." *Northeastern Logger* 6 (July 1957), 34-35, 54-56. Changes since the 1930s.

S279 Simmons, Fred C. "Logging—Yesterday, Today, and Tomorrow." *Southern Lumberman* 199 (Dec. 15, 1959), 195-98. On changes in logging technology and the transition to mechanization.

S280 Simmons, Fred C. "Reminiscences of a Forester." *Northern Logger and Timber Processor* 18 (Nov. 1969), 16-17, 33-34; 18 (Apr. 1970), 31, 40-42. On his career with the USFS in various parts of the country, including the Northeast, since the 1920s.

S281 Simmons, Fred C. "Howard A. Hanlon, November 19, 1896—October 20, 1973: Lumberman, Author, Ecologist, Humanitarian, Agronomist, Leader of Men and Friend." *Northern Logger and Timber Processor* 22 (Feb. 1974), 18-26. Hanlon was a New York lumberman and author of historical novels on forest themes.

S282 Simmons, James R. *The Historic Trees of Massachusetts*. Boston: Marshall Jones Company, 1919. xxi + 139 pp. Illus.

S283 Simmons, James R. *Cabin in the Woods: An Anthology of the Cabin Articles by the Keeper of the Cabin*. Boston: Meador Publishing Company, 1957. 181 pp. Illus., map, tables. Reminiscences of the author's life and experiences as a forester in various parts of the United States since 1930.

S284 Simmons, Perez, and Davies, Alfred H., eds. *Twentieth Engineers, France, 1917-1918-1919.* Portland: Twentieth Engineers Publishing Association, 1920. 218 pp. Illus., maps. World War I forestry regiment.

S285 Simmons, Roger E.; Holmes, John S.; Sackett, Homer S. *Wood-Using Industries of North Carolina*. North Carolina Economic and Geological Survey, Economic Paper No. 20. Raleigh, 1910. 74 pp. Illus. This and the following entries under Roger E. Simmons contain incidental history.

S286 Simmons, Roger E.; Rankin, M.C.; and Sackett, Homer S. *A Study of the Wood-Using Industries of Kentucky*. Frankfort: USFS and Kentucky State Board of Agriculture, Forestry, and Immigration, 1910. 74 pp. Illus.

S287 Simmons, Roger E.; Blair, T.C.; Sackett, Homer S. *The Wood-Using Industries of Illinois*. Urbana: USFS and University of Illinois, Department of Agriculture, 1910. 164 pp.

S288 Simmons, Roger E. *Wood-Using Industries of New Hampshire*. Concord: New Hampshire Forestry Commission, 1912. 111 pp. Illus.

S289 Simmons, Roger E. *Wood-Using Industries of Virginia*. Richmond: Virginia Department of Agriculture and Immigration, 1912. 88 pp. Illus.

S290 Simmons, Roger E. *Wood-Using Industries of Pennsylvania*. Pennsylvania Department of Forestry, Bulletin No. 9. Harrisburg, 1914. 204 pp. Illus.

S291 Simmons, Virginia McConnell. "Wheeler: 'Enchanted City' of the San Juans." *National Parks & Conservation Magazine* 48 (July 1974), 14-19. Includes a historical sketch of the Wheeler Geologic Area, designated as Wheeler National Monument until 1950 when reunited with the Rio Grande National Forest, Colorado.

S292 Simms, Denton Harper. *The Soil Conservation Service*. Praeger Library of U.S. Government Departments and Agencies. New York: Praeger Publishers, 1970. ix + 238 pp. Illus., charts, tables, apps., bib., index. Includes some history of the federal agency since 1935, and of its relation to the conservation movement.

S293 Simpson, Albert. "Sawmills Have Come a Long Way since That First One in 1623." *Wood-Worker* 69 (June 1950), 58-60.

S294 Simpson, Charles D., and Jackman, E.R. *Blazing Forest Trails,* Caldwell, Idaho: Caxton Printers, 1967. xiv + 384 pp. Illus., apps., index. History and reminiscence of the USFS in the West. Simpson was supervisor of four national forests in three regions and held many other posts with the USFS in his long career.

S295 Simpson, Thomas. "The Early Government Land Survey in Minnesota West of the Mississippi River." *Collections of the Minnesota Historical Society* 10, part 1 (1905), 57-67.

S296 Simpson Logging Company. *Railroads in Mason County, 1884-1959: 75th Anniversary.* Everett, Washington, n.d. 7 pp. Illus.

S297 Sims, Robert C. "James P. Pope, Senator from Idaho." *Idaho Yesterdays* 15 (Fall 1971), 9-15. Senator Pope (1933-1939) was involved in the politics of natural resources, including the proposed executive department reorganization, and was a director of the Tennessee Valley Authority until 1951.

S298 Sims, Robert C. "Idaho's Criminal Syndicalism Act: One State's Response to Radical Labor." *Labor History* 15 (Fall 1974), 511-27. On a campaign against radical lumberjacks and miners, mostly members of the Industrial Workers of the World, 1917-1920s.

S299 Sinclair, J.D. "Watershed Management Research in Southern California's Brush Covered Mountains." *Journal of Forestry* 58 (Apr. 1960), 266-68. On the USFS's San Dimas Experimental Forest in the San Gabriel Mountains, since 1933.

S300 Sipe, Henry. "To Prosecute, or Not to Prosecute." *Journal of Forestry* 47 (Oct. 1949), 796-801. Problems of incendiary fires and law enforcement on the Cumberland (now Daniel Boone) National Forest, Kentucky, since 1930.

Sirotkin, Phillip. See Stratton, Owen, #S688

S301 Sisam, J.W.B. "The Canadian Institute of Forestry—Historical Highlights." *Journal of Forestry* 57 (Jan. 1959), 4-8. Since 1908.

S302 Sisam, J.W.B. *Forestry Education at Toronto.* Toronto: University of Toronto Press, 1961. 116 pp. Illus., notes, bib. At the University of Toronto since 1907.

S303 Sisson, George W. "The Story of the Racquette River Paper Company, One of the North Country's Most Stable Industries." *North Country Life* 5 (Summer 1951), 15-17. Of Potsdam, New York, since 1892.

S304 Sizemore, W.R. "Timber Resources of the Alabama Piedmont." *Journal of the Alabama Academy of Science* 31 (Apr. 1960), 295-301.

S305 Skelton, O.D. "Wood-Pulp and the Tariff." *Journal of Political Economics* 14 (Dec. 1906), 632-36. Incidental history.

S306 Skolmen, Roger G. "Forests and Forest Products in Hawaii—Past, Present, and Future." *Southern Lumberman* 203 (Dec. 15, 1961), 158-61.

S307 Skuce, Thomas W. "West Virginia Forests and Forestry." *Journal of Forestry* 23 (July-Aug. 1925), 654-61. Since the 19th century.

S308 Skuce, Thomas W. "A Constructive Forestry Program." *West Virginia Review* 17 (Sept. 1940), 340-44, 354. A chronological history of forestry and forest protection in West Virginia since 1906.

S309 Sleicher, Charles A. *The Adirondacks: American Playground.* New York: Exposition Press, 1960. 287 pp. Illus., map, notes, bib. History, description, and lore of New York's Adirondack Mountains since the 16th century.

S310 Slichter, Gertrude. "Franklin D. Roosevelt's Farm Policy as Governor of New York State, 1928-1932." *Agricultural History* 33 (Oct. 1959), 167-76. Includes his programs for land use planning and reforestation in New York.

S311 Sloane, Eric. *A Reverence for Wood.* New York: Wilfred Funk, 1965. Reprint. New York: Ballantine Books, 1973. 111 pp. Illus. Includes much history of early America's dependence on forests and a wide variety of wood products.

S312 Sloan, Gordon M. *The Forest Resources of British Columbia.* Victoria, 1957. Incidental history.

S313 Slocomb, Jack T. "The Position of the Town Forest in New England's Forest Economy." Master's thesis, Yale Univ., 1940.

S314 Slocomb, Jack T. "The Wood Preserving Industry—A Brief History." *Journal of Forestry* 65 (Mar. 1967), 198-99. Especially on the preservation of wooden crossties since the 19th century.

S315 Slonaker, L.V. "Apache National Forest, 1916: Report on the Baseline Wireless Station." *Journal of Forest History* 18 (Apr. 1974), 23-27. A USFS telephone engineer's contemporary account of the installation in Arizona of the USFS's first wireless station.

S316 Slotkin, Richard. *Regeneration through Violence: The Mythology of the American Frontier, 1600-1860.* Middletown, Connecticut: Wesleyan University Press, 1973. 670 pp. Notes, bib., index. An interpretive history which deals in part with American attitudes toward wilderness.

S317 Slotnick, Herman. "The Ballinger-Pinchot Affair in Alaska." *Journal of the West* 10 (Apr. 1971), 337-47. On the disposition of Alaskan coal lands and other issues related to the Ballinger-Pinchot affair, 1908-1912.

S318 Small, Henry B. *Canadian Forests, Forest Trees, Timber and Forest Products.* Montreal, 1884. 64 pp.

S319 Smalley, Brian H. "Some Aspects of the Maine to San Francisco Trade, 1849-1852." *Journal of the West* 6 (Oct. 1967), 593-603. Includes references to the Pacific lumber trade.

S320 Smallwood, Johnny B., Jr. "George W. Norris and the Concept of a Planned Region." Ph.D. diss., Univ. of North Carolina, 1963. 458 pp. Includes some history of the conservation and regional planning movements prior to the establishment of the Tennessee Valley Authority, with special attention to the role of the Nebraska senator.

Smallwood, Mabel. See Smallwood, William M., #S321

S321 Smallwood, William M., and Smallwood, Mabel. *Natural History and the American Mind.* New York: Columbia University Press, 1941. xiii + 445 pp. On the history of naturalists in the United States, including their studies, philosophy, contributions, and recognition.

S322 Smith, A. Robert. *Tiger in the Senate: The Biography of Wayne Morse.* Garden City, New York: Doubleday, 1962. 455 pp. Illus. Morse, U.S. senator from Oregon, 1945-1969, was prominent in many political battles involving forests, conservation, and natural resources generally.

S323 Smith, Alice E. *Millstone and Saw: The Origins of Neenah-Menasha.* Madison: State Historical Society of Wisconsin, 1966. viii + 208 pp. Illus., maps, notes, bib., index. Neenah and Menasha, Wisconsin, to 1875, including some history of the lumber and woodworking industries.

S324 Smith, Alice E. *The History of Wisconsin: Volume 1, From Exploration to Statehood.* Madison: State Historical Society of Wisconsin, 1973. xiv + 753 pp. Illus., maps, notes, app., essay on sources, index. Includes a chapter on public land policy and incidental history of the lumber industry in territorial Wisconsin to 1848.

S325 Smith, Arthur Dwight. "The Status of Federal Land Grant Lands in Utah and Proposals for Their Management." Ph.D. diss., Univ. of Michigan, 1957. 173 pp. Since 1896.

S326 Smith, B.F. "Forestry at Elizabeth, Louisiana." *Journal of Forestry* 30 (Mar. 1932), 312-16. By the Industrial Lumber Company since 1912.

S327 Smith, C.E. "The History of the Development of Naval Stores Inspection and Standards." *Naval Stores Review* (Feb. 11, 1933), 16ff; (Feb. 18, 1933), 10 ff.

S328 Smith, Charles D. "The Movement for Eastern National Forests—1899-1911." Ph.D. diss., Harvard Univ., 1956.

S329 Smith, Charles D. "The Appalachian National Park Movement, 1885-1901." *North Carolina Historical Review* 37 (Jan. 1960), 38-65. On the movement to create a national park in western North Carolina, abandoned in 1901 in favor of national forest reserves.

S330 Smith, Charles D. "The Mountain Lover Mourns: Origins of the Movement for a White Mountain National Forest, 1880-1903." *New England Quarterly* 33 (Mar. 1960), 37-56. On the movement which culminated in the establishment of White Mountain National Forest, New Hampshire-Maine.

S331 Smith, Charles D. "Gentlemen, You Have My Scalp." *American Forests* 68 (Feb. 1962), 16-19. On the struggle for eastern national forests, 1899-1911, and the ultimate capitulation of House Speaker Joseph Cannon to the conservationist forces leading to the Weeks Act of 1911.

S332 Smith, Mrs. Chester W. "Sawmilling Days in Winneconne." *Wisconsin Magazine of History* 8 (Sept. 1924), 71-73. Reminiscence of Winneconne, Wisconsin, 1877-1890.

S333 Smith, Darrell H. *The Forest Service; Its History, Activities and Organization.* Institute for Government Research, Service Monographs of the United States Government, No. 58. Washington: Brookings Institution, 1930. Reprint. New York: AMS Press, 1973. xi + 268 pp. Notes, apps., bib., index. General history of the USFS, its predecessor agencies, and the forest conservation movement.

S334 Smith, David C. "Wood Pulp and Newspapers, 1867-1900." *Business History Review* 38 (Autumn 1964), 328-45. On the correlation between innovations in newsprint manufacture and increased circulation of newspapers.

S335 Smith, David C. "A History of Lumbering in Maine, 1860-1930." Ph.D. diss., Cornell Univ., 1965. 782 pp.

S336 Smith, David C. "Wood Pulp Paper Comes to the Northeast, 1865-1900." *Forest History* 10 (Apr. 1966), 12-25. Focuses on the growth of the Denison mills, the S.D. Warren Company, the International Paper Company, and the Great Northern Paper Company, primarily in Maine.

S337 Smith, David C. "Middle Range Farming in the Civil War Era: Life on a Farm in Seneca County." *New York History* 48 (Oct. 1967). Includes reference to farm-forest relationships.

S338 Smith, David C. "Forest History Research and Writing at the University of Maine." *Forest History* 12 (July 1968), 27-31.

S339 Smith, David C. "A Look at the United States Paper Industry in Its 19th Century Growing Years." *Paper Trade Journal* 152 (Sept. 9, 1968), 60-63.

Smith, David C. See Pride, Fleetwood, #P325

S340 Smith, David C. "Paper Mill Problems in Middle of Last Century Resemble Ours." *Paper Trade Journal* 153 (Jan. 13, 1969), 32-36.

S341 Smith, David C. "Bangor—The Shipping and Lumber Trade." In *A History of Bangor,* ed. by James Vickery. Bangor, Maine: Forbush-Roberts, 1969. Pp. 23-27.

S342 Smith, David C. "Maine and Its Public Domain: Land Disposal on the Northeastern Frontier." In *The Frontier in American Development: Essays in Honor of Paul Wallace Gates,* ed. by David M. Ellis. Ithaca, New York: Cornell University Press, 1969. Pp. 113-37.

S343 Smith, David C. *History of Papermaking in the United States (1691-1969).* New York: Lockwood, 1970. 693 pp. Illus., bib. A general history covering the entire United States, with treatment of changing raw materials, changing technology, institutional development, and labor-management relations.

S344 Smith, David C. "Toward a Theory of Maine History—Maine's Resources and the State." In *Explorations in Maine History, Miscellaneous Papers,* ed. by Arthur Johnson. Orono: University of Maine, 1970. Pp. 45-64.

Smith, David C. See Wood, Richard G., #W433

S345 Smith, David C., comp. *Lumbering and the Maine Woods: A Bibliographical Guide.* Portland: Maine Historical Society, 1971. 35 pp.

S346 Smith, David C. *A History of Lumbering in Maine, 1861-1960.* University of Maine Studies, No. 93. Orono: University of Maine Press, 1972. xvi + 469 pp. Illus., map, tables, apps., notes, bib., index. A history of forest industries in Maine which indicates the relative decline of small-scale lumber manufacturing after the 1870s and the trend toward corporate ownership of timberlands, industrial forestry, pulp and paper production, and conservation.

S347 Smith, David C. "The Logging Frontier." *Journal of Forest History* 18 (Oct. 1974), 96-106. On the concept of a logging frontier in American history, with special attention to the pivotal role of Maine lumbermen and logging technology, since ca. 1830s.

S348 Smith, David C. "Pulp, Paper, and Alaska." *Pacific Northwest Quarterly* 66 (Apr. 1975), 61-70. On the hopes, plans, and disappointments of the pulp and paper industry in Alaska in the 20th century, including the promotional role of the USFS.

S349 Smith, David M. "The Eli Whitney Forest." *Northeastern Logger* 5 (Jan. 1957), 19, 38-40. Brief history and description of the New Haven Water Company's forest in Connecticut.

S350 Smith, David M. "Connecticut Parks and Forests in the 20th Century." *Connecticut Woodlands* 35 (Spring 1970), 31-40.

S351 Smith, David M., and Hibbard, John E. "A History of the Connecticut Forest and Park Association." *Connecticut Woodlands* 35 (Spring 1970), 3-22. Since its founding as the Connecticut Forestry Association in 1895.

S352 Smith, Emma A. "Splendid Wooded Wealth of New Brunswick." *Canada Lumberman* 39 (May 1, 1919), 156-57. Incidental history.

S353 Smith, E.A. "Sawmill Railroading." *Railroad Magazine* 70 (Oct. 1959), 26-30. Reminiscences of railroads serving sawmills in southern Georgia and northern Florida since 1901.

S354 Smith, E.A. "Wild Hoggers and No Brakes." *Railroad Magazine* 70 (Oct. 1959), 26-30. Reminiscences of logging railroads in Florida and Georgia, 1902-1918.

S355 Smith, E.R. "History of Grazing Industry and Range Conservation Developments in the Rio Grande Basin." *Journal of Range Management* 6 (Nov. 1953), 405-09. New Mexico.

S356 Smith, Earl. *The Days of My Years.* Portland: Oregon Historical Society, 1968. 322 pp. Illus. Includes some reminiscences of the lumber industry in Oregon.

S357 Smith, Edmund Ware. "Maine's Wilderness and Mr. Baxter." *Down East* 4 (Sept. 1957), 22-25, 42-43. On Percival P. Baxter's acquisition of nearly 200,000 acres of wilderness land and his gift of it to Maine as Baxter State Park, 1931-1953.

S358 Smith, Edward. "Le Commerce du bois carré." *Société Géographique Québec, Bulletin* 5 (Oct. 1911), 335-47. Square timber trade in Quebec.

S359 Smith, Elizabeth. *A History of the Salmon National Forest.* Salmon, Idaho: USFS, Salmon National Forest, 1973. 174 pp. Notes, apps., tables, bib. History of the region since 1805.

S360 Smith, Elizur Y. "First Ground wood Mill in the United States." *News Print Service Bureau Bulletin* (July 14, 1942), 2-3.

S361 Smith, Esther R. *The History of Del Norte County, California, Including the Story of Its Pioneers with Many of Their Personal Narratives.* Oakland: Holmes Book Company, 1953. 224 pp. Illus., map, bib. Includes some history of logging and the lumber industry since the mid-19th century.

S362 Smith, Frank E. *Congressman from Mississippi.* New York: Pantheon Books, 1964. ix + 338 pp. Smith, a U.S. representative from Mississippi, 1951-1962, was involved in natural resource legislation as a means of promoting the economy of the South. He has since been a director of the Tennessee Valley Authority and has written on conservation topics.

S363 Smith, Frank E. *The Politics of Conservation.* New York: Random House, Pantheon Books, 1966. xii + 338 pp. Bib., index. This general history of conservation and resource developments demonstrates that many successes have been attributable to pork barrel politics.

S364 Smith, Frank E. "Theodore Roosevelt, Conservationist." *American History Illustrated* 2 (No. 8, 1967), 36-42. On his conservation policies and his close relationship with USFS chief Gifford Pinchot.

S365 Smith, Frank E. *Land Between the Lakes: Experiment in Recreation.* Lexington: University Press of Kentucky, 1971. xiv + 124 pp. Illus., map. History and description of Land Between the Lakes, an extensive recreation area conceived and administered by the Tennessee Valley Authority on forested lands between reservoirs of the Cumberland and Tennessee rivers in western Kentucky and Tennessee.

S366 Smith, Frank E.; Foss, Phillip O.; Doherty, William T., Jr.; and Dworsky, Leonard B., eds. *Conservation in the United States: A Documentary History.* 5 Volumes. New York: Chelsea House Publishers, 1971. Index. Two volumes on "Land and Water" conservation, edited by Smith, contain many documents pertinent to forest history. The volume on recreation, edited by Foss, also contains relevant material.

S367 Smith, Franklin H. *A Study of the Wisconsin Wood-Using Industries.* Madison, 1910. 68 pp.

S368 Smith, Franklin H. "Significant Trends in Lumber Production in the U.S." *American Forestry* 26 (Mar. 1920), 143-47.

S369 Smith, Guy Harold, ed. *Conservation of Natural Resources.* 1950. Fourth edition. New York: John Wiley and Sons, 1971. 685 pp. Illus. This text contains a historical review and evaluation of the conservation movement.

S370 Smith, Helen. "Crooked Journey on the 'Skunk'." *Westways* 48 (June 1956), 28-29. On the California Western Railroad in redwood country between Fort Bragg and Willits, California, a logging railroad since 1911.

S371 Smith, Henry Nash. "Rain Follows the Plow: The Notion of Increased Rainfall for the Great Plains, 1844-1880." *Huntington Library Quarterly* 10 (Feb. 1947), 169-93. Includes references to tree planting.

S372 Smith, Henry Nash. "Clarence King, John Wesley Powell, and the Establishment of the United States Geological Survey." *Mississippi Valley Historical Review* 34 (June 1947), 37-58.

S373 Smith, Henry Nash. *Virgin Land: The American West as Symbol and Myth.* Cambridge: Harvard University Press, 1950. xiv + 305 pp. Illus., notes. A landmark intellectual history of American attitudes toward the West, including the social implications of the American's exposure to a hostile environment or wilderness conditions, primarily as revealed in literature.

S374 Smith, Herbert A. "The Old Order Changes." *Journal of Forestry* 18 (Mar. 1920), 203-10. On the accomplishments of Henry S. Graves as chief of the USFS, 1910-1920.

S375 Smith, Herbert A. "William B. Greeley: An Appreciation and Interpretation." *Journal of Forestry* 26 (Apr. 1928), 423-29. Greeley was chief of the USFS, 1920-1928.

S376 Smith, Herbert A. "Robert Young Stuart." *Journal of Forestry* 29 (Dec. 1933), 885-90. Stuart was chief of the USFS from 1928 to 1933.

S377 Smith, Herbert A. "Forest Education Before 1898." *Journal of Forestry* 32 (Oct. 1934), 684-89. At Massachusetts Agricultural College, Cornell University, University of Michigan, University of Pennsylvania, and other colleges since the 1870s.

S378 Smith, Herbert A. "The Early Forestry Movement in the United States." *Agricultural History* 12 (Oct. 1938), 326-46. On the origins of the movement, 1860s to 1900.

S379 Smith, Herbert A. "State Forestry under Public Regulation." *Journal of Forestry* 39 (Feb. 1941), 99-103. Historical sketch of the regulation issue since 1905.

S380 Smith, Herbert E. "Cost Accounting Problems in the Lumber Industry." *Timberman* 25 (Nov. 1923), 194-200. Incidental history.

S381 Smith, Herbert F. *John Muir.* New York: Twayne Publishers, 1965. 158 pp. Muir (1838-1914) was a naturalist, writer, wilderness philosopher, and founder of the Sierra Club.

S382 Smith, Ian. *Vancouver Island: Unknown Wilderness.* Seattle: University of Washington Press, 1973. 186 pp. Illus. An illustrated natural history of heavily forested Vancouver Island, British Columbia.

S383 Smith, J. Harry G. "New Goals, Criteria, and Standards Are Urgently Needed for Improved Management of Canada's Forest Land Resources." *Forestry Chronicle* 50 (June 1974), 90-92. Includes some reminiscences of the editor of *Forestry Chronicle,* 1960-1966.

S384 Smith, James A. "The Structure of Wages in the Pacific Northwest Lumber Industry, 1939-1964." Ph.D. diss., Washington State Univ., 1967. 408 pp.

S385 Smith, James B. "The Movements for Diversified Industry in Eau Claire, Wisconsin, 1879-1907: Boosterism and Urban Development Strategy in a Declining Lumber Town." Master's thesis, Univ. of Wisconsin, 1967.

S386 Smith, James B. "Lumbertowns in the Cutover: A Comparative Study of the Stage Hypothesis of Urban Growth." Ph.D. diss., Univ. of Wisconsin, 1973. 365 pp. A quantitative study of 94 Wisconsin and Michigan lumber towns which reached maturity as centers of lumber production between 1880 and 1910.

S387 Smith, James L. "Early Lumbering Days in the Chemung Valley." *Northeastern Logger* 6 (May 1958), 20-21, 39. New York, 19th century.

S388 Smith, James L. "Saw-Mills of Erin." *Chemung County Historical Journal* 8 (Dec. 1964), 1328-32. In the township of Erin, Chemung County, New York, since 1824.

S389 Smith, John H.G., et al. *Economics of Reforestation of Douglas Fir, Western Hemlock, and Western Red Cedar in the Vancouver Forest District.* University of British Columbia, Forestry Bulletin No. 3. Vancouver, 1961. 144 pp. Incidental history.

S390 Smith, John Jay. *Reminiscences of John Jay Smith.* N.p., n.d. 41 pp. The author (b. 1856) was a Washington logger.

S391 Smith, Kenneth George. "Impact of the Sawmilling Industry on the Economy of New Brunswick." Master's thesis, Univ. of New Brunswick, 1970.

S392 Smith, L.W., and Wood, L.W. *History of Yard Lumber Size Standards.* Madison: USFS, Forest Products Laboratory, 1964. 56 pp. Summarizes the development of size standards for yard lumber in the United States.

Smith, Lloyd F. See Ware, E.R., #W68

S393 Smith, Mowry, Jr., and Clark, Giles. *One Third Crew, One Third Boat, One Third Luck: The Menasha Corporation (Menasha Wooden Ware Company) Story, 1849-1974.* Neenah, Wisconsin: Menasha Corporation, 1974. xi + 177 pp. Illus., app. History of a wood products firm established by Elisha D. Smith in Neenah, Wisconsin, 1849. Once known as the world's largest wooden ware manufacturer, it later became a fully integrated corporation with container mills, paperboard mills, wood flour plants, and extensive timberlands in the Great Lakes states, the Pacific Northwest, and Saskatchewan.

S394 Smith, Norman F. "Plantation Harvest." *Michigan Conservation* 23 (Mar.-Apr. 1954), 31-32. On the forest plantations of the Forestry Division of the Michigan Department of Conservation since 1904.

S395 Smith, Norman F. "Forestry Ahead." *Michigan Conservation* 28 (May-June 1959), 2-7. On public forest management in Michigan since 1909.

Smith, Paul R. See Peterson, Virgil G., #P157

Smith, Paul R. See Plant, Charles, #P244

S396 Smith, Paul R. *Paul R. Smith Views the Western Red Cedar Industry 1910 to the Present.* OHI by Elwood R. Maunder. Santa Cruz, California: Forest History Society, 1975. vii + 106 pp. Illus., map, apps., index. Processed. On Smith's career in the red cedar shingle and shake industry and its trade association in Washington.

S397 Smith, R.C. *Taxation of Forest Land in South Missouri.* University of Missouri, Agricultural Experiment Station, Research Bulletin No. 624. Columbia, 1957. 44 pp. Incidental history.

S398 Smith, Ralph H. "This Was the Forest Primeval: As Revealed by Pollen Analysis and Writings of Early Travelers and Surveyors." *New York University Bulletin to the Schools* 40 (Feb. 1954), 138-43.

S399 Smith, Richard G. "A Forest Recovers." *American Forests* 66 (Mar. 1960), 29-30, 53-54. On a forest fire on Mount Desert Island, Maine, 1947, and the reforestation of the damaged area.

S400 Smith, Rufus M. "Day of the Shay." *Forest & People* 21 (Second quarter, 1971), 15-17, 30. On the use of the Shay locomotive by the Rapides Lumber Company, Louisiana, late 19th century.

S401 Smith, T. Lynn, and Fry, Martha R. *The Population of a Selected 'Cut-Over' Area in Louisiana.* Baton Rouge: Louisiana State University and Agricultural and Mechanical College, 1936. On the social impact of logging; also appeared as Louisiana State University and Agricultural and Mechanical College, *Bulletin* No. 268 (Jan. 1936).

Smith, W.H. See Rowland, H.B., #R341

S402 Smith, Walker C. *The Everett Massacre: A History of the Class Struggle in the Lumber Industry.* Chicago: Industrial Workers of the World Publishing Bureau, 1918. Reprint. New York: Da Capo Press, 1971. 302 pp. Illus., table, notes. An IWW account of the violent conflict between Wobblies and the sheriff's posse in Everett, Washington, 1916.

S403 Smith, Willard S. "History and Analysis of Joint Job Analysis in the Pulp and Paper Industry of the Pacific Coast." Master's thesis, Princeton Univ., 1948.

S404 Smith, William F. "Reminiscences." *Michigan History Magazine* 16 (Autumn 1932), 504-11. On establishing a sawmill in Chicago, 1841.

S405 Smith, William H. *A History of the Cabinet of the United States, From President Washington to President Coolidge.* Baltimore: Industrial Printing Company, 1925. 537 pp. Index. See for information on interior and agriculture secretaries.

S406 Smith, Wyman. "The Largest Community Forest." *American Forests* 46 (Jan. 1940), 22-25. Includes some history of Seattle's 66,000-acre Cedar River Watershed and the practice of forestry thereon since 1900.

S407 Smith & Winchester Manufacturing Company. *A Century of Pioneering in the Paper Industry, 1828-1928.* South Windham, Connecticut: Providence, Livermore and Knight, 1928. 45 pp. Centennial history of Smith & Winchester Manufacturing Company.

Smyth, Mary W. See Eckstorm, Fannie H., #E20

S408 Smythe, Limen Towers. "The Lumber and Sawmill Workers Union in British Columbia." Master's thesis, Univ. of Washington, 1937.

S409 Snell, Donald W. "An Introduction to a History of Lumbering in Minnesota." Master's thesis, Univ. of Minnesota.

S410 Snell, Ralph M. "Pioneer Pulp and Paper Making at Niagara Falls." *Paper Maker* 1 (No. 2, 1932), 24-26; 1 (No. 3, 1932), 40-41; 2 (No. 1, 1933), 16-18; 2 (No. 3, 1933), 51-54. New York.

S411 Snell, Ralph M., and McBain, B.T. "Early Pulp and Paper Mills of the Pacific Coast." *Paper Trade Journal* 99 (Oct. 11, 1934), 42-50.

S412 Snider, E.W.B. "Waterloo County Forests and Primitive Economics." *Report of the Waterloo County Historical Society* 6 (1918), 14-36. Ontario.

S413 Snow, Albert G. "Maple Sugaring and Research." *Journal of Forestry* 62 (Feb. 1964), 83-88. Some history of the industry since the 19th century, with emphasis on changing technology.

S414 Snow, Russell. "Nursery Practice and Watershed Planting at Quabbin Resevoir." *Journal of the New England Water Works Association* 66 (Mar. 1952), 111-16. Massachusetts, since 1934.

S415 Snow, Sinclair. "Naval Stores in Colonial Virginia." *Virginia Magazine of History and Biography* 72 (Jan. 1964), 75-93. Especially tar, pitch, and turpentine.

S416 Snyder, Arnold P. *Wilderness Area Management: An Administrative Study of a Portion of the High Sierra Wilderness Area.* Washington: USFS, 1960. 63 pp. Sierra National Forest, California.

S417 Snyder, Thomas E.; Middleton, William; and Keen, Frederick Paul. "The Progress of Forest Entomology in the United States." *Journal of Economic Entomology* 16 (Oct. 1923), 413-20.

S418 Soady, Fred W., Jr. "The Making of the Shawnee." *Forest History* 9 (July 1965), 10-23. On the creation of Shawnee National Forest in southern Illinois since 1933.

Sochen, June. See Frederick, Duke, #F223

S419 Social Science Research Council. *A Survey of Research in Forest Land Ownership: Report of a Special Committee on Research in Forest Economics.* New York, 1939. 93 pp. Notes. Includes a historiographical review of the question of forest ownership.

S420 Society of American Foresters. *A Survey of State Forestry Administration in Kentucky.* Washington, 1946. 49 pp. Incidental history.

S421 Society of American Foresters. *A Survey of State Forestry Administration in North Carolina.* Washington, 1946. 38 pp. Incidental history.

S422 Society of American Foresters. *A Survey of State Forestry Administration in Tennessee.* Washington: Society of American Foresters and the Charles Lathrop Pack Forestry Foundation, 1947. 40 pp. Illus., bib. Incidental history.

S423 Society of American Foresters. *A Survey of State Forestry Administration in Colorado.* Washington, 1948. 38 pp. Processed. Incidental history.

S424 Society of American Foresters. *A Survey of State Forestry Administration in Idaho.* Washington, 1948. 35 pp. Processed. Incidental history.

S425 Society of American Foresters. *Forest Practices Developments in the United States, 1940 to 1955.* Washington, 1956. 39 pp. Bib. Summarizes state legislation on cutting practices.

S426 Soeriaatmadja, Roehajat E. "Fire History of the Ponderosa Pine Forests of the Warm Springs Indian Reservation, Oregon." Ph.D. diss., Oregon State Univ., 1966. 132 pp.

S427 Soffar, Allan J. "Conservation Controversies between the Department of Agriculture and the Department of the Interior, 1898-1910." Master's thesis, Univ. of Houston, 1967.

S428 Soffar, Allan J. "Differing Views on the Gospel of Efficiency: Conservation Controversies between Agriculture and Interior, 1898-1938." Ph.D. diss., Texas Technological Univ., 1974. 423 pp. Emphasizes inter-governmental disputes over the national forests and the USFS.

S429 Soffar, Allan J. "The Forest Shelterbelt Project, 1934-1944." *Journal of the West* 14 (July 1975), 95-107.

S430 Solberg, Erling D. *New Laws for New Forests: Wisconsin's Forest-Fire, Tax, Zoning, and County-Forest Laws in Operation.* Madison: University of Wisconsin Press, 1961. xxiv + 611 pp. Maps, tables, apps., notes, index. On conditions left at the end of logging operations in Wisconsin in the 1920s, and the legislation designed to restore forest resources to the cutover lands, such as the Forest Crop Law of 1927.

S431 Solin, Lawrence. *A Study of Farm Woodland Cooperatives in the United States.* New York State College of Forestry, Technical Paper No. 48. Syracuse, 1940. 117 pp. Incidental history.

S432 Sommarstrom, Allan R. "The Impact of Human Use on Recreational Quality: The Example of the Olympic National Park Backcountry User." Master's thesis, Univ. of Washington, 1967.

S433 Sommarstrom, Allan R. "Wild Land Preservation Crisis: The North Cascades Controversy." Ph.D. diss., Univ. of Washington, 1970. 185 pp. On the movement to create Washington's North Cascades National Park.

S434 Somrock, John W. "Incredible Ely." *American Forests* 80 (Sept. 1974), 8-11, 54-55. On Ely, a logging town in northeastern Minnesota, since the 1880s.

S435 Sonne, Conway Ballantyne. *Knight of the Kingdom: The Story of Richard Ballantyne.* Salt Lake City: Deseret Book Company, 1949. xiv + 230 pp. Illus. Biography of a Mormon leader and evangelist, with reference to his career as a Utah lumber dealer in the 19th century.

S436 Sorden, Leland George, and Ebert, Isabel J. *Logger's Words of Yesteryears.* Madison, Wisconsin: By the authors, 1956. v + 44 pp. Bib. Terminology used in the Great Lakes states between the 1850s and 1920s.

S437 Sorden, Leland George. *Lumberjack Lingo.* Madison: Wisconsin House, 1969. 150 pp. Illus. A revised version of the above entry; it includes logging terms from the New England and Great Lake states, ca. 1850-1920.

S438 Sorensen, Willis C. "The Kansas National Forest, 1905-1915." *Kansas Historical Quarterly* 35 (Winter 1969), 386-95. On the unsuccessful experiment to plant a forest on the sand hills of western Kansas.

S439 Sorg Paper Company. *Since 1852: 100 Years of Progress with America.* Middletown, Ohio, 1952.

S440 Sosin, Jack M. *The Revolutionary Frontier, 1763-1783.* New York: Holt, Rinehart & Winston, 1967. xiii + 241 pp. Includes some history of Eastern forests in relation to settlement.

S441 Soth, Lauren. "Mr. Hoover's Department of Agriculture." *Journal of Farm Economics* 31 (May 1949), 201-12.

S442 Soucie, Gary. "Congaree—Great Trees or Coffee Tables?" *Audubon* 77 (July 1975), 60-80. Includes some history of South Carolina's Congaree Swamp; early cypress lumbering in it by the owners, the Francis Beidler family; and recent efforts to preserve the swamp as a national monument.

S443 *Southern Lumber Journal.* "49 Years of Cypress Logging in Florida." *Southern Lumber Journal* 42 (Jan. 1938), 22-23.

S444 *Southern Lumberman.* "Choctaw Lumber & Shingle Co." *Southern Lumberman.* (May 2, 1908), 1, 32-34. Hulbert, Arkansas; incidental history.

S445 *Southern Lumberman.* "Green River Lumber Company, Memphis, Tennessee." *Southern Lumberman* (June 13, 1908), 42-44. Incidental history.

S446 *Southern Lumberman.* "Meridian as a Lumber Center." *Southern Lumberman* (Aug. 8, 1908), 31-36. Meridian, Mississippi.

S447 *Southern Lumberman.* "History of a Successful Retail Organization." *Southern Lumberman* (Oct. 9, 1909), 31-32. On a network of western Pennsylvania credit associations of retail lumbermen and building supply dealers, since 1903.

S448 *Southern Lumberman.* "Avoyelles Cypress Co. Ltd., Cottonport, La." *Southern Lumberman* (Dec. 25, 1909), 51-58. Incidental history.

S449 *Southern Lumberman.* "A Bit of History in Connection with Forest Industries." *Southern Lumberman* (Aug. 15, 1915), 37-39. On the James D. Lacey Company, international brokers in timberlands.

S450 *Southern Lumberman.* "Early History and Review of the Lumber Business in Eastern Kentucky." *Southern Lumberman* (Dec. 18, 1920), 180-81.

S451 *Southern Lumberman.* "The Allison Activities in Alabama." *Southern Lumberman* (Aug. 2, 1924), 43-58. Includes some history of the timber holdings, sawmills, and conservation activities of the Allison Lumber Company, Bellamy, Alabama.

S452 *Southern Lumberman*. "The Story of the 'Southern Lumberman'." *Southern Lumberman* (Dec. 15, 1931), 51-58. An important trade journal since 1881.

S453 *Southern Lumberman*. "Southern Hardwood Development." *Southern Lumberman* (Dec. 15, 1931), 73-76. Includes some history of early hardwood sawmills in the South.

S454 *Southern Lumberman*. "A Pioneer of the Southwest." *Southern Lumberman* (Dec. 15, 1931), 81. Reminiscences of Jasper Peavy, lumberman with operations throughout the South.

S455 *Southern Lumberman*. "Fifty Years of Advancement in Lumbering Equipment; Sawmill Machinery; Saw Manufacture." *Southern Lumberman* (Dec. 15, 1931), 82-85.

S456 *Southern Lumberman*. "A Century-Old Sawmill Operation." *Southern Lumberman* (Dec. 15, 1931), 99-100. R.F. Learned & Son, Natchez, Mississippi.

S457 *Southern Lumberman*. "The Oldest Southern Sawmill Is Operated by Negroes." *Southern Lumberman* (Dec. 15, 1931), 106. J.J. Sulton & Sons, Orangeburg, South Carolina, since 1825.

S458 *Southern Lumberman*. "Roll Call of the Markets." *Southern Lumberman* (Dec. 15, 1931), 131-46. Brief histories of fourteen market centers in the South and East. Some individual citations follow.

S459 *Southern Lumberman*. "Nashville, Former Hardwood Center of the World, Is Still Important." *Southern Lumberman* (Dec. 15, 1931), 131-32.

S460 *Southern Lumberman*. "Cincinnati—Once the Greatest Lumber Market, Still an Important Factor." *Southern Lumberman* (Dec. 15, 1931), 132-33.

S461 *Southern Lumberman*. "Louisville: No Longer a Sawmill Center, but Still Highly Important." *Southern Lumberman* (Dec. 15, 1931), 133-34.

S462 *Southern Lumberman*. "Memphis: Some Recollections of Early Days in the Hardwood Metropolis." *Southern Lumberman* (Dec. 15, 1931), 134-35.

S463 *Southern Lumberman*. "Houston: One of the Southwest's Great Distributing Centers." *Southern Lumberman* (Dec. 15, 1931), 135.

S464 *Southern Lumberman*. "Birmingham: Established Only Sixty Years Ago, Has Grown to be Great Market." *Southern Lumberman* (Dec. 1931), 135-36.

S465 *Southern Lumberman*. "Savannah: Since 1744 It Has Been of Importance in the Industry." *Southern Lumberman* (Dec. 15, 1931), 136-37.

S466 *Southern Lumberman*. "New Orleans: Important Lumber Center Specializing in the Export Trade." *Southern Lumberman* (Dec. 15, 1931), 137-38.

S467 *Southern Lumberman*. "Chicago: 'World's Greatest Lumber Market' Ships Billions of Feet Annually." *Southern Lumberman* (Dec. 15, 1931), 138-39.

S468 *Southern Lumberman*. "Norfolk: A Great Port and a Great Lumber Center, Past, Present and Future." *Southern Lumberman* (Dec. 15, 1931), 139-40.

S469 *Southern Lumberman*. "Baltimore Retains Strategic Importance as a Lumber Market." *Southern Lumberman* (Dec. 15, 1931), 140-41.

S470 *Southern Lumberman*. "Philadelphia: Trade Association Work Has Featured Its Lumber Activities." *Southern Lumberman* (Dec. 15, 1931), 141-42.

S471 *Southern Lumberman*. "New York: After 50 Years It Is Still the Greatest Lumber Market." *Southern Lumberman* (Dec. 15, 1931), 142-43.

S472 *Southern Lumberman*. "Boston: Fifty Years of Progress in the New England Field." *Southern Lumberman* (Dec. 15, 1931), 143-46.

S473 *Southern Lumberman*. "The History of Lumber Exporting." *Southern Lumberman* (Dec. 15, 1931), 163. From Southern ports since the colonial period.

S474 *Southern Lumberman*. "Forestry Rejuvenates Lumber Company Nearing End of Cut." *Southern Lumberman* (Dec. 15, 1944), 121-22. Louisiana Long Leaf Lumber Company, Fisher, Louisiana.

S475 *Southern Lumberman*. "The Country's Oldest Living Lumberman." *Southern Lumberman* (Dec. 15, 1944), 198. James H. Frazier, a pioneer Kentucky lumberman.

S476 *Southern Lumberman*. "Portable Gang Saw Mill." *Southern Lumberman* (Mar. 1, 1949), 45-46. Describes an invention of Simon Willard patented in 1830.

S447 *Southern Lumberman*. "Celebrate 100th Birthday: Centennial of Pope and Talbot, Inc., Marks a Highlight in West Coast Lumbering." *Southern Lumberman* (Dec. 15, 1949), 224-26. On the lumber and shipping operations of Pope & Talbot, Inc., since 1849.

S478 *Southern Lumberman*. "100 Years of Progress." *Southern Lumberman* (Dec. 15, 1951), 326-27. On the Coe Manufacturing Company, Painesville, Ohio, makers of woodworking and veneer machinery since 1852.

S479 *Southern Lumberman*. " 'America's No. 1 Lumberman.' " *Southern Lumberman* (Dec. 15, 1956), 249. On Marc Leonard Fleishel's association with various lumber companies in Louisiana and Florida since 1893, and on his presidency of the National Lumber Manufacturers Association.

S480 *Southern Lumberman*. "The History of the 'Southern Lumberman'." *Southern Lumberman* (Dec. 15, 1956), 110-15. Since 1881.

S481 *Southern Lumberman*. "Now and Then—Seventy-Five Years of Progress." *Southern Lumberman* (Dec. 15, 1956), 115-18. On the *Southern Lumberman's* influence on forest management and methods and concepts of lumber production in the South since its founding in 1881.

S482 *Southern Lumberman*. "Arkansas Group's Promotion Profitable." *Southern Lumberman* (Dec. 15, 1957), 136-37. On the Arkansas Soft Pine Bureau since 1912.

S483 *Southern Lumberman*. "Acquisition of R.F. Learned Records." *Southern Lumberman* (Nov. 15, 1958), 33. Records of the Natchez lumber company were deposited at the University of Mississippi Library.

S484 *Southern Lumberman.* "Last of the Plank Roads." *Southern Lumberman* (Dec. 15, 1958), 133-34. On a plank road built in 1920 on Graham Island of the Queen Charlotte Islands of British Columbia.

S485 *Southern Lumberman.* "History of the 'Southern Lumberman'." *Southern Lumberman* (Dec. 15, 1971), 58-61. Since 1881.

S486 Southern Pine Association. *Economic Conditions in the Southern Pine Industry.* New Orleans: Southern Pine Association, 1931. 136 pp. Illus., maps, tables. A report to the U.S. Timber Conservation Board containing some historical information.

S487 Southern West Virginia Forest Fire Protective Association. *History, Accomplishments and Aims: 8 Years of Forest Fire Protection in West Virginia, 1916-1924.* Charleston, 1924. 58 pp.

S488 *Southern Workman.* "The Beginning of Forestry in the United States." *Southern Workman* 39 (Feb. 1910), 111-12.

S489 Southworth, Constant. "The American-Canadian Newsprint Paper Industry and the Tariff." *Journal of Political Economy* 30 (Oct. 1922), 681-97. Incidental history.

S490 Southworth, Thomas. "Ontario's Progress Towards a National Forestry System." *Canadian Forestry Journal* 3 (Dec. 1907), 157-63.

S491 Sowinski, Edward S. "Rail Transportation of Lumber in the U.S.." Master's thesis, Yale Univ., 1949.

S492 Space, Ralph S. *The Clearwater Story: A History of the Clearwater National Forest.* Missoula, Montana: USFS, Northern Region, 1964. 163 pp. Illus., map, tables, notes. Idaho, since 1805.

S493 Space, Ralph S. "The Race for Clearwater Timber." *Idaho Yesterdays* 17 (Winter 1974), 2-5. On an expedition headed by C.O. Brown which laid claim in 1900 to a vast acreage of white pine timberlands in northern Idaho for Frederick Weyerhaeuser, John Humbird, and John Glover, lumbermen of the Great Lakes states.

Spada, B. See Larson, Robert W., #L75

Sparhawk, William N. See Zon, Raphael, #Z26

S494 Sparhawk, William N., and Brush, Warren D. *The Economic Aspects of Forest Destruction in Northern Michigan.* USDA, Technical Bulletin No. 92. Washington: GPO, 1929. 129 pp. Incidental history.

S495 Sparhawk, William N. "The History of Forestry in America." In *Trees: The Yearbook of Agriculture, 1949.* Washington: USDA and GPO, 1949. Pp. 702-14. Since the 17th century. This *Yearbook* contains historical information on nearly every aspect of forestry and forests in the United States. Other USDA *Yearbooks* should also be consulted for articles pertinent to forest history.

S496 Sparks, Theodore A. "Early Lumbering History of Winnipeg and District." *Canada Lumberman* 50 (Aug. 1, 1930), 97. On the founding of Theodore A. Burrows Lumber Company, Winnipeg, Manitoba, and the lumber industry of the region.

S497 Spears, Borden, ed. *Wilderness Canada.* Toronto: Clarke, Irwin, 1970. 174 pp. Incidental history.

S498 Speer, J.B. "A Saw Mill in the Panhandle of Texas." *Panhandle-Plains Historical Review* 9 (1936), 52-60. Built by J.D. Lard near Pampa, Texas, 1903.

S499 Speers, Ron. "Revolution in the Blue Ridge." *American Woodsman* 2 (July-Aug. 1952), 12-15, 30. On the establishment and management of the George Washington and Jefferson national forests in Virginia since 1912.

S500 Spence, Benjamin Arthur. "The National Career of John Wingate Weeks (1904-1925)." Ph.D. diss., Univ. of Wisconsin, 1971. 423 pp. The Massachusetts congressman sponsored important forest and conservation legislation, especially the Weeks Act of 1911.

S501 Spence, Vernon Charles. "A History of the Redwood Lumber Industry in Sonoma County." Master's thesis, Chico State College, 1962.

S502 Spencer, Betty Goodwin. *The Big Blowup.* Caldwell, Idaho: Caxton Printers, 1956. 286 pp. Illus., maps, bib. On the forest fires of 1910 which burned over three million acres in northern Idaho, northeastern Washington, and northwestern Montana.

Spencer, John S. See Choate, Grover A., #C307

S503 Spencer, Morris N. "The Union Pacific Railroad Company's Utilization of Its Land Grant, with Emphasis on Its Colonization Program." Ph.D. diss., Univ. of Nebraska, 1950.

S504 Spencer, P.L. "Ship and Shanty in the Early Fifties." *Ontario Historical Society, Papers and Records* 18 (1920), 25-31. Includes reminiscences of logging at Owen Sound on Georgian Bay, Ontario.

S505 Spencer, William A. "The Cooperative Shingle Mills of Western Washington." Master's thesis, Univ. of Washington, 1922.

S506 Spero, Sterling D., and Harris, Abram L. *The Black Worker—The Negro and the Labor Movement.* New York: Columbia University Press, 1931. x + 509 pp. Includes some history of labor conditions in the Southern lumber industry.

S507 Sprague, John F. "Making History in the Maine Woods." *Sprague's Journal of Maine History* 9 (July-Sept. 1921), 126-30. Reminiscences of Maine lumberjacks, the Great Northern Paper Company's Spruce Woods Department, and its monthly magazine, *The Northern,* 1860s to 1920.

S508 Sprague, George C. "The Land System of Colonial New York." Ph.D. diss., New York Univ., 1908.

S509 Sprague, Richard S. "Carl Sprinchorn in the Maine Woods." *Forest History* 14 (July 1970), 6-15. Sprinchorn painted forest and logging scenes of Maine, 1910s-1950s.

Sperry, James R. See Newton, Craig A., #N99

S510 Spicer, Stanley T. *Masters of Sail.* Toronto: Ryerson Press, 1968. 278 pp. Illus., maps. Includes some history of shipbuilding in the Maritime provinces, 19th century.

S511 Spillers, A.R., and Eldredge, Inman F. *Georgia Forest Resources and Industries.* USDA, Miscellaneous Publications No. 501 M. Washington: GPO, 1943. 70 pp. Incidental history.

Spoehr, H.A. See Bailey, Irving Widmer, #B15

S512 Spooner, Harry L. *Lumbering in Newaygo County.* White Cloud, Michigan: Cooper Press, 1948. Michigan.

S513 Spring, Ida R. "White Pine Portraits: Genial Don McLeod." *Michigan History Magazine* 30 (Jan.-Mar. 1946), 59-72. A lumberman of Newberry on Michigan's Upper Peninsula.

S514 Spring, Ida R. "White Pine Portraits: Big Dave Ranson." *Michigan History* 31 (Sept. 1947), 314-21. A lumberman from the Tahquamenon River region of Michigan's Upper Peninsula, ca. 1880s.

S515 Spring, Ida R. "White Pine Portraits: Con Culhane." *Michigan History* 31 (Dec. 1947), 437-42. A Michigan lumberman.

S516 Spring, Ida R. "White Pine Portraits: Norwegian Jack Ryland." *Michigan History* 32 (Sept. 1948), 295-300. A lumberman in Michigan's Upper Peninsula, 1890s.

Spring, Samuel N. See Recknagel, Arthur B., #R71

S517 Spring, Samuel N. "The Development of Forestry in New York State." *Empire Forester* 20 (1934), 7-12.

S518 Spring, Samuel N. "Fifty Years of Conservation." *American Forests* 41 (Aug. 1935), 355-57, 394. Forest conservation in New York since 1885.

S519 Spring, Samuel N. "Twenty-Five Years of Forestry, 1910-1935." *Memoirs of the Brooklyn Botanical Garden* 4 (May 7, 1936), 71-79. New York.

S520 Spring, Samuel N., ed. *The First Half-Century of the Yale School of Forestry.* New Haven: Yale University, School of Forestry, 1950. 211 pp. Illus., tables, bib. Contains many articles on forestry education at Yale University and its alumni.

S521 Springer, John S. *Forest Life and Forest Trees: Comprising Winter Camp-Life among the Loggers and Wild-Wood Adventure, with Descriptions of Lumbering Operations on the Various Rivers of Maine and New Brunswick.* New York: Harper and Brothers, 1851. Reprint. Somersworth, New Hampshire: New Hampshire Publishing Company, 1971. 292 pp. Illus. Springer's classic account is based in part on his own experiences as a lumberjack in Maine in the 1820s and 1830s.

S522 Spurr, Stephen H. *Aerial Photographs in Forestry.* New York: Ronald Press, 1948. xi + 340 pp. Illus. Incidental history.

S523 Spurr, Stephen H. "George Washington, Surveyor and Ecological Observer." *Ecology* 32 (Oct. 1951), 544-59.

S524 Spurr, Stephen H. "Origin of the Concept of Forest Succession." *Ecology* 33 (July 1952), 426-27. Development of the concept from 1792 to 1888.

S525 Spurr, Stephen H. *Forest Inventory.* New York: Ronald Press, 1952. xii + 476 pp. Illus. Incidental history.

S526 Spurr, Stephen H. "The Forests of Itasca in the Nineteenth Century as Related to Fire." *Ecology* 35 (Jan. 1954), 21-25. Itasca State Park, Minnesota, 1802-1917.

S527 Spurr, Stephen H. "Natural Restocking of Forests Following the 1938 Hurricane in Central New England." *Ecology* 37 (July 1956), 443-51. From 1938 to 1948.

S528 Spurr, Stephen H. "Plantation Success in the Harvard Forest as Related to Planting Site and Cleaning, 1907-1947." *Journal of Forestry* 54 (Sept. 1956), 577-79. Harvard Forest, Petersham, Massachusetts.

S529 Spurr, Stephen H. "Nine Successive Thinnings in a Michigan White Pine Plantation." *Journal of Forestry* 55 (Jan. 1957), 7-13. On forest management since 1915.

S530 Spurr, Stephen H., and Barnes, Burton V. *Forest Ecology.* 1964. Second edition. New York: Ronald Press, 1973. vii + 571 pp. Illus., maps, tables, diags., notes, bib., index. This text includes chapters on the "Historical Development of Forests" and "The American Forest Since 1600."

S531 Spurr, Stephen H. "Sam Dana." *American Forests* 79 (May 1973), 20-22. On Samuel Trask Dana's career in forestry since 1904.

Spurr, Stephen H. See Seaton, Fred A., #S157

S532 Squires, J.W. "Burning on Private Lands in Mississippi." *Tall Timbers Fire Ecology Conference Proceedings* 3 (1964), 1-9. Includes some history of the subject.

S533 Stabler, Herman. "Rise and Fall of the Public Domain." *Civil Engineering* 2 (1932), 541-46. Brief history of federal land policies.

S534 Stafford, Howard A. "Factors in the Location of the Paperboard Container Industry." *Economic Geography* 36 (July 1960), 260-66. Incidental history.

S535 Stahelin, R. *Thirty-Five Years of Planting on the National Forests of Colorado.* Fort Collins, Colorado: USFS, Rocky Mountain Forest and Range Experiment Station, 1941. 182 pp.

S536 Stahl, Rose M. *The Ballinger-Pinchot Controversy.* Smith College Studies in History, Volume 11, Part 2. Northampton, Massachusetts: Smith College, 1926. Pp. 69-138. On the disruption of the Taft administration and the conservation movement over controversy between Secretary of the Interior Richard A. Ballinger and USFS chief Gifford Pinchot, 1909-1910.

Stahl, William J. See Peirce, Earl S., #P92

S537 Stahlman, James G. "Some History of Papermaking in the South and Coosa River Newsprint Company." *Southern Pulp and Paper Manufacturer* 23 (Mar. 1960), 80-86. Alabama.

S538 Staley, Lewis E. "Conservation and Land Use in State Forestry." *Journal of Forestry* 31 (Mar. 1933), 265-69. In Pennsylvania since the 1890s.

S539 Stalley, Marshall. "Perpetuating Penn's Woods West." *American Forests* 65 (Oct. 1959), 28-30, 54-59. On the forest conservation work of the Western Pennsylvania Conservancy and its predecessor groups since 1931.

S540 Stamm, Alfred J., and Harris, E.E. *The Chemical Processing of Wood.* New York: Chemical Publishing Company, 1953. 595 pp. Incidental history.

S541 Stanchfield, Daniel. *History of Lumbering in Minnesota.* Minneapolis: By the author, 1900. 70 pp.

S542 Stanchfield, Daniel. "The History of Pioneer Lumbering on the Upper Mississippi and Its Tributaries, with Biographic Sketches." *Collections of the Minnesota Historical Society* 9 (1901), 324-62.

S543 Stanford Research Institute. *The Newsprint Situation in the Western Region of North America: A Report to the California Newspaper Publishers Association.* Los Angeles, 1952. 115 pp. Incidental history.

S544 Stanford Research Institute. *America's Demand for Wood: A Report to Weyerhaeuser Timber Company, Tacoma, Washington.* Tacoma: Weyerhaeuser Timber Company, 1954. xxii + 404 pp. Diags., tables. See also *America's Demand for Wood, 1929-1975: Summary of a Report to Weyerhaeuser Timber Company, Tacoma, Washington* (Tacoma: Weyerhaeuser Timber Company, 1954), 94 pp. Excerpts also appeared in several Pacific Coast lumber trade journals.

S545 Stanger, Frank M. "The Saga of Grabtown and Whiskey Hill." *Westways* 46 (June 1954), 16-17. Lumber towns in San Mateo County, California, settled ca. 1854.

S546 Stanger, Frank M. *Sawmills in the Redwoods: Logging on the San Francisco Peninsula, 1849-1967.* San Mateo, California: San Mateo County Historical Association, 1967. 160 pp. Illus., maps, notes, bib., index. Logging and sawmill technology, especially in the 19th century.

Stanger, Frank M. See Brown, Alan K., #B476

S547 Staniford, Edward F. "Governor in the Middle: The Administration of George C. Pardee, Governor of California, 1903-1907." Ph.D. diss., Univ. of California, Berkeley, 1956. Includes some history of conservation issues in California.

S548 Stanley, G.F.G. "The Canadian Forestry Corps, 1940-1943." *Canadian Geographic Journal* 28 (Mar. 1944), 134-45. World War II.

S549 Stanley, G.W. "Forty Years of Timber Conservation." *Cross Tie Bulletin* 39 (Nov. 1958), 69-75.

S550 Stanley, Robert D. "The Rise of the Penobscot Lumber Industry to 1860." Master's thesis, Univ. of Maine, 1963.

S551 Stansbury, Karl E. *The First Seventy Years: A Chronology of Thilmany Pulp and Paper Co., 1883-1953.* Kaukana, Wisconsin: Thilmany Pulp and Paper Company, 1953. 53 pp. Illus., tables.

S552 Stark, Charles R., Jr. *The Bering Sea Eagle.* Caldwell, Idaho: Caxton Printers, 1957. 170 pp. Illus., maps. On Harry L. Blunt, a Washington rancher and Alaska bush pilot who worked for the USFS as an aerial fire spotter during the 1920s.

S553 Starling, Robert B. "The Plank Road Movement in North Carolina." *North Carolina Historical Review* 16 (Jan. 1939), 1-22; 16 (Apr. 1939), 147-73. 1840s-1850s.

Starnes, George T. See Berglund, Abraham, #B223

S554 Starobin, Robert S. *Industrial Slavery in the Old South.* New York: Oxford University Press, 1970. xiii + 320 pp. Illus., tables, app., notes, bib., index. Includes some history of slave labor in the lumber and naval stores industries, 1790-1861.

S555 Starobin, Robert S. "The Economics of Industrial Slavery in the Old South." *Business History Review* 44 (Summer 1970), 131-65. Includes references to the use of slaves in the lumber and other forest industries.

Starr, John. See Werner, Morris Robert, #W170

S556 Starr, Merrett. "General Horace Capron, 1804-1885." *Illinois State Historical Society Journal* 18 (July 1925), 259-349. Capron was Commissioner of Agriculture, 1867-1871.

S557 Starr, Mrs. Morton H. "Wisconsin Pastimes." *Journal of American Folklore* 67 (Apr.-June 1954), 184. Practical jokes of Wisconsin lumberjacks, ca. 1900.

S558 Starring, Charles R. "Singapore: Michigan's Imaginary Pompeii." *Inland Seas* 9 (Winter 1953), 231-39; 10 (Spring 1954), 21-25, 72. On a sawmill settlement established at the mouth of the Kalamazoo River on Lake Michigan, 1837-1875, later buried by dunes.

S559 Stauffer, J.M. "Forestry in Alabama." *Alabama Historical Quarterly* 10 (1948), 65-67. Since 1923.

S560 Stauffer, J.M. "The Timber Resource of 'the Southwest Alabama Forest Empire'." *Journal of the Alabama Academy of Science* 30 (Jan. 1959), 52-67. Since the 18th century.

S561 Stauffer, J.M. "Historical Aspects of Forests and Vegetation of the Black Belt." *Journal of the Alabama Academy of Science* 32 (Jan. 1961), 485-92.

S562 Stearn, William T. "From Medieval Park to Modern Arboretum: The Arnold Arboretum and Its Historic Background." *Arnoldia* 32 (Sept. 1972), 173-197. Some history of the arboretum concept and of Harvard University's Arnold Arboretum, Jamaica Plain, Massachusetts.

S563 Stearns, Forest W. "Ninety Years Change in a Northern Hardwood Forest in Wisconsin." *Ecology* 39 (July 1949), 350-58.

Stearns, Raymond Phineas. See Frick, George Frederick, #F234

S564 Stecher, Gilbert E. *Cork: Its Origin and Industrial Uses.* New York: Van Nostrand, 1914. 83 pp. Tables, charts.

S565 Steen, Harold K. "Forestry in Washington to 1925." Ph.D. diss., Univ. of Washington, 1969. 309 pp.

S566 Steen, Harold K. *The History of American Forestry: An Overview.* Asheville, North Carolina: USFS, Pisgah National Forest, 1971. 84 pp. Processed.

S567 Steen, Harold K., and Schrepfer, Susan R. "Why Oral History?" *Forest History* 16 (Oct. 1972), 4-5. On the growth of oral history and the program of the Forest History Society.

S568 Steen, Harold K. "Grazing and the Environment: A History of Forest Service Stock-Reduction Policy." *Agricultural History* 49 (Jan. 1975), 238-42.

S569 Steen, Harold K. "SAF and Its History." *Journal of Forestry* 73 (Aug. 1975), 458-59. The Society of American Foresters has exhibited interest in the history of forestry and of the organization since 1922.

S570 Steen, Herman. *The O.W. Fisher Heritage.* Seattle: Frank McCaffrey Publishers, 1961. 224 pp. Illus., notes, index. On the Fisher family, which had lumber operations in Missouri, Louisiana, Montana, and Washington.

S571 Steen, Judith A., comp. *A Guide to Unpublished Sources for a History of the United States Forest Service.* Santa Cruz, California: Forest History Society, 1973. vi + 67 pp. Processed. Lists published and unpublished materials on interior and agriculture secretaries, USFS chiefs, and other associations and individuals affiliated with USFS history.

S572 Steer, Henry B., comp. *Stumpage Prices of Privately Owned Timber in the U.S.* USDA, Technical Bulletin No. 626. Washington: GPO, 1938. 163 pp. Tables, diags. Stumpage prices for individual species and comparisons with log and lumber prices, 1900-1934.

Steer, Henry B. See Guthrie, John D., #G394

S573 Steer, Henry B., comp. *Lumber Production in the United States, 1799-1946.* USDA, Miscellaneous Publication No. 669. Washington: GPO, 1948. 233 pp. Illus., map, graphs, tables, bib. Consolidates information from approximately 50 statistical publications of the USFS and the Bureau of the Census.

S574 Stegner, Wallace E. *Beyond the Hundredth Meridian: John Wesley Powell and the Second Opening of the West.* Boston: Houghton Mifflin, 1954. xxiii + 438 pp. Maps, illus., notes. From the 1860s through the 1890s, Powell explored and surveyed the West, advocating land reform and conservation. He was director of the U.S. Geological Survey, 1880-1894.

Stegner, Wallace E. See Powell, John Wesley, #P293

S575 Stegner, Wallace E. *The Sound of Mountain Water.* Garden City, New York: Doubleday & Company, 1969. 286 pp. Contains some history and philosophy of the wilderness movement.

S576 Stegner, Wallace E. "De Voto's Western Adventures: A Great Historian's Search for the West of Lewis and Clark and How It Transformed Him into an Ardent Conservationist." *American West* 10 (Nov. 1973), 20-27. As a columnist for *Harper's,* De Voto gave wide publicity to conservation issues, 1947-1955.

S577 Stegner, Wallace E. *The Uneasy Chair: A Biography of Bernard De Voto.* Garden City, New York: Doubleday & Company, 1974. xvi + 464 pp. Illus., notes, index. Includes sections on De Voto's involvement in the preservation movement, 1940s and 1950s.

Stegner, Wallace E. See De Voto, Bernard, #D156

S578 Stein, Robert E. "A History of Social Legislation and Its Influence on the Forest Industry." Master's thesis, Univ. of Michigan, 1951.

S579 Steinhacker, Charles, and Flader, Susan L. *The Sand Country of Aldo Leopold.* San Francisco: Sierra Club, 1973. 94 pp. Illus. Flader relates Leopold's work as a naturalist and ecologist in Wisconsin, 1930s-1940s.

S580 Steinheimer, Richard. *Backwoods Railroads of the West.* Milwaukee: Kalmbach Publishing Company, 1963. 178 pp. Illus. Essentially a photographic display, part of the book concerns logging railroads of the Pacific states.

Steinhoff, H.W. See Wagar, J.V.K., #W9

S581 Stene, E.O. *The Development of Wildlife Conservation Policies in Kansas: A Study in Kansas Administrative History.* University of Kansas, Bureau of Government Research, Governmental Research Series, No. 3. Topeka, 1946. 39 pp.

S582 Stephens, G.A. "Determinants of Lumber Prices." *American Economic Review* 7 (June 1917), 289-305. Incidental history.

S583 Stephens, George R., and Waggoner, Paul E. *The Forests Anticipated from 40 Years of Natural Transitions in Mixed Hardwoods.* Bulletin of the Connecticut Agricultural Experiment Station, No. 707. New Haven, 1970. 58 pp. Illus., tables, bib. Vegetational and forest changes on four mixed hardwood forests of central Connecticut, 1927-1967.

S584 Stephens, Kent. *Michigan California Lumber Company and the Camino, Placerville & Lake Tahoe R.R.* San Mateo, California: Western Railroader, 1967. 34 pp. Illus., maps. Also treated are the logging railroads of the American River Land & Lumber Company and the El Dorado Lumber Company of California since the late 19th century.

S585 Stephens, Rockwell R. *One Man's Forest: Pleasure and Profit from Your Own Woods.* Brattleboro, Vermont: Stephen Greene Press, 1974. 128 pp. Illus. An account of the author's efforts during the 1960s to turn a piece of Vermont woodland into a productive tree farm.

S586 Stephens, Thomas. *The Rise and Fall of Potash in America.* London, 1758. 43 pp.

S587 Stephenson, George M. "The Political History of the Public Lands from 1840 to 1862." Ph.D. diss., Harvard Univ., 1914.

S588 Stephenson, George M. *Political History of the Public Lands from 1840 to 1862, from Preemption to Homestead.* Boston: R.G. Badger, 1917. 296 pp. Maps, bib., notes.

S589 Stephenson, Isaac. *Recollections of a Long Life, 1829-1915.* Chicago: Privately printed, 1915. 264 pp. Illus. Autobiography of New Brunswick-born Wisconsin lumberman and politician. Stephenson (1829-1918) began logging for his father in Maine and New Brunswick, moved to Wisconsin in 1845, and engaged in the lumber business there and in Michigan until the 20th century. He was a U.S. representative from Wisconsin, 1883-1889, and U.S. senator, 1907-1915.

S590 Stephenson, J. Newell. "Canadian Pulp and Paper Association Twenty Years Old." *Pulp and Paper Magazine of Canada* 34 (Feb. 1933), 69-71.

S591 Stephenson, J. Newell. "Forest Protection in Quebec." *Pulp and Paper Magazine of Canada* 37 (Apr. 1936), 270-76. Incidental history.

S592 Stephenson, J. Newell. "Canada's Pulp and Paper Industry." *Canadian Geographical Journal* 19 (Nov. 1939), 268-87. Incidental history.

S593 Stephenson, J. Newell. "Twenty-Five Years of the Technical Section." *Pulp and Paper Magazine of Canada* 41 (Feb. 1940), 71-74. Technical section of the Canadian Pulp and Paper Association.

S594 Stephenson, J. Newell. "Pulp and Paper Making in the Maritimes." *Pulp and Paper Magazine of Canada* 44 (Aug. 1943), 681-90. Incidental history.

S595 Stephenson, J. Newell. "Quebec's Pulp and Paper Industry since 1803." *Pulp and Paper Magazine of Canada* 51 (Nov. 1950), 95-103.

Stephenson, J. Newell. See Turner, John S., #T286

Stephenson, Malvina. See Kerr, Robert S., #K116

S596 Sterling, Ernest A. "A Definite State Forest Policy: New York State's Progress in Reforesting the Adirondacks." *American Forestry* 18 (July 1912), 421-30. Since 1872.

S597 Sterling, Ernest A. "Historical Developments of Wood Preserving in the United States." *Lumber World Review* (No. 9, 1912), 24-26.

S598 Sterling, Ernest A. "Wood Preservation as a Factor in Forest Conservation." *American Forestry* 18 (Oct. 1912), 627-34. Includes some history of wood preservation by railroads since the mid-19th century.

S599 Sterling, Ernest A. "The Development and Status of the Wood Preserving Industry." *Scientific American* 76 (1913), supplement, 24-27.

S600 Sterling, Ernest A. "Forest Management of the Delaware & Hudson Adirondack Forest." *Journal of Forestry* 30 (May 1932), 569-74. Emphasis on forest management and timber utilization since 1903 by the Delaware & Hudson Company in New York's Adirondack Mountains, with references to earlier logging and forest uses.

S601 Sterling, Everett W. "The Powell Irrigation Survey, 1888-1893." *Mississippi Valley Historical Review* 27 (Dec. 1940), 421-34. On the congressional controversy over John Wesley Powell's promotion of forest conservation and arid-land irrigation in the West.

S602 Sterling, Keir Brooks. *Last of the Naturalists: The Career of C. Hart Merriam.* Natural Sciences in America Series. New York: Arno Press, 1974. xv + 478 pp. Illus., notes, bib. Clinton Hart Merriam (1855-1942), natural scientist, mammalogist, and chief of the U.S. Biological Survey (1885-1910), influenced Theodore Roosevelt and the conservation movement through his work.

S603 Stern, Theodore. *The Klamath Tribe: A People and Their Reservation.* American Ethnological Society, Monograph No. 41. Seattle: University of Washington Press, 1965. x + 301 pp. Illus., maps, notes, bib., index. The heavily timbered Klamath Indian Reservation in south central Oregon has long been an important source of logs. Since its termination in 1958 large parts of it have been established as the Winema National Forest.

S604 Sternitzke, Herbert S. *Tennessee's Timber Economy.* USFS, Forest Resource Report No. 9. Washington: GPO, 1955. 56 pp. Illus.

Sternitzke, Herbert S. See Wheeler, Philip R., #W218

S605 Sternitzke, Herbert S., and Putnam, J.A. *Forests of the Mississippi Delta.* Forest Survey Release No. 78. New Orleans: USFS, Southern Forest Experiment Station, 1956. 42 pp.

S606 Sternitzke, Herbert S. *Arkansas Forests.* Forest Survey Release No. 84. New Orleans: USFS, Southern Forest Experiment Station, 1960. 58 pp.

S607 Sternitzke, Herbert S. *Tennessee Forests.* Forest Survey Release No. 86. New Orleans: USFS, Southern Forest Experiment Station, 1962. 29 pp.

S608 Stetson, Sarah P. "The Traffic in Seeds and Plants from England's Colonies in North America." *Agricultural History* 23 (Jan. 1949), 45-56. Includes references to naturalists Mark Catesby, William Bartram, André Michaux, John Clayton, etc.

S609 Stetson, Sarah P. "William Hamilton and His 'Woodlands'." *Pennsylvania Magazine of History and Biography* 73 (Jan. 1949), 26-33. On Hamilton as a botanical gardener near Philadelphia and as a collector of plants, 1786-1813.

S610 Steuber, William F., Jr. "The Problem at Peshtigo." *Wisconsin Magazine of History* 42 (Autumn 1958), 13-15. On controversies regarding the cemetery at Peshtigo, Wisconsin, in which are buried victims of the forest fire of 1871.

S611 Stevens, Alden B. "The Redirected Career of a Naturalist." *Natural History* 67 (Oct. 1958), 418-25. On Theodore Roosevelt's early interest in wildlife and natural history, and his later initiative in promoting and enforcing measures for conservation, 1866-1913.

S612 Stevens, Carl M. "The Forest Industries and the Income Tax." *Journal of Forestry* 18 (Apr. 1920), 329-37. On the evolution of approaches to taxation of the forest industries taken by the Internal Revenue Service.

Stevens, Carl M. See Mason, David T., #M263

Stevens, Horace. See Puter, Stephen Arnold Douglass, #P353

S613 Stevens, James F. *Paul Bunyan.* 1925. Second edition. New York: Alfred A. Knopf, 1948. ix + 245 pp. Includes a historical introduction to the Paul Bunyan legend.

S614 Stevens, James F. "Logger Talk." *American Speech* 1 (1935), 135-40.

S615 Stevens, James F. *Timber! The Way of Life in the Lumber Camps.* Evanston, Illinois: Row, Peterson and Company, 1942. 72 pp. Illus. Includes some historical references to lumberjacks, especially in the Pacific Northwest.

S616 Stevens, James F. "Log Cabin History." *Southern Lumberman* 173 (Dec. 15, 1946), 169. On the introduction of log cabins to the Delaware River region by Swedes in the 1630s.

Stevens, James F. See Felton, Harold W., #F53

S617 Stevens, James F. "Paul Bunyan Lives On." *Southern Lumberman* (Dec. 15, 1952), 131-35.

S618 Stevens, James F. "On Wings of Science—Forest Products Research in Pacific Northwest." *Southern Lumberman* 187 (Dec. 15, 1953), 107-11.

S619 Stevens, James F. "The Simpson Lookout." *American Forests* 60 (Feb. 1954), 19-24. On the Shelton Cooperative Yield Unit maintained by the Simpson Logging Company and the USFS in Mason and Grays Harbor Counties, Washington, since 1947.

S620 Stevens, James F. "Pacific Rim Forester." *American Forests* 60 (Sept. 1954), 21, 68-76. On Clyde Sayers Martin as a professional forester since 1907 in Washington and Oregon, mostly with the Weyerhaeuser Timber Company.

S621 Stevens, James F. "Farthest West in the Timber—Grays Harbor County of Washington State." *Southern Lumberman* 191 (Dec. 15, 1955), 153-57. On logging and the lumber industry since 1937.

S622 Stevens, James F. "William B. Greeley: 'The Business of Life is to Go Forward'." *American Forests* 62 (Jan. 1956), 18-19, 46-48. A general account of Greeley's career in forestry, especially with the West Coast Lumbermen's Association in the Pacific Northwest, 1928-1946.

S623 Stevens, James F. *Green Power: The Story of Public Law 273.* Seattle: Superior Publishing Company, 1958. 95 pp. Illus., index. On federal legislation regarding forest conservation since 1905, particularly Public Law 273, the Sustained Yield Forest Management Act of 1944. Includes some history of forest conservation in California, Oregon, and Washington.

S624 Stevens, James F. "The Loggers' 'Smithsonian'." *Southern Lumberman* 197 (Dec. 15, 1958), 104-08. On Collier Memorial State Park and Logging Museum, near Chiloquin, Oregon, established by Alfred D. Collier in 1944.

S625 Stevens, James F. "The Making of a Folklorist: An Interview with James Stevens." OHI by Elwood R. Maunder. *Forest History* 7 (Winter 1964), 2-19. Reminiscences by the author of many Paul Bunyan tales.

S626 Stevens, James F. "The Greenkeepers." *American Forests* 71 (Dec. 1965), 20-21, 44-45. On the "Keep America Green" program sponsored by the American Forest Products Industries since 1940.

S627 Stevens, Sylvester K. "When Timber Was King in Pennsylvania." *Pennsylvania History* 19 (Oct. 1952), 391-95. An overview of the lumber industry from 1840 to 1900 and an introduction to this special issue on Pennsylvania forest history.

S628 Stevens, T.D. "Tree Farms versus Regulation." *Annals of the American Academy of Political and Social Science* 281 (May 1952), 99-104. Includes some history of the tree farm movement.

S629 Stevens, Walter B. "When a Missourian Forced a Special Session of Congress." *Missouri Historical Review* 23 (Oct. 1928), 44-48. On Secretary of the Interior David R. Francis's recommendation to President Grover Cleveland that he withdraw over 21,000,000 acres from the public domain for the establishment of forest reserves, 1897.

Stevenson, Donald D. See Meyer, Hans Arthur, #M426

S630 Stevenson, Donald D., and Schores, D.D. "A Case History of Industrial Forest Management in North Central Florida." *Journal of Forestry* 59 (June 1961), 411-16. The Buckeye Cellulose Corporation.

S631 Stevenson, John A. "Plants, Problems, and Personalities: The Genesis of the Bureau of Plant Industry." *Agricultural History* 28 (Oct. 1954), 155-62. The bureau was organized in the USDA in 1902.

S632 Stevenson, Louis T. *The Background and Economics of American Papermaking.* New York: Harper, 1940. xiii + 249 pp. Bib. Includes some history of the paper industry, with emphasis on economic, technological, and political developments in the 1930s.

Stewart, Charles L. See Clawson, Marion, #C392

S633 Stewart, Donald M. "Factors Affecting Local Control of White Pine Blister Rust in Minnesota." *Journal of Forestry* 55 (Nov. 1955), 832-37. On efforts to control forest disease, 1933-1951.

S634 Stewart, E. "Forestry on Dominion Lands." *Canadian Forestry Journal* 2 (Jan. 1906), 35-40. Incidental history.

Stewart, Fleming K. See Byrne, J.J., #B625

S635 Stewart, G.I. "Of Fires and Machines." *Michigan Conservation* 24 (Jan.-Feb. 1955), 15-18. On the machine shops and equipment employed at the Forest Fire Experiment Station of the Michigan Department of Conservation, Roscommon, Michigan, since 1929.

Stewart, George. See Cottam, Walter P., #C632

S636 Stewart, George. "Historic Records Bearing on Agricultural and Grazing Ecology in Utah." *Journal of Forestry* 39 (Mar. 1941), 362-75. Travelers' accounts and other records of range conditions in Utah in the 19th century.

S637 Stewart, George, and Widtsoe, J.A. "Contributions of Forest Land Resources to the Settlement and Development of the Mormon-Occupied West." *Journal of Forestry* 41 (Sept. 1943), 633-40. On the many uses of "open" forests by Mormon pioneers in Utah, 1840s-1890s.

S638 Stewart, George W. *Big Trees of the Giant Forest, Sequoia National Park, in the Sierra Nevada of California.* San Francisco: A.M. Robertson, 1930. 105 pp. Illus.

S639 Stewart, George W. "Early Governmental Attempts at Forest Conservation." *Sierra Club Bulletin* 16 (1931), 16-26.

S640 Stewart, John. *Early Days at Fraser Mills, B.C., from 1889 to 1912.* N.p., n.d. 29 pp. Processed. Logging at Fraser Mills, British Columbia.

S641 Stewart, Lowell O. *Public Land Surveys—History, Instructions, Methods.* Ames, Iowa: Collegiate Press, 1935. 202 pp.

S642 Stewart, Max Douglas. "Some Economic Aspects of the Canadian Wooden Match Industry and Public Policy." Ph.D. diss., Michigan State Univ., 1960. 430 pp. Since 1927.

S643 Stewart, Murray C. "Christmas Tree Growers Organization—A History of National Growth." *American Christmas Tree Journal* (Aug. 1968). Some account of plantations since the 1910s.

S644 Stewart, Omer C. "Burning and Natural Vegetation in the United States." *Geographical Review* 41 (Apr. 1951), 317-20. On the long-range effect of fires set by Indians, especially on prairies, 1528-1936.

S645 Stewart, Omer C. "The Forgotten Side of Ethnogeography." In *Method and Perspective in Anthropology: Papers in Honor of Wilson D. Wallis,* ed. by Robert F. Spencer. Minneapolis: University of Minnesota Press, 1954. Pp. 211-48. On the role of fire in the ecology of prairies and forests, and the controversies among scientists over grass and woods-burning practices of Indians and whites since the 19th century.

S646 Stewart, Omer C. "Forest Fires with a Purpose." *Southwestern Lore* 20 (Dec. 1954), 42-46; 20 (Apr. 1955), 59-64; 21 (June 1955), 3-9. On the deliberate setting of forest fires by Indians and the growing advocacy of "controlled burning" by foresters.

S647 Stewart, Ronald Lee. "Wilderness Heritage of the Arid Southwest." Ph.D. diss., Univ. of New Mexico, 1970. 356 pp. History and description of formal and informal wilderness areas of Arizona and New Mexico and their management by the USFS.

S648 Stewart, W.J. "Reminiscences of Early Lumbering in Ontario." *Canada Lumberman* 66 (May 1, 1946), 59, 108-12.

S649 Steyermark, Julian Alfred. *Vegetational History of the Ozark Forest.* University of Missouri Studies, Volume 31. Columbia: University of Missouri, 1959. 138 pp. Maps, bib. Based on a wide variety of accounts dating to 1810 and concluding that the present forest goes back "thousands of years," in contradiction of rival theories that in early historic times the area was an open grassland.

Stillinger, John R. See Hoyle, Raymond J., #H579

Stilson, Harold L. See Plant, Charles, #P244

S650 Stock, Chester. "Memorial to John C. Merriam." *Proceedings of the Geological Society of America* (1946), 183-98. Merriam was active in California's Save-the-Redwoods League.

S651 Stoddard, Charles H. "Reconciling Forest Conservation with Forest Liquidations in the Lake State Region." Master's thesis, Univ. of Michigan, 1938.

S652 Stoddard, Charles H. *Forest Farming and Rural Employment.* Washington: Charles Lathrop Pack Foundation, 1949. 39 pp. Incidental history.

S653 Stoddard, Charles H. *Essentials of Forestry Practice.* Second edition. New York: Ronald Press, 1959. 362 pp. Illus., maps, tables, graphs, apps., bib., index. Includes a chapter on the history of forestry.

Stoddard, Charles H. See Clawson, Marion, #C388

S654 Stoddard, Charles H. *The Small Private Forest in the United States.* Baltimore: Johns Hopkins Press, for Resources for the Future, 1961. 184 pp. Illus. Incidental history.

S655 Stoddard, Herbert Lee, Sr. *Memoirs of a Naturalist.* Norman: University of Oklahoma Press, 1969. 303 pp. Illus., bib., index. Stoddard (1889-1970) was a specialist in ornithology, forest ecology, and forest-wildlife management.

Stoeckeler, Joseph H. See Munns, Edward N., #M637

Stokes, Evelyn. See Dinsdale, Evelyn M., #D182, D183

S656 Stokes, Evelyn. "Kauri and White Pine: A Comparison of New Zealand and American Lumbering." *Annals of the Association of American Geographers* 56 (No. 3, 1966), 440-50. A comparison of lumbering methods, organization, and development in northern New York and New Zealand's North Island.

S657 Stokes, George A. "Lumbering in Southwest Louisiana: A Study of the Industry as a Culturo-Geographic Factor." Ph.D. diss., Louisiana State Univ., 1954. From 1893 to 1935.

S658 Stokes, George A. "Notes on Western Louisiana Sawmill Towns." *Proceedings of the Louisiana Academy of Sciences* 18 (1955), 71-75.

S659 Stokes, George A. "Western Louisiana Logging Trams." *Proceedings of the Louisiana Academy of Sciences* 19 (1956), 47-48. Logging railroads, 1895-1943.

S660 Stokes, George A. "Lumbering and Western Louisiana Cultural Landscapes." *Annals of the Association of American Geographers* 47 (Sept. 1957), 250-66. On logging roads, sawmills, lumber towns, houses, etc., in the cutover lands and second-growth timber of western Louisiana's longleaf pine area, 1895-1937.

S661 Stokes, George A. "Log-Rafting in Louisiana." *Journal of Geography* 58 (Jan. 1959), 81-89. From 1716 to 1930.

S662 Stoller, Leo. *After Walden, Thoreau's Changing Views on Economic Man.* Stanford, California: Stanford University Press, 1957. 163 pp. Notes, bib. Includes reference to Thoreau's views of the lumber industry and forest conservation, 1846-1862.

Stoltenberg, C.H. See Webster, H.H., #W129

S663 Stone, Flora Pierce; Reed, Frank A.; and Wein, Bernard A. "Some Wood-Using Industries in Northern New York." *Northeastern Logger* 8 (May 1960), 20-21, 32, 44-47. Incidental history.

S664 Stone, J. Herbert. *A Regional Forester's View of Multiple Use.* OHI by Elwood R. Maunder. Santa Cruz, California: Forest History Society, 1972. xi + 245 pp. Illus., apps., bib., index. Processed. On Stone's career in the USFS, 1927-1967, in the South, East, and as regional forester in the Pacific Northwest.

S665 Stone, J. Herbert. "An Oregon Heritage: National Forests." *Oregon Historical Quarterly* 76 (Mar. 1975), 28-38. General history since 1892 by a former regional forester.

S666 Stone, Lois C. "The Institute of Forest Genetics: A Legacy of Good Breeding." *Forest History* 12 (Oct. 1968), 20-29. James G. Eddy, Lloyd Austin, and the origins of the USFS's Institute of Forest Genetics, Placerville, California, 1920s.

S667 Stone, Richard Cecil. *Life-Incidents.* St. Louis: Southwestern, 1874. 352 pp. The author was associated with his son's lumber and paper business in Missouri in the 1860s.

S668 Stone, Robert N. "Preliminary Report on the Survey of Forest Plantations in Northern Lower Michigan." *Papers of the Michigan Academy of Science, Arts, and Letters* 45 (1960), 93-102. On areas of public and private forest planting since 1905.

S669 Stone, Robert N., and Bagley, W.T. *The Forest Resource of Nebraska.* Forest Survey Release No. 4. Fort Collins, Colorado: USFS, Rocky Mountain Forest and Range Experiment Station, 1961. 45 pp. Incidental history.

S670 Stone, Robert N., and Thorne, H.W. *Wisconsin's Forest Resources.* Paper No. 90. St. Paul: USFS, Lake States Forest Experiment Station, 1961. 52 pp. Incidental history.

S671 Stone, Robert N. "A Comparison of Woodland Owner Intent with Woodland Practice in Michigan's Upper Peninsula." Ph.D. diss., Univ. of Minnesota, 1969. 121 pp. Incidental history.

S672 Stone, W.H. "The Royal Pine of New Hampshire." *New England Magazine* 15 (1896), 26-32. Includes some history of British forest and mast policy.

Storey, Herbert C. See Lull, Howard W., #L301

Story, Isabelle F. See Yard, Robert Sterling, #Y4

S673 Story, Isabelle F. *The National Parks and Emergency Conservation.* Washington: USDI and GPO, 1933. 32 pp. Illus., tables. Includes some history of the national parks movement, national monuments, and National Park Service.

S674 Story, Isabelle F. "Park of Peace." *American Forests* 55 (Mar. 1949), 16, 32. On the Waterton-Glacier International Peace Park, bisected by the Montana-Alberta boundary, since 1932.

Story, Isabelle F. See Chittenden, Hiram Martin, #C303

S675 Story, Isabelle F. *The National Park Story in Pictures.* Washington: GPO, 1957. 88 pp. Map, illus. Since the 1870s.

S676 Stott, Calvin B. "A Short History of Continuous Forest Inventory East of the Mississippi." *Journal of Forestry* 66 (Nov. 1968), 834-37. Since established by the USFS in the 1930s.

S677 Stouffer, A.L. *The Story of Shell Lake.* Shell Lake, Wisconsin: Washburn County Historical Society, 1961. 253 pp. Illus., maps, tables. Includes a chapter on the lumber industry and especially the Shell Lake Lumber Company since 1881.

S678 Stout, Benjamin B. "The Harvard Black Rock Forest." *Bulletin of the Garden Club of America* 45 (Jan. 1957), 48-50. On an area in the Hudson Highlands north of New York City since the 17th century.

S679 Stout, Joe A., Jr. "Cattlemen, Conservationists, and the Taylor Grazing Act." *New Mexico Historical Review* 45 (Oct. 1970), 311-32. On the opponents and proponents of bills to establish grazing policy on the public lands, culminating in the passage of the Taylor Grazing Act of 1934.

S680 Stout, Ray L. "Over the Brush and through the Trees: Surveying, 1900-1909." *Oregon Historical Quarterly* 73 (Dec. 1972), 332-58. Reminiscences of land surveying in Oregon and Washington for the U.S. General Land Office.

S681 Stout, Wilbur. "The Charcoal Iron Industry of the Hanging Rock Iron District—Its Influence on the Early Development of the Ohio Valley." *Ohio Archaeological and Historical Quarterly* 42 (Jan. 1933), 72-104. Includes some history of wood-cutting and the charcoal industry in 19th-century Ohio and Kentucky.

S682 Stoveken, Ruth. "The Pine Lumberjacks in Wisconsin." *Wisconsin Magazine of History* 30 (Mar. 1947), 322-34. General history and description of life and customs.

S683 Strait, Robert F. "The Old McShane Tie Camp and the Rockwood Fire." As told to C.C. Rawlings. *Annals of Wyoming* 32 (Oct. 1960), 144-63. On the author's work for the McShane Tie Company on the Tongue River in Wyoming's Big Horn Mountains, including an account of its flumes, mills, crews, headquarters at Rockwood, and the forest fire that almost destroyed it, 1898-1899.

S684 Stratton, David H. "Albert B. Fall and the Teapot Dome Affair." Ph.D. diss., Univ. of Colorado, 1955. 537 pp. Fall was Warren Harding's Secretary of the Interior, 1921-1923.

S685 Stratton, David H. "New Mexico Machiavellian? The Story of Albert B. Fall." *Montana, Magazine of Western History* 7 (Oct. 1957), 2-14.

S686 Stratton, David H. "Behind Teapot Dome: Some Political Insights." *Business History Review* 31 (Winter 1957), 385-402. Concerns Albert B. Fall, the Teapot Dome scandals, and the political consequences of the scandals, 1920s.

Stratton, David H. See Fall, Albert Bacon, #F13

S687 Stratton, David H. "Two Western Senators and Teapot Dome: Thomas J. Walsh and Albert B. Fall." *Pacific Northwest Quarterly* 65 (Apr. 1974), 57-65. Includes the relationships of both to the conservation movement in the 1910s and 1920s.

S688 Stratton, Owen, and Sirotkin, Phillip. *The Echo Park Controversy.* Cases in Public Administration and Policy Formation, No. 46. University, Alabama: Inter-University Case Program, 1959. xii + 100 pp. Illus., maps, notes. Concerns the successful efforts of conservationists and preservationists to protect the canyons of Dinosaur National Monument, Utah-Colorado, from inundation by reservoir. The Echo Park area was the focus of a preservationist movement with broad implications for wilderness and scenic areas; it showed the influence of private interest groups on legislative decision-making in the 1940s-1950s.

S689 Strausberg, Stephen F. "The Administration and Sale of Public Land in Indiana, 1800-1860." Ph.D. diss., Cornell Univ., 1970. 452 pp.

S690 Strenge, F.A. "Dr. Alfred J. Stamm." *Journal of Forestry* 46 (June 1948), 461-62. On Stamm's career as a research chemist, especially with the Forest Products Laboratory, Madison, Wisconsin, since 1922.

S691 Strite, Daniel D. "Up the Kilchis." *Oregon Historical Quarterly* 72 (Dec. 1971), 293-314; 73 (Mar. 1972), 4-30; 73 (June 1972), 171-92; 73 (Sept. 1972), 212-27. Account of logging along the Kilchis River, Oregon, 1910s-1920s.

S692 Stroebel, Ralph W. *Tittabawassee River Log Marks.* Eddy Historical Series, No. 2. Saginaw, Michigan: Saginaw Public Libraries, 1967. 45 pp. Illus. On the Tittabawassee Boom Company and log marks used by lumber companies operating along Michigan's Tittabawassee River, 1890-1893.

S693 Strong, Anna Louise. *My Native Land.* New York: Viking Press, 1940. 299 pp. Includes a chapter on the "Westwood War," labor turmoil of the 1930s between the management and employees of the Red River Lumber Company, Westwood, California.

S694 Strong, Clarence C., and Webb, Clyde S. *White Pine: King of Many Waters.* Missoula, Montana: Mountain Press Publishing Company, 1970. xii + 212 pp. Illus., map, apps., index. A history of logging and of the lumber and shingle industries in the northern Idaho counties of Shoshone, Benewah, and Kootenai, since the 1870s. Appendixes include rosters of sawmills, shingle mills, and prominent lumbermen of the Coeur d'Alene region.

S695 Strong, Douglas H. "A History of Sequoia National Park." D.S.S. diss., Syracuse Univ., 1964. 350 pp. On the origins and history of the California park to 1926.

S696 Strong, Douglas H. "The History of Sequoia National Park, 1876-1926." *Southern California Quarterly* 48 (June 1966), 137-67; 48 (Sept. 1966), 265-88; 48 (Dec. 1966), 369-99. On the movement to establish the California park and the struggles to preserve and enlarge it.

S697 Strong, Douglas H. "Sequoia National Park: Discovery and Exploration." *Western Explorer* 4 (Sept. 1966), 9-27. History of the Sierra Nevada region prior to its establishment as Sequoia National Park in 1890; includes reference to logging and grazing in the area.

S698 Strong, Douglas H. "The Sierra Forest Reserve: The Movement to Preserve the San Joaquin Valley Watershed." *California Historical Society Quarterly* 46 (Mar. 1967), 3-17. On efforts to protect this forested region, 1889-1905.

S699 Strong, Douglas H. *Trees—or Timber? The Story of Sequoia and Kings Canyon National Parks.* Three Rivers, California: Sequoia Natural History Association, in cooperation with the U.S. National Park Service, 1968. 62 pp. Illus., maps, bib., index. History of the adjacent parks since their discovery.

S700 Strong, Douglas H. "The Man Who 'Owned' Grand Canyon." *American West* 6 (Sept. 1969), 33-40. Ralph Cameron, Arizona entrepreneur and U.S. senator, battled the General Land Office, USFS, and National Park Service from 1883 to 1926 in a losing effort to monopolize tourist facilities and other developments at Grand Canyon National Park.

S701 Strong, Douglas H. "The Rise of American Aesthetic Conservation: Muir, Mather and Udall." *National Parks & Conservation Magazine* 44 (Feb. 1970), 4-9. John Muir, Stephen T. Mather, and Stewart Udall.

S702 Strong, Douglas H. *The Conservationists.* Menlo Park, California: Addison-Wesley Publishing Company, 1971. 196 pp. Illus., maps, notes, bib., index. A history of the American conservation movement as interpreted through the careers of its leaders, including Henry David Thoreau, Frederick Law Olmsted, George Perkins Marsh, John Wesley Powell, Gifford Pinchot, John Muir, Stephen T. Mather, Aldo Leopold, Franklin D. Roosevelt, and Stewart Udall.

S703 Strong, Douglas H. "Lassen Volcanic National Park's Manzanita Lake: a Brief History." *Pacific Historian* 15 (Fall 1971), 68-82. Logging, shingle and shake making, and park history since the late 19th century.

S704 Strong, Douglas H. "High Sierra Wilderness." *Naturalist* 23 (Spring 1972), 38-44. Incidental history.

S705 Strong, Douglas H. *'These Happy Grounds': A History of the Lassen Region.* Red Bluff, California: Loomis Museum Association, in cooperation with the U.S. National Park Service, 1973. 101 pp. Illus., maps, notes, index. Lassen Volcanic National Park and vicinity, California, including the movement to preserve the park.

S706 Strong, Douglas H. "Teaching American Environmental History." *Social Studies* 65 (Oct. 1974), 196-200. A survey of teaching approaches and suggested reading for courses in environmental history.

S707 Strong, Douglas H. "To Save the Big Trees." *National Parks & Conservation Magazine* 49 (Mar. 1975), 10-14. A brief history of the discovery, exploitation, and preservation of sequoia groves in California's Sierra Nevada since the 1850s.

S708 Stuart, Robert Y. "The Relation of the Forest Service to the Mining Industry." *American Forestry* 19 (Mar. 1913), 154-63. And the latter's effort to obtain mine timbers from the national forests.

S709 Stuart, Robert Y. "The Father of Pennsylvania Forestry." *Journal of Forestry* 21 (Nov. 1923), 673-76. Joseph Trimble Rothrock (1839-1922).

Stuart, Robert Y. See Drinker, Henry S., #D262

S710 Stucker, Gilbert F. "Lake Superior's Island Wilderness." *National Parks Magazine* 43 (Mar. 1969), 4-9. Includes some history of Isle Royale National Park, Michigan.

S711 Studhalter, R.A. "Tree Growth: Some Historical Chapters in the Study of Diameter Growth." *Botanical Review* 29 (No. 3, 1963), 245-365.

S712 Studley, James D. *United States Pulp and Paper Industry.* U.S. Department of Commerce, Bureau of Foreign and Domestic Commerce, Trade Promotion Series, No. 182. Washington: GPO, 1938. 99 pp. Illus., tables, apps. Includes some history of papermaking.

S713 Stumbaugh, Thomas R. "Resource Development of North Lincoln County, Oregon." Master's thesis, Oregon State Univ., 1961.

S714 Stupka, Arthur. *Great Smoky Mountains National Park, North Carolina and Tennessee.* Natural History Handbook Series, No. 5. Washington: U.S. National Park Service, 1960. 79 pp. Illus., map, bib. Includes some history of the park since 1923.

S715 Sturdivant, Fred R. "The W.A. Woodward Lumber Company: A Case Study in 'Rugged Individualism'." Master's thesis, Univ. of Oregon, 1961. Cottage Grove, Oregon, since 1900.

S716 Sturgeon, Edward Earl. "Trends in Land Use and Ownership in Cheboygan County, Michigan, as Affected by Socio-Economic Development and the Land Disposal Policy of the Michigan Department of Conservation, with Emphasis on Lands Sold to Private Owners by the State of Michigan." Ph.D. diss., Univ. of Michigan, 1954. 355 pp.

S717 Styffe, Oscar. "Bunkhouses of Thirty Years Ago and Today." *Canada Lumberman* 58 (Jan. 15, 1938), 10-11. Improvements in logging camp quarters, 1910s-1930s.

Sudarsky, Peter. See Parson, Dudley L., #P46

S718 Suddath, John. "Forest Service History Began with Tree Survey." *Texas Forestry* 14 (Apr. 1974), 38-39. On the Texas Department of Forestry and Texas Forest Service since ca. 1914.

S719 Sudworth, George B. "The Origin and Development of Forest Work in the U.S." *Publications of the Michigan Political Science Association* 4 (July 1902), 54-84.

S720 Sullivan, Edward T. *The Wooden Container Industry in Minnesota.* Scientific Journal Series, Paper No. 4044. St. Paul: University of Minnesota, Agricultural Experiment Station, 1958. 42 pp. Maps, diags., tables, bib. From 1939 to 1954.

Summers, Clarence. See Jettmar, Karen, #J79

S721 Summers, Floyd G. "Norman J. Colman, First Secretary of Agriculture." *Missouri Historical Review* 19 (Apr. 1925), 404-08. Colman was Grover Cleveland's Commissioner of Agriculture, 1885-1889, and briefly in 1889, Secretary of Agriculture.

S722 Sumner, Mrs. George. "Sandalwood." *Bulletin of the Garden Club of America* 41 (Sept. 1953), 19-21. On the former abundance and later scarcity of sandalwood in the Hawaiian Islands, since 1780.

Sumner, Lowell. See Collins, George L., #C530

S723 Sundquist, James L. *Politics and Policy: The Eisenhower, Kennedy, and Johnson Years.* Washington: Brookings Institution, 1968. xi + 560 pp. Tables, notes, index. Chapter 8 ("For All, a Better Outdoor Environment") examines the legislative and administrative efforts to upgrade the national parks, to initiate a wilderness system, and to provide for outdoor recreation and highway beautification, 1950s-1960s.

S724 Surrey, N.M. Miller. *Commerce of Louisiana During the French Regime, 1699-1763.* New York: Columbia University Press, 1916. 476 pp. Includes incidental history of the lumber trade during this era.

S725 Susanik, R. "The Changes in the British Market for British Columbia's Lumber since 1935." *Forestry Chronicle* 30 (Sept. 1954), 328-30.

S726 Sutermeister, Edwin. *The Story of Papermaking.* Boston: S.D. Warren Company, 1954. 209 pp. Incidental history.

S727 Sutherland, A.D. *Sixty Years Afield and Observations on Conservation.* New York: Carlton Press, 1968. 77 pp. Reminiscences of a Wisconsin wildlife conservationist and leader in the Izaak Walton League of America.

S728 Sutherland, Charles F. "Consumption and Marketing of Forest Products in the Automobile: A Case Study of the Ford Motor Company." Ph.D. diss., Univ. of Michigan, 1961. 289 pp. From 1920 to 1956.

S729 Sutherland, Charles F. "Ford's Forest." *American Forests* 69 (Sept. 1963), 18-19, 50-51. Henry Ford's lumber operations in Michigan's Upper Peninsula provided wood materials for the Ford Motor Company, 1920s-1950s.

S730 Sutherland, Charles F. "Tin Lizzy's Wooden Heart." *Forest History* 16 (July 1972), 22-25. Ford Motor Company's lumber operations provided materials for automobile manufacturing, 1920s-1950s.

Sutherland, Jon N. See Detweiler, Robert, #D144

S731 Sutton, Ann, and Sutton, Myron. "The Man from Yosemite." *National Parks Magazine* 28 (July-Sept. 1954), 102-05, 131-32, 140. On Harold Child Bryant's career as vertebrate zoologist in California, lecturer on natural history, official of the National Park Service, and superintendent of the Grand Canyon National Park, 1908-1954.

S732 Sutton, Ann, and Sutton Myron. *Guarding the Treasured Lands: The Story of the National Park Service.* Philadelphia: J.B. Lippincott, 1965. 160 pp. Illus.

S733 Sutton, Ann, and Sutton, Myron. *The Wilderness World of the Grand Canyon: 'Leave It as It Is'.* Philadelphia: J.B. Lippincott, 1971. xii + 241 pp. Illus., map, apps., bib., index. Includes some incidental history of Grand Canyon National Park, Arizona.

S734 Sutton, Ann, and Sutton, Myron. *Yellowstone: A Century of the Wilderness Idea.* New York: Macmillan and Yellowstone Library and Museum Association, 1972. 219 pp. Illus., bib., index. Includes some history of the national park from the period of discovery and exploration.

S735 Sutton, Imre. "Land Tenure in the West: Continuity and Change." *Journal of the West* 9 (Jan. 1970), 1-23.

Sutton, Myron. See Sutton, Ann, #S732, S733, S734

S736 Sutton, S.B. *Charles Sprague Sargent and the Arnold Arboretum.* Cambridge: Harvard University Press, 1970. xvii + 382 pp. Illus., notes, index. Sargent (1841-1927) became the first director of Harvard's Arnold Arboretum in 1873. As a scientist, journalist, and publicist, he did much to advance forestry and preserve American forests.

S737 Sutton, S.B. "The Arboretum Administrators: An Opinionated History." *Arnoldia* 32 (Jan. 1972), 2-21. Harvard University's Arnold Arboretum, Jamaica Plain, Massachusetts.

S738 Swain, Donald C. "The Role of the Federal Government in the Conservation of Natural Resources, 1921-1933." Ph.D. diss., Univ. of California, Berkeley, 1961.

S739 Swain, Donald C. *Federal Conservation Policy, 1921-1933.* University of California Publications in History, Volume 76. Berkeley: University of California Press, 1963. x + 221 pp. Illus., notes, bib., note, index. The national conservation movement expanded during the 1920s, thanks to intelligent and effective administration of federal agencies responsible for natural resources. This thorough study covers national forests, national parks, wildlife, reclamation, minerals, soil conservation, and water and hydroelectric power.

S740 Swain, Donald C. "Harold Ickes, Horace Albright, and the Hundred Days: A Study in Conservation Administration." *Pacific Historical Review* 34 (Nov. 1965), 455-65. Horace M. Albright was director of the National Park Service under Secretary of the Interior Harold L. Ickes during the early months of the Franklin D. Roosevelt administration in 1933. Albright decisively influenced Ickes and his emerging conservation policies.

S741 Swain, Donald C. "The Passage of the National Park Service Act of 1916." *Wisconsin Magazine of History* 50 (Autumn 1966), 4-17. Documents the efforts of J. Horace McFarland, Stephen T. Mather, Horace M. Albright, and others in working for the National Park Service Act. The law represented the political emergence of esthetic conservationists, as opposed to utilitarian conservationists.

S742 Swain, Donald C. "The Founding of the National Park Service." *American West* 6 (Sept. 1969), 6-9. On the movement leading to passage of the National Park Service Act in 1916.

S743 Swain, Donald C. "The Bureau of Reclamation and the New Deal, 1933-1940." *Pacific Northwest Quarterly* 61 (July 1970), 137-46. Includes much general history of the conservation movement.

S744 Swain, Donald C. *Wilderness Defender: Horace M. Albright and Conservation*. Chicago: University of Chicago Press, 1970. 347 pp. Illus., bib., index. Albright was assistant to National Park Service director Stephen T. Mather, 1916-1919, superintendent of Yellowstone National Park, 1919-1929, and director of the National Park Service, 1929-1933. This biography of Albright also serves as a history of the National Park Service for the period.

S745 Swain, Donald C. "The National Park Service and the New Deal, 1933-1940." *Pacific Historical Review* 41 (Aug. 1972), 312-32.

S746 Swaine, J.M. *Forest Entomology and Its Development in Canada.* Department of Agriculture, Pamphlet No. 97. Ottawa, 1928. 20 pp.

S747 Swaine, J.M. "Progress in Forest Insect Control in Canada." *Forestry Chronicle* 4 (Mar. 1928), 35-40.

S748 Swaine, J.M. "The Forest Insect Situation in the Province of Quebec." *Forestry Chronicle* 9 (June 1933), 49-59. Since the 1910s.

S749 Swan, Kenneth D. *Splendid Was the Trail*. Missoula, Montana: Mountain Press Publishing Company, 1968. 170 pp. Illus. Reminiscences of his USFS career, primarily as a photographer in Montana and northern Idaho, 1911-1948.

S750 Swank, James M. *The Department of Agriculture: Its History and Objects.* Washington: GPO, 1872. 64 pp.

S751 Swank, James M. *History of the Manufacture of Iron in All Ages, and Particularly in the United States from Colonial Times to 1891. Also a Short History of Early Coal Mining in the United States and a Full Account of the Influences Which Long Delayed the Development of All American Manufacturing Industries.* Second edition. Philadelphia: American Iron and Steel Association, 1892. xix + 554 pp. Tables. Includes some history of the charcoal iron industry by area and state.

S752 Swank, Roy. *Trail to Marked Tree*. Ed. by Nolan Porterfield. San Antonio: Naylor Company, 1968. xvi + 165 pp. Map. Reminiscences of pioneering in the wooded delta lowlands of northeastern Arkansas, including some account of logging along the St. Francis River, 1919-1926.

S753 Swanke, A.E. "Forty Years of Northern Lumber Business." *Southern Lumberman* 181 (Dec. 15, 1950), 177-78. Survey of changes by the president of the Northern Hemlock and Hardwood Manufacturers Association.

S754 Swanson, Evadene B. "The Use and Conservation of Minnesota Game, 1850-1900." Ph.D. diss., Univ. of Minnesota, 1940. 321 pp.

S755 Swanson, Robert W. "A History of Railroad Logging." *Truck Logger* 17 (Jan. 1961), 40-49. British Columbia.

S756 Swanson, Robert E. "A History of Railroad Logging." *Loggers Handbook* 33 (1973), 21-23, 90-92. An account of early logging in British Columbia and especially the use of locomotives and railroads since 1900.

S757 Swanson, Robert W. "A History of Logging and Lumbering on the Palouse River, 1870-1905." Master's thesis, Washington State Univ., 1958. Of eastern Washington and northern Idaho.

S758 Sweezy, R.O. "Life in a Logging Camp." *Paper Trade Journal* 54 (No. 7, 1912), 225-31. Incidental history.

Swenson, Robert W. See Gates, Paul W., #G63

S759 Swift, Ernest F. *A History of Wisconsin Deer.* Publication No. 323. Madison: Wisconsin Conservation Department, 1946. 96 pp. Includes relationship to forest cover and forest growth.

S760 Swift, Ernest F. "The Conservation Movement—Men and Machines at Work." *Wisconsin Magazine of History* 37 (Autumn 1953), 3-6, 48-50. Historical sketch by a director of the Wisconsin Conservation Department.

S761 Swift, Ernest F. *The Glory Trail: The Great American Migration and Its Impact on Natural Resources.* Washington: National Wildlife Foundation, 1958. 50 pp. Includes some history of wildlife conservation.

S762 Swift, Ernest F. "They Laughed at Us." *American Forests* 67 (May 1961), 20-22, 54. On the problems of deer management in relation to forestry and the forest industries in Wisconsin since 1900.

S763 Swift, Ernest F. "When Wisconsin Pine Went to College." *American Forests* 72 (Jan. 1966), 8-9, 49-51. On Cornell University's timberlands in Wisconsin, 1860s-1900s.

S764 Swift, Ernest F. *A Conservation Saga*. Washington: National Wildlife Federation, 1967. vii + 264 pp. Illus. His personal account of the conservation movement. Swift (1897-1968) was director of the Wisconsin Conservation Department, 1926-1954, served as assistant director of the U.S. Fish and Wildlife Service, 1954-1955, and capped his career as executive director of the National Wildlife Federation, 1955-1960.

S765 Swift, Lon L. "Land Tenure in Oregon; Including the Topography, Disposition of Public Lands, Landlordism, Mortgages, Farm Output, and Practical Workings of Tenant Farming of the State, Together with Tables and Copies of Land Leases." *Oregon Historical Society Quarterly* 10 (June 1909), 31-135.

S766 Swigert, E.A. "History of Logging Machinery: Development of Logging." *Loggers Handbook* 9 (1949), 62-63. Since the 19th century.

S767 Swingler, William S. "History of the Forest Lands Around Eaglesmere, Pennsylvania." *Forest Leaves* 22 (Aug. 1930), 146-48.

S768 Swingler, William S. "Forest Service Work in New England and a Look Ahead." *Journal of Forestry* 48 (June 1950), 422-24. Since 1911.

S769 Swingler, William S. "The Forest Service and Private Forestry." *Journal of Forestry* 52 (July 1954), 484-86. On federal cooperation with private forestry since 1892.

S770 Swingler, William S. "Services to the Public Through Cooperation with States and Private Owners." *Journal of Forestry* 53 (Feb. 1955), 120-25. On the cooperative forestry work of the USFS since 1905.

S771 Swisher, Jacob A. *Iowa, Land of Many Mills.* Iowa City: Iowa State Historical Society, 1940. 317 pp. Illus., notes, index. Includes sawmills and the lumber industry since the 19th century.

S772 Switzler, William F. *Report on the Internal Commerce of the United States. Part II of Commerce and Navigation: Special Report on the Commerce of the Mississippi, Ohio, and Other Rivers, and of the Bridges Which Cross Them.* Washington: GPO, 1888. 591 pp. Includes statistics on the forest products trade in Mississippi Valley.

S773 Symington, D.F. *Forest Resources in Saskatchewan.* Conservation Bulletin No. 8. Regina: Saskatchewan Department of Natural Resources, 1960. 22 pp. Incidental history.

Sziklai, Oszkar. See Adamovich, Laszlo, #A16

T1 Taber, Edward O., and Thompson, Stith. "Paul Bunyan in 1910." *Journal of American Folklore* 59 (Apr.-June 1946), 134-35. On the origins of the Bunyan legend among lumbermen.

T2 Taber, Thomas T., and Casler, Walter. *Climax—An Unusual Steam Locomotive.* Rahway, New Jersey: Railroadians of America, 1960. 97 pp. Illus., tables. Built between 1888 and 1928 and widely used in logging operations.

T3 Taber, Thomas T. "Logging Railroads and Logging Locomotives in Eastern Pennsylvania." *Now and Then* 12 (Jan. 1960), 225-36. From 1875 to 1932.

T4 Taber, Thomas T. "The Railroading Era of Lumbering in Pennsylvania." *Pennsylvania Forests* 51 (Jan.-June 1961), 97-98, 134.

T5 Taber, Thomas T. *Ghost Lumber Towns of Central Pennsylvania.* Williamsport, Pennsylvania: Lycoming Printing Company, 1970.

T6 Taber, Thomas T. "Lumbering in Penn's Woods." *Southern Lumberman* 223 (Dec. 15, 1971), 99-101. Brief history of the lumber industry in Pennsylvania.

T7 Taber, Thomas T. "Williamsport—A Part of the Heritage of Pennsylvania Lumbering." *Northern Logger and Timber Processor* 20 (May 1972), 10-11, 28-30. The lumber industry in Williamsport and log drives on the West Branch of the Susquehanna River, 1838-1919.

T8 Tacoma Lumberman's Club. *Tacoma, Washington, the Lumber Capital of America.* Tacoma, 1923. 58 pp. A directory of the city's forest product industries, with many incidental historical references.

Taft, Philip. See Commons, John R., #C543

T9 Taft, Philip. *Organized Labor in American History.* New York: Harper and Row, 1964. xxi + 819 pp. Notes, index. See for reference to the lumber, pulp and paper, and related forest industries.

T10 Talbot, Murrell Williams. "Buffalo Bill's Top Hand." *American Forests* 66 (May 1960), 37, 61. Jesse W. Nelson, former cowboy and stunt rider, was employed by the USFS in the West, 1900-1944.

T11 Talbot, William S. "American Visions of Wilderness." *Living Wilderness* 33 (Winter 1969), 14-25. Includes some history of landscape painting and the artistic reaction to wilderness.

T12 Tanasoca, Donald. "Six Months in Garden Valley." Ed. by Elmo R. Richardson. *Idaho Yesterdays* 11 (Summer 1967), 16-24. Reminiscence of life in a Civilian Conservation Corps camp near Banks on Idaho's Payette National Forest, 1939.

T13 Tanner, N.E. " 'Hot Logging' Has Become Alberta's Custom." *Canada Lumberman* 60 (Aug. 1, 1940), 70-71, 117. On the evolution of logging methods in Alberta.

T14 Tassé, J. *Philemon Wright, ou colonisation et commerce de bois.* Montreal: Vapeur, 1871. Wright was an Ottawa Valley lumberman of the early 19th century.

T15 Tatter, Henry. "The Preferential Treatment of the Actual Settler in the Primary Disposition of the Vacant Lands in the United States to 1841—Preemption: Prelude to Homesteadism." Ph.D. diss., Northwestern Univ., 1933.

T16 Tatter, Henry. "State and Federal Land Policy During the Confederation Period." *Agricultural History* 9 (Oct. 1935), 176-86. 1780s.

T17 Tattersall, James N. "The Economic Development of the Pacific Northwest to 1920." Ph.D. diss., Univ. of Washington, 1960. 314 pp. Includes some history of the lumber industry, especially from 1880-1920.

T18 Taylor, Carl C., et al. *Rural Life in the United States.* New York: Alfred A. Knopf, 1949. xvii + 549 pp. Bib. See for reference to forest life and industries.

T19 Taylor, Dale L. "Forest Fires in Yellowstone National Park." *Journal of Forest History* 18 (July 1974), 68-77. A history of forest fire control and fire management policies in the park, with comments on their effect upon forest ecology.

Taylor, Frank J. See Tillotson, Miner R., #T112

Taylor, Frank J. See Albright, Horace M., #A49

T20 Taylor, Harve A. "More About the Old 'Long Bell' Planing Mill." *Heritage* (Crawford County [Arkansas] Historical Society) 1 (Jan. 1958), 3-5. On a mill of the Long-Bell Lumber Company, Van Buren, Arkansas, 1881-1895.

T21 Taylor, J. Crow. "Historical and Status Data on Veneering." *Veneers* 11 (No. 9, 1917), 15-16.

T22 Taylor, J. Crow. "Reminiscences of a Retail Secretary; A Review of the Changes Wrought by the Last Twenty Years." *Southern Lumberman* (Dec. 19, 1925), 246-48. By a secretary of the Kentucky Retail Lumber Dealers Association.

T23 Taylor, Paul S. "Reclamation: The Rise and Fall of an American Idea." *American West* 7 (July 1970), 27-33, 63. On the irrigation congresses held from 1891 to 1902.

T24 Taylor, Ray W. *Hetch Hetchy: The Story of San Francisco's Struggle to Provide a Water Supply for Her Future Needs.* San Francisco: R.J. Orozco, 1926. xii + 199 pp. Controversy over the inundation of Yosemite National Park's Hetch Hetchy Valley divided the conservation movement, 1910s.

T25 Taylor, W.S. "Different Stages in the Evolution of Overhead System of Logging." *Timberman* 15 (Jan. 1914), 30-31.

T26 Teagle, Ernest. *Out of the Woods: The Story of McCleary.* Shelton, Washington: Simpson Logging Company, 1956. 47 pp. Illus. The logging town of McCleary, Washington, since the 1890s.

T27 Teale, Edwin Way. *The Lost Woods: Adventures of a Naturalist.* New York: Dodd, Mead, and Company, 1945. xii + 326 pp.

Teale, Edwin Way. See Muir, John, #M618

T28 Tebeau, Charlton W. *Florida's Last Frontier: The History of Collier County.* Coral Gables, Florida: University of Miami Press, 1966. 278 pp. Illus., index. Includes some history of the cypress and pine lumber industry and of the Big Cypress Swamp since the 1920s.

T29 Tebeau, Charlton W. *Man in the Everglades: 2,000 Years of Human History in the Everglades National Park.* Coral Gables, Florida: Everglades Natural History Association and University of Miami Press, 1968. 192 pp.

Teclaff, Eileen M. See Haden-Guest, Stephen, #H7

T30 Teeguarden, Dennis; Casamajor, Paul; and Zivnuska, John. *Timber Marketing and Land Ownership in the Central Sierra Nevada Region.* University of California, Agricultural Experiment Station, Bulletin No. 774. Berkeley, 1960. 72 pp. Illus., map, tables, diags., notes, bib.

Teeguarden, Dennis. See Casamajor, Paul, #C207

T31 Teesdale, Clyde H. "History of Treated Wood Block Pavements in the United States." *Proceedings of the American Wood Preservers' Association* 11 (1915), 325-71.

T32 Telling, Irving. "Coolidge and Thoreau: Forgotten Frontier Towns." *New Mexico Historical Review* 29 (July 1954), 210-23. Includes some history of logging and sawmills in and near these New Mexico lumber towns, 1880s-1900s.

T33 Tennessee, University of. *Progress in Agricultural Development and Watershed Protection in the Tennessee Valley Area of Tennessee.* Knoxville, 1949. 41 pp. Map, tables. Since 1937.

T34 Terasmae, J. "A Discussion of Deglaciation and Boreal Forest History in Northern Great Lakes Region." *Proceedings of the Entomological Society of Ontario* 99 (1968), 31-43.

T35 Ter Bush, Frank A. "Southern Pine Prices and Business Cycles, 1904-1941." Master's thesis, State Univ. of New York, College of Forestry, 1951.

T36 Terrell, John U. *The United States Department of the Interior: A Story of Rangeland, Wildlife, and Dams.* New York: Duell, Sloan, and Pearce, 1963. 117 pp. Illus.

T37 Terrell, John U. *The United States Department of Agriculture: A Story of Food, Farms, and Forests.* New York: Duell, Sloan and Pearce, 1966. 130 pp.

T38 Terrell, John U. *The Man Who Rediscovered America: A Biography of John Wesley Powell.* New York: Weybright and Talley, 1969. 281 pp. Map, bib., index. Powell (1834-1902) was an explorer, director of U.S. Geological Survey, and advocate of irrigation, land reform, and forest conservation in the West.

T39 Terrell, John U. *Land Grab: The Truth About 'The Winning of the West'.* New York: Dial Press, 1972. viii + 277 pp. Includes some history of federal land policy in the West.

T40 Terres, John K. "W.L. McAtee, 1883-1962." *Journal of Wildlife Management* 27 (June 1963), 494-99. Employed by the U.S. Biological Survey of the USDA and its successor agency, the Fish and Wildlife Service of the USDI, 1904-1947, Waldo Lee McAtee also helped establish the *Journal of Wildlife Management.*

T41 Terrien, Norma Colburn. *Station in the Forest.* Escanaba, Michigan: Photo Offset Printing, 1969. 106 pp. Illus. History of Cornell, a lumber town of Michigan's Upper Peninsula, since ca. 1880s.

T42 Terry, William Z. "Native Trees of Utah and the Uses Made of Them by Early Settlers." *Proceedings of the Utah Academy of Science* (1941), 115-17.

T43 *Texas Forestry.* "Siecke Gave Time, Talents to Texas Forestry." *Texas Forestry* 14 (Apr. 1974), 5. E.O. Siecke was state forester of Texas, 1917-1942, and a leader in the Texas Forestry Association.

T44 *Texas Forestry.* "Jones Founded Texas Forestry Association." *Texas Forestry* 14 (Apr. 1974), 6. On the career of W. Goodrich Jones, sometimes called the "Father of Forestry in Texas."

T45 Texas. General Land Office. *History of Texas Land.* With a preface by Bill Allcorn. Austin, 1958. x + 22 pp. Illus., tables. Since 1836.

T46 Thackwell, Rhys G. "The Grand Old Man of the Lumber World." *Southern Lumberman* (Dec. 23, 1922), 127. On George W. Hotchkiss, lumberman and journalist of Chicago and author of the *History of the Lumber and Forest Industry of the Northwest* (1898).

Thane, Eric. See Henry, Ralph Chester, #H296

T47 Thatcher, Theodore O. *Forest Entomology.* Minneapolis: Burgess Publishing Company, 1961. 225 pp. Incidental history.

T48 Thayer, Alford F., and Thayer, Eliene F. *The Forest Empire, a Story of the Land Frauds in the West.* New York: International Company, 1909. 216 pp.

Thayer, Eliene F. See Thayer, Alford F., #T48

Thayer, Theodore. See Pendleton, Nathaniel, #P98

T49 Theiss, Lewis Edwin. "Muncy Dam and the Days of the Lumber Industry." *Now and Then* 9 (Jan. 1951), 272-77; 9 (Apr. 1951), 301-07. In Pennsylvania's Susquehanna River Valley, 1830-1900.

T50 Theiss, Lewis Edwin. "Lumbering in Penn's Woods." *Pennsylvania History* 19 (Oct. 1952), 396-412. Reminiscences and lore of Pennsylvania logging and log rafting in the 19th century.

T51 Theiss, Lewis Edwin. "The Last Raft: As It Appeared to a Contemporary." *Pennsylvania History* 19 (Oct. 1952), 464-75. On the voyage of a log raft down the West Branch and the main Susquehanna River, Pennsylvania, 1938.

T52 Theiss, Lewis Edwin. "The Pioneer and the Forest." *Pennsylvania History* 23 (Oct. 1956), 487-503. In the 18th and 19th centuries, especially in Pennsylvania.

T53 Thelen, Rolf. *The Substitution of Other Materials for Wood*. USDA, Report No. 117. Washington: GPO, 1917. 78 pp. Incidental history of the use of wood substitutes in the building industry and of its effect on the lumber industry.

T54 Thevenon, Michel Jean. "An Economic Analysis of Pulp, Paper and Board Exports from the Pacific Northwest." Ph.D. diss., Oregon State Univ., 1972. 176 pp. Especially in the 1960s.

T55 Thiesmeyer, L.R. "Research and Pulpwood Production." *Journal of Forestry* 50 (Oct. 1952), 723-28. Since 1914.

T56 Thirgood, J.V. "The Tillamook Burn and Its Rehabilitation." *Empire Forestry Review* 40 (Dec. 1961), 342-49. Oregon, since the 1930s.

T57 Thomas, Charles J. "Birth Pangs of a National Park." *Manitoba Pageant* 15 (Autumn 1969), 2-17. On the role of the Riding Mountain Association in establishing Riding Mountain National Park, Manitoba, 1927-1933, as told by a member of that organization.

T58 Thomas, David N. "Early History of the North Carolina Furniture Industry, 1880-1921." Ph.D. diss., Univ. of North Carolina, 1964. 412 pp.

T59 Thomas, David N. "Getting Started in High Point." *Forest History* 11 (July 1967), 22-32. The proximity of excellent hardwood timber was a factor in locating the furniture industry in High Point, North Carolina, ca. 1900.

T60 Thomas, Howard. *Black River in the North Country*. Prospect, New York: Prospect Books, 1963. vi + 213 pp. Illus., map. Includes chapters on the history of logging, the lumber industry, and the pulp and paper industry along the Black River of northern New York.

T61 Thomas, James C. "The Log Houses of Kentucky." *Antiques* (Apr. 1974), 791-98.

T62 Thomas, Pearl Edwin. *Cork Insulation: A Complete Illustrated Textbook on Cork Insulation, the Origin of Corks and History of Its Use for Insulation*. Chicago: Nickerson and Collins, 1928. 534 pp. Illus.

T63 Thomas, Richard J. "Caleb Blood Smith: Whig Orator and Politician, Lincoln's Secretary of Interior." Ph.D. diss., Indiana Univ., 1969. 249 pp.

T64 Thomas, William L., Jr., ed. *Man's Role in Changing the Face of the Earth*. Chicago: University of Chicago Press, 1956. 1232 pp. An anthology of essays on man's impact on the environment.

T65 Thompson, Allen E. "Forest Management on the Cedar River Watershed." *Journal of Forestry* 49 (Mar. 1951), 201-05. Seattle drew its water supply from the Cedar River as early as 1901 and began a program of forest management on the watershed in 1924.

T66 Thompson, Allen E. "Timber and Water—Twin Harvest on Seattle's Cedar River Watershed." *Journal of Forestry* 58 (Apr. 1960), 299-302. Since 1905.

Thompson, Ben. See Thompson, George B., #T69

T67 Thompson, C.D. "The Conservation Movement in Relation to Public Ownership." *Public Utilities Fortnightly* 18 (Sept. 24, 1936), 393-401.

Thompson, Esther Katherine. See Gray, Lewis C., #G274

T68 Thompson, Frederick W. *The I. W. W., Its First Fifty Years, 1905-1955: The History of an Effort to Organize the Working Class*. Chicago: Industrial Workers of the World, 1955. 203 pp. Notes, index. Including lumberjacks and millworkers of the South and Pacific Coast, 1910s.

T69 Thompson, George B., and Thompson, Ben. *A History of the Lumber Business at Davis, West Virginia, 1885-1924*. Parsons, West Virginia: McClain Printing, 1974. Illus. On a firm at Davis established by J.L. Rumbarger and later operated under the following names: Blackwater Boom and Lumber Company, 1887-1893; Blackwater Lumber Company, 1893-1905; Thompson Lumber Company, 1905-1907; and Babcock Lumber and Boom Company, 1907-1924.

T70 Thompson, George S. *Up to Date; or, the Life of a Lumberman*. Peterborough, Ontario: N.p., 1895. 126 pp. Illus. On the traditions and customs of Canadian lumbermen in the 19th century.

T71 Thompson, Henry A. "The Life of a Vermont Farmer and Lumberman: The Diaries of Henry A. Thompson of Grafton and Saxtons River." Ed. by Stuart F. Heinritz. Intro. by Walter H. Thompson. *Vermont History* 42 (Spring 1974), 89-139. Thompson's diary covers from 1864 to 1932.

T72 Thompson, Huston. "The NPA and Stephen Mather." *National Parks Magazine* 33 (May 1959), 12. On Mather's role in the establishment of the National Parks Association, 1915-1919.

T73 Thompson, Jack M. "James R. Garfield: The Career of a Rooseveltian Progressive, 1895-1916." Ph.D. diss., Univ. of South Carolina, 1958. 292 pp. Garfield was Secretary of the Interior, 1907-1909.

T74 Thompson, Kenneth. "The Australian Fever Tree in California: Eucalyptus and Malaria Prophylaxis." *Annals of the Association of American Geographers* 60 (June 1970), 230-44. Medical theory and the spread of eucalyptus culture after the 1870s.

T75 Thompson, M.W. "Uncle Sam's First Timber Sale and How It Has Written Permanence and Prosperity into a Mining Community of the Black Hills." *American Forests and Forest Life* 34 (Jan. 1928), 17-19. On the first USFS timber sale, made to the Homestake Mining Company of South Dakota in 1899, and its impact upon the Black Hills National Forest and the sawmill town of Nemo.

T76 Thompson, Mark J. *85 Years of Farming in the Northern Coniferous Forest Areas of Minnesota, Wisconsin and Michigan*. Minnesota Agricultural Experiment Station, Miscellaneous Report No. 35. St. Paul: University of Minnesota, 1959. 58 pp. Map, tables. Includes some history of agricultural use of the forests in Michigan's Upper Peninsula, twelve counties in Wisconsin, and fifteen counties of Minnesota, since 1870.

T77 Thompson, Michael E. "The Challenge of Unionization: Pacific Northwest Lumber Workers during the Depression." Master's thesis, Washington State Univ., 1968.

T78 Thompson, R.A.H. "What's Happening to Timber." *Harper's* 191 (Aug. 1945), 125-33. Includes some history of the lumber industry, especially in the Pacific Northwest.

T79 Thompson, Roger C. "The Doctrine of Wilderness: A Study of the Policy and Politics of the Adirondack Preserve-Park." Ph.D. diss., State Univ. of New York, College of Forestry, 1962. 526 pp.

T80 Thompson, Roger C. "Politics in the Wilderness: New York's Adirondack Forest Preserve." *Forest History* 6 (Winter 1963), 14-23. On the history of the Adirondack Forest Preserve and the problems of administering this wilderness area since the late 19th century.

Thompson, Stith. See Taber, Edward O., #T1

Thompson, Walter H. See Thompson, Henry A., #T71

T81 Thomson, Betty Flanders. "The Woods Around Us." *American Heritage* 9 (Aug. 1958), 50-53, 97-101. A general account of forest use in New England since the beginning of white settlement.

T82 Thomson, Betty Flanders. *The Changing Face of New England*. New York: Macmillan, 1958. ix + 188 pp. Maps, table. Includes reference to the changes in landscape and flora since the 16th century, resulting in part from human occupation.

Thomson, C.C. See Place, I.C.M., #P236

T83 Thomson, David D. "Destruction of American Forests and the Consequences." *Baptist Review* 2 (Oct.-Dec. 1880), 485-513. A survey of forest resources and devastation.

T84 Thomson, Don W. *Men and Meridians: The History of Surveying and Mapping Canada*. 2 Volumes. Ottawa: Department of Mines and Technical Surveys, 1966-1967. Illus., maps, notes, bib., index. Volume 1 covers to 1867; Volume 2 from 1867 to 1917.

T85 Thomson, John P. "Tie Hacking Now a Lost Art." *Pacific Northwesterner* 19 (Winter 1975), 8-13. History and description of the manufacture of railroad crossties in northeastern Washington, ca. 1900-1920s.

T86 Thomson, Roy B. *An Examination of Basic Principles of Comparative Forest Valuation*. Duke University, School of Forestry, Bulletin No. 6. Durham, North Carolina, 1942. 99 pp. Tables, bib. A chapter on the "Historical Development of Forest Valuation" emphasizes German thought in forest economics.

T87 Thomson, Sheila M. "Four Contibutors to the American Nature Movement." Master's thesis, Cornell Univ., 1940.

T88 Thoreau, Henry David. *The Maine Woods*. Boston: Ticknor & Fields, 1864. New York: Thomas Y. Crowell, 1909. x + 423 pp. Illus., apps. A classic account of wilderness appreciation, including some description of logging and lumberjacks. There are several modern reprints.

T89 Thoreau, Henry David. *Thoreau: Walden & Other Writings*. Ed. and with intro. by Joseph Wood Krutch. New York: Bantam Books, 1965. 433 pp. Thoreau's writings, especially *Walden*, have been particularly important to preservationists and nature lovers. There are many other editions and collections of Thoreau's writings.

Thorne, H.W. See Stone, Robert N., #S670

T90 Thorne, Wynne, ed. *Land and Water Use*. Washington: American Association for the Advancement of Science, 1963. 364 pp. An anthology of papers on resource utilization, some with historical information.

T91 Thornton, Henry Worth. "The Forest Policy of the Canadian National Railways." *American Forests and Forest Life* 32 (Feb. 1926), 69-72. Incidental history, especially of fire prevention.

T92 Thornton, P.L., and Morgan, J.T. *The Forest Resources of Iowa*. Forest Survey Release No. 22. Columbus, Ohio: USFS, Central States Forest Experiment Station, 1959. 46 pp. Maps.

T93 Thorpe, Lloyd. "Tillamook Comes Back." *American Forests* 74 (July 1968), 10-13, 48. On Oregon's Tillamook Burn, site of devastating forest fires in 1933, 1939, and 1945, and subsequent reforestation efforts.

T94 Thorpe, Lloyd. *Men to Match the Mountains*. Seattle: By the author, 1972. xi + 268 pp. Illus., apps. On California's Conservation Camp Program (and its predecessors since ca. 1915), which used adult felons and juvenile delinquents in many natural resource management operations, especially fighting forest fires since World War II.

T95 Throckmorton, Arthur L. "George Abernethy, Pioneer Merchant." *Pacific Northwest Quarterly* 48 (July 1957), 76-88. Abernethy (1807-1877), Oregon's first provisional governor and a leading merchant, owned a sawmill and sold lumber in Oregon City and Portland.

T96 Throckmorton, Arthur L. *Oregon Argonauts: Merchant Adventurers on the Western Frontier*. Portland: Oregon Historical Society, 1961. 372 pp. Illus., maps. Includes some history of the lumber industry and lumber dealers in Portland and its hinterland, 1839-1869.

T97 Thruelsen, Richard. *Boss Logger*. New York: Curtis Publishing, 1948. 22 pp. Illus. On the logging career of Jim Bates, foreman of a Weyerhaeuser Timber Company camp at Vail, Washington. Reprinted from the *Saturday Evening Post*.

T98 Thurmond, A.K. "Trends of the Paper Industry." Master's thesis, Yale Univ., 1931.

T99 Tidwell, Ed. "Fort Smith Furniture Industry." *Guild Ticker* 11 (May 1949), 12-13, 36. Fort Smith, Arkansas, 1880-1945.

T100 Tiemann, Harry D. *The Kiln Drying of Lumber*. 1917. Third edition. Philadelphia: J.B. Lippincott, 1920. xi + 318 pp. Incidental history.

T101 Tiemann, Harry D. "History of Artificial Seasoning." *Timberman* 27 (June 1926), 184-88.

T102 Tiemann, Harry D. *Wood Technology, Constitution, Properties and Uses*. 1942. Third edition. New York: Pitman Publishing Corporation, 1951. 390 pp. Graphs, charts, tables, index. Incidental history.

T103 Tiemann, Harry D. "The Lamented Chestnut: Possibility of Its Resuscitation in Wisconsin." *Southern Lumberman* 179 (Dec. 15, 1949), 239-42. On the destruction of the American chestnut by blight since 1904.

T104 Tiemann, Harry D. "The Development of Kiln-Drying during the Last 75 Years." *Southern Lumberman* 193 (Dec. 15, 1956), 230-34. Technological developments since the 1860s, as revealed by patent records.

T105 Tiemann, Harry D. "An Oral History Interview with Harry Donald Tiemann. OHI by Donald G. Coleman. *Southern Lumberman* 203 (Dec. 15, 1961), 137-40. Tiemann was chief of the Division of Timber Physics, USFS, Forest Products Laboratory, Madison, Wisconsin.

T106 Tiemann, Harry D. *An Oral History Interview with Harry Donald Tiemann*. OHI by Donald G. Coleman. Madison, Wisconsin: USFS, Forest Products Laboratory, 1961, 13 pp.

T107 Tilden, Freeman. *The National Parks, What They Mean to You and Me*. Intro. by Newton B. Drury. New York: Alfred A. Knopf, 1951. xvii + 417 + xxi pp. Illus., maps, apps., index. Intended primarily as a guide to national parks and monuments, this work also contains history of individual parks and their administration by the U.S. National Park Service.

T108 Tilden, Freeman. *Interpreting Our Heritage: Principles and Practices for Visitor Services in Parks, Museums, and Historic Places*. 1957. Revised edition. Chapel Hill: University of North Carolina Press, 1967. xviii + 120 pp. Illus., index. Includes some account of National Park Service interpretive programs.

T109 Tilden, Freeman. *The State Parks: Their Meaning in American Life*. New York: Alfred A. Knopf, 1962. xvi + 494 + xi pp. Illus., maps, apps., index. A descriptive guide to outstanding state parks, including some history of the state parks movement and incidental references to the history of individual parks.

T110 Tilden, Freeman. "The National Park Concept." *National Parks Magazine* 33 (May 1959), 2-5. On the establishment of Yosemite National Park (as a state park), and Yellowstone National Park, 1864-1872.

T111 Tilden, Freeman. "Riches of Being: The Century Since Yellowstone." *National Parks & Conservation Magazine* 46 (Jan. 1972), 4-9. Reflections on the national parks movement since 1872.

T112 Tillotson, Miner R., and Taylor, Frank J. *Grand Canyon Country*. 1929. Revised edition. Stanford, California: Stanford University Press, 1935. viii + 108 pp.

T113 Tilt, C.R. "Provincial Parks in Ontario." *Canadian Geographical Journal* 58 (Feb. 1959), 36-55. On the history of their establishment and some historical geography of individual parks.

T114 Tilton, Frank. *Sketch of the Great Fires in Wisconsin at Peshtigo, the Sugar Bush, Menekaune, Williamsonville, and Generally on the Shores of Green Bay, with Thrilling and Truthful Incidents by Eye Witnesses*. Green Bay, Wisconsin: Robinson & Kutterman, 1871. 116 pp. Illus. A nearly contemporary account of the forest fires that ravaged Peshtigo and northern Wisconsin in 1871. Also published in the *Green Bay Historical Bulletin* 7 (Jan.-June 1931).

T115 *Timber of Canada*. "Quebec and Lumbering History." *Timber of Canada* 4 (Aug. 1944), 23-25.

T116 *Timber of Canada*. "Tree Farm Story." *Timber of Canada* 14 (Nov. 1946), 38-42.

T117 *Timberman*. "Ocean Log Towing on the Pacific." *Timberman* 23 (Aug. 1922), 39. On the Davies log raft and its use at Grays Harbor, Washington.

T118 *Timberman*. "Early Logging on Puget Sound." *Timberman* 23 (Aug. 1922), 39. Reminiscences of Ed English from 1885 to 1904.

T119 *Timberman*. "History of the Northwest Broom Industry." *Timberman* 23 (Aug. 1922), 148.

T120 *Timberman*. "Lumbering in the Black Hills of South Dakota." *Timberman* 26 (Nov. 1924), 122. By the Homestake Mining Company.

T121 *Timberman*. "Oregon-Washington Furniture Industry." *Timberman* 26 (May 1925). Incidental history.

T122 *Timberman*. "Tacoma's First Sawmill." *Timberman* 27 (Nov. 1925), 208. Built by Nikolaus Delin in 1853.

T123 *Timberman*. "Pioneer California Planing Mill." *Timberman* 27 (May 1926), 82-86. On the Oakland Planing Mills Company since 1869.

T124 *Timberman*. "Pioneer Mahogany Lumber and Veneer Firm." *Timberman* 27 (May 1926), 37-39. On Ichabod T. Williams & Sons, New York City, dealers in Spanish cedar and mahogany since 1838.

T125 *Timberman*. "Ocean Log Rafts." *Timberman* 27 (July 1926). Incidental history.

T126 *Timberman*. "Forty Years of Saw Filing." *Timberman* 27 (Aug. 1926), 60. On Dan McKinnon, a saw filer since 1886 in Michigan, Wisconsin, Minnesota, Montana, and Washington.

T127 *Timberman*. "Lumber Manufacture Commences in America's Greatest Remaining White Pine Timber Stand, an Illustrated Description of the Plant Facilities and the Timber Resources of the Clearwater Timber Company, Lewiston, Idaho." *Timberman* 28 (July 1927), 50-63. Incidental history.

T128 *Timberman.* "The Oldest Business in the World." *Timberman* 28 (Sept. 1927), 174-75. On the Archers Company, Pinehurst, North Carolina, manufacturers of bows and arrows.

T129 *Timberman.* "Arizona's Largest Lumber Operation." *Timberman* 29 (Mar. 1928), 166-72. Some history and description of the woods and mill operations of Cady Lumber Company, McNary, Arizona.

T130 *Timberman.* "Selling Lumber in Packages." *Timberman* 29 (Apr. 1928), 44, 46. On the origins of Weyerhaeuser Timber Company's "4 Square" packaged lumber.

T131 *Timberman.* "Lumber Operations in Linn and Lane Counties, Developments in Two Outstanding Oregon Timber Districts." *Timberman* 29 (Apr. 1928), 48-50. Incidental history.

T132 *Timberman.* "Pioneer Lumbering in Klamath, History of Early Sawmill Experiences." *Timberman* 29 (July 1928), 42-49. Since establishment in 1878 by the Moore brothers of the first successful sawmill at Klamath Falls, Oregon.

T133 *Timberman.* "A Pioneer Michigan Lumber City." *Timberman* 29 (June 1928), 60-61. Saginaw.

T134 *Timberman.* "Pioneer Lumber Concern Re-established." *Timberman* 29 (Oct. 1928), 49, 54. Includes some history of the lumber enterprises of Asa Mead Simpson and descendants at Coos Bay, Oregon, since the 1850s.

T135 *Timberman.* "History of the Robertson Log Raft, Persistence and Enterprise of the Inventor Finally Triumph." *Timberman* 30 (Jan. 1929), 37-40. Includes a compilation of patent literature on log rafting since 1818.

T136 *Timberman.* "History of Log Rafting. II. Campbell Patents of 1905." *Timberman* 30 (Mar. 1929), 156-58.

T137 *Timberman.* "An Oregon Pioneer Woodcraftsman: Early History of Portland Planing Mill Industry." *Timberman* 30 (Apr. 1929), 68-72. On the Niccolai-Neppach Company of Portland since 1866.

T138 *Timberman.* "Evolution of the 'Tommy Moore' Block: Early Days on the Columbia River." *Timberman* 30 (May 1929), 46, 172. A piece of logging equipment invented by Henry Hoeck of Portland in 1893.

T139 *Timberman.* "Pioneer California Pine Company, History of Madera Sugar Pine Company." *Timberman* 30 (June 1929), 180-82. Madera County, California, since 1874.

T140 *Timberman.* "Handling Lumber on America's Greatest Tramway." *Timberman* 30 (July 1929), 37-39. By the El Dorado Lumber Company in California's American River Canyon since 1901.

T141 *Timberman.* "Pioneer Days in Tahoe Region." *Timberman* 30 (Aug. 1929), 178-79. J.H. Gardner's recollections of logging near Lake Tahoe, Nevada-California, 1865-1880.

T142 *Timberman.* "Rafting Coast Logs to Market." *Timberman* 30 (Sept. 1929). Incidental history.

T143 *Timberman.* "Pioneer Lumber Operation of Mendocino Coast." *Timberman* 30 (Oct. 1929), 186. On the Mendocino Lumber Company of California.

T144 *Timberman.* "Historic California Lumber Flume." *Timberman* 31 (May 1930), 124. Of the Hume-Bennett Lumber Company in California's San Joaquin Valley.

T145 *Timberman.* "Alaska's Modern Sawmills, Important Factors in Lumber Trade." *Timberman* 31 (Oct. 1930), 40-41. Incidental history.

T146 *Timberman.* "Old Ox-Team Hauling Contract." *Timberman* 32 (Jan. 1931), 60. Signed by the founder of Schafer Brothers Logging Company, Aberdeen, Washington, in 1895.

T147 *Timberman.* "Early Day Sawmills on Russian River: An Account of the 'Inexhaustible' Redwood Forests That Once Lay Within Fifty Miles of San Francisco." *Timberman* 32 (Feb. 1931), 28-29, 60.

T148 *Timberman.* "Experiences of a Veteran Mill Builder: James A. Allen Relates Interesting Experiences in Long Career as a Pacific Coast Sawmill Erecting Engineer." *Timberman* 32 (Mar. 1931), 38, 60.

T149 *Timberman.* "Development of Veneer Cutting Lathe, History of Early Patents and the Building of Plants Which Forecast Growth of the Industry." *Timberman* 32 (Apr. 1931), 85-86. Pacific Coast.

T150 *Timberman.* "Early Day Pacific Coast Veneer Plants: History of the First Manufacturers of Fruit and Vegetable Containers and Douglas Fir Plywood." *Timberman* 32 (May 1931), 83-84. On the first veneer plant in California (1868) and the first Douglas fir plywood plant in Portland, Oregon (1905).

T151 *Timberman.* "Veneer Industry of the Pacific Greater West; Manufacturing and Merchandizing Supplement to the 'Timberman'." *Timberman* 33 (Dec. 1931), 17-56H. Includes articles on the history of plywood and veneer manufacturing in the Pacific states.

T152 *Timberman.* "History of Pacific Hardwood Trade." *Timberman* 33 (Nov. 1931); 33 (Dec. 1931); 33 (Jan. 1932).

T153 *Timberman.* "Fifty Years of Service to an Industry: Washington Iron Works Observes Golden Anniversary of the Pioneer Enterprise Established by J.M. Frink." *Timberman* 33 (Jan. 1932), 21, 46. Seattle manufacturers of logging machinery since 1889.

T154 *Timberman.* "Forty Years of Shingle Making." *Timberman* 33 (Mar. 1932), 45. By Will S. Dippold in Oregon, 1890-1930.

T155 *Timberman.* "History of Willapa Harbor Lumbering." *Timberman* 33 (June 1932). Pacific County, Washington.

T156 *Timberman.* "Wood Preservation, a Record of Growth and Progress." *Timberman* 34 (Feb. 1933), 9-35. Incidental history of the wood preservation industry.

T157 *Timberman.* "Pioneer Eastern Oregon Lumber Firm." *Timberman* 34 (Sept. 1933), 48-49. On David Eccles, who cut ties for the Oregon Short Line Railroad in 1887 and founded the Oregon Lumber Company at Baker in 1889.

T158 *Timberman.* "Forty Years of Raft Building: Career of John A. Fastabend." *Timberman* 34 (Oct. 1933), 16-17, 24-

26. On the construction of cigar-shaped Benson log rafts in Wallace Slough of the Columbia River, Oregon. Includes some early history of ocean log rafting.

T159 *Timberman.* "Resettlement Administration Program." *Timberman* 37 (Jan. 1936), 12-13, 23. Includes some history of a land-use demonstration project in northeastern Washington, where submarginal farm land was converted to timber growing under a New Deal program.

T160 *Timberman.* "Robert Polson: Master Logger, 1863-1936." *Timberman* 37 (Mar. 1936), 34-36. Sketch of a pioneer logger in Grays Harbor County, Washington.

T161 *Timberman.* "Evolution of Gerlinger Lumber Carrier." *Timberman* 37 (July 1936), supplement, 48-49.

T162 *Timberman.* "Pioneer Truck Logger." *Timberman* 37 (Aug. 1936), 98. Fred S. Buck, Vancouver, British Columbia.

T163 *Timberman.* Lumber Export issue. *Timberman* 38 (Aug. 1937). Includes historical articles on the Pacific Lumber Inspection Bureau, the Douglas Fir Export Company, and other topics related to the export of logs and lumber.

T164 *Timberman.* "Defeat of Log Piracy." *Timberman* 39 (Dec. 1937), 16-18. On the use of patrols to halt the theft of logs from booms and log rafts on Washington's Puget Sound.

T165 *Timberman.* "Timber Preservation." *Timberman* 41 (Feb. 1940), 11-32. Includes some history of the wood preservation industry.

T166 *Timberman.* "Lumbering in the Southwest." *Timberman* 41 (Feb. 1940); 41 (Mar. 1940). Incidental history.

T167 *Timberman.* "Barging and Rafting." *Timberman* 41 (May 1940). Includes some history of the Pacific Coast lumber trade.

T168 *Timberman.* "Fifty Years with Saw and Axe." *Timberman* 41 (July 1940), 34-35. Reminiscences of Bill Carlile, a logger since the 1890s in Coos County, Oregon.

T169 *Timberman.* "It Does Come Back: Anaconda Copper Mining Company Takes Stock of Its New Pine Crop." *Timberman* 42 (Feb. 1941), 14-15. On Montana land logged in 1885, 1905, 1924, and 1934.

T170 *Timberman.* "Thirty-Year Record of Lumber Production." *Timberman* 42 (Apr. 1941), 58-59. Pacific Coast.

T171 *Timberman.* "Caspar President Tells of Early Days." *Timberman* 42 (July 1941), 60-62. On the early use of steam donkey engines and logging railroads in redwood logging by the Caspar Lumber Company, California, as told by C.E. DeCamp.

T172 *Timberman.* "Crown-Zellerbach's Sustained Yield Plan." *Timberman* 43 (Feb. 1942), 13-17. Forest management in the Pacific Northwest since 1906.

T173 *Timberman.* "Wooden Box Review." *Timberman* 43 (July 1942), 11-64. A special issue with some history of the box industry in the Pacific states since 1904.

T174 *Timberman.* "Balm for Paul Bunyan's Aching Muscles." *Timberman* 43 (Sept. 1942), 10-13, 28. On the development and increasing use of power saws.

T175 *Timberman.* "Simon Benson—Pioneer Logger." *Timberman* 43 (Sept. 1942). Of Oregon.

T176 *Timberman.* "Eighteen Years of Logging." *Timberman* 43 (Oct. 1942), 10-11, 32. At Headquarters, Idaho.

T177 *Timberman.* "Pioneer Mills of Willapa and the South Bend Branch." *Timberman* 43 (Oct. 1942), 12-14, 52-53. On the lumber industry in Pacific County, Washington, since 1862.

T178 *Timberman.* "Oregon Cruisers' Marks." *Timberman* 45 (Feb. 1944), 90-91. On Percy N. Pratt's collection of 60 cruisers' marks.

T179 *Timberman.* "Ox Team Logging in Jackson Hole Country." *Timberman* 45 (May 1944). Wyoming.

T180 *Timberman.* "Hinsdales Came in '52." *Timberman* 47 (Mar. 1946), 34ff. Gardiner Lumber Company, Gardiner, Oregon.

T181 *Timberman.* "Pilot Rock Vigorous Member of Western Pine Family." *Timberman* 47 (Aug. 1946), 42 ff. Pilot Rock Lumber Company, Pilot Rock, Oregon.

T182 *Timberman.* "Fifty Years at White River." *Timberman* 47 (Sept. 1946), 34ff. White River Lumber Company, Washington.

T183 *Timberman.* "Long Range Planning at Port Alberni." *Timberman* 48 (Sept. 1947), 84ff. History of sawmills at Port Alberni, British Columbia, since 1935.

T184 *Timberman.* "Who Was West Coast's First Woodworker?" *Timberman* 49 (Mar. 1948). On the E.C. Miller Cedar Lumber Company, Aberdeen, Washington.

T185 *Timberman.* "Surveyors of the Wilderness." *Timberman* 49 (July 1948), 52-53, 92-98. On 19th-century surveyors in Oregon.

T186 *Timberman.* "Oregon's Logger's Museum." *Timberman* 49 (July 1948), 62-64. Collier Memorial State Park Logging Museum near Chiloquin, Oregon.

T187 *Timberman.* "Yaquina Bay: Rising New Lumber Port on the Oregon Coast." *Timberman* 50 (July 1949), 102-04.

T188 *Timberman.* Fiftieth Anniversary Number. *Timberman* 50 (Oct. 1949), complete. Articles of historical interest are entered below or elsewhere by author.

T189 *Timberman.* "Lidgerwood Versus Mundy: Origin of Steel Skidder." *Timberman* 50 (Oct. 1949), 54-55. On the rival builders of steam logging equipment on the Pacific Coast, 1890s-1910s.

T190 *Timberman.* "Oxen, Horses, and Steam." *Timberman* 50 (Oct. 1949), 56. Logging operations in Washington of F.F. Williamson, 1884-1907.

T191 *Timberman.* "Evolution of the Logging Arch." *Timberman* 50 (Oct. 1949), 57-62. On log hauling and skidding devices used on the Pacific Coast during the early 20th century.

T192 *Timberman.* "Railroad Logging in the West." *Timberman* 50 (Oct. 1949), 63-78. On the rise and decline of railroad logging since the 1890s. Includes list of surviving logging railroads in the West and British Columbia.

T193 *Timberman.* "Birthplace of Fir Plywood." *Timberman* 50 (Oct. 1949), 86-91. Portland Manufacturing Company, St. Johns, Oregon, 1904-1905, and subsequent development of the industry on the Pacific Coast.

T194 *Timberman.* "Klamathon Chute." *Timberman* 50 (Oct. 1949), 98. A chute logging operation in Klamath County, Oregon, and Siskiyou County, California, 1890s.

T195 *Timberman.* "Pioneer Days Along the Rugged Mendocino Coast." *Timberman* 50 (Oct. 1949), 92-96. Photographs of redwood logging in the 19th century, Mendocino County, California.

T196 *Timberman.* "Early Days Along the Southern Sierras." *Timberman* 50 (Oct. 1949), 104-10. Logging in the sequoia regions of California's Sierra Nevada, and the use of giant flumes to transport logs to valley sawmills, 1854-1930s.

T197 *Timberman.* "Lumbering in Alaska, 1811-1949." *Timberman* 50 (Oct. 1949), 112, 114. From Russian logging operations near Sitka to contemporary logging in the Tongass National Forest.

T198 *Timberman.* "We Make a Start to Save an Empire." *Timberman* 50 (Oct. 1949), 140-46. On efforts in forest fire prevention and suppression in the Pacific Northwest since 1864, particularly by the Western Forestry and Conservation Association since 1909.

T199 *Timberman.* "Power Saws Come of Age: Chronicle of Heartaches and Ultimate Triumph." *Timberman* 50 (Oct. 1949), 150-63. On the evolution of mechanical saws since 1879.

T200 *Timberman.* "Beginning of Western Truck Logging." *Timberman* 50 (Oct. 1949), 168-70. Its history in the Pacific Northwest since 1914, and particularly on the career of J.T. McDonald, Lakeview, Oregon, who pioneered in truck logging.

T201 *Timberman.* "Billion Feet of Logs by Motor Truck." *Timberman* 50 (Oct. 1949), 172-80, 228. On the evolution of log trucks and the career of Lloyd Christensen, particularly as truck supervisor for Crown Zellerbach Corporation in Oregon and Washington, since 1914.

T202 *Timberman.* "Evolution of Lumber Handling Methods." *Timberman* 50 (Oct. 1949), 182-88. Lumber carriers and lumber handling equipment in the Pacific Northwest since 1905.

T203 *Timberman.* "Serving Logging Trade since 1881." *Timberman* 50 (Oct. 1949), 190-192. Washington Iron Works, Seattle, Washington.

T204 *Timberman.* "Farm Tractor Grandsire of Powerful Yarders." *Timberman* 50 (Oct. 1949), 198-202. Fordson tractors were adapted for logging and yarding uses in the 1920s.

T205 *Timberman.* "Sumner Now in 57th Year." *Timberman* 50 (Oct. 1949), 204-06. On the Sumner Iron Works of Everett, Washington, manufacturers of logging machinery.

T206 *Timberman.* "Progress in Planers and Matchers." *Timberman* 50 (Oct. 1949), 208-12. By George W. Stetson and Harry B. Ross in Washington since 1900.

T207 *Timberman.* "Growth of an Industry Five-Fold in Fifty Years." *Timberman* 50 (Oct. 1949), 214-16. Annual lumber production figures for eleven Western states and British Columbia, 1899-1947.

T208 *Timberman.* "Half a Century of Box Making." *Timberman* 50 (Oct. 1949), 218-26. Advances in the Pacific Coast wooden box industry since the 1850s.

T209 *Timberman.* "United We Stand." *Timberman* 50 (Oct. 1949), 236-46. A brief outline of Pacific Coast trade associations since the 1890s.

T210 *Timberman.* "One Hundred Years in the Lumber Trade." *Timberman* 51 (Dec. 1949), 36-37, 80. On the lumber and shipping firm of Pope & Talbot, Inc., a San Francisco firm with major operations on Puget Sound, since 1849.

T211 *Timberman.* "Diamond Donkeys and Other Machines: Interesting Chapter in Diamond Match History." *Timberman* 52 (Apr. 1951), 52-53, 80. Idaho.

T212 *Timberman.* "The Prince George Area." *Timberman* 59 (Dec. 1958), 36-42. The lumber industry in central British Columbia.

T213 Timmons, John F., and Murray, William G., eds. *Land Problems and Policies.* Ames: Iowa State College, 1950. Reprint. New York: Arno Press, 1972. viii + 298 pp. Illus. Includes some incidental history of land problems in relation to forests, range, wildlife, and recreation.

T214 Tininenko, Robert D. "Middle Snake River Development: The Controversy over Hells Canyon, 1947-1955." Master's thesis, Washington State Univ., 1967. Dams and hydroelectric power policy in the canyon dividing Oregon and Idaho.

T215 Tinker, E.W. "What's Happened to the Shelterbelt?" *American Forests* 44 (Jan. 1938), 6-10, 48. Includes some history of the USFS's Shelterbelt Project, the Prairie States Forestry Project, and the tree planting experiments which preceded them.

T216 Tinkey, Amos R. "Arcata & Mad River." *Railroad Magazine* 61 (June 1953), 34-53. On a logging railroad between Arcata and Eureka, California, since 1881.

Tippett, Maria. See Cole, Douglas, #C505

T217 Tippett, Maria. "Emily Carr's Forest." *Journal of Forest History* 18 (Oct. 1974), 132-37. On Emily Carr's artistic interpretations of the British Columbia forests.

T218 Titus, William A., ed. *History of the Fox River Valley, Lake Winnebago, and the Green Bay Regions.* 3 Volumes. Chicago, 1930. A center of the Wisconsin lumber industry.

T219 Titus, William A. "Two Decades of Wisconsin Forestry, 1905-1925." *Wisconsin Magazine of History* 30 (Dec. 1946), 187-91.

T220 Tobie, Harvey E. "Oregon Labor Disputes, 1919-1923." *Oregon Historical Quarterly* 48 (Mar. 1947), 7-24; 48 (Sept. 1947), 195-213; 48 (Dec. 1947), 309-21. Includes mention of labor conditions in the forest industries.

T221 Tobie, John. "One Way to Make History." *Timber Producer* (Apr. 1975), 20-23. Brief history of hydraulic loaders used in pulpwood logging, manufactured by Prentice Hydraulics Company, Prentice, Wisconsin, since 1955.

T222 Todd, A.S., and Yoho, James G. "Forestry in the Southern Economy." *Journal of Forestry* 60 (Oct. 1962), 694-704.

T223 Todd, William. *Todds of the St. Croix Valley.* Mount Carmel, Connecticut: Privately printed, 1943. 24 pp. On a family lumber business in Washington County, Maine.

T224 Todes, Charlotte. *Labor and Lumber.* New York: International Publishers, 1931. Reprint. New York: Arno Press, 1974. 208 pp. Illus., tables, notes. A historical survey which emphasizes labor organization and the grievances of workers.

T225 Toepperwein, Fritz A. *Charcoal and Charcoal Burners.* Boerne, Texas: Highland Press, 1950. 61 pp. Illus., map. Incidental history.

T226 Tolbert, Caroline Leona. *History of Mount Rainier National Park.* Seattle: Lowman and Hanford Company, 1933. 60 pp. Illus., map. Also issued as a master's thesis at the University of Washington, 1933.

T227 Toll, Roger. "Wilderness and Wildlife Administration in Yellowstone National Park." *American Planning and Civic Annual* (1936), 65-72. Incidental history.

T228 Tombs, G. "Recollections of a One-Time Freight Agent." *Pulp & Paper Magazine of Canada* 54 (May 1953), 157-60.

T229 Tomeraasen, Rollin W. "The Development of the Lumber Industry in the Northwest, 1870-1890." Master's thesis, Univ. of Illinois, 1933.

T230 Tomkin, R.D. "Old Water-Driven Sawmill Has Operated 125 Years." *Southern Lumberman* (Jan. 15, 1941), 49-50. A Pennsylvania sawmill.

T231 Tompkins, Charles R. *A History of the Planing Mill . . . with Practical Suggestions for the Construction, Care and Management of Wood Working Machinery.* New York: John Wiley, 1889. 222 pp.

T232 Tonkin, Joseph Dudley. *The Last Raft.* Harrisburg, Pennsylvania: By the author, 1940. 146 pp. Illus. On log driving and rafting on the West Branch of the Susquehanna River of Pennsylvania, from its beginnings to 1938.

T233 Tonkin, R. Dudley. "The Inside Story." *Pennsylvania History* 19 (Oct. 1952), 476-95. Reminiscence of the last log raft to go down Pennsylvania's Susquehanna River, 1938, by the lumberman who built it.

T234 Tonkin, R. Dudley. *My Partner, The River: The White Pine Story on the Susquehanna.* Pittsburgh: University of Pittsburgh Press, 1958. xii + 276 pp. Illus., diags. Logging, the lumber industry, and rafting on Pennsylvania's Susquehanna River, 1835-1915, including the author's reminiscences of the more recent period.

T235 Toole, Horace G. "The Land System of Colonial New York." Ph.D. diss., Univ. of Pennsylvania, 1932.

T236 Toole, K. Ross, and Butcher, Edward B. "Timber Depredations on the Montana Public Domain, 1885-1918." *Journal of the West* 7 (July 1968), 351-62. On the Free Timber Act of 1878 and subsequent struggles between mining and railroad interests and the federal government over Montana timber. Some of the suits against trespassers were not settled until 1918.

T237 Torrent, Lewis. *Muskegon County Log Marks.* Detroit: Great Lakes Model Shipbuilders Guild, 1956. Michigan.

T238 Torrey, Raymond H. *State Parks and Recreational Uses of State Forests in the United States.* Washington: National Conference on State Parks, 1926. 259 pp. Illus., tables.

T239 Torrey, Raymond H. "John Boyd Thacher State Park." *Scenic and Historic America* 4 (1935), 3-28. New York.

T240 Tosi, Joseph A. "The Land Use Study of Windham County, Vermont." Master's thesis, Yale Univ., 1948.

Toumey, James W. See Yale Forest School, #Y1

T241 Toumey, James W. "Forests Indispensable in War." *American Forestry* 24 (Jan. 1918), 16-20. On the importance of forest resources to the American effort in World War I, with incidental historical references.

T242 Toumey, James W. "Recent Progress and Trends in Forestry in the United States." *Journal of Forestry* 23 (Jan. 1925), 1-9. In national, state, and private forestry.

T243 Towne, Arthur W. *Pioneers in Paper: The Story of Blake, Moffitt and Towne.* San Francisco: Blake, Moffitt and Towne Company, 1930. 49 pp. Illus. A 75th anniversary company history.

T244 Towle, Jerry Charles. "Woodland in the Willamette Valley: An Historical Geography." Ph.D. diss., Univ. of Oregon, 1974. 210 pp. On changes in woodland areas since 1854.

T245 Tracey, C.H. "Forest Fire Protection in the Southern Portion of West Virginia." *Journal of Forestry* 30 (Jan. 1932), 58-61. On the Southern West Virginia Forest Fire Protective Association since 1916.

T246 Tracy, R.P. "Roughneck Pioneers: The Case-hardened Crew That Helped Organize the National Forests, as Seen by One of Them." *Sunset Magazine* 48 (No. 2, 1922), 32-35, 66.

T247 Trager, Martelle W. *National Parks of the Northwest.* New York: Dodd, Mead, 1939. 216 pp. Illus. Incidental history.

T248 Trani, Eugene P. "Conflict or Compromise: Harold L. Ickes and Franklin D. Roosevelt." *North Dakota Quarterly* 36 (Winter 1968), 20-29. On Ickes as Roosevelt's Secretary of the Interior and his unsuccessful efforts to reorganize the USDI into a Department of Conservation, 1930s.

T249 Trani, Eugene P. "Hubert Work and the Department of the Interior, 1923-1928." *Pacific Northwest Quarterly* 61 (Jan. 1970), 31-40. Work served Warren G. Harding and Calvin Coolidge as Secretary of the Interior.

T250 Traxler, Ralph N., Jr. "The Land Grants for the Thirty-Second Parallel Railroad from the Mississippi to the Pacific." Ph.D. diss., Univ. of Chicago, 1953.

T251 Trayer, George W. *Wood in Aircraft Construction: Supply, Suitability, Handling, Fabrication, Design.* Washington: National Lumber Manufacturers Association, 1930. 276 pp. Illus., tables, bib. Incidental history.

T252 Treat, Payson Jackson. "Origin of the National Land System under the Confederation." *Annual Report of the American Historical Association,* 1905 (1906), 231-39. On Virginia's cession of western lands and the question of their disposal by Congress in 1784.

T253 Treat, Payson Jackson. *The National Land System, 1785-1820.* New York: E.B. Treat Company, 1910. Reprint. New York: Russell & Russell, 1967. xiii + 426 pp. Maps, tables, notes, bib., index. Also accepted by Stanford University as a doctoral dissertation in 1910, this work is a study of the public domain and its disposal.

T254 Treen, Edward B. *100 Years in the Lumber Industry.* Buffalo: Mixer & Company, 1957. 64 pp. Illus. Mixer and Company, a wholesale lumber business in Buffalo, New York, since 1857.

T255 Treen, Edward W. "The Niagara Frontier and Lumber." *Niagara Frontier* 13-16 (1966-1969). A ten-part history; New York.

T256 *Tree Ring Bulletin.* "Andrew Ellicott Douglass (1867-1962)." *Tree Ring Bulletin* 24 (Nos. 3 & 4, 1962), 3-10. Biographical sketch and list of his writings on dendrochronology.

T257 Trefethen, James B. "Carl Schurz: Forestry's Forgotten Pioneer." *American Forests* 67 (Sept. 1961), 24-27. As Secretary of the Interior, 1877-1881, the German-born Schurz attempted reforms of the public land system and encouraged the practice of forestry.

T258 Trefethen, James B. *Crusade for Wildlife: Highlights in Conservation Progress.* Harrisburg, Pennsylvania: Stackpole Company and Boone and Crockett Club, 1961. 337 pp. Bib. A history of wildlife conservation in North America and also of the Boone and Crockett Club.

T259 Trefethen, James B. "The 1928 'ORRRC'." *American Forests* 68 (Mar. 1962), 8, 38-39. On the establishment and work of the National Conference on Outdoor Recreation, 1924-1928.

T260 Trefethen, James B. *An American Crusade for Wildlife.* New York: Winchester Press, 1975. 384 pp. Illus. A general history of evolving attitudes and legislation concerning wildlife in North America.

T261 Trenk, Fred B. "Forests Return to the Peshtigo." *American Forests* 54 (Oct. 1948), 444-45, 474-76, 478-79. On a forest fire near Peshtigo, Wisconsin, 1871, and subsequent rehabilitation of the damaged area.

T262 Trenk, Fred B. "Evolution of Modern Tree Planting Machines." *Journal of Forestry* 61 (Oct. 1963), 726-30. Since the 1880s.

T263 Trimble, W.J. "The Influence of the Passing of the Public Lands." *Atlantic Monthly* 113 (June 1914), 755-67.

T264 Trippensee, Reuben E. "Wildlife Management in the United States: Past, Present and Future." *Forestry Chronicle* 12 (Dec. 1936), 375-81.

T265 Trippensee, Reuben E. *Wildlife Management.* New York: McGraw-Hill, 1948. x + 479 pp. Illus. Incidental history.

T266 Troetschel, Henry, Jr. "John Clayton Gifford: An Appreciation." *Tequesta* 10 (1950), 35-47. On Gifford's life in Florida as a forestry educator at the University of Miami, 1900s-1940s.

T267 Trotter, John E. "State Park System in Illinois." Ph.D. diss., Univ. of Chicago, 1962.

T268 Trotter, Spencer. "The Atlantic Forest Region of North America." *Popular Science Monthly* 75 (Oct. 1909), 370-92. And its history.

T269 Troubetzkoy, Dorothy U. "Virginia Paper Industry." *Virginia and the Virginia County* 7 (July 1953), 24-28, 40-43. Since 1744.

T270 Trower, Peter. "From the Hill to the Spill . . . ." *Raincoast Chronicles* 1 (Winter 1973), 4-14. History of logging and logging technology along the British Columbia coast since the 19th century.

T271 True, Alfred C. "The United States Department of Agriculture." *Annals of the American Academy of Political and Social Science* 40 (Mar. 1912), 100-09. Its history and work.

T272 True, Alfred C. *A History of Agricultural Extension Work in the United States, 1785-1923.* USDA, Miscellaneous Publication No. 15. Washington: GPO, 1928. iv + 220 pp.

T273 True, Alfred C. *A History of Agricultural Education in the United States, 1785-1925.* USDA, Miscellaneous Publication No. 36. Washington: GPO, 1929. 436 pp. Illus., notes, bib. Includes some history of forestry education.

T274 True, Alfred C. *A History of Agricultural Experimentation and Research in the United States, 1607-1925, Including a History of the United States Department of Agriculture.* USDA, Miscellaneous Publication No. 251. Washington: GPO, 1937. vi + 321 pp. Illus., notes, bib.

T275 True, Rodney H. "François André Michaux, the Botanist and Explorer." *Proceedings of the American Philosophical Society* 78 (1937), 313-27. Michaux made a bequest in 1855 to further the study of silviculture in America.

T276 Tryon, Henry H. *Forest Fires in South Carolina.* Extension Service Circular No. 77. Clemson, South Carolina: Clemson Agricultural College, 1926. 11 pp. Illus. Incidental history.

T277 Tryon, Henry H. *Forests and Forestry in South Carolina.* Extension Bulletin No. 81. Clemson, South Carolina: Clemson Agricultural College, 1926. 40 pp. Incidental history.

T278 Tryon, Henry H. *Practical Forestry in the Hudson Highlands.* Black Rock Forest, Bulletin No. 12. Cornwall, New York, 1934. 50 pp. Incidental history.

T279 Tryon, Rolla M. *Household Manufactures in the United States, 1640-1860.* Chicago: University of Chicago Press, 1917. xii + 413 pp. Tables, bib. Includes some history of forest products industries. There are several modern reprints.

T280 Tucker, Edwin A., and Fitzpatrick, George. *Men Who Matched the Mountains: The Forest Service in the Southwest.* Washington: GPO, 1972. 293 pp. Illus., maps. On the USFS in Arizona and New Mexico, compiled from oral histories and reminiscences. The emphasis is on individual rangers and events rather than the institution.

T281 Tugwell, Rexford G. "Forester's Heart." *New Republic* (Mar. 4, 1940), 304-05. Brief biographical sketch and appreciation of Ferdinand A. Silcox, USFS chief, 1933-1939.

T282 Turchen, Michael A. "The National Park Movement: A Study of the Impact of Temporal Change on Rhetoric." Ph.D. diss., Purdue Univ., 1972. 178 pp. A comparison of the periods 1890-1915 and 1950-1970.

T283 Turley, T.J. "Changes in the Cross Tie Industry during the Last Forty Years." *Cross Tie Bulletin* 39 (Nov. 1958), 67-69.

T284 Turner, George. *Narrow Gauge Nostalgia: A Compendium of California Short Lines.* 1965. Second edition. Corona del Mar, California: Trans-Anglo Books, 1971. 160 pp. Illus. Includes some history of the Brookings Lumber and Box Railroad, Highland; Bodie and Benton Railroad, Bodie; Diamond and Caldor Railway, Diamond Springs.

T285 Turner, Jerry M. "Rafting on the Mississippi." *Wisconsin Magazine of History* 23 (Dec. 1939), 163-76; 23 (Mar. 1940), 313-27; 23 (June 1940), 430-38; 24 (Sept. 1940), 56-65. Reminiscences of a lumber raft captain, 1850s-1880s.

T286 Turner, John S., and Stephenson, J. Newell. "Milestones: Pulp and Paper Industry's Growth." *Pulp and Paper of Canada* 54 (June 1953), 102-10; 54 (July 1953), 76-91; 54 (Aug. 1953), 93-108; 54 (Sept. 1953), 109-15; 54 (Oct. 1953), 121-25. On developments in the industry since 1903. Appeared also in *Tappi* 37 (May 1954), 18A-68A; 37 (June 1954), 14A-50A, under the title, "Fifty Years with the Canadian Pulp and Paper Industry."

T287 Turner, Robert D. *Vancouver Island Railroads.* San Marino, California: Golden West Books, 1973. 170 pp. Illus. Includes some history of logging railroads on Vancouver Island, British Columbia.

T288 Turpin, David Howard. "Pulpwood Price Trends in the Southeast." Master's thesis, Duke Univ., 1950.

T289 Turrentine, John W. *Potash: A Review, Estimate and Forecast.* New York: John Wiley, 1926. 188 pp. Illus.

T290 Turrentine, John W. *Potash in North America.* American Chemical Society, Monograph Series, No. 91. New York: Reinhold, 1943. 186 pp. Illus.

T291 Turrentine, John W. "The American Potash Industry." *Scientific Monthly* 70 (Jan. 1950), 41-47. Since 1910.

T292 Tutt, George T. "Lumber Wholesaling in the Pacific Northwest." Master's thesis, Univ. of Montana, 1967.

T293 Tuttle, J.M. "The Minnesota Pineries." *Harper's* 36 (Mar. 1868), 409-23.

T294 Tveten, John L. "The Big Thicket: A Texas Treasure in Trouble." *National Parks & Conservation Magazine* 48 (Jan. 1974), 4-8. Includes some history of efforts to preserve Big Thicket as a national park since 1927.

T295 Tweed, William. "Sequoia National Park Concessions, 1898-1926." *Pacific Historian* 16 (Spring 1972), 36-60. Also includes an account of the California park's origins.

T296 Tweton, D. Jerome. "Theodore Roosevelt and Land Law Reform." *Mid-America* 49 (Jan. 1967), 44-54. On President Theodore Roosevelt, the work and recommendations of the Public Lands Commission, and subsequent action by Congress 1902-1909. Among particular topics treated here are the Forest Homestead Act of 1906, the repeal of the Forest Lieu Act of 1897 in 1905, the failure to repeal the Timber and Stone and Desert Land acts, and federal grazing policy.

T297 Tweton, D. Jerome. "Theodore Roosevelt and the Arid Lands." *North Dakota Quarterly* 36 (Spring 1968).

T298 Twiford, Ormand H. "Life of an Arkansas Logger in 1901." Ed. by Walter L. Brown. *Arkansas Historical Quarterly* 21 (Spring 1962), 44-74. Twiford kept a journal while logging near Mena, Arkansas, in 1901; includes a biographical sketch of the author.

T299 Twight, Ben W. "The Tenacity of Value Commitment: The Forest Service and the Olympic National Park." Ph.D. diss., Univ. of Washington, 1971. 202 pp. Includes some history of rivalry between the USFS and National Park Service over the forested Olympic Peninsula.

T300 Twining, Charles E. "Lumbering and the Chippewa River." Master's thesis, Univ. of Wisconsin, 1963.

T301 Twining, Charles E. "Plunder and Progress: The Lumbering Industry in Perspective." *Wisconsin Magazine of History* 42 (Winter 1963), 116-24. Argues that most interpretations of 19th-century lumbering are overburdened by the shadow of 20th-century conservation ideas.

T302 Twining, Charles E. "Orrin Ingram: Wisconsin Lumberman." Ph.D. diss., Univ. of Wisconsin, 1970. 383 pp.

T303 Twining, Charles E. *Downriver: Orrin H. Ingram and The Empire Lumber Company.* Madison: State Historical Society of Wisconsin, 1975. ix + 309 pp. Illus., map, notes, app., tables, bib., index. Ingram, whose lumbering experience began in New York and Ontario, became the foremost lumberman of Eau Claire, Wisconsin, 1857-1918. This work emphasizes marketing and Ingram's conflict and later accommodation with Frederick Weyerhaeuser.

T304 Tyler, Robert L. "Rebels of the Woods and Fields: A Study of the I.W.W. in the Pacific Northwest." Ph.D. diss., Univ. of Oregon, 1953. 318 pp.

T305 Tyler, Robert L. "The Everett Free Speech Fight." *Pacific Historical Review* 23 (Feb. 1954), 19-30. On efforts of the Industrial Workers of the World to organize laborers in Everett, Washington, during a lumber mill strike in 1916.

T306 Tyler, Robert L. "I.W.W. in the Pacific Northwest: Rebels of the Woods." *Oregon Historical Quarterly* 55 (Mar. 1954), 3-44. The Industrial Workers of the World and its methods of circulating its doctrines in the lumber camps of Washington and Oregon, 1906-1933.

T307 Tyler, Robert L. "Violence at Centralia, 1919." *Pacific Northwest Quarterly* 45 (Oct. 1954), 116-24. On the events leading to the Armistice Day riot between American Legionnaires and members of the Industrial Workers of the

World, in Centralia, Washington, 1919. The IWW was active among workers in the lumber industry.

T308 Tyler, Robert L. "The Rise and Fall of an American Radicalism: The I.W.W." *Historian* 19 (Nov. 1956), 48-65. Includes references to lumberjacks and millworkers, 1905-1929.

T309 Tyler, Robert L. "The I.W.W. and the West." *American Quarterly* 12 (Summer 1960), 175-87. Including the role of lumberjacks and millworkers, 1900s-1920s.

T310 Tyler, Robert L. "The United States Government As Union Organizer: The Loyal Legion of Loggers and Lumbermen." *Mississippi Valley Historical Review* 47 (Dec. 1960), 434-51. On an organization established to counteract a strike of the Industrial Workers of the World against the lumber operators of the Pacific Northwest during World War I, 1917-1918, continued under Colonel Brice P. Disque for the U.S. War Department, and disbanded in 1935.

T311 Tyler, Robert L. *Rebels in the Woods: The I. W. W. in the Pacific Northwest.* Eugene: University of Oregon Books, 1967. 230 pp. Notes. A general study of working conditions in the lumber industry, and the strikes and other activities of the Industrial Workers of the World, 1900s-1920s.

T312 Tyler, Ronnie C. "Robert T. Hill and the Big Bend: An 1899 Expedition That Helped Establish a Great National Park." *American West* 10 (Sept. 1973), 36-43. On the Hill expedition as a step in the movement to create Big Bend National Park, Texas.

T313 Tyndall, James H. "Forestry in a Great Metropolitan Area." *Journal of Forestry* 47 (Jan. 1949), 29-35. On the Forest Preserve District of Cook County, Illinois, with emphasis on recreation policy, since 1903.

T314 Tyrrell, George F. "Background and Development of Cadastral Surveys." *Surveying and Mapping* 17 (Jan.-Mar. 1957), 33-41. Since the 17th century.

T315 Tyson, Willie K. "History of the Utilization of Longleaf Pine (*Pinus palustris* Mill.) in Florida from 1513 until the Twentieth Century." Master's thesis, Univ. of Florida, 1956.

U1 Udall, Stewart L. "National Parks for the Future." *Atlantic Monthly* 207 (June 1961), 81-84. On the deterioration of the national parks and the progress of the Mission 66 program.

U2 Udall, Stewart L. *The Conservation Challenge of the Sixties.* Horace M. Albright Conservation Lectures, No. 3. Berkeley: University of California, School of Forestry, 1963. 22 pp. Incidental history.

U3 Udall, Stewart L. "Pinchot and the Foresters." *American Forests* 69 (Nov. 1963), 24-27, 53-54. On Gifford Pinchot's role in the conservation movement.

U4 Udall, Stewart L. *The Quiet Crisis.* New York: Holt, Rinehart and Winston, 1963. xiii + 209 pp. Illus., bib., index. A popular history of the conservation movement and its leaders, including Henry David Thoreau, George Perkins Marsh, Carl Schurz, John Wesley Powell, Frederick Law Olmsted, Gifford Pinchot, John Muir, Theodore Roosevelt, and Franklin D. Roosevelt. The author, himself a leader of the movement, was Secretary of the Interior, 1961-1969.

U5 Udall, Stewart L. *The National Parks of America.* New York: G.P. Putnam's Sons and Country Beautiful Foundation, 1966. 225 pp. Illus. Incidental history.

U6 Udall, Stewart L, and the editors of Country Beautiful Corporation. *National Parks of the U.S.* Waukesha, Wisconsin: Country Beautiful Corporation, 1972. 226 pp. Illus. Incidental history.

U7 Uhl, Harry G. "Progress of Great Movement: Standardization of Lumber Manufacture." *Canada Lumberman* 44 (Sept. 15, 1924), 116-17.

U8 Ulm, Amanda. "Remember the Chestnut!" *American Forests* 54 (Apr. 1948), 156-59, 190, 192. On the chestnut blight in the United States, the introduction of blight-resistant Chinese and Japanese types, and their crossing with American types, since 1904.

U9 Ulmer, Grace. "Economic and Social Development of Calcasieu Parish, Louisiana, 1840-1912." *Louisiana Historical Quarterly* 32 (July 1950), 519-630. Includes some history of the lumber and other forest industries.

Underhill, Lonnie E. See Littlefield, Daniel F., Jr., #L212, L213

U10 Underwood, Marsh. *The Log of a Logger.* Portland: By the author, 1938. 62 pp. Illus. A river driver's account of loggers, logging, and logging camps from New England to Wisconsin, Oregon, and Washington, since the 1880s.

U11 U.S. Army. Spruce Production Division. *History of the Spruce Production Division, United States Army and United States Spruce Production Corporation.* Washington: U.S. Army, 1920. ix + 126 pp. Illus., maps. On the government agency organized to assure production of spruce lumber for aircraft manufacture during World War I. Colonel Brice P. Disque headed the operation in the Pacific Northwest.

U12 U.S. Bureau of Corporations. *The Lumber Industry.* 3 Volumes. Washington: GPO, 1913-1914. Entries to individual volumes follow.

U13 U.S. Bureau of Corporations. *Report of the Commissioner of Corporations, Part 1, Standing Timber.* Washington: GPO, 1913. xxiii + 301 pp. A general survey of standing timber in the United States, the concentration of its ownership, and the effect of federal and state land policies in furthering that concentration.

U14 U.S. Bureau of Corporations. *The Lumber Industry; Part II. Concentration of Timber Ownership in Important Selected Regions; Part III. Land Holdings of Large Timber Owners (With Ownership Maps).* Washington: GPO, 1914. xx + 264 pp. Maps, tables, index. Includes much history of timberland grants and ownership in the Pacific states, Louisiana, Florida, and Michigan. The large timber owners include the Southern Pacific Company, Northern Pacific Railway, and Weyerhaeuser Timber Company.

U15 U.S. Bureau of Corporations. *The Lumber Industry; Part IV. Conditions in Production and Wholesale Distribution Including Wholesale Prices.* Washington: GPO, 1914. xxi + 933 pp. Graphs, tables, index. Contains some history of the trade associations and of production of lumber and other forest products by species.

U16 U.S. Bureau of Corporations. *Special Report on Present and Past Conditions in the Lumber and Shingle Industry in the State of Washington.* Washington: GPO, 1914. 43 pp. Incidental history; excerpted from above.

U17 U.S. Bureau of Labor Statistics. *Wages and Hours of Labor in the Lumber, Millwork, and Furniture Industries, 1890 to 1912.* Bulletin No. 129. Washington: GPO, 1913. 178 pp. Tables.

U18 U.S. Bureau of Labor Statistics. *Wages and Hours of Labor in the Lumber, Mill Work, and Furniture Industries, 1907-1913.* Bulletin No. 153. Washington: GPO, 1914. Tables. Similar bulletins were issued in later years.

U19 U.S. Bureau of Labor Statistics. *History of Wages in the United States from Colonial Times to 1928.* Bulletin No. 499. Washington: GPO, 1929. 527 pp. Tables. Includes a chapter on wages in the lumber and woodworking industries since 1845.

U20 U.S. Bureau of Labor Statistics. *Impact of Technological Change and Automation in the Pulp and Paper Industry.* Bulletin No. 1347. Washington: GPO, 1962. 92 pp. Processed.

U21 U.S. Bureau of Land Management. *The Public Land Records . . . Footnotes to American History.* Washington: GPO, 1959. 36 pp. Illus., diags., maps. On the records maintained by the federal government since 1785.

U22 U.S. Bureau of Land Management. *The Public Lands: A Brief Sketch in United States History.* Washington: GPO, 1960. ii + 98 pp. Diag., maps, tables, notes. Since 1780.

U23 U.S. Bureau of Land Management. *Historical Highlights of Public Land Management.* Washington: GPO, 1962. 91 pp. Index. A chronological outline of legislation and events, many pertaining to forests and forestry.

U24 U.S. Bureau of Land Management. *Landmarks in Public Land Management.* Washington: GPO, 1962. 44 pp. Illus., tables. Historical sketch of the General Land Office and the Bureau of Land Management.

U25 U.S. Bureau of Statistics. *Imports and Exports of Wood and Manufactures of Wood by Articles and Countries, 1894-1904.* Foreign Commerce and Navigation of the U.S., Volume 2. Washington: GPO, 1904.

U26 U.S. Bureau of Statistics. *Foreign Trade of the United States in Forest Products, 1851-1908.* Bulletin No. 51. Washington: GPO, 1909. 32 pp.

U27 U.S. Civilian Conservation Corps. *Reforestation by the CCC.* Washington: GPO, 1941. 13 pp. Illus.

U28 U.S. Congress. Senate. *Report of the National Conservation Commission.* 3 Volumes. Ed. by Henry Gannett. Senate Document No. 676, 60th Congress, 2nd Session. Washington: GPO, 1909. Maps, tables, graphs, notes, index. Contains many historical references to forest resources and conservation.

U29 U.S. Congress. Senate. Committee on Commerce. *Problems of the Softwood Lumber Industry: Hearings, April 16-July 18, 1962, on Impact of Lumber Exports on the United States Softwood Lumber Industry.* 87th Congress, 2nd Session. Washington: GPO, 1962. 976 pp. Incidental history.

U30 U.S. Congress. Senate. Committee on Manufactures. *Newsprint Paper Industry Investigation: Report.* Senate Report 662, 66th Congress, 2nd Session. Washington: GPO, 1920. Incidental history.

U31 U.S. Congress. Joint Committee on Forestry. *Forest Lands of the United States.* Senate Document 32, 77th Congress, 1st Session. Washington: GPO, 1941. 44 pp. Incidental history.

U32 U.S. Congress. Joint Committee to Investigate Interior Department and Forestry Service. *Investigation of the Department of the Interior and of the Bureau of Forestry.* 13 Volumes. Senate Document No. 719, 61st Congress, 3rd Session. Washington: GPO, 1911. The hearings held to investigate the charges in the Ballinger-Pinchot controversy of 1910. They contain some history of the federal agencies involved and are a rich contemporary source for historians of the conservation movement.

U33 U.S. Council of National Defense. Advisory Commission. *The Douglas Fir Lumber Industry; an Inter-Departmental Study.* Washington: GPO, 1941. 98 pp. Map. Often referred to as the Keezer report; it contains a chapter on the financial history of the industry.

U34 U.S. Crop Reporting Board. *Naval Stores Statistics, 1900-1954.* USDA, Statistical Bulletin No. 181. Washington: GPO, 1956. 32 pp. Tables, diags., map. Statistics on turpentine and rosin production since 1900.

U35 USDA. *Message from the President of the United States Transmitting A Report of the Secretary of Agriculture in Relation to the Forests, Rivers, and Mountains of the Southern Appalachian Region.* Washington: GPO, 1902, 210 pp. Illus., index. Contains some history of this forested region and its forest industries.

U36 USDA. *The Western Range: Letter from the Secretary of Agriculture Transmitting in Response to Senate Resolution No. 289 a Report on the Western Range—A Great but Neglected Natural Resource.* Senate Document 199, 74th Congress, 2nd Session. Washington: GPO, 1936. 620 pp. Illus., maps, tables, charts, index. Includes much history of forest rangelands and USFS range policies.

U37 USDA. *Early American Soil Conservationists.* USDA, Miscellaneous Publication No. 449. Washington: GPO, 1941.

U38 USDA. Agricultural Research Administration. *Research and Related Services in the United States Department of Agriculture.* 4 Volumes. Washington: GPO, 1951. There are chapters on the history of all research and related activities within the USDA and by the USDA in association with the state land-grant colleges and other state agencies since 1839.

U39 U.S. Department of Commerce. *By-Products of the Lumber Industry.* Washington: GPO, 1913. 68 pp. Incidental history.

U40 USDI. *Years of Progress, 1945-1952.* Washington: GPO, 1953. 195 pp. Illus., diags., tables. On the administration of natural resources by the USDI since World War II.

U41 USDI. *Highlights in the History of Forest and Related Natural Resource Conservation.* Conservation Bulletin No. 41. Washington: GPO, 1959. 35 pp. Illus. Since 1783.

U42 U.S. Excess Profits Tax Council. *Lumber Manufacturing Industry, with Special Emphasis on Southern and Ponderosa Pine.* Washington: GPO, 1949. 74 pp. Tables, diags. Investigation of the production and profits cycles of the industry for the purpose of establishing equitable tax policies.

U43 U.S. Federal Trade Commission. *Report of the Federal Trade Commission on the News-Print Paper Industry.* Washington: GPO, 1917. 162 pp. Incidental history.

U44 U.S. Federal Trade Commission. *Report of the Federal Trade Commission on Lumber Manufacturers' Trade Associations.* Washington: GPO, 1922. x + 150 pp. Tables, diags. Includes some history of national and regional trade associations, their influence on legislation, trade and pricing practices, etc.

U45 USFS. *The Timber Supply of the United States.* Circular No. 166. Washington: USFS, 1909. 24 pp.

U46 USFS. "Wood-Using Industries of Georgia." *Lumber Trade Journal* 67 (May 15, 1915), 19-29.

U47 USFS. *Timber Depletion, Lumber Prices, Lumber Exports, and Concentration of Timber Ownership.* Report on Senate Resolution 311, 66th Congress, 2nd Session. Washington: GPO, 1920. 71 pp. Maps, tables, charts. Known generally as the Capper Report, it contains some history of the lumber industry and attributes high lumber prices to the cumulative effects of forest depletion.

U48 USFS. *The National Forests of Arizona.* USDA, Circular No. 318. Washington: GPO, 1924. 18 pp. Illus., map.

U49 USFS. *The National Forests of Idaho.* USDA, Miscellaneous Circular No. 61. Washington: GPO, 1926. 34 pp. Illus., tables, maps.

U50 USFS. *National Forests of Wyoming.* USDA, Miscellaneous Circular No. 82. Washington: GPO, 1927. 26 pp. Illus., maps, table.

U51 USFS. *American Forests and Forest Products.* Statistical Bulletin No. 21. Washington: GPO, 1928. 324 pp. Maps, graphs, tables. Contains historical statistics.

U52 USFS. *A National Plan for American Forestry, Letter from the Secretary of Agriculture, Transmitting in Response to Senate Resolution 175 the Report of the Forest Service . . . on the Forest Problem of the United States.* 2 Volumes. Senate Document 12, 73rd Congress, 1st Session. Washington: GPO, 1933. Maps, graphs, charts, index. Known generally as the Copeland Report, it includes some history of forest resources and their utilization.

U53 USFS. *Florida National Forests.* Washington: GPO, 1939. 46 pp. Illus., map.

U54 USFS. *Sawtooth National Forest.* Washington: GPO, 1939. 14 pp. Illus., map. Idaho.

U55 USFS. *Wyoming National Forest, Wyoming.* Washington: GPO, 1940. 26 pp. Now part of the Bridger National Forest.

U56 USFS. *White River National Forest, Colorado.* Washington: GPO, 1940.

U57 USFS. *Tongass National Forest, Alaska.* Washington: GPO, 1940. 46 pp.

U58 USFS. *Idaho National Forest, Idaho.* Washington: GPO, 1940. 20 pp. Illus., map. Now part of the Payette National Forest.

U59 USFS. *Caribbean National Forest of Puerto Rico.* Washington: GPO, 1940. 29 pp. Illus., map.

U60 USFS. *National Forests in the Southern Appalachians.* Washington: GPO, 1940. 46 pp. Map.

U61 USFS. *Apache National Forest.* Washington: GPO, 1941. 22 pp. Illus., map. Arizona-New Mexico.

U62 USFS. *Golden Anniversary: Shoshone National Forest, Wyoming.* Washington: GPO, 1941. 28 pp. Illus.

U63 USFS. *Superior National Forest, Minnesota.* Washington: GPO, 1941. 36 pp. Illus., map.

U64 USFS. *National Forests of Michigan.* Washington: GPO, 1941. 46 pp.

U65 USFS. *Challis National Forest, Idaho.* Washington: GPO, 1941. 22 pp.

U66 USFS. *San Juan National Forest, Colorado.* Washington: GPO, 1942. 22 pp. Illus., maps.

U67 USFS. *Chippewa National Forest, Minnesota.* Washington: GPO, 1942. 30 pp. Illus., map.

U68 USFS. *Cache National Forest.* Washington: GPO, 1943. 20 pp. Illus., map. Utah-Idaho.

U69 USFS. *Forests and National Prosperity: A Reappraisal of the Forest Situation in the United States.* Miscellaneous Publication No. 668. Washington: GPO, 1948. 99 pp. Illus. Known as the Reappraisal Report, it contains some forest history.

U70 USFS. *Black Hills National Forest, 50th Anniversary.* Washington: GPO, 1948. ii + 43 pp. Illus., tables. History of the Black Hills, Harney, and Sundance national forests of South Dakota and Wyoming since 1898.

U71 USFS. *Payette National Forest.* Washington: GPO, 1950. 20 pp. Illus., map. Idaho.

U72 USFS. *Nebraska National Forest.* Washington: GPO, 1952. 30 pp. Illus., map.

U73 USFS. *Highlights in the History of Forest Conservation.* USDA, Agricultural Information Bulletin No. 83. Washington: GPO, 1952. 30 pp. Tables. A chronology of the forest conservation movement since the colonial period, including dates of establishment and location of national forests. Revised in 1958, 1961, and 1964.

U74 USFS. *Price Trends and Relationships for Forest Products: Letter from Assistant Secretary of Agriculture, Transmitting a Report Pursuant to Section 402 of the Agricultural Act of 1956.* House Document No. 195, 85th Congress, 1st Session. Washington: GPO, 1957. 58 pp. Bib. Statistical trends since 1910 in the lumber industry and since the 1920s and 1930s in other forest products industries.

U75 USFS. *Timber Resources for America's Future.* Forest Resource Report No. 14. Washington: GPO, 1958. 713 pp. Illus., maps, tables, graphs, index. Known as the Timber Review Report, it includes a section on earlier forest surveys and statistics.

U76 USFS. *The John Weeks Story: A Chapter in the History of American Forestry.* Washington: GPO, 1961. 15 pp. Weeks was the sponsor of forestry legislation.

U77 USFS. *The National Forest Reservation Commission: A Report on Progress in Establishing National Forests, Published on the Occasion of the 50th Anniversary of the Weeks Law, 1961.* Washington: GPO, 1961. 27 pp. Map, apps. Includes some history of federal forestry since the 1870s.

U78 USFS. *The Forest Products Laboratory: A Brief Account of Its Work and Aims.* USDA, Agricultural Information Bulletin No. 105. Washington: GPO, 1962. 29 pp.

U79 USFS. *The Story of Green Mountain National Forest in Vermont.* Washington: GPO, 1962. 20 pp. Illus., tables.

U80 USFS. *The Story of Jefferson National Forest.* Washington: GPO, n.d. 29 pp. Illus., graphs, tables. Virginia.

U81 USFS. *This is the Monongahela National Forest in West Virginia.* Washington: GPO, 1964. 30 pp. Illus., graphs, tables.

U82 USFS. *Court Cases Related to Administration of the Range Resource on Lands Administered by the Forest Service.* Washington: GPO, 1964. 121 pp.

U83 USFS. *Timber Trends in the United States.* Forest Resource Report No. 17. Washington: GPO. 1965. viii + 235 pp. Illus., tables, graphs, index. Includes some statistics on timber supply and demand since 1905.

U84 USFS. *The Outlook for Timber in the United States.* Forest Resource Report No. 20. Washington: GPO, 1973. xiii + 367 pp. Illus., charts, tables, notes. An analysis of the contemporary timber situation including some incidental historical references.

U85 USFS. *The Principal Laws Relating to Forest Service Activities.* USDA Handbook No. 453. Washington: GPO, 1974. ix + 265 pp. Public laws since 1891 are arranged by subject and chronologically. There are earlier editions under different titles.

U86 USFS. Alabama National Forests. "The Alabama National Forests." *Alabama Historical Quarterly* 14 (1952), 47-64. On the William B. Bankhead, Conecuh, and Talladega national forests since 1918.

U87 USFS. Division of Engineering. *Engineering in the Forest Service: (A Compilation of History and Memoirs), 1905-1969.* Washington: USFS, 1970. Illus. Processed.

U88 USFS. Division of Forest Economics. *Materials Survey: Lumber, Railroad Ties, Veneer and Plywood, Poles, and Piles.* Washington, 1950. xii + 253 pp. Diags., map, tables, bib. Statistics on production, imports, exports, stocks, and consumption since 1935.

U89 USFS. Division of Range Research. "The History of Western Range Research." *Agricultural History* 18 (July 1944), 127-43. Federal and other range research since 1817.

U90 USFS. Forest Products Laboratory. *The Forest Products Laboratory: A Decennial Record, 1910-1920.* Madison, 1921. 196 pp. Illus., app. Madison, Wisconsin.

U91 USFS. Forest Products Laboratory. *Toward Wiser Use of Wood: A Mid-Century Tribute to the Men and Women Who Conceived and Founded the U.S. Forest Products Laboratory and Guided It Through Its First 40 Years.* Madison: By the employees of the Laboratory, 1952. 36 pp.

U92 USFS. Forest Products Laboratory. *50 Years of Service Through Wood Research, 1910-1960; Golden Anniversary, Forest Products Laboratory.* Miscellaneous Publication No. 820. Washington: GPO, 1960. 16 pp.

U93 USFS. Intermountain Region. *History, Payette National Forest.* Ogden, Utah, 1968. 175 pp. Map, tables, bib. Idaho.

U94 USFS. Lake State Forest Experiment Station. *Possibilities of Shelterbelt Planting in the Plains Region: A Study of Tree Planting for Protective and Ameliorative Purposes as Recently Begun in the Shelterbelt Zone of North and South Dakota, Nebraska, Kansas, Oklahoma, and Texas by the Forest Service; Together with Information as to Climate, Soils, and other Conditions Affecting Land Use and Tree Growth in the Region.* Washington: GPO, 1935. 201 pp. Illus., maps, diags. Includes some history of tree planting in the region.

U95 USFS. National Forests in Mississippi. *The 40th Year for the National Forests in Mississippi.* Jackson, 1973. On the recovery of forests from barren, eroded, and burned-over lands, 1933-1970s.

U96 USFS. North Central Forest Experiment Station. *Rebuilding a Resource: 50 Years of Forestry Research.* St. Paul, 1973. 15 pp. Illus. Highlights of the station's history since 1923.

U97 USFS. North Pacific Region. *Industrial Development of Olympic Peninsula Counties.* Portland, 1937. On the effects of forest devastation upon the lumber towns of northwestern Washington.

U98 USFS. Northern Region. *Early Days in the Forest Service.* 3 Volumes. Missoula, Montana, 1944-1962. Illus., Processed. Reminiscences of foresters and other USFS employees in Montana, northern Idaho, and eastern Washington.

U99 USFS. Northern Region, Division of Fire Control. *History of Smokejumping.* Missoula, Montana, 1972. 35 pp. Illus., table. Some history of smokejumping and aerial fire control, especially on the national forests of the West, since the 1920s.

U100 USFS. Ouachita National Forest. *Ouachita National Forest: Arkansas-Oklahoma, 40th Anniversary, 1907-1947.* Hot Springs, Arkansas, 1947. 23 pp. Graphs, tables. Processed.

U101 USFS. Rogue River National Forest. *History of the Rogue River National Forest.* 2 Volumes. Medford, Oregon, 1960, 1969.

U102 USFS. Southeastern Forest Experiment Station. *Anniversary Report, 1921-1946: Twenty-Five Years of Forest Research at the Appalachian Forest Experiment Station.* Asheville, North Carolina, 1946. 71 pp.

U103 USFS. Southern Region. *National Forests of the Southern Region: A Report of Progress, 1934-1954.* Atlanta, 1954. 23 pp. Illus., graphs.

U104 USFS. Uinta National Forest. *Utah's First Forest's First 75 Years.* Provo, Utah, 1972. 72 pp. Illus., bib., index. Since 1897.

U105 U.S. General Land Office. *Public Land System of the United States: Historical Outline.* Washington: GPO, 1924. 18 pp.

U106 U.S. National Conservation Commission. *Report of the National Conservation Commission and a Chronological History of the Conservation Movement.* National Conservation Commission, Bulletin No. 4. Washington: Joint Committee on Conservation, 1909. 52 pp.

U107 U.S. National Forest Reservation Commission. *Review of the Work of the National Forest Reservation Commission, 1911-1933.* Washington: GPO, 1933. 21 pp. Illus.

U108 U.S. National Forest Reservation Commission. *Report on Land Planning.* Washington: GPO, 1938. Includes a section on the history of recreation on public lands.

U109 U.S. National Forest Reservation Commission. *A Report on Progress in Establishing National Forests.* Washington: GPO, 1961. 27 pp. Illus., map, tables. Since established by the Weeks Act in 1911.

U110 U.S. National Park Service. *A Bibliography of National Parks and Monuments West of the Mississippi River.* 2 Volumes. Berkeley, 1941. Processed.

U111 U.S. National Park Service. *Study of the Park and Recreation Problems of the U.S.* Washington: GPO, 1941. Incidental history.

U112 U.S. National Park Service. *Grand Teton National Park and Jackson Hole National Monument, Wyoming.* Washington: GPO, 1948. 16 pp. Illus., map. Incidental history.

U113 U.S. National Park Service. *Yellowstone National Park, Wyoming.* Washington: GPO, 1940. 15 pp. Illus., map. Incidental history.

U114 U.S. National Park Service. *Your National Parks: Brief History, with Questions and Answers.* Washington, 1961. 18 pp. Illus.

U115 U.S. National Park Service. *The Redwoods, A National Opportunity for Conservation and Alternatives for Action.* Washington, 1964. 52 pp. Includes some history of redwood logging and preservation.

U116 U.S. National Park Service. *Proposed Voyageurs National Park, Minnesota.* Omaha, 1964. 43 pp. Illus., maps. Incidental history.

U117 U.S. National Recovery Administration. Division of Economic Research. *Statistical and Economic Material Bearing on the Lumber and Timber Producing Industries.* Washington: GPO, 1933. 115 pp. Illus.

U118 U.S. Outdoor Recreation Resources Review Commission. *Outdoor Recreation for America: A Report to the President and to the Congress.* Washington: GPO, 1962. 245 pp. Illus., tables, diags. Incidental history.

U119 U.S. Outdoor Recreation Resources Review Commission. *Wilderness and Recreation—A Report on Resources, Values, and Problems.* ORRRC Study Report No. 3. Washington: GPO, 1962. 352 pp. Contains an inventory of sixty-four wilderness areas, with some historical references.

U.S. Public Land Law Review Commission. See Gates, Paul W., #G63

U120 U.S. Public Land Law Review Commission. *One Third of the Nation's Land: A Report to the President and to the Congress.* Washington, 1970. xiii + 342 pp. Illus., maps, apps., index. Includes some history of federal land policy.

U121 U.S. Public Lands Commission. *Report of the Public Lands Commission, with Appendix.* Senate Document No. 189, 58th Congress, 3rd Session. Washington: GPO, 1905. On the operation and results of public land laws, including state-by-state information on the disposition of the public lands to 1904.

U122 U.S. Soil Conservation Service. *The CCC at Work, A Story of 2,500,000 Young Men.* Washington: GPO, 1942. 103 pp. Prepared jointly with the USFS, this contains some history of the Civilian Conservation Corps.

U123 U.S. Tennessee Valley Authority. *Tennessee Valley Resources, Their Development and Use.* Knoxville, 1947. 145 pp. Includes a section on the restoration of forest resources in the Tennessee Valley area.

U124 U.S. Tennessee Valley Authority. Division of Forestry Development. *TVA and Forestry, Twentieth Century Goals.* Norris, Tennessee, 1962. 42 pp. On the development of forests, fish, and wildlife in the Tennessee Valley.

U125 U.S. Work Projects Administration, Michigan. *Michigan Log Marks, Their Function and Use During the Great Michigan Pine Harvest.* Michigan Agricultural Experiment Station, Memoir Bulletin No. 4. Lansing, 1941. 89 pp. Illus. On the use of log marks, especially in the 19th century.

U126 U.S. Work Projects Administration, Minnesota. *Logging Town: The Story of Grand Rapids, Minnesota.* Grand Rapids: Village of Grand Rapids, 1941. 77 pp. Illus., tables, index. Since the 1870s.

U127 U.S. Work Projects Administration, Utah. *Forest Conservation in Utah, Weber County.* Ogden, Utah, 1942. 95 pp. Illus., map, bib. Processed. Includes some history of the subject.

U128 United States Pulp Producers Association. *Wood Pulp: A Basic Fiber; the Story of the Origin, Development and Economic Status of a Great American Industry.* New York, 1955. 48 pp.

U129 University of the South. Forestry Department. *The University of the South Forest: A Demonstration of Practical Applied Forestry for Student Instruction and Regional Use.* Sewanee, Tennessee, 1953. 124 pp. On management of the 8,220-acre tract since 1857.

U130 Urquhart, R.G. "Nova Scotia Domain of the Portable Sawmill; Seven Hundred Portable Mills in Nova Scotia Keep Lumbering an Important Industry." *Canada Lumberman* 50 (Aug. 1, 1930), 131-32. Incidental history.

U131 Usborne, John. "Forests for the News: Relationship between a Large Industrial Corporation and Two Environments in Which It Operates: Newfoundland, Tennessee." *Geographical Magazine* 30 (May 1957), 44-54. On the Bowaters Newfoundland Pulp & Paper Mills, Ltd., Corner Brook, Newfoundland, and Bowaters Southern Paper Corporation, Calhoun, Tennessee.

U132 Usher, Ellis B. *Wisconsin: Its History and Biography, 1848-1913.* 8 Volumes. Chicago: Lewis Publishing Company,

1914. Includes some history of lumbermen and the lumber industry.

U133 Usher, Ellis B. "Cyrus Woodman, A Character Sketch." *Wisconsin Magazine of History* 2 (June 1919), 393-412. Woodman (1814-1889) was a dealer in Wisconsin pine lands.

U134 Utley, Robert M. "Historic Preservation and the Environment." *Colorado Magazine* 51 (Winter 1974), 1-12.

V1 Vagnarelli, Adelaide N. "A History of Lumbering in the Ausable Valley, 1800-1900: A Critical Essay." Master's thesis, Cornell Univ., 1944. New York.

Vail, William C. See Walls, Sara, #W49

V2 Vale, Robert B. "The Return of Penn's Woods." *American Forests and Forest Life* 31 (Aug. 1925), 463-65. Includes some history of Pennsylvania's state forests.

V3 Valentine, W. Cullen. "Fire-Fighting Equipment Was Made, Not Born." *Forests & People* 2 (Second quarter, 1954), 23-26, 42-43. On the evolution of fire fighting equipment in Louisiana.

V4 Valentine, W. Cullen. "From Fire Flaps to Superplows." *Forests & People* 13 (First quarter, 1963), 40-41, 102-03. Fire fighting equipment in Louisiana since 1913.

Van Arnam, Lewis S. See Van Arnam, Ralph N., #V5

V5 Van Arnam, Ralph N., and Van Arnam, Lewis C. "Dannatburg: Lewis County Ghost Town." *North Country Life* 4 (Spring 1950), 6-11. On a northern New York lumber town, originally named Crandallville, 1855-1900.

V6 Van Atta, Robert B. "Jefferson County's First Settlers and First Community." *Western Pennsylvania Historical Magazine* 56 (No. 2, 1973), 129-39. The lumbering operations of Joseph Barnett and family made Port Barnett a business center in western Pennsylvania in the 18th and 19th centuries.

V7 Van Blaricom, G.B. "Past and Present: Some Milestones Which Mark Fifty Years' Service of 'Canada Lumberman' to Canadian Forest Products Industry." *Canada Lumberman* 50 (Aug. 1, 1930), 87-88.

V8 Van Brocklin, Ralph M. "The Movement for the Conservation of Natural Resources in the United States before 1901." Ph.D. diss., Univ. of Michigan, 1953. 269 pp. A survey of conservation since the 17th century.

V9 Van Camp, J.L. "Fighting Forest Fires for Forty Years." *Forest and Outdoors* 51 (Feb. 1955), 11-13. Canada.

V10 Vance, J.A. "Story Behind the Canadian Tree Farm Movement, How It Started, What It Is." *Forest and Outdoors* 51 (Feb. 1955), 18, 22.

V11 Vance, Maurice M. "Charles Richard Van Hise: A Biography." Ph.D. diss., Univ. of Wisconsin, 1953. 318 pp.

V12 Vance, Maurice M. *Charles Richard Van Hise: Scientist Progressive.* Madison: State Historical Society of Wisconsin 1960. 246 pp. Illus., notes, bib. Van Hise (1857-1918), a Wisconsin geologist and educator, wrote the first general text on conservation and was otherwise involved in the natural resource field.

V13 Vance, Rupert B. *Human Geography of the South; a Study in Regional Resources and Human Adequacy.* 1932. Second Edition. Chapel Hill: University of North Carolina Press, 1935. xviii + 596 pp. Illus. Includes some history of forests and forest industries.

Van Den Berge, J. See Van Groenou, H. Broese, #V22

V14 Vanderhill, Burke G. "Settlement in the Forest Lands of Manitoba, Saskatchewan, and Alberta: A Geographic Analysis." Ph.D. diss., Univ. of Michigan, 1956. 315 pp. Since 1670.

V15 Vandermillen, Edmund J. "Forestry in Connecticut: Some History, Progress and Problems." *Connecticut Woodlands* 38 (Summer 1973), 3-7.

V16 Van Dersal, William R. See Graham, Edward H., *History and Its Uses.* New York: Oxford University Press, 1943. xvi + 215 pp. Includes some history of forests and other natural resources.

V17 Van Dersal, William R., and Graham, Edward H. *The Land Renewed: The Story of Soil Conservation.* New York: Oxford University Press, 1946. 109 pp.

Van Dersal, William R. See Graham, Edward H., #G231

Vandersluis, Charles. See Bourgeois, Euclid J., #B372

Van der Zee, John. See Wilkerson, Hugh, #W296

Van Doren, Charles. See McHenry, Robert, #M97

Van Doren, Mark. See Bartram, William, #B117

V18 Van Dresser, Cleveland. "Harvesting Timber on Wildlife Refuge." *American Forests* 59 (Nov. 1953), 18-19, 91. On the timber program of the U.S. Fish and Wildlife Service on Georgia's Piedmont National Wildlife Refuge since 1944.

V19 Van Duers, George. "The Russians Logged the Redwoods First." *American Forests* 65 (Jan. 1959), 29-31, 61-62. On Fort Ross and adjacent buildings near Bodega Bay, California, a trading post of the Russian American Company, 1812-1842.

V20 Vanek, Jaroslav. *The Natural Resource Content of United States Foreign Trade, 1870-1955.* Cambridge: M.I.T. Press, 1963. xvi + 142 pp. Tables, graphs, notes, index. A theoretical and statistical analysis.

V21 Van Epps, Percy M. " 'Jones's Boys They Built a Mill': Folk History of the Evas Kill." *New York Folklore Quarterly* 3 (Autumn 1947), 231-36. On the establishment of a sawmill near Schenectady, New York, in ca. 1700, including personal reminiscences of the author.

V22 Van Groenou, H. Broese; Rischen, H.W.L.; and Van Den Berge, J. *Wood Preservation during the Last 50 Years.* Leiden, Netherlands: A.W. Sijthoff's, 1951. xii + 318 pp. Illus., tables, graphs, bib., index. Including in the United States and Canada.

V23 Vangsness, G.A. "The Lumberjack." *Southern Lumberman* 169 (Dec. 15, 1944), 179-80. On working conditions and logging methods in the 19th century.

V24 Van Hise, Charles Richard. *The Conservation of Natural Resources in the United States.* New York: Macmillan, 1910. xiv + 413 pp. Illus., maps, notes, apps., index. The first general text on conservation; includes a large section on forest conservation.

V25 Van Holmes, Jeanne. "The Big Cypress." *American Forests* 61 (Oct. 1955), 28-31. On the Lee Tidewater Cypress Company in southwest Florida since ca. 1880, and the preservation of several hundred acres of its former holdings as a "museum stand."

V26 Van Kirk, Sylvia. "The Development of National Park Policy in Canada's Mountain National Parks, 1885 to 1930." Master's thesis, Univ. of Alberta, 1969. Banff, Jasper, Yoho, Kootenay, Glacier, and Waterton Lakes national parks, Alberta-British Columbia.

V27 Van Name, Willard Gibbs. *Vanishing Forest Reserves: Problems of the National Forests and National Parks.* Boston: Richard G. Badger, 1929. 190 pp. Illus.

Van Noppen, Ina Woestemeyer. See Van Noppen, John, #V28

V28 Van Noppen, John, and Van Noppen, Ina Woestemeyer. "The Genesis of Forestry in the Southern Appalachians: A Brief History." *Appalachian Journal* 1 (Autumn 1972), 63-71. On the lumber industry and the beginnings of forestry at George Vanderbilt's Biltmore Estate near Asheville, North Carolina, 1890s-1900s. Includes some account of the roles of Gifford Pinchot and Carl Alwin Schenck.

Van Ravenswaay, Charles. See Whitehill, John, #W255

V29 Van Ravenswaay, Charles. "America's Age of Wood." *Proceedings of the American Antiquarian Society* 80 (1970), 49-66. On the importance of wood and wood technology in preindustrial America.

V30 Van Ravenswaay, Charles. "A Historical Checklist of the Pines of Eastern North America." *Winterthur Portfolio* 7 (1971), 175-215. Includes some history of pines used in the furniture, shipbuilding, and other wood-using industries.

V31 Van Tassel, Alfred J., and Bluestone, David W. *Mechanization in the Lumber Industry: A Study of Technology in Relation to Resources and Employment Opportunity.* National Research Project Report, No. M-5. Philadelphia: Work Projects Administration, 1940. 201 pp. Illus., maps, tables, charts, graphs, notes, apps. Especially in the South and the Pacific states.

V32 Van Tassel, Alfred J. "The Influence of Changing Technology and Resources on Employment in the Lumber Industry of the Pacific Northwest." Ph.D. diss., Columbia Univ., 1964. 411 pp.

V33 Van Zandt, Roland. "The Catskills and the Rise of American Landscape Painting." *New-York Historical Society Quarterly* 49 (July 1965), 257-81. Thomas Cole and the Hudson River School of American landscape painters derived inspiration from the forested Catskill Mountains of New York, 1820s and later.

V34 Varney, Charles B. "Economic and Historical Geography of the Gulf Coast of Florida: Cedar Keys to St. Marks." Ph.D. diss., Clark Univ., 1963. 340 pp. Includes some history of forest industries.

V35 Varossieau, W.W. *Forest Products Research and Industries in the U.S.* Amsterdam: J.M. Meulenhoff, 1954. 796 pp. A report on American research and methods commissioned by the Technical Assistance Branch, European Recovery Administration.

V36 Varvel, Carl D. "Forestry Legislation in Ohio, 1885-1929." *Ohio Social Science Journal* 2 (May 1930), 58-64.

V37 Vassault, F.I. "Lumbering in Washington." *Overland Monthly* 20 (1892), 23-32. Incidental history.

V38 Vastokas, Joan M. "Architecture and Environment: The Importance of the Forest to the Northwest Coast Indian." *Forest History* 13 (Oct. 1969), 12-21. Pacific Northwest, British Columbia, and Alaska.

V39 Vatter, Barbara Amy. "A Forest History of Douglas County, Oregon, to 1900." Ph.D. diss., Univ. of Minnesota, 1971. 454 pp.

V40 Vaughan, Thomas J. "Life of the Wisconsin Lumberjack 1850-1880." Master's thesis, Univ. of Wisconsin, 1951.

V41 Vaughan, Thomas J., and Ferriday, Virginia Guest, eds. *Space, Style and Structure: Building in Northwest America.* Portland: Oregon Historical Society, 1974. 2 Volumes. Illus., maps, bib., index. Historical essays on building construction and architecture in the Pacific Northwest. The central theme is the tradition of wood in Northwest buildings.

Vaux, Henry J. See Duerr, William A., #D284, D286

V42 Vaux, Henry J. "How Resource Decisions Are Made." *Forest History* 11 (Apr. 1967), 32-38. Reflections on the process of making forest and resource policies.

V43 Veitschegger, Rodney Dean. "Forest Taxation in Washington State." Master's thesis, Univ. of Washington, 1959.

Verardo, Denzil. See Lowry, Alexander, #L284

Verner, William K. See Masten, Arthur H., #M287

V44 Verner, William K. "Wilderness and the Adirondacks—An Historical View." *Living Wilderness* 33 (Winter 1969), 27-47. Includes some history of the movement to preserve Adirondack State Park in northern New York.

V45 Vernon, Charles C. "A History of the San Gabriel Mountains, Part IV." *Historical Society of Southern California Quarterly* 38 (Dec. 1956), 373-84. On forestry and recreational development in the area since 1892.

V46 Vessel, Matt Frank. "A Study of Conservation Education in the Rural Areas of the U.S." Ph.D. diss., Cornell Univ., 1941. 256 pp.

Vetleson, Jack. See Lutz, George, #L320

V47 Vezina, Rosemarie. *Nevers Dam: The Lumbermen's Dam.* St. Croix Falls, Wisconsin: Standard-Press, 1965. 38 pp. On the St. Croix River, Wisconsin-Minnesota.

V48 Victor, Arthur E. "Fred Herrick and Bill Grotte: Idaho's Paul Bunyan and His Bull of the Woods." *Pacific Northwesterner* 16 (Summer 1972), 33-48. Reminiscence of logging for Herrick and Grotte near St. Maries and in the St. Joe Valley of northern Idaho, 1920s.

V49 Victor, Arthur E. "The Ubiquitous Bulldozer." *Pacific Northwesterner* 19 (Winter 1975), 1-8. Reminiscence of USFS and CCC use of bulldozers in Washington, Idaho, and Montana, 1930s.

V50 Vincent, Joseph J. *Streak o'Lean and a Streak o'Fat.* Tampa: Southern Historical Associates, 1953. ix + 109 pp. Illus., map. Includes some reminiscences of the lumber industry in southeastern Texas and southwestern Louisiana, late 19th century.

V51 Vincent, Paul Y. "Grazing on Texas National Forests." *Journal of Forestry* 50 (Mar. 1952), 214-15. Since 1937.

V52 Vincent, Paul Y. "Conservation of Timber—and Game." *Southern Lumberman* 193 (Dec. 15, 1956), 252-53. By the USFS in the South since 1933.

V53 Vinette, Bruno. "Early Lumbering on the Chippewa." *Wisconsin Magazine of History* 9 (June 1926), 442-47.

V54 Vining, Charles. *Newsprint Prorating: An Account of Government Policy in Quebec and Ontario.* Montreal, 1940.

V55 Vinnedge, Robert W. *The Pacific Northwest Lumber Industry and Its Development.* Lumber Industry Series, No. 4. New Haven: Yale University, School of Forestry, 1923. 26 pp. A summary with emphasis on the struggle for markets since the 1850s; also appeared in *Lumber World Review* 45 (Dec. 25, 1923).

V56 Vinnedge, Robert W. "The Genesis of the Pacific Northwest Lumber Industry and Its Development." *Timberman* 35 (Dec. 1933).

V57 Vinson, Frank B. "Conservation and the South, 1890-1920." Ph.D. diss., Univ. of Georgia, 1971. 378 pp. On individuals and organizations which advocated forest conservation, wildlife protection, swamp drainage, and multiple-purpose waterways development.

Vinten, C.R. See Beard, Daniel B., #B147

Viorst, Milton. See Anderson, Clinton P., #A427

V58 Virginia. Commission on Forest Resources. *Forest Resources of Virginia.* Richmond, 1955. 74 pp. Incidental history.

V59 Virginia Academy of Science. James River Project Committee. *The James River Basin, Past, Present, and Future.* Richmond, 1950. xxxi + 843 pp. Illus., maps, tables. There are chapters on wildlife, conservation, and forests and forestry.

V60 Virtue, G.O. "The Co-operative Coopers of Minneapolis." *Quarterly Journal of Economics* 19 (Aug. 1905), 527-44. On a movement for cooperation in production, 1886-1905.

V61 Visher, Stephen S. "Conservation Progress in Indiana." *Proceedings of the Indiana Academy of Science* 65 (1956), 198-99. Since 1908.

V62 Voelker, Keith E. "The History of the International Brotherhood of Pulp, Sulphite, and Paper Mill Workers from 1906 to 1929: A Case Study of Industrial Unionism before the Great Depression." Ph.D. diss., Univ. of Wisconsin, 1960. 406 pp.

V63 Vogel, John H. "The Pulp and Paper Industry." In *The Development of American Industries: Their Economic Significance,* ed. by John G. Glover and Rudolph L. Lagai. Fourth edition. New York: Simmons-Boardman, 1959. Pp. 376-93. Sponsored by the American Pulp and Paper Association, this article concerns developments in the industry since 1690.

Vogenberger, Ralph A. See Lehman, John W., #L139

V64 Vogt, William. *Road to Survival.* New York: William Sloane Associates, 1948. xvi + 335 pp. Notes, index. A general account of world resource problems, including some historical references to forests.

V65 Voigt, William, Jr. "The Izaak Walton League of America." *Journal of Forestry* 44 (June 1946), 424-25. On the history and activities of this conservation organization since its founding in 1922.

V66 Volkober, John Anton, Jr. "Factors Affecting the Competitive Position of the Pacific Northwest and British Columbia in United States Lumber Markets: Regional Advantage." Master's thesis, Reed College, 1967.

von Schrenk, Hermann. See Schrenk, Hermann von, #S116, S117

V67 Voorhis, Manning Curlee. "The Land Grant Policy of Colonial Virginia, 1607-1774." Ph.D. diss., Univ. of Virginia, 1940. 114 pp.

V68 Voss, Ray. "It's Your Land." *Michigan Conservation* 21 (Mar.-Apr. 1952), 12-16. On recreation areas in Michigan owned by the state and federal governments, since 1909.

V69 Voss, Walter A. "Colorado and Forest Conservation." Master's thesis, Univ. of Colorado, 1931.

W1 Wacker, Peter O. "Forest, Forge, and Farm: An Historical Geography of the Musconetcong Valley, New Jersey." Ph.D. diss., Louisiana State Univ., 1966. 364 pp.

W2 Wacker, Peter O. *The Musconetcong Valley of New Jersey: A Historical Geography.* New Brunswick, New Jersey: Rutgers University Press, 1968. xi + 207 pp. Illus., maps, apps., notes, bib., index. Includes a chapter on the charcoal iron industry and other historical references to forest industries in northwestern New Jersey.

W3 Waddell, R.D. "Early Day Wood Pipe." *Timberman* 29 (Aug. 1928), 184. As used in the Chicago water supply system in the 1830s.

Wade, B. See Sharp, S.J., #S187

W4 Wade, Curtis. "Story of the Shingle." *Portland Shingle News* 1 (May-Dec. 1948). On the shingle industry in the Pacific Northwest.

W5 Wadsworth, Frank H. "The Development of the Forest Land Resources of the Luquillo Mountains, Puerto Rico." Ph.D. diss., Univ. of Michigan, 1949.

W6 Wadsworth, Frank H. "Notes on the Climax Forests of Puerto Rico and Their Destruction and Conservation Prior to 1900." *Caribbean Forester* 11 (1950), 38-47.

W7 Wagar, J.V.K. "Some Major Principles in Recreation Land-Use Planning." *Journal of Forestry* 49 (June 1951),

431-35. Based largely on the history and literature of American forestry since the 1890s.

W8 Wagar, J.V.K. "An Analysis of the Wilderness and Natural Area Concept." *Journal of Forestry* 51 (Mar. 1953), 178-83. Since the 19th century.

W9 Wagar, J.V.K., and Steinhoff, H.W. "Often Overlooked Foundations of Conservation Education." *Transactions of the North American Wildlife Conference* 21 (1956), 567-75. On conservation legislation from 1872 to 1935.

W10 Wagener, Willis W. "Past Fire Incidence in Sierra Nevada Forests." *Journal of Forestry* 59 (Oct. 1961), 739-48. Since the early 19th century, based on decay studies and other historical and anthropological sources.

W11 Wager, Paul W., and Hayman, Donald B. *Resource Management in North Carolina: A Study in Public Administration.* Chapel Hill: University of North Carolina, Institute for Research in Social Sciences, 1947. x + 192 pp. Maps, tables, notes, index. Includes some history and description of federal, state, and local administration of natural resources, including forests.

W12 Waggener, Thomas R. "The Federal Land Grant Endowments: A Problem in Forest Resource Management." Ph.D. diss., Univ. of Washington, 1966. 258 pp. Includes some history of these lands in Washington.

Waggoner, Paul E. See Stephens, George R., #S583

W13 Wagner, Jack R. "California Western." *American Forests* 54 (Sept. 1948), 408-09, 418-19. On the California Western Railroad, a logging line between Fort Bragg and Willits, since 1911.

W14 Wagner, Jack R. *McCloud River Railroad.* San Mateo, California: Francis A. Guido, 1953. 16 pp. Illus., map. A logging railroad between Mount Shasta City and Hambone, California, 1897-1938. Reprinted from *Western Railroader* 16 (Aug. 1953).

W15 Wagner, Jack R. "Railroad Through the Redwoods." *Railroad Magazine* 61 (Aug. 1953), 32-49. On the Northwestern Pacific Railroad, California, since 1862.

W16 Wagner, Jack R. *Short Line Junction: A Collection of California-Nevada Railroads.* Fresno: Academy Library Guild, 1956. 266 pp. Illus. Includes some history of logging railroads.

W17 Wagner, Joseph B. *Cooperage: A Treatise on Modern Shop Practice and Methods; from the Tree to the Finished Article.* Yonkers, New York, 1910. 396 pp. Illus. Incidental history.

W18 Wagoner, Ed. "The Green Gold of Texas." *American Forests* 70 (Sept. 1964), 26-29, 61-63. A brief history of forestry in Texas with emphasis on the role of the Texas Forestry Association since 1914.

W19 Wainwright, K.B. "A Comprehensive Study of Nova Scotian Rural Economy, 1788-1872." *Collections of the Nova Scotia Historical Society* 30 (1954), 78-119. Including forest industries.

W20 Wakeley, Philip C. "Forest Tree-Improvement Work in the South." *Southern Lumberman* 195 (Dec. 15, 1957), 126-29. Since 1920.

W21 Wakeley, Philip C. "The Ups and Downs of Pioneer Planting Research in Louisiana – Problems in the Design and Analysis of Planting Trials." *Forestry Chronicle* 43 (June 1967), 135-44.

W22 Wakeley, Philip C. "The South's First Big Plantation." *Forests & People* 23 (Second quarter, 1973), 26-28, 30-31. Of the Great Southern Lumber Company, Bogalusa, Louisiana, since 1920.

W23 Waldron, Richard S. "A Forest Enterprise." *Northeastern Logger* 4 (June 1956), 18-19, 46-47, 52. On the Chadbourne Lumber Company of East Waterford, Maine, a family-owned business since 1900.

W24 Wales, H.B. "National-Forest Land Management in Michigan." *Papers of the Michigan Academy of Science, Arts, and Letters* 26 (1941), 143-52.

Walker, C.E. See Barraclough, K.E., #B94

W25 Walker, Clyde M. "A Forest Restored." *Journal of Forestry* 71 (Oct. 1973), 648-49. On the Tillamook Burn of western Oregon, a vast region burned by forest fires, 1933-1945. Rehabilitated by the Oregon Department of Forestry, the area was dedicated in 1973 as Tillamook State Forest.

W26 Walker, Helen M. *CCC Through the Eyes of 272 Boys: A Summary of a Group Study of the Reactions of 272 Cleveland Boys to Their Experiences in the Civilian Conservation Corps.* Cleveland: Western Reserve University Press, 1938. 94 pp. Tables.

W27 Walker, Herman, Jr., and Parks, W. Robert. "Soil Conservation Districts: Local Democracy in a National Program." *Journal of Politics* 8 (Nov. 1946), 538-49. Incidental history of soil conservation districts.

W28 Walker, James Herbert, ed. *Rafting Days in Pennsylvania.* Comp. by John C. French, John H. Chatham, Mahlon J. Colcord, Albert D. Karstetter, and others. Altoona, Pennsylvania: Times-Tribune Co., 1922. 122 pp. Illus. Rafting of logs and lumber.

Walker, James Herbert. See Korson, George G., #K236

K29 Walker, John B. "Three Centuries of North Carolina Pine." *American Lumberman* (May 4-Oct. 26, 1907). Twenty weekly installments.

W30 Walker, Joseph E. "Hopewell Village: Some Aspects of the Social and Economic History of an Iron-Making Community with Special Emphasis upon the Period 1800-1850." Ed. D. diss., Temple Univ., 1964. 512 pp.

W31 Walker, Joseph E. *Hopewell Village: A Social and Economic History of an Iron-Making Community.* Philadelphia: University of Pennsylvania Press, 1966. 526 pp. Illus., notes, tables, apps., bib., index. The charcoal iron industry in Hopewell Village, Pennsylvania, especially during the early 19th century.

W32 Walker, Joseph E. "Negro Labor in the Charcoal Iron Industry of Southeastern Pennsylvania." *Pennsylvania*

Magazine of History and Biography 93 (Oct. 1969), 466-86. In the 18th and early 19th centuries.

W33 Walker, Laurence C. *Ecology and Our Forests.* Cranbury, New Jersey: A.S. Barnes, 1972. 175 pp. Illus. Includes some history of forest ecology.

W34 Walker, Platt B., comp. *Sketches of the Life of Honorable T.B. Walker: A Compilation of Biographical Sketches by Many Authors.* Minneapolis: Lumberman Publishing Company, 1907. 131 pp. Illus. Thomas Barlow Walker was a prominent lumberman in Minneapolis. His Red River Lumber Company operated large mills at Akeley, Minnesota, and Westwood, California.

W35 Walker, Thomas. "The Chief Foresters and American Forest Policy, 1898-1928." Master's thesis, Univ. of Houston, 1967.

W36 Wall, Brian R. *Log Production in Washington and Oregon: An Historical Perspective.* Resource Bulletin No. 42. Portland: USFS, Pacific Northwest Forest and Range Experiment Station, 1972. 89 pp. Illus. Examines trends in log production and forest land ownership in Oregon and Washington (by region, state, and county) from 1925 to 1970.

W37 Wall, Brian R. *Relationship of Log Production in Oregon and Washington to Economic Conditions.* Research Paper No. 147. Portland: USFS, Pacific Northwest Forest and Range Experiment Station, 1972. 13 pp. On the relationship between national demand for wood products (especially as shown in housing starts) and log production in Oregon and Washington, 1949-1969.

W38 Wall, Geoffrey. "Pioneer Settlement in Muskoka." *Agricultural History* 44 (Oct. 1970), 393-400. Includes some history of lumbering and farming in the Muskoka region of Ontario during the late 19th century.

Wall, Louise Herrick. See Lane, Franklin K., #L36

W39 Wall, Ralph T. "Milestones in Louisiana Forestry." *Forest Farmer* 26 (Apr. 1967), 8-9, 24-26, 36-40.

W40 Wallace, Andrew. "To the Memory of Will Croft Barnes, 1858-1936." *Arizona and the West* 2 (Autumn 1960), 203-04. On his life in Arizona as a soldier, cattleman, writer, and grazing inspector for the USFS. From 1915 to 1928 he directed the USFS Branch of Grazing.

W41 Wallace, Frederick W. "Ships of the Timber Trade." *Canadian Geographical Journal* 11 (1935), 3-14. On the timber trade and the Canadian merchant marine.

W42 Wallace, Joseph H. "Old Timer's Reminiscences." *Paper Trade Journal* 92 (Jan. 29, 1931-Apr. 16, 1931). Wallace was a paper mill engineer in Holyoke, Massachusetts, in the 1880s, and designed the first sulphate mill in North Carolina.

W43 Wallace, Joseph H. "The Beginning of the Kraft Paper Industry in the Southern States." *Paper Trade Journal* 103 (July 23, 1936), 21, 24. North Carolina.

W44 Wallace, Tom. "A State That Abandoned Its Forests." *American Forestry* 39 (May 1923), 277-82. Includes some historic references to forest uses in Kentucky.

W45 Wallen, Arnold F. "Jackson Forest." *American Forests* 53 (Mar. 1947), 122-23, 142. On Jackson State Forest in Mendocino County, California, including some history of logging by the Caspar Lumber Company on the site since 1861.

W46 Waller, Robert Alfred. "Business Reactions to the Teapot Dome Affair: 1922 to 1925." Master's thesis, Univ. of Illinois, 1958.

W47 Wallis, Hugh M. "James Wallis, Founder of Fenelon Falls and Pioneer in the Early Development of Peterborough." *Ontario History* 53 (Dec. 1961), 257-71. Wallis was an Ontario lumberman.

W48 Wallner, Richard L. "The Role of the Newspaper in the Conservation Movement." Master's thesis, Univ. of Michigan, 1952.

W49 Walls, Sara, and Vail, William C. "45 Years of Papermaking." *Forest Farmer* 34 (Oct. 1974), 10-11, 14-16. Brief history of the pulp and paper industry in Alabama since 1928, especially the Gulf States Paper Corporation, Tuscaloosa, and International Paper Company, Mobile.

W50 Walsh, Margaret. "The Manufacturing Frontier: Pioneer Industry in Antebellum Wisconsin, 1830-1860." Ph.D. diss., Univ. of Wisconsin, 1969. 585 pp.

W51 Walsh, Margaret. *The Manufacturing Frontier: Pioneer Industry in Antebellum Wisconsin, 1830-1860.* Madison: State Historical Society of Wisconsin, 1972. xvi + 263 pp. Maps, tables, apps., notes, bib., index. Lumber milling constituted a significant minority of the manufacturing enterprises during the 1850s. It was the pioneering industry in most Wisconsin counties.

W52 Walsh, Thomas R. "Charles E. Bessey: Land-Grant College Professor." Ph.D. diss., Univ. of Nebraska, 1972. 245 pp. Includes a chapter on Bessey's conservation activities in Nebraska and elsewhere.

W53 Walsh, Thomas R. "The American Green of Charles Bessey." *Nebraska History* 53 (Spring 1972), 35-57. Bessey (1845-1915), a botanist at the University of Nebraska, advocated conservation and reforestation policies in his state and elsewhere. Known as the "father" of the Nebraska National Forest, Bessey also was involved in efforts to save the Calaveras Big Trees in California and to promote Appalachian forest reserves from Maine to Georgia.

W54 Walsh, William D. "The Diffusion of Technological Change in the Pennsylvania Pig Iron Industry, 1850-1870." Ph.D. diss., Yale Univ., 1967. Includes treatment of the industry's reliance on charcoal. Published under the same title by Arno Press in 1975.

W55 Walster, Harlow Leslie. "George Francis Will, 1884-1955: Archeologist, Anthropologist, Ethnologist, Naturalist, Nurseryman, Seedsman, Historian: A Biography." *North Dakota History* 23 (Jan. 1956), 5-25.

W56 Walters, C.S. *The Illinois Veneer Container Industry.* Illinois Agricultural Experiment Station, Bulletin No. 534. Urbana, 1948. Pp. 385-432. Since 1883.

W57 Walton, Perry. *Two Related Industries; An Account of Papermaking, and of Papermakers' Felts as Manufactured at the Kenwood Mills, Rensselaer, New York, U.S.A., and Arn-*

prior, Ontario, Canada. Boston: F.C. Huyck & Sons, 1920. F.C. Huyck & Sons.

W58 Wambold, L.D. "25 Years of Good Will & Good Logging." *Loggers Handbook* 34 (1974), Section III, 8-13. History and recollection of the Sierra-Cascade Logging Conference since 1949.

Wampole, John H. See Mohler, Levi L., #M512

Wand, Ben. See Richardson, F.M., #R188

W59 Wand, Ben. "Fifty Years of Wholesaling Lumber." *Southern Lumber Journal* 42 (Jan. 1938), 32.

Wandesforde-Smith, Geoffrey. See Cooley, Richard A., #C586

W60 Wandrey, Clarence J. "History and Development of Farm Woodlands." Master's thesis, Colorado Agricultural and Mining College, 1947.

W61 Wangaard, Frederick F., ed. *Proceedings of Wood Symposium on 100 Years of Engineering Progress with Wood.* Washington: Timber Engineering Company, 1953. Collected papers, some on the history of wood engineering.

W62 Wangaard, Frederick F. "Forest Products Industries of the United States." *Mechanical Engineering* 75 (Jan. 1953), 9-15. Since the 17th century.

W63 Wanlass, William L. *The United States Department of Agriculture, a Study in Administration.* Johns Hopkins University Studies, Series 38, No. 1. Baltimore: Johns Hopkins Press, 1920. 131 pp. Includes some incidental references to the USFS.

W64 Ward, David. *The Autobiography of David Ward.* New York: Privately printed, 1912. 194 pp. Ward (1822-1900) was a Michigan lumberman who owned timberlands in the South and West.

Ward, G.B. See Winters, Robert K., #W377

W65 Ward, Gerald R. "An Intelligence Report on Sandalwood." *Journal of Pacific History* 3 (1968), 178-80. A brief history of New England sandalwood traders in the South Pacific during the early 19th century, including a manuscript report from 1817 which refers to the Hawaiian Islands.

W66 Ward, Richard T. "The Beech Forest of Wisconsin—Changes in Forest Composition and the Nature of the Beech Border." *Ecology* 37 (July 1956), 407-19. Since 1834.

W67 Ward, Willis C. "Reminiscences of Michigan's Logging Days." *Michigan History Magazine* 20 (Autumn 1936), 301-12. Reminiscences of the David Ward estate, the last large pine tract to be logged in lower Michigan, 1880s-1912.

W68 Ware, E.R., and Smith, Lloyd F. *Woodlands of Kansas.* Kansas Agricultural Experiment Station, Bulletin No. 285. Manhattan, 1939. 42 pp. Illus., tables. Includes a brief history of forestry and tree planting in Kansas.

W69 Ware, Lamar M. *History and Contributions of the Forestry Program of the Alabama Polytechnic Institute.* Auburn: Alabama Polytechnic Institute, 1947. 75 pp. Bib.

W70 Warken, Phillip W. "A History of the National Resources Planning Board, 1933-1943." Ph.D. diss., Ohio State Univ., 1969. 296 pp.

W71 Warne, H.T. "Even with Beans Twenty-One Times a Week, We Thrived Fifty Years Ago." *Canada Lumberman* 50 (Aug. 1, 1930), 92. Logging camps in the Maritime provinces, 1880s.

W72 Warne, H.T. "How Export Business Was Handled Years Ago." *Canada Lumberman* 54 (May 1, 1934), 31. Lumber exporting from Nova Scotia.

W73 Warne, William E. *The Bureau of Reclamation.* Praeger Library of U.S. Government Departments and Agencies. New York: Praeger Publishers, 1973. x + 270 pp. Illus., maps, charts, apps., bib., index. Includes some organizational history of the agency since its founding in 1902, accounts of its major irrigation and reclamation projects in the West, and assessments of its relations with other natural resource agencies.

W74 Warner, Claude Kent. "Conservation and Science Education in Teacher Training Institutions." Ph.D. diss., Cornell Univ., 1949. 80 pp.

W75 Warner, John R. "History and Financial Results of a Cooperative Forest Products Market Operated through Farmers Mutual Inc. of Durham, N.C." Ph.D. diss., Duke Univ., 1953. 198 pp.

W76 Warner, John R., and Chase, C.D. *The Timber Resource of North Dakota.* Paper No. 36. St. Paul: USFS, Lake States Forest Experiment Station, 1956. 39 pp.

W77 Warner, Richard E. "A Forest Dies on Mauna Kea." *Pacific Discovery* 13 (Mar.-Apr. 1960), 6-14. On the destruction of a forest on the island of Hawaii by overgrazing of sheep, soil erosion, and unfavorable government policies, since 1834.

W78 Warren, Edna. "Forests and Parks in the Old Line State." *American Forests* 62 (Oct. 1956), 13-26, 56-58, 62-72. On forests in Maryland since 1633 and the growth of a state forestry program since 1906.

W79 Warren, George H. *The Pioneer Woodsman As He Is Related to Lumbering in the Northwest.* Minneapolis: Hahn & Harmon Company, 1914. 184 pp. Illus. Reminiscences of a lumberjack and timber cruiser in northern Wisconsin and Minnesota, 1871-1910.

W80 Warren, Judith Ann. "The Lumber Industry in the Plumas-Lassen Area of the Northern Sierra-Southern Cascades in California: An Historical Geography." Master's thesis, Univ. of California, Berkeley, 1968.

W81 Warren, Lawrence R. "The Development of the Memorial Grove System in Humboldt Redwoods State Park." Master's thesis, Sacramento State College, 1968.

W82 Warren, Viola Lockhart. "The Eucalyptus Crusade." *Historical Society of Southern California Quarterly* 44 (Mar. 1962), 31-42. On the widespread planting of eucalyptus trees in California, ca. 1850s-1900s, and their failure as a commercial species.

W83 Warren Manufacturing Company. *Fifty Years of Paper Making: A Brief History of the Origin, Development and Present Status of the Warren Manufacturing Company, 1873-1923.* Boston: Walton Advertising and Printing Company, 1923. 39 pp. Illus. A New Jersey firm.

W84 Warren, S.D., Company. *A History of S.D. Warren Company, 1854-1954.* Westbrook, Maine, 1954. vi + 120 pp. Illus. A Maine pulp and paper firm.

W85 Warshow, H.T., ed. *Representative Industries in the United States.* New York: Henry Holt and Company, 1928. xiii + 702 pp. Illus., tables, charts. Includes a chapter on the lumber industry.

W86 Washburn, Bradford. *A Tourist Guide to Mount McKinley.* Anchorage: Alaska Northwest Publishing Company, 1971. 79 pp. Illus., maps, bib., index. Incidental history of Mount McKinley National Park.

W87 Washburn, Charles G. *The Life of John W. Weeks.* Boston: Houghton Mifflin, 1928. xix + 349 pp. Illus., apps. As a Massachussets congressman Weeks (1860-1926) sponsored forestry legislation—most notably the Weeks Act of 1911 which provided for national forests in the East.

W88 Washburn, C. Marshall. "Bangor, Gateway to Forest Wealth." *Northern Logger and Timber Processor.* 18 (May 1970), 20-21, 49. On Bangor, Maine, and the Penobscot Valley lumber industry in the mid-19th century.

W89 Wasson, George S. *Sailing Days on the Penobscot: The Story of the River and the Bay in the Old Days.* 1932. Revised edition. Intro. and epilogue by Walter Muir Whitehill. New York: W.W. Norton, 1949. 246 pp. Illus., map. Includes some history of the lumber port of Bangor, Maine, and of the shipping and shipbuilding industries, 1860-1926.

W90 Watkins, T.H. *The Grand Colorado: The Story of a River and Its Canyons.* Palo Alto, California: American West Publishing Company, 1969. 310 pp. Illus., maps, index. Includes some history of Grand Canyon National Park, Arizona.

W91 Watkins, T.H. "Pilgrims' Pride." *American West* 6 (Sept. 1969), 48-54. On Arizona's Grand Canyon, 1870s-1910s.

W92 Watson, E. "Fifty Years of Tree Planting." *Country Life in America* 11 (1906), 47-50.

W93 Watson, Hilton. "Alabama's Sawmill Industry." *Southern Lumberman* 193 (Dec. 15, 1956), 158-61. Since 1814.

W94 Watson, Judge. "The Economic and Cultural Development of Eastern Kentucky from 1900 to the Present." Ph.D. diss., Indiana Univ., 1963. 341 pp. Includes some history of forest industries.

W95 Watson, Judy. "The Red River Raft." *Texana* 5 (No. 1, 1967), 68-76. On the extensive log jam in Louisiana's Red River and its removal by U.S. Army engineers in the 19th century.

W96 Watson, Russell. "Forest Devastation in Michigan: A Study of Some of Its Deleterious Economic Effects." *Journal of Forestry* 21 (May 1923), 425-51. Includes some history of logging and forest devastation in general since ca. 1830s.

W97 Watson, Russell. "Historic Sources of Congressional Trouble in Conservation." *Journal of Forestry* 22 (May 1924), 480-89. Since 1777, with emphasis on philosophical influences of Alexander Hamilton and Adam Smith.

Watson, Russell. See Dana, Samuel Trask, #D19

W98 Wattenburger, Ralph T. "The Redwood Lumbering Industry on the Northern California Coast, 1850-1900." Master's thesis, Univ. of California, Berkeley, 1931.

W99 Wattles, Marshall D. "Observations on Foreign Trade in Lumber." *Land Economics* 34 (May 1958), 168-74. On lumber exports from Oregon and Washington, 1923 to 1952.

W100 Watts, Lyle F. "'The Greatest Good of the Greatest Number'." *Journal of Forestry* 48 (Feb. 1950), 81-83. On Gifford Pinchot and his conservation philosophy; an address delivered by the USFS chief at the dedication of the Gifford Pinchot National Forest, Washington, 1949.

W101 Watts, Lyle F. "Edward Norfolk Munns." *Journal of Forestry* 49 (Nov. 1951), 798-99. Sketch of Munns's career with the USFS, 1912-1951, especially in forestry research and as a bibliographer of the literature of forestry.

W102 Watts, Lyle F. "They Never Retire." *American Forests* 60 (July 1954), 13, 52-54. Reminiscences of his experiences in the USFS, 1914-1952.

W103 Waugh, Frank A. "What is a Forest?" *Journal of Forestry* 20 (Mar. 1922), 209-14. On he etymology of the word *forest* in the English language.

W104 Way, Ronald L. *Ontario's Niagara Parks.* Toronto: Niagara Parks Commission, 1946. 349 pp.

W105 Weatherby, H. "Preventing the Fires." *Timber of Canada* 19 (No. 4, 1958), 65-74. Includes some history of the forest protection activities of the British Columbia Forest Service.

W106 Weatherby, James B. "The Hells Canyon Controversy: A Study of the Hells Canyon Associations and Their View of Comprehensive River Basin Development." Master's thesis, Univ. of Idaho, 1968. Dam and hydroelectric power policy in the Snake River Canyon, Oregon-Idaho.

W107 Weaver, Alexander. *Paper, Wasps and Packages: The Romantic Story of Paper and Its Influence on the Course of History.* Chicago: Container Corporation of America, 1937. 80 pp.

W108 Weaver, Harold. "Fire As an Ecological Factor in the Southwestern Ponderosa Pine Forests." *Journal of Forestry* 49 (Feb. 1951), 93-98. Conclusions from study of fire-scarred stumps in Arizona, dating from 1708.

W109 Weaver, Harold. "Effects of Prescribed Burning in Ponderosa Pine." *Journal of Forestry* 55 (Feb. 1957), 133-38. On fire ecology studies conducted on Washington's Colville Indian Reservation, 1942-1955.

W110 Weaver, Harold. "Effects of Prescribed Burning in Second Growth Ponderosa Pine." *Journal of Forestry* 55 (Nov. 1957), 823-26. On fire ecology studies conducted on Oregon's Klamath Indian Reservation since 1939.

W111 Weaver, Harold. "Ecological Changes in the Ponderosa Pine Forest of the Warm Springs Indian Reservation in Oregon." *Journal of Forestry* 57 (Jan. 1959), 15-20. Based on fire ecology studies made since 1903.

W112 Weaver, Harry. "Labor Practices in the East Texas Lumber Industry to 1930." Master's thesis, Stephen F. Austin State College, 1961.

W113 Webb, Bernice Larson. "Company Town—Louisiana Style." *Louisiana History* 9 (Fall 1968), 325-39. History of Elizabeth, Louisiana, site of a Calcasieu Paper Company mill since 1926. Includes some account of industrial strikes and attendant violence in the 1950s.

W114 Webb, Burdine. "Old Times in Kentucky." *Southern Lumberman* 187 (Dec. 15, 1953), 163-64. On the lumber industry in eastern Kentucky, 1870s-1880s.

W115 Webb, Burdine. "Old Times in Eastern Kentucky." *Southern Lumberman* 193 (Dec. 15, 1956), 178G-178J. On the lumber industry since the 1870s.

W116 Webb, Charles G. "Tioga County Formed to Help Accelerate Logging Sales to Speculators' Wilderness Land." Pennsylvania Department of Internal Affairs, *Monthly Bulletin* 22 (Oct. 1954), 6-10, 25-26. Historical sketch of Tioga County since 1787.

W117 Webb, Clyde S. "Truck Hauling in the Inland Empire." *Timberman* 38 (Mar. 1937), 120-22. Reminiscences of 20 years of change in truck logging, mainly in northern Idaho.

Webb, Clyde S. See Strong, Clarence C., #S694

W118 Webb, F.T. "The Canadian Pulp and Paper Industry." *Canadian Geographical Journal* 31 (Dec. 1945), 285-99.

W119 Webb, George W. "The Resources of the Cumberland Plateau as Exemplified by Cumberland County, Tennessee: A Geographic Analysis." Ph.D. diss., Univ. of Tennessee, 1956. 301 pp. Includes some history of forest resources and industries.

W120 Webb, George W. "The Hardwood Lumber Industry of the Eastern Highland Rim." *Journal of the Tennessee Academy of Sciences* 32 (July 1957), 216-27. In central Tennessee and south central Kentucky, 1940s-1950s.

W121 Webb, W.B. "Eastern Kentucky Timber Development: A History of the Past." *Southern Lumberman* 86 (Dec. 22, 1917), 101.

W122 Webb, W.B. "Retrospect of the Lumber Industry in Eastern Kentucky: Story of Fifty Years of Progress." *Southern Lumberman* 113 (Dec. 22, 1923), 175-76.

W123 Webb, W.B. "Early Days in Kentucky Hardwoods." *Southern Lumberman* 117 (Dec. 20, 1924), 183, 187.

W124 Webb, William J. "United States Government War Agencies and Texas-Louisiana Pine Manufacturers in World War I." Master's thesis, Univ. of Houston, 1968.

W125 Webber, G. Pierce. "The Webber Timberlands, an Horatio Alger Story." *Northeastern Logger* 11 (Dec. 1962), 22, 68-72. The Webber family timberlands in Maine since 1850.

W126 Weber, Arnold N. "Redwood Cut-Over Lands and Their Use." Master's thesis, Univ. of California, Berkeley, 1926.

W127 Weber, Daniel Barr. "John Muir: The Function of Wilderness in an Industrial Society." Ph.D. diss., Univ. of Minnesota, 1964. 285 pp. On Muir (1838-1914) as a naturalist, conservationist, and philosopher, and his political activities in behalf of preserving Yosemite National Park.

W128 Weber, Mary Bond. "A Lumberman's Dream." *Glades Star* (Garrett County [Maryland] Historical Society) 2 (Mar. 1958), 429-31, 433-35. On the Du Bois and Bond Brothers Lumber Corporation, their timberlands in Maryland's Savage River Valley, their logging railroad, and the lumber town of Bond, 1900-1910.

W129 Webster, H.H., and Stoltenberg, C.H. *The Timber Resources of New Jersey.* Upper Darby, Pennsylvania: USFS, Northeastern Forest Experiment Station, 1958. 40 pp. Map.

W130 Webster, Paul, and editors of *American West. The Mighty Sierra: Portrait of a Mountain World.* Great West Series. Palo Alto, California: American West Publishing Company, 1972. 287 pp. Illus., maps, apps., bib., index. Includes historical chapters on the discovery and exploration of the Sierra Nevada, the lumber industry, and forest conservation.

W131 Weddell, D.J. "Who Should Control the Forests? An Historical Study of Government Regulation." *Journal of Forestry* 42 (Dec. 1944), 861-70. Surveys the issue of public regulation of forests to 1937; includes a bibliography of fifty references.

W132 Weddle, Ferris M. "Wilderness Champion—Olaus J. Murie." *Audubon* 52 (July-Aug. 1950), 224-33. Murie (1889-1963) was a field naturalist for the U.S. Biological Survey, a wildlife conservationist, and a director of the Wilderness Society.

W133 Weeks, David; Wieslander, A.E.; and Hill, C.L. *The Utilization of El Dorado County Land.* Bulletin No. 572. Berkeley: University of California, Agricultural Experiment Station, 1934. 115 pp. Some history of forest industries in El Dorado County, California.

W134 Weeks, David; Wieslander, A.E.; Josephson, H.R.; and Hill, C.L. *Land Utilization in the Northern Sierra Nevada.* Berkeley: University of California, Agricultural Experiment Station, 1943. 127 pp. Illus., maps, tables, diags. Includes some history of forest industries.

W135 Weeks, Lyman H. *A History of Paper-Manufacturing in the United States, 1690-1916.* New York: Lockwood Trade Journal Company, 1916. xv + 352 pp. Illus., notes, index. A standard history with much information on leaders and firms in the industry; first appeared serially in the *Paper Trade Journal.*

W136 Wehrwein, George S. "A Social and Economic Program for the Sub-Marginal Areas of the Lake States." *Journal of Forestry* 29 (Oct. 1931), 915-24. Includes some history of the cutover lands of the Great Lakes states since 1860.

W137 Wehrwein, George S. "Second Conservation Movement." *Journal of Land and Public Utility Economics* 12 (Nov. 1936), 421-22. New Deal.

Wehrwein, George S. See Ely, Richard T., #E87

W138 Wehrwein, George S., and Barlowe, Raleigh. *The Forest Crop Law and Private Forest Taxation in Wisconsin.* Bulletin No. 519. Madison: Wisconsin Conservation Commission, 1945. Wisconsin's Forest Crop Law of 1927.

W139 Weidberg, Dorothy. "The History of John Kentfield & Company, 1854-1925." Master's thesis, Univ. of California, Berkeley, 1940. A San Francisco lumber and shipping firm.

W140 Weigle, W.G. "The Great Idaho Fire of 1910." *Timberman* 35 (July 1934), 16-20. Recollections of a supervisor of Idaho's Coeur d'Alene National Forest.

Wein, Bernard A. See Stone, Flora Pierce, #S663

Weinmayr, V. Michael. See Fabos, Julius Gy., #F1

W141 Weinstein, Robert A. "Lumber Ships at Puget Sound." *American West* 2 (Fall 1965), 50-63. On lumber ships and crews at Port Blakeley, Washington, 1890s-1900s, with photographs by Wilhelm Hester.

W142 Weinstein, Robert A. "Grays Harbor Country, 1880-1920." *The Record, 1974* (Washington State University, Friends of the Library) 35 (1974), 5-44. History and pictorial essay on Aberdeen, Hoquiam, and Cosmopolis, lumber towns on Grays Harbor, Washington; photographs by Charles R. Pratsch and Colin MacKenzie.

W143 Weintraub, Sidney. "Price-Making in Forest Service Timber Sales." *American Economic Review* 49 (Sept. 1959), 628-37. Incidental history.

W144 Weir, Richard S. "The Work of the Department of Agriculture During the Incumbency of Secretary James Wilson." Master's thesis, Univ. of Washington, 1926. 93 pp. From 1897 to 1913.

W145 Weisberger, Bernard A. "Here Come the Wobblies!" *American Heritage* 13 (June 1967), 30-35, 87-93. On the labor organizing activities of the Industrial Workers of the World, especially among lumberjacks and millworkers in the Pacific Northwest, ca. 1905-1920.

W146 Weiskittel, Ralph. "The Cincinnati Lumber Market." *Southern Lumberman* 193 (Dec. 15, 1956), 150-52. From 1840 to 1900.

Weiss, Grace M. See Weiss, Harry B., #W147, W148, W149

W147 Weiss, Harry B., and Weiss, Grace M. *Forgotten Mills of Early New Jersey: Oil, Plaster, Bark, Indigo, Fanning, Tilt, Rolling, and Slitting Mills, Nail and Screw Making.* Trenton: New Jersey Agricultural Society, 1960. 94 pp. Illus., diags., notes, bib. From 1770 to 1882.

W148 Weiss, Harry B., and Weiss, Grace M. *Some Early Industries of New Jersey: Cedar Mining, Tar, Pitch, Turpentine, Salt Hay.* Trenton: New Jersey Agricultural Society, 1965. 70 pp.

W149 Weiss, Harry B., and Weiss, Grace M. *The Early Sawmills of New Jersey.* Trenton: New Jersey Agricultural Society, 1968. v + 98 pp. Illus., notes, bib., index. A general account of sawmill technology throughout the East, with emphasis on New Jersey. Chapter 5 identifies 250 sawmills existing in New Jersey between 1666 and 1881.

Weiss, Harry B. See Sim, Robert J., #S275

W150 Weiss, Howard F. *The Preservation of Structural Timber.* 1914. Second edition. New York: McGraw-Hill, 1916. xx + 361 pp. Illus., tables, graphs, apps., index. Includes some history of wood preservation.

W151 Weiss, Howard F. "Forest Products Investigations." *Journal of Forestry* 23 (July-Aug. 1925), 565-73. A general account of forest products research since 1900.

W152 Welch, Fay. "Sharon J. Mauhs—A Dedicated Conservationist." *Conservation Council Comments* 4 (May-June 1965), 6-7. Also appeared in *Ad-i-ron-dac* 29 (Sept.-Oct. 1965), 66-69.

W153 Welker, Robert H. *Birds & Men: American Birds in Science, Art, Literature, and Conservation, 1800-1900.* Cambridge: Harvard University Press, 1955. Reprint. New York: Atheneum, 1966. 230 pp. Illus., bib., index.

W154 Welling, J.C. *The Land Politics of the United States.* New York: New York Historical Society, 1888. 40 pp.

W155 Wellington, Raynor G. *The Political and Sectional Influence of the Public Lands, 1828-1842.* Cambridge: Harvard University Press, 1914.

Wellons, John C. See Anderson, Darlene G., #A428

W156 Wells, Donald T. *The TVA Tributary Development Program.* University: University of Alabama, Bureau of Public Administration, 1964. ix + 153 pp. Illus. Includes some history of the Tennessee Valley Authority's small watershed programs and conflicts with the Soil Conservation Service.

W157 Wells, George S. "The Forest That Blew Away." *American Forests* 60 (Mar. 1954), 14-16, 51-52. On the National Park Service's program to preserve forests along North Carolina's coast from burial by dunes, since 1936.

W158 Wells, Philip P. "Philip P. Wells in the Forest Service Law Office." *Forest History* 16 (Apr. 1972), 22-29. His memoir of 1913 pertains to legal questions, conservation issues, and Gifford Pinchot, 1906-1911.

W159 Wells, Robert W. *Fire at Peshtigo.* Englewood Cliffs, New Jersey: Prentice-Hall, 1968. 243 pp. On the forest fire which destroyed Peshtigo, Wisconsin, in 1871.

W160 Wells, S.D. "History and Development of Semi-Chemical Pulping." *Paper Trade Journal* 133 (Dec. 15, 1951), 32, 36, 40.

W161 Welsh, Peter C. "A. Cardon and Company, Brandywine Tanners, 1815-1826, with Notes on the Early History of Tanning in Delaware." *Delaware History* 8 (Sept. 1958), 121-47.

W162 Welsh, Peter C. *Tanning in the United States to 1850: A Brief History.* Bulletin No. 242. Washington: Smithsonian Institution, Museum of History and Technology, 1964.

vii + 99 pp. Illus., notes, apps., bib., index. Emphasizes the tanning industry in Delaware.

W163 Welsh, Thomas J. "Logging on the Northwest Angle." *Minnesota History* 34 (Spring 1954), 1-8. Reminiscences of the author's logging camp on the northwest side of Minnesota's Lake of the Woods, 1925-1931.

W164 Wengert, Norman I. "TVA—Symbol and Reality." *Journal of Politics* 13 (Aug. 1951), 369-92. On the Tennessee Valley Authority's importance as a symbol of regional development, "grass roots democracy," and conservation of resources, with incidental history.

W165 Wengert, Norman I., and Honey, John C. "Program Planning in the U.S. Department of the Interior, 1946-53." *Public Administration Review* 14 (Summer 1954), 193-201.

W166 Wengert, Norman I. *Natural Resources and the Political Struggle.* New York: Random House, 1955. 71 pp. Notes. Includes some history of natural resource policies and conservation concepts, with emphasis on intragovernmental relations and the competing bureaucracies.

W167 Wengert, Norman I. "The Ideological Basis of Conservation and Natural Resource Policies and Programs." *Annals of the American Academy of Political and Social Science* 344 (Nov. 1962), 65-75. On the influence of conservation ideology on the formation of natural resource policies.

W168 Wengert, Norman I. "Changing Relations Between Government and the Forest Products Industries: An Exploration of Policy Processes." In *First National Colloquium on the History of Forest Products Industries, Proceedings,* ed. by Elwood R. Maunder and Margaret G. Davidson. New Haven: Forest History Society, 1967. Pp. 28-48. Since 1876.

W169 Wenkham, Robert. *Kauai and the Park Country of Hawaii.* San Francisco: Sierra Club Books, 1969. 160 pp. Illus. Incidental history.

Wentling, J.P. See Cheyney, Edward G., #C283

W170 Werner, Morris R., and Starr, John. *Teapot Dome.* New York: Viking Press, 1959. x + 306 pp. Illus., notes. On the Teapot Dome scandals of the 1920s, including their impact on the conservation movement.

Werthman, Michael S. See Detweiler, Robert, #D144

W171 Wertz, Daniel. "Early Days as an Indiana Hardwood Pioneer." *Southern Lumberman* 144 (Dec. 15, 1931), 80.

W172 Wescott, Richard R. "Economic, Social and Governmental Aspects of the Development of Maine's Vacation Industry, 1850-1920." Master's thesis, Univ. of Maine, 1959.

W173 Weslager, Clinton A. "Log Houses in Pennsylvania during the Seventeenth Century." *Pennsylvania History* 22 (July 1955), 256-66.

W174 Weslager, Clinton A. *The Log Cabin in America: From Pioneer Days to the Present.* New Brunswick, New Jersey: Rutgers University Press, 1969. xxv + 382 pp. Illus., notes, index. From log cabins on the frontier to log cabins as a symbol in American politics.

W175 Wessel, Thomas R. "Prologue to the Shelterbelt, 1870-1934." *Journal of the West* 6 (Jan. 1967), 119-34. Tree planting in the Great Plains states.

W176 Wessel, Thomas R. "Roosevelt and the Great Plains Shelterbelt." *Great Plains Journal* 8 (Spring 1969), 57-74. On Franklin D. Roosevelt's interest in the Shelterbelt Project (Prairie States Forestry Project) and its results, 1934-1942.

W177 West, Bruce. *The Firebirds.* Toronto: Ontario Ministry of Natural Resources, 1974. 258 pp. A history of the Ontario Provincial Air Service, aerial fire spotters since 1924.

W178 West, Oswald. "Oregon and California Railroad Land Grant Management." *Oregon Historical Quarterly* 53 (Sept. 1952), 177-80. Returned to the federal government in 1916, these forested lands in Oregon have been jointly administered by the USDI and USFS.

W179 *West Coast Lumberman.* "Lumber Shipped by Water from Northwest in Past Thirty Years." *West Coast Lumberman* (May 1, 1924), 106-09.

W180 *West Coast Lumberman.* "First Log Band Mill in America." *West Coast Lumberman* (Sept. 1930), 41ff.

W181 *West Coast Lumberman.* "Production in Washington, Oregon and British Columbia by Year from 1908 to 1930." *West Coast Lumberman* (May 20, 1931), 41. Shingle production.

W182 *West Coast Lumberman.* "Since 1863." *West Coast Lumberman.* 66 (June 1939), 12-14, 57. Brief history of Dolbeer and Carson Lumber Company, a California redwood firm.

W183 *West Coast Lumberman.* "Westwood, California: The Home of Paul Bunyan and the Red River Lumber Company." *West Coast Lumberman* 68 (Dec. 1941), 10-15.

W184 *West Coast Lumberman.* "For Forty-nine Years." *West Coast Lumberman* 69 (Aug. 1942), 18-24. Union Lumber Company, Fort Bragg, California, since the 1880s.

W185 *West Coast Lumberman.* "Champion of Good Forestry." *West Coast Lumberman* 75 (Dec. 1948), 58-60, 62, 64-65, 97-98. Western Forestry and Conservation Association, Portland, since 1909.

W186 West Coast Lumbermen's Association. *Uniform Cost Accounting System.* Seattle: West Coast Lumbermen's Association, 1919. 214 pp. Incidental history.

W187 West Coast Lumbermen's Association. *The Story of West Coast Lumber.* Portland, 1956. 14 pp.

W188 Westcott, Linn H. "Shay Locomotive." *Trains* 1 (July 1941), 28-33. Widely used in logging operations since the late 19th century.

W189 *Western Conservation Journal.* "St. Regis Celebrates Its 75th Anniversary." *Western Conservation Journal* 20 (Oct. 1963), 62-65. A brief history of its Tacoma predecessor, St. Paul and Tacoma Lumber Company, since 1888.

W190 *Western Conservation Journal.* "Deer Park Firm Pioneers Logging Methods." *Western Conservation Journal* 21 (Feb.-Mar. 1964), 24-26, 39. Deer Park Pine Industry, a subsidiary of Potlatch Corporation in eastern Washington, since 1914.

W191 Western Forestry and Conservation Association. *Permanent Forest Production: An Analysis of Sustained Yield Management in Western Forests.* Portland, 1948. 63 pp. Illus. Incidental history.

Western Forestry and Conservation Association. See Hamilton, Eloise, #H63

W192 omitted.

W193 Western Forestry and Conservation Association. *Forest Laws of California: History of Forest and Fire Laws—Summary of Forest Practice Act.* Portland, 1950. 29 pp.

W194 Western Forestry and Conservation Association. *Forest Laws of Oregon; History of Forest and Fire Laws; Summary of Forest Conservation Act.* Portland, 1956. 36 pp. Map. Since 1864.

W195 *Western Railroader.* "The Pickering Lumber Company." *Western Railroader* 12 (Oct. 1949), 3-6. On two logging railroads owned by this California company: The Sugar Pine Railway and the Empire City Railway, 1902-1929.

W196 Weston, Byron. "History of Paper Making in Berkshire County, Mass." *Collections of the Berkshire Historical and Scientific Society* 2 (1895).

W197 Weston, Harry E. "Sulphate Industry of the U.S." *Paper Industry and Paper World* 20 (Mar. 1939), 1249-58.

W198 Weston, Harry E. "Chronology of Papermaking in the United States." *Paper Maker* (Sept. 1944), 9-11; (Feb. 1945), 8-10.

W199 Weston, Harry E. "When the U.S. Wood Pulp Industry was Young: Random Selections from the Trade Press of the Time." *Paper Maker* (Feb. 1946), 7-11. From 1880 to 1904.

W200 Weston, Harry E. "Paper Mill Memories: News Snatches Gathered from the Trade Press of Years Ago." *Paper Maker* 17 (Feb. 1947), 9-15. From 1880 to 1904.

W201 Weston, Nathan A. "A History of the Land System of the State of New York with Special Reference to Financial Legislation." Ph.D. diss., Cornell Univ., 1901. 61 pp.

W202 Westphall, Victor. "The Public Domain in New Mexico, 1854-1891." Ph.D. diss., Univ. of New Mexico, 1956. 344 pp.

W203 Westphall, Victor. "The Public Domain in New Mexico, 1854-1891." *New Mexico Historical Review* 33 (Jan. 1958), 24-52; (Apr. 1958), 128-43.

W204 Westphall, Victor. *The Public Domain in New Mexico, 1854-1891.* Albuquerque: University of New Mexico Press, 1965. xv + 212 pp. Maps, tables, notes, apps., bib., index. Includes a chapter on the Timber Culture Act of 1873, as well as other references to forested lands and federal land policy.

W205 Westveld, R.H. *Applied Silviculture in the United States.* 1935. Second edition. New York: John Wiley and Sons, 1949. xi + 590 pp. Includes historical introductions to chapters on each forest region.

W206 Westveld, R.H. *Forest Restoration in Missouri.* Bulletin No. 392. Columbia: University of Missouri, Agricultural Experiment Station, 1937. 153 pp. Incidental history.

W207 Westveld, R.H. *Forestry in Farm Management.* 1941. Second edition. New York: John Wiley and Sons, 1951. ix + 339 pp. Incidental history.

W208 Westveld, R.H. "Forest Research in Colleges and Universities Offering Forestry Education." *Journal of Forestry* 52 (Feb. 1954), 85-89. Includes some history of forest research at educational institutions since 1899.

W209 West Virginia Pulp and Paper Company. *Fifty Years of Paper Making.* New York, 1937. 39 pp.

W210 Westwood, Richard W. "The American Nature Association." *Journal of Forestry* 44 (Oct. 1946), 722-23. On the conservation organization founded in 1922 by Charles Lathrop Pack and Arthur Newton Pack.

W211 Weyerhaeuser, Frederick K. *Trees and Men.* New York: Newcomen Society in North America, 1951. 32 pp. Illus. On the Weyerhaeuser Timber Company, Tacoma, Washington, and its origins in the Upper Midwest.

Weyerhaeuser, Louise L. See Hill, William Bancroft, #H375

W212 Weyerhaeuser Timber Company. *Men, Mills and Timber: Fifty Years of Progress in the Forest Industry.* Tacoma, Washington: Weyerhaeuser Timber Company, 1950. 50 pp. Illus. Brief history of the Tacoma-based firm and its Pacific Northwest operations.

W213 Whaley, Ross S. "Structure of the Sawmill Industry in Colorado." Master's thesis, Colorado State Univ., 1961.

W214 Wharton, Don. "Douglas Fir: Our Most Valuable Tree." *American Forests* 80 (Jan. 1974), 30-33, 58-60. Includes some history of its identification and use since 1825.

W215 Wharton, Mel. "Oriental Lumber Trade." *Export and Shipping Journal* 3 (Mar. 1922), 7-10. Incidental history.

W216 Wheatley, R.A. "An Industry of National Importance: The Origin and Development of the Great Lakes Paper Company Limited." *Pulp & Paper Magazine of Canada* 44 (July 1943), 559-78. At Fort William [Thunder Bay], Ontario.

W217 Wheeler, Charles F. *A Sketch of the Original Distribution of White Pine on the Lower Peninsula.* Bulletin No. 162. Lansing: Michigan Agricultural College, Experiment Station, 1898.

Wheeler, Lilla C. See Wheeler, William F., #W223

W218 Wheeler, Philip R., and Sternitzke, Herbert S. "Timber Trends in the Mid-South." *Southern Lumberman* 193 (Dec. 15, 1956), 179-81. Tennessee, Alabama, Mississippi, Louisiana, Arkansas, Texas, and Oklahoma since the 1870s.

W219 Wheeler, Philip R. "The Coming of Forest Research." *Forests & People* 13 (First quarter, 1963), 66-67, 96-101, 110-11. On forestry research in Louisiana since 1902, with emphasis on the work of the USFS's Southern Forest Experiment Station in New Orleans and of its branches.

W220 Wheeler, Raymond H. *Booms, Depressions, and Tree Rings.* Crystal Lake, Illinois: Weather Forecasts for the

Weather Science Foundation, 1951. 15 pp. Diags. On the relations between business cycles and the major sequoia tree ring maxima and minima, since 1794.

W221 Wheeler, W. Reginald. "N.P. Wheeler: Lumberman, Congressman, Christian." *Pennsylvania History* 19 (Oct. 1952), 421-34. Nelson P. Wheeler (1841-1920) owned the Wheeler and Dusenbury Lumber Company in Pennsylvania.

W222 Wheeler, W. Reginald. *Pine Knots and Bark Peelers: The Story of Five Generations of American Lumbermen.* La Jolla, California: By the author, 1960. 252 pp. Illus. The Wheeler family was involved in the lumber industry since the late 18th century, primarily in New York, Pennsylvania, and Michigan, with timberlands on the Pacific Coast. The bulk of the book concerns Nelson P. Wheeler (1841-1920), a U.S. representative from Pennsylvania, and his role in the Wheeler and Dusenbury Lumber Company, whose timberlands now constitute part of Pennsylvania's Allegheny National Forest.

W223 Wheeler, William F. *Autobiographic Sketch of the Hon. William F. Wheeler, of Portville, N.Y.* Ed. by Lilla C. Wheeler. Portville, New York: By the author, 1890. Wiilliam French Wheeler (1811-1892) was a New York lumberman. His dictated reminiscences include accounts of lumber rafting on the Allegheny River of New York and Pennsylvania.

W224 Wheelwright, William B. "Two and One-Half Centuries of American Papermaking." *Paper Trade Journal* 106 (Feb. 24, 1938), 202-04.

W225 Wheelwright, William B. "Pioneers in Wood Pulp." *Paper Maker* (No. 3, 1941), 4-7. On the introduction of wood pulp to the United States by the Pagenstecher brothers, Curtisville, Massachusetts, 1866, and subsequent use of wood in papermaking.

W226 Whetsler, Bill. "Log Drives on the Priest River." *The Big Smoke 1974* (Pend Oreille County [Washington] Historical Society) (1974), 2-17. Reminiscences of logging and log driving on the northern Idaho river, 1916-1949.

W227 Whipple, Gurth. *A History of Half a Century of the Management of the Natural Resources of the Empire State, 1885-1935.* Albany: New York Conservation Department and New York State College of Forestry, 1935. 199 pp. Illus., tables. The cover title is *Fifty Years of Conservation in New York State, 1885-1935.* The emphasis is on forest conservation and parks.

W228 Whisenhunt, Donald W. *The Environment and the American Experience: A Historian Looks at the Ecological Crisis.* Port Washington, New York: Kennikat Press, 1974. vi + 136 pp. Notes, index.

W229 Whisler, Ezra Leroy. *Shepherd of the Cowlitz: The Autobiography of Ezra Leroy Whisler.* With exposition by Geraldine Crill Eller. Elgin, Illinois: Brethren Publishing House, 1957. 169 pp. Illus., notes. Includes some reminiscences of the author's life as owner of a sawmill in Centralia, Washington, during the early 20th century.

Whisnant, Archie. See Holbrook, Stewart H., #H442

W230 Whitaker, Howard E. *Humane Enterprise: An Account of the Mead Corporation (1846-1963).* New York: Newcomen Society in North America, 1963. 24 pp. Illus. A paper company with headquarters in Dayton, Ohio, and operations in the East and South.

Whitaker, J. Russell. See Parkins, Almon E., #P41

W231 Whitaker, J. Russell, and Ackerman, Edward A. *American Resources, Their Management and Conservation.* New York: Harcourt, Brace, 1951. Reprint. New York: Arno Press, 1972. 497 pp. Maps, graphs, tables, index. Incidental history.

W232 Whitbeck, Ray H. "The Industries of Wisconsin and Their Geographic Basis." *Annals of the Association of American Geographers* 2 (1912), 55-64. Includes some history of the lumber industry.

W233 White, Arthur S. "Grand Rapids Furniture Centennial." *Michigan History Magazine* 12 (Apr. 1928), 267-79.

W234 White, Aubrey. *A History of Crown Timber Regulations from the Date of the French Occupation to the Year 1899.* 1899. Reprint. Toronto: Ontario Department of Lands and Forests, 1957. 282 pp. Issued originally as part of the *Annual Report* of the Ontario Clerk of Forestry.

W235 White, Aubrey. "History and Results of the Fire Ranging System in Ontario." Canadian Forestry Association, *Annual Report* 4 (1903), 31-41.

W236 White, C.H. "Origin and History of Hardwoods: Description of Species Used on the Pacific Coast." *Timberman* 25 (Feb. 1924), 72-76.

W237 White, C.H. "History of Pacific Hardwood Trade." *Timberman* 33 (Nov. 1931), 90-92; 33 (Dec. 1931), 68-70; 33 (Jan. 1932), 34-35.

White, Dale. See Place, Marian T., #P237, P238, P239

W238 White, David E. "The Economic Problems of the Lumber Industry in New York State." Ph.D. diss., State Univ. of New York, College of Forestry, 1965. 179 pp. Incidental history.

W239 White, David G. "The Trend in Chestnut Production and Consumption." *Southern Lumberman* (Apr. 15, 1930), 46-48. Since 1899.

W240 White, E.N. "Old Days of Lumbering in the Ottawa Valley." *Canada Lumberman* 51 (Dec. 15, 1931), 20ff.

W241 White, Francis Harding. "The Administration of the General Land Office, 1812-1911." Ph.D. diss., Harvard Univ., 1912.

W242 White, G. Edward. *The Eastern Establishment and the Western Experience: The West of Frederic Remington, Theodore Roosevelt, and Owen Wister.* Yale Publications in American Studies, No. 14. New Haven: Yale University Press, 1968. 238 pp. Notes, bib., index. Chapter 8 is entitled "Technocracy and Arcadia: Conservation under Roosevelt."

W243 White, George O. "The Forest Service and the States." *Journal of Forestry* 53 (Feb. 1955), 133-35. Includes some history of cooperation between state forestry agencies and the USFS.

W244 White, Helen M. *A Survey of Forest History Resources for the St. Croix River Valley, Minnesota-Wisconsin.* St. Paul: Forest History Foundation, 1955. 18 pp. Processed.

W245 White, Henry G. "Public Forest Homesteads." *Papers of the Michigan Academy of Science, Arts, and Letters* 27 (1941), 169-80. Includes some history of the Forest Homestead Act of 1906 and other projects providing for settlement of cutover lands in the Great Lakes states.

W246 White, Henry G. "Forest Regulation: A Study of Public Control of Cutting Practices on Private Forest Land in the U.S." Master's thesis, Univ. of Minnesota, 1948.

W247 White, Henry G. "Forest Ownership Research in Historical Perspective." *Journal of Forestry* 48 (Apr. 1950), 261-64. On forest ownership patterns since the 1860s.

White, J.H. See Howe, Clifton D., #H569

W248 White, J. Wesley, comp. *Historical Sketches of the Quetico-Superior.* 10 Volumes. Duluth: USFS, Superior National Forest, 1967-1972. Processed. On the Superior National Forest of northeastern Minnesota, with some account of the adjacent Quetico Provincial Park in Ontario. Volume 8 is a bibliography.

White, James A. See Guthrie, John D., #G394

White, John R. See Fry, Walter, #F275

W249 White, John R., and Pusateri, Samuel J. *Sequoia and Kings Canyon National Parks.* Stanford, California: Stanford University Press, 1949. xviii + 212 pp. Illus., maps, bib., index. A guidebook which includes some history of these two California national parks.

W250 White, Roy R. "Austin Cary and Forestry in the South." Ph.D. diss., Univ. of Florida, 1960. 277 pp.

W251 White, Roy R. "Austin Cary, the Father of Southern Forestry." *Forest History* 5 (Spring 1961), 2-5. Cary (1865-1936), a pioneer in industrial forestry in New England, began with the USFS in 1910 and after 1917 worked wholly in the South, where he had much success persuading lumbermen to practice forestry.

W252 White, Tom. "Moving Logs in B.C. Waters." *Loggers Handbook* 33 (1973), 8-11, 148. Log transportation in coastal waters of British Columbia since the mid-19th century, including mention of "Davis rafts" and more recent log barges.

W253 White, William C. *Adirondack Country.* New York: Duell, Sloan & Pearce, 1954. 315 pp. Map. Includes some forest history of the New York region since the late 18th century. A revised edition was published by Alfred A. Knopf in 1967.

W254 White, Zebulon W. "Growth of Industrial Forestry." In *First National Colloquium on the History of the Forest Products Industries, Proceedings,* ed. by Elwood R. Maunder and Margaret G. Davidson. New Haven: Forest History Society, 1967. Pp. 93-106.

W255 Whitehill, John. "John Whitehill, Carpenter." Ed. by Charles Van Ravenswaay. *Bulletin of the Missouri Historical Society* 16 (July 1960), 342-44. Whitehill (1794-1871) writes of his life as a carpenter and lumber yard operator in St. Louis, 1819-1866.

Whitehill, Walter Muir. See Wasson, George S., #W89

W256 Whitford, Harry N., and Craig, Roland D. *Forests of British Columbia.* Ottawa: Commission of Conservation, 1918. viii + 409 pp. Illus., maps, tables, apps., index. Includes some history of forest resources, management, and policy.

W257 Whitford, Kathryn. "Thoreau and the Woodlots of Concord." *New England Quarterly* 23 (Sept. 1950), 291-306. Henry David Thoreau as a forest ecologist, 1845-1862.

W258 Whitford, Philip, and Whitford, Kathryn. "Thoreau: Pioneer Ecologist and Conservationist" *Scientific Monthly* 73 (Nov. 1951), 291-96. On Henry David Thoreau, 1847-1862.

W259 Whitham, William B. "L'Industrie canadienne de pâtes et papiers." *Actualité Economique* 45 (No. 2, 1969), 267-98. Proposes a framework for a history of the Canadian pulp and paper industry, including a chronology of its development and suggested areas for research.

W260 Whiting, Perry. *Autobiography of Perry Whiting, Pioneer Building Material Merchant of Los Angeles.* Los Angeles: Smith-Barnes Corporation, 1930. 334 pp.

Whitlock, Harry T. See Guthrie, John D., #G394

W261 Whitnall, Rolfe. "Forty Years of Logging at Cascade." *Timberman* 44 (Nov. 1942), 14-17, 42-44. On the Cascade Lumber Company, Yakima, Washington, and logging in central Washington since 1901.

W262 Whitnall, Rolfe. "The Saga of St. Maries." *Timberman* 47 (Apr. 1946), 34ff. St. Maries Lumber Company, St. Maries, Idaho.

W263 Whitney, Stephen R. "The Sierra Club Foundation: A Strong Right Arm." *Sierra Club Bulletin* 57 (Sept. 1972), 34-36. Since its establishment in 1960.

W264 Whitson, J.F. "Ontario's Forest Fires." *Canadian Forestry Journal* 6 (Dec. 1910), 107-09. History of fires since the 1850s.

W265 Whittaker, Jack. "Jack Whittaker." OHI by Sue Baptie. *Sound Heritage* 3 (No. 2, 1974), 13-24. Reminiscences of a sawmill superintendent for Industrial Timber Mills, Limited, Youbou, British Columbia, 1930s.

W266 Whittaker, Robert H. "Recent Evolution of Ecological Concepts in Relation to the Eastern Forests of North America." *American Journal of Botany* 44 (Feb. 1957), 197-206. Since 1905.

W267 Whittemore, Laurence F. *Industry in Canada and in the United States of America: Friends of Freedom.* New York: Newcomen Society in North America, 1950. 32 pp. On the Brown Corporation of Canada, formerly the Quebec and St. Maurice Industrial Company, and its wood-pulp paper mill at La Tuque, Quebec, since 1905.

W268 Whittles, Thomas D. *The Lumberjack Sky Pilot.* Chicago. Winona Publishing Company, 1908. 233 pp. Illus. On Frank E. Higgins, minister to Minnesota lumberjacks since the 1890s.

W269 Whittles, Thomas D. *The Parish of the Pines: The Story of Frank Higgins, the Lumberjacks' Sky Pilot.* New York, Chicago: Fleming H. Revell Company, 1912. 247 pp.

W270 Whittles, Thomas D. *Frank Higgins, Trail Blazer.* New York: Interchurch Press, 1920. 148 pp. Illus. Higgins was a minister in Minnesota lumber camps.

W271 Whitton, Charlotte. *A Hundred Years A-Fellin': Some Passages from the Timber Saga of the Ottawa in the Century in Which the Gillies Have Been Cutting in the Valley, 1842-1942.* Ottawa: Runge Press, 1942. xiv + 172 pp. Illus., maps. Company history of Gillies Brothers, Limited, Braeside, Ontario, with logging operations in the Ottawa Valley.

W272 Wible, Ralph C. "Re-Building Penn's Woods." *American Forests* 65 (Oct. 1959), 22-23, 69-73. On state forests in Pennsylvania since 1898.

W273 Wickman, Kenneth P. "Historical and Locational Aspects of Economic Decline of New England Furniture Industry." Ph.D. diss., Syracuse Univ., 1962. 487 pp.

W274 Wickman, Marjorie. "Conservation Comes of Age: An Analysis of Select Conservation Policies Adopted by President Theodore Roosevelt." Master's thesis, George Mason College, 1974.

W275 Wickstrom, George W., and Ainsworth, Charles P. *Always Lumber: The Story of Dimock, Gould & Co., 1852-1952.* Rock Island, Illinois: Augustana Book Concern, 1953. 136 pp. Illus. On DeWitt Clinton Dimock, John Maxfield Gould, and Charles Rodney Ainsworth, founders of the lumber firm in Moline, Illinois. The company history emphasizes the period to 1900.

W276 Widner, Ralph R., ed. *Forests and Forestry in the American States: A Reference Anthology.* Washington: National Association of State Foresters, 1968. xx + 594 pp. Illus., map, tables, index. Contains many historical articles (some published elsewhere) on logging, forest conservation, state forestry, and state forestry organizations. Nearly every state is represented.

Widtsoe, J.A. See Stewart, George, #S637

W277 Wiebe, Mrs. Oscar. "The J.G. Wiebe Lumber Company." *Mennonite Life* 8 (July 1953), 127-28. Beatrice, Nebraska, since 1877.

W278 Wiegman, Carl. *Trees to News: A Chronicle of the Ontario Paper Company's Origin and Development.* Toronto: McClelland and Stewart, 1953. xii + 364 pp. Illus., maps, tables, index. Established in 1912 to supply newsprint for Robert R. McCormick's *Chicago Tribune,* the company held timberlands in Ontario and Quebec.

W279 Wienker, Curtis W. "McNary: A Predominantly Black Company Town in Arizona." *Negro History Bulletin* 37 (Aug.-Sept. 1974), 282-85. The blacks were transferred from Louisiana when the Cady Lumber Company (later McNary Lumber Company and Southwest Forest Industries) moved to Arizona in the 1920s.

Wieslander, A.E. See Weeks, David, #W133, W134

W280 Wiest, Edward. *Agricultural Organization in the United States.* Kentucky University Studies in Economics and Sociology, Volume 2. Lexington: University of Kentucky, 1923. 618 pp. Includes a chapter on the USFS and Bureau of Plant Industry.

Wight, Charles L. See Bourgeois, Euclid J., #B372

W281 Wight, D.B. *The Wild River Wilderness: A Saga of Northern New England.* Littleton, New Hampshire: Courier Printing Company, 1971. 158 pp. Illus., index. History of lumbering and other forest use in the Wild River Valley, White Mountain National Forest, New Hampshire-Maine.

W282 Wilbur, Ray Lyman, and DuPuy, William A. *Conservation in the Department of the Interior.* Washington: GPO, 1931. 252 pp. Illus. A popular history of conservation in the USDI, including chapters on the national parks, the U.S. Geological Survey, and U.S. General Land Office. Wilbur was Secretary of the Interior, 1929-1933.

W283 Wilbur, Ray Lyman, and Hyde, Arthur M. *The Hoover Policies.* New York: Charles Scribner's Sons, 1937. viii + 667 pp. Index. Wilbur was Secretary of the Interior and Hyde Secretary of Agriculture in the Hoover administration, 1929-1933. Their work includes a chapter on "Policies in Conservation."

W284 Wilbur, Ray Lyman. *The Memoirs of Ray Lyman Wilbur, 1875-1949.* Ed. by Edgar Eugene Robinson and Paul Carroll Edwards. Stanford, California: Stanford University Press, 1960. xiv + 687 pp. Illus., notes, bib. Wilbur was Secretary of the Interior, 1929-1933.

W285 Wilcox, Arthur N. "The Development of the Cedar Creek Forest Natural History Area." *Proceedings of the Minnesota Academy of Science* 18 (1950), 163-68. In Anoka and Isanti counties, Minnesota, since 1940.

W286 Wilcox, Earley V., and Wilson, Flora H. *Tama Jim.* Boston: Stratford Company, 1930. 196 pp. Illus., index. James Wilson was Secretary of Agriculture under Presidents McKinley, Roosevelt, and Taft, 1897-1913.

W287 Wilcox, R.F. "John Fensel's Woodlot." *American Forests and Forest Life* 35 (Nov. 1929), 698-99. Farm forestry on a twenty-acre woodlot at Montpelier, Indiana, since 1881.

W288 Wildavsky, Aaron. "Aesthetic Power or the Triumph of the Sensitive Minority over the Vulgar Mass: A Political Analysis of the New Economics." *Daedalus* 96 (Fall 1967), 1115-28. The "new" resources economics, esthetic conservationists, and natural resources problems.

W289 Wilde, Margaret F. "History of the Public Land Policy of Maine, 1620-1820." Master's thesis, Univ. of Maine, 1932.

W290 Wilder, Judith C. "The Years of a Desert Laboratory." *Journal of Arizona History* 8 (No. 3, 1967), 179-99. On a multipurpose laboratory near Tucson, Arizona, operated by the USFS, 1940-1960.

W291 Wildman, Edward E. *Penn's Woods, 1682-1932.* Philadelphia: Christopher Sower Company, 1933. viii + 201 pp. Illus. Historic trees and forests of Pennsylvania.

W292 Wiles, Ralph W. *Pipe Dreams and Memories: An Autobiography.* New York: William-Frederick Press, 1953. 312 pp. Reminiscences of the lumber industry in Nova Scotia and New England, 1882-1940s.

Wiley, Farida A. See Burroughs, John, #B588

Wiley, Farida A. See Roosevelt, Theodore, #R290

W293 Wiley, Richard T. "Ship and Brig Building on the Ohio and Its Tributaries." *Ohio Archaeological and Historical Quarterly* 22 (Jan. 1913), 54-64. An enterprise of the late 1790s and early 1800s.

W294 Wilhelm, Samuel A. "The Wheeler and Dusenbury Lumber Company of Forest and Warren Counties." *Pennsylvania History* 19 (Oct. 1952), 413-20. History of a Pennsylvania firm, 1834-1922.

W295 Wilhelm, Samuel A. "History of the Lumber Industry of the Upper Allegheny River Basin during the Nineteenth Century." Ph.D. diss., Univ. of Pittsburgh, 1953. 264 pp. On logging, rafting, and the lumber industry in six northwestern Pennsylvania counties.

W296 Wilkerson, Hugh, and Van der Zee, John. *Life in the Peace Zone: An American Company Town.* New York: Macmillan, 1971. 158 pp. Scotia, California, a company town of the Pacific Lumber Company.

W297 Wilkes, George C. "Ground Rent for Provincial Forest Land in Ontario." *Canadian Journal of Economics and Political Science* 22 (Feb. 1956), 63-72. On the history of forest ground rent as a source of provincial revenue.

Wilkie, David. See Hearn, George, #H260

W298 Wilkins, Austin H. *The Forests of Maine: Their Extent, Character, Ownership and Products.* Bulletin No. 8 Augusta: Maine Forest Service, 1932. 107 pp. Illus. Incidental history.

W299 Wilkins, Austin H. "The Story of the Maine Forest Fire Disaster." *Journal of Forestry* 46 (Aug. 1948), 568-73. Fires burned 220,000 acres in 1947.

W300 Wilkins, Thurman. *Clarence King: A Biography.* New York: Macmillan, 1958. ix + 441 pp. Illus., bib. King (1842-1901), a geologist, mining engineer, and writer, was also the first director of the U.S. Geological Survey.

W301 Wilkins, Thurman. *Thomas Moran: Artist of the Mountains.* Norman: University of Oklahoma Press, 1966. xvi + 315 pp. Illus., notes, app., bib., index. Moran (1837-1926) was with the Hayden surveying expedition in the Yellowstone area in the 1870s and subsequently painted most of the natural wonders of the West. His work interested many in conservation and the national parks.

W302 Wilkinson, Norman B. "Land Policy and Speculation in Pennsylvania, 1779-1800." Ph.D. diss., Univ. of Pennsylvania, 1958. 375 pp.

W303 Wilkinson, William. *Memorials of the Minnesota Forest Fires in the Year 1894, with a Chapter on the Forest Fires in Wisconsin in the Same Year.* Minneapolis: By the author, 1895. 412 pp. + 67 pp. Illus., app.

W304 Will, Emery Lewis. "A Study of Conservation Education in the Secondary Schools of the U.S." Ph.D. diss., Cornell Univ., 1949. 80 pp.

W305 Will, Thomas Elmer. "A Chapter of Conservation History." *Conservation* 15 (Aug. 1909), 495-97.

W306 Willey, Day Allen. "The Forest's Guardian: Gifford Pinchot, National Forester and Head of the Conservation Commission." *Putnam's Magazine* 7 (Nov. 1909), 161-71.

W307 Williams, A.W. "Changes in the Louisville Lumber Area." *Southern Lumberman* 193 (Dec. 15, 1956), 178J-178M. Since 1870.

W308 Williams, Arthur B. *The Native Forests of Cuyahoga County, Ohio.* Cleveland: Holden Arboretum, Cleveland Museum of Natural History, 1949. 90 pp. A theoretical reconstruction of the original vegetation based on historical sources and evidence.

W309 Williams, Asa S. "Logging by Steam." *Forestry Quarterly* 6 (Mar. 1908), 1-33. Includes some history of steam logging technology.

W310 Williams, Ben Ames. "The Return of the Forest." *American Forests and Forest Life* 33 (Aug. 1927), 451-54. To a former agricultural region between Belfast and Augusta, Maine, since 1870.

W311 Williams, Burton J. "Trees But No Timber: The Nebraska Prelude to the Timber Culture Act." *Nebraska History* 53 (Spring 1972), 77-86. On tree planting experiments, largely unsuccessful, in several decades prior to 1873.

W312 Williams, Charles E. *The Life of Abner Coburn: A Review of the Political and Private Career of the Late Ex-Governor of Maine.* Bangor: Press of T.W. Burr, 1885. Coburn was also a large landowner and lumberman.

W313 Williams, Charles Fredrick. "William M. Jardine and the Development of Republican Farm Policy, 1925-1929." Ph.D. diss., Univ. of Oklahoma, 1970. 271 pp. Incidental reference to USFS affairs.

W314 Williams, Edgar Leon. "A Study of the Sawmills of Northern Idaho: Their Facilities, Production, Log Procurement, and Lumber Sales." Master's thesis, Univ. of Idaho, 1962.

W315 Williams, Edward C. "The First Redwood Operations in California." *Pioneer Western Lumberman* 58 (July 15, 1912), 9-13. On the first sawmill and foreign shipment of redwood lumber.

W316 Williams, Ellis T. "Tax Assessment of Forest Land and Timber Shows Improvement." *Journal of Forestry* 54 (Mar. 1956), 172-76. On federal income tax assessments on private forests since 1941.

W317 Williams, George H. *Wilderness and Paradise in Christian Thought.* New York: Harper & Row, 1962. 245 pp. Includes some history of man's conception of wilderness, beginning with the Hebraic tradition and continuing to the New World. Although wilderness is considered largely in the spiritual context, the background is useful for modern thought about wild nature.

W318 Williams, Guy. *Logger Talk.* University of Washington Chapbooks, No. 41. Seattle: University of Washington Bookstore, 1930.

W319 Williams, John A. "Davis and Elkins of West Virginia: Businessmen in Politics." Ph.D. diss., Yale Univ., 1967. 347 pp. Henry G. Davis and Stephen B. Elkins were U.S. senators from West Virginia in the era of the 1870s to the 1910s. Both had interests in the lumber business.

W320 Williams, John E. "Recollections of Commercial Forestry." *Lumber* 62 (Sept. 23, 1918-Jan. 20, 1919). A serialized account of the white pine and Southern pine industries.

W321 Williams, Justin. "England's Colonial Naval Stores Policy, 1588-1776." *Iowa University Studies in the Social Sciences* 10 (1934), 32-45. An abstract of the author's doctoral dissertation, University of Iowa, 1933.

W322 Williams, Justin. "English Mercantilism and Carolina Naval Stores, 1705-1776." *Journal of Southern History* 1 (May 1935), 169-85. On the establishment of the naval stores industry in the Carolinas.

W323 Williams, Marion, ed. "Tales from Tug Hill." *New York Folklore Quarterly* 13 (Spring 1957), 23-31. Reminiscences of the lumber industry in New York's Tug Hill plateau, collected from oral sources.

W324 Williams, R.S. "Forest Conditions in the Klondike." *Journal of the New York Botanical Garden* 1 (1900), 44-46. Incidental history.

W325 Williams, Roy. "Evolution of the Tie-Hack." *American Forests* 46 (Dec. 1940), 550-52. On changes in the railroad crosstie industry and among tie hacks since 1913 on the Washakie (Shoshone) National Forest of Wyoming.

W326 Williams, Thomas R. "Recollections of My 73 Years in the Mahogany Business." *Southern Lumberman* 223 (Dec. 15, 1971), 104-07. By the president of the New York importing and manufacturing firm, Ichabod T. Williams & Sons.

W327 Williams, W.R. "John Island's Stolen Sawmill." *Inland Seas* 8 (Summer 1952), 139-40. Transported illegally from De Tour, Michigan, to John Island, Ontario, in Georgian Bay, 1890-1918.

W328 Williamson, Harold F., ed. *The Growth of the American Economy: An Introduction to the Economic History of the United States.* New York: Prentice-Hall, 1944. xiii + 804 pp. Tables, charts, notes. See for reference to forest industries.

W329 Williamson, Harold F. "Prophecies of Scarcity or Exhaustion of Natural Resources in the U.S." *American Economic Review* 35 (May 1945), supplement, 97-109. On the accuracy of estimates made between 1860 and the 1930s.

Williamson, Joe. See Newell, Gordon R., #N84

W330 Willins, Henry H. "The History of Oak Flooring." *Southern Lumberman* 193 (Dec. 15, 1956), 200-03. Since the 1880s.

W331 Willson, Lillian M. *Forest Conservation in Colonial Times.* Forest Products History Foundation Series, No. 3. St. Paul: Forest Products History Foundation, Minnesota Historical Society, 1948. 36 pp. Illus., notes, bib. From 1607 to 1775.

W332 Wilm, H.G. "The Status of Watershed Management Concepts." *Journal of Forestry* 44 (Nov. 1946), 968-71. A review of research and experimentation on the stream-flow controversy since 1909.

W333 Wilson, A.K. *Timber Resources of Idaho.* Forest Survey Release No. 3. Ogden, Utah: USFS, Intermountain Forest and Range Experiment Station, 1962. 42 pp.

W334 Wilson, Alan. *The Clergy Reserves of Upper Canada: A Canadian Mortmain.* Canadian Studies in History and Government, No. 8. Toronto: University of Toronto Press, 1968. Including forested lands of Ontario on which commercial lumbering took place.

W335 Wilson, Ben Hur. "Plank Road Fever." *Palimpsest* 15 (Sept. 1934), 289-318. History of plank roads in Iowa, mid-19th century.

W336 Wilson, Bethany (Lovell). "It's an Ill Wind." *Michigan Alumnus Quarterly Review* 63 (Dec. 1956), 56-62. On a forest fire that burned much of several counties in Michigan's "Thumb" in 1881, and the organization of disaster relief by the American Red Cross.

W337 Wilson, Charles Morrow. *Backwoods America.* Chapel Hill: University of North Carolina Press, 1934. 209 pp. A history of frontier life, this work treats pioneer uses of the forest.

W338 Wilson, Charles Morrow. *Aroostook: Our Last Frontier, Maine's Picturesque Potato Empire.* Brattleboro, Vermont: Stephen Daye Press, 1937. 240 pp. Illus. Includes several chapters on the lumber industry in Aroostook County, Maine, since the early 19th century.

W339 Wilson, David A. "An Analysis of Lumber Exports from the Coast Region of British Columbia to the United Kingdom and United States, 1920-1950." Ph.D. diss., Univ. of California, Berkeley, 1955. 105 pp.

W340 Wilson, Ellwood. "A Forester's Work in a Northern Forest." *Forestry Quarterly* 7 (Mar. 1909), 2-14. On the author's efforts to organize a forestry division of the Laurentide Paper Company, Quebec.

W341 Wilson, Ellwood. "Paper Mills and Forestry in Canada." *American Forestry* 18 (Dec. 1912), 769-76. On the adoption of forestry practices by the Laurentide Company of Quebec since 1905.

W342 Wilson, Ellwood. "Planting Forests for Profit: The Interesting Story of the Laurentide Company's Success on Waste Lands near Grand Mere, P.Q." *Canadian Forestry Journal* 11 (Dec. 1915), 284-88.

W343 Wilson, Ellwood. "The Use of Seaplanes in Forest Mapping." *Journal of Forestry* 18 (Jan. 1920), 1-5. The author pioneered in the use of airplanes for mapping and fire detection while a forester for the Laurentide Paper Company, Quebec.

W344 Wilson, Ellwood. "Reforestation in Canada." *Forestry Chronicle* 5 (June 1929), 14-18. Includes some history of nurseries and seeding experiments.

W345 Wilson, Ellwood. "Reviewing 40 Years of Forest Conservation." *Pulp and Paper Magazine of Canada* 43 (July 1942), 631-32.

W346 Wilson, Ellwood, "Through Canadian Wilds: Three Sketches of Early Forestry in Quebec." *Forest History* 11 (Jan. 1968), 16-25. Selected writings (1909, 1912, 1920) by a forester employed by the Laurentide Paper Company.

Wilson, Flora H. See Wilcox, Earley V., #W286

W347 Wilson, Frederick G. "Wood at Work in Wisconsin." *American Forests* 63 (Oct. 1957), 27-34, 84-94. On the lumber industry and forestry since the early 19th century.

W348 Wilson, Frederick G. "Zoning for Forestry and Recreation: Wisconsin's Pioneer Role." *Wisconsin Magazine of History* 41 (Winter 1958), 102-06. Reminiscences of zoning and nonfarm rural land use in Wisconsin since 1927.

Wilson, G.C. See Wilson, R.L., #W358

W349 Wilson, Harold F. *The Hill Country of Northern New England: Its Social and Economic History, 1790-1830.* Columbia University Studies in the History of American Agriculture, No. 3. New York: Columbia University Press, 1936. xiv + 455 pp. Illus., bib. Includes references to logging and other forest uses.

W350 Wilson, Henry. "Cherry Tree Joe McCreery." *Pennsylvania History* 19 (Oct. 1952), 461-63. Verses and lore about a lumberjack of Pennsylvania's Susquehanna River, 1805-1895.

W351 Wilson, Herbert E. *The Lore and the Lure of Sequoia; the Sequoia Gigantea, Its History and Description.* Los Angeles, 1928. 132 pp. Illus.

W352 Wilson, James. "Politics of Pollution: The Case of Maine." Ph.D. diss., Syracuse Univ., 1963. 270 pp. Includes some history of pollution and Maine's forest industries.

W353 Wilson, James S. *Alexander Wilson: Poet-Naturalist.* New York: Neale Publishing Company, 1906. 179 pp. Wilson (1766-1813) glorified nature and wilderness in his writings.

W354 Wilson, Minter Lowther. *The Light of Other Days.* Boston: Christopher Publishing House, 1959. 283 pp. Essays by a West Virginia judge, including reference to his early career as a lumberjack.

W355 Wilson, Neill C., ed. *Deep Roots: The History of Blake, Moffitt, & Towne, Pioneers in Paper since 1855.* San Francisco: Privately printed, 1955. 112 pp. Illus., chronological chart. On a firm of San Francisco papermakers founded by Francis Blake, James Moffitt, and James W. Towne, with branches in several Western states.

W356 Wilson, R.C. "Early Day Lumber Operations in the Santa Cruz Redwood Region." *Timberman* 38 (May 1937), 12-15.

W357 Wilson, R.C. "Redwood of the Santa Cruz—A Logging Saga." *American Forests* 43 (Oct. 1937), 478-81, 510-11. On the discovery of redwoods by Spaniards in 1769 and the lumber industry in California's Santa Cruz Mountains throughout the 19th century.

W358 Wilson, R.L., and Wilson, G.C. *Theodore Roosevelt, Outdoorsman.* New York: Winchester Press, 1971. 278 pp. Illus., bib., app., index. There are chapters on Roosevelt as a naturalist and conservationist, as well as other sections pertinent to his interest in natural resources and the outdoor life.

W359 Wilson, Ralph R. "Controversy over Water Power Rights in the National Forests and National Parks in the 20th Century." Master's thesis, Univ. of Washington, 1934.

W360 Wilson, Robert, and Cobb, Francis E. *Development of Cooperative Shelterbelt Demonstrations on the Northern Great Plains.* USDA, Bulletin No. 1113. Washington: GPO, 1923. 28 pp.

Wilson, Robert F. See Crowell, Benedict, #C732

W361 Winegar, B. "Logging Operations in the Province of Quebec." *Forestry Quarterly* 8 (Sept. 1910), 294-98. Incidental history.

W362 Winer, Herbert I. "History of the Great Mountain Forest, Litchfield County, Connecticut." Ph.D. diss., Yale Univ., 1956.

W363 Wing, David L. "A Michigan Lumber Village." Master's thesis, Massachusetts Institute of Technology, 1898.

W364 Wingate, George Wood. "Early Days of the N.R.A." *American Rifleman* 99 (May 1951), 32-35; 99 (June 1951), 39-41, 46. Reminiscences of the National Rifle Association, an organization sometimes involved in conservation, 1871-1900.

W365 Winkenwerder, Hugo A. *Forestry in the Pacific Northwest.* Washington: American Tree Association, 1928. 48 pp. Includes some incidental history of forestry, forest resources, and the forest industries.

W366 Winkenwerder, Hugo A. "Forests and American History." *California University Chronicle* 14 (Apr. 1912), 218-45. A survey of changing forest use.

W367 Winkley, John W. *John Muir: A Concise Biography of the Great Naturalist.* Martinez, California: Contra Costa County Historical Society of California, 1959. 141 pp. Illus. On the life of Muir (1838-1914) and especially his efforts to preserve the mountains and natural beauty of the Pacific Coast states.

W368 Winn, Fred. "Coronado's Journey and the Forests and Wildlife of 1540." *American Forests* 46 (July 1940), 296-99. Coronado's expedition to the Southwest and impressions of forests and wildlife.

W369 Winslow, Carlile P. "The Development of the Forest Products Industries in the South." *Southern Lumber Journal* 48 (May 15, 1922), 18-22.

W370 Winslow, Carlile P. "The Forest Products Laboratory: A Twenty-First Anniversary Retrospect." *Southern Lumberman* (Dec. 15, 1931), 115-17. Madison, Wisconsin.

W371 Winslow, Carlile P. "Wood in the National Economy." *Proceedings of the American Philosophical Society* 89 (July 18, 1945), 403-07. On economic trends in the forest products industries.

W372 Winter, Charles E. *Four Hundred Million Acres: The Public Lands and Resources. History, Acquisition, Disposition, Proposals, Memorials, Briefs, Status.* Casper, Wyoming: Overland Publishing Company, 1932. 349 pp. Illus., tables, maps.

W373 Winters, Donald L. "Henry Cantwell Wallace and the Farm Crisis of the Early Twenties." Ph.D. diss., Univ. of Wisconsin, 1966. 487 pp.

W374 Winters, Donald L. *Henry Cantwell Wallace as Secretary of Agriculture: 1921-1924.* Urbana: University of Illinois Press, 1970. x + 313 pp. Notes, bib., essay, index. Under Presidents Harding and Coolidge, Wallace resisted efforts to transfer the USFS to the USDI.

W375 Winters, Robert K. "Trends in the Production of Southern Hardwood Lumber, 1889-1929." *Southern Lumberman* 144 (June 15, 1931), 85.

W376 Winters, Robert K.; Putnam, J.A.; and Eldredge, Inman F. *Forest Resources of North-Louisiana Delta*. USDA, Miscellaneous Publications No. 309. Washington: GPO, 1938. 49 pp. Incidental history.

Winters, Robert K. See Hutchison, S. Blair, #H676

W377 Winters, Robert K.; Ward, G.B.; and Eldredge, Inman F. *Louisiana Forest Resources and Industries*. USDA, Miscellaneous Publications No. 519. Washington: GPO, 1943. 44 pp. Incidental history.

W378 Winters, Robert K., ed. *Fifty Years of Forestry in the U.S.A.* Washington: Society of American Foresters, 1950. x + 385 pp. Illus., apps., bib., index. Nineteen chapters by different authors provide outlines of progress in the technical and institutional aspects of American forestry.

Winters, Robert K. See King, D.B., #K150

Winters, Robert K. See Hutchison, O. Keith, #H671

W379 Winters, Robert K. "How Forestry Became a Part of FAO." *Journal of Forestry* 69 (Sept. 1971), 574-77. Background to international forestry since 1905.

W380 Winters, Robert K. *The Forest and Man*. New York: Vantage Press, 1974. 393 pp. Illus., notes, bib., index. A broad survey of the world including several chapters on the growth of forest conservation and the development of the science of forestry in North America.

W381 Winther, Oscar Osburn. *The Great American Northwest: A History*. 1947. Second edition. New York: Alfred A. Knopf, 1950. xvii + 491 + xxx pp. Illus., maps, notes, bib., index. Includes a chapter on the lumber industry and other references to forests.

W382 Wirt, George H. "Joseph Trimble Rothrock: The Father of Forestry in Pennsylvania." *Journal of Forestry* 37 (May 1939), 361-63. Rothrock (1839-1922) was the first commissioner of forestry in Pennsylvania and did much to promote forest conservation in the state.

W383 Wirt, George H. "Joseph Trimble Rothrock, Father of Forestry in Pennsylvania." *American-German Review* 8 (Feb. 1942), 5-8.

W384 Wirt, George H. "A Half Century of Forestry in Pennsylvania." *Journal of Forestry* 41 (Oct. 1943), 730-34. An appraisal of Pennsylvania state forestry since 1893, with some reference to earlier developments in the forestry movement.

W385 Wirt, George H. "The State Forest Academy." *Pennsylvania Forests* 38 (Spring 1953), 43-45, 54, 57. Reminiscences and history of forestry in Pennsylvania since 1877 and of the origins of the Pennsylvania State Forest School at Mont Alto in 1903.

W386 Wirth, Conrad L. *Civilian Conservation Corps Program of the United States Department of the Interior, March 1933 to June 30, 1943.* Washington: GPO, 1945. Illus. tables.

W387 Wirth, Fremont Philip. "The Disposition of the Iron Lands in Minnesota." Ph.D. diss., Univ. of Chicago, 1925. 205 pp.

W388 Wirth, Fremont Philip. "The Operation of the Land Laws in the Minnesota Iron District." *Mississippi Valley Historical Review* 13 (Mar. 1927), 483-98. At the time of the mining boom (1880s-1890s), these northern Minnesota lands were also valued for their timber.

W389 Wirth, Fremont Philip. *The Discovery and Exploitation of the Minnesota Iron Lands*. Cedar Rapids, Iowa: Torch Press, 1937. viii + 247 pp. See for reference to timber resources.

W390 Wisconsin. Committee on Land Use and Forestry. *Forest Land Use in Wisconsin: Report of the Committee on Land Use and Forestry*. Madison: Executive Office, 1932. vii + 156 pp. Maps, tables, graphs, app., bib., index. Includes some history of forested lands, forestry, and the forest products industries.

W391 Wisconsin. Conservation Department. *Wisconsin State Forests: A Report on Their Origin, Development, Public Usefulness, and Potentialities*. Madison, 1955. 129 pp. Illus., maps, tables.

W392 Wisconsin Commercial Forestry Conference. *Forestry in Wisconsin: A New Outlook.* Milwaukee: Wisconsin Commercial Forestry Conference, 1928. 194 pp. Illus. Includes some articles on the history of forests and forestry.

W393 *Wisconsin Magazine of History*. "Historic Trees in Wisconsin." *Wisconsin Magazine of History* 2 (Sept. 1918), 92-98.

W394 *Wisconsin Magazine of History*. "The Wisconsin State Forest Reserve." *Wisconsin Magazine of History* 2 (June 1919), 461-63.

W395 *Wisconsin Magazine of History*. "Question Box: Knapp, Stout & Co." *Wisconsin Magazine of History* 3 (June 1920), 469-71. Knapp, Stout, and Company, a lumber firm of Menomonie, Wisconsin, since 1846.

W396 *Wisconsin Magazine of History*. "Early Lumbering and Lumber Kings of Wisconsin." *Wisconsin Magazine of History* 4 (Sept. 1920), 102-04.

W397 *Wisconsin Then and Now*. "The Battle of Beef Slough." *Wisconsin Then and Now* 10 (No. 7, 1964), 4-5. An episode in 19th-century Wisconsin lumbering history.

W398 *Wisconsin Then and Now*. "Reforestation Began Here as Fire Threat Subsided." *Wisconsin Then and Now* 19 (Mar. 1973).

W399 Wise, Erma (Clement). *Pioneering Days in Oregon*. New York: Vantage Press, 1955. 104 pp. Includes some account of the author's life in logging camps near Sacramento, California, 1920s-1940s.

Wiser, Vivian. See Baker, Gladys L., #B27

W400 Wiswall, Clarence A., and Crafts, Eleanor Bait. *One Hundred Years of Paper Making: A History of the Industry on the Charles River at Newton Lower Falls, Massachusetts*. Reading, Massachusetts: Reading Chronicle Press, 1938. 115 pp.

W401 Witherspoon, John C. "The Oldest Planting." *Southern Lumberman* (Dec. 15, 1948), 242. On a shortleaf-loblolly pine plantation set out in Sumter County, South Carolina, 1873.

W402 Witherspoon, John C. "A Wildcat and a Paper Mill: The Story of the Birthplace of IP's Southern Kraft Division." *Forests & People* 23 (Second quarter, 1973), 16-17, 38-40. On the origins of the Bastrop Pulp and Paper Company mill in Bastrop, Louisiana, 1916-1920. It is now owned by International Paper Company.

Witte, E.E. See Fleming, Robben Wright, #F123

W403 Witten, James W. *Report on the Agricultural Prospects, Natives, Salmon Fisheries, Coal Prospects and Development of Timber and Lumber Interests of Alaska, 1903*. Washington: GPO, 1904. Illus. Includes a historical review of these resources.

W404 Wittke, Carl F. "Carl Schurz and Rutherford B. Hayes." *Ohio Historical Quarterly* 65 (Oct. 1956), 337-55. Schurz was Secretary of the Interior in the Hayes administration, 1877-1881; the article contains an account of his selection and his performance in the post.

W405 Wittwer, Charlotte. "The 1908 White House Governor's Conference." *Environmental Education* 1 (Summer 1970), 142-45. A landmark in the conservation movement.

W406 Wohlenberg, E.T.F. "Western Forestry and Conservation Association." *Journal of Forestry* 44 (July 1946), 505-06. Includes some history of this industrial conservation association, founded in 1909.

W407 Wohlenberg, E.T.F. "Income Tax Regulations and Long Term Forestry Programs." *Journal of Forestry* 48 (May 1950), 360-61. On timber taxation legislation since 1913.

Woike, L.R. See Silen, R.R., #S271

W408 Wojta, J.F. "Town of Two Creeks: From Forest to Dairy Farms." *Wisconsin Magazine of History* 27 (June 1944), 420-35.

W409 Wolcott, Frank H. "Monarch of Grand County." *Westerners Brand Book* (Denver) 10 (1955), 139-53. Monarch, Colorado, an abandoned lumber town, 1903-1908.

Wolcott, Leon O. See Gaus, John M., #G72

W410 Wold, Albert N. "My Father Was A 'Tree-Climber'." *North Dakota History* 26 (Oct. 1959), 171-80. On the efforts of the author's father to plant trees in 1878 in North Dakota; includes comments on the purposes, effects, and repeal of the Timber Culture Acts and related legislation, 1873-1891.

W411 Wolf, R.E. "A Partial Survey of Forest Legislation in New York State." Master's thesis, State Univ. of New York, College of Forestry, 1948.

W412 Wolf, Roy O. "A History of Oregon School Lands, 1849-1900." Master's thesis, Univ. of Oregon, 1940.

W413 Wolfanger, Louis A. "Economic Regions of Alaska." *Economic Geography* 2 (Oct. 1926), 508-36. Includes some history of forest resources and industries.

W414 Wolfe, Charles J. "Hannah, Lay & Company—A Study in Michigan Lumber Industry." Master's thesis, Wayne State Univ., 1938. 88 pp.

W415 Wolfe, Linnie Marsh, ed. "An Unpublished Journal of John Muir." *North American Review* 245 (1938), 24-51.

Wolfe, Linnie Marsh. See Muir, John, #M615

W416 Wolfe, Linnie Marsh. *Son of the Wilderness: The Life of John Muir*. New York: Alfred A. Knopf, 1945. 364 pp. This is the standard biography of Muir (1838-1914), the naturalist, wilderness philosopher, and founder of the Sierra Club.

W417 Wolfe, S.L. *Wood-Using Industries of South Carolina*. Columbia: USFS and South Carolina Department of Agriculture, 1913. 53 pp. Incidental history.

W418 Wolff, A.E. "A Study of the American Lumber Trade with Central America." Master's thesis, State Univ. of New York, College of Forestry, 1936.

W419 Wolff, Julius F., Jr. "Minnesota Conservation, or Resource Management in Minnesota." Ph.D. diss., Univ. of Minnesota, 1950. 227 pp. Includes some legal and administrative history of the Minnesota Department of Conservation.

W420 Wolff, Julius F., Jr. "Some Major Forest Fires in the Sawbill Country." *Minnesota History* 36 (Dec. 1958), 131-38. Northeastern Minnesota.

W421 Wolff, Julius F., Jr. "Of Fire, Forests and People." *Minnesota Volunteer* 37 (Mar.-Apr. 1974), 12-20. Forest fires in northern Minnesota, 1890-1910, and 1970-1971.

W422 Wollenberg, R.P., and Cooper, E.N. "Labor in the Pacific Coast Paper Industry: A Case in Collective Bargaining." *Harvard Business Review* 16 (1938), 366-72.

W423 Wolman, William. "The Development of Manufacturing Industry in the State of Washington, 1899-1947." Ph.D. diss., Stanford Univ., 1957. 297 pp.

W424 Wolman, William. *The Development of Manufacturing Industry in the State of Washington*. Bureau of Economic and Business Research, Economic and Business Studies, Bulletin No. 31. Pullman: Washington State College, 1958. 208 pp. Diags., tables, notes, bib. Includes forest products industries, 20th century.

W425 Wood, Andrew D., and Linn, Thomas G. *Plywoods: Their Development, Manufacture and Application*. New York: Chemical Publishing Company, 1943. xxi + 373 pp. Illus., tables, glossary, index. Includes some incidental history of the manufacture and use of plywood and veneer.

Wood, Dorothy. See Wood, Frances., #W426

W426 Wood, Frances, and Wood, Dorothy. *Forests Are For People: The Heritage of Our National Forests*. New York: Dodd, Mead & Company, 1971.

W427 Wood, Jack. "Jacob Babler: His Contribution to the State Park Movement in Missouri." *Bulletin of the Missouri*

Historical Society 15 (July 1959), 285-95. Jacob L. Babler (1870-1945) established and endowed the Dr. Edmund A. Babler Memorial State Park near St. Louis, 1934-1938; also includes some history of the state park system of Missouri.

W428 Wood, Leslie C. *Rafting on the Delaware River*. Livingston Manor, New York: Livingston Manor Times, 1935. 272 pp.

W429 Wood, Leslie C. *Holt! T'Other Way! . . . Raftsmen . . . in the Delaware Country*. Middletown, New York: By the author, 1950. 252 pp.

Wood, L.W. See Smith, L.W., #S392

W430 Wood, Nancy. *Clearcut: The Deforestation of America*. Battlebook Series, No. 3. San Francisco: Sierra Club, 1971. 151 pp. Illus., tables, graphs. Although essentially an argument against clearcutting, this book contains some history of the USFS, the forest industries, and their cutting practices.

W431 Wood, Nancy. "Of Trees and Men." *Living Wilderness* 37 (Winter 1973-1974), 15-21. Includes some reminiscences of logging by Bob Ziak of Knappa, Oregon.

W432 Wood, Richard G. "A History of Lumbering in Maine, 1820-1861." Ph.D. diss., Harvard Univ., 1934.

W433 Wood, Richard G. *A History of Lumbering in Maine, 1820-1861*. University of Maine Studies in History and Government, Second Series, No. 33. 1935. Reprint. Intro. by David C. Smith. Orono: University of Maine Press, 1971. 267 pp. Illus., maps, tables, notes, bib., index. This standard study plots the lumbering frontier's advance into northern Maine. There are many references to individual firms and lumbermen.

Wood, Richard G. See Lowe, Vyron D., #L268

W434 Woodard, C.S. "The Public Domain, Its Surveys and Surveyors." *Michigan Pioneer and Historical Collections* 27 (1896), 306-23.

W435 Woodbury, Angus M. *A History of Southern Utah and Its National Parks*. N.p., 1950. 123 pp. Illus. Reprinted with revisions from *Utah Historical Quarterly* (July-Oct. 1944).

W436 Woodbury, Robert. "William Kent: Progressive Gadfly, 1864-1928." Ph.D. diss., Yale Univ., 1967. 372 pp. Kent (1864-1928) was a California congressman active in the conservation movement. He donated the land which became Muir Woods National Monument.

W437 Woodbury, T.D. "Development of Silvicultural Practices in the California National Forests." *Journal of Forestry* 28 (May 1930), 693-700. Since 1908.

W438 Woodruff, George W., comp. *Federal and State Forest Laws*. USDA, Bureau of Forestry, Bulletin No. 57. Washington: GPO, 1904. 259 pp. Apps., index. A state-by-state account of legislation pertaining to forest reserves, timber trespass, forest fires, and other forestry matters.

W439 Woodruff, George W. "The Disposal of Public Lands." *Proceedings of the Society of American Foresters* 1 (Nov. 1905), 53-61. Brief history of federal land policy, especially as applied to forested lands in the 19th century.

W440 Woods, James. "The Legend and the Legacy of Franklin D. Roosevelt and the Civilian Conservation Corps (CCC)." D.S.S. diss., Syracuse Univ., 1964. 453 pp. On his personal interest in the origin and management of the CCC, 1933-1942.

W441 Woods, John B. "Moves on the Oregon Checkerboard." *American Forests* 46 (Oct. 1940), 451-53, 478-79. Includes some recent history of the revested Oregon and California Railroad grant lands in western Oregon, especially on the legislation providing for sustained-yield forest management since 1937.

W442 Woods, John B. "Forty Years of Private Forest Ownership." *Journal of Forestry* 39 (Feb. 1941), 106-10.

Woods, John B. See Hastings, Alfred B., #H194

W443 Woods, John B. "Woodlands of Kansas." *American Forests* 54 (Apr. 1948), 170-72. On forests and forestry in Kansas since 1865.

W444 Woods, John B. "Revolt in the O & C Timberlands." *American Forests* 54 (May 1948), 205-07, 218, 231, 233, 235. On the revested Oregon and California Railroad grant lands in western Oregon since 1916.

W445 Woods, John B. "The Forests of Maine." *American Forests* 54 (June 1948), 266-68, 285, 287. Since 1631.

W446 Woods, John B. "Taxes and Tree Growing." *American Forests* 54 (Aug. 1948), 354-55, 384. On timber taxation by states since 1861.

W447 Woods, John B. "Forests of Utah." *American Forests* 54 (Sept. 1948), 406-07, 428-29. On forests and forestry in Utah since 1847.

W448 Woods, John B. "Biltmore Days." *American Forests* 54 (Oct. 1948), 452-53, 469-70. Reminiscences of Carl A. Schenck and Biltmore Forest School, near Asheville, North Carolina, 1912-1913.

W449 Woods, John B. "Forests of New Mexico." *American Forests* 55 (Jan. 1949), 26-27, 46-48. Incidental history of forests and forestry since the 1930s.

W450 Woods, John B. "The Rehabilitation of the Tillamook Burn." *Journal of Forestry* 48 (May 1950), 362-64. Oregon, since 1933.

W451 Woods, John B. "Those Recent Moves on the Oregon Checkerboard." *American Forests* 57 (Aug. 1951), 6-10. Logging on the revested Oregon and California Railroad grant lands and reconveyed Coos Bay Wagon Road grant lands since 1916.

W452 Woodson, Carter G. *The Rural Negro*. Washington: Association for the Study of Negro Life and History, 1930. xvi + 265 pp. Includes some history of labor conditions in the Southern lumber industry.

Woodson, Carter G. See Greene, Lorenzo J., #G321

W453 Woodson, Weldon D. "Eucalyptus Boom and Bust." *Railroad Magazine* 53 (Oct. 1950), 74-79. On the introduction and uses of eucalyptus in California since 1856.

W454 Woodward, Carol H. "The Elms of America: What Is to Be Their Fate?" *Journal of the New York Botanical Garden* 49 (Mar. 1948), 46-69. On the ravages of *phloem necrosis* and Dutch elm disease since 1918.

W455 Woodward, Earl F. "Hon. Albert B. Fall: The Frontier's Fallen Star of Teapot Dome." *Montana, Magazine of Western History* 23 (Winter 1973), 14-23. Fall was Secretary of the Interior, 1921-1923. This sketch emphasizes his earlier career in New Mexico.

W456 Woodward, Karl W. "Philip W. Ayres, Forester." *American Forests* 42 (Feb. 1936), 72, 92. Ayres served since 1900 as the forester for the Society for the Protection of New Hampshire Forests and was also an advocate of the White Mountain National Forest.

W457 Woodward, Karl W. "Land Use in New Hampshire." *Journal of Forestry* 34 (Nov. 1936), 975-82. Since the colonial period.

W458 Woodwell, William H. "The Woodwell Shipyard, 1759-1852." *Bulletin of the Business Historical Society* 21 (June 1947), 58-74. Includes excerpts from the account books of this shipyard in Newburyport, Massachusetts, indicating purchases and costs of timber.

W459 Woolley, H.E. "What Has Been Accomplished in Fire Protection on the National Forests." *American Forestry* 19 (Nov. 1913), 760-68. Since 1905.

W460 Woolley, Mary E. "The Development of the Love of Romantic Scenery in America." *American Historical Review* 3 (Oct. 1897), 56-66. In the 18th century.

W461 Woolsey, Theodore S., Jr. "Scaling Government Timber." *Forestry Quarterly* 5 (June 1907), 166-73. Reminiscences of work on national forests of the West.

W462 Woolsey, Theodore S., Jr. "Early Arizona Problems." *Journal of Forestry* 18 (Feb. 1920), 135-42. Correspondence of Fred S. Breen, supervisor of the Coconino National Forest in 1906, reveals problems of administering timber sales on Arizona national forests.

W463 Woolsey, Theodore S., Jr., and Chapman, Herman Haupt. *American Forest Regulation*. New Haven: Tuttle, Morehouse and Taylor Company, 1922. xv + 217 pp. Notes, tables, apps., index. Includes some history of forest regulation policies.

W464 Woolsey, Theodore S., Jr. "Battell Forest: An Epoch-Making Project Among American Colleges." *American Forests and Forest Life* 31 (Nov. 1925), 678-81. Brief history of a Vermont forest belonging to Middlebury College, since 1916.

W465 Woolsey, Theodore S., Jr. *Riding the Chuck Line: A Forester in Peace and War*. New Haven, Connecticut: Tuttle, Morehouse and Taylor Company, 1930. 116 pp. On the author's experiences with the USFS and with the forestry regiment in France, 1903-1918.

W466 Wooten, H.H. *Major Uses of Land in the United States*. USDA, Technical Bulletin No. 1082. Washington: GPO, 1953. 100 pp. On changes in land use since 1880.

W467 Worcester, Hugh M. *Hunting the Lawless*. Berkeley: American Wildlife Associates, 1955. 298 pp. Illus. Autobiography of a game management agent for the U.S. Fish and Wildlife Service, California-Nevada.

W468 Work, Herman, "Forestry in the Paper Industry." *Journal of Forestry* 39 (Feb. 1941), 110-14. Since the 19th century.

W469 Work, Herman. "Forests on the Old Fields." *Paper Maker* 26 (Sept. 1957), 5-16. Incidental history of forest use, forest products industries, forestry, and reforestation.

W470 Workman, Gilbert. "Only Lovers Can Be Sound Critics: Bernard De Voto and American Conservation." Master's thesis, San Jose State College, 1967.

W471 Workman, R.M. "Bull-Cooks of the Logging Camps." *Timberman* 48 (July 1947), 48ff. Pacific Northwest.

Worley, Lillian. See Ray, Joseph M., #R62

W472 Worrell, Albert C. *Economics of American Forestry*. New York: John Wiley & Sons, 1959. 441 pp. Tables, graphs, index. Incidental history.

W473 Worrell, Albert C. "Pests, Pesticides and People." *American Forests* 66 (July 1960), 39-81. A broad discussion of the social factors involved in policy decisions concerning use of pesticides; includes a section on the history of the gypsy moth campaign since 1890.

W474 Worster, Donald, ed. *American Environmentalism: The Formative Period, 1860-1915*. New York: John Wiley & Sons, 1973. vi + 234 pp. A collection of documents illuminating the origin and development of the environmental movement.

Worth, Harold E. See Kuenzel, John G., #K273

W475 Wright, Benjamin C. *San Francisco's Ocean Trade, Past and Future: A Story of the Deep Water Service of San Francisco, 1848 to 1911. Effect the Panama Canal Will Have Upon It*. San Francisco: A. Carlisle and Company, 1911. 212 pp. Illus. Includes some history of the Pacific lumber trade.

W476 Wright, Carl C. "The Mesquite Tree: From Nature's Boon to Aggressive Invader." *Southwestern Historical Quarterly* 69 (July 1965), 38-43. Observations on its use in Texas since the early 19th century.

W477 Wright, Chester W. *Economic History of the United States*. New York: McGraw-Hill, 1941. xxviii + 1120 pp. Maps, charts, bib. See for reference to forest industries.

W478 Wright, E.W., ed. *Lewis and Dryden's Marine History of the Pacific Northwest: An Illustrated Review of the Growth and Development of the Maritime Industry, from the Advent of the Earliest Navigators to the Present Time, with Sketches and Portraits of a Number of Well Known Marine Men*. Portland: Lewis and Dryden Printing Company, 1895. Reprint. Seattle: Superior Publishing Company, 1971. Includes some history of the Pacific lumber trade.

W479 Wright, Elizur, and Wright, Ellen. *Elizur Wright's Appeals for the Middlesex Fells and the Forests, with a Sketch of What He Did for Both*. Medford, Massachusetts, 1893. 156 pp.

Wright, Ellen. See Wright, Elizur, #W479

W480 Wright, J.G. "Progress in Forest-Fire Research." *Journal of Forestry* 36 (Oct. 1938), 1092-94. Canada, since 1914.

Wright, John K. See Haden-Guest, Stephen, #H7

W481 Wright, Jonathan W. "A Survey of Forest Genetics Research." *Journal of Forestry* 51 (May 1953), 330-33. In the 20th century.

W482 Wright, Livingston. "New England Logging Days—the Old and the New." *Southern Lumberman* (Dec. 23, 1922), 157-58.

W483 Wright, Martin. "Log Culture in Hill Louisiana." Ph.D. diss., Louisiana State Univ., 1956. 195 pp. Log construction distinguishes a cultural area of Louisiana.

W484 Wright, Sam. "To Jump Three Thousand Years." *American West* 8 (Mar. 1971), 24-27. On Robert Marshall's tree-planting experiments in Alaska's Brooks Range in the 1930s, and similar experiments by the author in the 1960s.

W485 Wright, W. Gilchrist. "Research Work of the Dominion Forest Service." *Journal of Forestry* 20 (Jan. 1922), 62-66. Canada, since 1917.

W486 Wright, W. Gilchrist. "How Price Brothers and Company Guard Timber." *Canadian Forest and Outdoors* 20 (Aug.-Sept. 1924), 528-29. Price Brothers and Company, Quebec, Quebec.

W487 Wroten, William H., Jr. "The Railroad Tie Industry in the Central Rocky Mountain Region, 1867-1900." Ph.D. diss., Univ. of Colorado, 1956. 350 pp. Colorado and Wyoming.

W488 Wroten, William H., Jr. "Cross Tie Industry: 1866-1900, in Central Rocky Mountain Area." *Cross Tie Bulletin* 38 (Mar. 1957), 9-12, 14.

Wulpi, Meinrad. See Knight, E. Vernow, #K212

W489 Wurm, Theodore G., and Graves, Alvin C. *The Crookedest Railroad in the World: A History of the Mt. Tamalpais and Muir Woods Railroads of California.* Fresno: Academy Library Guild, 1954. Revised edition. Berkeley: Howell-North, 1960. 123 pp. Illus., maps, bib. On a line originally named the Mill Valley and Mt. Tamalpais Scenic Railway, patronized mainly by sightseers, 1896-1930.

W490 Wurm, Theodore G. *Hetch Hetchy and Its Dam Railroad.* Berkeley: Howell-North, 1973. 290 pp. Illus., maps, apps., bib., index. Includes some history of logging along the Hetch Hetchy Railroad, built by San Francisco to facilitate construction of a dam and reservoir in Yosemite National Park, 1910s-1930s.

W491 Wurster, L. Rodman. "Memories of the Susquehanna Boom: The Museum of the Lycoming County Historical Society." *Pennsylvania History* 19 (Oct. 1952), 496-99. On relics of logging in Pennsylvania's Susquehanna Valley.

W492 Wurtele, D.J. "Mossum Boyd: Lumber King of the Trent Valley." *Ontario History* 50 (Autumn 1958), 177-88. Trent Valley, Ontario.

W493 Wyckoff, J.M. "History of Lumbering in Southeastern Alaska." *Timberman* 31 (June 1930), 38-40. Near Sitka and elsewhere on the Tongass National Forest since 1878.

W494 Wyman, Donald. "The Arboretums and Botanical Gardens of North America." *Chronica Botanica* 10 (Summer 1947), 395-482. Incidental history.

W495 Wyman, Edgar L. "Where Did All the Sawmills Go?" *Northern Logger* 13 (Jan. 1965), 18-19, 28-29. On the decline of the lumber industry in Connecticut.

W496 Wyman, Edgar P. "The White Mountain Museum of Forest History." *Northern Logger and Timber Processor* 23 (June 1975), 6-7. On the movement to establish the museum near Bethlehem, New Hampshire, 1960s-1970s.

Wyman, Walker D. See Blanchard, Louis, #B308

W497 Wynn, Graeme Clifford. "The Assault on the New Brunswick Forest, 1780-1850." Ph.D. diss., Univ. of Toronto, 1974. On the lumber industry and attitudes toward forest exploitation.

W498 Wyoming. State Planning Board. *A Study of Forest Lands in Wyoming.* Cheyenne, 1939. 343 pp. Processed. Incidental history.

W499 Wyss, Max Albert. *Magic of the Wilderness.* New York: Viking Press, 1973. 86 pp. Illus. A heavily illustrated work on the influence of wilderness areas throughout the world; incidental history.

Y1 Yale Forest School. *Biographical Record of the Graduates and Former Students of the Yale Forest School, with Introductory Papers on Yale in the Forestry Movement and the History of the Yale Forest School.* New Haven: Yale Forest School, 1913. 350 pp. Illus., index. Includes a brief history of the school by James W. Toumey and 402 biographical sketches of former students and graduates.

Y2 Yale University. School of Forestry. *The First Thirty Years of the Yale School of Forestry: In Commemoration of the Third Decennial Reunion at New Haven, Conn., Feb. 21-22, 1930.* New Haven, 1930. 59 pp. Includes Henry S. Graves, "The Evolution of a Forest School," pp. 1-26; Thomas H. Gill, "Men Who Made Yale Foresters," pp. 27-46; and C. Edward Behre, "The Alumni of the Yale School of Forestry," pp. 47-59.

Y3 *Yale Law Journal.* "Management of Public Land Resources." *Yale Law Journal* 60 (Mar. 1951), 455-82. On the policies, activities, and comparative records of the USFS and the Bureau of Land Management.

Y4 Yard, Robert Sterling. *The National Parks Portfolio.* 1916. Sixth edition. Revised by Isabelle F. Story. Washington: GPO, 1931. 274 pp. Illus., map. Includes some historical references to national parks and monuments.

Y5 Yard, Robert Sterling. *The Book of the National Parks.* New York: Charles Scribner's Sons, 1919. xvi + 436 pp. Illus., maps, tables, index. Description and history of national parks and monuments.

Y6 Yard, Robert Sterling. "Economic Aspects of Our National Parks Policy." *Scientific Monthly* 16 (Apr. 1923), 380-87. Incidental history.

Y7 Yard, Robert Sterling. *Our Federal Lands: A Romance of American Development.* New York: Charles Scribner's Sons, 1928. xvi + 360 pp. Illus., map, tables, index. Outlines the historical problems connected with the public domain in the

West, including chapters on national forests, national parks and monuments, and the conservation movement.

Y8 Yard, Robert Sterling. "Organizing the National Parks." *American Forests and Forest Life* 35 (Aug. 1929), 462-63, 516. On the movement to create a National Park Service since ca. 1900.

Y9 Yard, Robert Sterling. "The Unforgotten Story of Hetch Hetchy." *American Forests* 40 (Dec. 1934), 566-69, 595-96. On the long and involved political controversy over the Hetch Hetchy Valley and reservoir in Yosemite National Park, 1890s-1910s.

Y10 Yarnall, Ira T. "John Arden Ferguson: Forester, Educator, and Builder of Men." *Journal of Forestry* 47 (Dec. 1949), 985-86. On Ferguson's career as a forestry educator, especially at Pennsylvania State College, 1913-1937.

Y11 Yates, Bowling C. "Glimpse into the Early Life of Gifford Pinchot." *Journal of Forestry* 66 (Sept. 1968), 666-69. On Pinchot's forestry work for George Vanderbilt at Biltmore, his North Carolina estate, 1892-1895.

Y12 Yates, Bowling C. "On the Trail of Dr. Schenck." *American Forests* 79 (Dec. 1973), 28-31, 60-61. On the German background and life of Carl Alwin Schenck (1868-1955), who directed the Biltmore Forest School near Asheville, North Carolina, 1898-1913.

Y13 Yates, Harry O., III. "The First Fifteen Years of Southern Forest Insect Work Conference—1956-1970." *Journal of the Georgia Entomological Society* 8 (Jan. 1973), 1-16.

Y14 Yeager, Dorr G. *Your Western National Parks: A Guide.* New York: Dodd, Mead, 1947. 264 pp. Incidental history.

Y15 Yeager, Dorr G. *National Parks in California.* Menlo Park, California: Lane Publishing Company, 1959. 96 pp. Illus., maps. Includes some history of Yosemite, Sequoia, Kings Canyon, and Lassen Volcanic national parks, with an appendix on national monuments of California.

Y16 Yerburgh, Richard E.M. "An Economic History of Forestry in British Columbia." Master's thesis, Univ. of British Columbia, 1931.

Ylvisaker, Paul N. See Bedard, Paul W., #B168

Y17 Yoho, James G. "Private Forest Ownership and Management in Thirty-One Counties of the Northern Portion of the Lower Peninsula of Michigan." Ph.D. diss., Michigan State Univ., 1956. 343 pp. Incidental history.

Yoho, James G. See James, Lee Morton, #J38

Yoho, James G. See Todd, A.S., #T222

Yoho, James G. See Pomeroy, Kenneth B., #P270

Y18 Yoho, James G. "Unity and Diversity in Marketing." In *First National Colloquium on the History of the Forest Products Industries, Proceedings,* ed. by Elwood R. Maunder and Margaret G. Davidson. New Haven: Forest History Society, 1967. Pp. 107-30. On the evolution of marketing and trade associations within the forest products industries since the 19th century.

Y19 Yonce, Frederick J. "Public Land Disposal in Washington." Ph.D. diss., Univ. of Washington, 1969. 299 pp.

Y20 Yonce, Frederick J. "The Public Land Surveys in Washington." *Pacific Northwest Quarterly* 63 (Oct. 1972), 129-41. Conducted by the General Land Office, 1850s to 1920s.

Y21 York, Harlan H., and Littlefield, Edward W. "The Naturalization of Scotch Pine, Northeastern Oneida County, N.Y." *Journal of Forestry* 40 (July 1942), 552-59. On the establishment of Scotch pine plantations in this region, beginning ca. 1870, and their subsequent development.

Y22 Yoshpe, Harry B., and Brower, Philip P. *Preliminary Inventory of the Land-Entry Papers of the General Land Office.* Preliminary Inventory No. 22. Washington: U.S. National Archives, 1949. iii + 77 pp. Entries named by state and name of district land office, 1800-1925.

Y23 Yost, L. Morgan. "American Houses from Victorian to Modern." *American Lumberman* (Sept. 11, 1948), 210-28.

Y24 Young, Carl Henry. "Life as a Forest Ranger in 1904." Ed. by Inez Long Fortt. *Lane County Historian* 16 (Winter 1971), 63-74. From the field notes of a ranger on Oregon's Cascade (Umpqua and Willamette) National Forest.

Y25 Young, Erma Thurston. *Yesterday and Today.* Manchester, New Hampshire: N.p., 1953. 109 pp. Illus. Memories of rural life in eastern New Hampshire and of a sawmill near the head of Bear River, 1892-1907.

Y26 Young, Ewart. *Grand Falls, Bishop's Falls, Badger, Millertown, Terra Nova, Newfoundland: Paper and Pulpwood Towns of the Interior.* Montreal: Guardian Associates Ltd., 1951. 48 pp. Incidental history.

Y27 Young, Gerald James. "Tree Distribution in St. Clair and Macomb Co., Michigan, Based on Original Land Survey Records." Master's thesis, Univ. of Michigan, 1952.

Y28 Young, Gerald Loren. "The South Umpqua Ranger District: A Case Study in Multiple Use." Master's thesis, Oregon State Univ., 1962.

Y29 Young, Leigh Jarvis. "The Saginaw Forest—Some Historical Highlights." *Washtenaw Impressions* 13 (Nov. 1957), 1-4. On an experimental forest of the University of Michigan near Ann Arbor since 1903.

Y30 Young, Leigh Jarvis. "Pioneering in Forestry at Michigan." *Washtenaw Impressions* 14 (June 1960), 6-13. History and reminiscences of forestry education at the University of Michigan since 1881.

Y31 Young, Mary E. "Congress Looks West: Liberal Ideology and Public Land Policy in the Nineteenth Century." In *The Frontier in American Development: Essays in Honor of Paul Wallace Gates,* ed. by David M. Ellis. Ithaca, New York: Cornell University Press, 1969. Pp. 74-112.

Young, Robert J. See Marshall, Hubert, #M229

Y32 Young, Vertrees. *Forestry and the Paper Industry.* New York: American Paper and Pulp Association, 1950. 19 pp. On advances in forestry made by the industry since the 1920s.

Y33 Young, W. Gordon. "Four Generations of Fords Have Made Paper in Canada." *Pulp and Paper Magazine of Canada* 33 (Mar. 1932), 10-13. Joseph Ford & Company of Quebec.

Y34 Youtsler, James S. "Collective Bargaining Accomplishments in the Paper Industry." *Southern Economic Journal* 21 (Apr. 1955), 441-52. The International Brotherhood of Pulp and Sulphite and Paper Workers and the International Brotherhood of Paper Makers, especially in the South since 1901.

Z1 Zahniser, Howard. "Wildlife Returns to Seney." *American Forests* 49 (Oct. 1943), 483-85, 496. Includes some references to logging near Seney on Michigan's Upper Peninsula since the 1880s.

Z2 Zaremba, Joseph M. "The Trend of Lumber Prices." *Journal of Forestry* 56 (Mar. 1958), 179-81. Since 1860.

Z3 Zaremba, Joseph M. *Economics of the American Lumber Industry.* New York: Robert Speller & Sons, 1963. xii + 232 pp. Tables, notes, references, index. Incidental history.

Z4 Zavitz, Edmund J. "Reforestation in Ontario." *Journal of Forestry* 20 (Jan. 1922), 18-24. Reviews progress of reforestation efforts on southern Ontario wastelands and northern cutover lands.

Z5 Zavitz, Edmund J. "The Development of Forestry in Ontario." *Forestry Chronicle* 15 (Mar. 1939), 36-43.

Z6 Zavitz, Edmund J. *50 Years of Reforestation in Ontario.* Toronto: Ontario Department of Lands and Forests, 1961.

Z7 Zavitz, Edmund J. *Recollections: 1875-1964.* Toronto: Ontario Department of Lands and Forests, 1965. 26 pp. Illus. Zavitz was a provincial forester of Ontario.

Z8 Zavitz, C. Harold, comp. *A History of the Lake Erie Forest District.* Toronto: Ontario Department of Lands and Forests, 1963. 85 pp. Tables. In southern Ontario bordering Lakes Erie, Ontario, and Huron.

Z9 Zelinsky, Wilbur. "The Log House in Georgia." *Georgia Review* 43 (Apr. 1953), 173-93. On its construction, distribution, and chronology in Georgia after 1780.

Z10 Zellerbach Corporation. *Zellerbach, the House of Paper.* San Francisco, 1927. 51 pp. A historical sketch of the Zellerbach Corporation and its pulp and paper operations in California since 1868.

Z11 Zellerbach Paper Company. *History and Description of Paper Making. A Brief History of Zellerbach Paper Company.* San Francisco, 1931. 36 pp.

Z12 Ziegler, E.A. "Dr. John C. Gifford." *Journal of Forestry* 45 (June 1947), 455-56. Gifford (1870-1949) was a forestry educator at the University of Miami, a specialist in tropical forestry, and a prolific writer on forestry and conservation topics.

Z13 Ziegler, E.A. "Mont Alto in Early Forestry Education—1903-1929." *Pennsylvania Forests* 38 (Fall 1953), 95-97, 114-16. On the Pennsylvania State Forest School. See other short articles on the history of Mont Alto in the same issue.

Z14 Ziewacz, Lawrence E. "Thomas W. Palmer: A Political Biography." Ph.D. diss., Michigan State Univ., 1971. 313 pp. Palmer, a Michigan lumberman, held several state and federal government posts, including U.S. senator from 1883 to 1889.

Z15 Zigler, Joe T. "Fifty Years of Logging." *The Big Smoke 1974* (Pend Oreille County [Washington] Historical Society) (1974), 23-26. In northern Idaho and northeastern Washington, ca. 1910-1960.

Z16 Zimmerman, W.E., and Lutz, E.A. *The Pulp and Paper Industries of New York State: An Economic Analysis.* Albany: New York State Division of Commerce, 1942. 111 pp. Illus., maps, Incidental history.

Z17 Zimmet, Steven. *The Public Domain: Its Place in Our History. A Brief Historical Sketch of the Great Land Inheritance and How It Has Enhanced a Nation.* Washington: U.S. Bureau of Land Management, 1966. Illus., maps. maps.

Zinn, Donald. See Seaton, Fred A., #S157

Z18 Zivnuska, John A. "Business Cycles, Building Cycles, and the Development of Commercial Forestry in the United States." Ph.D. diss., Univ. of Minnesota, 1948.

Z19 Zivnuska, John A. "Commercial Forestry in an Unstable Economy." *Journal of Forestry* 47 (Jan. 1949), 4-13. On the relation of business cycles to commercial sustained-yield forestry; supported by data from 1904 to 1946.

Z20 Zivnuska, John A. *Business Cycles, Building Cycles, and Commercial Forestry.* New York: Institute of Public Administration, 1952. vii + 254 pp. Tables, diags., notes, bib. Includes a chapter on the history of the lumber industry and many other historical references to related economic and business problems.

Z21 Zivnuska, John A. "Supply, Demand and the Lumber Market." *Journal of Forestry* 53 (Aug. 1955), 547-53. On trends since the 1920s.

Z22 Zivnuska, John A. "Lumber Markets and the Law of Supply and Demand." *Southern Lumberman* 191 (Dec. 15, 1955), 196-200. From 1919 to 1953.

Zivnuska, John S. See Casamajor, Paul, #C207

Zivnuska, John A. See Teeguarden, Dennis, #T30

Z23 Zivnuska, John A. *U.S. Timber Resources in a World Economy.* Washington: Resources for the Future, 1967. xii + 125 pp. Graphs, tables, notes. Incidental history.

Z24 Zobler, L. "Economic-Historical View of Natural Resource Use and Conservation." *Economic Geography* 38 (July 1962), 189-94.

Z25 Zollinger, James P. *Sutter: The Man and His Empire.* New York: Oxford University Press, 1939. xv + 374 pp. Illus. John Augustus Sutter (1803-1880) had lumber interests in California. His sawmill was the site of a gold discovery in 1848.

Zon, Raphael. See Bates, Carlos G., #B121

Z26 Zon, Raphael, and Sparhawk, William N. *Forest Resources of the World.* 2 Volumes. New York: McGraw-Hill, 1923. Maps, tables, bib., index. Chapter 4 of Volume 2 covers the "Forest Situation in Northern North America," including some historical references.

Z27 Zon, Raphael. "The Search for Forest Facts." *American Forests and Forest Life* 36 (July 1930), 421-23, 482. On the origins and development of USFS research and experiment stations.

Z28 Zon, Raphael. "The Society Comes of Age." *Journal of Forestry* 29 (Mar. 1931), 308-15. On the Society of American Foresters and the growth of the profession since 1900.

Z29 Zon, Raphael, and Duerr, William A. *Farm Forestry in the Lake States: An Economic Problem.* USDA, Circular No. 661. Washington: GPO, 1942. 34 pp. Includes incidental references to the history of farm forestry.

Z30 Zon, Raphael. "Public Good Comes First." American Forests 52 (Nov. 1946), 544, 557. A brief evaluation of Gifford Pinchot's career.

Z31 Zon, Raphael. "Forestry Mistakes and What They Have Taught Us." *Journal of Forestry* 49 (Feb. 1951), 179-83. Concerns eight major errors in policy since 1900.

Z32 Zumwalt, Eugene Vernon. *Taxation and Other Factors Affecting Private Forestry in Connecticut.* Bulletin No. 58. New Haven: Yale University, School of Forestry, 1953. 142 pp. Diags., maps, tables, bib. Since the 17th century.

Subject Index

Subject Index
Introduction

This is a subject index. The main body of the bibliography, arranged alphabetically by author, already serves one indexing purpose. Authors of autobiographical works are also listed as subjects in this index. Titles are not indexed.

Geographical and political place-names do not appear as main indexing entries. The exceptions to this rule include named forests and parks ("Shawnee National Forest, Illinois," "Yoho National Park, British Columbia," "Bernheim Forest, Kentucky," and "Adirondack State Park, New York") and a few mountain ranges, swamps, and other geographical features marked by their forested character. Place-names, however, assume a large importance within the hierarchy of subentries. Under a broad subject entry such as "lumber industry," "state forestry," or "conservation movement," place-name subentries are preceded only by period designations ("colonial" and "20th c.") and common terms ("records," "accounting," "railroads," and "white pine"). Place-name subentries are followed by personal and corporate names for which geographical references are unknown, irrelevant, or so diverse as to be misleading.

Place-name designations are often very general and overlapping. The user seeking historical references to the lumber industry in northeastern Oregon, for example, should consult titles found under "lumber industry—Oregon," "lumber industry—Pacific Northwest," "lumber industry—Inland Empire," "lumber industry—Pacific states," and "lumber industry—West," as well as under related subject entries such as "logging," "forest industries," and "sawmills."

For books and articles examined by the compiler, index terms are based on the known subject contents rather than on key words in titles. Indexing together the contents of lengthy books and brief articles, however, results in more precise references to the shorter and usually more specific work—a bias which may mislead the inexperienced user. Those seeking information on a narrow topic should always begin under the most specific indexing term but should also examine titles under more general entries for further leads. For example, many more books and articles contain information on Gifford Pinchot than is suggested by the sixty-three entry numbers following his name. The user should examine titles under such related entries as "Ballinger-Pinchot controversy," "USFS," "conservation movement," "forestry movement," "Theodore Roosevelt," and "Taft administration." "See also" references assist with this task, but, as with any index, the user must bring patience and imagination to the search.

An asterisk indicates a work that the compiler knows to be a thorough or scholarly historical treatment of a subject. Asterisks will help the user in search of a quick reference to discern important titles from among large blocks of numbers behind a general indexing entry. Most users, however, will want to examine all listed numbers and determine for themselves which are most important to their research.

Subject Index

Warren, S.D., Company, Maine. *S336, W84